*Cover Image Credit/Copyright Attribution: IIIerlok_Xolms/Shutterstock*

Bruce L. Kutter • Majid T. Manzari
Mourad Zeghal
Editors

# Model Tests and Numerical Simulations of Liquefaction and Lateral Spreading

LEAP-UCD-2017

*Editors*
Bruce L. Kutter
Department of Civil and Environmental Engineering
University of California, Davis
Davis, CA, USA

Majid T. Manzari
Department of Civil and Environmental Engineering
George Washington University
Washington, DC, USA

Mourad Zeghal
Department of Civil and Environmental Engineering
Rensselaer Polytechnic Institute
Troy, NY, USA

https://doi.org/10.1007/978-3-030-22818-7

# Preface

Liquefaction of saturated sands during earthquakes and associated ground deformations continue to be a major concern for engineers involved with design and evaluation of civil infrastructure, such as buildings, transportation facilities (e.g., bridges, roads, and ports), and water systems (e.g., dams, levees, pipelines, and aqueducts). Numerous constitutive models and numerical simulation platforms have been developed and are being used to predict the behavior of saturated sands during liquefaction. These simulation tools are now being used in practice to help evaluate civil infrastructure even though there is no currently viable formal process available for validation of these tools.

**L**iquefaction **E**xperiments and **A**nalysis **P**rojects (LEAP) is a series of collaborative research projects with the main goal of producing reliable experimental data to be used for the assessment, calibration, and validation of constitutive models and numerical modeling techniques available for analysis of soil liquefaction and its effects. LEAP was initially envisioned as a follow on from the pioneering VELACS (Verification of Liquefaction Analysis by Centrifuge Studies) project spearheaded by K. Arulanandan and R.F. Scott in the 1990s.

This volume presents results from one LEAP project (LEAP-UCD-2017) that culminated in a workshop at the UC Davis in December 2017. It was preceded by LEAP projects hosted by Professor Iai at the University of Kyoto and LEAP-GWU-2015 hosted at the George Washington University and was followed in 2019 by LEAP-ASIA, at Kansai University, hosted by Professors Tobita and Ueda.

LEAP-UCD-2017 addresses the repeatability, variability, and sensitivity of centrifuge tests in modeling lateral spreading of mildly sloping liquefiable soils. To achieve this goal, 24 centrifuge tests were conducted at 9 different centrifuge facilities around the world. The results of these centrifuge tests allowed for defining a response surface and enabled an assessment of the sensitivity and variability of the tests. For the first time, a sufficient number of experiments were conducted on the same test configuration to enable the assessment of the test-to-test and facility-facility variability of the centrifuge test results.

The experimental data obtained in this project provided a unique opportunity to assess the capabilities of a number of numerical modeling techniques currently available for the analysis of soil liquefaction. Nine of the 24 centrifuge tests were selected for a type B prediction exercise in which 12 numerical simulation teams participated. To assist the numerical modeling effort, a large number of laboratory tests were performed to characterize the physical and mechanical response of Ottawa F-65 Sand and to define the stress-strain-strength response of this soil in cyclic loading. These tests along with other available experimental data on Ottawa sand were used by the numerical simulation teams to calibrate the constitutive models embedded in their numerical simulation software. The results of both the experiments and numerical simulations were discussed in an international workshop that was held in December 2017 on the campus of the University of California, Davis.

The background, observations, and lessons learned in the course of LEAP-UCD-2017 project are described throughout the volume, organized in four main sections:

1. The *Overview Papers* section includes papers that describe the specifications for experiments, material properties for the tested sand, and comparisons between all of the experiments and all of the simulations.
2. The *Centrifuge Experiment Papers* section includes one paper from each centrifuge experiment team.
3. The *Numerical Simulation Papers* section includes one paper from each team that participated in type B and type C simulations.
4. Finally, the *Workshop Essays* section includes short essays on topics related to LEAP that were submitted at the time of the LEAP-UCD-2017 workshop.

The papers in this volume contain many conclusions and important observations. In this preface, one or more of the editors want to highlight some important points:

1. The data produced by LEAP-UCD-2017 is readily available to the general public through the NHERI Cyberinfrastructure Center's DesignSafe at https://doi.org/10.9517603/DS2N10S. We envision that the future researchers could use the data and different metrics or techniques for type C simulations and as calibration/validation benchmarks.
2. The LEAP-UCD-2017 numerical simulation exercise demonstrated the role of experience and careful peer review in achieving reasonable simulations. Modelers with significant experience in numerical simulation of geotechnical engineering problems predicted some key aspects of the response (e.g., magnitude of lateral spreading) with reasonable accuracies. This was particularly interesting as the constitutive models used by the same modelers were not necessarily the most efficient in capturing the stress-strain response of the soil in element tests.
3. The results of centrifuge model tests are dependent on model initial conditions and experimental setup. Replication and repeatability of the model tests are essential to interpret the variability in measurements and observations.
4. The interpretation of variability of the initial state was improved significantly by using a newly designed LEAP cone penetration test in almost all of the centrifuge tests. Due to errors associated with direct measurement of density from mass and

volume, the direct measurement of density by current procedures appears to be a less reliable indicator of state than the cone penetration resistance.

5. There still is a significant scatter observed in the measurements of minimum and maximum dry density for Ottawa F-65 sand; this scatter led to difficulty in the proper selection of relative density by the numerical modelers.
6. The use of high-speed cameras in LEAP-UCD-2017 for measuring the lateral displacements of ground surface during lateral spreading proved to be particularly valuable and complimentary to measurements typically made after the test.
7. In addition to the results of monotonic and cyclic triaxial and direct simple shear tests that were made available to the simulation teams in LEAP-2017 project, a larger database of element tests that include other relevant stress paths (e.g., hollow cylinder torsional shear) and cover a wider range of stress states (smaller confining stresses that are comparable to average stresses in the centrifuge specimen) will enhance the ability of the numerical modelers to further calibrate and fine-tune the parameters of their constitutive models.
8. Close coordination of the experimental efforts and continuous communication of the research teams was a key in obtaining high-quality and consistent results. Numerous conference calls and web-based meeting with the participants across the United States, Europe, and Asia were held to develop and disseminate standards of practice regarding the method of sample preparation, density measurement, cone penetration testing, and displacement measurement.
9. Although the readers may begin to draw conclusions regarding which simulation platforms and constitutive models are best for a given type of analysis, it is clear that different individuals will be able to draw different conclusions in this endeavor. Furthermore, acceptable numerical tools may be sensitive to errors in the input and to the accuracy of the model calibration and calibration data. Even one individual could draw different conclusions regarding the acceptability of simulation tools depending on the metrics used for assessment. The bases for assessment of numerical models could involve multiple aspects of behavior such as:

   (a) Ability to predict element test results, such as the effective stress friction angle and liquefaction triggering curves (e.g., stress ratio vs. number of cycles to 3% strain)
   (b) Ability to predict the shear modulus reduction and damping curves in laboratory element tests
   (c) Respect for perceived fundamentals of soil behavior, such as critical state soil mechanics and stress-dilatancy relationships
   (d) Ability to predict the ultimate residual displacement observed in a series of centrifuge model tests
   (e) Ability to predict the cyclic strain amplitudes leading up to and following "triggering" in the centrifuge tests
   (f) Ability to predict the time history traces from all of the nuances pore pressure and acceleration sensors in the centrifuge tests
   (g) Ability of the numerical tools to predict the evolution of density and soil behavior in a series of shaking events

Some engineers may consider a numerical tool to be disqualified if it cannot pass stage (a) of validation. Others may focus only on the end result (e.g., item (d) residual displacement). Others may put different weights on each of the above items considered in the assessment. It is clear that significant work remains in establishing a formal widely accepted framework for the assessment of liquefaction simulation tools.

Davis, CA, USA — Bruce L. Kutter
Washington, DC, USA — Majid T. Manzari
Troy, NY, USA — Mourad Zeghal

Participants in the LEAP-UCD-2017 Workshop

| Attendee | Affiliation |
|---|---|
| Tarek Abdoun | Rensselaer Polytechnic Institute |
| Pedro Arduino | University of Washington |
| Richard Armstrong | California State University, Sacramento |
| Arul Arulmoli | Earth Mechanics, Inc. |
| Andres Barrero | The University of British Columbia |
| Michael Beaty | Beaty Engineering, Portland, Oregon |
| Emilio Bilotta | University of Naples Federico II |
| Ross Boulanger | University of California at Davis |
| Jonathan Bray | University of California, Berkeley |
| Trevor Carey | University of California at Davis |
| Long Chen | University of Washington |
| Zhao Cheng | Itasca Consulting Group, Inc. |
| Yannis Dafalias | University of California at Davis |
| Kathleen Darby | University of California at Davis |
| Shideh Dashti | University of Colorado, Boulder |
| Jason DeJong | University of California at Davis |
| Ricardo Dobry | Rensselaer Polytechnic Institute |
| Maya El Kortbawi | University of California at Davis |
| Mohamed El Ghoraiby | George Washington University |
| Richard Fragaszy | US, National Science Foundation |
| David Frost | Georgia Institute of Technology |
| Kiyoshi Fukutake | Shimizu Corporation |
| Andreas Gavras | University of California at Davis |
| Alborz Ghofrani | University of Washington |
| Nithyagopal Goswami | Rensselaer Polytechnic Institute |
| Jeong-Gon Ha | KAIST (Korea Advanced Institute of Science and Technology) |
| Stuart Haigh | Cambridge University |
| Timothy Haynes | University of California at Davis |
| Gabby Hernandez | University of California at Davis |
| Francisco Humire | University of California at Davis |
| Susumu Iai | FLIP Consortium |
| Koji Ichii | Kansai University |
| Boris Jeremic | University of California at Davis and LBNL |
| Ali Khosravi | University of California at Davis |
| Mohammad Khosravi | University of California at Davis |
| Dongsoo Kim | KAIST (Korea Advanced Institute of Science and Technology) |
| Seongnam Kim | KAIST (Korea Advanced Institute of Science and Technology) |
| Takatoshi Kiriyama | Shimizu Corporation |
| Evangelia Korre | Rensselaer Polytechnic Institute |
| Steve Kramer | University of Washington |
| Kevin Kuei | University of California at Davis |
| Bruce L. Kutter | University of California at Davis |
| Hoe Ling | Columbia University |
| Kai Liu | Zhejiang University |
| Jorge Macedo | University of California, Berkeley |

(continued)

| | |
|---|---|
| Gopal SP Madabhushi | University of Cambridge |
| Srikanth SC Madabhushi | University of Cambridge |
| Andrew Makdisi | University of Washington |
| Ian Maki | California Division of Safety of Dams |
| Erik Malvick | California Division of Safety of Dams |
| Majid T. Manzari | The George Washington University |
| Alejandro Martinez | University of California at Davis |
| Frank McKenna | University of California, Berkeley |
| Lelio Mejia | Geosyntec Consultants, Inc. |
| Jack Montgomery | Auburn University |
| Tyler Oathes | University of California at Davis |
| Kyle O'Hara | University of California at Davis |
| Mitsu Okamura | Ehime University |
| Osamu Ozutsumi | Meisosha Corporation |
| DongSoon Park | K-water Research Institute |
| Nicholas Paull | University of California at Davis |
| Renmin Pretell | University of California at Davis |
| Adam Price | University of California at Davis |
| Zhijian Qiu | University of California, San Diego |
| Ellen Rathje | University of Texas |
| Inthuorn Sasanakul | University of South Carolina |
| Sumeet Kumar Sinha | University of California at Davis |
| Asri Nurani Sjafruddin | Ehime University |
| Nicholas Stone | University of California at Davis |
| Alex Sturm | University of California at Davis |
| Mahdi Taiebat | The University of British Columbia |
| Jiro Takemura | Tokyo Institute of Technology |
| Liao Ting-Wei | National Central University, Taiwan |
| Tetsuo Tobita | Kansai University |
| Dimitra Tsiaousi | Fugro, Walnut Creek, California |
| Kyohei Ueda | Kyoto University |
| Jose Ugalde | Fugro, Walnut Creek, California |
| Ryosuke Uzuoka | Kyoto University |
| Ruben Rodrigo Vargas Tapia | Kyoto University |
| Toma Wada | Kyoto University |
| Rui Wang | Tsinghua University |
| Hung Wen-Yi | National Central University, Taiwan |
| Dan Wilson | University of California at Davis |
| Tianlong Xu | Georgia Institute of Technology |
| Ming Yang | The University of British Columbia |
| Mourad Zeghal | Rensselaer Polytechnic Institute |
| Barry Zheng | University of California at Davis |
| Yanguo Zhou | Zhejiang University |
| Katerina Ziotopoulou | University of California at Davis |

# Contents

## Part III Numerical Simulation Papers



## Part IV Workshop Essays

# Contributors

**Tarek Abdoun** Department of Civil and Environmental Engineering, Rensselaer Polytechnic Institute, Troy, NY, USA

**Pedro Arduino** Department of Civil and Environmental Engineering, University of Washington, Seattle, WA, USA

**Arul K. Arulmoli** Earth Mechanics, Inc., Fountain Valley, CA, USA

**Philippe Audrain** IFSTTAR, GERS, GMG, Bouguenais, France

**Andres R. Barrero** Department of Civil Engineering, University of British Columbia, Vancouver, BC, Canada

**R. Beber** Department of Civil, Environmental and Mechanical Engineering, University of Trento, Trento, Italy

**Emilio Bilotta** Department of Civil, Architectural and Environmental Engineering, University of Napoli Federico II, Naples, Italy

**Masoud Hajialilue Bonab** Department of Civil Engineering, University of Tabriz, Tabriz, Iran

**Trevor J. Carey** Department of Civil and Environmental Engineering, University of California, Davis, CA, USA

**Long Chen** Department of Civil and Environmental Engineering, University of Washington, Seattle, WA, USA

**Renren Chen** Department of Hydraulic Engineering, Tsinghua University, Beijing, China

**Yun-Min Chen** Department of Civil Engineering, Zhejiang University, Hangzhou, China

**Zhao Cheng** Itasca Consulting Group, Inc., Minneapolis, MN, USA

**Anna Chiaradonna** Department of Civil, Architectural and Environmental Engineering, University of Napoli Federico II, Naples, Italy

**Yannis F. Dafalias** Department of Civil and Environmental Engineering, University of California, Davis, CA, USA
Department of Mechanics, School of Applied Mathematical and Physical Sciences, National Technical University of Athens, Athens, Greece

**A. Dobrisan** Department of Engineering, Cambridge University, Cambridge, UK

**Ahmed Elgamal** Department of Structural Engineering, Cambridge University, La Jolla, CA, USA

**Sandra Escoffier** IFSTTAR, GERS, SV, Bouguenais, France

**Gianluca Fasano** Department of Civil, Architectural and Environmental Engineering, University of Napoli Federico II, Naples, Italy

**Pengcheng Fu** Atmospheric, Earth, and Energy Division, Lawrence Livermore National Laboratory, Livermore, CA, USA

**William Fuentes** Universidad del Norte, Barranquilla, Colombia

**Kiyoshi Fukutake** Institute of Technology, Shimizu Corporation, Tokyo, Japan

**Andreas Gavras** Department of Civil and Environmental Engineering, University of California, Davis, CA, USA

**Alborz Ghofrani** Department of Civil and Environmental Engineering, University of Washington, Seattle, WA, USA

**Mohamed El Ghoraiby** The George Washington University, Washington, DC, USA

**Nithyagopal Goswami** Department of Civil and Environmental Engineering, Rensselaer Polytechnic Institute, Troy, NY, USA

**Jeong-Gon Ha** Korea Advanced Institute of Science and Technology, Daejeon, South Korea

**Stuart K. Haigh** Department of Engineering, Cambridge University, Cambridge, UK

**Jin-Shu Huang** Institute of Geotechnical Engineering, Zhejiang University, Hangzhou, China

**Wen-Yi Hung** Department of Civil Engineering, National Central University, Jhongli City, Taoyuan, Taiwan

**Junichi Hyodo** Tokyo Electric Power Service Co, Tokyo, Japan

**Susumu Iai** FLIP Consortium, Kyoto, Japan

**Koji Ichii** Faculty of Societal Safety Science, Kansai University (Formerly at Hiroshima University), Osaka, Japan

**Dong-Soo Kim** Department of Civil and Environmental Engineering, Korea Advanced Institute of Science and Technology, Daejeon, South Korea

**Nam Ryong Kim** K-water Research Institute, Korea Water Resources Corporation, Daejeon, South Korea

**Seong-Nam Kim** Water Management Department, Korea Advanced Institute of Science and Technology, Daejeon, South Korea

**Takatoshi Kiriyama** Institute of Technology, Shimizu Corporation, Tokyo, Japan

**Evangelia Korre** Department of Civil and Environmental Engineering, Rensselaer Polytechnic Institute, Troy, NY, USA

**Bruce L. Kutter** Department of Civil and Environmental Engineering, University of California, Davis, CA, USA

**Carlos Lascarro** University of British Columbia, Vancouver, BC, Canada

**Moon-Gyo Lee** Earthquake Research Center, Korea Institute of Geoscience and Mineral Resources, Daejeon, Republic of Korea

**Ting-Wei Liao** Department of Civil Engineering, National Central University, Taoyuan, Taiwan

**Kai Liu** Institute of Geotechnical Engineering, Zhejiang University, Hangzhou, China

**Gopal S. P. Madabhushi** Department of Engineering, Cambridge University, Cambridge, UK

**Srikanth S. C. Madabhushi** Department of Engineering, Cambridge University, Cambridge, UK

**Majid T. Manzari** Department of Civil and Environmental Engineering, George Washington University, Washington, DC, USA

**Lelio H. Mejia** Geosyntec Consultants, Oakland, CA, USA

**Di Meng** Institute of Geotechnical Engineering, Zhejiang University, Hangzhou, China

**Vicente Mercado** University of British Columbia, Vancouver, BC, Canada

**Jack Montgomery** Department of Civil Engineering, Auburn University, Auburn, AL, USA

**Mitsu Okamura** Department of Civil Engineering, Ehime University, Matsuyama, Japan

**Naoki Orai** Chuden Engineering Consultants, Hiroshima, Japan

**Osamu Ozutsumi** Meisosha Corporation, Tokyo, Japan

**Hanna Park** Department of Civil and Environmental Engineering, George Washington University, Washington, DC, USA

**Zhijian Qiu** Department of Structural Engineering, University of California, San Diego, La Jolla, CA, USA

**Inthuorn Sasanakul** Department of Civil and Environmental Engineering, University of South Carolina, Columbia, SC, USA

**Yu She** Institute of Geotechnical Engineering, Zhejiang University, Hangzhou, China

**Asri Nurani Sjafruddin** Department of Civil Engineering, Ehime University, Matsuyama, Japan

**Nicholas Stone** Department of Civil and Environmental Engineering, University of California, Davis, CA, USA

**Mahdi Taiebat** Department of Civil Engineering, University of British Columbia, Vancouver, BC, Canada

**Tetsuo Tobita** Department of Civil Engineering, Kansai University, Osaka, Japan

**Thaleia Travasarou** Fugro, Walnut Creek, CA, USA

**Dimitra Tsiaousi** Fugro, Walnut Creek, CA, USA

**Kyohei Ueda** Disaster Prevention Research Institute, Kyoto University, Kyoto, Japan

**Kazuaki Uemura** Core Laboratory, Oyo Corporation, Saitama, Japan

**Jose Ugalde** Fugro, Walnut Creek, CA, USA

**Ruben R. Vargas** Disaster Prevention Research Institute, Kyoto University, Kyoto, Japan

**Toma Wada** Department of Civil and Earth Resources Engineering, Kyoto University, Kyoto, Japan

**Rui Wang** Department of Hydraulic Engineering, Tsinghua University, Beijing, China

**Peng Xia** Institute of Geotechnical Engineering, Zhejiang University, Hangzhou, China

**Ming Yang** Department of Civil Engineering, University of British Columbia, Vancouver, BC, Canada

**Gang Yao** Institute of Geotechnical Engineering, Zhejiang University, Hangzhou, China

**Hikaru Yatsugi** Faculty of Environmental and Urban Engineering, Kansai University, Osaka, Japan

**Mourad Zeghal** Department of Civil and Environmental Engineering, Rensselaer Polytechnic Institute, Troy, NY, USA

**Jian-Min Zhang** Department of Hydraulic Engineering, Tsinghua University, Beijing, China

**Bao Li Zheng** Department of Civil and Environmental Engineering, University of California, Davis, CA, USA

**Yan-Guo Zhou** Department of Civil Engineering, Zhejiang University, Hangzhou, China

**Katerina Ziotopoulou** Department of Civil and Environmental Engineering, University of California, Davis, CA, USA

# Part I
# Overview Papers

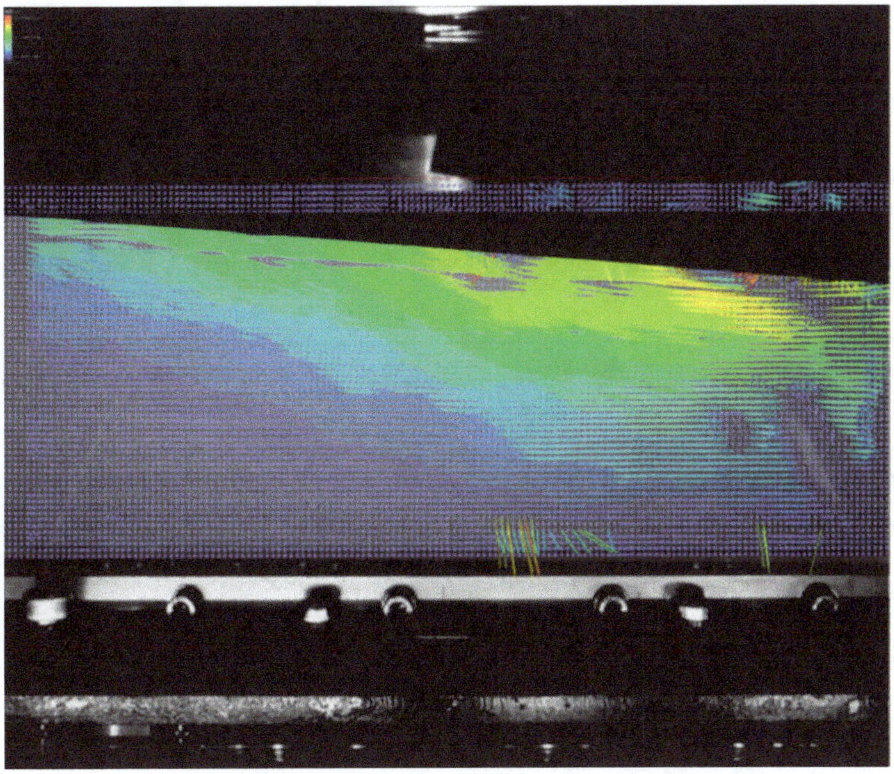

Image credit M. Okamura and A. N. Sjafruddin

# Chapter 1
# LEAP-UCD-2017 V. 1.01 Model Specifications

**Bruce L. Kutter, Trevor J. Carey, Nicholas Stone, Masoud Hajialilue Bonab, Majid T. Manzari, Mourad Zeghal, Sandra Escoffier, Stuart K. Haigh, Gopal S. P. Madabhushi, Wen-Yi Hung, Dong-Soo Kim, Nam Ryong Kim, Mitsu Okamura, Tetsuo Tobita, Kyohei Ueda, and Yan-Guo Zhou**

B. L. Kutter (✉) · T. J. Carey · N. Stone
Department of Civil and Environmental Engineering, University of California, Davis, CA, USA
e-mail: blkutter@ucdavis.edu

M. H. Bonab
Department of Civil Engineering, University of Tabriz, Tabriz, Iran

M. T. Manzari
Department of Civil and Environmental Engineering, George Washington University, Washington, DC, USA

M. Zeghal
Department of Civil and Environmental Engineering, Rensselaer Polytechnic Institute, Troy, NY, USA

S. Escoffier
IFSTTAR, GERS, SV, Bouguenais, France

S. K. Haigh · G. S. P. Madabhushi
Department of Engineering, Cambridge University, Cambridge, UK

W.-Y. Hung
Department of Civil Engineering, National Central University, Jhongli City, Taoyuan, Taiwan

D.-S. Kim
Department of Civil and Environmental Engineering, Korea Advanced Institute of Science and Technology, Yuseong, South Korea

N. R. Kim
K-water Research Institute, Korea Water Resources Corporation, Daejeon, South Korea

M. Okamura
Department of Civil Engineering, Ehime University, Matsuyama, Japan

T. Tobita
Department of Civil Engineering, Kansai University, Osaka, Japan

K. Ueda
Disaster Prevention Research Institute, Kyoto University, Kyoto, Japan

Y.-G. Zhou
Department of Civil Engineering, Zhejiang University, Hangzhou, China



B. Kutter et al. (eds.), *Model Tests and Numerical Simulations of Liquefaction and Lateral Spreading*, https://doi.org/10.1007/978-3-030-22818-7_1

**Abstract** This paper describes the specifications developed by and distributed to all of the centrifuge test facilities involved in LEAP-UCD-2017. The specified experiment consisted of a submerged medium dense clean sand with a 5-degree slope subjected to 1 Hz ramped sine wave base motion in a rigid container. This document describes the detailed geometry, sensor locations, methods of preparation, quality control, shaking motions, surface markers, and surface survey techniques.

## 1.1 Introduction

### *1.1.1 Differences Between This Paper and Pre-test Specifications*

This paper documents the specifications for LEAP-UCD-2017 centrifuge model tests as they existed just prior to the testing for LEAP-UCD-2017. The primary goal of the specifications was to improve the specimen preparation, the testing procedures, and the accuracy of the data and in addition to provide data to quantify the uncertainties associated with the experiments. The specifications were first drafted for the LEAP-GWU-2015 project described by Kutter et al. (2017). Based upon experience in 2014–2015, the specifications were updated and improved for the LEAP-UCD-2017 exercise.

While some of the data in this paper (e.g., the maximum and minimum densities) were subsequently superseded, it was decided to maintain this document as it was prior to centrifuge testing as a record of the specifications. The remainder of this paper is therefore nearly a verbatim copy the specifications developed prior to the centrifuge testing.

### *1.1.2 Goals and Overview*

The goals of this LEAP are to perform a sufficient number of experiments to characterize the median response and the uncertainty of the median response of a specific sloping deposit of sand to a specified ground motion. To put the uncertainty in context, it was considered critical to also determine the sensitivity of response to relative density, the sensitivity of the response to the ground motion intensity, and the sensitivity of the response to unspecified components of the ground motion that are superimposed on the specified ground motion.

The specific median soil deposit to be tested at each centrifuge facility is a 4-m-deep, 20-m-long deposit of Ottawa F-65 sand with a dry density of about 1650 kg/m$^3$ and a ground slope of 5°. The specified median ground motion is a ramped sine wave input motion similar to the target motion for LEAP-GWU-2015. The primary response quantity of interest is the displacement and deformed shape of the soil deposit. Important secondary response quantities include time series data from acceleration, pore pressure, and displacement sensors.

## 1.2 Scaling Laws

The length scale factor, $L^*$, is defined as $L^* = L_{model}/L_{prototype}$. According to the conventional centrifuge scaling laws, gravity will be scaled according to $g^* = 1/L^*$. The gravity in the model $g_{model} = \omega^2 R_{ref}$, where $R_{ref}$ is measured at 1/3 the depth of sand in the middle of the plan area of the soil deposit. Viscous pore fluid will be used in all the experiments and the viscosity should be scaled according to $\mu = \mu_{water}/L^*$. Scaling laws are used in accordance with recommendations by Garnier et al. (2007).

## 1.3 Description of the Model Construction and Instrumentation

### *1.3.1 Soil Material: Ottawa F-65 Sand*

Ottawa F-65 sand was chosen as the standard sand for LEAP-GWU-2015 and LEAP-UCD-2017. Ottawa F-65 sand is a clean (less than 0.5% fines), sub-rounded to sub-angular whole grain silica sand, provided by US Silica, in Ottawa, Illinois. Specific gravity and grain size parameters determined during LEAP-GWU-2015 are summarized in Table 1.1; results from maximum and minimum dry density tests are given in Table 1.2. Additional material properties of Ottawa F-65 sand, including triaxial, simple shear, and permeability test data, may be found in the LEAP Soil Properties and Element Test Database (Carey et al. 2017).

It appears from standard deviations given in Table 1.1 that the grain sizes of the tested sand were very consistent. (It should be noted that many sites used 0.25 and 0.125 mm sieve sizes without intermediate sieve sizes, so the values of $D_{10}$–$D_{60}$ were interpolated between the percentages retained on those two sieves; this interpolation is estimated to produce errors on the order of 0.01 mm for $D_{30}$ and $D_{50}$.)

In late 2016, however, a different batch of sand delivered from US Silica to UC Davis had quite different grain sizes; $D_{50}$ was approximately 0.28 mm and some variability was noticed from bag to bag. The 2016 batch of material was considered

**Table 1.1** Specific gravity and grain size characteristics of Ottawa F-65 sand used in LEAP-GWU-2017 (Kutter et al. 2017)

| | Num of Tests | Average | Stand. Dev. |
|---|---|---|---|
| $G_s$ | 4 | 2.665 | 0.012 |
| $D_{10}$ (mm) | 5 | 0.133 | 0.005 |
| $D_{30}$ (mm) | 5 | 0.173 | 0.009 |
| $D_{50}$ (mm) | 5 | 0.203 | 0.013 |
| $D_{60}$ (mm) | 5 | 0.215 | 0.016 |
| Fines (%) | 5 | 0.154 | 0.133 |

**Table 1.2** Maximum and minimum dry density for Ottawa F-65 sand (Kutter et al. 2017)

| Data source | Test method | Min. density (kg/m$^3$) | Max. density (kg/m$^3$) |
|---|---|---|---|
| Cooper Labs (UCD) | ASTM D4254 and D4253 | 1515 | 1736 |
| GeoTesting Express (RPI) | ASTM D4254 and D4253 | 1494 | 1758 |
| Andrew Vasco (GWU) | ASTM D4254 and D4253 | 1538 | 1793 |
| Andrew Vasco (GWU) | Lade et al. (1998)(using graduated cylinder) | 1521 | 1774 |
| Cerna Alvarez (UCD) | Lade et al. (1998) (using graduated cylinder) | 1415 | 1720 |
| Cerna Alvarez (UCD) | Modified ASTM D4254(a) | 1406 | |
| Parra Bastidas (UCD) | ASTM D4254, JIS A 1224 | 1455 | 1759 |
| Wen-Yi Hung (NCU) | — | 1482 | 1781 |
| Yan-Guo Zhou (ZJU) | DL/T5355-2006[a] | 1456 | 1733 |
| Average of tests | | 1475 | 1756 |
| Stand. dev. of all methods | | 46 | 25 |

Please see Carey et al. paper in this proceedings describing more recent index density tests
[a]Chinese "Code for soil tests for hydropower and water conservancy engineering," 2006

to be non-satisfactory for LEAP. All of the sand shipped from Davis in early 2017 to LEAP participants was therefore taken from the original batch of sand shipped to UC Davis in 2014. Unfortunately, the inconsistency in the later delivery has raised concerns about the quality control of the sand in the future. Therefore, it is important for each LEAP site to perform quality control checks on the grain characteristics of the Ottawa F-65 sand. A sufficient number of bags to prepare an entire model should be mixed together first, and then the grain size analysis and maximum and minimum tests should be conducted on this mixture. If the grain size is significantly different from that in Table 1.1, researchers should scalp the material so that $D_{50}$ and $D_{10}$ conform to be within one standard deviation of the data reported in Table 1.1. In addition, at least one maximum and minimum density test, as described below, should be performed on the mixed material used for each centrifuge model.

The sand placement is to be prescribed based on a target density (as opposed to relative density) to avoid the small conversion error associated with uncertainty in specific gravity and larger uncertainty associated with maximum and minimum densities. From Table 1.3, it can be seen that considerable scatter is observed in the values of maximum and minimum density. The two experiments performed by professional laboratories in the USA according to ASTM D4253 and D4254 were quite consistent with each other. But use of a 1000 ml graduated cylinder (modified Lade et al. approach) typically produced significantly lower minimum densities.

$$D_r = \frac{\rho_{d\max}(\rho_d - \rho_{d\min})}{\rho_d(\rho_{d\max} - \rho_{d\min})} \tag{1.1}$$

**Table 1.3** Grain size parameters of Ottawa F-65 sand published by US Silica

| USA STD SIEVE SIZE | | TYPICAL VALUES | | |
|---|---|---|---|---|
| | | % RETAINED | | % PASSING |
| MESH | MILLIMETERS | INDIVIDUAL | CUMULATIVE | CUMULATIVE |
| 30 | 0.600 | 0.0 | 0.0 | 100.0 |
| 40 | 0.425 | 0.1 | 0.1 | 99.9 |
| 50 | 0.300 | 4.0 | 4.1 | 95.9 |
| 70 | 0.212 | 37.0 | 41.1 | 58.9 |
| 100 | 0.150 | 45.8 | 86.9 | 13.1 |
| 140 | 0.106 | 11.4 | 98.3 | 1.7 |
| 200 | 0.075 | 1.6 | 99.9 | 0.1 |
| 270 | 0.053 | 0.1 | 100.0 | 0.0 |

To facilitate frequent quality control, a simple and consistent method of measuring maximum and minimum densities is required. Researchers are requested to measure maximum and minimum densities at least once for each model test. The methods described below should be used (modified from ASTM procedures). In addition, researchers may decide to measure the index densities using common standard procedures used in their country. A new section of the data reporting template for reporting the results of the grain size and maximum/minimum densities will be developed soon.

Independent measurement of any of the soil properties should be reported using a spreadsheet with format consistent with the formats of existing soil properties in the LEAP Soil Properties and Element Test Database (https://nees.org/resources/13689/).

## Modified ASTM D4254 Method C for Minimum Dry Density

A glass graduated cylinder of 1000 ml will be used for measurement of the minimum density. The humidity of the "dry" sand source should be measured by burying a humidity sensor into the sand until the reading stabilizes. The humidity of the room should also be measured. About 500 g of dry sand should be carefully weighed and placed inside a 1000 ml graduated cylinder. The top of the graduated cylinder should be covered with a sheet of latex and held by hand to seal the top of the cylinder. The sample should then be turned upside down and steadily rotated upright within about 60 s. The volume of the loose sample can then read from the graduated cylinder. Data to be recorded include date, researcher, mass of sand, volume of loose sand, humidity of the laboratory, humidity of the sand source, temperature of the room, and the minimum dry density. The calibration of the volume marks on the side of the graduated cylinder should be checked by weighing it with a known volume of water and assuming the density of water is 998 kg/m$^3$.

**Modified Lade et al. (1998) Method for Maximum Density**

Place approximately 500 g of soil in approximately 50 g increments into a plastic 1000 ml graduated cylinder. After placing each 50 g increment, the side of the graduated cylinder should be firmly tapped eight times (two times each on the north, south, east, and west sides of the cylinder) with the plastic handle of a screwdriver. The mass of the screwdriver should be approximately 140 g and total length of about 250 mm. To consistently apply the firm taps, the operator should hold the screwdriver by the metal part, and the plastic handle should be about 250–300 mm away from the cylinder between taps to produce consistency. After the last 50 g increment of soil is placed and tapped, each of four sides of the graduated cylinder should be lightly tapped six times (24 total) up and down the sample. To level the top surface for purposes of accurate reading, five or ten very light taps are made while the cylinder is tilted. The volume of the sand may be read from the graduated cylinder and the maximum density calculated. Data to be recorded include date, researcher, mass of sand, volume of loose sand, humidity of the laboratory, humidity of the sand source, temperature of the room, and the maximum density. A video of the recommended procedure for checking maximum and minimum density is posted in the General Report for the data archive for LEAP-UCD-2017 (https://doi.org/10.17603/DS2N10S).

## *1.3.2 Placement of the Sand by Pluviation*

The sand should be pluviated through a screen with opening size approximately 1.20 mm. The screen will be partially blocked to restrict the flow (see Fig. 1.1). Three arrangements with different parts of the screen masked off should be used to achieve different sample densities as indicated in Fig. 1.2. Small modifications to the opening patterns are allowable if necessary. In the tests used to create Fig. 1.3, 100-mm-long slots in a US standard sieve were used. If the screens need to be placed inside the model container during pluviation, a non-circular sieve may be fabricated to facilitate pouring near the edges and corner of the container. The 1.2 mm screen may be cut from a standard sieve, and the screen may be placed in a rectangular-shaped custom sieve. To avoid cutting a good sieve, the screen may be available from rolls from soil mechanics laboratory suppliers. The slot lengths in a rectangular screen should be greater than or equal to 100 mm.

If the screen or the geometry of the open parts of the screen is significantly modified, calibration curve (density as a function of drop height), mass flow rate during deposition, as well as the dimensions of the container used to perform the calibration will be reported in the data report.

The elevation of the screen above the container during pluviation should be continually adjusted to maintain a constant vertical spacing between the sand surface and the screen; the drop height should not change by more than 5% during

**Fig. 1.1** (**a**) Standard sieve used for calibration tests in this specification. (**b**) Side view of sand falling for 7.1 mm slots spaced at 17 mm. (**c**) Duct tape used to mask off slots for pluviation; the 7.1 mm slots correspond to three mesh widths of the standard sieve. (**d**) Ad hoc arrangement of sieve attached to hopper used to feed sand to the sieve during pluviation

deposition. The sieve should be steadily and continuously moved (by robot or by hand) to avoid local mounds with side slopes that affect the deposition. Some tapping of the sieve is necessary if/when the sieves become clogged.

Calibration curves for the above three screen masking patterns are presented in Fig. 1.3. If researchers notice a significant difference in the calibration curve, they should report their observed calibration curve (density vs drop height).

It is suspected that humidity of the sand and electrostatic forces developed during repeated handling of the sand could affect the results. To investigate these issues, a sideview photograph of the sand flowing should be included in the sample preparation report. A humidity sensor should be embedded into the sand after deposition to measure the temperature and humidity of the sand after placement. The humidity of the room should also be measured. Suitable humidity sensors, for example, are

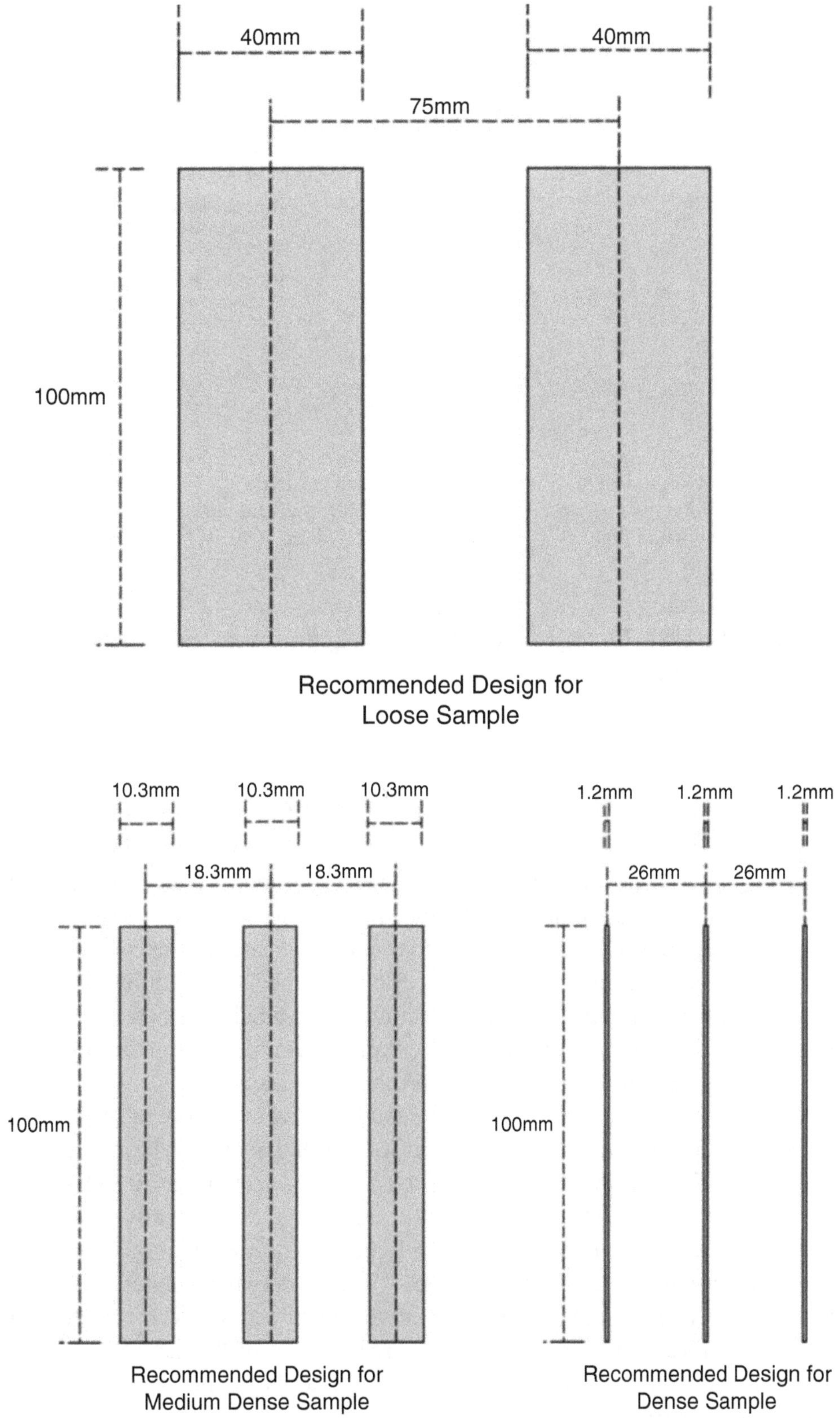


**Fig. 1.2** Recommended opening dimensions to be masked off over a mesh of 1.18–1.22 mm opening to achieve different densities of samples by dry pluviation. The mass flow rates for these designs were 3.6, 2.5, and 1 kg/min with flow rates increasing as slot width increases

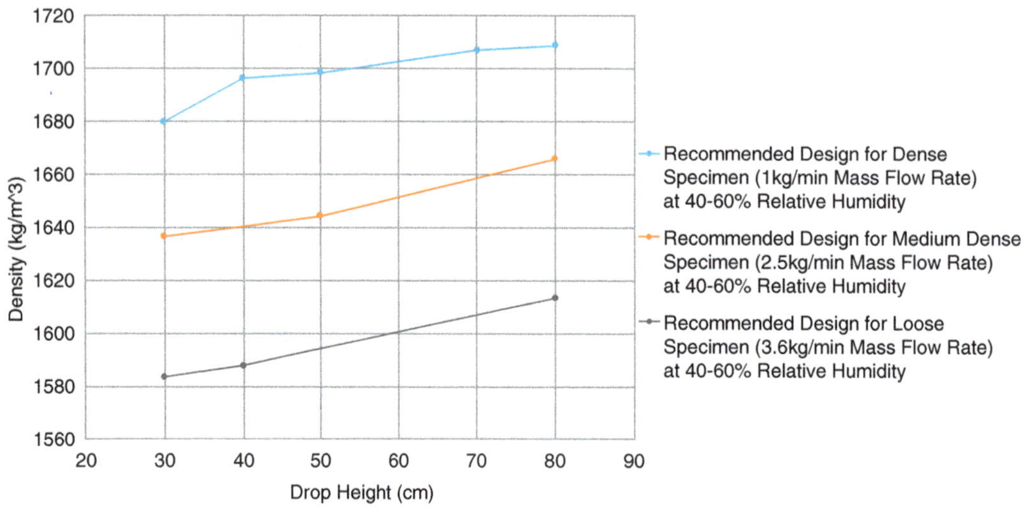


**Fig. 1.3** Density vs drop height (from tests at UCD in Jan to March 2017) for three recommended sieve designs tested at 40–60% relative humidity. The blue, red, and black lines correspond to the slot dimensions shown in Fig. 1.2

available in the USA for approximately $75 (http://www.testequipmentdepot.com/extech/pdf/rh300.pdf).

### 1.3.3 Measurement of Density of the Sand

To measure the as-deposited density of the sand in the model container, an accurate method of measuring the container dimensions and the volume of the sand in the container is required. Container dimensions may be measured accurately using a rigid steel ruler and then the volume of the container checked by filling the container with water, covering it with a flat plate, eliminating air bubbles, and measuring the mass of water that it holds. Confirm that the volumes computed from dimensions and water mass are consistent; check and report their repeatability.

A suggested method for measuring the sand height is to smooth the surface of the sand using a 25-mm-thick, approximately 70-mm-diameter, acrylic plastic disk as shown in Fig. 1.4 followed by placement of about 15 or more 20-mm-diameter, 6-mm-thick PVC or HDPE disks gently on the surface as uniformly spaced representative grid locations. The disks may be lightly twisted without pushing down to make their base rest evenly on the sand. Using a Vernier caliper, measure the location of the disks relative to a stiff smooth beam across the top of the model container as indicated in Fig. 1.4c. The average height of the sand must be determined to an accuracy better than 0.5%. Superior tools for measuring volume are acceptable if the process and accuracy are documented.

The mass of the sand should also be measured to better than 0.5%. The accuracy and the repeatability of the measuring method should be verified.

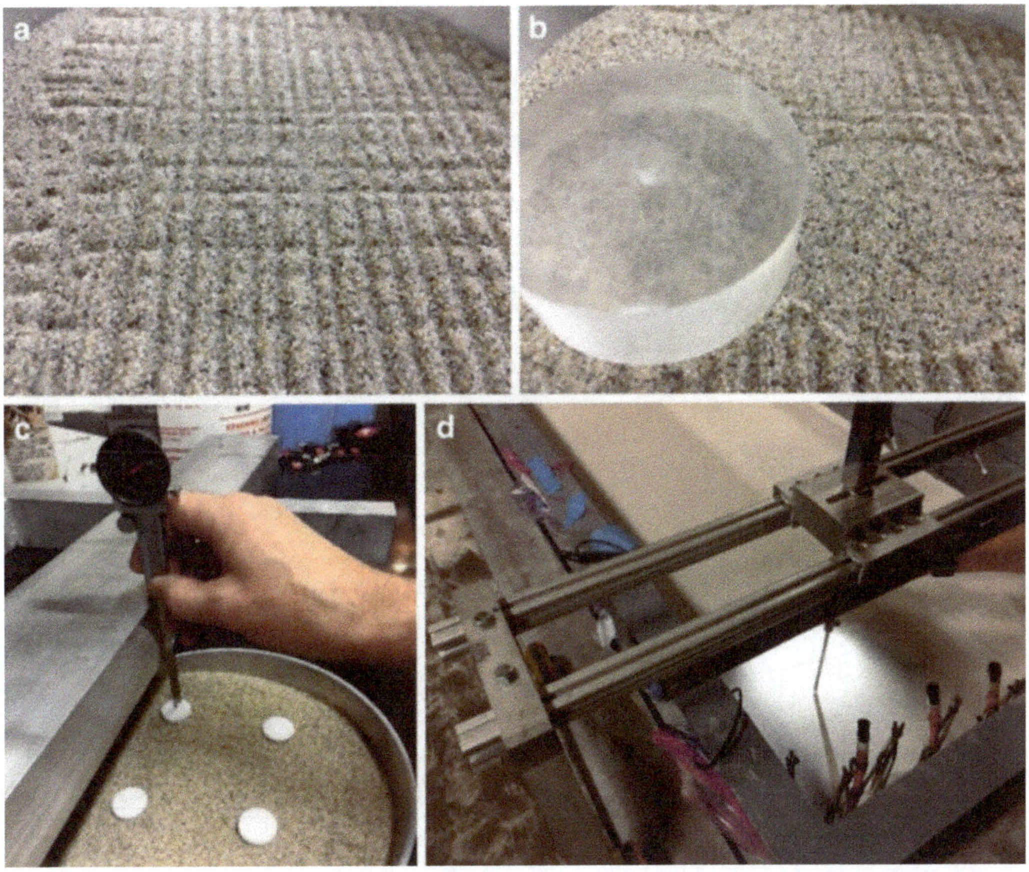

**Fig. 1.4** Illustration of recommended method to measure the volume of sand specimen. (**a**) Rough surface of sand after shaping with vacuum. (**b**) 25 × 70 mm acrylic disk lightly set down on surface, slightly twisted as necessary to smooth the surface. (**c**) 5-mm-thick × 18-mm-diameter PVC or HDPE plastic disks gently placed on the smoothed sand in a grid. X, Y, and Z coordinates of the disk to be measured using caliper. (**d**) Measuring rails used on the large centrifuge at UCD. Good quality of light with a distinct shadow aids in resolving the location of the surface

### *1.3.4 Geometry of the Model*

The required shape of soil surface depends on direction of shaking relative to the curved g-field as shown in Fig. 1.5. In addition, if the centrifuge bucket is not freely swinging (e.g., for the Cambridge centrifuge where the bucket swing seats against stops), then the surface of the sand should be modified to eliminate unintended slope angles in any direction.

### *1.3.5 Saturation of the Model*

To facilitate dissolution of gas bubbles that are trapped in the sand, the model container will be repeatedly evacuated and flooded with $CO_2$ to replace 98% of

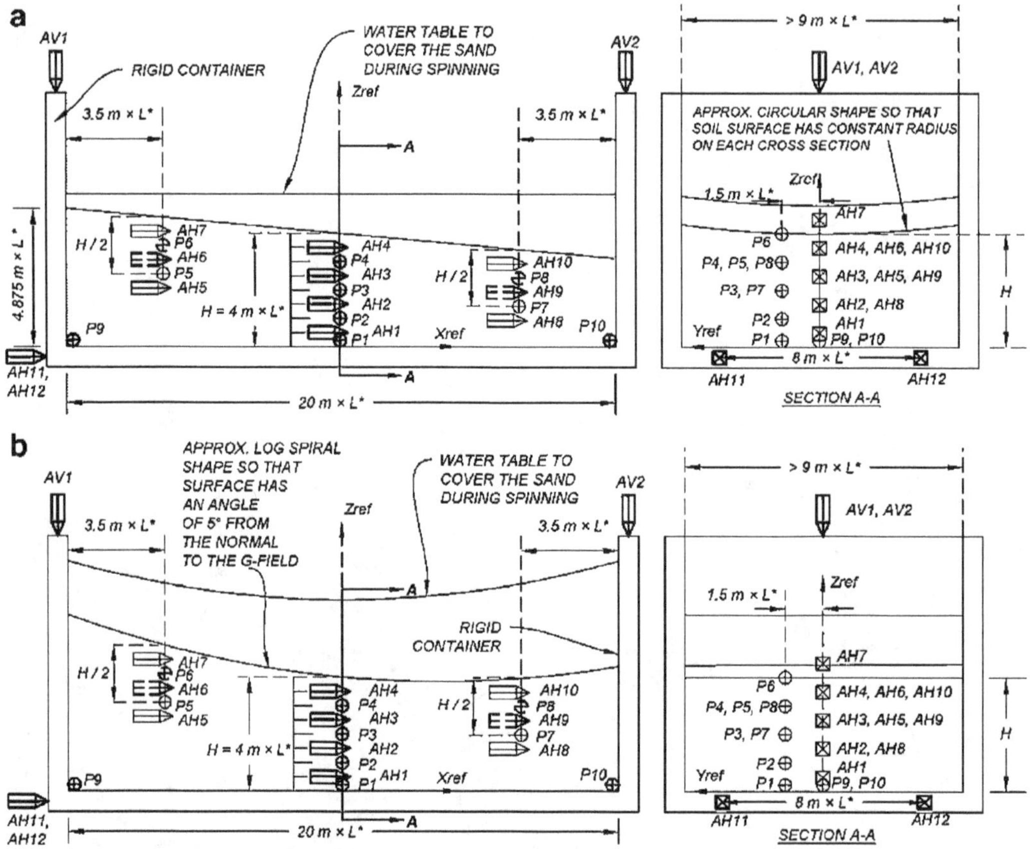


**Fig. 1.5** (**a**) Baseline schematic for LEAP-UCD-2017 experiment for shaking parallel to the axis of the centrifuge. Bold sensors are required. Bold-dashed sensors are highly recommended. Thin-lined sensors are recommended. (**b**) Baseline schematic for LEAP-UCD-2017 experiment for shaking in the circumferential direction of the centrifuge. Bold sensors are required. Bold-dashed sensors are highly recommended. Thin-lined sensors are recommended

the air in the chamber by pure $CO_2$ prior to saturation with de-aired viscous fluid (Kutter 2013). If 90% vacuum is applied, then the process must be repeated at least twice to remove more than 98% of the air. If 50% vacuum is applied, then the vacuum and $CO_2$ flooding should be repeated at least five times. After 98% replacement of air by $CO_2$, de-aired viscous pore fluid should be dripped into the low end of the model slope followed by infiltration with de-aired water while under vacuum as shown in Fig. 1.6. The de-aired viscous fluid may be prepared by letting it sit under a vacuum of at least 80 kPa (absolute pressure is 21 kPa or less). The vacuum should be continuously applied to the water supply reservoir while it is being introduced to the specimen. If this is not possible, steps should be taken to prevent gas from re-dissolving in the fluid prior to infiltration. Flow of viscous fluid to the model from the reservoir may be driven by gravity feed or peristaltic pump.

Documentation that the method of saturation is successful either by the measurement of p-wave velocity or by measuring the raising and lowering of the water level due to applying vacuum to the entire container (as described by Okamura and Inoue 2012) is required.

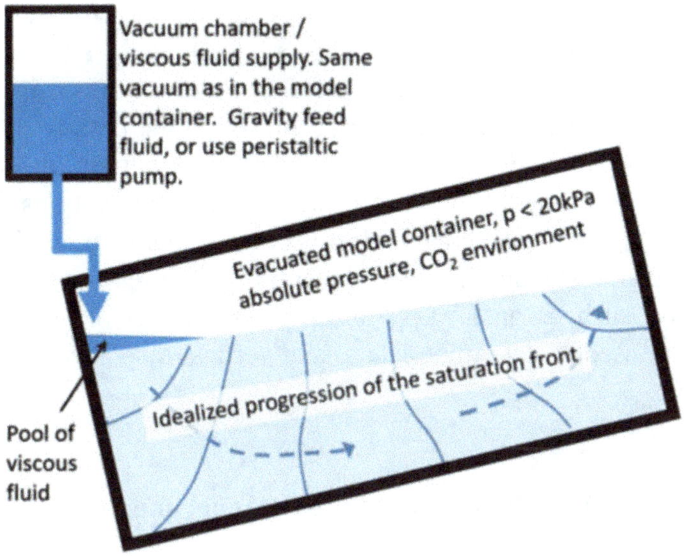


**Fig. 1.6** Saturation by dripping de-aired viscous fluid under vacuum into the container. Once a pool has formed at the low end of the soil surface, the pool should be continuously maintained with some free pore fluid. The elevation of the pool may gradually be increased to cover the outcrop of the saturation front. A sponge may be placed on the surface of the sand to protect the sand from the impact of fluid dropping from the vacuum chamber. The tilting of the box by a few degrees is suggested to help reduce likelihood of trapping an air pocket in the bottom corner of the box

## 1.4 Instrumentation of the Model

### *1.4.1 Required Instrumentation*

The required instrumentation is very similar to that for LEAP-GWU-2015 as indicated in Fig. 1.5a, b. AH1–AH4, AH11, AH12, AV1, AV2, P1–4, P9, and P10 are required sensors. AH6, AH9, P6, and P8 are highly recommended. AH5, AH7, AH8, AH10, P5, and P7 are recommended sensors. Bender elements are also recommended be used to monitor the evolution of the shear wave velocity during the centrifuge testing. To the extent possible, cables from the sensors should run in the transverse direction toward the side walls of the model container to minimize the reinforcing effect of the sensor cables. The routing of cables should avoid regions were CPT tests could intersect the cables.

### *1.4.2 Displacement Measurements*

As described by Kutter et al. (2017), it is possible to determine dynamic displacements by integration of accelerometer records. Acceleration records should therefore be reported with as little analog and digital signal processing as possible; if filtering is necessary, the characteristics of the filter must be indicated. The corner frequencies of the data acquisition system should be reported, especially for the accelerometer sensors.

There is clearly a fundamental difficulty associated with measuring residual or permanent displacements from contact sensors founded in submerged liquefied, laterally spreading ground. A concerted effort is required to obtain more accurate displacement measurements than has been achieved in the past.

At least four techniques to be pursued for determining displacements are described below. Experimenters are encouraged to share innovations that will enable improved displacement measurements. If it is not possible to assess lateral displacements before and after shaking from the cameras or pore pressure sensors, then some alternate method such as an LVDT, linear potentiometer, or laser displacement sensor should be used to determine lateral and vertical displacements as accurately as possible.

## Careful Before and After Photographs of the Model and Surface Markers

Prior to spinning the centrifuge, after stopping after the first destructive motion, and after all of the destructive shaking events, photographs must be taken. The photographs should use a camera looking vertically down at the center of the specimen at the same distance and in the same lighting conditions. The location of the camera should be carefully placed above the top center of the model, and good lighting should be used to enable these reference photos to be used to determine the lateral deformation at various stages of the testing. These reference photographs should be taken using the same camera, with the same lens with the same amount of water on top of the specimen and with similar lighting. Make sure that the surface markers are clearly visible in the photographs.

Surface markers can be manufactured by 3D printing. The images in Fig. 1.7 are from an order placed by UCD. The shape of the markers is designed to anchor to the soil and provide minimal restriction to pore pressure drainage, with a taper to facilitate insertion with reduced disturbance. The sand in the middle of the markers may also be colored to improve edge contrast for image analysis of marker locations. The markers were made from a plastic material with specific gravity of approximately 1.5. The file used to produce the markers, Surface_Marker_ImprovedDesign.stl, is in an industry standard format for 3D printing. The file may be obtained by contacting either of the first two authors of this paper. Each site should manufacture their own markers. A permanent marker (e.g., a Sharpie) or paint/stain should be used to improve contrast for photography.

If 3D printing is not possible, surface markers should be made of tapered PVC (polyvinyl chloride) approximately 25 mm outside diameter thick-walled (approx. 2.5 mm wall thickness) common water pipe. It may be tapered using a lathe to approximate drawing in Fig. 1.7. The PVC can be stained with dark paint or primer prior to installation in the model to improve contrast with your sand and lighting conditions. If PVC is not available, the markers should be made from plastic with specific gravity approximately 1.5 to mitigate potential for uplifting out of the liquefying sand (Fig. 1.8).

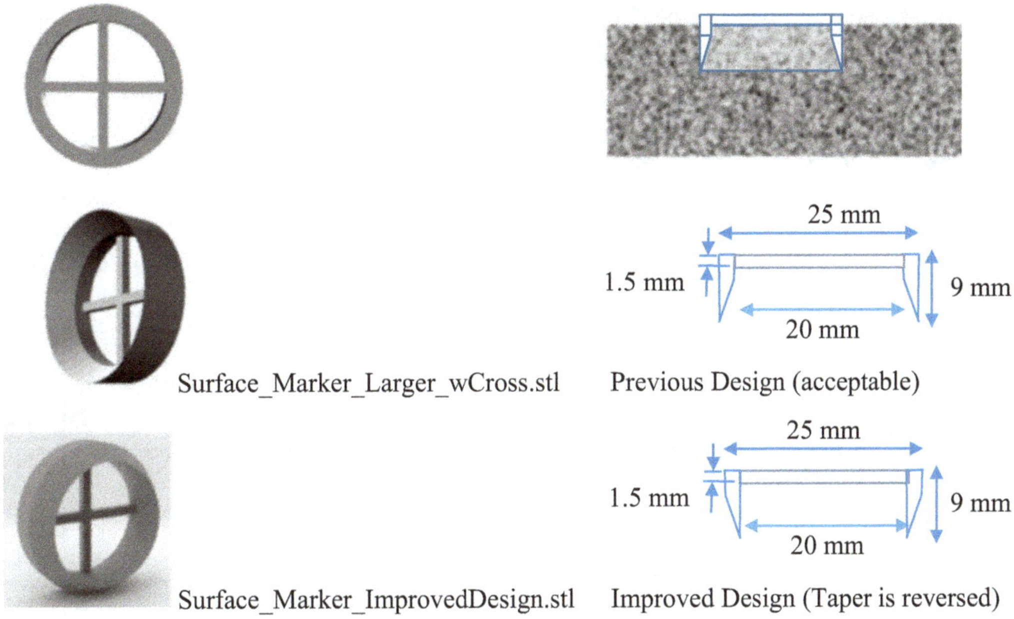


**Fig. 1.7** Unstained 3D printed surface markers. Top right shows installation with 1.5 mm of marker above the soil surface. Dimensions are not critical but are recommended for consistency

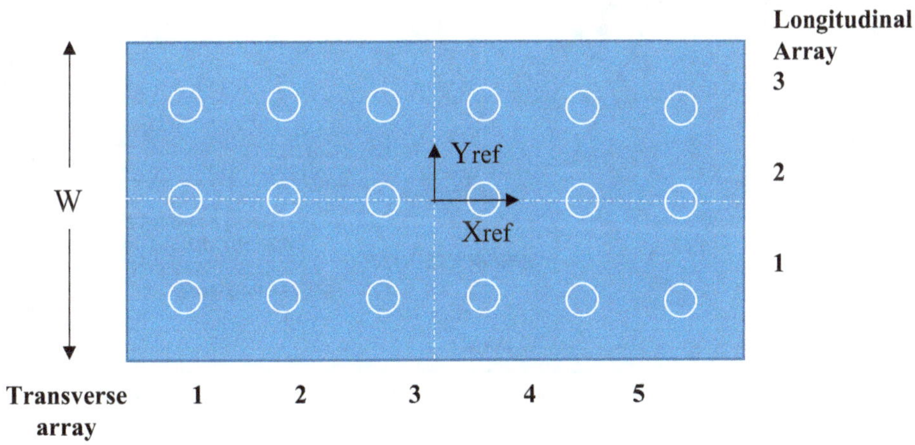


**Fig. 1.8** Eighteen required surface markers at $X_{ref} = (-7.5, -3.5, -1.5, +1.5, +3.5, +7.5)$ m and $Y_{ref} = (0.3\ W, 0, -0.3\ W)$. Coordinate system was defined in Fig. 1.5 with the origin in the middle of the container, with $Z_{ref} = 0$ at the top surface of the base of the container

## Lateral Displacements from Cameras Mounted on the Centrifuge

Cameras (preferably high speed) should be mounted to measure the plan view lateral displacements of surface markers during spinning. Camera with good resolution of horizontal movement should be used. UCD is presently planning to use a thick rigid plastic cover plate on top of the specimen to prevent surface waves. Four GoPro

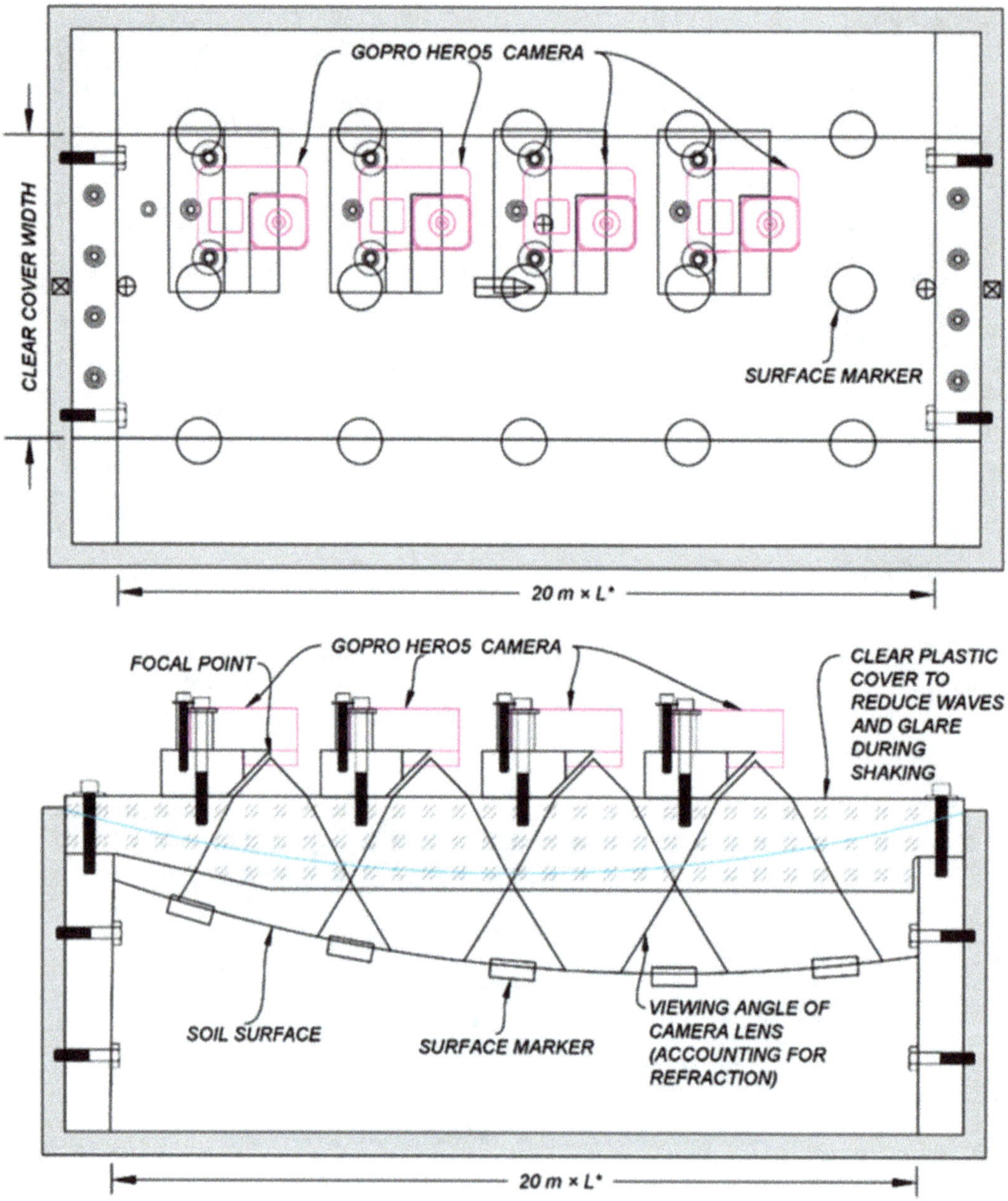


**Fig. 1.9** UCD plan for cameras mounted on top of a thick plastic sheet to eliminate waves in images

cameras (running at 240 frames per second in 700p resolution) will be attached to the top of the cover plate to view different regions of the specimen (see Fig. 1.9). LED lights will also be attached to the plastic cover plate.

### Residual Settlements from Pore Pressure Sensors

Residual pore pressures from pore pressure sensors will be used to determine the settlement of the pore pressure transducers before and after shaking. This requires the sensors to be embedded in the soil (not attached to the model container) and very accurate recording data for a period of time after the pore water pressures are completely dissipated. Three main sources of error have been considered to make this strategy successful: (1) electronic noise, (2) incomplete pore pressure dissipation, and (3) drifting of the pore pressures due to residual tilting of the model container (perhaps due to sporadic overcoming of swing bearing friction during spinup and shaking) or due to changes in the water pressure associated with evaporation, changes in centrifuge g-level, or the degree of saturation.

1. The effect of electric noise on pore pressure sensor records can be dealt with by taking many data points over a period of time longer than the period of the noise. For example, if the noise is 50 or 60 Hz or higher, we could take the Residual Pore Pressure Average (RPPA) reading of data recorded at 2000 data points per second for a full second. This RPPA recording must be repeated before spinning the centrifuge, just before each shaking event, after complete dissipation of pore pressure from each shaking event, and after stopping the centrifuge.
2. As we are trying to resolve very small residual pore pressures to accurately measure residual settlements, it is crucial to ensure complete pore pressure dissipation. A good way to do this is to visually inspect the pore pressure record, find the time $t_{99}$ required for pore pressures to drop to within 1% of the initial absolute pore pressure, and then wait ten times longer than $t_{99}$ before recording the residual pore pressures. This might take about 10 min total between shaking events.
3. Sensors P9 and P10 accurately located in the bottom corner of the model containers should read the same RPPA increase during spinup from 1 g to the test acceleration, and ideally, they should return to the same RPPA after shaking and the same 1-g RPPA after stopping the centrifuge. Reasons for the difference between P9 and P10 include friction in the bucket hinge, the container base not being normal to the resultant g-vector, leakage or evaporation from the model container, or drifting of the centrifuge acceleration. Thus, the data from P9 to P10 sensors are crucial to allow us to compensate for small changes in g-level and water table elevation. Report RPM to an accuracy better than 0.5% before and after shaking. All of the RPPA values for all of the pore pressure sensors should be reported on one worksheet of the template.

Many facilities zeroed out the pore pressure sensors prior to each shaking event in LEAP-GWU-2015. This should not be done because it will defeat our effort to determine sensor settlements from residual pore pressures. If it is necessary to offset the electronic zero of the pore pressure sensors (e.g., for some electronic instrumentation limitation), RPPA values should be recorded before and after each electronic offset.

## Direct Measurements of Sensor and Surface Marker Locations

The X, Y, and Z coordinates of the 15 surface markers shall be surveyed. Record the surface survey data in the spread sheet template showing the X, Y, and Z coordinates of the top center of each marker. The required accuracy of the measurement is 1 mm for horizontal displacements and 0.5 mm for vertical settlements. The markers may be located using photogrammetric techniques, a laser scanner, or a depth gauge with a Vernier caliper that measures the locations relative to a rigid guide frame such as that pictured below. Suggested surface markers to facilitate analysis of lateral movements are specified in section (circular disks described above).

### Colored Sand Layers, Noodles, and Sensor Locations During Dissection

Horizontal colored sand layers will be placed at the elevations of the central array of pore pressure sensors. The elevation of the sand relative to the model container reference coordinate system and relative to the pore pressure sensors should be measured before and after the tests.

Vertical markers will consist of spaghetti noodles placed vertically in the dry sand. The noodles should be placed in the sand after about half of the slope has been placed. An array of noodles along the window and along the longitudinal centerline should be placed before the test. The length of the noodles should be adjusted so they stick out about 1 or 2 mm above the ground surface before saturation. The deformed shape of the noodles should be exposed by excavation and photographed at the end of the shaking events.

### Settlement Gage Sensors

Modified pore pressure transducers may be used to measure the surface settlement of the deposit (if possible at three locations along the surface). The concept of the modified pore pressure sensors is explained in Fig. 1.10. Traditional pore pressure transducers are sealed and connected through a flexible thin tubing to a constant head water reservoir. When the model is in equilibrium, the modified sensor reads the pressure imposed on it by the constant water head. As the soil settles during shaking and the sensor moves with it, the sensor reads an additional pressure that is directly proportional to the soil settlement. These modified were used in RPI test for LEAP-GWU-2015. Details of the sensors development are given in Antonaki et al. (2014). Additional details are given by Kokkali et al. (2018).

### Tactile Pressure Sensors

Tactile pressure sensors may be attached to the container boundaries to measure the transient and residual soil pressure at the soil-container interface. Such sensors were used by RPI for LEAP-GWU-2015. The RPI group used tactile pressure sensors manufactured by Tekscan, Inc. (Cambridge, MA). The sensors need conditioning, equilibration, and calibration. Since the sensors are not waterproof, they were laminated, prepared for the specific soil-container interfaces, and then installed in

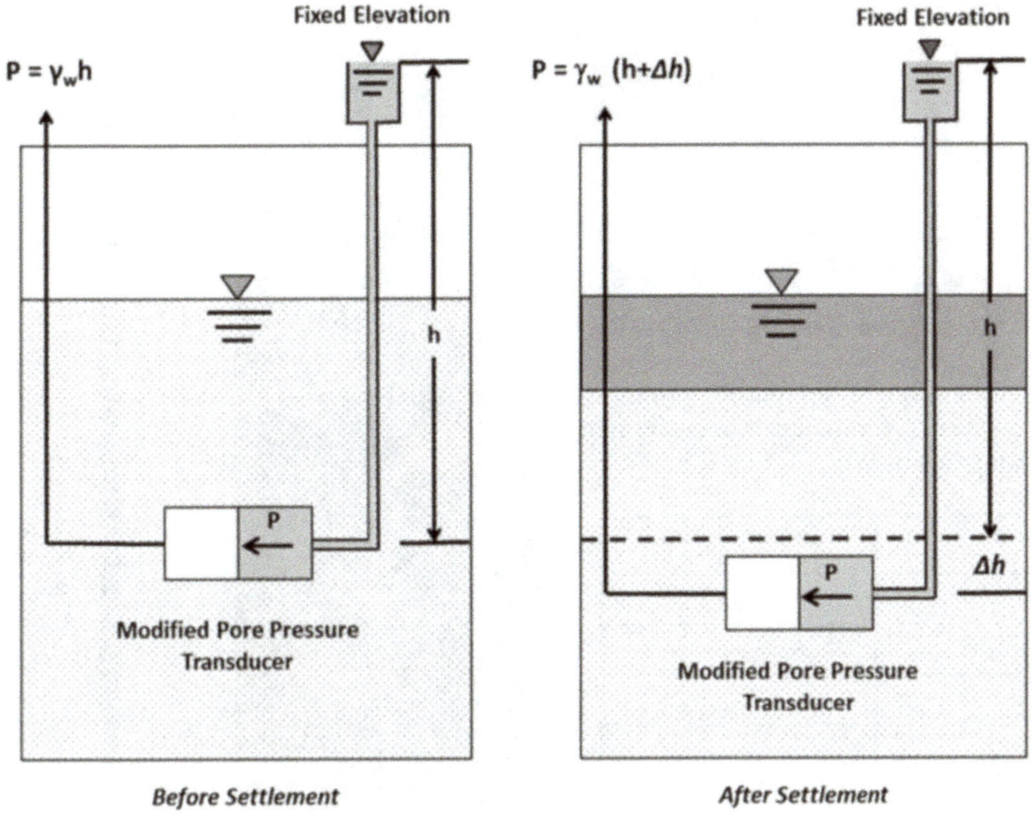


**Fig. 1.10** Concept of the pressure transducer settlement gauges

the model container prior to the model construction (Figs. 3.3.1 and 3.3.2). The preparation procedure is described in detail in El Ganainy et al. (2014) and Kokkali et al. (2018).

## 1.5 Cone Penetration Testing

A new cone has been designed with a top load cell and no custom strain gauges at the cone tip as indicated in Fig. 1.11. The drawings are in a separate file in the shared box.com folder. Estimated cost for fabrication of this cone at UCD is approximately $1800 which includes a $500 load cell.

Cone penetration tests should be performed before and after every destructive ground motion. The rate of cone penetration, in model scale, should be scaled depending on the pore fluid viscosity in the model:

$$V_{\mathrm{cpt,\,modelScale}} = (100\ \mathrm{mm/s})/\mu^*$$

The spacing of the four M6x1mm bolts is 25.4 mm center to center in a square pattern.

| Cone | | Ref. Length (mm) (not counting conical tip) |
|---|---|---|
| 1 | Ehime | 124 |
| 2 | Kwater | 202 |
| 3 | Kaist | 202 |
| 4 | Kyoto | 194 |
| 5 | NCU | 294 |
| 6 | NCU | 294 |
| 7 | Zhejiang | 694 |
| 8 | Zhejiang | 644 |

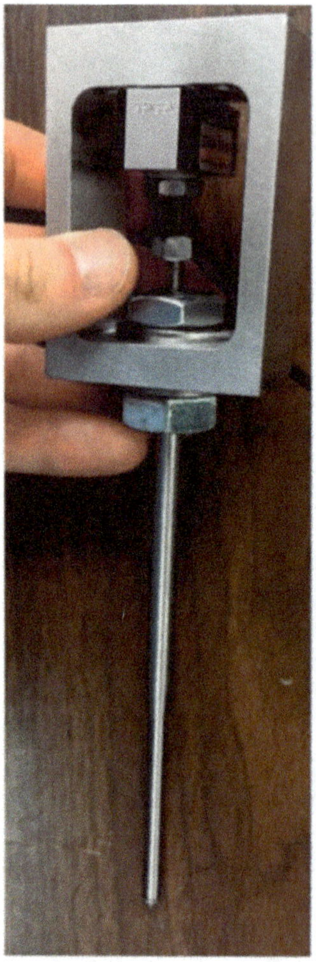

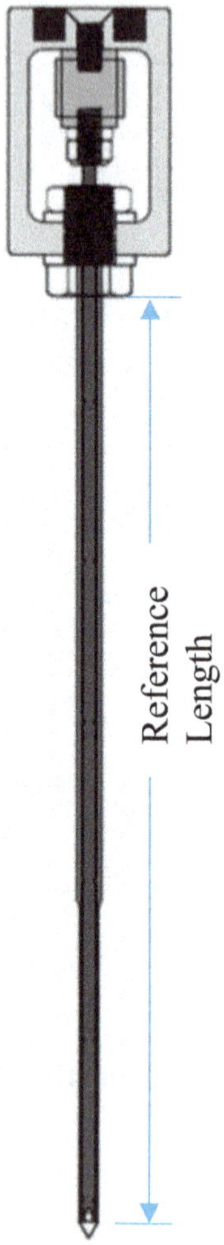


**Fig. 1.11** Images of a new design for LEAP-2017 CPT device and rod lengths of the devices used at different facilities

Thus if the test is done at 1/20 scale and $\mu^* = 20$, the velocity of penetration would be 5 mm/s, model scale. The data reported includes time, force, and actuator displacement in model units and qc vs depth below ground surface in prototype units. The tip resistance should be reported at depth intervals of 1 cm or less, prototype scale.

## 1.6 Shear Wave Velocity

Free bender elements, 12.7 mm × 8 mm in area as described by Brandenberg et al. (2006), could be used for measurement of shear wave velocity profiles. The methods used for interpretation of the data (see, for example methods described by Brandenberg et al. (2008), Lee and Santamarina (2005), and Montoya et al. (2012)) are to be determined by those doing the experiments. The sensors will be placed at mid-depth in the soil layer as indicated in Fig. 1.12 in a horizontal plane

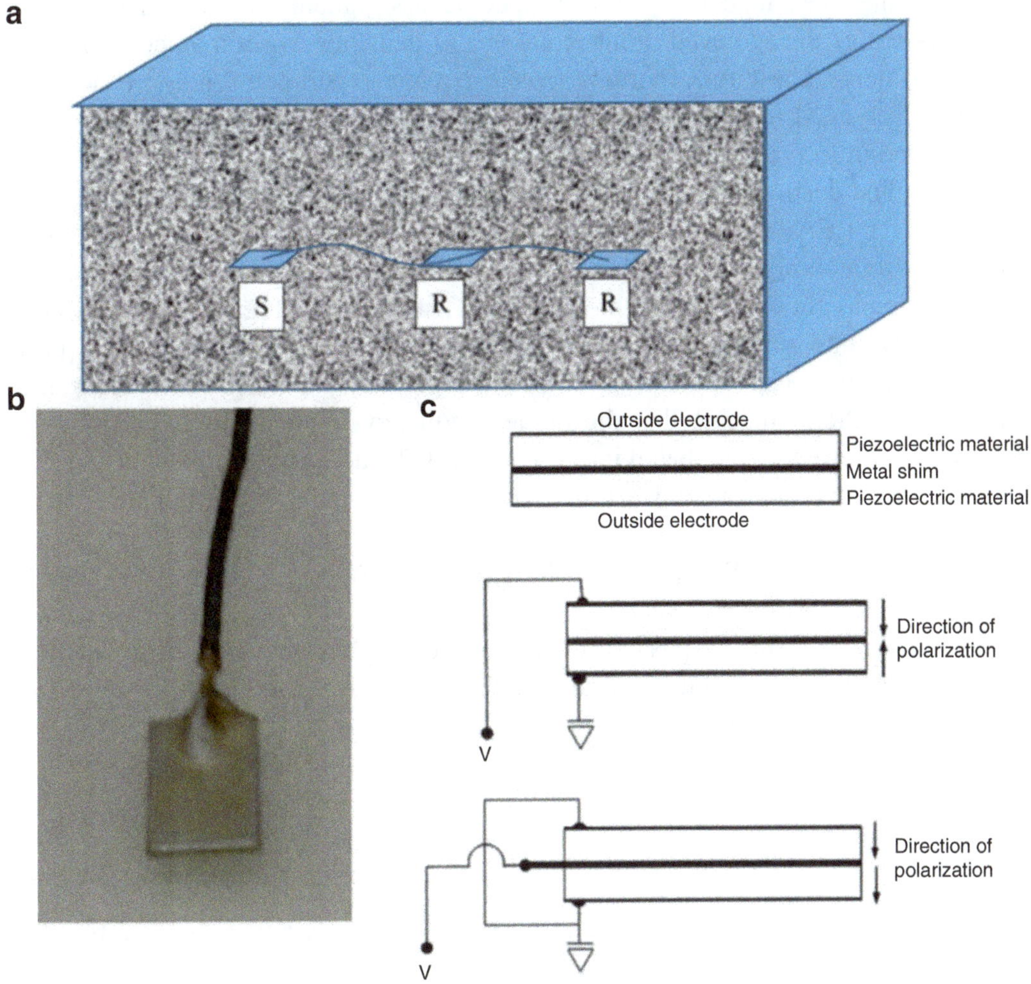


**Fig. 1.12** (**a**) SV wave produced by source bender (S) and recorded by one or more receivers. (**b**) "Free" (not fixed to a block) bender element ready for embedment in soil. (**c**) Parallel and series wiring of benders (Brandenberg et al. 2006)

and measure vertically polarized horizontally travelling SV shear wave velocity. If capability is available, additional sensors may be placed at the quarter depths.

Shear wave velocity measurements should be made before and after every destructive shake. About half of the test sites are expected to retrieve shear wave velocity data in some LEAP-2017 experiments.

## 1.7 Ground Motions

### *1.7.1 Destructive Ground Motions*

An excel worksheet that summarizes the target ground motions and the actual achieved ground motions posted in the Box.com folder and each site should refer to the detailed test matrix to confirm specific assignments for their facility and promptly report the achieved input motions and densities as soon as possible after the experiment. The target motions in subsequent experiments may be adjusted based on achieved motions in previous experiments. The target motions are also summarized in Fig. 1.13.

For the first destructive motion, we will target a ramped sine wave very similar to that used for LEAP-GWU-2015. In that exercise we learned that many facilities shakers introduce high-frequency noise superimposed on the smooth ramped sine wave motion as shown, for example, in Figs. 1.14 and 1.15. Figure 1.16 shows that the velocity time series is much less affected by the high frequency that is so apparent in the acceleration time series of Fig. 1.14. Studies performed since LEAP-GWU-2015 indicate that the higher-frequency components have some but relatively less effect on the behavior of the model. To account for the reduced effect of high frequency, we have a working hypothesis that the effective PGA is

$$PGA_{effective} = PGA_{1Hz} + 0.5^{*}PGA_{hf} \tag{1.2}$$

where $PGA_{hf}$ represents the peak acceleration of the higher frequency components of the ground motion. The method of determining the $PGA_{1Hz}$ and $PGA_{hf}$ is described below and summarized in Fig. 1.15.

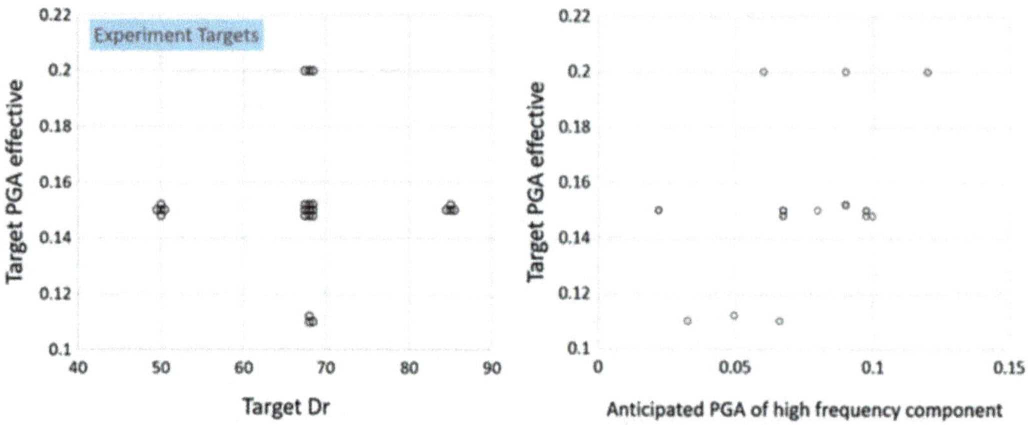


**Fig. 1.13** Anticipated input motions for the experimental results

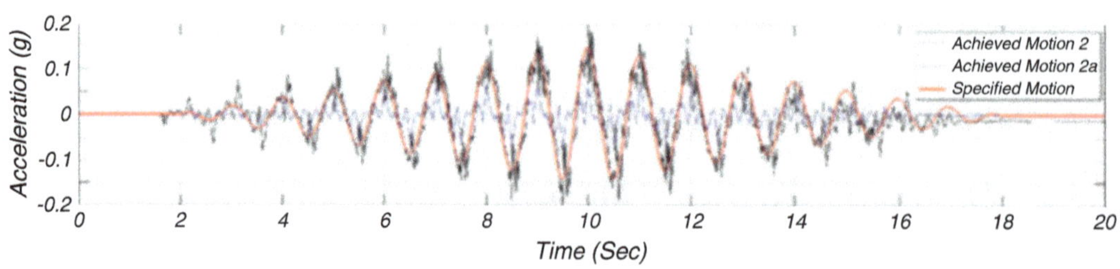


**Fig. 1.14** Achieved and specified acceleration time history for Motions #2 and #2a of the ground motion sequence

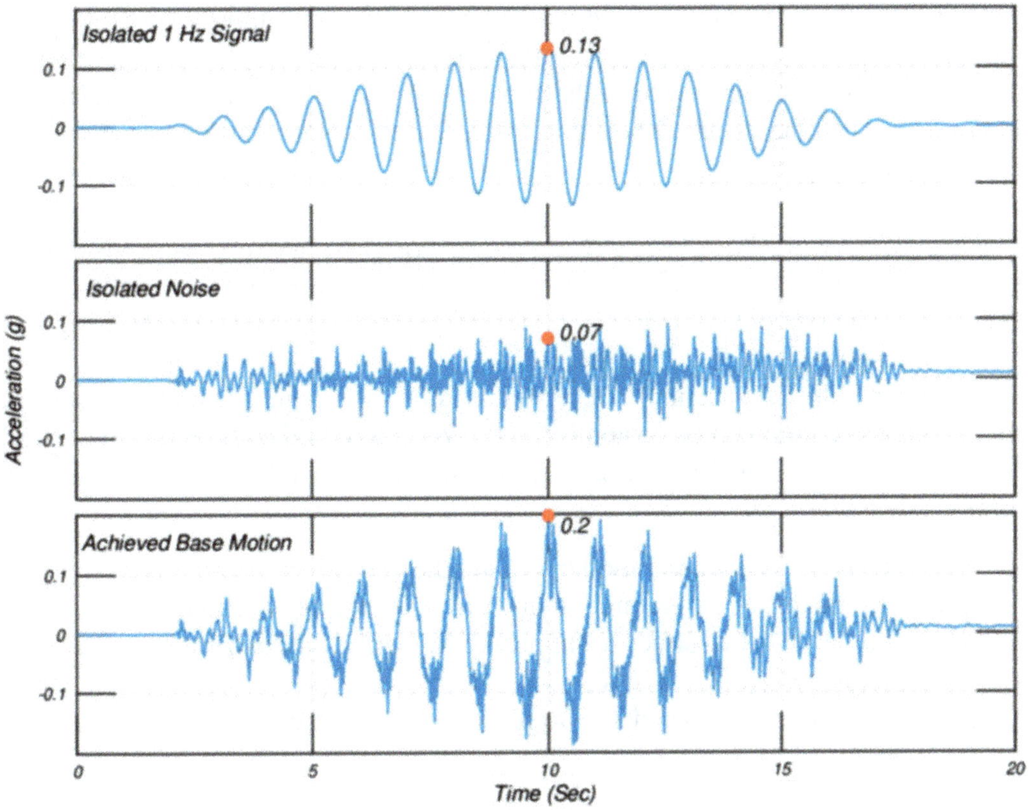


**Fig. 1.15** Achieved base acceleration time history (bottom), high-frequency noise isolated from achieved signal (middle), and 1 Hz signal (top)

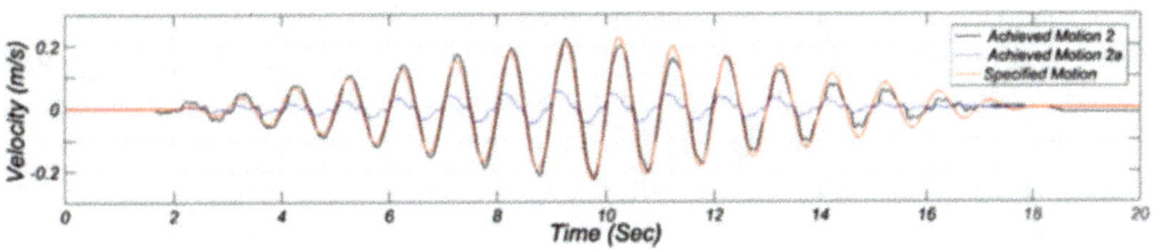


**Fig. 1.16** Achieved and specified velocity time history for Motions #2 and #2a of the ground motion sequence

For the second destructive motion, pending additional discussion, discretion is given to the researchers at the test site. If, for example, the target for Destructive Motion #1 was missed in the first destructive motion, then Destructive Motion #2 would attempt to more accurately hit or to bracket the target. Another option for Destructive Motion #2 would be to attempt to impose the target motion for Destructive Motion #1 in a subsequent test for purposes of calibrating the shaking equipment. This would be a "practice" to enable more accurate performance in the subsequent test, and would also allow comparison of performance with different shaking histories. Other possibilities for Destructive Motion #2 are to try to mimic more realistic ground motions, to intentionally introduce noise in the input motion to explore the validity of Eq. 1.2. Some researchers may also precisely repeat the first destructive motion to more rigorously evaluate the evolution of the behavior of the model due to the previous shaking event. Others may decide to vary the centrifuge acceleration, thereby increasing the prototype depth of the model, prior to the second destructive shaking event.

The experiment should be terminated before deformations are catastrophic because this will allow meaningful interpretation of post-test sensor location measurements, photographs of colored sand, surface marker locations, and spaghetti noodle markers.

However, if the deformations are very small after the second destructive motion, a third destructive motion may be applied at the discretion of the researcher. The third destructive motion should be the final ground motion. Possible options for the third motion are:

- Duplicate an achieved input motions of another LEAP site.
- Investigate effects of higher predominant frequency input motion.
- Investigate stress level effect by imposing higher/lower centrifuge g-level (different prototype layer thickness).
- Compare response in tapered sine to response in realistic earthquakes.

### *1.7.2 Nondestructive Ground Motions*

The small amplitude 1 Hz motions used as nondestructive motions in the 2015 LEAP were found to be not very useful for characterizing the stiffness of the soils. It is possible that a high-frequency motion (with wavelengths comparable to the model thickness will be more useful). If $Vs = 200$ m/s and the wavelength is $\lambda = 4$ H (first mode resonance), the required frequency would be 12.5 Hz in prototype scale for the 4-m-deep deposit. It may be possible for the existing shakers to produce a very small vibration at this frequency Hz; if so, this could be useful for characterizing the soil between shakes. Caution must be exercised to avoid damaging or changing the state of stress of the model by shaking it too intensely. Banded white noise in the frequency range 6–20 Hz (or higher) prototype scale could be useful. RPI may use a Ricker wavelet to produce a short pulse containing these frequencies (RPI now in trials with practice test).

### *1.7.3 Assessment of Tapered Sine Wave (TSW) Ground Motions*

The achieved base motion will be compared to the specified acceleration in a variety of ways as illustrated in Figs. 1.14, 1.15, and 1.16. The predominant frequency component of the motion (e.g., 1 Hz component for Motion #1) will be isolated by use of a notched band pass filter with corner frequencies of approximately 0.9 and 1.1 times the predominant frequency. The predominant frequency component will be subtracted from the achieved motion in the time domain, producing a separate record of the high-frequency components as shown in Fig. 1.15 (middle). The record from UCD Motion#2 in LEAP-GWU-2015 was processed to produce the comparisons in Figs. 1.14, 1.15, and 1.16. The 0.2 g PGA of Motion #2, apparent in 14, has significant contributions from higher frequencies; the 1 Hz component of 0.13 g (Fig. 1.15) reasonably matches the specified 0.15 g for Motion #2. The achieved velocity obtained by integration of the acceleration record for Motion #2 follows that for the specified motion as illustrated in Fig. 1.16. A MATLAB script designed to process the signals (as shown in Fig. 1.14 to 1.16) has been posted for LEAP researchers in the Box.com folder.

## 1.8 Data Reporting Anticipated Plan/Concept

A specification for centrifuge test data will be detailed in a separate document; this will include templates for data submission and methods of uploading and sharing data.

### *1.8.1 New Leap Database*

Results from LEAP-GWU-2015 were archived in a NEEShub Database. As part of this project, the database may be migrated to the new tools developed by the US NHERI DesignSafe Cyber Infrastructure Center. Experimenters should expect to fill out an excel workbook template for each model tested. The excel workbook data template will be similar to that used for LEAP-GWU-2015. In the meantime, we will share information using shared folders on box.com. Participating researchers should contact Bruce Kutter and Trevor Carey if they do not yet have access the LEAP-UCD-2017 Box.com files.

### *1.8.2 Dynamic Shaking Sensor Data*

The dynamic sensor data must be recorded and reported at greater than 50 Hz prototype scale. Ideally, the data acquisition rate should be 200 samples per second (prototype time).

Acceleration data during shaking events should be reported in two formats: (1) as absolute acceleration without any baseline correction and (2) absolute acceleration with baseline correction. Baseline correction parameters must be reported. A spreadsheet template will be provided.

### 1.8.3 Pore Pressure Long-Term Time Series Data

Residual pore pressure changes from before and after the earthquakes will be used as explained above to determine the residual settlement of the pore pressure transducers during the earthquakes. A separate tab in the spreadsheet workbook will be used to report the sensor data (at approximately one sample per second) obtained during the entire spin. Alternatively, Residual Pore Pressure Averages (RPPAs) will be recorded and reported at several stages of the experiment as described above.

### 1.8.4 Summary of Other Anticipated Report Requirements to Be Detailed in a Separate Document

A new template is under design for recording various items during model construction. It will likely involve another spreadsheet, with sections designed to take the following data.

- Sensor locations and marker locations.
- Description of the saturation process and documentation.
- Density calculations, photograph, and/or drawing of tools used to measure mass and volume of model.
- Quality control checks (grain size analysis and $e_{max}$, $e_{min}$).
- Description of any deviations from the specifications.
- Dimensions, mass, and part numbers of sensors.
- Description of how the model surface was curved.
- Photographic record of the top view of the surface markers at every stage of the test, and table of surface marker locations at various stages of the test.
- Photographic record of dissection and post-test sensor location measurement, and table showing locations of sensors, colored sand layers, and noodles before and after the testing.
- Results of inspection of the model and description of any special features of the deformed model surface, e.g., presence of sand boils, cracks at the boundaries, non-symmetric deformation, etc.
- Signal processing (analog filters, baseline correction, other corrections) to sensor data. If custom software is used, provide the software if possible.

**Acknowledgments** The experimental work on LEAP-UCD-2017 was supported by different funds depending mainly on the location of the work. The work by the US PIs (Manzari, Kutter, and Zeghal) is funded by National Science Foundation grants: CMMI 1635524, CMMI 1635307, and CMMI 1635040. The work at Ehime U. was supported by JSPS KAKENHI Grant Number 17H00846. The work at Kyoto U. was supported by JSPS KAKENHI Grant Numbers 26282103, 5420502, and 17H00846. The work at Kansai U. was supported by JSPS KAKENHI Grant Number 17H00846. The work at Zhejiang University was supported by the National Natural Science Foundation of China, Nos. 51578501 and 51778573; Zhejiang Provincial Natural Science Foundation of China, LR15E080001; and National Basic Research Program of China (973 Project), 2014CB047005. The work at KAIST was part of a project titled "Development of performance-based seismic design," funded by the Ministry of Oceans and Fisheries, Korea. The work at NCU was supported by MOST: 106-2628-E-008-004-MY3.

## References

Antonaki, N., Sasanakul, I., Abdoun, T., Sanin, M. V., Puebla, H., & Ubilla, J. (2014). Centrifuge modeling of deposition and consolidation of fine-grained mine tailings. *Geo-Congress Technical Papers Geo-characterization and Modeling for Sustainability*, 3223–3232.

Brandenberg, S., Kutter, B., & Wilson, D. (2008). Fast stacking and phase corrections of shear wave signals in a noisy environment. *Journal of Geotechnical and Geoenvironmental Engineering, 134*(8), 1154–1165. https://doi.org/10.1061/(ASCE)1090-0241(2008)134:8(1154).

Brandenberg, S. J., Choi, S., Kutter, B. L., Wilson, D. W., & Santamarina, J. C. (2006). A bender element system for measuring shear wave velocities in centrifuge models. *Proceedings, 6th International Conference on Physical Modeling in Geotechnics, 1*, 165–170. https://nees.org/data/get/NEES-2006-0149/Documentation/References/Brandenberg_2006.pdf.

Carey, T. J., Kutter, B. L., Manzari, M. T., Zeghal, M., & Vasko, A. (2017). *LEAP soil properties and element test data*. https://doi.org/10.17603/DS2WC7W.

El Ganainy, H., Tessari, A., Abdoun, T., & Sasanakul, I. (2014). Tactile pressure sensors in centrifuge testing. *ASTM Geotechnical Testing Journal, 37*(1). https://doi.org/10.1520/GTJ20120061.

Garnier, J., Gaudin, C., Springman, S. M., Culligan, P. J., Goodings, D. J., Konig, D., Kutter, B. L., Phillips, R., Randolph, M. F., & Thorel, L. (2007). Catalogue of scaling laws and similitude questions in geotechnical centrifuge modelling. *International Journal of Physical Modelling in Geotechnics, 8*(3), 1–23.

Kokkali, P., Abdoun, T., & Zeghal, M. (2018). Physical modeling of soil liquefaction: Overview of LEAP Production Test 1 at Rensselaer Polytechnic Institute. *International Journal of Soil Dynamics and Earthquake Engineering, 113*, 629–649, https://doi.org/10.1016/j.soildyn.2017.01.036.

Kutter, B., Carey, T., Hashimoto, T., Zeghal, M., Abdoun, T., Kokalli, P., Madabhushi, G., Haigh, S., Hung, W.-Y., Lee, C.-J., Iai, S., Tobita, T., Zhou, Y. G., Chen, Y., & Manzari, M. T. (2017). LEAP-GWU-2015 experiment specifications, results, and comparisons. *International Journal of Soil Dynamics and Earthquake Engineering*. https://doi.org/10.1016/j.soildyn.2017.05.018.

Kutter, B. L. (2013). Effects of capillary number, bond number, and gas solubility on water saturation of sand specimens. *Canadian Geotechnical Journal, 50*(2), 133–144.

Lade, P. V., Liggio, C. D., & Yamamuro, J. A. (1998). Effects of non-plastic fines on minimum and maximum void ratios of sand. *Geotechnical Testing Journal*, 21(4), 336.

Lee, J. S., & Santamarina, J. C. (2005). Bender elements: Performance and signal interpretation. *Journal of Geotechnical and Geoenvironmental Engineering, 131*(9), 1063–1070.

Montoya, B. M., Gerhard, R., DeJong, J. T., Wilson, D. W., Weil, M. H., Martinez, B. C., & Pederson, L. (2012). Fabrication, operation, and health monitoring of bender elements for aggressive environments. *Geotechnical Testing Journal, 35*(5), 1–15. https://doi.org/10.1520/GTJ103300. ISSN 0149-6115.

Okamura, M., & Inoue, T. (2012). Preparation of fully saturated models for liquefaction study. *International Journal of Physical Modeling in Geotechnics, 12*(1), 39–46. https://doi.org/10.1680/ijpmg.2012.12.1.39.

# Chapter 2
# Grain Size Analysis and Maximum and Minimum Dry Density Testing of Ottawa F-65 Sand for LEAP-UCD-2017

**Trevor J. Carey, Nicholas Stone, and Bruce L. Kutter**

**Abstract** Ottawa F-65 sand (supplied by US Silica, Ottawa, Illinois) was selected as the standard sand for LEAP-UCD-2017. Between December 2017 and February 2018, each LEAP research team sent 500 g samples of sand to UC Davis for grain size analysis and minimum and maximum dry density testing. The purpose of this testing was to confirm the consistency of the sand used at various test sites and to provide updated minimum and maximum density index values. The variation of measured properties among the different samples is similar to the variation measured during repeat testing of the same sample. Modified LEAP procedures to measure index densities are used to confirm consistency of the sands, and the results from these procedures are compared to results from ASTM procedures. The LEAP procedures give repeatable results with median index densities of $\rho_{min} = 1457$ kg/m$^3$, $\rho_{max} = 1754$ kg/m$^3$. Relative densities calculated with facility-specific index densities varied by less than 4%, so we conclude that average index densities from all the sites may be used for analysis of the results. The LEAP procedures are easier to perform than the ASTM procedures and do not require specialized equipment; therefore, continued use of the LEAP procedure for frequent quality control purposes is recommended. However, the values from ASTM procedures are expected to be more consistent with values adopted in liquefaction literature in the past; therefore, we recommend using the median ASTM values for analysis of LEAP data. Index densities from ASTM procedures ($\rho_{min} = 1490.5$ kg/m$^3$, $\rho_{max} = 1757.0$ kg/m$^3$) produce relative densities that are 4 –10% smaller than the index densities from the LEAP procedures.

## 2.1 Background and Introduction

The standard sand selected for the LEAP-UCD-2017 exercise is Ottawa F-65, a clean, poorly graded, whole grain silica sand, with less than 0.5% fines by mass, supplied by US Silica, Ottawa, Illinois. The LEAP-GWU-2015 exercise also used Ottawa F-65 as the standard sand; therefore, material characterization and element

T. J. Carey (✉) · N. Stone · B. L. Kutter
Department of Civil and Environmental Engineering, University of California, Davis, CA, USA
e-mail: tjcarey@ucdavis.edu



B. Kutter et al. (eds.), *Model Tests and Numerical Simulations of Liquefaction and Lateral Spreading*, https://doi.org/10.1007/978-3-030-22818-7_2

test data from the LEAP-GWU-2015 exercise are applicable to the current LEAP exercise (Kutter et al. 2017; Vasko 2015). Recently, several researchers have performed additional element strength tests of Ottawa F-65 for different initial densities, confinements, and stress paths (Parra Bastidas 2016; Ziotopoulou et al. 2018).

The primary batch of Ottawa F-65 sand used in the LEAP-UCD-2017 exercise was delivered to UC Davis in March of 2013 as a 20-ton single-batch shipment of several pallets loaded with 25 kg bags. Being from the same shipment does not guarantee identical samples in each bag because US Silica does not mix the mined sand prior to shipment. Recognizing that the potential variability in soil properties among the 25 kg bags from a single batch may be smaller than the variability between different batches, the UC Davis team shipped sand to National Central University (Taiwan), Zhejiang University (China), Kyoto University, University of Cambridge, IFSTTAR, Ehime University, KAIST University, and K-Water Corporation. George Washington University (USA) ordered sand directly from US Silica in a single shipment; RPI ordered at least two shipments of sand from US Silica. Following centrifuge testing, between December 2017 and February 2018, all nine centrifuge facilities and George Washington University sent 500 g samples to UC Davis for grain size analysis and minimum and maximum index density testing using a modified LEAP procedure. However, the sand sent to Davis from RPI was from their most recent shipment of sand, while their LEAP experiments were conducted using a previous shipment of sand.

In August 2017, when instructions were released to the numerical simulation teams predicting the LEAP-UCD-2017 centrifuge experiments, Kutter et al. (2019) stated that based on tests conducted by many laboratories, the average minimum and maximum index dry densities for Ottawa F-65 were 1476 and 1765 kg/m$^3$, with standard deviations of 46 and 25 kg/m$^3$, respectively. Table 2.1 summarizes these and other index densities for Ottawa F-65 sand that the LEAP project has previously used. Updated recommendations for index densities are presented later.

**Table 2.1** Previously used minimum and maximum densities for Ottawa F-65 sand

| Phase | Minimum density (kg/m$^3$) | Maximum density (kg/m$^3$) |
|---|---|---|
| LEAP-GWU-2015 | 1519 | 1736 |
| LEAP-UCD-2017, (Kutter et al. 2017) | 1475 | 1756 |
| LEAP-UCD-2017, August 2017, instructions to numerical modelers | 1476 | 1765 |

## 2.2 Grain Size Analysis

This section presents grain size analysis results of the 500 g samples that were sent from the experiment sites to UC Davis following the LEAP-UCD-2017 workshop. For each sample, a dry sieve analysis was performed, following ASTM C136 procedures. The US standard sieve numbers of 20, 30, 40, 50, 60, 70, 100, and 200 were selected to describe the grain size distribution of Ottawa F-65 sand, with a specified mean grain size of 0.23 mm, equal to the number 65 sieve opening. The grain size distribution (GSD) curves for the ten 500 g samples are shown in Fig. 2.1. Shown in Fig. 2.2 is the envelope of the gradation curves from Fig. 2.1. The specified gradation curve for Ottawa F-65 is shown with a dark black line in Fig. 2.1 and a dashed line in Fig. 2.2.

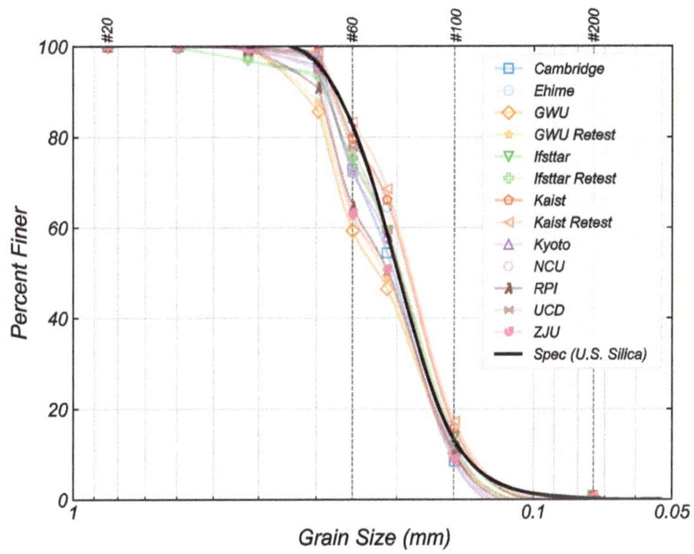


**Fig. 2.1** Grain size distribution based on testing of 500 g samples sent from each experimental site and tested at Davis in spring 2018

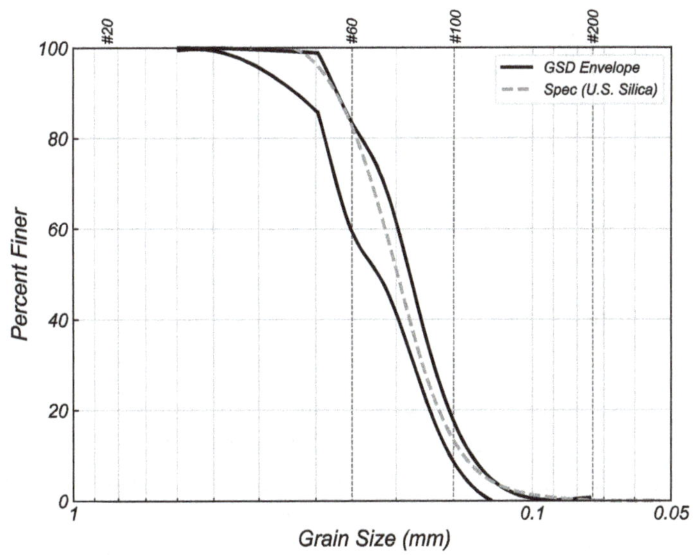


**Fig. 2.2** Envelope of the grain size distribution shown in Fig. 2.1

**Table 2.2** Reported grain size distribution properties from sieve analyses performed at select experimental sites

| Facility (test) | $D_{10}$ (mm) | $D_{30}$ (mm) | $D_{50}$ (mm) | $D_{60}$ (mm) |
|---|---|---|---|---|
| University of Cambridge (1) | 0.16 | 0.19 | 0.22 | 0.24 |
| University of Cambridge (2) | 0.18 | 0.25 | 0.29 | 0.32 |
| University of California Davis | 0.13 | 0.16 | 0.19 | 0.20 |
| George Washington University | 0.14 | 0.17 | 0.21 | 0.23 |
| KAIST University (1) | 0.13 | 0.17 | 0.20 | 0.21 |
| KAIST University (2) | 0.13 | 0.17 | 0.20 | 0.21 |
| KAIST University (supplement 1) | 0.12 | 0.17 | 0.20 | 0.21 |
| KAIST University (supplement 2) | 0.14 | 0.18 | 0.20 | 0.22 |
| RPI | 0.14 | 0.18 | 0.22 | – |
| Zhejiang University (1) | 0.10 | 0.14 | 0.17 | 0.19 |
| Zhejiang University (2) | 0.10 | 0.14 | 0.17 | 0.19 |
| Zhejiang University (3) | 0.10 | 0.14 | 0.17 | 0.19 |
| IFSTTAR (1) | 0.14 | 0.20 | 0.22 | 0.26 |
| IFSTTAR (2) | 0.13 | 0.19 | 0.22 | 0.24 |
| IFSTTAR (3) | 0.15 | 0.21 | 0.23 | 0.24 |
| Average | 0.13 | 0.18 | 0.21 | 0.23 |
| Standard deviation | 0.02 | 0.03 | 0.03 | 0.04 |

**Table 2.3** Grain size distribution properties (from Fig. 2.1) based on testing of 500 g samples sent from each test site to UC Davis for testing following the 2017 workshop

| Facility (test) | $D_{10}$ (mm) | $D_{30}$ (mm) | $D_{50}$ (mm) | $D_{60}$(mm) |
|---|---|---|---|---|
| University of Cambridge | 0.15 | 0.18 | 0.2 | 0.22 |
| Ehime University | 0.14 | 0.17 | 0.19 | 0.2 |
| George Washington University | 0.15 | 0.18 | 0.22 | 0.25 |
| George Washington University (retest) | 0.15 | 0.18 | 0.21 | 0.24 |
| IFSTTAR | 0.14 | 0.17 | 0.2 | 0.21 |
| IFSTTAR (retest) | 0.14 | 0.17 | 0.2 | 0.21 |
| KAIST University | 0.14 | 0.17 | 0.19 | 0.2 |
| KAIST University (retest) | 0.14 | 0.16 | 0.19 | 0.2 |
| Kyoto University | 0.15 | 0.17 | 0.2 | 0.21 |
| National Central University | 0.15 | 0.17 | 0.2 | 0.21 |
| RPI | 0.15 | 0.18 | 0.21 | 0.24 |
| University of California Davis | 0.15 | 0.17 | 0.2 | 0.21 |
| Zhejiang University | 0.15 | 0.18 | 0.21 | 0.24 |
| Average | 0.15 | 0.17 | 0.20 | 0.22 |
| Standard deviation | 0.01 | 0.01 | 0.01 | 0.02 |

Tabulated in Table 2.3 are the $D_{10}$, $D_{30}$, $D_{50}$, and $D_{60}$ values for the curves shown in Fig. 2.1. Several research teams performed independent dry sieve testing and reported the $D_{10}$, $D_{30}$, $D_{50}$, and $D_{60}$ values from their analyses; the values obtained from their data templates are reported in Table 2.2. Also in Table 2.2 is the value reported for GWU by El Ghoraiby et al. (2017).

### 2.2.1 Discussion of Grain Size Analyses

Figures 2.1 and 2.2 show the GSDs of the ten samples clustered around the specified distribution for Ottawa F-65. The GWU, RPI, and ZJU samples deviate from the specified distribution at grain sizes between 0.25 and 0.2 mm. The GWU, RPI, and ZJU $D_{60}$ values reported in Table 2.3 are about 20% larger than the test average, but the $D_{50}$ values are within about 5% of the average. The sand from GWU and RPI came from separate batches/shipments than the eight centrifuge facilities, and this small variability may be attributed to the different batches. ZJU used sand provided by UC Davis. The sand from the seven other facilities that received sand from UCD had almost identical gradations. Based upon all of the GSD analyses, especially those shown in Table 2.3 (tests done by one operator using the same set of sieves and sieve shaker), it appears that the sands used at various facilities are similar enough for practical purposes.

Assuming the permeability of the sand follows Hazen's empirical relationship, ($k = D_{10}^2$), the uncertainty associated with the $D_{10}$ values in Table 2.3 will propagate to the calculated permeability of the sand. The factor by which the permeability will change due to the uncertainty of $D_{10}$ can be expressed as $D_{10(Avg\ \pm\ 1\sigma)}^{2}/D_{10(Avg)}^{2}$. For example, using the average $D_{10}$ of 0.15 mm and standard deviation of 0.01 mm from Table 2.3 results in a change in permeability by a factor of $(0.16/0.15)^2 = 1.14$.

All sieve analyses were performed by the same operator, using the same equipment, eliminating uncertainty due to operator and equipment variation. Duplicate tests were conducted for the KAIST, IFSTTAR, and GWU samples to determine the repeatability of results and better understand the variability of the testing protocol. Table 2.4 summarizes the variation, in percent passing, of the three duplicate tests. The largest variation is observed on the No. 60 sieve, but the average range is still less than 3%. From Table 2.3, the $D_{30}$, $D_{60}$, and $D_{10}$ values of the retest data were all within 0.01 mm of each other. The variation detailed in Table 2.4 is consistent with the 0–4% average variation found by Tiedemann (1973), who using a different sand studied variability of duplicate grain size gradation testing.

**Table 2.4** Summary of grain size gradation tests percent passing variations of duplicate tests

| | No. 20 | No. 30 | No. 40 | No. 50 | No. 60 | No. 70 | No. 100 | No. 200 |
|---|---|---|---|---|---|---|---|---|
| Minimum range | 0.0% | 0.0% | 0.0% | 0.1% | 0.9% | 0.1% | 0.4% | 0.1% |
| Maximum range | 0.0% | 0.0% | 0.0% | 1.9% | 3.9% | 2.4% | 1.4% | 0.2% |
| Average range | 0.0% | 0.0% | 0.0% | 0.9% | 2.6% | 1.7% | 1.1% | 0.1% |
| Standard deviation | 0.0 | 0.0 | 0.0 | 0.0 | 0.0 | 0.0 | 0.0 | 0.0 |

## 2.3 Minimum and Maximum Index Dry Density

Minimum and maximum index dry density tests on the ten 500 g sand samples sent to Davis following the workshop are presented here. The detailed procedure to determine the index densities are described by Kutter et al. (2019), which uses a modified ASTM D4254 Method C and Modified Lade (1988) procedure, respectively. For the remainder of this publication, the Kutter et al. (2019) methods to measure density will be referred to as the LEAP method. The ASTM international standard was not ideal for quality control purposes at all facilities because it requires specialized equipment and it is more time-consuming. Furthermore, different standards are used in other countries. The LEAP method to measure index dry densities are relatively quick and reliable and were thought to be more practical for repeated quality control checks. However, several tests were also done using ASTM 4254 Method A (using a funnel device) and 4253 Method 1B (wet soil, vertically vibrating table) procedures by three different private laboratories and two university researchers, and these results are described later.

### *2.3.1 LEAP Minimum Density Procedure*

Initially, 500 g of sand is placed in a 1000 ml glass graduated cylinder. With the top of the cylinder sealed, the sample is turned upside down, then steadily rotated back upright at a constant rate, taking approximately 30–60 s to reach vertical. The volume of sand is then measured using the gradations on the graduated cylinder. The mass of the sand was measured, and the density calculated. For each sample, the test was repeated three times with the same sample of sand. This procedure is similar to the ASTM 4254 Method C procedure except the size of the graduated cylinder is reduced from 2000 to 1000 ml, which is considered to be large enough for the fine sand used in the LEAP exercise.

### *2.3.2 LEAP Maximum Density Procedure*

Maximum density is found by adding ten 50 g increments of sand to a 1000 ml plastic graduated cylinder. After each increment of sand is added, the side of the cylinder is tapped two times with the plastic handle of a screwdriver at the level of the sand and rotated 90 degrees and tapped again, for a total of eight taps per 50 g increment of sand. The striking distance the screwdriver is swung prior to contacting the plastic cylinder is 250 to 300 mm, with a target distance of 275 mm for this study. The mass of the screwdriver is approximately 140 g. Following the eight taps for the final 50 g increment of sand, six additional lighter taps are made on each 90-degree face (24 total taps). To level the top surface for purposes of accurate reading of the cylinder gradation, five to ten very light taps are made while the cylinder is tilted. The volume is read from the graduated cylinder, and the mass of the soil is measured.

### 2.3.3 Results of Index Dry Density Testing

The index density testing was performed and repeated by three operators, following the LEAP procedures. All three individuals used the same graduated cylinders and screwdrivers, minimizing systematic variability from different equipment. Graduated cylinder volume measurements were calibrated by filling the cylinder with a measured mass of water, and the necessary correction (about 1%) was applied to all volume measurements. Prior to testing, the samples were set out in grounded metal trays for 1 week to equilibrate to ambient moisture and limit differences of static electrical charge. To minimize the effect of differential relative humidity on results, much of the testing occurred on the same day, in the same laboratory with near-constant relative humidity. Following each density test, the sample of sand was placed back in the grounded metal tray to dissipate electrostatic charge that may have accumulated during the movement of the soil. A humidity sensor was also buried in the sand to check if the humidity in the soil pore space was similar to the ambient room humidity. The humidity measurements varied from 32 to 48%.

Table 2.5 summarizes the results from the three operators of this study and provides the site-specific average values. The minimum dry density reported for each operator is the average of three trials of tilting the cylinder and recording the volume and mass. Trial-to-trial standard deviations of these three trials for each individual operator, 1, 2, and 3, are indicated by $STD_1$, $STD_2$, and $STD_3$. $STD_O$ is the operator-to-operator standard deviation, calculated from the operator average values ($Avg_1$, $Avg_2$, $Avg_3$) for each sample. For example, for CU, 14.4 kg/m$^3$ is the standard deviation of 1465, 1448, and 1436 kg/m$^3$; the minimum density site-specific average (Min $Avg_{ss}$) from all operators (1450 kg/m$^3$ for CU) is the average of averages from each operator 1465, 1448, and 1436 kg/m$^3$. $STD_S$ shown in the bottom row is the standard deviations of all of the $Avg_{SS}$ values. Since $STD_S = 7.7$ kg/m$^3$ is less than the average $STD_O = 12.2$ kg/m$^3$, it seems that the site-to-site variability of minimum density is smaller than the operator-to-operator variability of the LEAP method.

While three trials per operator were done for minimum density tests, one trial of the maximum density test was done by each of three operators. Therefore, standard deviation of the maximum densities (STD Max) includes trial-to-trial and operator-to-operator variability.

Illustrated in Fig. 2.3a, b are the facility-specific averages from Table 2.5. Figure 2.3b shows the entire range of data, whereas Fig. 2.3a uses a split vertical scale to exaggerate the differences. The vertical lines in Fig. 2.3a, b represent plus and minus one standard deviation ($STD_O$). The mean minimum and maximum index densities determined with the LEAP procedures are 1451 and 1753 kg/m$^3$, respectively.

**Table 2.5** Summary of index dry densities (kg/m$^3$) of the ten 500 g samples measured at UCD after the LEAP-UCD-2017 workshop measured using the LEAP method

| | Operator 1 | | | Operator 2 | | | Operator 3 | | | Site-specific statistics | | | |
|---|---|---|---|---|---|---|---|---|---|---|---|---|---|
| Sample | Min $Avg_1$ | $STD_1$ Min | $Max_1$ | Min $Avg_2$ | $STD_2$ Min | $Max_2$ | Min $Avg_3$ | $STD_3$ Min | $Max_3$ | Min $Avg_{SS}$ | $STD_O$ Min | Max $Avg_{SS}$ | STD Max |
| CU | 1465 | 9.9 | 1784 | 1448 | 5.2 | 1754 | 1436 | 7.8 | 1742 | 1450 | 14.4 | 1760 | 21.3 |
| Ehime | 1454 | 3.4 | 1750 | 1441 | 6.0 | 1753 | 1445 | 6.8 | 1747 | 1447 | 6.3 | 1750 | 3.1 |
| GWU | 1472 | 5.5 | 1779 | 1466 | 1.3 | 1748 | 1441 | 2.9 | 1757 | 1460 | 16.3 | 1761 | 15.9 |
| IFSTTAR | 1477 | 4.2 | 1776 | 1459 | 7.2 | 1754 | 1459 | 8.1 | 1754 | 1465 | 10.5 | 1761 | 12.7 |
| KAIST | 1444 | 7.1 | 1772 | 1446 | 1.6 | 1735 | 1428 | 12.5 | 1744 | 1440 | 9.9 | 1751 | 19.4 |
| Kyoto | 1471 | 5.6 | 1754 | 1459 | 3.7 | 1739 | 1447 | 8.2 | 1732 | 1459 | 12.4 | 1742 | 10.9 |
| NCU | 1451 | 1.6 | 1755 | 1467 | 3.6 | 1745 | 1432 | 5.6 | 1736 | 1450 | 17.6 | 1745 | 9.2 |
| RPI | 1452 | 2.7 | 1759 | 1457 | 3.2 | 1746 | 1439 | 0.5 | 1746 | 1449 | 9.1 | 1750 | 7.2 |
| UCD | 1469 | 4.6 | 1758 | 1446 | 1.5 | 1744 | 1434 | 4.7 | 1756 | 1450 | 18.0 | 1753 | 7.7 |
| ZJU | 1445 | 1.8 | 1762 | 1444 | 3.7 | 1748 | 1433 | 7.1 | 1754 | 1441 | 7.1 | 1755 | 7.0 |
| Average | 1460 | 4.6 | 1765 | 1453 | 3.7 | 1747 | 1439 | 6.4 | 1747 | 1451 | 12.2 | 1753 | 11.4 |
| Site-to-site $STD_s$: | | | | | | | | | | 7.7 | | 6.3 | |

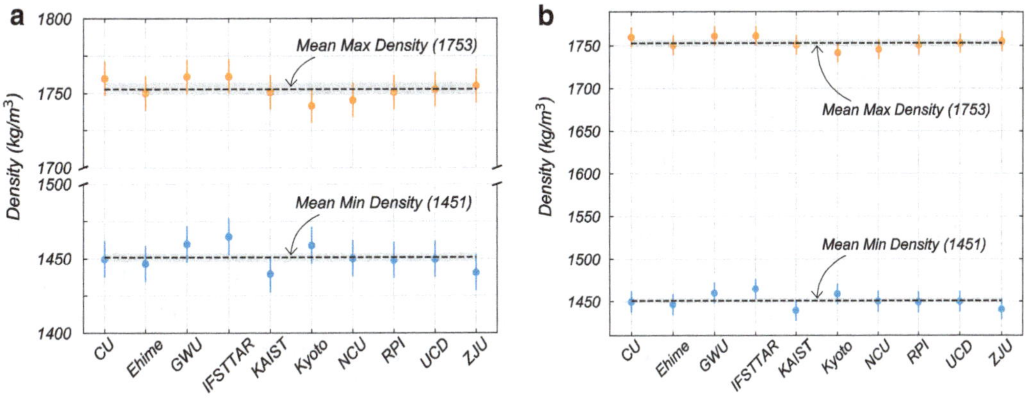


**Fig. 2.3** Index dry density values of the sand from each test facility, with the mean value and 95% confidence of the mean (gray region) shown with, (**a**) split scale and (**b**) linear scale. The error bars in both figures represent the average operator-to-operator standard deviation, $STD_O = 12.2$ kg/m$^3$ of the minimum densities and the average STD = 11.4 kg/m$^3$ of the maximum densities

## 2.3.4 *Discussion of Minimum Density*

Overall, the minimum densities are closely grouped around the average of 1451 kg/m$^3$. The trial-to-trial standard deviation was about 5 kg/m$^3$, the operator-to-operator standard deviation was about 12 kg/m$^3$, and the site-to-site variation was about 8 kg/m$^3$. The standard deviation for the entire data set (all 90 tests) is 14 kg/m$^3$ (not provided in Table 2.5). In contrast, a standard deviation of 46 kg/m$^3$ was provided to the numerical simulation teams by Kutter et al. (2019), which in addition to material variability includes the variability of different procedures to measure density, equipment, and different operators.

Using a two-way analysis of variance framework, the comparison interval, or the vertical length of the standard deviation bar, for each facility can be evaluated to determine if they overlap the mean minimum density of 1451 kg/m$^3$. The IFSTTAR sample comparison interval does not overlap the mean minimum density line. If the 95% confidence interval of the mean is considered, which is shown as the gray region surrounding the mean value line, the IFSTTAR comparison interval overlaps with the mean region. The facility-specific minimum densities vary up to 1% from the mean. The differences appear to be small but statistically distinguishable.

## 2.3.5 *Discussion of Maximum Density*

Kutter et al. (2019) calculated a standard deviation of 25 kg/m$^3$ using data aggregated from many laboratories performing different methods to measure maximum density. The standard deviation for the entire data produced for the present paper (all

30 tests) is only 13 kg/m$^3$, which is 0.7% of the maximum density (1753 kg/m$^3$). This includes trial-to-trial, operator-to-operator, and sample-to-sample variability.

The variability of the STD of the maximum density in Table 2.5 is due to the limited sample size (three samples per STD evaluation). Furthermore, the STD of the maximum density includes trial-to-trial and operator-to-operator variability. For the minimum density test, much of the trial-to-trial variability was removed by averaging the three trials before obtaining $STD_O$. The STD bars in Fig. 2.3 include operator-to-operator and trial-to-trial variability of all ten maximum density samples which overlap the mean value of the maximum density 1753 kg/m$^3$. Thus, based on maximum density tests, it is not possible to statistically distinguish soil specimens from each other.

## 2.4 Testing Results Effect on Relative Density

Relative density can be expressed as:

$$D_r = \frac{\rho_{max}(\rho_d - \rho_{min})}{\rho_d(\rho_{max} - \rho_{min})} \tag{2.1}$$

where $\rho_{max}$ is the maximum dry density, $\rho_{min}$ is the minimum dry density, and $\rho_d$ is the dry density. Equation 2.1 illustrates the sensitivity of relative density to changes in $\rho_{max}$, $\rho_{min}$, and $\rho_d$. For example, if $\rho_{max} = 1752$ kg/m$^3$, $\rho_d = 1600$ kg/m$^3$, and $\rho_{min}$ is decreased by 1% from 1486 to 1471 kg/m$^3$, the calculated relative density would increase 3% from 47 to 50%; similarly, if $\rho_d$ is increased by 1% from 1600 to 1616 kg/m$^3$, the calculated relative density would increase 6% from 47 to 53%.

Table 2.6 compares relative densities that would be obtained from the reported densities for each LEAP experiment, calculated by mass and volume of sand, using the average index density values determined from the present study ($\rho_{min} = 1451$ kg/m$^3$, $\rho_{max} = 1753$ kg/m$^3$), site-specific average values by the LEAP method ($Avg_{ss}$ values in Table 2.5), and the average values from the ASTM method ($\rho_{min} = 1490.5$ kg/m$^3$, $\rho_{max} = 1757.0$ kg/m$^3$).

## 2.5 Measurements by ASTM Method

Tabulated in Table 2.7 are index densities using the ASTM procedures 4253 Method 1B and 4254 Method A performed by four different professional laboratories and two researchers on three different batches of Ottawa F-65 sand. The results from Cooper Labs (located in Palo Alto, CA) tested material from the March 2013 UC Davis shipment. The material tested was taken from a single bag from the batch. Material sent to Gulf Shores Exploration (located in Rancho Cordova, CA) was

**Table 2.6** Relative densities obtained by using an average, site-specific, and ASTM index density values

| Facility | Density (kg/m$^3$) | $D_r$ (%) ($\rho_{min}$ = 1451 $\rho_{max}$ = 1753 LEAP method average) | $D_r$ (%) ($\rho_{min}$ and $\rho_{max}$ using site-specific, Avg$_{ss}$, average values from Table 2.5) | $D_r$ (%) ($\rho_{min}$ = 1490.5 $\rho_{max}$ = 1757 ASTM average from Table 2.7) |
|---|---|---|---|---|
| CU1 | 1656 | 72% | 71% | 66% |
| CU2 | 1606 | 56% | 55% | 47% |
| Ehime1 | 1649 | 70% | 71% | 63% |
| Ehime2 | 1657 | 72% | 73% | 66% |
| Ehime3 | 1693 | 83% | 84% | 79% |
| IFSTTAR1 | 1696 | 84% | 81% | 80% |
| IFSTTAR2 | 1624 | 62% | 58% | 54% |
| KAIST1 | 1701 | 85% | 87% | 82% |
| KAIST2 | 1593 | 52% | 54% | 42% |
| KyU1 | 1683 | 80% | 82% | 75% |
| KyU2 | 1659 | 73% | 74% | 67% |
| KyU3 | 1637 | 66% | 67% | 59% |
| NCU1 | 1652 | 71% | 72% | 64% |
| NCU2 | 1652 | 71% | 72% | 64% |
| NCU3 | 1652 | 71% | 72% | 64% |
| RPI1 | 1650 | 70% | 71% | 64% |
| RPI2 | 1659 | 73% | 73% | 67% |
| RPI3 | 1623 | 62% | 62% | 54% |
| UCD1 | 1665 | 75% | 75% | 69% |
| UCD2 | 1648 | 69% | 70% | 63% |
| UCD3 | 1658 | 72% | 73% | 67% |
| ZJU1 | 1651 | 70% | 71% | 64% |
| ZJU2 | 1599 | 54% | 55% | 45% |
| ZJU3 | 1703 | 86% | 86% | 82% |
| AVG | – | 71% | 71% | 65% |

mixed from eight different bags, from four different pallets of the March 2013 shipment. The sand was mixed and split into two identical samples without telling the laboratory that the samples were identical. The results from GeoComp Express (NY) used sand from a different shipment. Considering only the four tests done by commercial laboratories (tests 1, 2, 5, and 6), the standard deviations (13.7 and 10.2 kg/m$^3$) are much less than the standard deviations from all six tests (28.3 and 20.8 kg/m$^3$).

**Table 2.7** ASTM index dry densities for Ottawa F-65 sand from commercial laboratories and researchers

| Test no: | Testing company | Date | Soil batch | Minimum density (D4254 A) (kg/m$^3$) | Maximum density (D4253-1B) (kg/m$^3$) |
|---|---|---|---|---|---|
| 1 | Cooper Labs | Apr 13 | UC Davis March 2013 | 1515 | 1736 |
| 2 | GeoComp Express | Dec 14 | RPI shipment | 1494 | 1758 |
| 3 | Vasko (2015) | Dec 14 | GWU shipment | 1538 | 1793 |
| 4 | Parra Bastidas (2016) | Jun 15 | UC Davis March 2013 | 1455 | Not measured |
| 5 | Gulf Shore Exploration and Testing (test 1 of 2) | May 18 | UC Davis March 2013 | 1485 | 1752 |
| 6 | Gulf Shore Exploration and Testing (test 2 of 2) | May 18 | UC Davis March 2013 | 1487 | 1757 |
| | Average (tests 1,2,5,6) | | | 1485.0 | 1744.0 |
| | Median (tests 1,2,5,6) | | | 1490.5 | 1754.5 |
| | STD (tests 1,2,5,6) | | | 13.7 | 10.2 |
| | Average (tests 1–6) | | | 1495.7 | 1759.2 |
| | STD (tests 1–6) | | | 28.3 | 20.8 |
| | Median (tests 1–6) | | | 1490.5 | 1757.0 |

## 2.6 Conclusions

Ottawa F-65, an unprocessed mined sand, was chosen as the standard sand for LEAP-UCD-2017. Most of the LEAP-UCD-2017 experiments were performed with sand shipped from one large shipment to UCD in March 2013. GWU and RPI ordered sand independently from different shipments. Between December 2017 and February 2018, the nine centrifuge facilities participating in LEAP-UCD-2017 and GWU sent 500 g samples of their Ottawa F-65 sand to UC Davis for quality control testing. Index dry densities using the LEAP method and grain size analysis were performed to determine how consistent the sand was across the facilities.

The sand used for the LEAP-UCD-2017 exercise at different facilities was reasonably consistent, but it does seem from this suite of testing that there are small but detectable differences in the index dry densities; the differences might affect the computed relative densities by up to 4%. However, the measured site-to-site variability is less than the operator-to-operator variability. The LEAP procedure to measure index dry densities produces consistent results, lending credence to the procedure's use as a quality control measure. The LEAP procedures give median values $\rho_{\min} = 1451$ kg/m$^3$, $\rho_{\max} = 1753$ kg/m$^3$.

The ASTM procedures, including several tests by commercial laboratories, produced different values. The relative densities using ASTM procedures are 4–10% smaller than the relative densities that would be calculated using the LEAP procedures. It is believed that ASTM procedures are more likely to be representative of procedures used in liquefaction literature. We therefore recommend using the median ASTM values for future analysis of LEAP data: $\rho_{\min} = 1490.5$ kg/m$^3$, $\rho_{\max} = 1757.0$ kg/m$^3$.

**Acknowledgments** This study was supported by NSF CMMI Grant Number 1635307. The authors appreciate the assistance of the researchers at all of the LEAP-UCD-2017 experiment sites for providing their grain size analysis, index density tests, and for sending the required samples back to UC Davis.

## References

El Ghoraiby, M. A., Park, H., Manzari, M. T. (2017). *LEAP2017: Soil characterization and element tests for Ottawa F65 Sand*. Report by George Washington University, Washington, DC, 14th March 2017, 37p.

Kutter, B. L., Carey, T. J., Bonab, M. H., Stone, N., Manzari M., Zeghal M., Escoffier, S., Haigh, S., Madabhushi, G., Hung, W., Kim, D., Kim, N., Okamura, M., Tobita, T., Ueda, K., Zhou, Y. (2019). LEAP UCD 2017. 1.01 model specifications. In B. Kutter et al. (Eds.), *Model tests and numerical simulations of liquefaction and lateral spreading: LEAP-UCD-2017*. New York: Springer.

Kutter, B. L., Carey, T. J., Hashimoto, T., Zeghal, M., Abdoun, T., Kokkali, P., Madabhushi, G., Haigh, S. K., Burali d'Arezzo, F., Madabhushi, S., Hung, W.-Y., Lee, C.-J., Cheng, H.-C., Iai, S., Tobita, T., Ashino, T., Ren, J., Zhou, Y.-G., Chen, Y.-M., Sun, Z.-B., & Manzari, M. T. (2017). LEAP-GWU-2015 experiment specifications, results, and comparisons. *Soil Dynamics and Earthquake Engineering, 113*, 616–628. https://doi.org/10.1016/j.soildyn.2017.05.018.

Lade, P. V. (1988). Double hardening constitutive model for soils, parameter determination and predictions for two sands. *Constitutive Equations for Granular Non-Cohesive Soils*, 367–382.

Parra Bastidas, A. M. (2016). *Ottawa F-65 Sand Characterization*. PhD Dissertation, University of California, Davis.

Tiedemann, D. A. (1973). Variability of laboratory relative density test results. In *Evaluation of relative density and its role in geotechnical projects involving Cohesionless soils*. ASTM International.

Vasko, A. (2015). *An investigation into the behavior of Ottawa sand through monotonic and cyclic shear tests*. MS Thesis, George Washington University.

Ziotopoulou, K., Montgomery, J., Parra Bastidas, A. M., Morales, B. (2018). *Cyclic Strength of Ottawa F-65 Sand: Laboratory Testing and Constitutive Model Calibration: Geotechnical Earthquake Engineering and Soil Dynamics V*. The University of Texas at Austin, 10–13 June 2018.

# Chapter 3
# Physical and Mechanical Properties of Ottawa F65 Sand

**Mohamed El Ghoraiby, Hanna Park, and Majid T. Manzari**

**Abstract** This paper presents the results of soil characterization and element tests of Ottawa F65 sand. The data presented is intended to be used as calibration material for the prediction exercise conducted as part of the Liquefaction Experiments and Analysis Project (LEAP 2017). The databank generated includes soil specific gravity tests, particle size analysis, hydraulic conductivity tests, maximum and minimum void ratio tests, and cyclic triaxial stress-controlled tests. An effort was made to ensure the consistency and repeatability of the test results by reducing the sources of variability in the sample preparations and increasing the number of tests. The uniformity of the soil was evaluated by conducting tests on samples from five different batches. The results showed that the sand is uniform among the five batches. Due to significant variability in previously reported maximum and minimum void ratio results, the effects of the test operator were studied by comparing test results obtained from three different operators. For the triaxial tests, a constant height dry pluviation method was used for sample preparation. To eliminate the effect of the human error in maintaining a constant drop height and to ensure consistency of the sand fabric between different samples, a device was developed to facilitate the sample preparation. The cyclic triaxial experiments were performed using three different soil densities, and a liquefaction strength curve was obtained for each density based on a 2.5% single amplitude axial strain criteria. The developed databank in this study was made publicly available for the community through DesignSafe.

## 3.1 Introduction

The Liquefaction Experiments and Analysis Project (LEAP) is a combined effort between several multinational research institutions to investigate the liquefaction phenomenon. As the project name suggests, the main objectives of LEAP is to

M. El Ghoraiby
The George Washington University, Washington, DC, USA

H. Park · M. T. Manzari (✉)
Department of Civil and Environmental Engineering, George Washington University, Washington, DC, USA
e-mail: manzari@gwu.edu



B. Kutter et al. (eds.), *Model Tests and Numerical Simulations of Liquefaction and Lateral Spreading*, https://doi.org/10.1007/978-3-030-22818-7_3

generate a large database of centrifuge experiments which model the consequences of liquefaction on different geo-structures and to assess the capability of the current state of the art analytical tools to predict the response of liquefiable soil. A reliable assessment of the analytical tools is achieved through a blind prediction exercise (Type-B). In a Type-B prediction exercise, the modelers are given the centrifuge experiment geometry and the achieved density and base motion, and then they are required to model the soil response without prior knowledge of the experiment results. A databank for soil characterization and element tests is generated in order to aid the modelers in their model calibrations. Details of this databank are presented here.

Previous experimental studies were conducted for characterizing Ottawa F65 sand including cyclic triaxial as well as cyclic direct simple shear tests (Bastidas 2016; Vasko 2015; Vasko et al. 2018). Vasko (2015) conducted a series of triaxial experiments for the planning phase of LEAP (PLEAP). The data was used in a prediction exercise of centrifuge experiments performed during that phase (Manzari et al. 2017). In this database the cyclic triaxial tests were performed on a single soil density which is the test density. The sample preparation technique used was dry pluviation with minor tapping on the mold. Bastidas (2016) produced a database of monotonic and cyclic direct simple shear experiments. The performed experiments were conducted on two densities. Dry funnel deposition was the sample preparation method for the loose specimen, while air pluviation was used for dense soil.

The databank presented in this paper is intended for the LEAP 2017 phase of the project. The databank includes soil characterization tests such as specific gravity, particle size distribution, permeability, and maximum and minimum void ratio tests. Additionally, a cyclic triaxial stress-controlled testing program is conducted to obtain the liquefaction strength of the soil at various densities. The theme of the 2017 phase is to assess the sensitivity of the soil response to variations in soil density and input motion. As a result of this study, the centrifuge experiments would yield a range of response instead of a single point. In order to properly model these centrifuge experiments, the prediction exercise in LEAP 2017 dictates that the databank provided for model calibration to be of high quality.

The LEAP 2017 Laboratory experiment databank is intended to address the needs of the project. Maintaining a high level of consistency and repeatability was one of the main objectives. The consistency of the sample among different sand batches was addressed by obtaining samples from five different batches. In obtaining the soil permeability, a large number of experiments were conducted to produce a trend between the soil density and the permeability. Previously obtained maximum and minimum void ratios of the Ottawa F65 sand have shown large variation among the reporting research groups (Kutter et al. 2017). To address the source of variability, the maximum and minimum void ratio tests were conducted by three different operators to quantify the variation in the results and eliminate the operator effect.

For the triaxial testing program of LEAP 2017, a series of cyclic stress-controlled tests were conducted on Ottawa F65 sand samples at three different densities. The centrifuge experiments performed in the LEAP 2017 project were prepared using dry pluviation technique. To obtain a similar fabric as the centrifuge tests, a constant height dry pluviation sample preparation technique was used. The effect of human error in maintaining a constant drop height and achieving consistent fabric was eliminated by using a device developed to facilitate the sample preparation. The liquefaction strength curves based on a 2.5% single amplitude axial strain criterion were produced from the performed tests.

The remainder of this paper is divided into two sections. The first addresses the soil characterization tests including the soil specific gravity, particle size distribution, hydraulic conductivity, and the maximum and minimum void ratio tests. For each test, the procedures are first presented followed by the results. The second section covers the cyclic triaxial testing program. The procedures for running the tests are first discussed followed by the sample preparation technique. The consistency of the samples produced is addressed by presenting the statistics of the sample measurement. Finally, the results obtained from the testing program are presented, followed by concluding remarks.

## 3.2 Ottawa F65 Soil Characterization

This section covers the characterization tests performed to obtain the Ottawa F65 properties. The tests include soil specific gravity tests, particle size analysis, soil permeability, and maximum and minimum void ratio tests. The tests were performed as per the ASTM standard testing procedures, and any additional steps taken or modifications are further discussed.

### 3.2.1 *Specific Gravity Tests*

The specific gravity tests were conducted in accordance with the ASTM D854 testing standard procedures. The sample de-aeration was achieved by subjecting the sample to a vacuum for at least 2 h. The results of the specific gravity were used as one way to confirm the sample uniformity and consistency among different sand batches. The tests were performed on samples obtained from five different sand batches. For each batch a series of six tests were conducted. Table 3.1 shows a summary of the results including the mean and coefficient of variation (COV).

An average specific gravity for Ottawa F65 sand of 2.65 was obtained with a mean coefficient of variation of 0.78%. Figure 3.1 shows a box plot of the spread of the specific gravity measurements from the average value for each soil batch.

### 3.2.2 *Particle Size Distribution Analysis*

A sieve analysis was performed to obtain the particle size distribution of Ottawa F65 sand. The analysis follows the ASTM D422 standard procedures. Five tests were performed on samples from different sand batches. Figure 3.2 shows the plot of the particle size distributions, while Table 3.2 shows a summary of test results.

**Table 3.1** Specific gravity of Ottawa F65

| Test | Batch #1 | Batch #2 | Batch #3 | Batch #4 | Batch #5 |
|---|---|---|---|---|---|
| 1 | 2.63 | 2.67 | 2.63 | 2.66 | 2.67 |
| 2 | 2.65 | 2.65 | 2.64 | 2.64 | 2.62 |
| 3 | 2.64 | 2.68 | 2.66 | 2.65 | 2.63 |
| 4 | 2.66 | 2.58 | 2.66 | 2.65 | 2.67 |
| 5 | 2.65 | 2.68 | 2.65 | 2.64 | 2.65 |
| 6 | 2.65 | 2.66 | 2.68 | 2.65 | 2.64 |
| Mean | 2.65 | 2.65 | 2.65 | 2.65 | 2.65 |
| COV (%) | 0.40 | 1.47 | 0.73 | 0.30 | 0.71 |

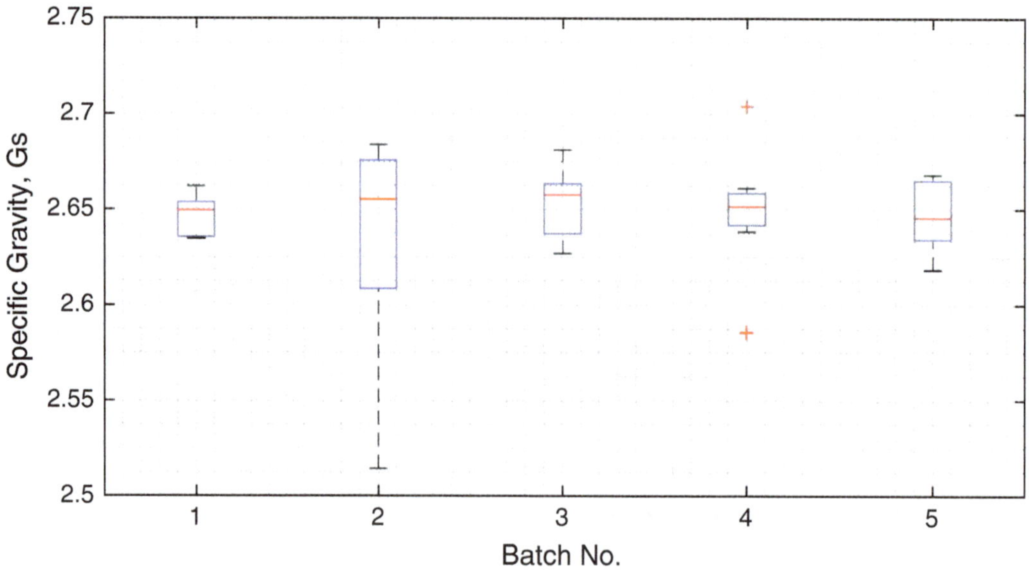


**Fig. 3.1** Ottawa F65 specific gravity test results

Figure 3.3 shows a comparison between the average distribution and the one obtained by Vasko in 2015. $D_{10}$, $D_{30}$, and $D_{60}$ are indicated in Fig. 3.3, while the $D_{50}$ values are presented in Table 3.3. The soil coefficient of uniformity, $C_u$, and coefficient of curvature, $C_c$, were 1.728 and 0.947, respectively. According to the Unified Soil Classification System, the soil is classified as poorly graded sand with designation SP.

### *3.2.3 Hydraulic Conductivity*

A series of constant head hydraulic conductivity tests were conducted to determine the permeability of Ottawa F65 sand at different densities. The tests followed the ASTM D 2434 standard testing procedures. Table 3.4 shows the hydraulic conductivity for different initial void ratios. The hydraulic conductivity values presented are

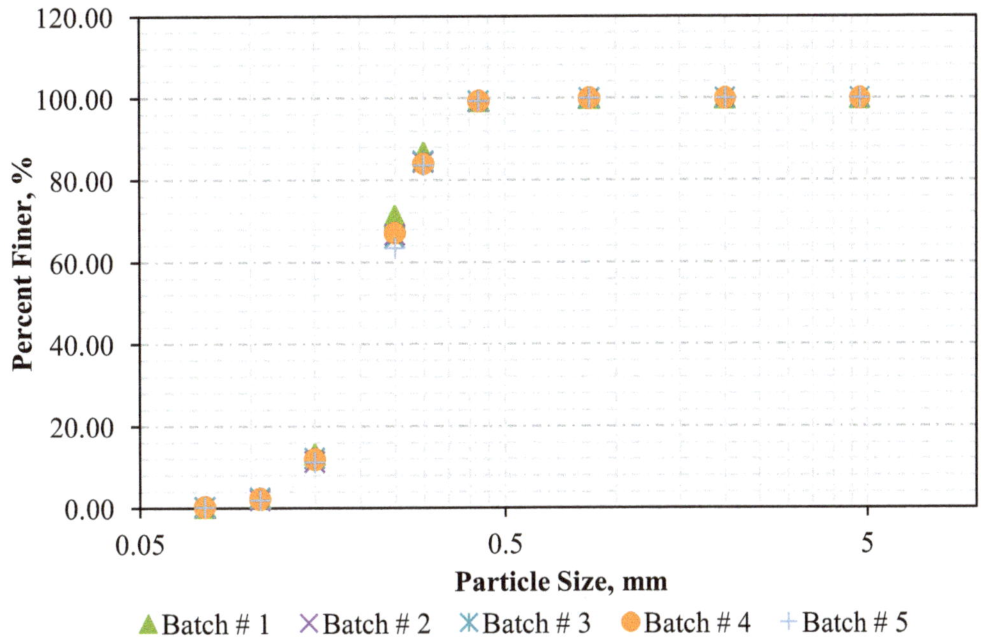


**Fig. 3.2** Ottawa F65 particle size distribution

corrected to a temperature of 20 °C. The test samples were prepared using a constant height dry pluviation technique. The same technique was used for preparation of the soil samples for the triaxial experiments presented in the following section. Figure 3.4 shows a plot of the hydraulic conductivity versus the initial void ratio. The permeability reported here is in the same order of magnitude as the results reported by Vasko (2015) and Bastidas (2016). While the results obtained by Vasko (2015) fall within the range of values reported herein, the permeability reported by Bastidas (2016) was slightly higher.

### 3.2.4 *Maximum and Minimum Void Ratios*

Maximum and minimum void ratio tests were conducted on Ottawa F65 sand by different researchers. Table 3.5 shows the average values from the tests performed by the different groups (Kutter et al. 2017). Two different testing procedures were used in obtaining the maximum and minimum void ratios. The first is based on the ASTM standard testing procedures D4253 and D4254, while the second method is the one proposed by Lade et al. (1998). The results show a wide range of variation among the different reporting groups. The sources of variations are either inherent in the Ottawa F65 sand or caused by the testing procedures.

In order to improve our understanding of the causes of the variability observed in the maximum and minimum void ratio results, additional tests were conducted at GWU. The tests were performed using the method proposed by Lade et al. (1998). The tests were conducted by different researchers to address the human factor in the

**Table 3.2** Particle size distribution of Ottawa F65

| Sieve number | Diameter (mm) | Percent passing – P#1 (%) | Percent passing – P#2 (%) | Percent passing – P#3 (%) | Percent passing – P#4 (%) | Percent passing – P#5 (%) | Average percent passing (%) | Standard deviation (%) |
|---|---|---|---|---|---|---|---|---|
| 4 | 4.75 | 100.00 | 100.00 | 100.00 | 100.00 | 100.00 | 100. | 0 |
| 10 | 2 | 100.00 | 100.00 | 100.00 | 100.00 | 100.00 | 100. | 0 |
| 20 | 0.85 | 100.00 | 100.00 | 100.00 | 100.00 | 100.00 | 100. | 0 |
| 40 | 0.425 | 99.39 | 99.27 | 99.32 | 99.29 | 99.34 | 99.32 | 0.05 |
| 50 | 0.3 | 86.81 | 84.44 | 84.73 | 84.07 | 83.55 | 84.72 | 1.25 |
| 60 | 0.25 | 71.36 | 66.81 | 66.04 | 67.20 | 63.50 | 66.98 | 2.84 |
| 100 | 0.15 | 13.19 | 11.33 | 12.25 | 11.92 | 11.16 | 11.97 | 0.81 |
| 140 | 0.106 | 2.63 | 2.10 | 2.60 | 2.28 | 1.99 | 2.32 | 0.29 |
| 200 | 0.075 | 0.37 | 0.29 | 0.40 | 0.34 | 0.26 | 0.33 | 0.06 |

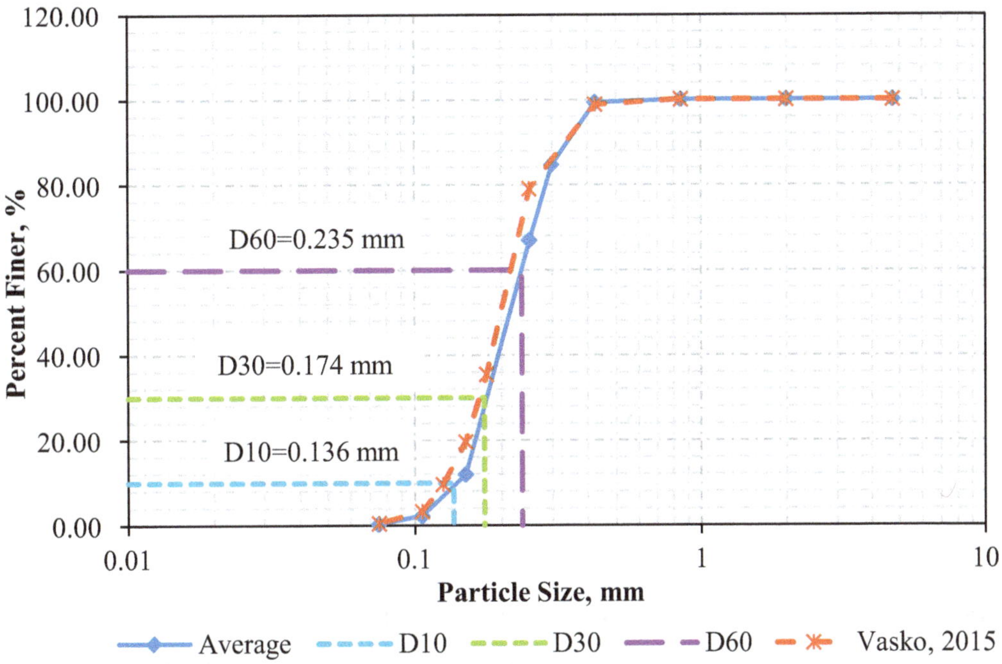


**Fig. 3.3** Ottawa F65 average particle size distribution

**Table 3.3** D50 of Ottawa F65 sand

| | Sample 1 | Sample 2 | Sample 3 | Sample 4 | Sample 5 | Vasko (2015) |
|---|---|---|---|---|---|---|
| D50, mm | 0.21 | 0.21 | 0.21 | 0.21 | 0.22 | 0.2 |

**Table 3.4** Hydraulic conductivity of Ottawa F65

| Void ratio, $e_o$ | Permeability, $T = 20$ °C, cm/s | Void ratio, $e_o$ | Permeability, $T = 20$ °C, cm/s |
|---|---|---|---|
| 0.571 | 0.010 | 0.702 | 0.012 |
| 0.697 | 0.011 | 0.706 | 0.014 |
| 0.668 | 0.012 | 0.713 | 0.016 |
| 0.590 | 0.010 | 0.697 | 0.012 |
| 0.476 | 0.009 | 0.644 | 0.010 |
| 0.766 | 0.015 | 0.650 | 0.015 |
| 0.699 | 0.012 | 0.653 | 0.014 |
| 0.486 | 0.008 | 0.646 | 0.014 |
| 0.674 | 0.013 | 0.642 | 0.016 |
| 0.721 | 0.015 | 0.622 | 0.014 |
| 0.694 | 0.014 | 0.494 | 0.011 |

results. Tables 3.6 and 3.7 show the mean, range, and coefficient of variation for a series of tests performed. The data sets 1 and 2 were obtained at GWU in 2017 by the authors, while data set 3 was the results of experiments by Vasko in 2015. Set 4 considers the range, mean, and coefficient of variation of the results presented in Table 3.5.

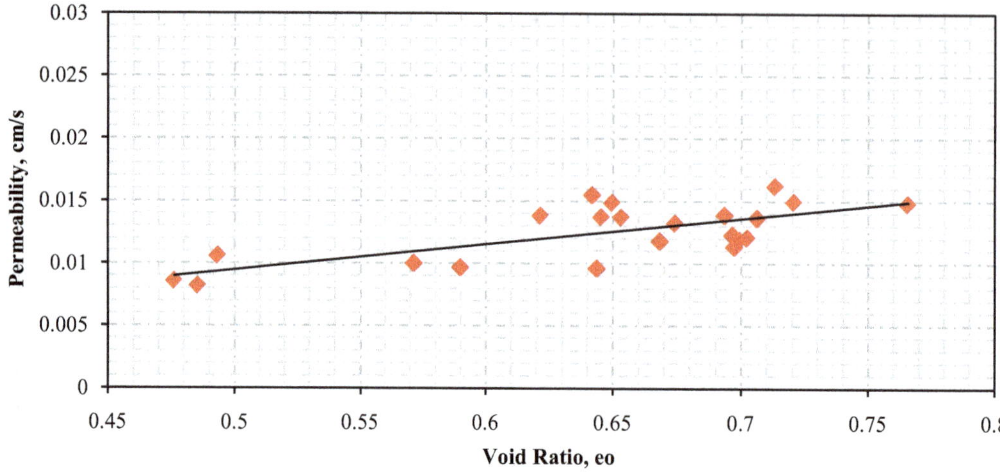


**Fig. 3.4** Hydraulic conductivity, $k$, vs. void ratio, $e_o$

**Table 3.5** Maximum and minimum void ratio obtained by different researchers (Kutter et al. 2017)

| | | $(\rho dry)_{min}$ kg/m$^3$ | $(\rho dry)_{max}$ kg/m$^3$ | $e_{max}$ | $e_{min}$ |
|---|---|---|---|---|---|
| Cooper Labs (UCD) | ASTM D4254 & ASTM D4253 | 1515 | 1736 | 0.75 | 0.53 |
| GeoComp (RPI) | ASTM D4254 & ASTM D4253 | 1494 | 1758 | 0.77 | 0.51 |
| Andrew Vasco (GWU) | ASTM D4253 | 1538 | 1793 | 0.72 | 0.48 |
| Andrew Vasco (GWU) | Lade et al. (1998) | 1521 | 1774 | 0.74 | 0.49 |
| Eduardo Cerna (UCD) | Lade et al. (1998) | 1415 | 1720 | 0.87 | 0.54 |
| Wen-Yi Hung (NCU) | N/A | 1482 | 1781 | 0.79 | 0.49 |
| Yan-Guo Zhou (ZJU) | N/A | 1456 | 1733 | 0.82 | 0.53 |

**Table 3.6** Minimum void ratio

| $e_{min}$ | 1 | 2 | 3 | 4 |
|---|---|---|---|---|
| Max | 0.519 | 0.521 | 0.522 | 0.581 |
| Min | 0.479 | 0.475 | 0.469 | 0.478 |
| Average | 0.500 | 0.501 | 0.491 | 0.515 |
| COV (%) | 2.7 | 2.8 | 3.7 | 6.8 |

**Table 3.7** Maximum void ratio

| $e_{max}$ | 1 | 2 | 3 | 4 |
|---|---|---|---|---|
| Max | 0.786 | 0.779 | 0.777 | 0.825 |
| Min | 0.712 | 0.729 | 0.702 | 0.723 |
| Average | 0.740 | 0.763 | 0.739 | 0.774 |
| COV (%) | 3. | 2. | 3.3 | 5. |

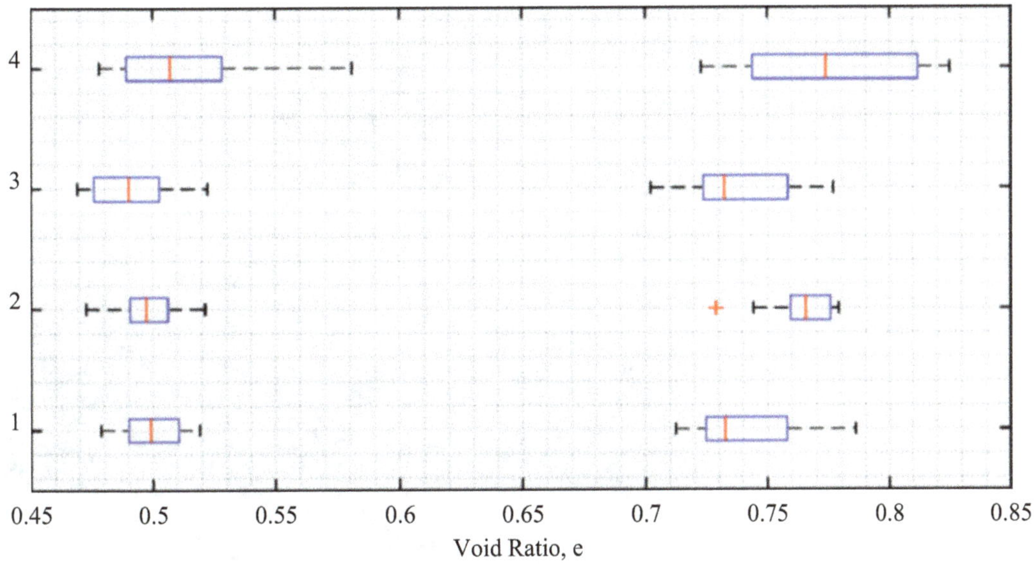


**Fig. 3.5** Scatter of the maximum and minimum void ratios obtained by different experimenters

The scatter of the results is represented by a box plot in Fig. 3.5. It can be seen from the results that data sets 1–3 have a consistent range of variation while it is higher for data set 4. It can also be seen that there is a larger variation in the maximum void ratio than in the minimum void ratio.

Carey et al. (2019) performed additional tests to obtain maximum and minimum void ratios on the samples obtained from the testing facilities that participated in LEAP 2017 project. These tests were also conducted using the same procedures as the ones employed in the tests reported in this paper. The results showed similar ranges of variation in the $e_{min}$ and $e_{max}$. As a compromise, the median $e_{max}$ and $e_{min}$ values obtained by using the ASTM procedures were adopted for future analysis of LEAP data, where $e_{max} = 0.78$ and $e_{min} = 0.51$. These agreed upon values were adopted after the LEAP 2017 prediction exercise.

## 3.3 Cyclic Triaxial Tests

Stress-controlled tests were conducted on three different soil densities to evaluate the soil's liquefaction strength. The three soil densities have average initial void ratios of 0.585, 0.547, and 0.515. The experiment setup and procedures and the sample preparation technique are first discussed. Then, the results obtained are presented.

In this study the testing equipment used was the CKC e/p cyclic triaxial testing system manufactured by C.K. Chan at the Soil Engineering Equipment Company. The setup of the machine is shown in Fig. 3.6. The system is an electro-pneumatic system, which relies on in-house pressure for operation. Different testing conditions are permitted in this system including consolidation testing, monotonic stress/strain-controlled testing, and cyclic stress/strain-controlled testing among other tests.

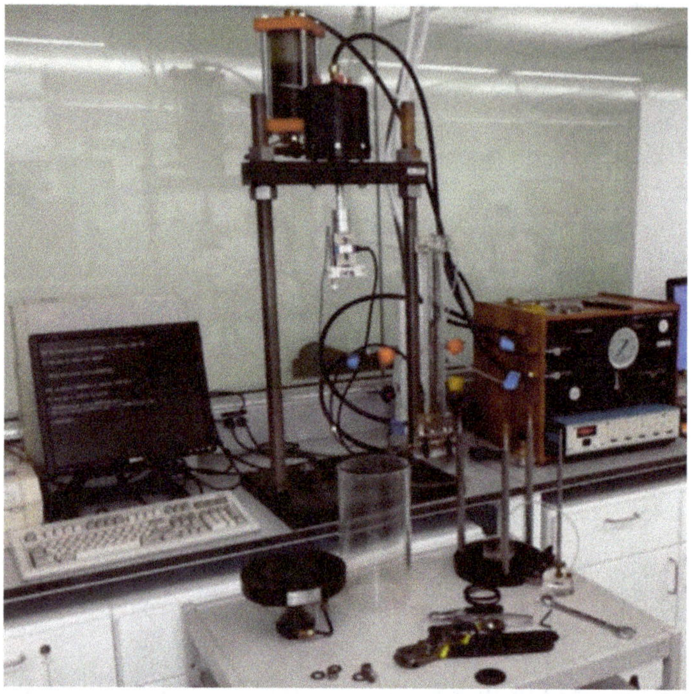

**Fig. 3.6** CKC e/p cyclic triaxial testing machine

### *3.3.1 Experiment Procedures*

A cyclic triaxial experiment consists of sample preparation, saturation, consolidation, and shearing. In sample preparation the soil specimen is set up to match the target density. This step is the most critical to the consistency among experiments. Careful measurements have to be taken to reduce the variability in the results.

For the cyclic triaxial experiments performed in this study, the following steps were followed. First, the triaxial cell was cleaned and then dried using pressurized air. Then, a rubber membrane was placed around the bottom cap. The thickness of the membrane was measured from its top and bottom. Afterwards the triaxial mold was placed around the bottom cap and the membrane was stretched over the mold. Vacuum was then applied to the mold to hold the membrane onto it. At this point, the weight of the cell base, the mold, and the porous stones used was measured. Next, the sample was prepared using the *constant height dry pluviation* method to achieve a target density. Once, the sample was prepared, vacuum was applied to the soil specimen to hold its shape. The weight of the soil specimen, cell base, the mold, and porous stones was then measured. The difference between the two measurements of the weight yields the dry weight of the soil specimen. Then the dimensions of the soil specimen were taken including the diameter and the sample height at five different locations. From the measurement obtained, the volume of the specimen, the dry density, and the initial void ratio were computed.

Figure 3.7 shows the prepared soil specimen subjected to vacuum and attached to the triaxial base and top cap. The statistics obtained on the measurements of the sample dimensions and weight are shown in Table 3.8. It can be seen from the results

**Fig. 3.7** Prepared sample subject to vacuum

**Table 3.8** Statistics of soil specimens

| Measurement | Mean | COV (%) |
|---|---|---|
| Height, mm | 164.64 | 0.24 |
| Diameter, mm | 71.14 | 0.18 |
| Weight ($e_o = 0.515$), g | 1142.3 | 0.30 |
| Weight ($e_o = 0.547$), g | 1120.3 | 0.42 |
| Weight ($e_o = 0.585$), g | 1089.0 | 0.51 |

in the table that the coefficient of variations for the specimen measurements was below 1%. Figures 3.8, 3.9, and 3.10 show the cumulative distribution for the sample weight, height, and diameter. The distribution of the measured data can be fitted using a normal distribution as can be seen in the figures. While Fig. 3.8 shows the distribution of the sample weight for each void ratio separately, Figs. 3.9 and 3.10 combines the measurements of all the tests specimen for the sample height and diameter.

Once the sample was prepared, then it was enclosed with the cell chamber. The sample was then saturated with de-aerated water. Figure 3.11 shows the saturation

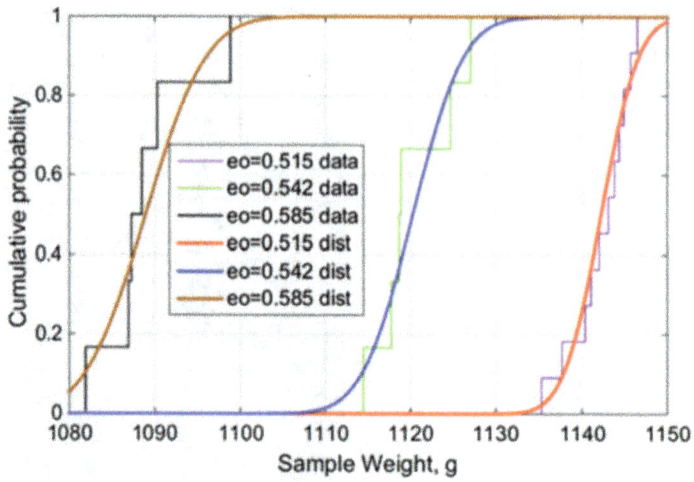


**Fig. 3.8** Experimental and fitted cumulative distribution of the sample weight data

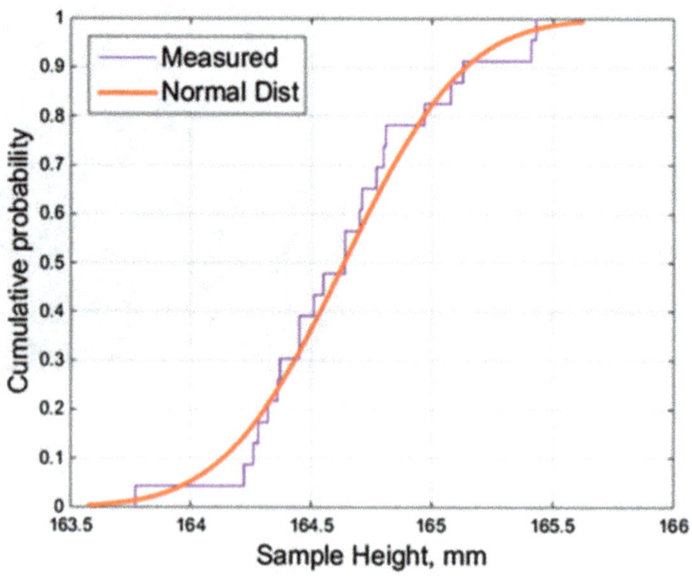


**Fig. 3.9** Experimental and fitted cumulative distribution of the sample height data

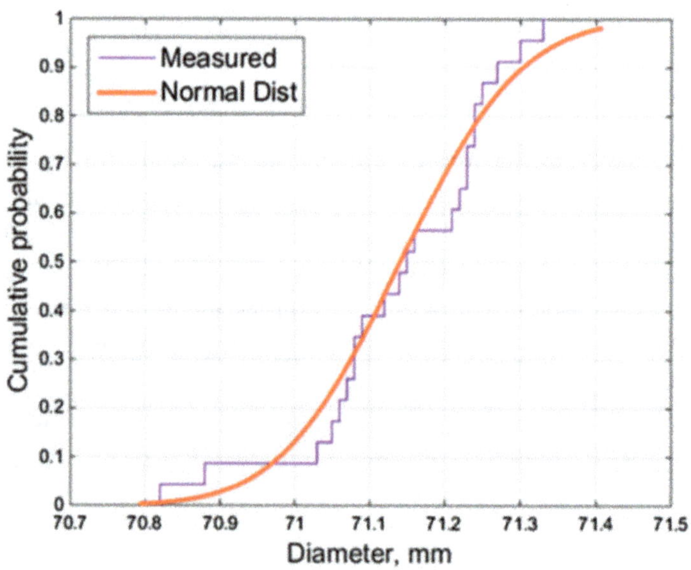


**Fig. 3.10** Experimental and fitted cumulative distribution of the sample diameter data

**Fig. 3.11** Saturation of test specimen

setup which was composed of two containers. The first container was filled with the de-aerated water and connected to the bottom of the soil specimen. The second one was subjected to vacuum and connected to the top of the specimen. The saturation step continued until there were no air bubbles coming in to the vacuumed container. After this step was completed, the triaxial cell was attached to the load frame and the pressure tubes were connected to the soil specimen and the chamber.

Once the specimen was connected to the machine, the testing sequence began. First the specimen was subjected to back pressure saturation. Pressure increments of 10 kPa were gradually applied until a B value greater than 0.95 was achieved. Next, the specimen was consolidated to an effective confining pressure of 100 kPa for 45 min. At the end of the consolidation, the volumetric strain developed was obtained and the void ratio was updated. Subsequently, the specimen was subjected to the cyclic shear stress that matches the target cyclic stress ratio (CSR). In the shearing step, the load was applied at a period of 1 min/cycle for the denser specimens with void ratios of 0.515 and 0.542, while a frequency of 3 min/cycle was used for the looser specimens with a void ratio of 0.585.

### 3.3.2 Sample Preparation

A *constant height dry pluviation* method was used for the sample preparation. In such method the target density of the sample is achieved by pouring the sand into the mold from a specific height. In order to create a more uniform sample, a constant height of drop should be maintained between the sample surface and hopper. In order to produce samples with similar fabric and consistency, two types of sand pluviators were considered.

The first pluviator, Fig. 3.12, was used for preparing the sample with void ratio of 0.515 ($\rho_d = 1774.2$ kg/m$^3$). The pluviator is composed of a sand hopper with a top diameter of 24 cm and total depth of 21 cm. The exit diameter of the hopper is 7.5 cm. The exit of the bucket is covered with a shutter plate with 13 openings. Each opening has a diameter of 0.5 cm. The end of the pluviator has a number 8 sieve with a mesh opening of 2.36 mm. The distance between the sieve and the shutter plate is 15.0 cm. The drop height of the sand (distance from the sieve to the surface of the deposited sand) is 19.8 cm. In order to control the flow rate of the sand and the uniformity of the sand placed in the pluviator, a dispersion cup hanging from the top of the pluviator is used to place the sand in the hopper with a consistent and uniform density.

The second pluviator, Fig. 3.13, was used for preparation of specimen with initial void ratios 0.585 ($\rho_d = 1665.6$ kg/m$^3$) and 0.542 ($\rho_d = 1712.6$ kg/m$^3$). The pluviator is composed of a hopper with the same dimensions as pluviator 1 with a diameter of 24.0 cm and depth of 21 cm. The distance from the exit of the hopper to the end of

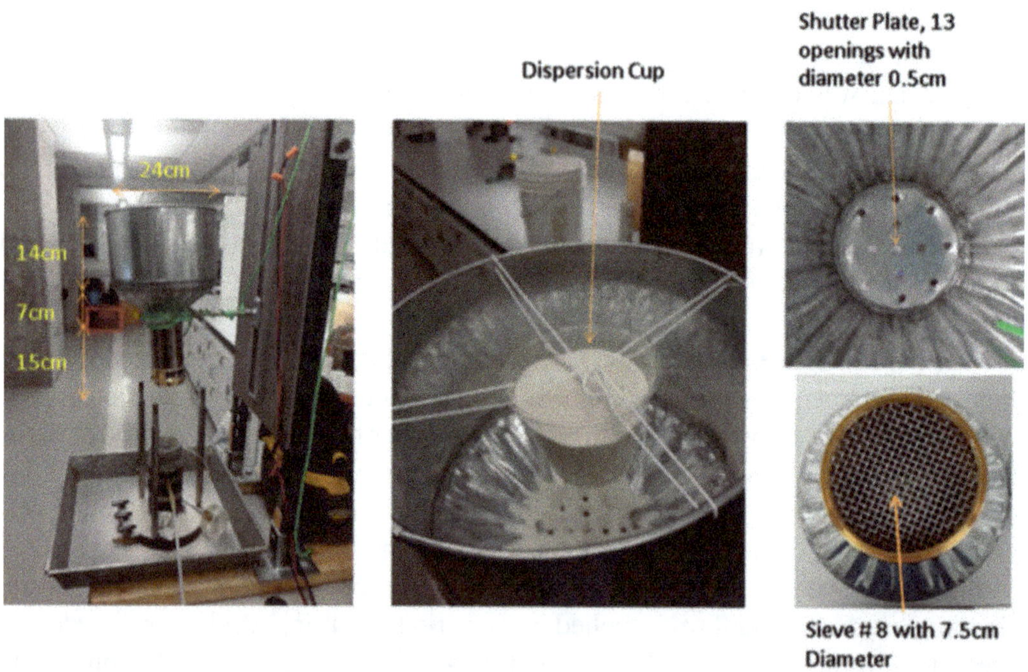


**Fig. 3.12** Sample pluviator 1

**Fig. 3.13** Sample pluviator 2

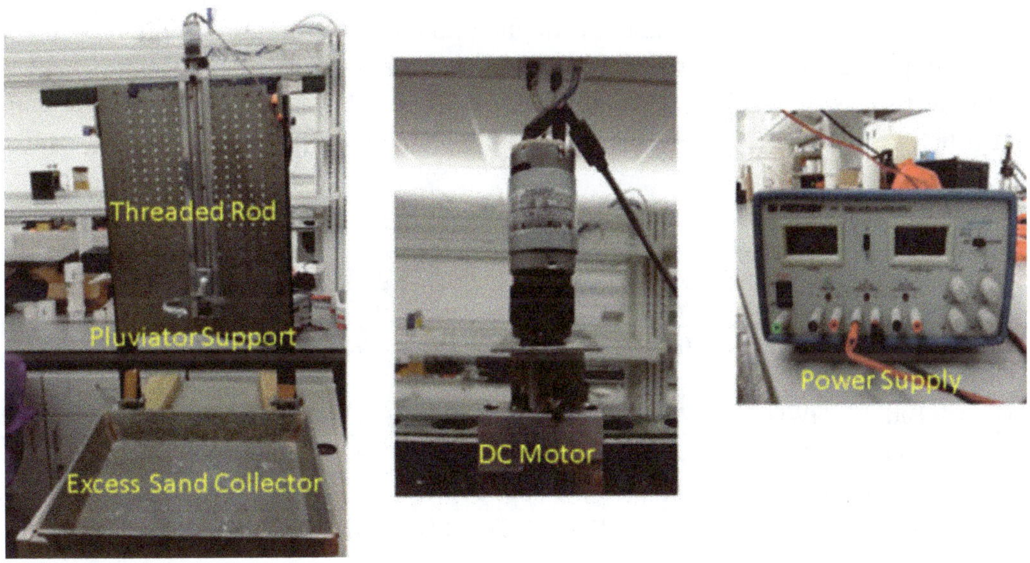


**Fig. 3.14** Pluviator lift

the pluviator is 30 cm. The end of the pluviator has an opening of 2.0 cm, and it is covered with a number 8 sieve. The drop height is 2.0 cm and 4.0 cm for samples with void ratio of 0.585 and 0.542, respectively. A dispersion cup is used for sample uniformity and flow rate control.

In order to maintain a constant height during the sample preparation, a pluviator lift, Fig. 3.14, was used. The pluviator lift is composed of a support for the sample pluviator. The support is attached to a steel frame, and it is moved upwards and downwards through a threaded rod. The movement of the threaded rod is controlled by a DC motor. The speed of the motor is controlled by a power supply. During the sample preparation, the sand is gradually deposited until it flows from the surface of the mold. Afterwards the sample surface is leveled in order to ensure a consistent sample height. The excess sand is collected in the sand collector.

### 3.3.3 Summary of Experimental Results and Observations

Three sets of experiments were conducted with the following void ratios of 0.515, 0.542, and 0.585 ($\rho_d$ = 1744.2, 1712.6, 1665.6 kg/m$^3$). As it was shown in the previous section (Tables 3.5 and 3.6), the reported maximum and minimum void ratios for Ottawa F65 sand by various methods have significant variability that causes a significant uncertainty in the computed relative density for a given void ratio. Table 3.9 summarizes the computed relative density for each achieved void ratio based on the average reported maximum and minimum void ratios. It can be seen from the results in Table 3.9 that the relative density corresponding to the achieved void ratio can have a coefficient of variation of about 12%.

In Fig. 3.15 the results obtained from one of the cyclic triaxial tests are presented. The sample tested is a dense sample with an initial void ratio of 0.515 and a confining stress of 100 kPa. The specimen is subjected to a cyclic shear stress with a cyclic stress ratio (CSR = $\sigma_d/2\sigma'^3$) of 0.325. The effective stress path, the stress

**Table 3.9** Relative density of tested specimen according to the reported $e_{min}$ and $e_{max}$

| Tested by | Institute | $\rho_{min}$ kg/m$^3$ | $\rho_{max}$ kg/m$^3$ | $e_{max}$ | $e_{min}$ | $e$ = 0.515 Dr (%) | $e$ = 0.542 Dr (%) | $e$ = 0.585 Dr (%) |
|---|---|---|---|---|---|---|---|---|
| Mohamed El Ghoraiby | GWU | 1503 | 1765 | 0.76 | 0.5 | 94.8 | 84.5 | 68.1 |
| Hannah Park | GWU | 1523 | 1767 | 0.74 | 0.5 | 93.6 | 82.4 | 64.5 |
| Andrew Vasko | GWU | 1524 | 1777 | 0.74 | 0.49 | 90.4 | 79.5 | 62.1 |
| Andrew Vasko | GWU | 1538 | 1793 | 0.72 | 0.48 | 84.9 | 73.9 | 56.3 |
| Eduardo Cerna | UC Davis | 1415 | 1720 | 0.87 | 0.54 | 107.7 | 99.6 | 86.7 |
| Cooper Labs | UC Davis | 1515 | 1736 | 0.75 | 0.53 | 105.2 | 93.0 | 73.7 |
| Wen-Yi Hung | NCU | 1482 | 1781 | 0.79 | 0.49 | 91.0 | 82.0 | 67.7 |
| GeoComp | RPI | 1494 | 1758 | 0.77 | 0.51 | 97.1 | 87.0 | 70.9 |
| Yan-Guo (Eagle) ZHOU | ZJU | 1456 | 1733 | 0.82 | 0.53 | 104.9 | 95.6 | 80.8 |
| Ana Maria Parra Bastidas | UC Davis | 1455 | 1759 | 0.82 | 0.51 | 97.3 | 88.7 | 75.1 |
| Carey et al. (2019) | Various | 1491 | 1757 | 0.78 | 0.51 | 97.5 | 87.5 | 71.5 |
| Average | | | | | | 96.8 | 86.7 | 70.7 |
| COV | | | | | | 7.2 | 8.6 | 12 |

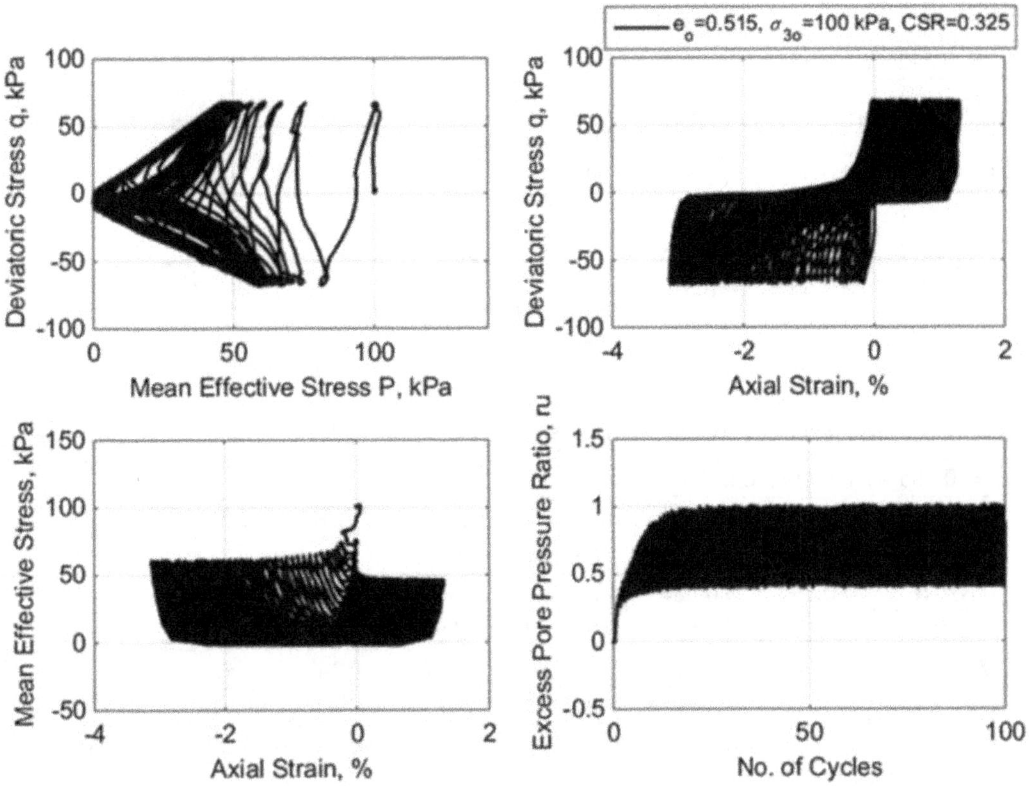


**Fig. 3.15** Triaxial experiment results of Ottawa F65 sand specimen with void ratio of 0.515 and CSR of 0.325

**Table 3.10** Summary of experiments on specimen with $e_o = 0.515$

| $e_o = 0.515 - \rho_d = 1744.2$ kg/m$^3$ | | | | | | | |
|---|---|---|---|---|---|---|---|
| Date | $e_a$ | $e_o$ | B | $e_s$ | CSR | 2.5% s.a. | $r_u = 1.0$ |
| 10/10/2016 | 0.522 | 0.515 | 0.971 | 0.499 | 0.600 | 14 | 15.53 |
| 10/6/2016 | 0.526 | 0.522 | 0.953 | 0.506 | 0.500 | 17 | 26 |
| 10/7/2016 | 0.517 | 0.512 | 0.977 | 0.497 | 0.450 | 19 | 15.55 |
| 9/16/2016 | 0.520 | 0.515 | 0.962 | 0.500 | 0.375 | 26 | 25 |
| 9/27/2016 | 0.518 | 0.514 | 0.952 | 0.501 | 0.365 | 29 | 16 |
| 9/30/2016 | 0.512 | 0.507 | 0.959 | 0.494 | 0.325 | 41 | 18 |
| 9/22/2016 | 0.518 | 0.514 | 0.950 | 0.502 | 0.315 | 46 | 37 |
| 9/28/2016 | 0.521 | 0.516 | 0.951 | 0.500 | 0.300 | 48 | 31 |
| 9/13/2016 | 0.522 | 0.517 | 0.961 | 0.505 | 0.275 | 60 | 35 |
| 9/29/2016 | 0.520 | 0.515 | 0.959 | 0.503 | 0.265 | 70 | 45 |
| 10/12/2016 | 0.517 | 0.513 | 0.951 | 0.501 | 0.225 | 191 | 140 |

strain response, the mean effective stress vs. strain response, and the excess pore pressure development are shown. It can be seen that the response of the soil specimen follows a typical cyclic mobility pattern. The liquefaction strength of the soil is measured based on the number of cycles to achieve a certain level of shear strain. In this study, the criterion is taken to be the number of cycles to develop a single amplitude shear strain of 2.5%. Tables 3.10, 3.11, and 3.12 show a summary

**Table 3.11** Summary of experiments on specimen with $e_o = 0.542$

| $e_o = 0.542 - \rho_d = 1712.6$ kg/m$^3$ | | | | | | | |
|---|---|---|---|---|---|---|---|
| Date | $e_a$ | $e_o$ | B | $e_s$ | 2.5% s.a. | $r_u = 1.0$ | CSR |
| 11/20/2016 | 0.556 | 0.550 | 0.956 | 0.539 | 16 | 12 | 0.28 |
| 11/18/2016 | 0.545 | 0.540 | 0.96 | 0.529 | 18 | 16 | 0.24 |
| 11/16/2016 | 0.540 | 0.535 | 0.958 | 0.523 | 22 | 14 | 0.22 |
| 11/16/2016 | 0.544 | 0.538 | 0.973 | 0.527 | 28 | 25 | 0.21 |
| 11/16/2016 | 0.555 | 0.550 | 0.971 | 0.533 | 41 | 36 | 0.2 |
| 11/21/2016 | 0.544 | 0.538 | 0.958 | 0.524 | 50 | 41 | 0.19 |

**Table 3.12** Summary of experiments on specimen with $e_o = 0.585$

| $e_o = 0.585 - \rho_d = 1665.6$ kg/m$^3$ | | | | | | | |
|---|---|---|---|---|---|---|---|
| Date | $e_a$ | $e_o$ | B | $e_s$ | CSR | 2.5% s.a. | $r_u = 1.0$ |
| 11/10/2016 | 0.589 | 0.581 | 0.955 | 0.57 | 0.2 | 9 | 9 |
| 11/1/2016 | 0.592 | 0.584 | 0.963 | 0.562 | 0.17 | 15 | 13 |
| 11/2/2016 | 0.592 | 0.587 | 0.953 | 0.567 | 0.16 | 17 | 16 |
| 11/4/2016 | 0.581 | 0.575 | 0.955 | 0.557 | 0.14 | 33 | 31 |
| 11/7/2016 | 0.605 | 0.598 | 0.958 | 0.581 | 0.12 | 59 | 58 |
| 11/14/2016 | 0.588 | 0.583 | 0.954 | 0.566 | 0.10 | 188 | 186 |

**Table 3.13** Statistics of experiments on specimen with $e_o = 0.515$

| | $e_a$ | $e_o$ | B | $e_s$ |
|---|---|---|---|---|
| Mean | 0.519 | 0.515 | 0.959 | 0.501 |
| SD | 0.004 | 0.004 | 0.009 | 0.003 |
| COV (%) | 0.696 | 0.703 | 0.918 | 0.681 |

**Table 3.14** Statistics of experiments on specimen with $e_o = 0.542$

| | $e_a$ | $e_o$ | B | $e_s$ |
|---|---|---|---|---|
| Mean | 0.547 | 0.542 | 0.963 | 0.529 |
| SD | 0.007 | 0.007 | 0.007 | 0.006 |
| COV (%) | 1.199 | 1.204 | 0.765 | 1.136 |

of the experiments performed for each initial void ratio. The tables show the achieved sample void ratio after pluviation ($e_a$), consolidation ($e_o$), and shearing ($e_s$) the test specimens. The number of cycles until 2.5% single amplitude (s.a.) of strain and until an excess pore pressure ratio of 1.0 are first achieved is shown. Tables 3.13, 3.14, and 3.15 show the statistics of the void ratios and the B values achieved. Figure 3.16 shows the liquefaction strength curves obtained for the three soil densities tested. The results of each experiment are archived in a databank on DesignSafe allowing for public access (El Ghoraiby et al. 2018).

While the curves shown in Fig. 3.16 have been used to describe the liquefaction strength of the soil, they only represent a single snapshot of the soil response at the

**Table 3.15** Statistics of experiments on specimen with $e_o = 0.585$

| | $e_a$ | $e_o$ | B | $e_s$ |
|---|---|---|---|---|
| Mean | 0.591 | 0.585 | 0.956 | 0.567 |
| SD | 0.008 | 0.008 | 0.004 | 0.008 |
| COV (%) | 1.334 | 1.310 | 0.384 | 1.434 |

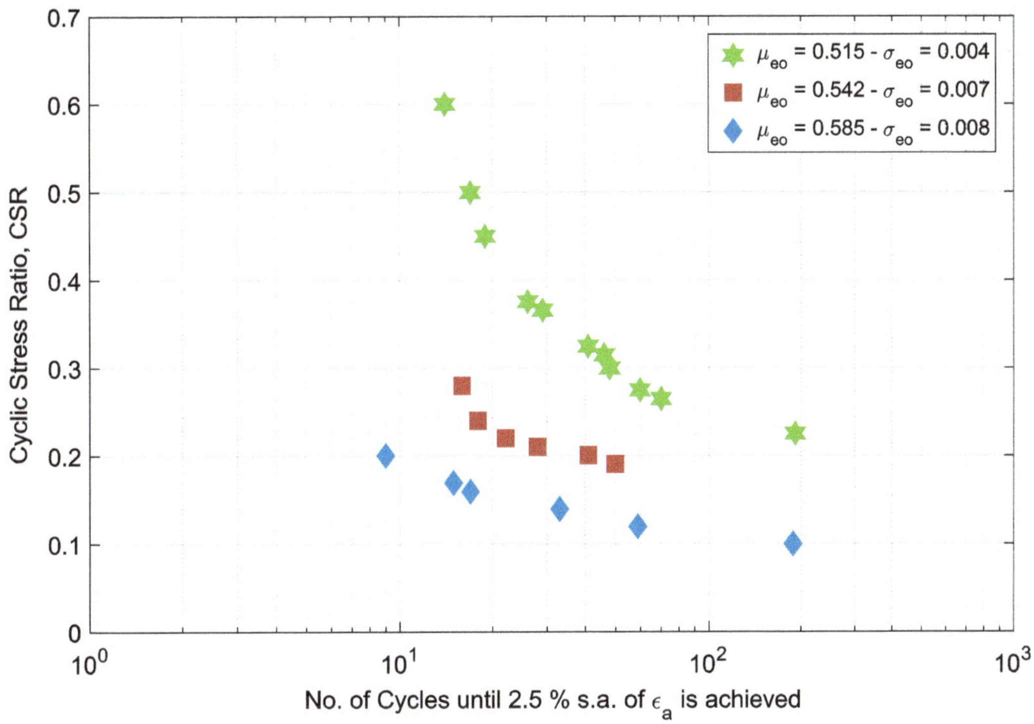


**Fig. 3.16** Liquefaction strength curves for Ottawa F65 sand

end of the tests. In order to gain more insight on the progression of the soil response, additional information were obtained by counting the number of cycles it took to reach different levels of shear strain. Figures 3.17, 3.18, and 3.19 show the strength curves corresponding to a range of strains (0.5–2.5%) for the soil specimens with initial void ratios of 0.585, 0.542, and 0.515, respectively. It can be seen that as the soil density increases, the number of shear cycles required to move from one shear strain level to the next level increases.

It is also useful to observe the progression of the soil response in terms of excess pore water pressure ratio, $r_u$. Figures 3.20, 3.21, and 3.22 show the variation of number of shear stress cycles required to reach certain levels of $r_u$ (in the range of 0.7–0.99) for the samples with initial void ratios of 0.585, 0.542, and 0.515, respectively. It is observed that as the soil density increases the number of shear stress cycles required to generate certain levels of excess pore pressure ratio increases. For the loosest sand, the number of cycles to reach $r_u = 0.7$ is slightly less than the number of cycles to reach $r_u = 0.99$. For the denser sand, the number of cycles to reach $r_u = 0.7$ is much less than the number of cycles to reach $r_u = 0.99$.

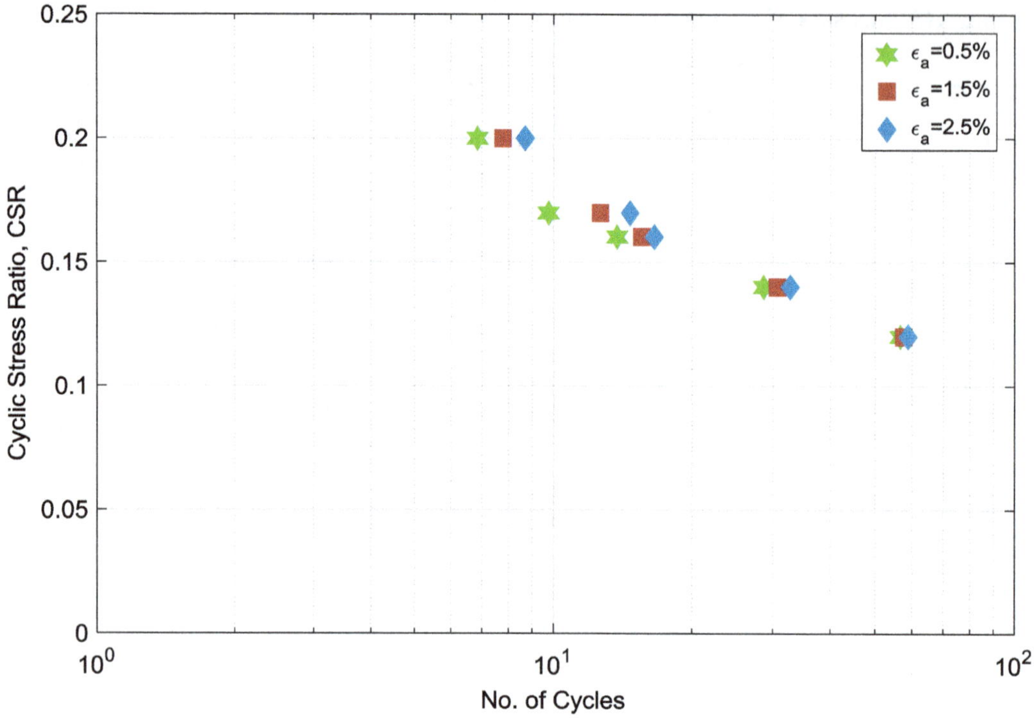


**Fig. 3.17** Strength curves for different axial strain amplitudes—$e_o = 0.585$

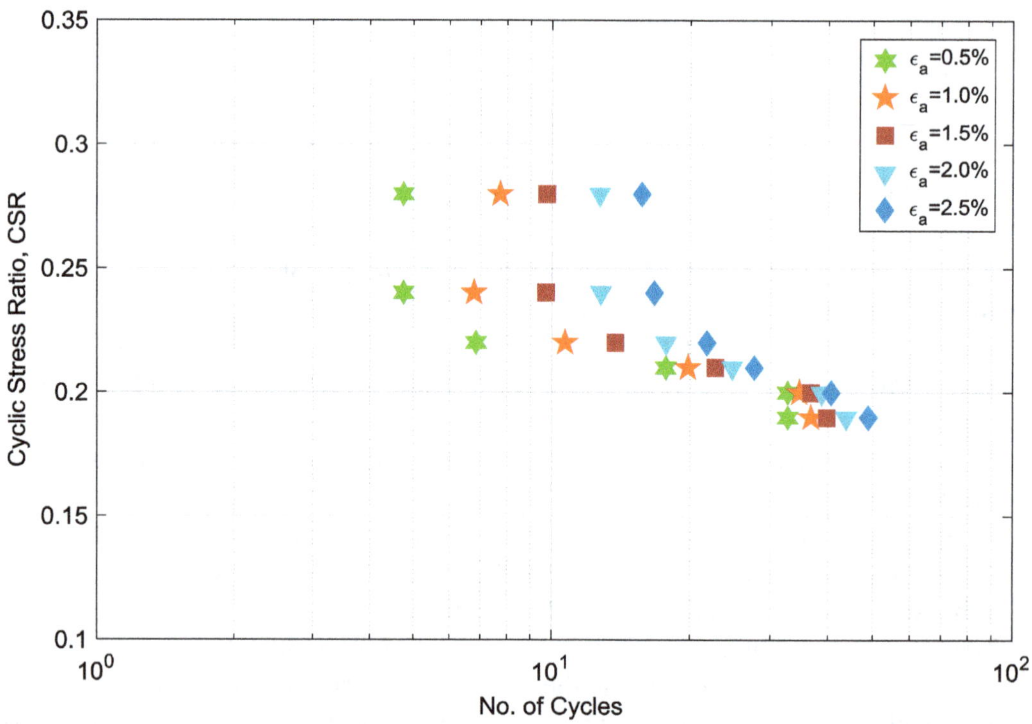


**Fig. 3.18** Strength curves for different axial strain amplitudes—$e_o = 0.542$

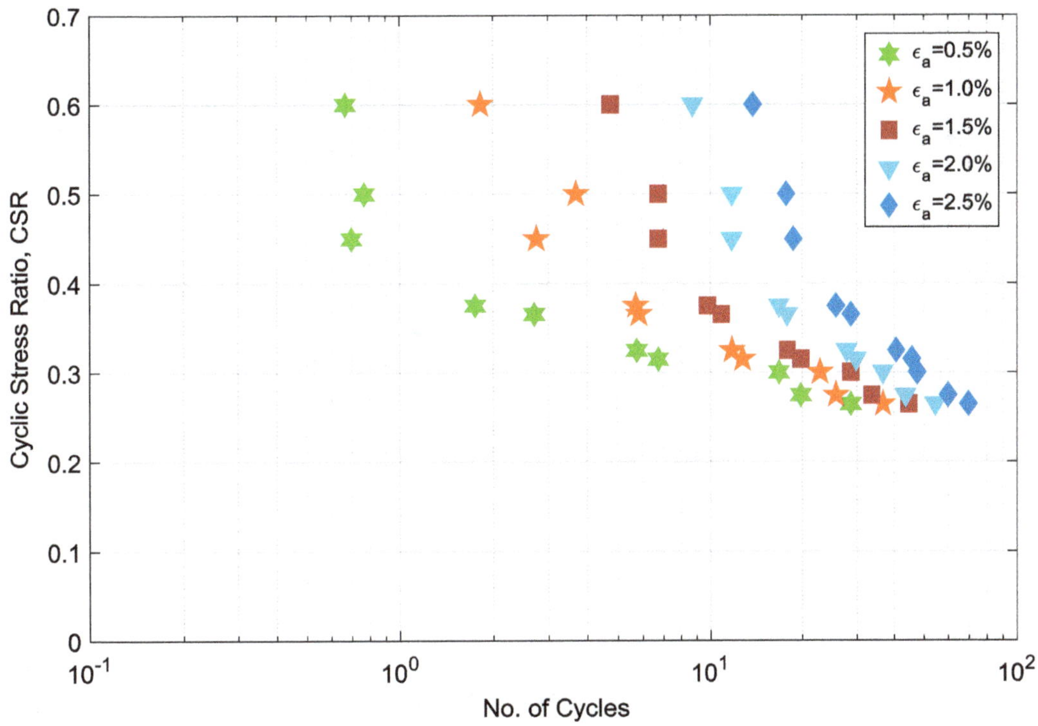


**Fig. 3.19** Strength curves for different axial strain amplitudes—$e_o = 0.515$

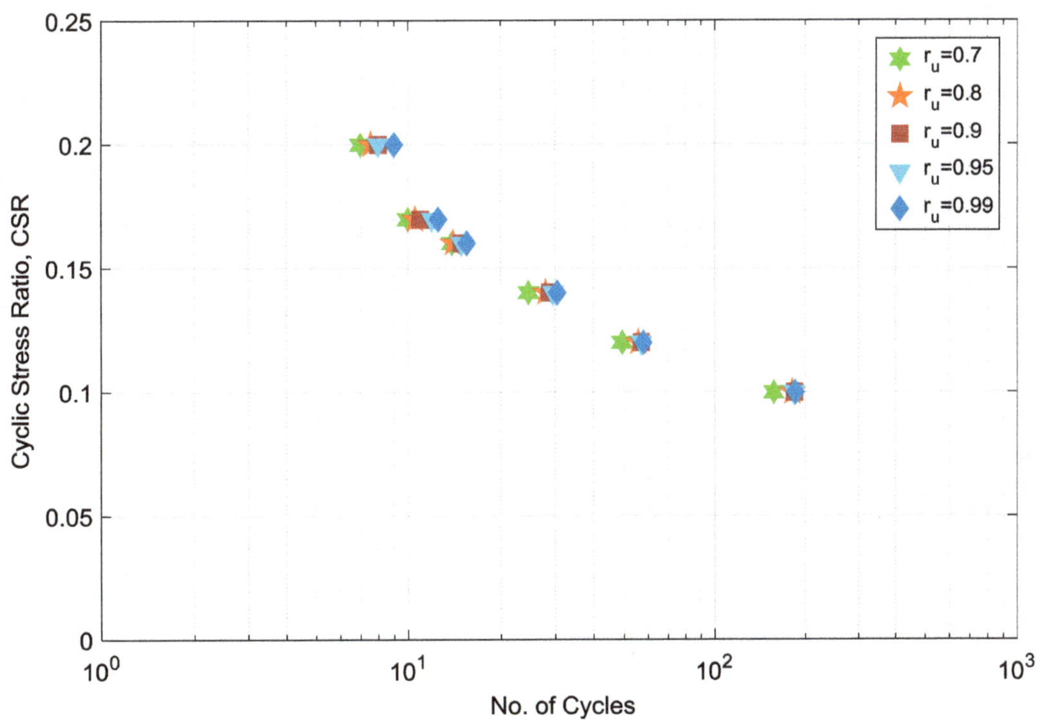


**Fig. 3.20** Strength curves for different excess pore pressure ratios—$e_o = 0.585$

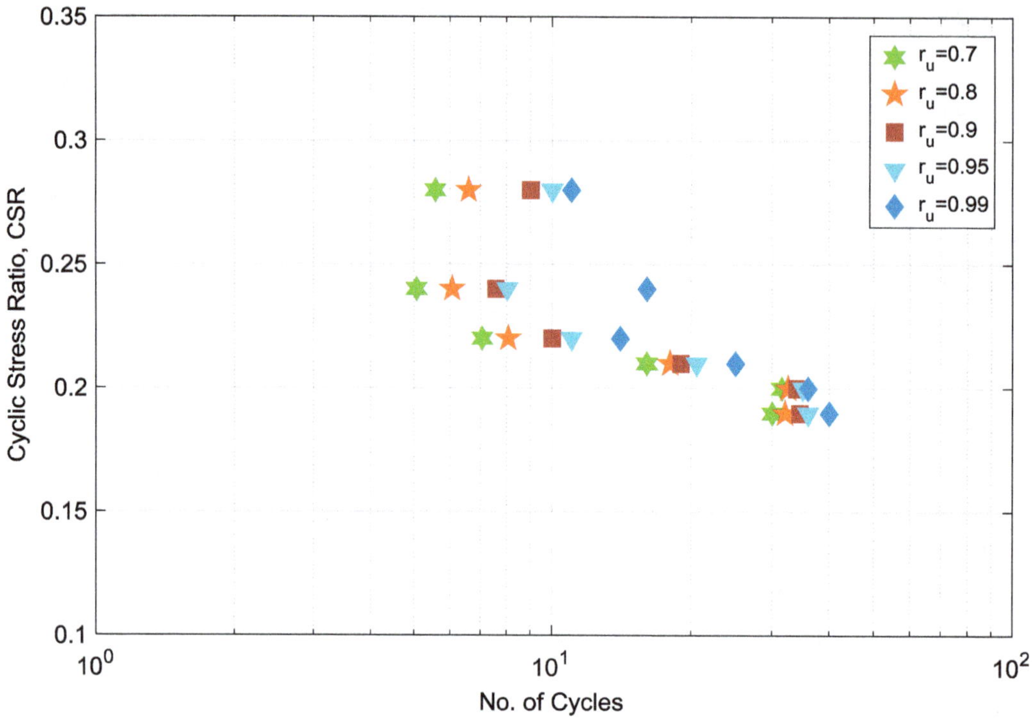


**Fig. 3.21** Strength curves for different excess pore pressure ratios—$e_o = 0.542$

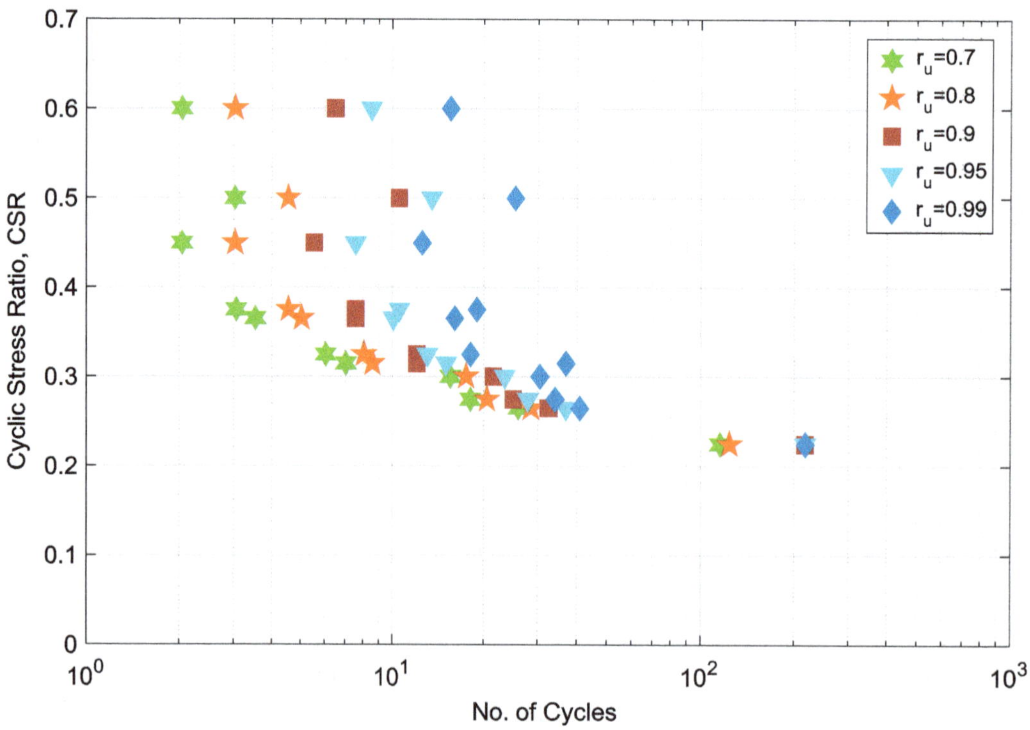


**Fig. 3.22** Strength curves for different excess pore pressure ratios—$e_o = 0.515$

## 3.4 Concluding Remarks

A series of laboratory experiments were conducted to characterize the physical and mechanical properties of Ottawa F65 sand. A large number of experiments were performed to obtain specific gravity, particle size distribution, maximum and minimum void ratios, hydraulic conductivity, and stress-strain-strength behavior of Ottawa sand. The results of 23 cyclic triaxial tests were reported for three different initial void ratios. The results of cyclic triaxial tests are presented in form of liquefaction strength curves. A databank of this experimental study has been created and uploaded on DesignSafe (El Ghoraiby et al. 2018).

## References

Bastidas, A. M. B. (2016). *Ottawa F-65 sand characterization*. Davis, CA: University of California.

Carey, T. J., Stone, N., & Kutter, B. L. (2019). Grain size analysis and maximum and minimum dry density of Ottawa F-65 sand for LEAP-UCD-2017. In B. Kutter et al. (Eds.), *Model tests and numerical simulations of liquefaction and lateral spreading: LEAP-UCD-2017*. New York: Springer.

El Ghoraiby, M. A., Park, H., & Manzari, M. (2018). *LEAP-2017 GWU Laboratory Tests*. DesignSafe-CI, Dataset. https://doi.org/10.17603/DS2210X.

Kutter, B. L., Carey, T. J., Hashimoto, T., Zeghal, M., Abdoun, T., Kokkali, P., Madabhushi, G., Haigh, S. K., Burali d'Arezzo, F., Madabhushi, S., Hung, W.-Y., Lee, C.-J., Cheng, H.-C., Iai, S., Tobita, T., Ashino, T., Ren, J., Zhou, Y.-G., Chen, Y.-M., Sun, Z.-B., & Manzari, M. T. (2017). LEAP-GWU-2015 experiment specifications, results, and comparisons. *Soil Dynamics and Earthquake Engineering, 4*, 1–31.

Lade, P. V., Liggio, C. D., & Yamamuro, J. A. (1998). Effects of non-plastic fines on minimum and maximum void ratios of sand. *Geotechnical Testing Journal, 21*(4), 336–347.

Manzari, M. T., ElGhoraiby, M., Kutter, B. L., Zeghal, M., Abdoun, T., Arduino, P., Armstrong, R. J., Beaty, M., Carey, T., Chen, Y., Ghofrani, A., Gutierrez, D., Goswami, N., Haigh, S. K., Hung, W.-Y., Iai, S., Kokkali, P., Lee, C.-J., Madabhushi, S. P. G., Mejia, L., Sharp, M., Tobita, T., Ueda, K., Zhou, Y., & Ziotopoulou, K. (2017). Liquefaction experiment and analysis projects (LEAP): Summary of observations from the planning phase. *Soil Dynamics and Earthquake Engineering*. https://doi.org/10.1016/j.soildyn.2017.05.015

Vasko, A. (2015). *An investigation into the behavior of Ottawa sand through monotonic and cyclic shear tests*. Washington: George Washington University.

Vasko, A., ElGhoraiby, M. A., & Manzari, M. T. (2018). *LEAP-GWU-2015 Laboratory tests*. DesignSafe-CI, Dataset. https://doi.org/10.17603/DS2TH7Q.

# Chapter 4
# LEAP-UCD-2017 Comparison of Centrifuge Test Results

**Bruce L. Kutter, Trevor J. Carey, Nicholas Stone, Bao Li Zheng, Andreas Gavras, Majid T. Manzari, Mourad Zeghal, Tarek Abdoun, Evangelia Korre, Sandra Escoffier, Stuart K. Haigh, Gopal S. P. Madabhushi, Srikanth S. C. Madabhushi, Wen-Yi Hung, Ting-Wei Liao, Dong-Soo Kim, Seong-Nam Kim, Jeong-Gon Ha, Nam Ryong Kim, Mitsu Okamura, Asri Nurani Sjafruddin, Tetsuo Tobita, Kyohei Ueda, Ruben Vargas, Yan-Guo Zhou, and Kai Liu**

B. L. Kutter (✉) · T. J. Carey · N. Stone · B. L. Zheng · A. Gavras
Department of Civil and Environmental Engineering, University of California, Davis, CA, USA
e-mail: blkutter@ucdavis.edu

M. T. Manzari
Department of Civil and Environmental Engineering, George Washington University, Washington, DC, USA

M. Zeghal · T. Abdoun · E. Korre
Department of Civil and Environmental Engineering, Rensselaer Polytechnic Institute, Troy, NY, USA

S. Escoffier
IFSTTAR, GERS, SV, Bouguenais, France

S. K. Haigh · G. S. P. Madabhushi · S. S. C. Madabhushi
Department of Engineering, Cambridge University, Cambridge, UK

W.-Y. Hung · T.-W. Liao
Department of Civil Engineering, National Central University, Taoyuan, Taiwan

D.-S. Kim
Department of Civil and Environmental Engineering, Korea Advanced Institute of Science and Technology, Daejeon, South Korea

S.-N. Kim
Water Management Department, Korea Advanced Institute of Science and Technology, Daejeon, South Korea

J.-G. Ha
Korea Advanced Institute of Science and Technology, Yuseong, South Korea

N. R. Kim
K-water Research Institute, Korea Water Resources Corporation, Daejeon, South Korea

M. Okamura · A. N. Sjafruddin
Department of Civil Engineering, Ehime University, Matsuyama, Japan



B. Kutter et al. (eds.), *Model Tests and Numerical Simulations of Liquefaction and Lateral Spreading*, https://doi.org/10.1007/978-3-030-22818-7_4

**Abstract** This paper compares experimental results from every facility for LEAP-UCD-2017. The specified experiment consisted of a submerged medium-dense clean sand with a 5-degree slope subjected to 1 Hz ramped sine wave base motion in a rigid container. The ground motions and soil density were intentionally varied from experiment to experiment in hopes of defining the slope of the relational trend between response (e.g., displacement, pore pressure), intensity of shaking, and density or relative density. This paper is also intended to serve as a useful starting point for overview of the experimental results and to help others find specific experiments if they want to select a subset for further analysis. The results of the experiments show significant differences between each other, but the responses show a significant correlation, $R^2 \sim 0.7–0.8$, to the known variation of the input parameters.

## 4.1 Introduction

Twenty-four separate model tests were conducted at nine different centrifuge facilities for this LEAP exercise. The first goal of this paper is to provide an overview of all the experimental data from the 24 experiments. This overview will allow readers to quickly scan through the key time series data and various performance measures to evaluate the extent of liquefaction in the different experiments. A second goal of this paper is to demonstrate that the experiments are consistent with each other and that they define a response function or trend between key input parameters and key liquefaction response parameters. From the comparison of the results to empirical response functions, it is possible to obtain meaningful assessments of the sensitivity of the results to variations of input parameters and to assess the variability of the results in terms of their deviation from the response functions.

All of the experiments were intended to model a 4 m-deep, 20 m-long prototype soil deposit of submerged uniform sand with a 5-degree surface slope instrumented as indicated in Fig. 4.1. This paper attempts to compare the results from all of the first shaking events of each model test. The same sand, Ottawa F65 from US Silica, was used in all of the experiments. Carey et al. (2019c) describe results of grain size and max/min density tests of the sand used at different facilities to determine the index properties and to check the consistency of the sand used at various facilities.

El Ghoraiby et al. (2017) and Para Bastidas et al. (2017) report results of laboratory testing including cyclic triaxial and DSS tests. Permeability tests using

T. Tobita
Institute of Geotechnical Engineering, Zhejiang University, Hangzhou, China

K. Ueda · R. Vargas
Disaster Prevention Research Institute, Kyoto University, Kyoto, Japan

Y.-G. Zhou
Department of Civil Engineering, Zhejiang University, Hangzhou, P. R. China

K. Liu
Department of Civil Engineering, Kansai University, Suita, Japan

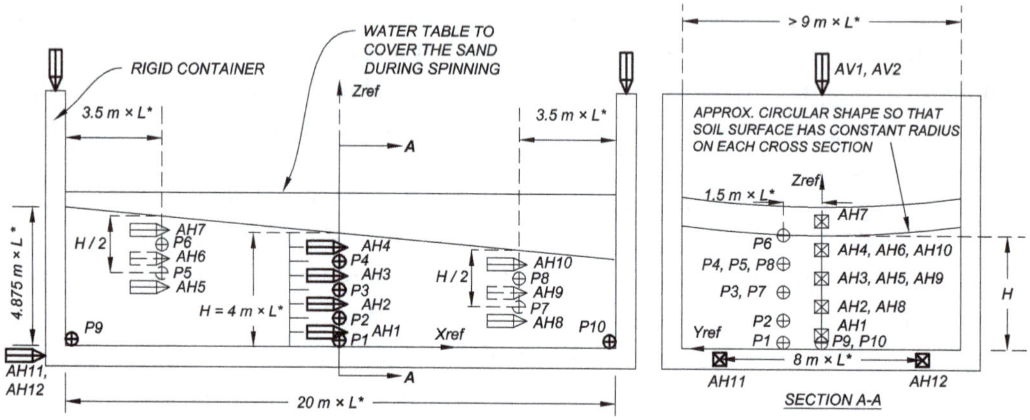


**Fig. 4.1** Configuration of all of the model tests as specified by Kutter et al. (2019)

**Table 4.1** Test facilities, length scale factor, shaking direction, radius of centrifuge, and model container length/width ratio for LEAP-UCD-2017

| Centrifuge facility institution | $L^*$ | Shaking direction | Radius (m) | Container length/ width |
|---|---|---|---|---|
| Cambridge University, UK | 1/40 | Tangential | 3.56 | 0.45 |
| Ehime University, Japan | 1/40 | Parallel to axis | 1.184 | 0.24 |
| IFSTTAR, France | 1/50 | Parallel to axis | 5.063 | 0.5 |
| KAIST, Rep. of Korea | 1/40 | Parallel to axis | 5 | 0.45 |
| Kyoto University, Japan | 1/44.4 | Tangential | 2.5 | 0.32 |
| National Central Univ., Taiwan | 1/26 | Parallel to axis | 2.716 | 0.45 |
| Rensselaer Poly. Inst., USA | 1/23 | Parallel to axis | 2.7 | 0.42 |
| Univ. of California, Davis, USA | 1/43.75 | Tangential | 1.094 | 0.63 |
| Zhejiang University, China | 1/30 | Parallel to axis | 4.315 | 0.59 |

water as the pore fluid as reported by El Ghoraiby et al. (2017) were fit with a linear regression line through data over the range of $e = 0.5–0.75$:

$$k = (0.0207(e) - 0.0009)\ \mathrm{cm/s} \tag{4.1}$$

As the pore fluid viscosity was scaled in the centrifuge tests, this measured permeability corresponds to the prototype permeability.

Some basic parameters including the institutions and direction of shaking for the nine centrifuge facilities are summarized in Table 4.1. The table also shows that the scale factor selected for the model tests ($L^* = L_{model}/L_{prototype}$) varied from 1/50 to 1/23, and the radii varied between 1 and 5 m. The models were all tested in a rigid model container to avoid the uncertainties associated with more complex flexible model containers. To control side boundary effects, Kutter et al. (2019) recommended 0.45 as a minimum desired width/length ratio of the model container; the actual width/length ratios are summarized in Table 4.1.

## 4.2 Densities and Penetration Resistances

Each experimental facility was given suggestions regarding target densities and target input motions for the first shaking event; each site was given some latitude in deciding what input motions to apply in subsequent shaking events, if any. The details of results for all the shaking events should be described in separate papers produced by each experimental facility. The density of the sand in each model was characterized by mass and volume measurements of each model. However, it is deceptively difficult to directly measure the mass and volume to the desired level of accuracy. Small errors due to sand mounding near the container side walls during pluviation, imperfect container rectangularity, and uneven (rough) surfaces at the base and top of the sand deposit, in combination with resolution and accuracy of the load cells used to measure the weight of the sand and the empty container, contribute to the uncertainty of the mass and volume measurement. Also note that the relative density is very sensitive to density; at $D_r = 60\%$, a 1% error in density results in a 6% error in relative density.

For an independent check on the density, new 6 mm-diameter cone penetrometers were developed and manufactured at UC Davis (Carey et al. 2018b, 2019a) and then distributed to the centrifuge sites. The penetrometers had different rod lengths but were otherwise identical. Cone penetration tests were to be performed at the test acceleration prior to each destructive shaking event for every test site. At Cambridge, the centrifuge was spun up, the penetration test was performed, the ground motion was triggered, and finally a second CPT was pushed in the same location of the model all during the same spin, without stopping the centrifuge. For all other sites, the centrifuge was spun up to the test acceleration ($g^* = 1/L^*$) for a penetration test and then stopped for removal of the penetrometer. Then the centrifuge was spun up to the test acceleration to apply the model earthquake. Subsequent penetration tests were done at different locations in subsequent spins. Figure 4.2 shows cone penetration tests for every model test. Blank figures indicate unsuccessful penetration tests.

Since the same cone design and the same sand were used at different centrifuge facilities, the results should be comparable. However, the length scale factor in the centrifuge tests varies between 1/23 and 1/50, so the prototype diameter of the cone varies between 138 and 300 mm. At mid depth of the 4 m-thick prototype layer, the depth/diameter ratio varied between 6.7 and 14.5. Bolton et al. (1999) indicated that for dense specimens ($D_r$~80%) the normalized penetration distance was not sensitive to depth if depth/diameter is greater than about 10. For LEAP, we may expect minor reductions of normalized penetration resistance for cases where depth/diameter < 10. Carey et al. (2019a) observed about 5–10% greater penetration resistance for low g tests (large depth/diameter) compared to high g tests.

Bolton et al. (1999) also identified an effect of the container width on the penetration resistance. Narrow containers produced about 15–20% increase in

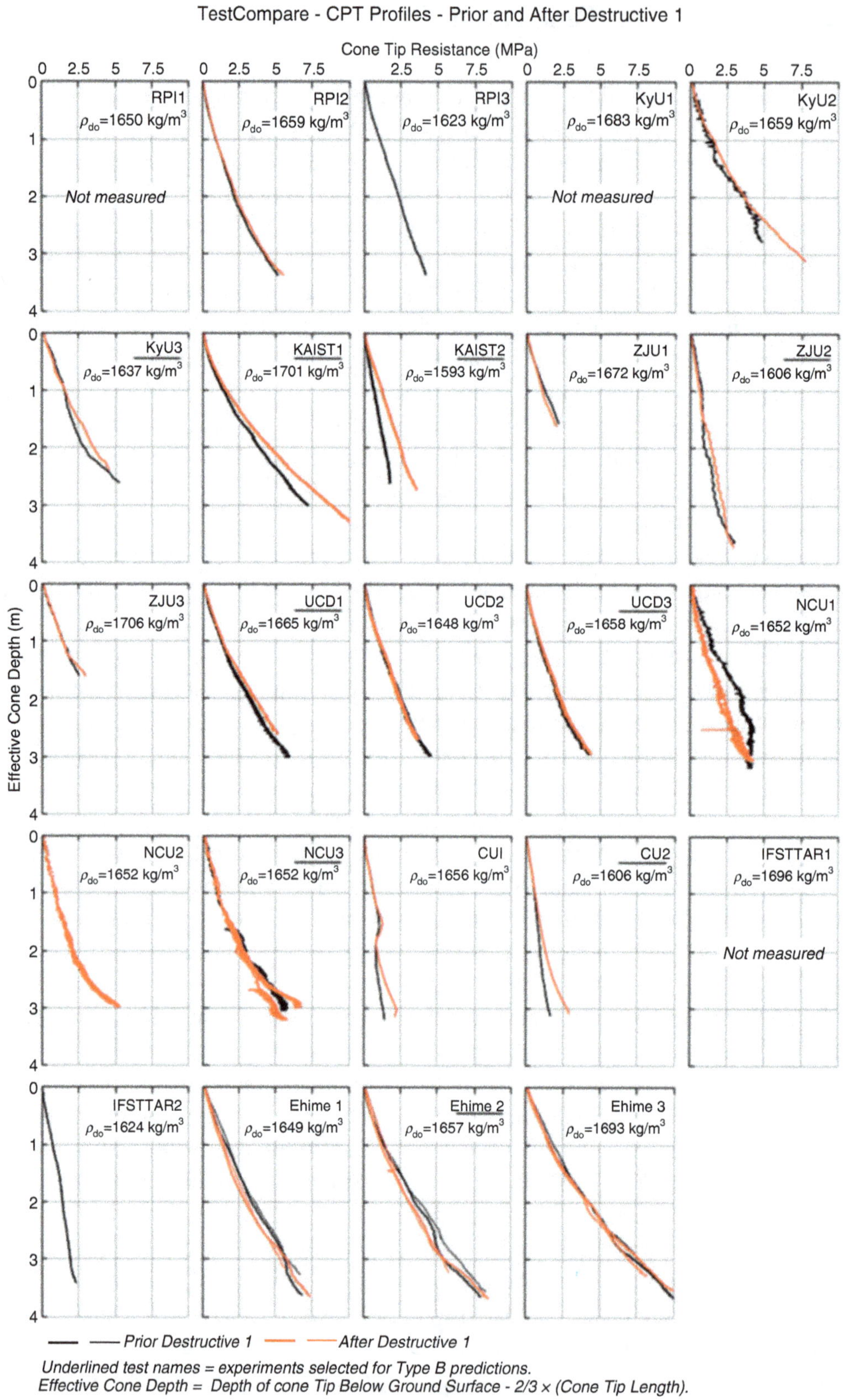


**Fig. 4.2** CPT profiles before the first destructive ground motion

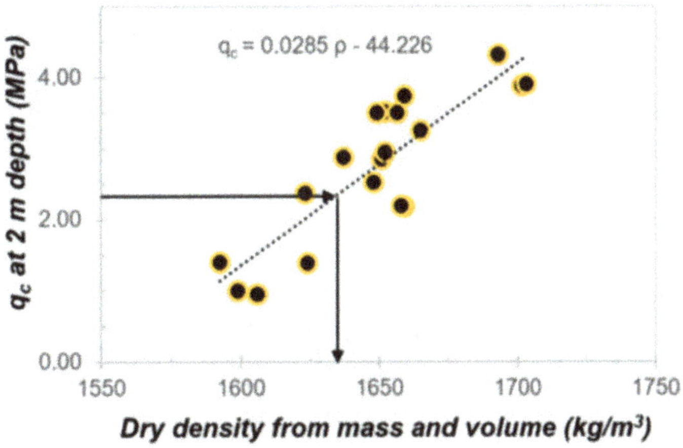


**Fig. 4.3** Correlation between $q_c$ and dry density from CPT tests

penetration resistance. In addition to producing an increased penetration resistance for a given density, wall friction from narrow containers would also restrict liquefaction deformations for a given density; these errors are expected to counteract each other to some extent. Effects of container width on penetration resistance were not accounted for in the correlations presented later in this paper.

As will be demonstrated later, the cone penetration resistance at a 2 m depth is more highly correlated to the liquefaction behavior than is the density determined by mass and volume measurements. For this paper, the dry density was determined by least squares fit to the data shown in Fig. 4.3. The inverse form of the equation of the regression line indicated in Fig. 4.3 is $\rho_d = a(q_c) + b$, with a = 35.1 kg/m$^3$/MPa and b = 1553 kg/m$^3$. As indicated by the arrows in the figure, one model was reported to have a density based on mass and volume measurements of 1623 kg/m$^3$ and a $q_c(2\text{ m}) = 2.37$ MPa. At the intersection of $q_c(2\text{ m}) = 2.37$ MPa and the regression line one finds that the dry density from $q_c$ is $\rho_d(q_c(2\text{ m})) = 1636$ kg/m$^3$.

## 4.3 Base Input Motions in First Destructive Motion

Figure 4.4 shows an example of a horizontal base input motion for test UCD3. Ideally the input motions would have been a smooth ramped sine wave similar to the top trace of Fig. 4.4. While every centrifuge shaker used in LEAP produced motions that contained the ramped sine wave, they also produced high-frequency accelerations superimposed on the ramp. In Fig. 4.4, the bottom trace is the actual achieved base acceleration recorded in the UCD3 test. The top trace of Fig. 4.4 shows the low-frequency (approx. 1 Hz) component of the achieved motion (the portion passed through a 0.5–1.2 Hz band-pass filter); the middle trace is the high-frequency component of the base motion determined by subtracting the low-frequency component from the achieved base motion.

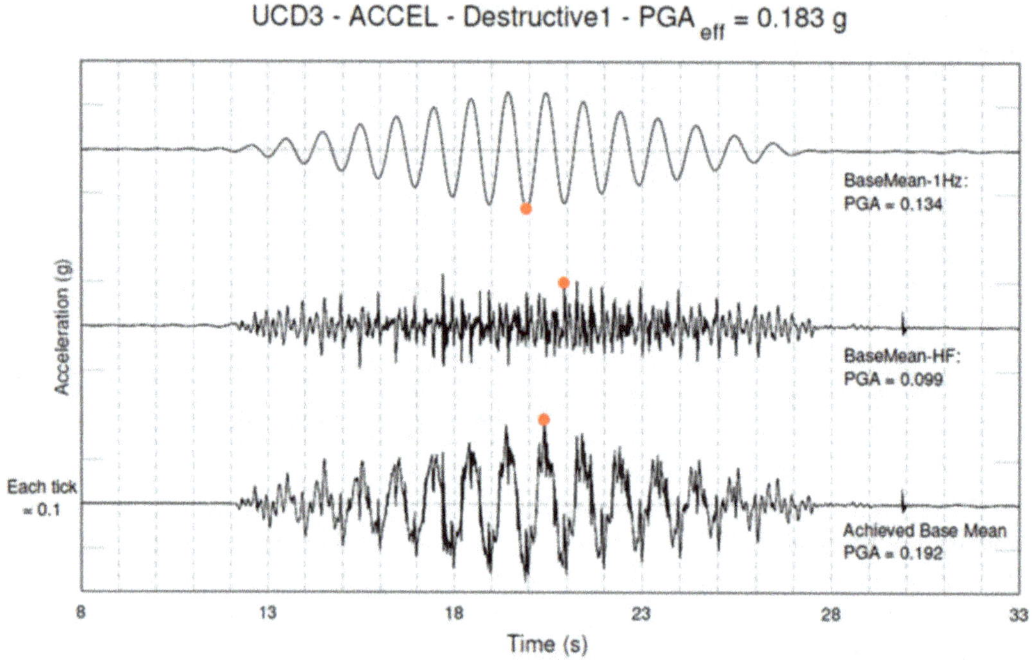


**Fig. 4.4** Example from test UCD3 of a horizontal base input motion and an illustration of the method to determine $PGA_{eff}$

As a measure of the intensity of the achieved input motion, one may use the PGA, the PGA of the low-frequency component ($PGA_{1Hz}$), the PGA of the high-frequency component, or traditional intensity measures such as the PGV, the Arias intensity ($I_a$), or cumulative absolute velocity ($CAV_5$). For the purposes of organizing and comparing the LEAP experiments, another parameter, $PGA_{eff}$, was found to be useful:

$$PGA_{eff} = PGA_{1Hz} + 0.5\ PGA_{HF} \tag{4.2}$$

where $PGA_{HF}$ is determined from the peak of the high-frequency component that occurs within 1 s of the peak of the $PGA_{1Hz}$. (Note that the larger peak of the high-frequency component at about 17.7 s in Fig. 4.4 was not used because it was not near the PGA of the 1 Hz component.)

Table 4.2b summarizes the intensity measures for each of the first destructive motions for each experiment. It should be emphasized that many of the LEAP experiments included a total of two or three destructive motions. This paper focuses on results from the first motion only. Papers by each experiment facility explain the results from subsequent destructive motions.

To allow for a qualitative comparison of the input motions, Fig. 4.5 presents time histories from the horizontal base motions AH11 and AH12 and from the vertical accelerometers AV1 and AV2. The top trace in each subplot shows the measured

**Table 4.2a** Summary of density measures for each of the models

| Test ID | Dry density from mass and volume ρ(M&V) [1] | $D_r$(M&V) assuming $\rho_{max}$ = 1757, $\rho_{min}$ = 1490.5 [2] | Pen. Resist. at 2 m depth $q_c$(2 m) [3] | $\rho(q_c(2\ m)) = a\ q_c + b$ a = 35.1; b = 1553 [4] | $D_r$ from $q_c$ ρ from [4] and $\rho_{max}$ = 1757 $\rho_{min}$ = 1490.5 [5] |
|---|---|---|---|---|---|
| | kg/m$^3$ | | MPa | kg/m$^3$ | |
| CU1 | 1656 | 0.66 | 0.81 | 1581 | 0.38 |
| CU2 | 1606 | 0.47 | 0.95 | 1586 | 0.40 |
| Ehime1 | 1649 | 0.63 | 3.50 | 1676 | 0.73 |
| Ehime2 | 1657 | 0.66 | 3.50 | 1676 | 0.73 |
| Ehime3 | 1693 | 0.79 | 4.31 | 1704 | 0.83 |
| IFSTTAR1 | 1696 | 0.80 | | | |
| IFSTTAR2 | 1624 | 0.56 | 1.38 | 1602 | 0.46 |
| KAIST1 | 1701 | 0.82 | 3.88 | 1689 | 0.77 |
| KAIST2 | 1593 | 0.42 | 1.40 | 1602 | 0.46 |
| KyU1 | 1683 | 0.75 | | | |
| KyU2 | 1659 | 0.67 | 3.74 | 1684 | 0.76 |
| KyU3 | 1637 | 0.59 | 2.88 | 1654 | 0.65 |
| NCU1 | 1652 | 0.64 | 3.51 | 1676 | 0.73 |
| NCU2 | 1652 | 0.64 | | | |
| NCU3 | 1652 | 0.64 | 2.95 | 1656 | 0.66 |
| RPI1 | 1650 | 0.64 | | | |
| RPI2 | 1659 | 0.67 | 2.18 | 1630 | 0.56 |
| RPI3 | 1623 | 0.54 | 2.37 | 1636 | 0.59 |
| UCD1 | 1665 | 0.69 | 3.26 | 1667 | 0.70 |
| UCD2 | 1648 | 0.63 | 2.52 | 1642 | 0.61 |
| UCD3 | 1658 | 0.67 | 2.19 | 1630 | 0.56 |
| ZJU1 | 1651 | 0.64 | 2.85 | 1653 | 0.65 |
| ZJU2 | 1599 | 0.45 | 1.00 | 1588 | 0.41 |
| ZJU3 | 1703 | 0.82 | 3.90 | 1690 | 0.78 |

horizontal base velocities obtained by time integration of the data from AH11 and AH12. The bottom trace in each subplot shows the high-frequency component of the base acceleration, and the second trace from the bottom shows the 1 Hz component of the base motion.

Figure 4.5 shows that the base motion for test UCD2 contains several large-amplitude sharp spikes, IFSTTAR1 has more continuous high-frequency components, and CU1, CU2, and RPI2 contain significant 3 Hz components superimposed on the motion. RPI2 motion was intentionally varied to allow emulation of the high-frequency component observed in the CU experiments. The first few and last few

**Table 4.2b** Summary of ground motion intensity measures for the first destructive shake

| Test ID | PGA | $PGA_{eff}$ | $PGA_{1Hz}$ | $PGA_{HF}$ | PGV | Cumulative abs. vel. $CAV_5$ | Arias intensity $I_a$ |
|---|---|---|---|---|---|---|---|
| | g | g | g | g | m/s | m/s | $m^2/s$ |
| CU1 | 0.190 | 0.186 | 0.123 | 0.125 | 0.253 | 7.75 | 1.20 |
| CU2 | 0.206 | 0.195 | 0.122 | 0.146 | 0.259 | 8.04 | 1.31 |
| Ehime1 | 0.169 | 0.158 | 0.135 | 0.045 | 0.202 | 8.26 | 1.07 |
| Ehime2 | 0.180 | 0.158 | 0.134 | 0.048 | 0.206 | 8.25 | 1.07 |
| Ehime3 | 0.168 | 0.155 | 0.136 | 0.039 | 0.200 | 8.24 | 1.07 |
| IFSTTAR1 | 0.214 | 0.165 | 0.119 | 0.106 | 0.184 | 7.44 | 0.98 |
| IFSTTAR2 | 0.135 | 0.129 | 0.095 | 0.069 | 0.166 | 5.68 | 0.56 |
| KAIST1 | 0.178 | 0.168 | 0.119 | 0.098 | 0.209 | 7.18 | 0.85 |
| KAIST2 | 0.185 | 0.166 | 0.120 | 0.092 | 0.210 | 7.30 | 0.86 |
| KyU1 | 0.071 | 0.064 | 0.047 | 0.034 | 0.084 | 2.63 | 0.13 |
| KyU2 | 0.119 | 0.111 | 0.098 | 0.026 | 0.155 | 5.72 | 0.50 |
| KyU3 | 0.143 | 0.133 | 0.116 | 0.033 | 0.185 | 6.74 | 0.72 |
| NCU1 | 0.292 | 0.237 | 0.180 | 0.114 | 0.291 | 8.93 | 1.73 |
| NCU2 | 0.224 | 0.202 | 0.151 | 0.101 | 0.247 | 7.43 | 1.20 |
| NCU3 | 0.217 | 0.176 | 0.125 | 0.102 | 0.205 | 5.84 | 0.83 |
| RPI1 | 0.150 | 0.146 | 0.135 | 0.021 | 0.221 | 7.02 | 0.82 |
| RPI2 | 0.144 | 0.148 | 0.106 | 0.085 | 0.208 | 6.67 | 0.74 |
| RPI3 | 0.170 | 0.162 | 0.144 | 0.036 | 0.233 | 7.25 | 0.92 |
| UCD1 | 0.165 | 0.149 | 0.119 | 0.060 | 0.197 | 6.28 | 0.64 |
| UCD2 | 0.339 | 0.210 | 0.149 | 0.122 | 0.249 | 8.25 | 1.13 |
| UCD3 | 0.192 | 0.183 | 0.134 | 0.099 | 0.228 | 7.31 | 0.90 |
| ZJU1 | 0.167 | 0.134 | 0.094 | 0.080 | 0.151 | 5.12 | 0.49 |
| ZJU2 | 0.191 | 0.148 | 0.099 | 0.098 | 0.160 | 5.33 | 0.54 |
| ZJU3 | 0.135 | 0.111 | 0.078 | 0.065 | 0.126 | 4.33 | 0.33 |
| Average | 0.181 | 0.158 | 0.120 | 0.077 | 0.201 | 6.791 | 0.859 |

cycles of the motion produced by the Ehime shaker are lower frequency than 1 Hz; this is a nuance of their mechanical shaker. The long period components did not much affect the PGA but did affect the cumulative absolute velocity and Arias intensity; Ehime motions were just below the median in terms of $PGA_{eff}$, but well above the median in terms of $CAV_5$ and $I_a$. From the highlighting in Table 4.2b, it is apparent that in most cases the intensity measures are highly correlated to each other; two apparent exceptions include the aforementioned effect of low-frequency components for the Ehime motions and weak correlation between $PGA_{HF}$ to the intensity measures other than PGA.

**Table 4.2c** Performance measures for each of the experiments during the first destructive motion

| Test ID | Integrated pos. rel. vel. | Peak dyn. rel. disp. | depth of liq., zliq | Duration of lique-faction at P4 | Ux mean of all markers | Ux St. Dev. of all markers | Ux mean of 8 markers | Ux St. Dev. of 8 markers | Ux mean of 2 markers | Ux St. Dev. of 2 markers |
|---|---|---|---|---|---|---|---|---|---|---|
| | m | m | m | s | mm | mm | Mm | mm | mm | mm |
| CU1 | 5.80 | 0.066 | 3 | 58 | 359 | 77 | 403 | 56 | 440 | 57 |
| CU2 | 5.52 | 0.056 | 3 | 28 | 359 | 96 | 428 | 65 | 490 | 42 |
| Ehime1 | 5.42 | 0.047 | 2 | 0 | -- | -- | -- | -- | -- | -- |
| Ehime2 | 7.18 | 0.061 | 1 | 0 | 89 | 48 | 103 | 39 | 100 | 28 |
| Ehime3 | 9.63 | 0.055 | 0 | 0 | 56 | 37 | 65 | 37 | 60 | 28 |
| IFSTTAR1 | 3.22 | 0.032 | 1 | 0 | 21 | 62 | 25 | 46 | 50 | 71 |
| IFSTTAR2 | 5.66 | -- | 2 | 31 | 272 | 160 | 297 | 227 | 438 | 53 |
| KAIST1 | 3.05 | 0.037 | 1 | 2 | 1 | 3 | 0 | 4 | 2 | 2 |
| KAIST2 | 4.58 | 0.063 | 1 | 20 | 0 | 57 | 0 | 37 | 0 | 28 |
| KyU1 | 0.20 | 0.001 | 0 | 0 | 64 | 216 | 117 | 283 | 377 | 283 |
| KyU2 | 4.61 | 0.040 | 3 | 20 | 108 | 44 | 141 | 43 | 150 | 61 |
| KyU3 | 4.84 | 0.044 | 1 | 12 | 12 | 110 | 11 | 62 | 0 | 63 |
| NCU1 | 8.73 | 0.113 | 2 | 13 | 197 | 70 | 248 | 40 | 287 | 1 |
| NCU2 | 6.64 | 0.079 | 1 | 11 | 188 | 83 | 239 | 48 | 256 | 0 |
| NCU3 | 4.61 | 0.054 | 2 | 14 | 233 | 79 | 270 | 59 | 279 | 56 |
| RPI1 | 5.28 | 0.045 | 2 | 16 | -- | -- | 101 | 25 | 94 | 11 |
| RPI2 | 5.73 | 0.061 | 2 | 14 | -- | -- | 128 | 22 | 134 | 12 |
| RPI3 | 7.16 | 0.072 | 2 | 32 | -- | -- | 123 | 18 | 126 | 10 |
| UCD1 | 0.63 | 0.004 | 0 | 0 | 12 | 30 | 0 | 28 | 0 | 28 |
| UCD2 | 4.48 | 0.047 | 2 | 25 | 109 | 48 | 131 | 25 | 125 | 9 |
| UCD3 | 5.54 | 0.051 | 2 | 19 | 116 | 80 | 134 | 35 | 160 | 3 |
| ZJU1 | 4.39 | 0.035 | 1 | 18 | 133 | 56 | 150 | 53 | 135 | 21 |
| ZJU2 | 5.41 | 0.046 | 1 | 25 | 221 | 86 | 283 | 33 | 263 | 53 |
| ZJU3 | 0.40 | 0.002 | 0 | 0 | 29 | 42 | 26 | 16 | 30 | 21 |

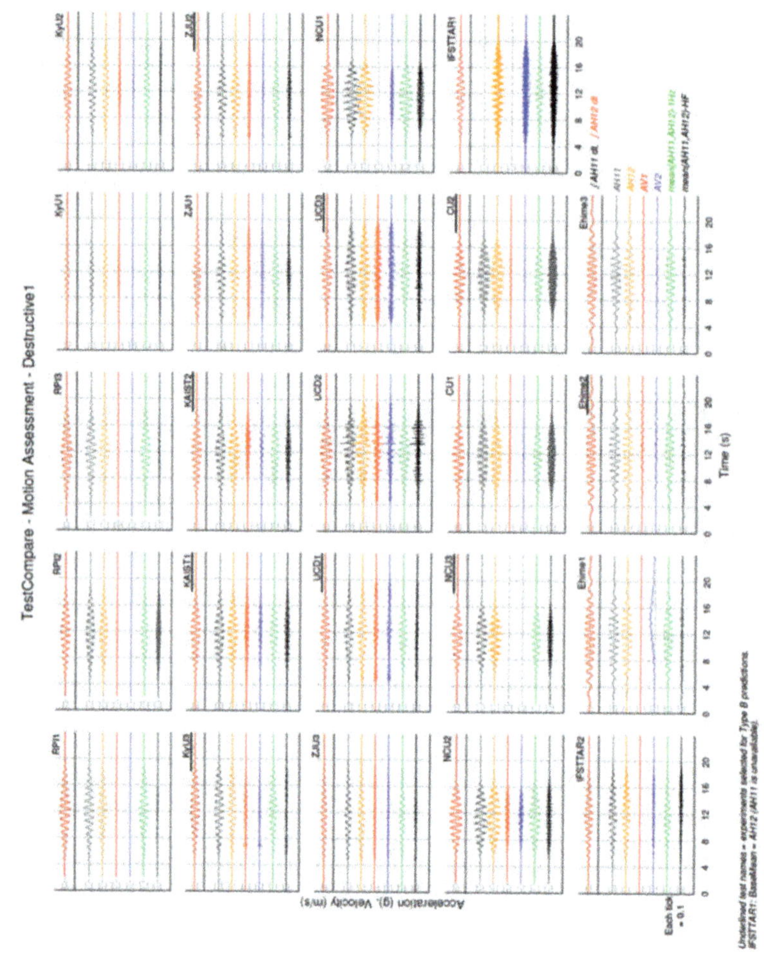


**Fig. 4.5** Velocity time series data, horizontal accelerations, vertical accelerations, 1 Hz component, and high-frequency component of horizontal base motion

## 4.4 Acceleration Response of Soil Layers in First Destructive Motion

To allow for qualitative comparison of the acceleration response of the models, Fig. 4.6 presents time series data for the base horizontal acceleration in the soil. From top to bottom, each subplot shows AH4, AH3, AH2, AH1, and the base input motion presented in the same sequence as their physical location defined in Fig. 4.1.

Three of the experiments (KyU1, ZJU3, and UCD1) show almost uniform acceleration behavior—in other words, the models behaved like a rigid body—a clear indication that liquefaction did not occur in these experiments. All of the other experiments showed significant evidence of nonlinear behavior and evidence of liquefaction. The sharp downward spikes, most significant in AH3 and AH4, we call "dilation spikes" because they are caused by the sudden increase in effective stress and hence increase in stiffness associated with negative pore water pressures produced by the tendency of the sand to dilate in response to the imposition of large shear strains. The spikes are larger in the downward direction because this corresponds to shearing in the downslope direction; strains tend to accumulate in the downslope direction.

Some aspects of the recorded data are obviously influenced by faulty instrumentation. For example, the data from AH3 and AH4 in UCD1 show almost uniform behavior, similar to the base acceleration, indicating very little deformation of the soil; therefore, it is clear that the offset seen in AH1 and to a lesser extent AH2 are anomalous and probably due to an instrumentation issue. AH1 is not reported for UCD3, and AH1 is not reported for IFSTTAR2. AH1 appears to be nonfunctional (flat) in CU1.

Based upon the response recorded by the upper accelerometers (AH3 and AH4), CU1 shows the most severe isolation of the ground surface motion associated with liquefaction; towards the end of the earthquake record, the surface motion is almost flat. Other surface records that show severe spikes or isolation are Kaist2, ZJU2, NCU1, NCU2, NCU3, CU2, IFSTTAR2, and Ehime1. Consistent with this, all of these events also produced permanent displacements larger than 250 mm (see Table 4.2c).

## 4.5 Displacement Response of the Soil Layers in First Destructive Motion

It is difficult to directly measure the potentially large multidirectional deformations of submerged slopes by conventional contact sensors. A reliable alternative approach for measuring permanent displacements is by surveying the location of the surface markers before and after liquefaction as described in the specifications for LEAP-UCD-2017 (Kutter et al. 2019). The surveys may be accomplished by direct measurement using rulers and calipers or by photography or surface scanners. High-

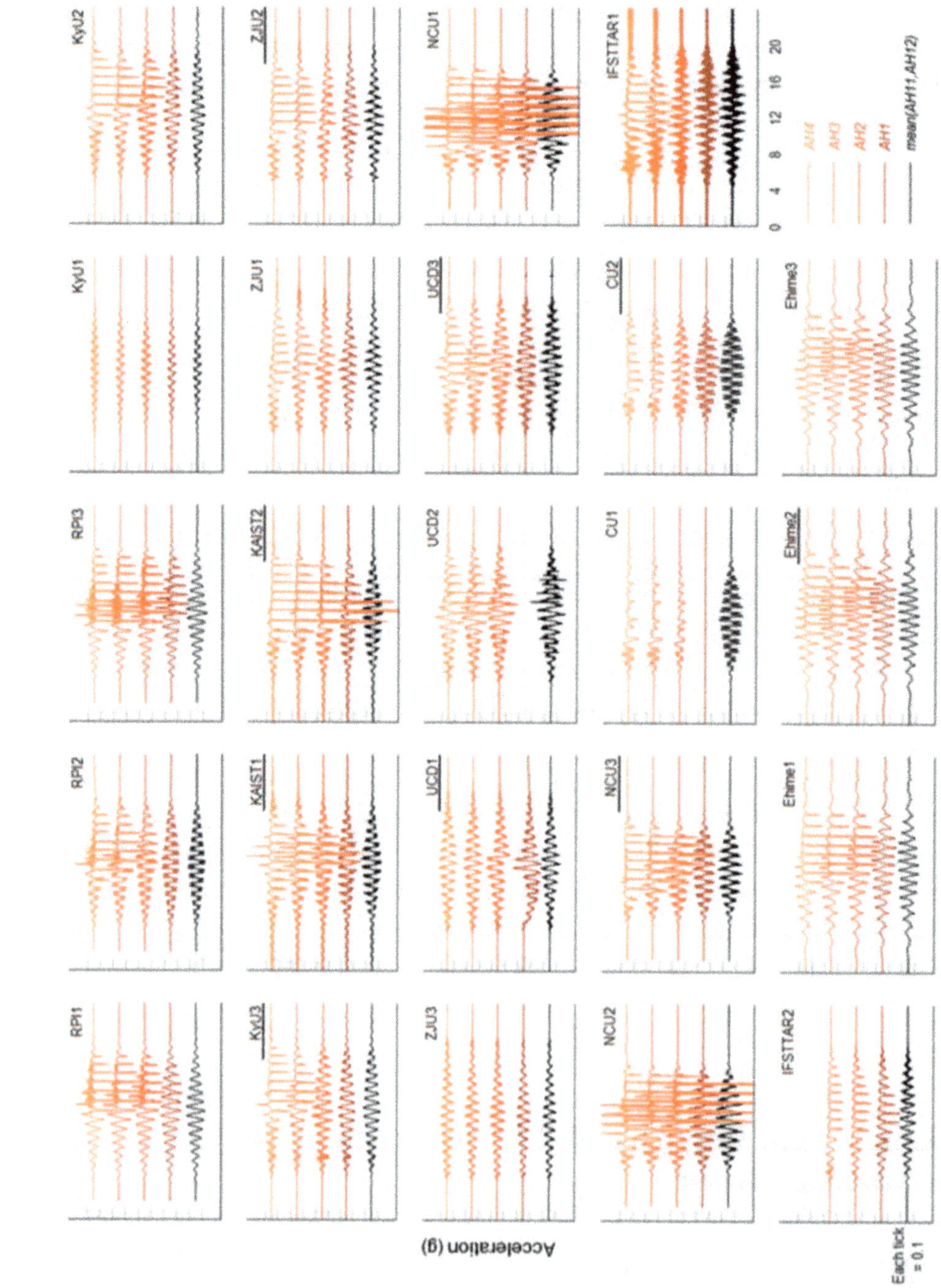


**Fig. 4.6** Acceleration response in the soil layers

speed photography at some sites allowed not only the determination of the residual deformations but also dynamic measurement of displacements during shaking. Dynamic displacements from photography are presented in the papers submitted by the experimenters from each site.

Vectors of the horizontal component of the displacement from before-and-after surveys of the surface markers are shown in Fig. 4.7. The data source for these plots is available in a spreadsheet document that is archived in the LEAP-UCD-2017 data archive in DesignSafe (Kutter et al. 2018b). The length of the displacement vectors is magnified by a factor of 6 compared to the geometry of the model. According to the specifications, markers were to be placed in a 6 × 3 grid across the surface. One site (RPI) reported displacement data from high-speed camera measurements of a different pattern of surface markers, so the patterns of the vectors are different. The data from KyU1 indicate substantial out-of-plane displacements that might be explained by a systematic error. Figure 4.7 also shows the locations of the model boundaries. All of the models were 20 m long in prototype scale, but some of the models (Ehime and Kyoto) were narrower than the others. In general, the displacements tended to be greatest near the middle of the container and tapered off near the ends of the container. (The ends of the container are defined by X-coordinate = ±10 m.) Settlements tended to be large on the upslope (left side of each subfigure) and smaller or negative on the downslope end of the sample container.

The X-components and Z-components of the displacement vectors are shown in contour plots in Figs. 4.8 and 4.9, respectively. From the contour plots, it is apparent that the experiments covered a wide range of displacements.

Although the dynamic component of the surface displacement cannot be determined from the before and after surface marker displacements, the dynamic component of the relative displacement, shown by the lighter lines in Fig. 4.10, was obtained by subtracting acceleration at AH4 from the acceleration at the base of the container and then double integrating with respect to time. Following each integration, a 0.2 Hz (prototype scale) high-pass filter was applied to remove drift error due to integration.

Also in Fig. 4.10, the curve labeled "scaled IPRV" represents a ramp, the shape of which is determined by the "integrated positive relative velocity" defined by Kutter et al. (2017):

$$
\begin{gathered}
\mathrm{IPRV} = \int_0^{\infty} \chi[v_{\mathrm{rel}}(t)]dt \\
\text{where } \chi = \begin{Bmatrix} 0 & \text{if} & v_{\mathrm{rel}}(t) < 0 \\ 1 & \text{if} & v_{\mathrm{rel}}(t) > 0 \end{Bmatrix}
\end{gathered}
\tag{4.3}
$$

where $v_{\mathrm{rel}}$ is the dynamic component of the relative velocity from single integration of the difference between AH4 and the base acceleration. This function produces a reasonably shaped ramp that should be representative of the accumulation of the permanent displacements. The IPRV ramp function is then scaled to make it agree with the displacement determined from surface marker surveys. The time series

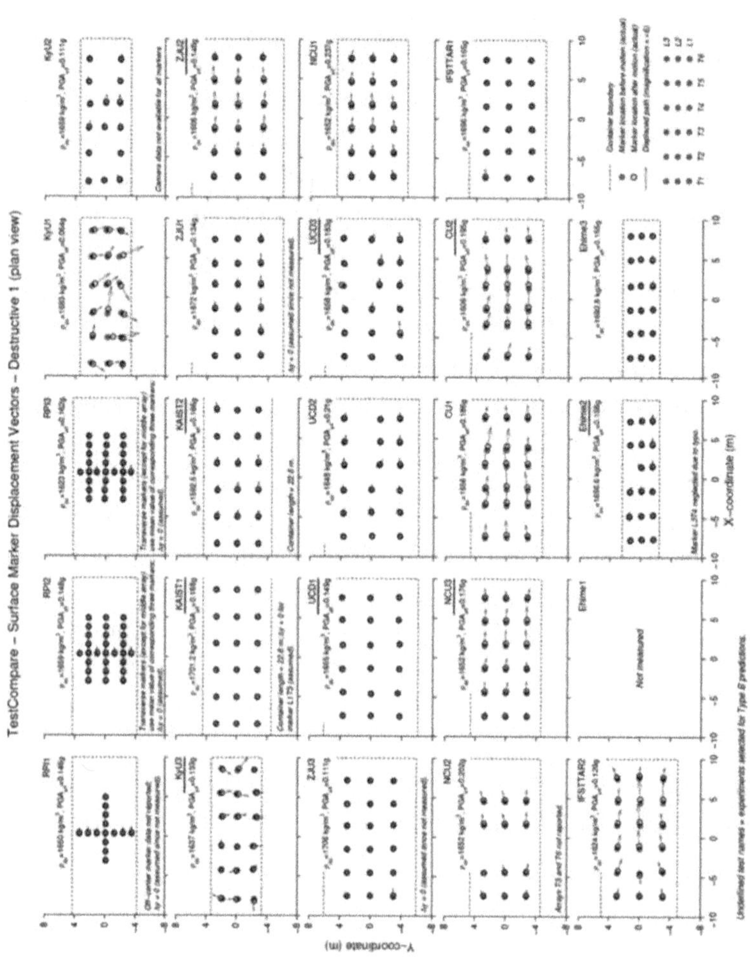


**Fig. 4.7** Horizontal component of the displacement vectors superimposed on plan view of the surface markers on the top of the sand

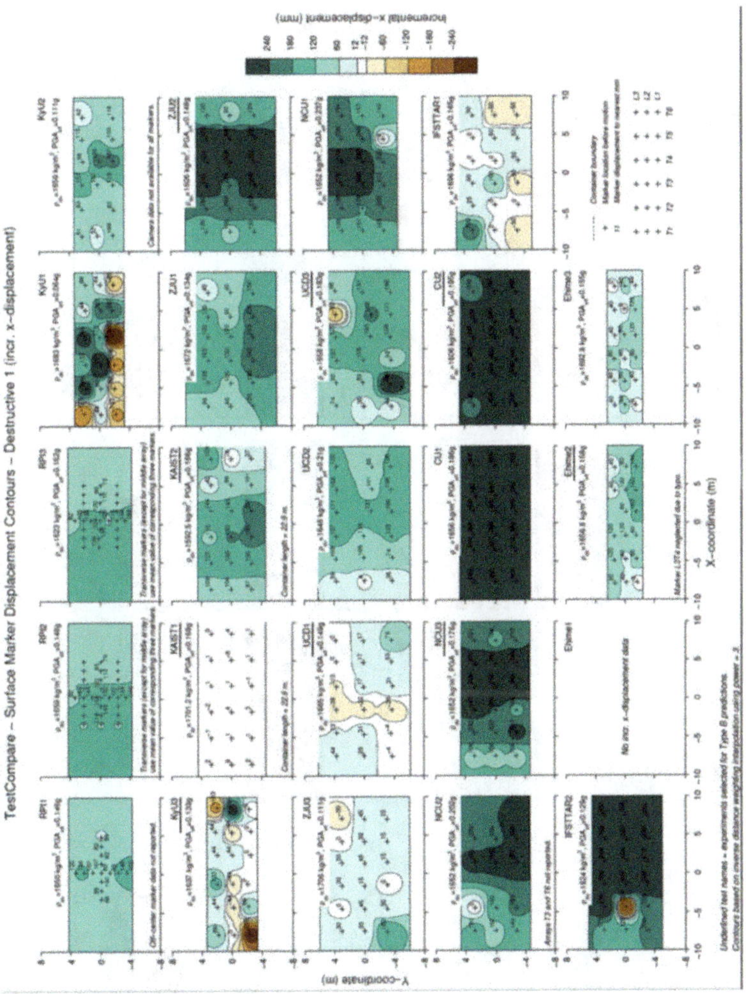


**Fig. 4.8** Contours of X-component of the surface marker displacement

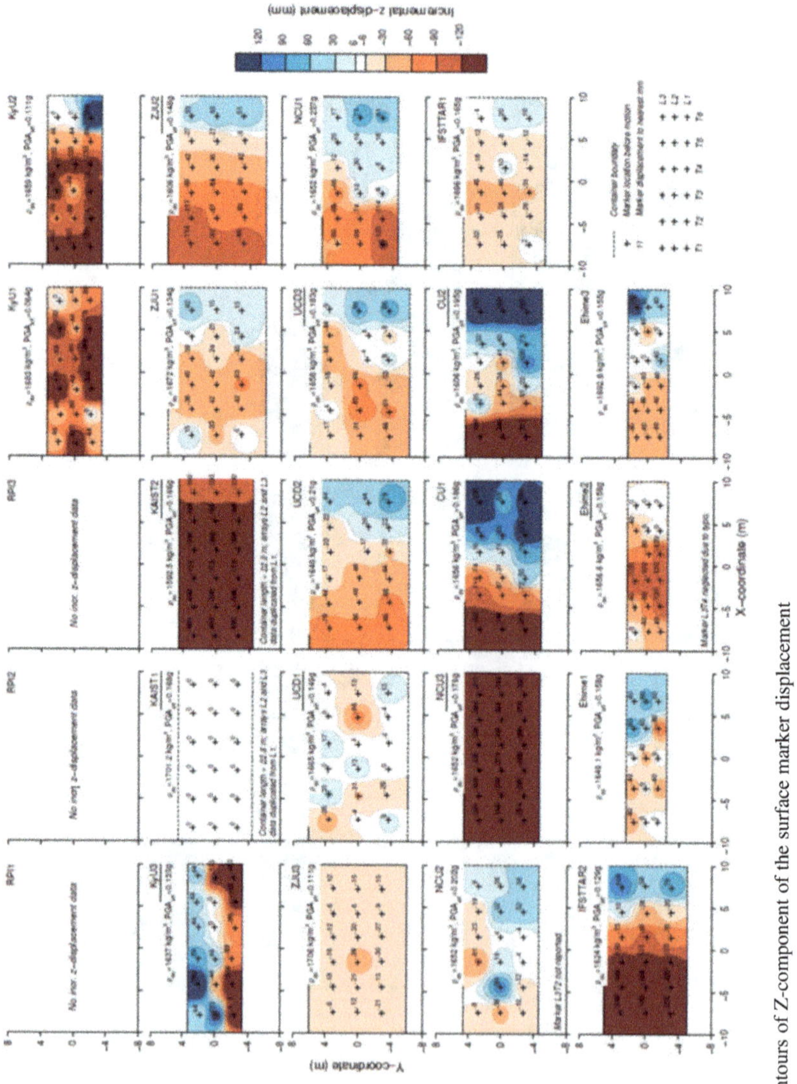

**Fig. 4.9** Contours of Z-component of the surface marker displacement

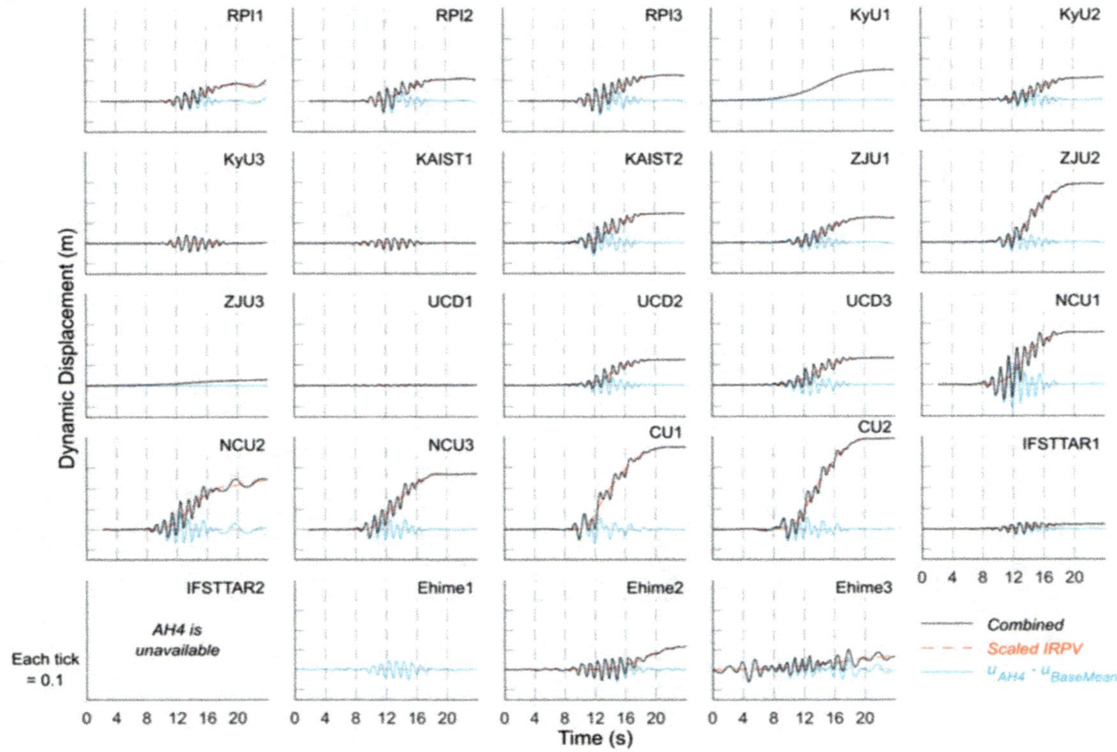


**Fig. 4.10** Dynamic component of the relative displacement $u_{AH4} - u_{BaseMean}$, the scaled integrated positive relative velocity, and the sum of the IPRV and the dynamic component of relative displacement

labeled "combined" is obtained by adding the scaled IPRV ramp to the dynamic component of the displacement. Carey et al. (2018a, 2019b) independently determined the displacement of the surface markers as a function of time using the high-speed cameras and demonstrated that superposition of the dynamic displacements from accelerometers on the IPRV ramp produces a reasonable approximation of the displacement time series. From Fig. 4.10, the cyclic components of the displacements are negligible for tests UCD1, ZJU3, and KyU1, as would be expected considering that these models did not liquefy. The fact that KyU1 showed a permanent displacement but no cyclic displacement may be explained by the earlier observation that there may have been a systematic error in the surface marker measurement for KyU1. The cyclic displacements were the largest for test NCU1. The IPRV and the amplitude of the cyclic component of relative displacements are considered to be meaningful and reliably quantified measures of the performance of the models that would not be affected by errors in surface marker measurements; therefore, these performance measures are listed in Table 4.2c.

## 4.6 Pore Pressure Response of Soil Layers in First Destructive Motion

Figure 4.11 compares the pore water pressure responses from the first destructive motions for the central array for all of the experiments. Note that the time scale is broken, with compressed time scale in the latter part of the graph to show the dissipation of pore pressures. The top four traces in each subplot show results from P4, P3, P2, and P1 (depth $\approx$ 1, 2, 3, and 4 m prototype scale; initial effective overburden stress $\approx$ 10, 20, 30, and 40 kPa, respectively). The buoyant unit weight of the saturated sand is approximately 10 kN/m$^3$. The tick marks on the side of each subplot correspond to 10 kPa. The last trace in each subplot shows results from P10, a sensor in the bottom corner of the containers. It is interesting to note that, especially near the beginning of shaking, the cyclic pore pressures at P10 tend to be greater than those in the central array, possibly in response to the cyclic total stress oscillations near the wall.

Consistent with the small cyclic relative displacements in UCD1, ZJU3, and KyU1 apparent from the accelerometer arrays, these three tests showed relatively small pore pressures throughout the layer. In UCD1 and ZJU3, the pore pressure approached the overburden of 10 kPa at P4 during shaking, but only at the peaks of the cycles, and dissipation began during shaking. For all of the other models, the pore pressures appeared to reach the effective overburden stress. The extent of liquefaction could be determined by pore pressure ratios, but small errors in the depth of the sensors could make the difference between pore pressure ratios of 100% and 90%, which is significant.

In search of a more robust measure of the extent of liquefaction, we systematically measured the duration of time over which the high pore pressures were

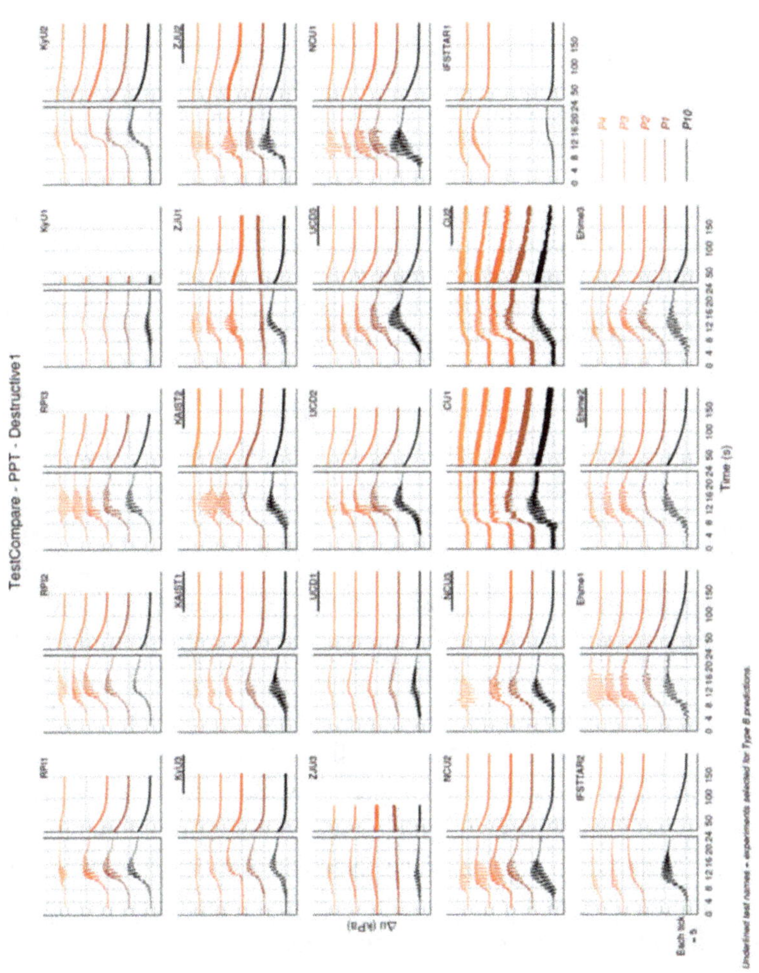


**Fig. 4.11** Pore pressure time series from the central array, first destructive ground motion

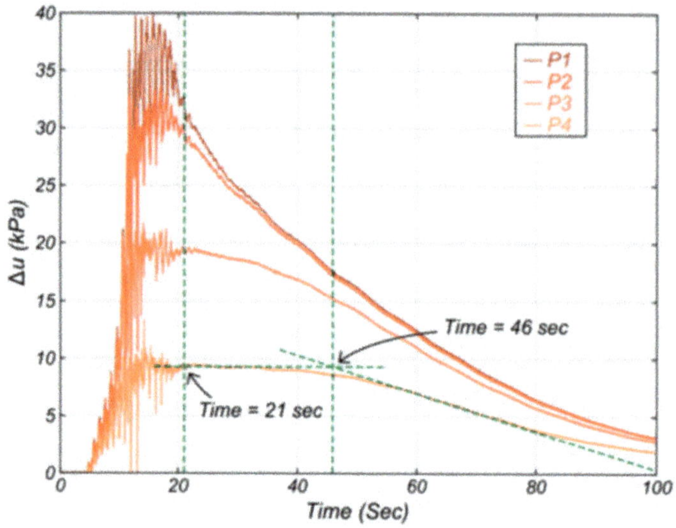


**Fig. 4.12** Pore pressures in the central array for UCD2 to illustrate the method of estimating the duration of liquefaction at P4. The second vertical line is drawn through the intersection of a near horizontal line through the time of sustained high pore pressure and a sloping line through the inflection of the dissipation curve. The end of shaking is indicated by the first vertical line and the beginning of dissipation is indicated by the second vertical line

sustained at sensor P4 after shaking stops as illustrated in Fig. 4.12. Sustained high pore pressures at P4 is an indication that large hydraulic gradients are uniform around that point. If hydraulic gradients are uniform at P4, then the upward flow toward P4 will be equal to the upward flow away from P4, and the pore pressure would be constant. The duration of sustained pore pressures at P4 is indicative of the duration of large exit gradients at the soil surface; hence, the duration of sustained pore pressures will be an indicator of the volumetric strains caused by the shaking event. If, for example, the duration of large exit gradients (approximately, $i \sim i_{\mathrm{crit}} \sim 1$) is 20s in prototype scale, and the permeability of the sand is $1.5 \times 10^{-4}$ m/s (El Ghoraiby et al. 2017), then the volume of water expelled would be (20 s) ($1.5 \times 10^{-4}$ m/s) = 3 mm in prototype scale. After some time a break in the dissipation curve is apparent in Fig. 4.12. Such a construction was repeated for each of the first shaking events for all 24 experiments. The duration of sustained pore pressures determined by this method is summarized in Table 4.2c.

Also apparent in Figs. 4.11 and 4.12 are large spikes of negative pore pressure that appear in many traces after liquefaction develops; these negative pore pressures are attributed to dilatancy and have been observed in many laboratory element tests as well as centrifuge tests in the past. As expected, these spikes of negative pore pressure increase the effective stress and stiffen the sand and tend to be aligned with corresponding spikes of ground acceleration. The large dilatancy spikes correspond to the arresting of downslope displacements. In some sensors for some experiments, positive spikes of pore pressure are also apparent; because effective stress in sand cannot be negative, the only mechanism for which pore pressures in a soil layer could be greater than the initial total vertical stress is if the total stress is momentarily increased by dynamic vertical accelerations. Anomalous positive pore pressures might also be recorded by sensors due to local pushing or pulling on sensor cables or other soil-sensor interactions.

## 4.7 Correlations Between Displacement, $D_r$, and IMs

Figure 4.13, reproduced from Kutter et al. (2018a), shows two views of 3D plots of correlations between the observed displacement from the average of the two central surface markers (Ux2 in units of mm) as a function of relative density from cone penetration resistance ($D_r(q_c(2\text{ m}))$) and $PGA_{eff}$ for 16 of the 19 tests that provided this information. All of the available data for the 24 tests used as a data source are listed in Tables 4.2a, 4.2b, and 4.2c. In addition, the data will also be available in a spreadsheet document available in the LEAP-UCD-2017 data archive in DesignSafe (https://www.designsafe-ci.org). Three of the 19 tests were excluded from the correlations because they were thought to be "outliers." With the outliers excluded, the coefficient of correlation $R^2 = 0.846$, indicating that 84.6% of the variation between these results could be explained by the two variables $PGA_{eff}$ and $D_r(q_c(2\text{ m}))$. Kutter et al. (2018a) also presented surface fits through using the same fitting function but for all 19 points without excluding outliers. Inclusion of the outliers reduced the correlation coefficient considerably to $R^2 = 0.578$.

The shape of the surface used to perform the regression was loosely based on curves presented by Yoshimine et al. (2006). Idriss and Boulanger (2008) approximated the Yoshimine et al. curves by:

$$\gamma_{max} = 0.035\left(2 - FS_{liq}\right)\frac{1 - F_\alpha}{FS_{liq} - F_\alpha} \tag{4.4}$$

where $FS_{liq}$ = CRR/CSR is the factor of safety with respect to triggering of liquefaction and $F_\alpha$ is a function of relative density. Note that Eq. 4.4 is not applicable if $FS_{liq}$ is greater than 2 and would return a strain potential, $\gamma_{max}$, of zero for $FS_{liq} = 2$. The curve fit equation used for displacement for this study was:

$$Ux = b_2\left\langle b_1 - \frac{(D_r - 0.125)^{n3} + 0.05}{1.3\frac{a_{max}}{g}}\right\rangle^{n1}\left(\frac{a_{max}}{g}\right)^{n2}(1 - D_r)^{n4} \tag{4.5}$$

where the second term inside the Macauley brackets $\langle\rangle$ is meant to be analogous to $FS_{liq}$ and the b1 term corresponds to the constant, 2, in Eq. 4.4. Note that $\langle x\rangle = x$ if $x > 0$; $\langle x\rangle = 0$ if $x < 0$. However, for the present study, coefficients $b_1$, $b_2$, n1, n2, n3, and n4 are determined by nonlinear regression. Inclusion of the term in Macauley brackets, with the restriction that $0.125 < D_r < 1$, produces a smooth function and prevents this function from producing not-physically realistic uphill residual displacements. As an example, the curve fit parameters determined using a nonlinear

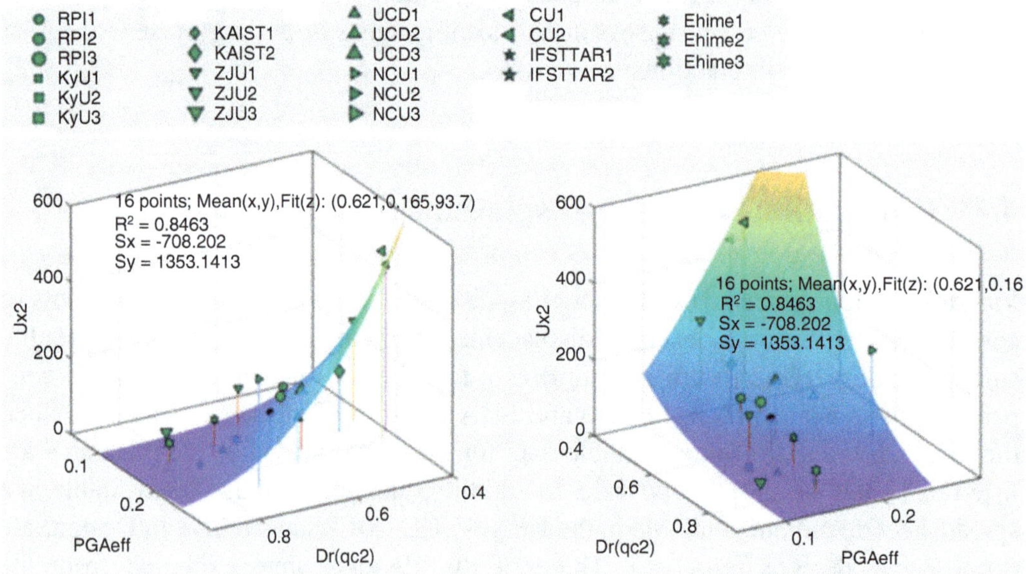


**Fig. 4.13** Two views of the correlation between the median displacement of the two central surface markers (U×2 mm, prototype scale), $PGA_{eff}$ (g), and relative density determined from $q_c$(2 m). This result is for Case 1/6 in Table 4.3 and was also presented by Kutter et al. (2018a). An excellent coefficient of correlation $R^2 = 0.846$ is obtained

regression algorithm in Matlab that produced the surface plotted in Fig. 4.13 are $b_1 = 12$, $b_2 = 0.0456$, $n1 = 4.57$, $n2 = 1.157$, $n3 = 1$, and $n4 = 2$.

The first four cases summarized in Table 4.3 are reproduced from Kutter et al. (2018a) using the six-parameter fitting equation (Eq. 4.5). The table summarizes the results of the surface fitting discussed in the previous paragraphs and in addition the surface fitting results of the analysis using $D_r$(Mass & Vol.) in place of $D_r(q_c(2\text{ m}))$ in case 3 and using PGA instead of $PGA_{eff}$ as a shaking intensity measure in case 4. The $R^2$ values summarized in Table 4.3 suggest that $PGA_{eff}$ is a better indicator of shaking intensity than PGA and that $D_r(q_c(2\text{ m}))$ is a better indicator of liquefaction resistance than is $D_r$(Mass & Vol.) for the present dataset. The mean density and shaking intensity measure (IM) of the data points analyzed are summarized in the last column of Table 4.3 along with the evaluation of the surface function at these means.

Case 5 shows a later analysis done after one more data point from test IFSTTAR2 became available. IFSTTAR2 produced relatively large deformations compared to the rest of the dataset; hence the $R^2$ value decreased from 0.846 to 0.718 by including this one point. Note however that evaluation of the curve fit at the median was not affected much by this point, and the computed sensitivities were not drastically changed by the addition of this data point. For the five cases summarized, the mean $D_r$ varied from 0.61 to 0.65; the mean IM varied from 0.161 to 0.185 g, and the surface fit at the median point varied from 94 to 154 mm. The table also summarizes the sensitivity of the displacement to variation of the $D_r$ and IM. The sensitivity of displacement to $D_r(q_c(2\text{ m}))$ varied by a factor of 1.29 (between −645 and −829 mm), and the sensitivity to $PGA_{eff}$ varied by a factor of 1.57 (between 1356 and 2125 mm/g) for the various cases. While there is

**Table 4.3** Results of nonlinear regression between displacement, relative density, and motion intensity for the first destructive motion in LEAP-UCD-2017

| Case/ (# pars) | Motion intensity measure (IM) (g) | Basis to determine $D_r$ | Data points used/ excluded outliers | Correlation coef. $R^2$ | Sensitivity to $D_r$ at mean (mm) | Sensitivity to IM at mean (mm/g) | Mean $D_r$, mean IM, and evaluation of curve fit at mean (1, g, mm) |
|---|---|---|---|---|---|---|---|
| 1/6 | $PGA_{eff}$ | $q_c$(2 m) | 16/3 | 0.846 | −708 | 1356 | 0.62, 0.165, 94 |
| 2/6 | $PGA_{eff}$ | $q_c$(2 m) | 19/0 | 0.578 | −645 | 2125 | 0.62, 0.161, 131 |
| 3/6 | $PGA_{eff}$ | Mass & Vol. | 19/4 | 0.603 | −492 | 1804 | 0.65, 0.166, 131 |
| 4/6 | PGA | $q_c$(2 m) | 19/0 | 0.485 | −829 | 611 | 0.62, 0.185, 154 |
| 5/6 | $PGA_{eff}$ | $q_c$(2 m) | 17/3 | 0.718 | −568 | 2339 | 0.60, 0.163, 106 |
| 6/4 | $PGA_{eff}$ | $q_c$(2 m) | 17/3 | 0.756 | −598 | 2681 | 0.60, 0.163, 106 |

variability in the sensitivities obtained by these methods, it is believed that the sensitivities are consistent enough to claim that the results are statistically significant. The compilation of sufficient data from centrifuge tests to enable quantification of the mean, sensitivities, and correlation is unprecedented.

Figure 4.14 (lower left) shows the same contour plot from Fig. 4.14 (upper left), with displacements normalized by the 4 m-layer thickness to produce an average shear strain using green dashed lines. Superimposed on Fig. 4.14 are contours from Eqs. 89–92 of Idriss and Boulanger (2008); their shear strain equations are based on a curve fit to cyclic stress-controlled triaxial tests reported by Yoshimine et al. (2006). Their equations relate cyclic stress ratio and relative density to shear strains. To map their CSR values onto the $PGA_{eff}$ vertical axis, the CSR values were divided by 1.3 for reasons explained below.

To revisit the assumptions of the nonlinear regression, the regression model was simplified from the six-parameter regression model (Eq. 4.5), to a four-parameter regression model (Eq. 4.6), and surprisingly, a better coefficient of correlation was obtained: $R^2 = 0.753$. The result (b1 = 5.985, b2 = 1.416, n1 = 4, n3 = 0.705) is plotted in Fig. 4.15. The values of the exponents n1 and n2 were arbitrarily limited to not exceed 4.0, and the converged value of n1 = 4 was fixed by this constraint.

$$Ux = \mathrm{b}_2 \left\langle \mathrm{b}_1 - \frac{(D_\mathrm{r} - 0.125)^{\mathrm{n3}} + 0.05}{1.3\frac{a_{\max}}{g}} \right\rangle^{\mathrm{n1}} \tag{4.6}$$

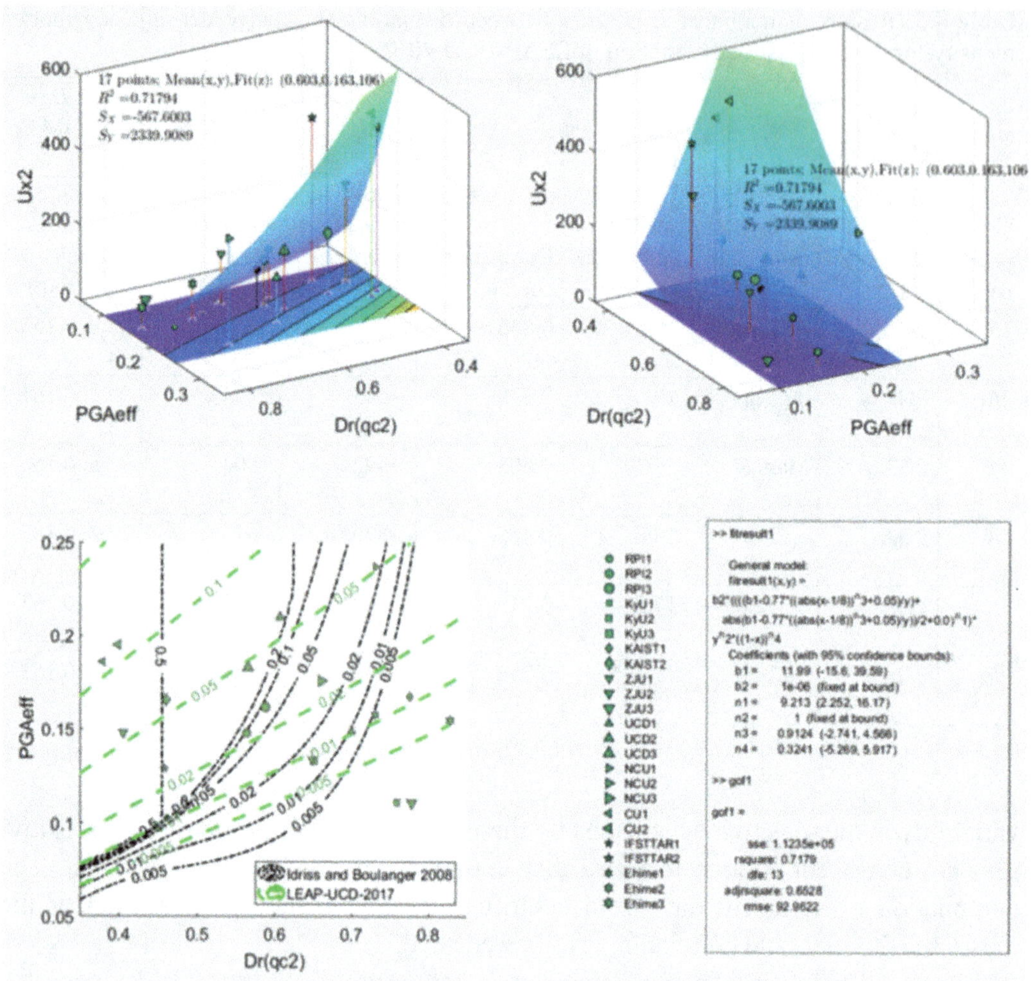


**Fig. 4.14** (**a**) Reevaluation of the six-parameter model surface fit after adding a new data point (IFSTTAR2) to the same dataset as used for Fig. 4.13. $R^2$ is reduced to 0.718. This result is for Case 5/6 in Table 4.3. The contour plot on the bottom left of the figure maps the same fitted surface displacement divided by the soil layer thickness (4000 mm) to convert the displacement to an average soil strain. These strains from the LEAP-UCD-2017 data are compared to a contour plot of Idriss and Boulanger (2008) (see Eq. 4.4)

In statistics, the "adjusted $R^2$" value is meant to compensate for the tendency for $R^2$ to decrease as additional parameters are introduced to the model. The "adjusted $R^2$" = 0.653 for the six-parameter model illustrated in Fig. 4.14. For the four-parameter model, the "adjusted $R^2$" = 0.722 is superior to that for the six-parameter model. It is interesting but certainly possible that introduction of additional parameters being fit by a nonlinear regression algorithm selected in Matlab could result in convergence to a different local minimum. Despite the simplification of the model, the resulting contours from the four-parameter model (Fig. 4.15) bear major resemblance to contours from the six-parameter model (Fig. 4.15). The results from regression to the same 17 data points for the four-parameter model are summarized as Case 6/4 in Table 4.3.

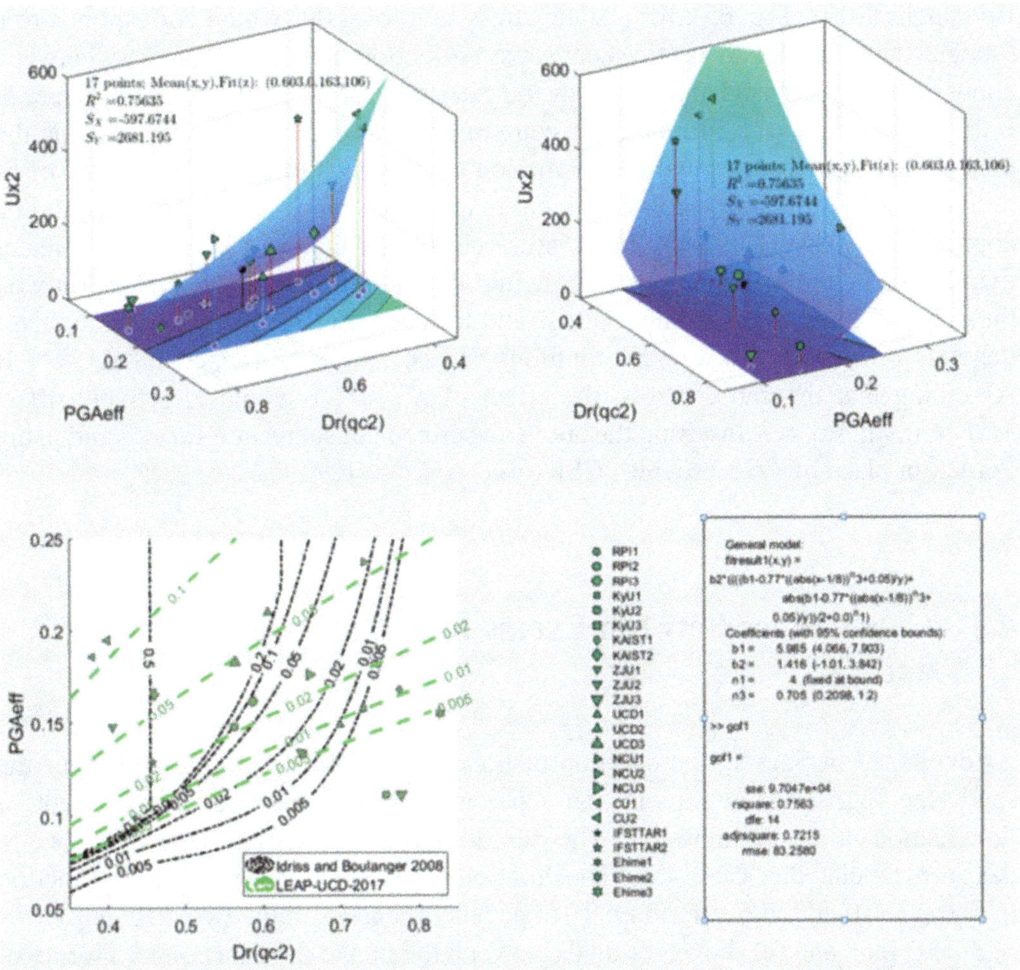


**Fig. 4.15** Results of linear regression using the simplified four-parameter model; $R^2 = 0.756$. This result is for Case 6/4 in Table 4.3. (**b**) The contour plot on the bottom left of the figure maps the same fitted surface displacement divided by the soil layer thickness (4000 mm) to convert the displacement to an average soil strain. These strains from the LEAP-UCD-2017 data are compared to a contour plot of Idriss and Boulanger (2008) (see Eq. 4.4)

## 4.7.1 *Rationale for Scaling Between PGA and CSR for Simplified Procedure*

According to the Idriss and Boulanger (2008) simplified procedure, the cyclic stress ratio scaled to a magnitude 7.5 earthquake with a vertical effective stress of 1 atm (101 kPa) is given by:

$$\mathrm{CSR}_{M=7.5,\ \sigma'_{vc}=1} = 0.65(\gamma/\gamma')(\mathrm{PGA}/g)(r_d)(1/\mathrm{MSF})(1/K_\sigma)(1/K_\alpha) \quad (4.7)$$

The ratio of total to buoyant densities of the soil $\gamma/\gamma' \approx 2$, and the depth reduction factor, $r_d$, is very close to 1.0 for a 4 m-deep deposit. According to Idriss and

Boulanger (2008, Fig. 65), for a static stress ratio of 0.09 (which corresponds to a 5-degree slope angle), the static shear stress correction factor, $K_\alpha$, may vary between about 0.8 for looser sand and 1.2 for the denser sand; for the purposes of this paper, it is assumed that $K_\alpha = 1$. From a cycle counting procedure, it was determined that the prescribed ramped sine wave LEAP motion corresponds to an earthquake of magnitude $M = 7.7$ to 7.9; the corresponding magnitude scaling factor, MSF, would be approximately 0.9. The overburden stress correction factor, $K_\sigma$, would be greater than 1 because the confining pressures at mid-depth of the liquefiable soil are only on the order of 20 kPa. According to Idriss and Boulanger (2008), the correction factor depends on density as well as confining stress, and it is capped at $K_\sigma \leq 1.1$. Assuming that the cap controls, the effect of $K_\sigma \approx 1.1$ would effectively offset $MSF \approx 0.9$ in Eq. 4.7. Inserting the above-described constants into Eq. 4.7 and using $\mathrm{PGA_{eff}}$ in place of $PGA$ provides $\mathrm{CSR}_{M=7.5,\ \sigma'_{vc}=1} \approx 1.3(\mathrm{PGA_{eff}})$.

## 4.8 Correlations Between Excess Pore Pressures, $D_r$, and IMs

As explained in Sect. 6, the duration of liquefaction near the top boundary of the sand (see Fig. 4.12) is explored as a potential robust measure of the extent of liquefaction in a centrifuge test. The durations of liquefaction at sensor P4 (1 m deep) are tabulated in Table 4.2c. The duration of liquefaction is plotted as a function of $D_r(q_c(2\ \mathrm{m}))$ and PGAeff in Fig. 4.16; panels (a) and (b) show two different views of the 3D plot, and (c) shows residuals between the fit and the data points. Twenty of the twenty-four centrifuge tests provided $q_c(2\ \mathrm{m})$. One of these 20 was considered to be an outlier and is excluded from Fig. 4.16. The coefficient of correlation for this regression was found to be $R^2 = 0.78$, indicating that 78% of the variation can be explained by the fitting function with the variables $\mathrm{PGA_{eff}}$ and $D_r(q_c(2\ \mathrm{m}))$. This is considered to be a clear indication that the LEAP centrifuge tests performed at different centrifuge facilities are very consistent from centrifuge to centrifuge. The thin dash-dot line in Fig. 4.16 shows the liquefaction triggering curve from Boulanger and Idriss based on the relative density at mid depth of the layer. The CRR (Cyclic Resistance Ratio) curves were scaled according to $\mathrm{CRR} = 1.3 \times \mathrm{PGA_{eff}}$ as explained in Sect. 7.1. The empirical triggering curve seems to be consistent with the LEAP-UCD-2017 data.

Figure 4.17 presents a similar set of plots, with $R^2$ reduced from 0.78 to 0.47 due to replacement of the density measure $D_r(q_c(2\ \mathrm{m}))$ by the density measure $D_r$(rho). $D_r$(rho) is determined from direct measurements of mass and volume, while $D_r(q_c(2\ \mathrm{m}))$ uses the density obtained by correlations with the cone penetration resistance at mid-depth. As was the case for the correlations to displacement, $D_r(q_c(2\ \mathrm{m}))$ produces a better correlation than relative density from mass and volume measurements. This is likely caused by cumulative errors in the direct measurement of mass and volume.

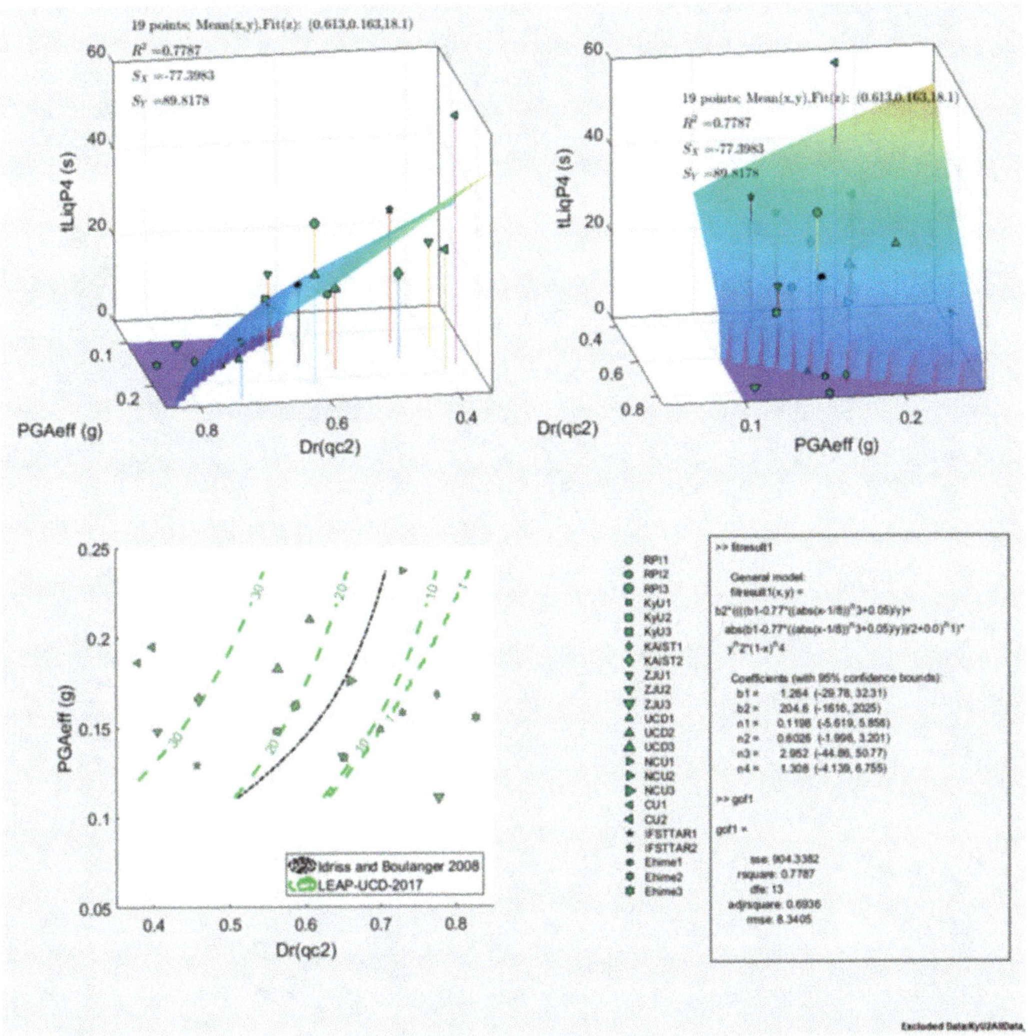


**Fig. 4.16** Correlation between duration of liquefaction at P4, $PGA_{eff}$, and $D_r(q_c(2\ m))$. 19 of 20 experiments with requisite data are plotted (one outlier was excluded), and $R^2 = 0.78$. Panels (**a**) and (**b**) show two views of the same surface fit to the data. Panel (**c**) shows a contour plot in green, overlaid on the Idriss and Boulanger (2008) triggering curve. The same six-parameter model was used (Eq. 4.5)

$R^2$ reduced slightly from 0.78 in Fig. 4.16 to 0.73 in Fig. 4.18 when $CAV_5$ is used in place of $PGA_{eff}$ as the IM. As was concluded by Kutter et al. (2018a), the $R^2$ value is best when IM = $PGA_{eff}$ and when relative density is based on the cone measurements. However, the correlation of the duration of liquefaction at P4 to $CAV_5$ is also very good.

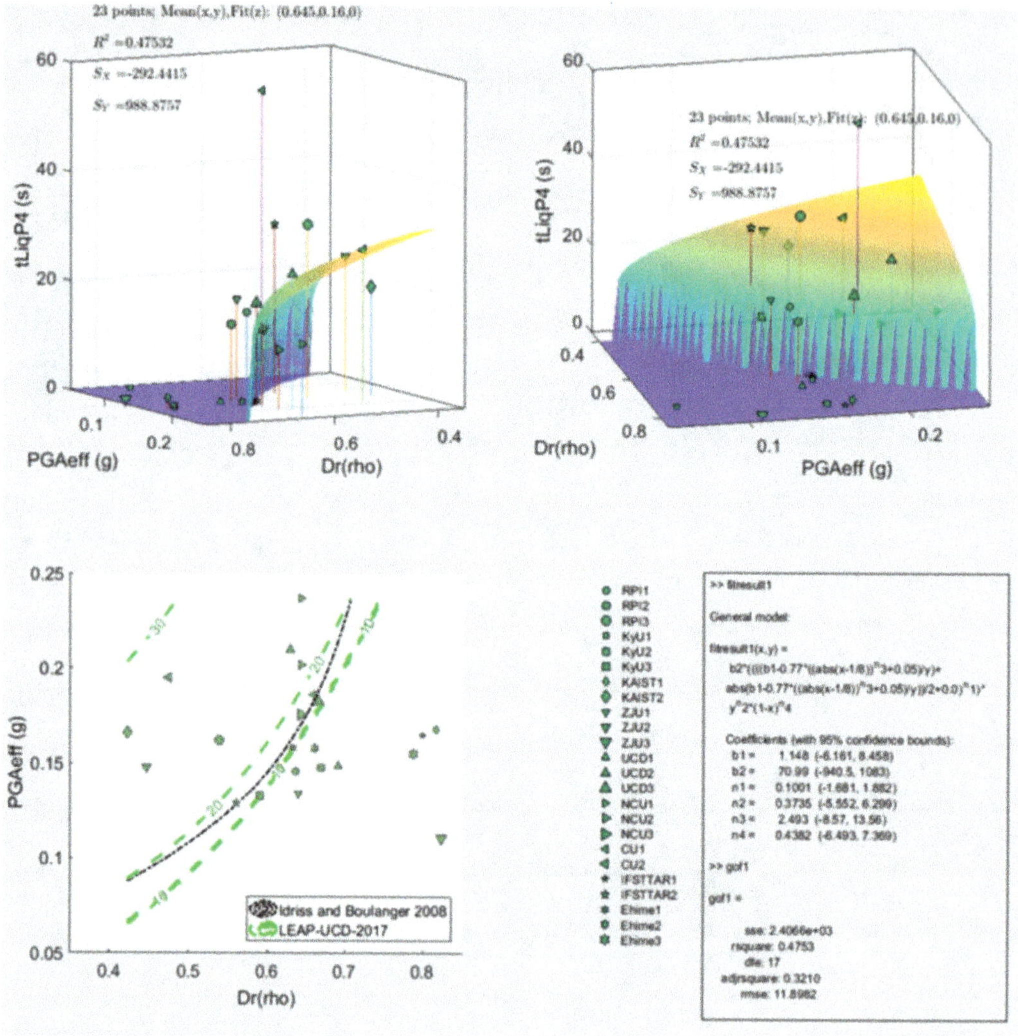


**Fig. 4.17** Correlation between duration of liquefaction at P4, $PGA_{eff}$, and $D_r$(rho), based on measurements of mass and volume. 23 of 24 experiments with requisite data are plotted (one outlier was excluded), and $R^2 = 0.48$. Panels (**a**) and (**b**) show two views of the same surface fit to the data. Panel (**c**) shows a contour plot in green, overlaid on the Idriss and Boulanger (2008) triggering curve. The same six-parameter model was used (Eq. 4.5)

## 4.9 Correlations Between Peak Cyclic Displacements, $D_r$, and IMs

Another easily measurable quantity thought to be indicative of the extent of liquefaction is the magnitude of the average cyclic component of the shear strains in the soil layer. This quantity can be reliably computed as described by Kutter et al. (2017). Briefly, it is obtained by subtracting the accelerations of the base of the container (average of accelerometers AH11 and AH12) from the accelerations measured at the surface of the soil layer (accelerometer AH4) and then double

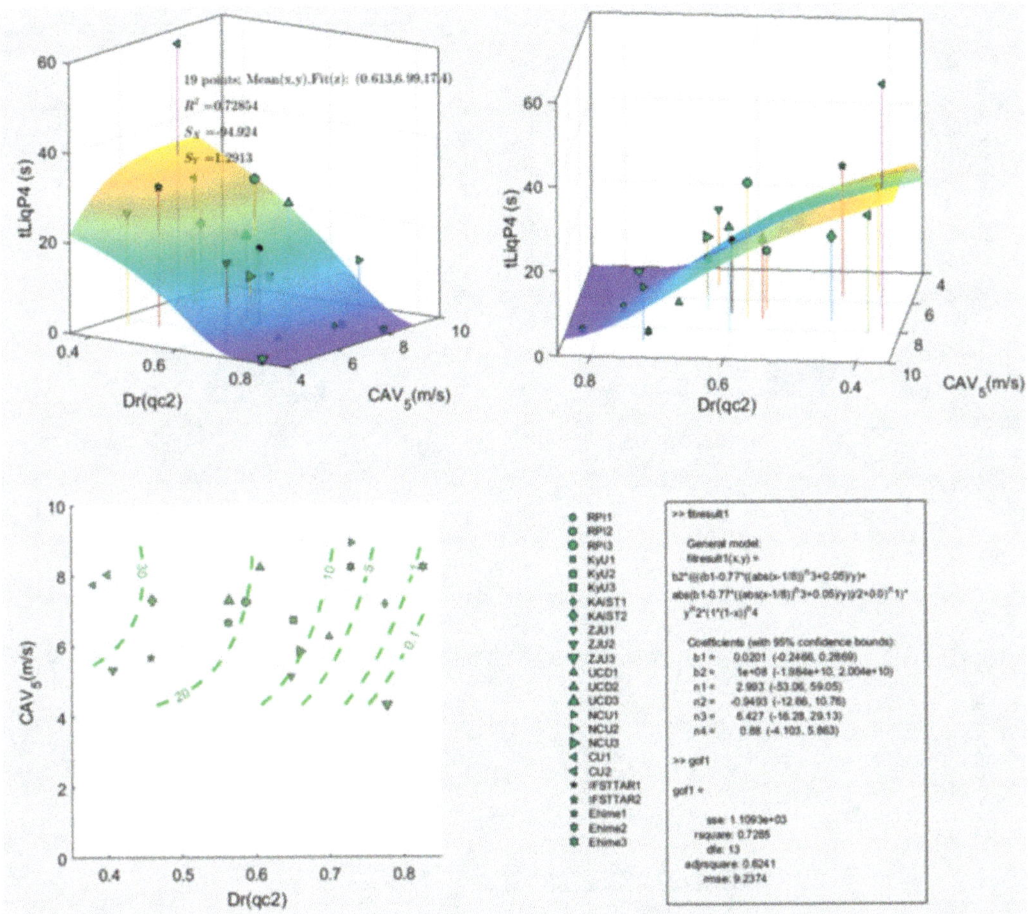


**Fig. 4.18** Correlation between duration of liquefaction at P4, CAV5, and $D_r(q_c(2\text{ m}))$. 19 of 20 experiments with requisite data are plotted (one outlier was excluded), and $R^2 = 0.73$. Panel (**a**) shows a contour plot; panel (**b**) shows a side view of the function. The same six-parameter model was used (Eq. 4.5)

integrating to determine the displacement as a function of time. A high-pass filter (about 0.3 Hz prototype scale) is used to remove the low-frequency components of the accelerations before integrating to obtain velocities and displacements. The low-frequency components, often a result of small electrical drift in the signal, become large during integration and are not reliably measured by accelerometers. Unfortunately, the filtering of the low-frequency component also removes the evidence of permanent displacements on the acceleration signal. Nevertheless, the amplitude of the cyclic component (higher than 0.3 Hz prototype scale) is also indicative of softening due to liquefaction. Figure 4.10 showed the cyclic time series of the cyclic component of displacement as a function of time. The peak of the cyclic displacement time series is summarized in Table 4.2c.

Figure 4.19 shows the relationship between the relative density, effective PGA, and the peak cyclic component of the relative displacement. The curve fit almost produced a step function; relative displacements are negligible if liquefaction is not

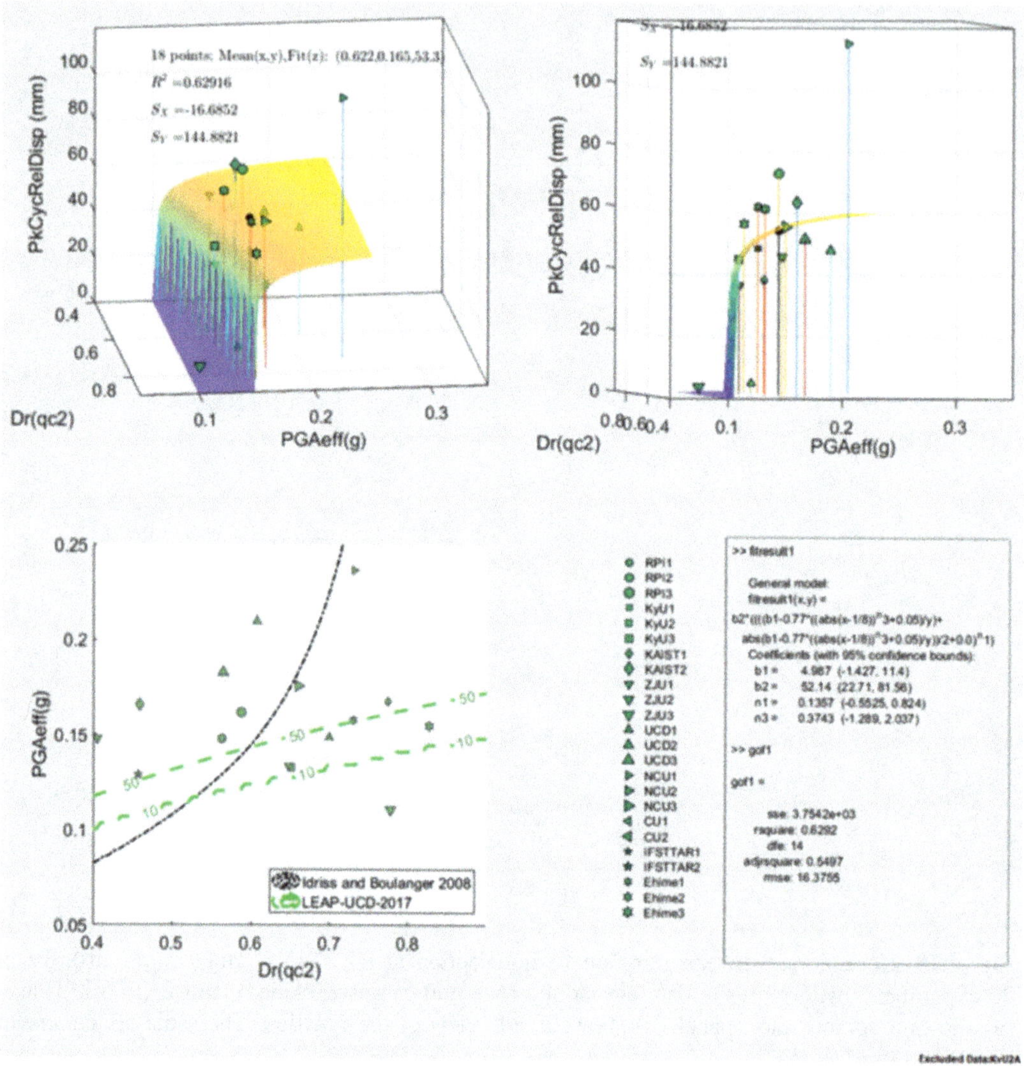


**Fig. 4.19** Correlation between peak cyclic relative displacement between surface accelerometers and the base as a function of $PGA_{eff}$ and $D_r(q_c(2\ m))$. 19 of 20 experiments with requisite data are plotted (one outlier was excluded), and $R^2 = 0.66$. Panel (**a**) shows a contour plot; panel (**b**) shows a side view of the function. The same six-parameter model was used (Eq. 4.5)

triggered and relative displacements are 40–80 mm (in most cases) where liquefaction was triggered.

In a theoretical special case of complete liquefaction, the base might be expected to move while the ground surface was isolated from the base motion; hence, the relative cyclic displacement would equal the base cyclic displacement. The average peak of the 1 Hz component of the input base motion listed in Table 4.2b is 0.12 g. This acceleration corresponds to a cyclic displacement of ±39 mm. As is apparent in Fig. 4.19, the measured cyclic relative displacements are typically in the range of 40–80 mm—significantly greater than the base displacement. The relative displacement could be greater than the input base displacement if the surface displacements are of opposite phase to the base displacements and/or if the displacements are amplified at the ground surface.

In one case cyclic displacements were significantly greater than 80 mm; NCU1 displayed relative displacements of 113 mm (see Fig. 4.19 and/or Table 4.2c); the large relative displacement in NCU1 is at least partly explained by the fact that the 1 Hz component of the input base acceleration for NCU1 0.18 g is about 50% greater than the average 1 Hz component (0.12 g) listed in Table 4.2b. The relatively large amplitude of the low-frequency component of the base displacement for NCU1 helps explain why the cyclic displacements are the greatest for NCU1.

It should also be recalled that, due to the mechanical nature of the shaker at Ehime University, the Ehime motions contained a significant lower-frequency displacement that is very apparent for Ehime3 in Fig. 4.10.

## 4.10 Summary and Conclusions

The first goal of this paper is to provide an overview of all the experimental data from centrifuge testing for LEAP-UCD-2017. This overview will allow readers to quickly scan through the key time series data and various performance measures to evaluate the extent of liquefaction in various experiments. The second goal of this paper is to demonstrate that the experiments are consistent with each other and that they define a response function or trend between key input parameters and key liquefaction response parameters.

Time series data from input accelerations, accelerations and pore pressures in the central array and relative cyclic displacements obtained by integration of accelerations in the time domain are qualitatively compared. Residual displacements are characterized by the measured displacement from surface markers. Contour plots of lateral displacement and settlements of the surface markers are presented. Key density measures, cone penetration data, ground motion intensity measures, and response parameters are tabulated for all 24 experiments in Tables 4.2a, 4.2b, and 4.2c. All of the data in Tables 4.2a, 4.2b, and 4.2c and more data not presented here are also available in a spreadsheet document archived in the LEAP-UCD-2017 data archive in the NHERI DesignSafe (Kutter et al. 2018b). The results from Tables 4.2a, 4.2b, and 4.2c are cross plotted in 3D plots along with nonlinear regression surfaces to show the trend and to estimate the degree of correlation of the data to the response surface.

The consistency of the centrifuge experiments performed for LEAP-UCD-2017 is demonstrated by showing that responses (permanent displacement, duration of liquefaction, and the amplitude of the cyclic displacements observed in different experiments) are highly correlated to key parameters describing the resistance to liquefaction (e.g., relative density and cone penetration resistance) and the base motion shaking intensity measures (e.g., PGA, $PGA_{effective}$, and $CAV_5$).

The extent of liquefaction in the experiments is more highly correlated to the dry density correlated with cone penetration resistance than the dry density determined by direct measurement of mass and volume of the models, partly due to uncertainties and errors in direct density measurement. Errors in volume measurement arise due to the fact that the surfaces of the model are rough, sloped, and curved. Errors in measurement of container dimensions, sand surface location, and resolution and accuracy of load cells used to measure the weight of the container combine to produce inaccurate density measurements.

The correlation between dry density and cone penetration resistance is obtained by linear regression between dry density determined from mass and volume measurements and $q_c$ at a depth of 2 m. Then, the densities of the models are obtained from the measured $q_c$ and the inverse linear regression line.

The PGA of the recorded base acceleration was found to be very sensitive to high-frequency components of the base motion, which varied significantly from facility to facility. The response of the model on the other hand was more sensitive to the lower frequency components of the input motion. Thus PGA was not a good parameter to use to describe the shaking intensity.

$PGA_{effective} = PGA_{1Hz} + 0.5\ PGA_{HighFrequency}$ was arbitrarily guessed as a trial function to help researchers at different facilities decide the appropriate input motions given the unique high-frequency noise produced by their centrifuge shakers. As it turned out, $PGA_{effective}$ is much more highly correlated to model response than is PGA. Another, less arbitrary intensity measure, cumulative absolute velocity ($CAV_5$), was also a good predictor of the duration of high excess pore pressures in the model.

Suggested measures of liquefaction response for this and future LEAP exercises are robust and easily and accurately measured and meaningful indicators of liquefaction phenomena. Until each facility can demonstrate that accurate measurements of surface markers using photographic or other scanning procedures, direct measurement of permanent displacements (especially lateral spreading and settlements) should be made using rulers, calipers, and surface markers. The displacements vary with position in the models; markers near the boundaries (end walls and side walls) are restricted by the boundaries. Displacements should be made near the boundaries and far from the boundaries to help assess boundary effects.

The duration of liquefaction of the top pore pressure sensor of a uniform soil deposit is proposed as a robust and useful indicator of the extent of liquefaction. This parameter is meaningful because the volume change of the soil deposit will be correlated to the duration of high pore pressures near the drainage boundary; the rate of volume change may be estimated using the water exit velocity from Darcy's law, $v = ki$. The maximum pore pressure ratio, $r_u = \Delta u / \sigma'_{vo}$, is theoretically an important pore pressure intensity measure, but it is less robust because it is sensitive to errors in the estimation of the depth of the sensor and corresponding initial effective stress.

Until non-contact methods such as photography and stereo photogrammetry are developed for more accurate measurement of time series of boundary displacements, the cyclic component relative displacement should be used as an indicator of the magnitude of cyclic strains induced by the shaking. The cyclic relative displacements of a layer may be obtained by subtracting displacements of the top and bottom of the layer obtained by integration of acceleration records at the top and bottom of the layer. The magnitude of cyclic strains has been shown to correlate well with liquefaction because strain levels change drastically during the onset of liquefaction.

LEAP-UCD-2017 produced an unprecedented quantity of model tests of sloping ground with intentionally varying input motions and soil density. Through data analysis summarized in this paper, we have shown a repeatable response function between liquefaction response and key input parameters including shaking intensity and relative density. The repeatability of this response function proves that the results are consistent with each other within a range of uncertainty. The matrix of test results is sufficient to not only quantify the median response but also the sensitivity of response to variations in the input parameters and the centrifuge-centrifuge variability of the experimental results. The credibility of the data provided by demonstrated interlaboratory consistency allows us to move forward with meaningful use of this data for assessment of the accuracy of numerical simulation procedures. Since we have mapped out an experimental response surface with some ability to quantify experiment-experiment variability, it is recommended that the numerical procedures also be required to map out the same response surfaces.

It should be emphasized that many of the LEAP experiments included a total of two or three destructive motions. This paper focuses on results from the first motion only. Papers by each experiment facility describe the results from subsequent destructive motions.

**Acknowledgments** The experimental work on LEAP-UCD-2017 was supported by different funds depending mainly on the location of the work. The work by the US PI's (Manzari et al. 2017) is funded by National Science Foundation grants: CMMI 1635524, CMMI 1635307, and CMMI 1635040. The work at Ehime U. was supported by JSPS KAKENHI Grant Number 17H00846. The work at Kyoto U. was supported by JSPS KAKENHI Grant Numbers 26282103, 5420502, and 17H00846. The work at Kansai U. was supported by JSPS KAKENHI Grant Number 17H00846. The work at Zhejiang University was supported by National Natural Science Foundation of China (Nos. 51578501 and 51778573), Zhejiang Provincial Natural Science Foundation of China (LR15E080001), and National Basic Research Program of China (973 Project) (2014CB047005). The work at KAIST was part of a project titled "Development of performance-based seismic design," funded by the Ministry of Oceans and Fisheries, Korea. The work at NCU was supported by MOST: 106-2628-E-008-004-MY3.

## References

Bolton, M. D., Gui, M. W., Garnier, J., Corte, J. F., Bagge, G., Laue, J., & Renzi, R. (1999). Centrifuge cone penetration tests in sand. *Geotechnique, 49*(4), 543–552.

Carey, T. J., Gavras, A., & Kutter, B. L. (2019a). Comparison of LEAP-UCD-2017 CPT results. In B. Kutter et al. (Eds.), *Model tests and numerical simulations of liquefaction and lateral spreading: LEAP-UCD-2017*. New York: Springer.

Carey, T. J., Stone, N., Bonab, M. H., & Kutter, B. L. (2019b). LEAP-UCD-2017 Centrifuge Test at University of California, Davis. In B. Kutter et al. (Eds.), *Model tests and numerical simulations of liquefaction and lateral spreading: LEAP-UCD-2017*. New York: Springer.

Carey, T. J., Stone, N., & Kutter, B. L. (2019c). Grain size analysis and maximum and minimum dry density of Ottawa F-65 sand for LEAP-UCD-2017. In B. Kutter et al. (Eds.), *Model tests and numerical simulations of liquefaction and lateral spreading: LEAP-UCD-2017*. New York: Springer.

Carey, T., Stone, N., Kutter, B., & Hajialilue-Bonab, M. (2018a). A new procedure for tracking displacements of submerged sloping ground in centrifuge testing. In *Proceedings of 9th International Conference on Physical Modelling in Geotechnics, ICPMG 2018* (Vol. 1, pp. 829–834). London: CRC Press/Balkema.

Carey, T., Gavras, A., Kutter, B., Haigh, S. K., Madabhushi, S. P. G., Okamura, M., Kim, D. S., Ueda, K., Hung, W.-Y., Zhou, Y.-G., Liu, K., Chen, Y.-M., Zeghal, M., Abdoun, T., Escoffier,

S., & Manzari, M. (2018b). A new shared miniature cone penetrometer for centrifuge testing. In *Proceedings of 9th International Conference on Physical Modelling in Geotechnics, ICPMG 2018* (Vol. 1, pp. 293–229). London: CRC Press/Balkema.

El Ghoraiby, M. A., Park, H., & Manzari, M. T. (2017). *LEAP 2017: Soil Characterization and Element Tests for Ottawa F65 Sand*. Washington, DC: George Washington University.

Idriss, I. M., & Boulanger, R. W. (2008). *Soil Liquefaction During Earthquakes, MNO-12* (p. 142). Oakland, CA: Earthquake Engineering Research Institute.

Kutter, B. L., Carey, T. J., Bonab, M. H., Stone, N., Manzari, M., Zeghal, M., Escoffier, S., Haigh, S., Madabhushi, G., Hung, W., Kim, D., Kim, N., Okamura, M., Tobita, T., Ueda, K., & Zhou, Y. (2019). LEAP-UCD-2017 V. 1.01 model specifications. In B. Kutter et al. (Eds.), *Model tests and numerical simulations of liquefaction and lateral spreading: LEAP-UCD-2017*. New York: Springer.

Kutter, B. L., Carey, T. J., Hashimoto, T., Zeghal, M., Abdoun, T., Kokkali, P., Madabhushi, S. P. G., Haigh, S. K., Burali d'Arezzo, F., Madabhushi, S. S. C., Hung, W.-Y., Lee, C.-J., Cheng, H.-C., Iai, S., Tobita, T., Ashino, T., Ren, J., Zhou, Y.-G., Chen, Y., Sun, Z.-B., & Manzari, M. T. (2017). LEAP-GWU-2015 experiment specifications, results, and comparisons. *International Journal of Soil Dynamics and Earthquake Engineering, Elsevier, 113*, 616. https://doi.org/10.1016/j.soildyn.2017.05.018.

Kutter, B. L., Carey, T. J., Zheng, B. L., Gavras, A., Stone, N., Zeghal, M., Abdoun, T., Korre, E., Manzari, M., Madabhushi, G. S., Haigh, S., Madabhushi, S. S., Okamura, M., Sjafuddin, A. N., Escoffier, S., Kim, D.-S., Kim, S.-N., Ha, J.-G., Tobita, T., Yatsugi, H., Ueda, K., Vargas, R. R., Hung, W.-Y., Liao, T.-W., Zhou, Y.-G., & Liu, K. (2018a). Twenty-four centrifuge tests to quantify sensitivity of lateral spreading to Dr and PGA. In S. J. Brandenberg & M. T. Manzari (Eds.), *Geotechnical Earthquake Engineering and Soil Dynamics V, GSP 293* (pp. 383–393). Alexandria, VA: ASCE. https://doi.org/10.1061/9780784481486.040.

Kutter, B., Zeghal, M., Manzari, M. (2018b). *LEAP-UCD-2017 Experiments (Liquefaction Experiments and Analysis Projects)*, DesignSafe-CI [publisher], Dataset. https://doi.org/10.17603/DS2N10S

Manzari, M., Ghoraiby, M. E., Kutter, B. L., Zeghal, M., Abdoun, T., Arduino, P., Armstrong, R. J., Beaty, M., Carey, T., Chen, Y.-M., Ghofrani, A., Gutierrez, D., Goswami, M., Haigh, S. K., Hung, W.-Y., Iai, S., Kokkali, P., Lee, C.-J., Madabhushi, S. P. G., Mejia, L., Sharp, M., Tobita, T., Ueda, K., Zhou, Y.-G., & Ziotopoulou, K. (2017). Liquefaction analysis and experiment projects (LEAP): Summary of observations from the planning phase. *International Journal of Soil Dynamics and Earthquake Engineering, 113*, 714. https://doi.org/10.1016/j.soildyn.2017.05.015.

Parra Bastidas A.M., Boulanger, R.W., Carey, T., DeJong, J. (2017) Ottawa F-65 Sand Data from Ana Maria Parra Bastidas, https://datacenterhub.org/resources/ottawa_f_65, doi: 10.17603/DS2MW2R.

Yoshimine, M., Nishizaki, H., Amano, K., & Hosono, Y. (2006). Flow deformation of liquefied sand under constant shear load and its application to analysis of flow slide in infinite slope. *Soil Dynamics and Earthquake Engineering, 26*, 253–264.

# Chapter 5
# Archiving of Experimental Data for LEAP-UCD-2017

**Bruce L. Kutter, Trevor J. Carey, Nicholas Stone, and Bao Li Zheng**

**Abstract** This paper describes how the LEAP-UCD-2017 data is organized in DesignSafe; it is intended to help users, archivers, and curators find or organize data of interest. Several key files, folders, and documents included in the archive are discussed: (1) an Excel format data template used to document much of the data and metadata for each model (sensor data, cone penetrometer data, and surface marker data as reported by the experimenters), (2) processed sensor data files with time offsets and zero and calibration corrections that facilitate comparison of consistently formatted data from various model tests, (3) plots of processed data for quick overview and comparison of results among experiments, and (4) photographs taken during construction and testing.

## 5.1 Introduction

Twenty-four separate model tests were conducted at nine different centrifuge facilities for this LEAP exercise. The first goal of this paper is to describe how the data is archived and organized to help future users of the data. The data are archived in a Project: PRJ-1843: LEAP-UCD-2017 Experiments (Liquefaction Experiments and Analysis Projects) in the NSF-funded DesignSafe (https://www.designsafe-ci.org/) developed by the NHERI cyberinfrastructure center.

The original basis of the archived data was an Excel workbook data template distributed to each of the experiment sites; each site reported their data from each model test using this data template. Different worksheets in the workbooks contain the following data: a summary of the key initial conditions required for blind prediction, the testing sequence, sensor data for each event, long-term sensor data, results of cone penetration tests, results of surface marker surveys, and other data. The completed worksheet templates were submitted to LEAP-UCD-2017 organizers in a Box.com folder. These completed worksheets are included in archives for every experiment. The data in these templates were then processed and plotted in a

B. L. Kutter (✉) · T. J. Carey · N. Stone · B. L. Zheng
Department of Civil and Environmental Engineering, University of California, Davis, CA, USA
e-mail: blkutter@ucdavis.edu



B. Kutter et al. (eds.), *Model Tests and Numerical Simulations of Liquefaction and Lateral Spreading*, https://doi.org/10.1007/978-3-030-22818-7_5

consistent format, mostly by researchers at UC Davis, to facilitate consistent cross comparisons between data collected at different facilities.

The authors of the data and experiment were asked to review the data uploaded to the Project in DesignSafe; many of them corrected some errors and added some key information such as photos taken during the sample preparation and testing processes. Each set of experiments from each experimental facility is to be published with a separate publication and different DOI (digital object identifier). So, the LEAP-UCD-2017 project is to include nine different Published Datasets describing two or three model tests in each dataset.

The projects in the DesignSafe Data Depot are presently organized such that experimenters may upload all of the data relevant to conducting the experiment into a "Working Directory". Then a subset of the data in the Working Directory (the portion of the data deemed to be of value to future users of the data) was categorized and published. Thus, there are two hierarchies of interest: (1) Working Directory and (2) Published Datasets. The data in the Working Directory is of interest to researchers conducting and documenting the experiments and data; the Published Datasets contain information deemed useful to numerical modelers and other future users of the data.

## 5.2 Accessing Published LEAP-UCD-2017 Data in DesignSafe

This section of this paper attempts to walk a future data user along a suggested path to understand and access the nine Published Datasets of LEAP-UCD-2017.

First it is necessary to navigate to the landing page of the published LEAP-UCD-2017 project in DesignSafe. It should look similar to the screenshots shown in Fig. 5.1a, b. Below the title and keywords, a short description of the project is provided and below that is a list of the nine "Experiments"; each experiment contains the two or three model tests performed at a given facility. They are listed in alphabetical order by the name of the test facility. Each experiment has its own author list and digital object identifier (DOI). Below the list of nine experiments, there is a list of Model Config details (one entry per model test) and a list of Sensor Info details (one entry per model test). A general report (describing all 24 experiments) may be also accessed from this page.

From the landing page, it is recommended that the user click on the General Experiment Report to view its contents. The contents of the general report are illustrated in the screenshot at the bottom of Fig. 5.1b. The files and folders have been numbered according to a suggested sequence for navigating through the data for the first time. Each item in the report is briefly described below.

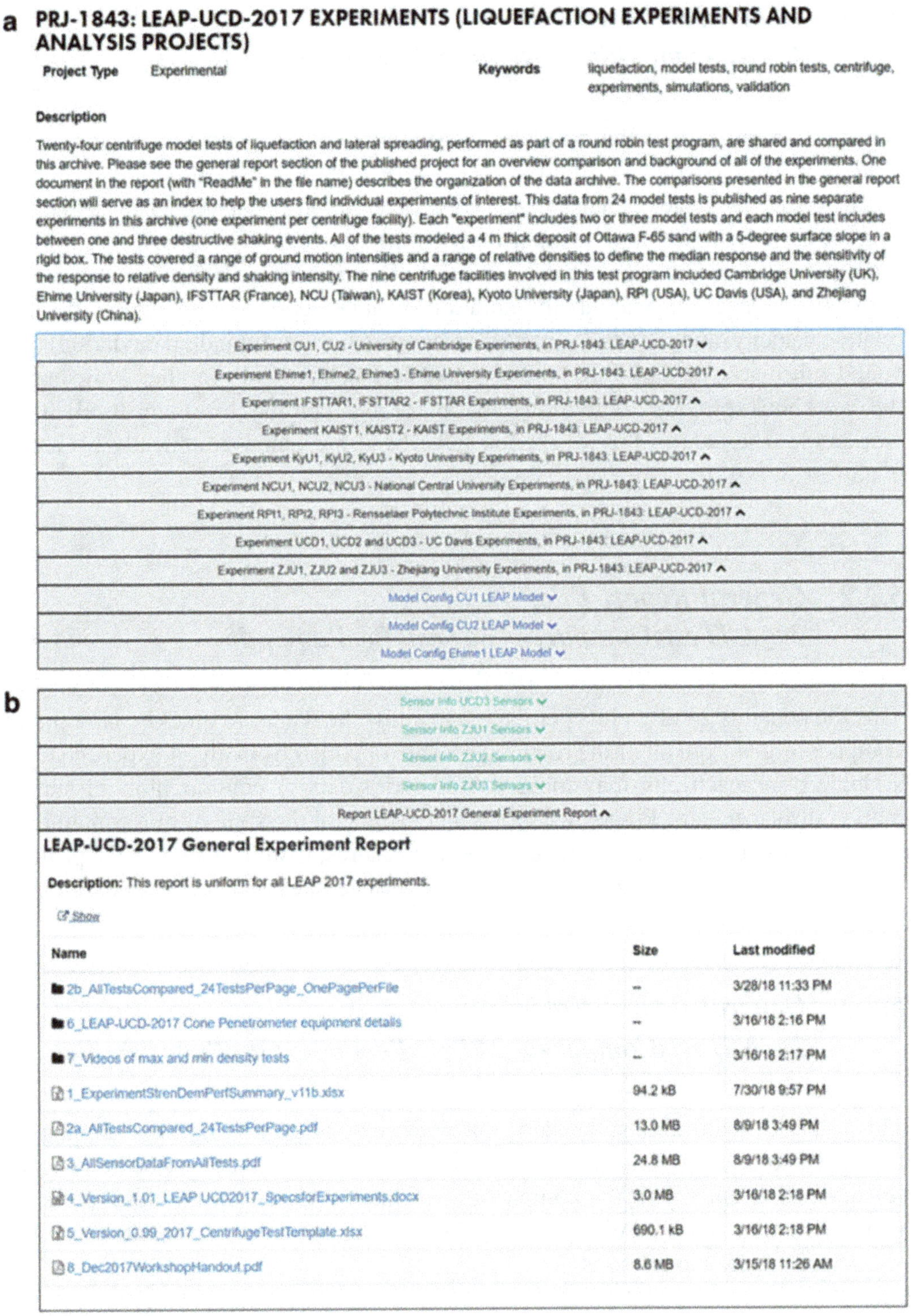


**Fig. 5.1** (**a**) Screenshot of LEAP-UCD-2017 project in DesignSafe: header, experiment list, and some of the Model Configuration lists. (The appearance of the user interface is subject to change.). (**b**) Screenshot of LEAP-UCD-2017 project in DesignSafe: part of the Sensor Info List and contents of the General Experiment Report. (The appearance of the user interface is subject to change)

### 5.2.1 General Report File: 1_ExperimentStrenDemPerfSummary_v11b.xlsx

This file is a spreadsheet that tabulates the strength, demand, and performance of the models in the first two destructive motions of each experiment. The strength is characterized by relative density and cone penetration resistance. The demand is characterized by input motion Intensity Measures such as PGA, $PGA_{effective}$, PGV, $CAV_5$, Arias Intensity, and other quantities. The performance is characterized by performance measures such as the duration of liquefaction, the integrated positive relative velocity, and statistical analysis (mean, median, and standard deviation) of x, y, and z displacements of different subsets of surface markers. Another worksheet in the workbook provides definitions for all of the column headings used in the workbook. The data in this workbook have been presented graphically by Kutter et al. (2018, 2019b).

### 5.2.2 General Report File: 2a_AllTestsCompared_24TestsPerPage.pdf

This file contains 24 tiny plots per page that allow the reader to quickly and roughly compare results from all of the required sensors obtained from all 24 experiments on a single page; each plot may include time series data or contour plots of surface marker displacements. Figure 5.2 shows an example of the style of plot contained in this file. Similar plots are provided for pore pressures, displacements, CPT profiles, and contours of x, y, and z displacements of surface markers.

### 5.2.3 General Report Folder: 2b_AllTestsCompared_24TestsPerPage_OnePagePerFile

This folder contains almost the same information as File 2a. However each page of the file is reproduced as a separate file in this folder. This format may be more useful for navigating quickly to the comparisons of interest based on the descriptive file names. In addition to data plotted in File 2a, the folder also contains plotted response spectra of the ground motions.

### 5.2.4 General Report File: 3_AllSensorDataFromAllTests.pdf

This file contains 151 pages of plots, presenting all of the sensor data from Destructive Motions #1 and #2 for all of the models in one large document. All of the data

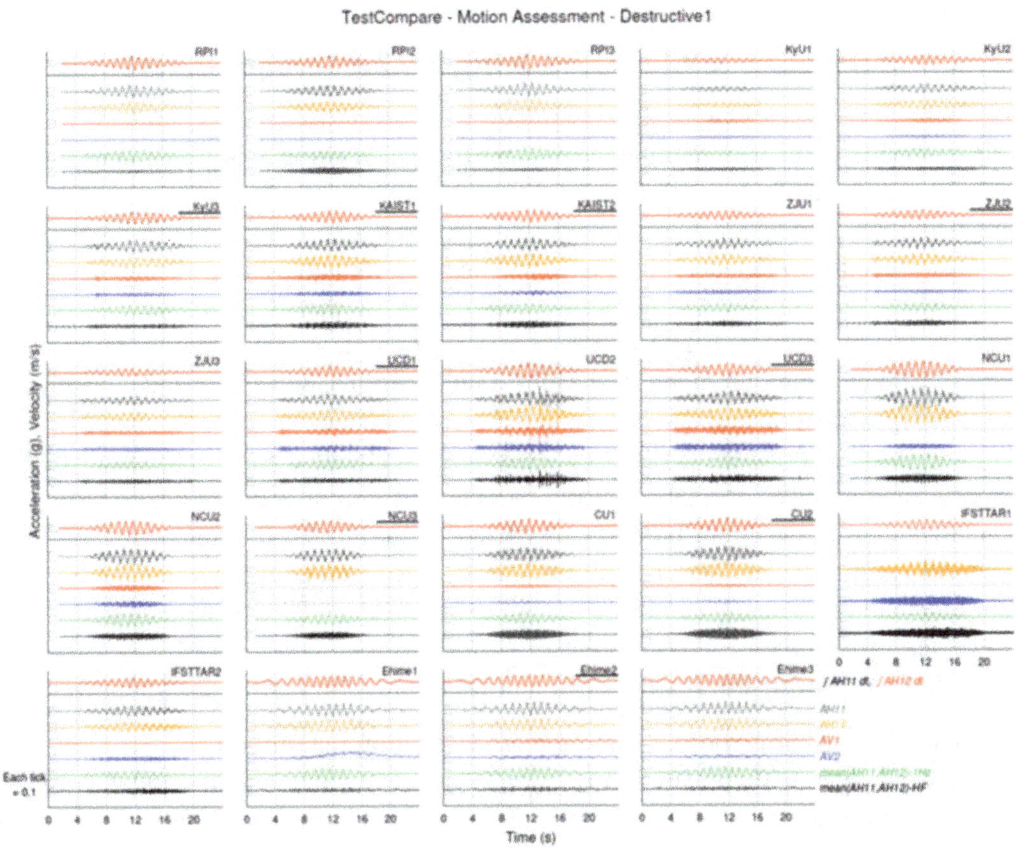


**Fig. 5.2** Input motions for Motion #1 for 24 experiments presented as an example of style of plots available in File: 2a_AllTestsCompared_24TestsPerPage.pdf

are processed consistently and plotted at consistent scale. The scale is large enough that the details of the motion can be seen. The experiments are presented in alphabetical order: CU1, CU2,. . ., ZJU3 in this document.

### *5.2.5 General Report File: 4_Version1.01_LEAP UCD2017_SpecsforExperiments.docx*

This file contains all the instructions provided to experimenters in preparation for their centrifuge tests. The previously provided specifications were cleaned up and presented in this proceedings document by Kutter et al. (2019a) because they would be of interest to future users of the data.

### 5.2.6 General Report File: 5_Version_0.99_2017_CentrifugeTestTemplate.xlsx

Each experimenter was required to submit their data using this blank spreadsheet workbook as a template. The first worksheet in this workbook contains instructions for using the template.

The second worksheet "(o)SummarySheet" summarizes key dimensions and properties essential for the numerical simulation of the experiments (density of soil, radius of the centrifuge, centrifuge acceleration, degree of saturation, viscosity of fluid, and the measurements of the elevation of the ground surface and comparison to the target locations).

The third worksheet "(a)TestLogSequence" is intended to document all of the events and measurements taken during model construction, saturation, and testing. It includes spaces for reporting measurements of grain size distribution, humidity, saturation, and density and asks the researchers to document when they measured surface markers. The time and date of spins, shakes, and CPT data collection are also to be reported in the TestLogSequence.

Other worksheets document:

- A worksheet for recording surface marker location measurements.
- A worksheet with sensor information such as serial number, model number, orientation, and x, y, z coordinates of the sensor during installation and during excavation of the model.
- A worksheet for reporting uncorrected CPT penetration resistance versus depth. Carey et al. (2019a) describes how the CPT data was corrected for more detailed analysis.
- A worksheet for recording residual pore pressure averages (RPPA). This sheet was intended to enable calculation of the settlement of pore pressure sensors by computing how much the water pressure changed during shaking.

Sensor time series data, sampled at high speed, is recorded on a separate sheet for each dynamic event (each ground motion). A separate sheet documents uninterrupted lower-speed sampling of the sensors to document the sequence of each entire spin; this is valuable for checking the response of pore pressure transducers during spin-up and slowdown of the centrifuge. Finally, separate custom sheets were used if researchers collected additional data such as shear wave velocity data or high-speed camera data.

Some of the first completed experiments were submitted using Version_0.96_2017 of this data template; Version 0.96 did not contain the "(o) SummarySheet." For cases where Version 0.96 was completed, a Version 0.99 Template summary sheet was added later to the data set.

### 5.2.7 *General Report Folder: 6_LEAP-UCD-2017 Cone Penetrometer Equipment Details*

This folder contains all of the design drawings, design calculations, and instructions for assembly and preloading for the CPT devices constructed at UC Davis and distributed to the LEAP facilities.

### 5.2.8 *General Report Folder: 7_Videos of Max and Min Density Tests*

This folder contains video instructions describing the procedure recommended for quality control tests on the sand used at each facility. It was requested that each facility perform maximum and minimum density tests using the same procedure to provide additional assurance that the properties of the Ottawa F-65 sand tested at each facility were the same. The purpose of the video was to improve consistency of test procedure at different sites. Further study of this important topic is described by Carey et al. (2019b).

### 5.2.9 *General Report File: 8_Dec2017WorkshopHandout.pdf*

This file, primarily of historical interest, contains the program for the December 2017 Workshop at UC Davis. It also contains some selected experimental results that were printed and distributed to the workshop participants.

## 5.3 Detailed Data for Each Model Test

### 5.3.1 *Selecting an Experiment Site*

After reviewing the information in the General Experiment Report, the interested user may decide to download specific experimental data or to try to figure out some details or metadata for a specific model test. It should be noted that each model test is also explained in some detail in a separate paper in these proceedings. To download or explore the data in DesignSafe, click on the experiment of interest from the menu illustrated in Fig. 5.1a. If you were to click on the Experiment titled “ZJU1, ZJU2 and ZJU3—Zhejiang University Experiments, in PRJ-1843: LEAP-UCD-2017,” you would see something like the screenshot shown in Fig. 5.3; this would show the list of authors, the “doi” for the selected experiments, and some basic information about their centrifuge. Below that, the event data for each of their experiments would

**Fig. 5.3** Screenshot that appears after selecting the Zhejiang University Experiments. (The appearance of the user interface is subject to change)

be visible. One of the items in the event data would be the Excel templates (.xlsx files), which could be downloaded to review the summary sheet, surface marker measurements, or other information described in Sect. 2.6. But, for analysis of the dynamic data from the destructive motions, it is recommended to use the processed data in the .csv files, instead of the data in the Excel spreadsheets. The processed data has been processed uniformly from experiment to experiment to offset the zeros, correct the polarity of sensors if necessary, and change the units of the measurement (e.g., g instead of $m/s^2$). The time range of the processed data has also been truncated consistently to include the most interesting duration of the dynamic sensor data, with $t = 0$ set so that the motions should be in phase if one experiment was plotted with data from another experiment from this LEAP exercise. The summary.pdf file shows all the sensor data for the first two destructive motions; this plotted data, also available in the general report in the large pdf file, is repeated for convenience. The data in the . csv files is the same data that is plotted in the pdf files. All of the above data is intended to be available for every model test of LEAP-UCD-2017 exercise.

In addition, experimenters were encouraged to upload photographs that document their experiments, raw data from other shaking events including non-destructive motions, and other relevant information.

## 5.3.2 *Model Configuration Data*

If one selects the link to "Model Configuration" in Fig. 5.3, one would find another link back to the Excel template. The Excel sheet contains the basic model configuration data in various worksheets. In addition, one would find photographs, reports, or powerpoint presentations that describe most of the experiments. It is worth repeating that the papers published in these proceedings prepared by the experimenters will also be a useful source of information about configuration of every model.

### 5.3.3 *Sensor Information*

If one selects the "Sensor List" link in Fig. 5.3, one would be provided again with a link back to the same Excel template. One of the worksheets in this workbook contains all the sensor information including serial number, model number, orientation, and location of the sensor (measured at the time of model construction and at the time of model excavation).

## 5.4 Working Directory for Data LEAP-UCD-2017

The Working Directory itself is not published, so it is accessible only to the members of the project. The Working Directory for LEAP is the storage location for all of the data to be published. All of the items included in the general report were uploaded to the root folder of the Working Directory. In addition, a folder called "Individual Centrifuge Facility Data" was created in the Working Directory. Inside this folder there are nine subfolders, one for each centrifuge facility as shown by screenshot in Fig. 5.4.

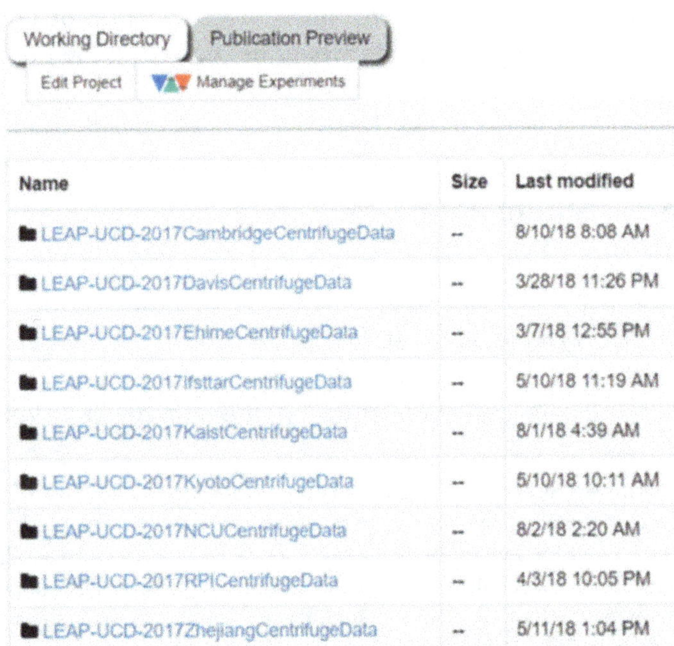


**Fig. 5.4** Screenshot of individual centrifuge facility data in the Working Directory. (The appearance of the user interface is subject to change)

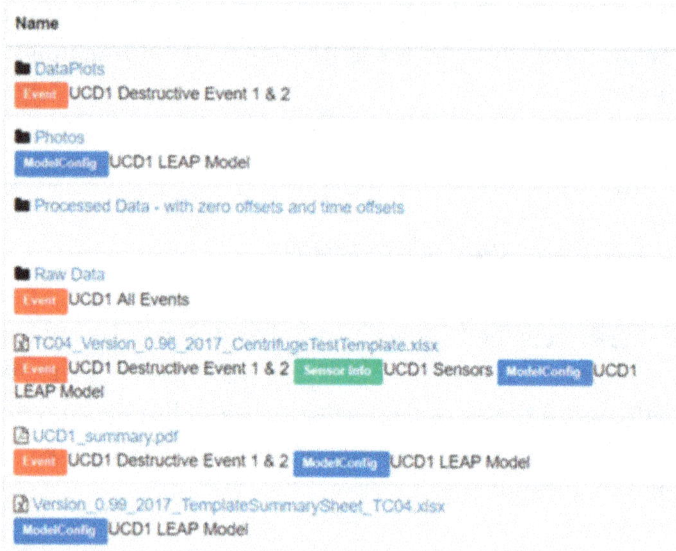


**Fig. 5.5** Tagged contents of subfolder LEAP-UCD-2017DavisCentrifugeData/UCD1_TC04. (The appearance of the user interface is subject to change)

In each of the folders of Fig. 5.4, all of the other important metadata for the experiments may be archived and tagged. An example of a subfolder in the LEAP-UCD-2017DavisCentrifugeData folder is shown by screenshot in Fig. 5.5.

The data is organized in a conventional folder hierarchy that varies to some extent from facility to facility. The example in Fig. 5.5 is a folder of data about experiment UCD1, with a more elaborate structure than those used for most experiments. Subfolders were created for data plots, photos, processed data, and raw data. Files or folders are then tagged as "Event," "ModelConfig," or "SensorInfo" as indicated by the red, blue, and green tags in Fig. 5.5. Some files such as the data in the Excel template have all three tags because some worksheets in the file contain event data, and others contain sensor information for model configuration data. The photos are generally tagged "ModelConfig." The selection of tags in the Working Directory determines where the data would be located in the published data set.

## 5.5 Summary

This paper is meant to help explain the organization of the data in the LEAP-UCD-2017 project archived in DesignSafe. Most of the paper explains how future users might access the published data. Some of the paper also explains how the producers of the data organized their data in the unpublished Working Directory.

**Acknowledgments** The experimental work on LEAP-UCD-2017 was supported by different funds depending mainly on the location of the work. The work by the US PI's (Manzari, Kutter, and Zeghal) is funded by National Science Foundation grants: CMMI 1635524, CMMI 1635307, and CMMI 1635040. The work at Ehime University was supported by JSPS KAKENHI Grant Number 17H00846. The work at Kyoto University was supported by JSPS KAKENHI Grant Numbers 26282103, 5420502, and 17H00846. The work at Kansai University was supported by

JSPS KAKENHI Grant Number 17H00846. The work at Zhejiang University was supported by National Natural Science Foundation of China (Nos. 51578501 and 51778573), Zhejiang Provincial Natural Science Foundation of China (LR15E080001), and National Basic Research Program of China (973 Project) (2014CB047005). The work at KAIST was part of a project titled "Development of performance-based seismic design," funded by the Ministry of Oceans and Fisheries, Korea. The work at NCU was supported by MOST: 106-2628-E-008 -004 -MY3.

## References

Carey, T. J., Stone, N., & Kutter, B. L. (2019a). Interpretation of CPT profiles for LEAP-UCD-2017. In B. Kutter et al. (Eds.), *Model tests and numerical simulations of liquefaction and lateral spreading: LEAP-UCD-2017*. New York: Springer.

Carey, T. J., Stone N., & Kutter, B. L. (2019b). Grain size analysis and maximum and minimum dry density of Ottawa F-65 sand for LEAP-UCD-2017. In B. Kutter et al. (Eds.), *Model tests and numerical simulations of liquefaction and lateral spreading: LEAP-UCD-2017*. New York: Springer.

Kutter, B. L., Carey, T. J., Zheng, B. L., Gavras, A., Stone, N., Zeghal, M., Abdoun, T., Korre, E., Manzari, M., Madabhushi, G. S., Haigh, S., Madabhushi, S. S., Okamura, M., Sjafuddin, A. N., Escoffier, S., Kim, D.-S., Kim, S.-N., Ha, J.-G., Tobita, T., Yatsugi, H., Ueda, K., Vargas, R. R., Hung, W.-Y., Liao, T.-W., Zhou, Y.-G., & Liu, K. (2018). Twenty-four centrifuge tests to quantify sensitivity of lateral spreading to Dr and PGA. In S. J. Brandenberg & M. T. Manzari (Eds.), *Geotechnical Earthquake Engineering and Soil Dynamics V, GSP 293* (pp. 383–393). Alexandria, VA: ASCE. https://doi.org/10.1061/9780784481486.040.

Kutter, B. L., Carey, T. J., Stone, N., Bonab, M. H., Manzari, M., Zeghal, M., Escoffier, S., Haigh, S., Madabhushi, G., Hung, W.-Y., Kim, D.-S., Kim, N.-R., Okamura, M., Tobita, T., Ueda, K., & Zhou, Y.-G. (2019a). LEAP-UCD-2017 V. 1.01 model specifications. In B. Kutter et al. (Eds.), *Model tests and numerical simulations of liquefaction and lateral spreading: LEAP-UCD-2017*. New York: Springer.

Kutter, B. L., Carey, T., Stone, N., Zheng, B. L., Gavras, A., Manzari, M., Zeghal, M., Abdoun, T., Korre, E., Escoffier, S., Haigh, S., Madabhushi, G, Madabhushi, S. S. C., Hung, W. -Y., Liao, T.-W., Kim, D.-S., Kim, S.-N., Ha, J.-G., Kim, N. R., Okamura, M., Sjafuddin, A. N., Tobita, T., Ueda, K., Vargas, R., Zhou, Y.-G., & Liu, K. (2019b). LEAP-UCD-2017 comparison of centrifuge test results. In B. Kutter et al. (Eds.), *Model tests and numerical simulations of liquefaction and lateral spreading: LEAP-UCD-2017*. New York: Springer.

# Chapter 6
# Comparison of LEAP-UCD-2017 CPT Results

**Trevor J. Carey, Andreas Gavras, and Bruce L. Kutter**

**Abstract** A cone penetrometer was specifically designed for the LEAP project to provide an assessment of centrifuge model densities independent from mass and volume measurements. This paper presents the design of the CPT and analyses of the results. Due to uncertainty in the specifications about how to define zero depth of penetration, about 20% of the CPT profiles were corrected to produce more accurate results. The procedure for depth correction is explained. After these corrections, penetration resistances at the representative depths of 1.5, 2, 2.5, and 3 m (prototype depth) are correlated to the reported specimen dry densities by linear regression. Using the inverse form of the linear regression equations, the density of each specimen could be estimated from the penetration resistance. Kutter et al. (LEAP-UCD-2017 comparison of centrifuge test results. In Model Tests and Numerical Simulations of Liquefaction and Lateral Spreading: LEAP-UCD-2017, 2019b) found that the density determined from penetration resistance was a more reliable predictor of liquefaction behavior than the reported density itself. Finally, the centrifuge tests at different LEAP facilities modeled the same prototype in different containers using different length scale factors (1/20 to 1/44); thus the ratio of layer thickness to cone diameter was different in each experiment. It appears that the penetration resistances are noticeably affected by container width and, to a lesser extent, resistance is affected by the length scale factor.

## 6.1 Introduction

One of the challenges encountered during LEAP-GWU-2015 was independently assessing the achieved density of the centrifuge models (Kutter et al. 2017). Uncertainty in the calculated density from measurements of mass and volume is affected by irregular or bumpy surfaces produced by sand pluviation and limited precision of the load cells used to measure mass. Cone penetrometers (CPTs) were selected as an independent method to assess the achieved specimen density.

T. J. Carey (✉) · A. Gavras · B. L. Kutter
Department of Civil and Environmental Engineering, University of California, Davis, CA, USA
e-mail: TJCarey@ucdavis.edu



B. Kutter et al. (eds.), *Model Tests and Numerical Simulations of Liquefaction and Lateral Spreading*, https://doi.org/10.1007/978-3-030-22818-7_6

To eliminate variability that would be encountered by using different CPT designs at each centrifuge laboratory, a new CPT was designed with the goal that it will be low cost and not require specialty machining (Carey et al. 2018). Total cost for the device, including the load cell, is less than $2000 US. The devices were fabricated at a machine shop at UC Davis and distributed to each of the centrifuge laboratories.

Carey et al. (2018) discussed the design and calibration and provided a cross and direct comparison of results using the LEAP CPT. This paper will briefly review the design of the device and provide additional analysis of the results from the LEAP-UCD-2017 exercise. Using a linear regression and penetration resistance at mid-depth of the model, densities are estimated for each experiment. The apparent effects of centrifuge length scale factor and model container width on CPT penetration resistance are also discussed.

## 6.2 Design

The LEAP CPT, sketched in Fig. 6.1, is a 60-degree cone, 6 mm in diameter, and is fabricated from stock stainless steel materials. The device uses a rod, protected by a hollow sleeve to transmit tip forces to a load cell, eliminating the need for a costly submerged tip strain gauge bridge. The load cell and the hollow sleeve are attached to a rigid aluminum block, which allows for simultaneous advancement of both the rod and sleeve. Carey et al. (2018) calculated for a vertical effective stress of 100 kPa and a maximum allowable tip force of 900 N, the device could safely penetrate sand with a relative density of about 100% to a depth of 4 m prototype scale without yielding the cone rod in compression.

Located 100 mm behind the cone shoulder is a 20-degree taper that transitions the 6 mm diameter sleeve to 8 mm. The 8 mm section increases the ruggedness of the device, and the location of the taper was chosen to be sufficient to limit increases in overburden stress at the cone tip from bearing stress generated at the taper.

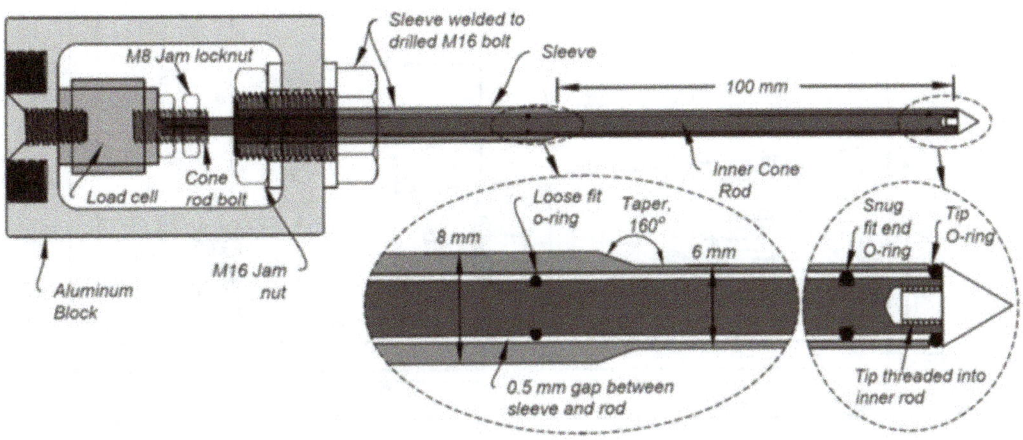


**Fig. 6.1** Sketch of LEAP CPT from Carey et al. (2018)

The internal rod is solid stainless steel with a diameter of 4 mm. O-rings that have a diameter slightly smaller than the 5 mm inner diameter of the sleeve are spaced every 100 mm to reduce the unsupported length of the rod. The O-rings rest in grooves that are cut in the rod. Located just above the cone tip is a slightly larger tip O-ring with a 4 mm inner diameter and 1 mm cross section to prevent sand from filling the gap between the rod and sleeve. Preloading the tip O-ring is performed using a procedure described by Carey et al. (2018).

Carey et al. (2018) performed a series of calibration experiments to test (1) load transfer from the tip O-ring into the sleeve, (2) the influence of lateral force on cone tip forces, and (3) cycling loading to check if the internal O-rings had changing hysteresis during repeat cycles of loading. The load transfer from the tip O-ring to the sleeve under full design load was roughly 3% of the applied load; this effect is accounted for in the calibration of the cone. Tests confirm that the load cell was insensitive to inadvertent lateral loading of the cone. During cyclic loading hysteresis was observed, but following the first cycle of loading and unloading, additional cycles followed the same hysteresis as the prior cycle; the width of the hysteresis curve suggests that the magnitude of error due to hysteresis would correspond to less than 31 kPa of tip resistance, which is about 1.1% of the average tip resistance at mid-depth of the LEAP experiments.

## 6.3 LEAP-UCD-2017 Experiment

The specifications for the LEAP-UCD-2017 experiment are presented by Kutter et al. (2019a). The experiment consists of a uniform profile of Ottawa F-65 sand, inclined at 5 degrees in a rigid container. Test geometry, sensor layout, and prototype dimensions for the experiment are shown in Fig. 6.2. Twenty-four experiments were

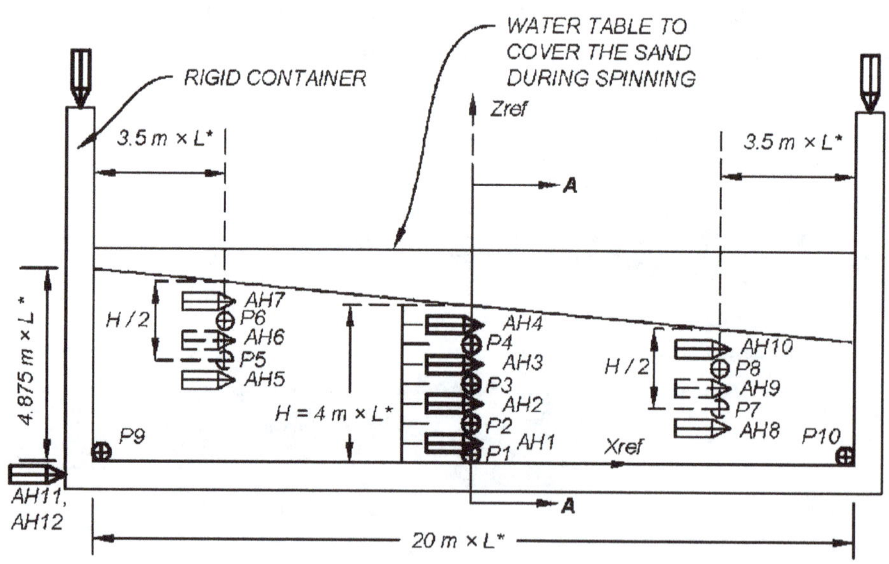


Fig. 6.2 Test geometry, sensor layout, and prototype dimensions of the LEAP-UCD-2017 experiment

conducted at nine participating facilities (National Central University (Taiwan), Zhejiang University (China), Kyoto University, University of Cambridge, IFSTTAR, UC Daivs, Ehime University, KAIST, Rensselaer Polytechnic Institute). The goal of the LEAP-UCD-2017 exercise was to perform experiments with relative densities ranging from about 50% (about 1599 kg/m$^3$) to about 80% (1703 kg/m$^3$) to determine the sensitivity of the model response to initial relative density.

## 6.4 Depth at Which the Cone Tip Touches the Surface (Depth of Zero Penetration)

When comparing results from different centrifuge tests, it is critical that the effective depth of the cone be consistently defined. A two-step process was used to determine the effective depth: (1) determine the depth of zero penetration and (2) adjust the depth to the 2/3 point on the cone tip. This section describes the first step in this process. Identifying when the tip of the cone touches the ground surface is complicated because the penetration resistance near the surface is very small and the surface area of the point of the cone tip is zero. Recognizing the complications in rectifying the point of zero penetration, the preferred practice was to accurately advance the tip of the cone to the sand surface before spinning the centrifuge. Unfortunately, some of the LEAP-UCD-2017 tests did not begin this way.

The proposed criteria compare the recorded penetration resistance ($q_c$) when the conical tip is just embedded into the soil with the ultimate bearing capacity ($q_{bL}$) of an equivalent prototype area square footing. It is assumed that the equivalent square footing will apply a similar magnitude of stress to the soil as would a fully embedded conical tip. Shown in Fig. 6.3 is an illustration of the point at which the conical tip is just embedded and an equivalent square footing. The bearing stress, $q_{bL,}$ of the square footing was determined to be 0.1 MPa for a dry density of 1668 kg/m$^3$, the median target density for the LEAP-UCD-2017 experiment. It was assumed that the depth at which the penetration resistance is 0.1 MPa corresponded approximately to the depth at which the conical tip is just embedded as shown in Fig. 6.3.

The different initial densities of the centrifuge models result in different penetration resistances at the stage depicted in Fig. 6.3. Therefore, when the recorded penetration resistance at full embedment of the conical tip was measured between $0.05 \leq q_c \leq 0.2$ MPa, it was deemed that no adjustment to the point of zero penetration (the location at which the point of the cone touches the surface) was

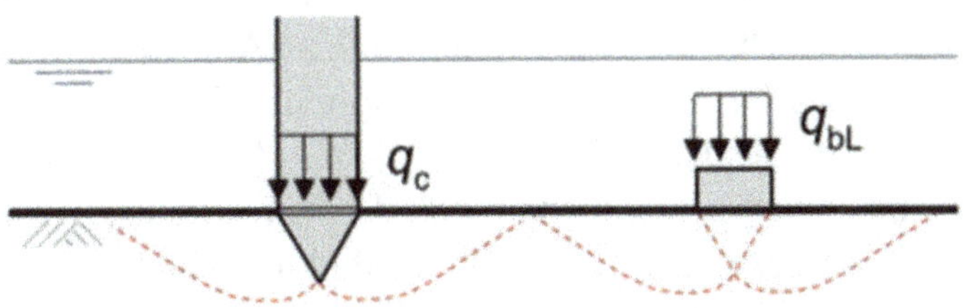


**Fig. 6.3** Conical tip just embedded into the soil and equivalent area square surface footing

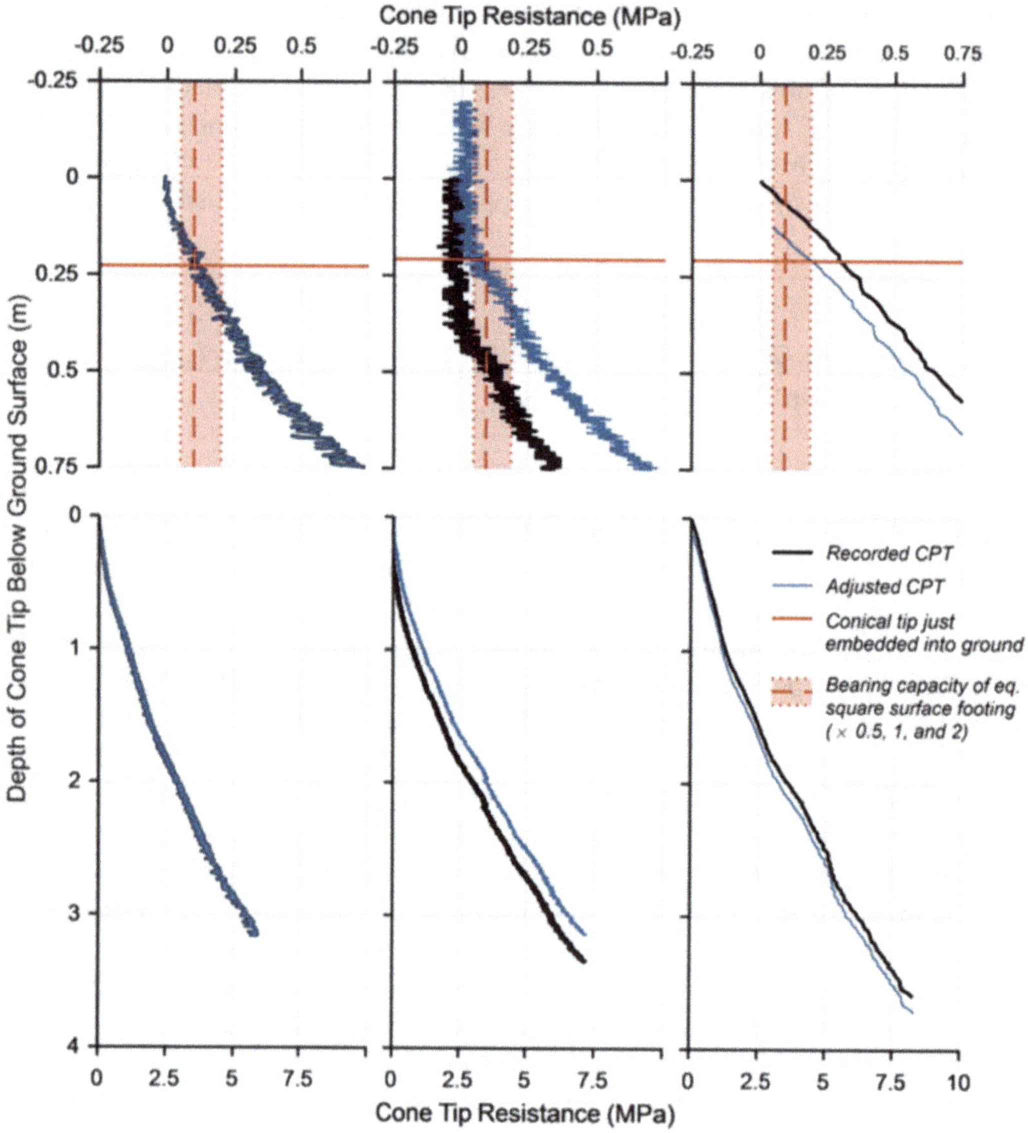


**Fig. 6.4** Examples of CPT profiles were (1) No adjustment was needed; (2) Adjustment I, CPT started above ground surface; and (3) Adjustment II, CPT started embedded in ground surface

required. Roughly 80% of the CPT profiles did not require depth correction by these criteria. If the penetration resistance was $q_c < 0.05$ MPa, the "real" point of zero penetration is located above the reported depth = 0. The depth of the point of zero penetration is then corrected by shifting the data upward by the prototype length of the conical tip ((3 mm)/(tan 30°) × length scale factor) from where $q_c$ is equal to 0.1 MPa. This scenario is referred to as Adjustment I and was applied to about 12% of the CPT soundings. If the reported point of zero penetration is located too far below the actual soil surface, $q_c > 0.2$ MPa. For this case, called Adjustment II, the point of zero penetration is shifted downward until $q_c$ is equal to 0.1 MPa. Figure 6.4 provides examples of a CPT sounding where adjustment was not required and soundings adjusted in accordance with Adjustment I and II.

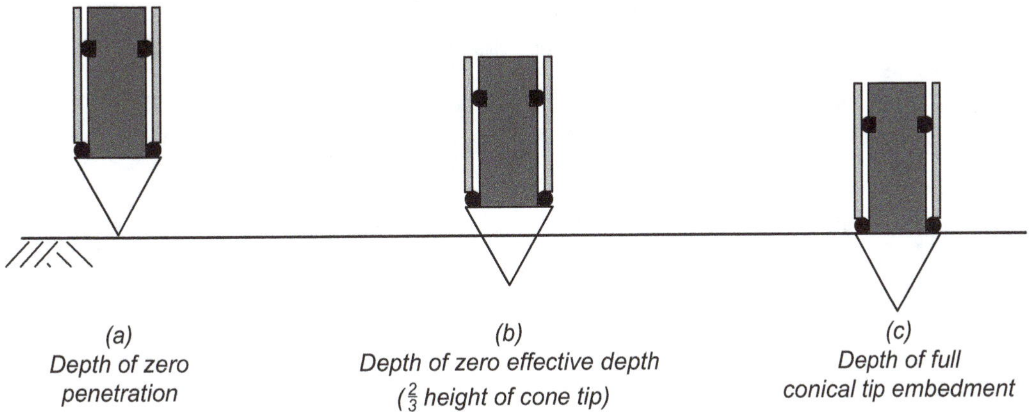


**Fig. 6.5** Definition of different reference depths. (**a**) Depth of zero penetration (tip located at group surface), (**b**) depth corresponding to zero effective depth (2/3 height of the cone tip), (**c**) depth at which the conical tip is fully embedded

After the correct points of zero penetration (the locations where the point of the cone tip touches the ground) were determined, all soundings were adjusted so the effective depth of the cone tip corresponded to the 2/3 point of the cone tip height (2/3)(5.19 mm), which is consistent with industry practice. Figure 6.5 shows the point at which the effective depth (z) of the cone tip is considered to be zero.

Eleven CPT devices were manufactured at the University of California Davis and shipped to seven of the centrifuge facilities. The University of Cambridge used their own cone design, which Carey et al. (2018) showed produced comparable penetration resistances as the LEAP design. Shown in Fig. 6.6 are all 59 CPT soundings and reported dry densities for each experiment. CPT measurements were performed prior to shaking for assessment of specimen density. Additional CPTs were performed prior to each destructive motion (i.e., second and third).

For interpretation of the CPT data, penetration tip resistances at representative depths of 1.5, 2.0, 2.5, and 3.0 m prototype were considered. In Fig. 6.7, the penetration resistances at the representative depths are plotted against the density reported from mass and volume measurements of the model specimen. General trends of increasing cone tip stress with depth and initial dry density can be observed. The red solid line is a linear mean regression, and the blue dashed lines are the 95% confidence interval of the linear fit. The confidence intervals become wider at the 2.5 and 3.0 m depths due to (1) larger variance in the data and (2) increased standard error from fewer data points. (Not all facilities were able to push to 2.5 and 3.0 m depths.)

The linear trend, with the form $q_c = \mathrm{a}\cdot\rho_d + \mathrm{b}$, has depth-dependent coefficients a and b, which are provided in Table 6.1. With the inverse trend, $\rho_d = (q_c - \mathrm{b})/\mathrm{a}$, the specimen density can be estimated from the penetration resistance, an alternative assessment of the reported densities.

Shown in Fig. 6.8 is an example of the density correction for points to the left and right of the trend line at 2.0 m depth. As an example, the experiment to the right of

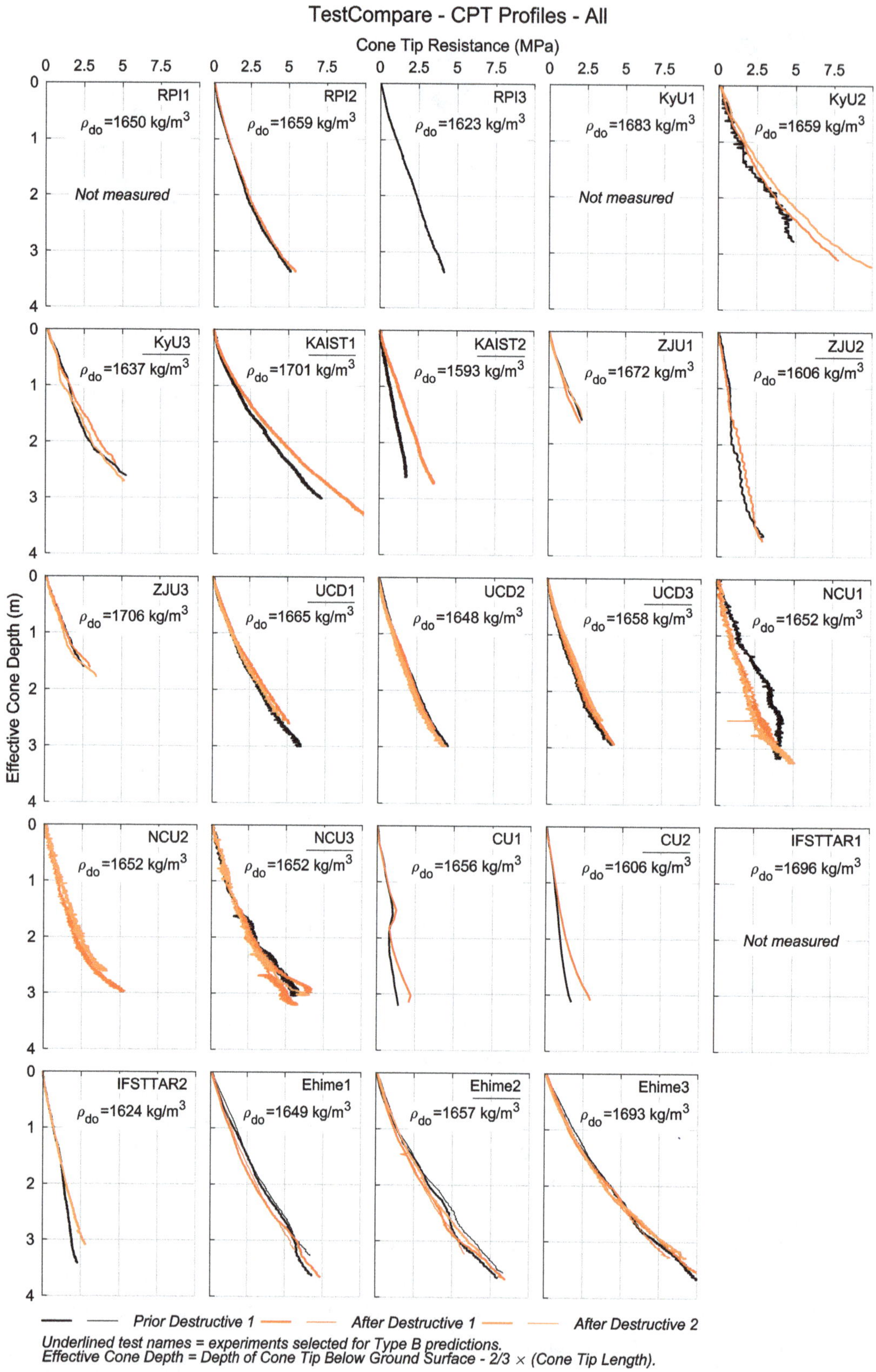


**Fig. 6.6** All CPT profiles for LEAP-UCD-2017

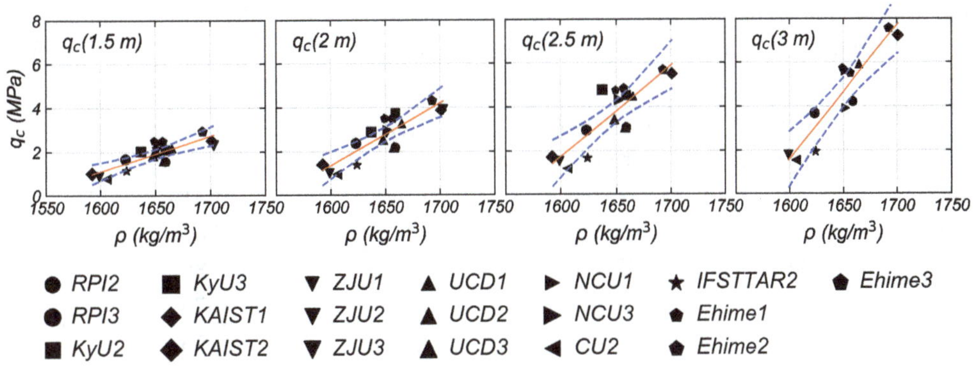


**Fig. 6.7** Penetration resistances at 1.5, 2.0, 2.5, and 3.0 m depths vs. reported specimen density. The red solid lines are the linear mean regressions fit, and the blue dashed lines are the 95% confidence interval of the linear fit

**Table 6.1** Coefficients of the linear trend lines ($q_c = a \cdot \rho_d + b$) for depths 1.5, 2.0, 2.5, and 3 m shown in Fig. 6.7

| Depth (m) | A | B |
|---|---|---|
| 1.5 | 0.0163 | −24.9 |
| 2.0 | 0.0285 | −44.3 |
| 2.5 | 0.0416 | −64.8 |
| 3.0 | 0.0611 | −96.1 |

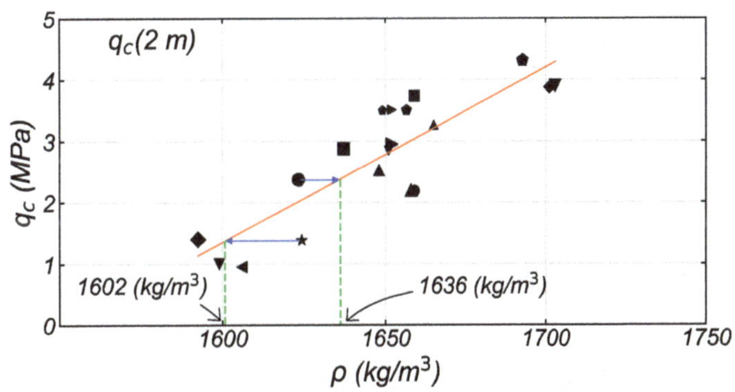


**Fig. 6.8** Density corrected using the linear trend at 2 m depth and penetration resistance

the line has a reported density of 1624 kg/m$^3$ and a penetration resistance of 1.38 MPa. When that point is corrected back to the trendline, the density decreases to 1602 kg/m$^3$. Kutter et al. (2019b) showed that at 2.0 m depth, densities determined from penetration resistances better correlate with liquefaction performance measurements than densities from mass and volume measurements.

Tabulated in Table 6.2 are the reported dry densities from mass and volume and densities calculated from the linear regression and penetration resistance. Relative densities of the mass and volume density measurements are also provided in Table 6.2 ($\rho_{min}$ = 1490.5 kg/m$^3$ and $\rho_{max}$ = 1757.0 kg/m$^3$ from Carey et al. 2019).

**Table 6.2** Reported dry density, relative density, cone penetration resistances at 2 m depth, and density using linear regression and penetration resistance at 2 m depth (NT = no CPT data available)

| Test ID | Dry density from mass and volume (kg/m$^3$) | Relative density from mass and volume (%) | Cone tip stress at 2 m depth $q_c$(2 m) (MPa) | Dry density from cone tip stress at 2.0 m depth $\rho(q_c(2))$ (kg/m$^3$) |
|---|---|---|---|---|
| CU1 | 1656 | 66 | 0.81 | 1581 |
| CU2 | 1606 | 47 | 0.95 | 1586 |
| Ehime1 | 1649 | 63 | 3.50 | 1676 |
| Ehime2 | 1657 | 66 | 3.50 | 1676 |
| Ehime3 | 1693 | 79 | 4.31 | 1704 |
| IFSTTAR1 | 1696 | 80 | NT | NT |
| IFSTTAR2 | 1624 | 56 | 1.38 | 1602 |
| KAIST1 | 1701 | 82 | 3.88 | 1689 |
| KAIST2 | 1593 | 42 | 1.40 | 1602 |
| KyU1 | 1683 | 75 | NT | NT |
| KyU2 | 1659 | 67 | 3.74 | 1684 |
| KyU3 | 1637 | 59 | 2.88 | 1654 |
| NCU1 | 1652 | 64 | 3.51 | 1676 |
| NCU2 | 1652 | 64 | NT | NT |
| NCU3 | 1652 | 64 | 2.95 | 1656 |
| RPI1 | 1650 | 64 | NT | NT |
| RPI2 | 1659 | 67 | 2.18 | 1630 |
| RPI3 | 1623 | 54 | 2.37 | 1636 |
| UCD1 | 1665 | 69 | 3.26 | 1667 |
| UCD2 | 1648 | 63 | 2.52 | 1642 |
| UCD3 | 1658 | 67 | 2.19 | 1630 |
| ZJU1 | 1651 | 64 | 2.85 | 1653 |
| ZJU2 | 1599 | 45 | 1.00 | 1588 |
| ZJU3 | 1703 | 82 | 3.90 | 1690 |

## 6.5 Effects of Scale Factor and Container Width

Since each facility used the same 6 mm diameter cone in model scale, the prototype size of the cone varied from 138 mm (RPI) to 300 mm (IFSTTAR). Bolton et al. (1999) demonstrated that if the normalized penetration depth to CPT diameter ratio was greater than 10 for relative densities of 80%, then the normalized penetration is not sensitive to depth. At a 2 m prototype depth in the IFSTTAR experiment, the depth to diameter ratio is 2000 mm/300 mm = 6.7, which is less than 10; therefore, reductions in penetration resistance may be expected in this case. In addition, cone penetration resistance for dilatant sand in rigid containers is known to increase as container width decreases.

To investigate container size and length scale factor effects further, the data for penetration resistance at a depth of 2 m were sorted, color coded, and replotted as

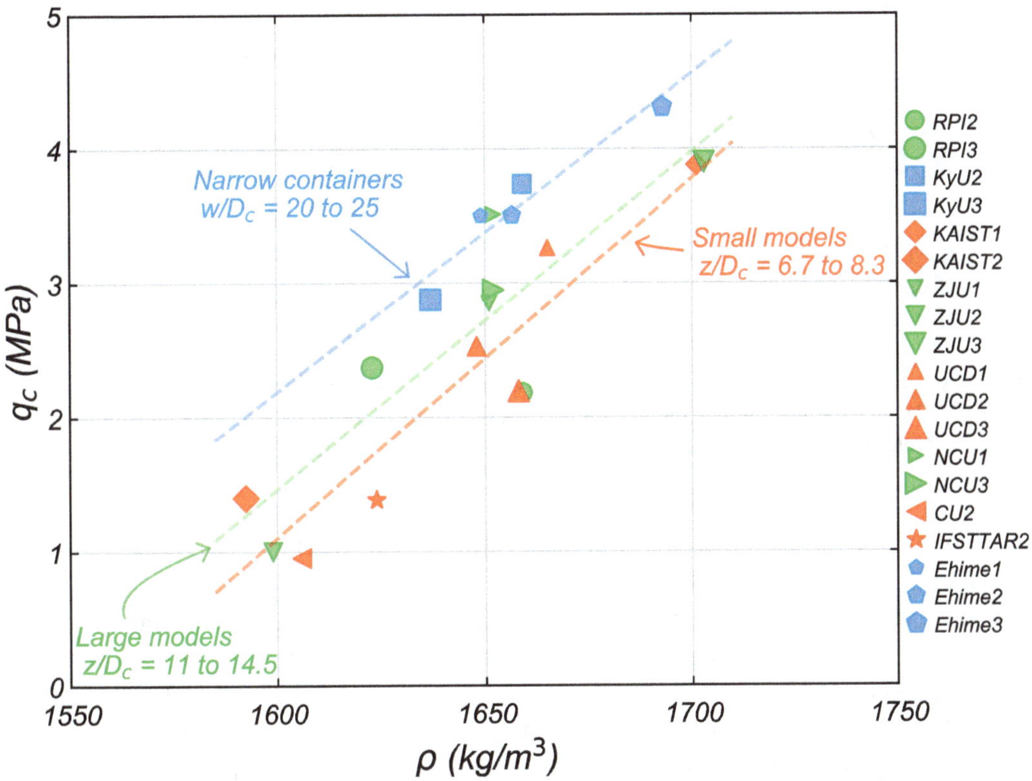


**Fig. 6.9** Experiments are separated by (1) large models tested at lower centrifuge acceleration ($z/D_c$ = 11 to 14.5) (green), (2) small models tested at high-g ($z$/Dc = 6.7 to 8.3) (red), and (3) narrow containers ($w$/Dc 20–25) (blue)

shown in Fig. 6.9. The experiments shown in green represent the larger models with wider containers tested at low g-levels (RPI, ZJU, and NCU). The number of normalized cone diameters ($z/D_c$ = depth of interest/cone diameter) to $z = 2$ m for these experiments ranged from $z/D_c$ = 11 to 14.5. Experiments represented by red points were the small models with wide containers tested at higher g-levels (KAIST, UCD, CU, and IFSTTAR); for these models, the ratio of depth to cone diameter at mid-depth of the model was only $z$/Dc = 6.7 to 8.3. Blue points represent experiments conducted in relatively narrow containers where the width of the container to cone diameter ratio was $w/D_c$ = 20–25 (KyU, Ehime). For experiments other than KyU and Ehime, the container width ranged from $w$/Dc ≈ 33 to 67 cone diameters.

Linear regressions were computed for each group of data points; interestingly, these regressions defined the nearly parallel lines shown in Fig. 6.9. At a density of 1662 kg/m$^3$ (relative density of 68%), using the regressions for the large and small models, the predicted penetration resistances are 2.73 and 2.5 MPa, respectively, an 8% difference. Reduction in penetration resistances for the high-g experiments is consistent at 1.5 and 2.5 m depths where adequate data is available. Shown in Fig. 6.10 are the four color-coded depths of interest.

The experiments performed with narrow containers consistently have elevated penetration resistances. Bolton et al. (1999) showed that if the number of normalized

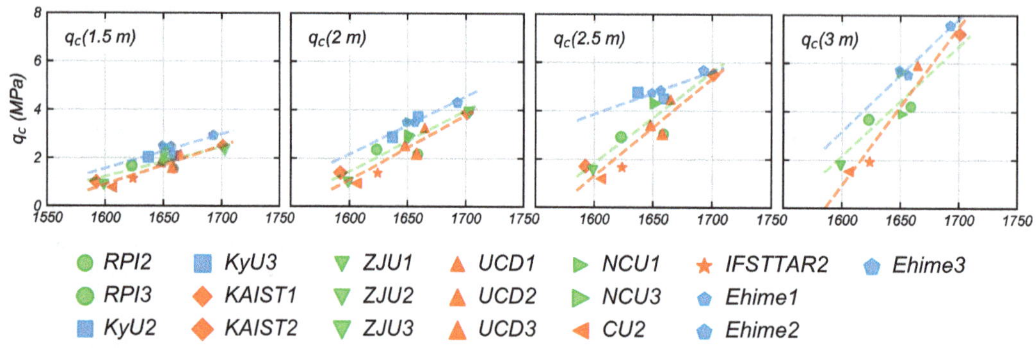


**Fig. 6.10** Experiments are separated by (1) large models tested at lower centrifuge acceleration (green), (2) small models tested at high-g (red), and (3) narrow containers (blue) for each depth of interest. Note: data presented for 2 m depth is consistent with data presented in Fig. 6.9

cone diameters to the container boundary wall was less than 10, then the penetration resistance would increase. CPTs performed along the longitudinal centerline of the narrow containers would be roughly 10–12 normalized cone diameters from the boundary wall at the most. Using the narrow container linear regression, the predicted penetration resistance at a relative density of 68% is 3.37 MPa, 23% larger than the low-g experiments. The elevated penetration resistance for the narrow container experiments is expected to range from 10 to 25% for all relative densities.

Additional work is required to confirm the trends observed in Fig. 6.9, but the results are encouraging and are generally consistent with expectations based on correlations presented by Bolton et al. (1999). Much of the variance seen in the correlation depicted in Fig. 6.7 may be explained by container width and model size. Kutter et al. (2019b) showed however that the resistance to liquefaction is better characterized by the cone penetration resistance than it is by the measured mass and volume. It is possible that the effects of model size on penetration resistance somehow compensate the effects of model size on liquefaction resistance (i.e., perhaps $q_c$ increases as container size decreases, and liquefaction resistance increases as container size increases.)

## 6.6 Conclusions

One of the challenges during LEAP-GWU-2015 was the lack of consistent measurements of the achieved specimen density. Using mass and volume to find density introduces uncertainty from both measurements; scales are typically only accurate to 0.1 kg, and point measurements of bumpy or irregular soil surfaces do not accurately represent the entire depth of soil. A low-cost CPT device was developed in an effort to maximize the likelihood of obtaining a useful correlation between penetration tests and density at different facilities. Carey et al. (2018) provides an overview of the design and calibration procedures.

Facilities reported their penetration tests as $q_c$ vs. depth, but there was no guarantee that the cone tip started at the ground surface. A procedure was presented to evaluate when the cone tip just touches the ground surface using the recorded penetration resistance ($q_c$) and a reference ultimate bearing capacity from an equivalent area square footing. Furthermore, the effective depth ($z$) was adjusted for all CPT profiles so that depth = 0 when the conical tip is embedded to 2/3 of its length (2/3 of the distance from the point of the cone-to-cone shoulder).

Penetration resistances at depths of 1.5, 2.0, 2.5, and 3.0 m depths were plotted against specimen density, computed from mass and volume. Linear trend lines were computed for each depth. Inverse equations from the linear regression lines were used then to determine dry density from penetration resistance. Kutter et al. (2019b) showed the densities calculated from penetration resistance at 2.0 m depth were better correlated to liquefaction performance measures than densities from mass and volume.

Additional investigation was done to explore the effects of container width and scale factor on penetration resistance. Consistent with work from Bolton et al. (1999), systematic sensitivity to the normalized container width (w/Dc) and the normalized depth ($z$/Dc = depth/cone diameter) seems apparent. It was argued that the effect of $w$/Dc on penetration resistance might be similar to the effect of w/Dc on the liquefaction resistance. If this is so, then this might contribute to explain why liquefaction resistance is better correlated to penetration resistance than to reported densities.

**Acknowledgments** Funding for this work is supported by NSF under CMMI Grant Number 1635307. The authors would like to the thank each of the participating LEAP facilities for providing CPT data. Thank you, Kate Darby, for providing data during the initial design process. Barry Zheng's help during data processing was much appreciated. The authors would also like to thank Andy Cobb at UCD's BAE shop for providing construction recommendations and manufacturing each device.

## References

Bolton, M. D., Gui, M. W., Garnier, J., Corte, J. F., Bagge, G., Laue, J., & Renzi, R. (1999). Centrifuge cone penetration tests in sand. *Geotechnique, 49*(4), 543–552.

Carey, T., Gavras, A., Kutter, B., Haigh, S. K., Madabhushi, S. P. G., Okamura, M., Kim, D. S., Ueda, K., Hung, W.-Y., Zhou, Y.-G., Liu, K., Chen, Y.-M., Zeghal, M., Abdoun, T., Escoffier, S., & Manzari, M. (2018). A new shared miniature cone penetrometer for centrifuge testing. In *Proceedings of 9th International Conference on Physical Modelling in Geotechnics, ICPMG 2018* (Vol. 1, pp. 293–229). London: CRC Press/Balkema.

Carey, T. J., Stone, N., & Kutter, B. L. (2019). Grain size analysis and maximum and minimum dry density testing of Ottawa F-65 sand for LEAP-UCD-2017. In B. Kutter et al. (Eds.), *Model tests and numerical simulations of liquefaction and lateral spreading: LEAP-UCD-2017*. New York: Springer.

Kutter, B. L., Carey, T. J., Bonab, M. H., Stone, N., Manzari, M., Zeghal, M., Escoffier, S., Haigh, S., Madabhushi, G., Hung, W., Kim, D., Kim, N., Okamura, M., Tobita, T., Ueda, K., & Zhou, Y. (2019a). LEAP UCD 2017 version 1.01 model specifications. In B. Kutter et al. (Eds.),

*Model tests and numerical simulations of liquefaction and lateral spreading: LEAP-UCD-2017.* New York: Springer.

Kutter, B. L., Carey, T., Stone, N., Zheng, B. L., Gavras, A., Manzari, M., Zeghal, M., Abdoun, T., Korre, E., Escoffier, S., Haigh, S., Madabhushi, G., Madabhushi, S. S. C, Hung, W.-Y., Liao, T.-W., Kim, D.-S., Kim, S.-N., Ha, J.-G., Kim, N. R., Okamura, M., Sjafuddin, A. N., Tobita, T., Ueda, K., Vargas, R., Zhou, Y.-G., & Liu, K. (2019b). LEAP-UCD-2017 comparison of centrifuge test results. In B. Kutter et al. (Eds.), *Model tests and numerical simulations of liquefaction and lateral spreading: LEAP-UCD-2017.* New York: Springer.

Kutter, B. L., Carey, T. J., Hashimoto, T., Zeghal, M., Abdoun, T., Kokkali, P., Madabhushi, S. P. G., Haigh, S. K., Burali d'Arezzo, F., Madabhushi, S. S. C., Hung, W.-Y., Lee, C.-J., Cheng, H.-C., Iai, S., Tobita, T., Ashino, T., Ren, J., Zhou, Y.-G., Chen, Y., Sun, Z.-B., & Manzari, M. T. (2017). LEAP-GWU-2015 experiment specifications, results, and comparisons. *International Journal of Soil Dynamics and Earthquake Engineering, Elsevier, 113*, 616. https://doi.org/10.1016/j.soildyn.2017.05.018.

# Chapter 7
# Difference and Sensitivity Analyses of the LEAP-2017 Experiments

**Nithyagopal Goswami, Mourad Zeghal, Bruce L. Kutter, Majid T. Manzari, Tarek Abdoun, Trevor Carey, Yun-Min Chen, Sandra Escoffier, Stuart K. Haigh, Wen-Yi Hung, Dong-Soo Kim, Seong-Nam Kim, Evangelia Korre, Ting-Wei Liao, Kai Liu, Gopal S. P. Madabhushi, Srikanth S. C. Madabhushi, Mitsu Okamura, Asri Nurani Sjafruddin, Tetsuo Tobita, Kyohei Ueda, Ruben Vargas, and Yan-Guo Zhou**

N. Goswami · M. Zeghal (✉) · T. Abdoun · E. Korre
Department of Civil and Environmental Engineering, Rensselaer Polytechnic Institute, Troy, NY, USA
e-mail: zeghal@rpi.edu

B. L. Kutter · T. Carey
Department of Civil and Environmental Engineering, University of California, Davis, CA, USA

M. T. Manzari
Department of Civil and Environmental Engineering, George Washington University, Washington, DC, USA

Y.-M. Chen · Y.-G. Zhou
Department of Civil Engineering, Zhejiang University, Hangzhou, China

S. Escoffier
IFSTTAR, GERS, SV, Bouguenais, France

S. K. Haigh · G. S. P. Madabhushi · S. S. C. Madabhushi
Department of Engineering, Cambridge University, Cambridge, UK

W.-Y. Hung · T.-W. Liao
Department of Civil Engineering, National Central University, Taoyuan, Taiwan

D.-S. Kim
Department of Civil and Environmental Engineering, Korea Advanced Institute of Science and Technology, Daejeon, South Korea

S.-N. Kim
Water Management Department, Korea Advanced Institute of Science and Technology, Daejeon, South Korea

K. Liu
Institute of Geotechnical Engineering, Zhejiang University, Hangzhou, China

M. Okamura · A. N. Sjafruddin
Department of Civil Engineering, Ehime University, Matsuyama, Japan

T. Tobita
Department of Civil Engineering, Kansai University, Osaka, Japan

K. Ueda · R. Vargas
Disaster Prevention Research Institute, Kyoto University, Kyoto, Japan



B. Kutter et al. (eds.), *Model Tests and Numerical Simulations of Liquefaction and Lateral Spreading*, https://doi.org/10.1007/978-3-030-22818-7_7

**Abstract** The experimental results of LEAP (Liquefaction Experiments and Analysis Projects) centrifuge test replicas of a saturated sloping deposit are used to assess the sensitivity of soil accelerations to variability in input motion and soil deposition. A difference metric is used to quantify the dissimilarities between recorded acceleration time histories. This metric is uniquely decomposed in terms of four difference component measures associated with phase, frequency shift, amplitude at 1 Hz, and amplitude of frequency components higher than 2 Hz (2 + Hz). The sensitivity of the deposit response accelerations to differences in input motion amplitude at 1 Hz and 2 + Hz and cone penetration resistance (used as a measure reflecting soil deposition and initial grain packing condition) was obtained using a Gaussian process-based kriging. These accelerations were found to be more sensitive to variations in cone penetration resistance values than to the amplitude of the input motion 1 Hz and 2 + Hz (frequency) components. The sensitivity functions associated with this resistance parameter were found to be substantially nonlinear.

## 7.1 Introduction

The Liquefaction Experiments and Analysis Projects (LEAP) are an ongoing series of international collaborations to produce high-quality (centrifuge) experimental data of saturated soil systems and to use this data to validate and assess the performance of constitutive models and numerical tools used in soil liquefaction analyses (Manzari et al. 2018). In 2017, the LEAP exercise involved repeating the same centrifuge test of a sloping deposit at nine centrifuge facilities in China (Zhejiang University, ZJU), France (Institut Français des Sciences et Technologies des Transports, de l'Aménagement et des Réseaux, IFSTTAR), Japan (Ehime University and Kyoto University, KyU), Korea (KAIST University), Taiwan (National Taiwan University, NCU), the UK (Cambridge University, CU), and the USA (UC Davis, UCD, and Rensselaer, RPI). In addition to the main goal of numerical model validation, the tests are aimed at assessing the repeatability, reproducibility, and sensitivity of experimental results among the different facilities.

Assessing the repeatability and reproducibility of the conducted centrifuge experiments requires metrics to quantify the similarities and differences among both the recorded input motions and the responses of the test replicas. A sensitivity analysis may then be used to evaluate how the different input motions and other uncertainties affect the observed soil response. This article proposes a new approach to identify and quantify the differences between time histories of input or response quantities, such as accelerations, velocities, and displacements, provided by experiments and centrifuge tests. The differences are decomposed in terms of measures associated with phase, frequency shift, amplitude at 1 Hz, and amplitude of frequency components higher than 2 Hz (2 + Hz). This approach is used herein to assess the differences and similarities between input accelerations achieved during 26 centrifuge test replicas of the same sloping deposit and evaluate the sensitivity of the experimental results to differences in input motion and deposition.

## 7.2 Experiment Overview

The LEAP-2017 centrifuge model is a deposit of Ottawa F-65 sand sloping at an angle of 5° to the horizontal and having a height of 4 m at mid-slope (Fig. 7.1). The sand was deposited through pluviation in a level rigid container to achieve a range of mass densities (with a reference mean value of 1652 kg/m$^3$). The corresponding relative densities varied from about 50 to 85% (evaluated using the minimal and maximal densities reported by Kutter et al. (2018, 2019), as displayed in Fig. 7.2).

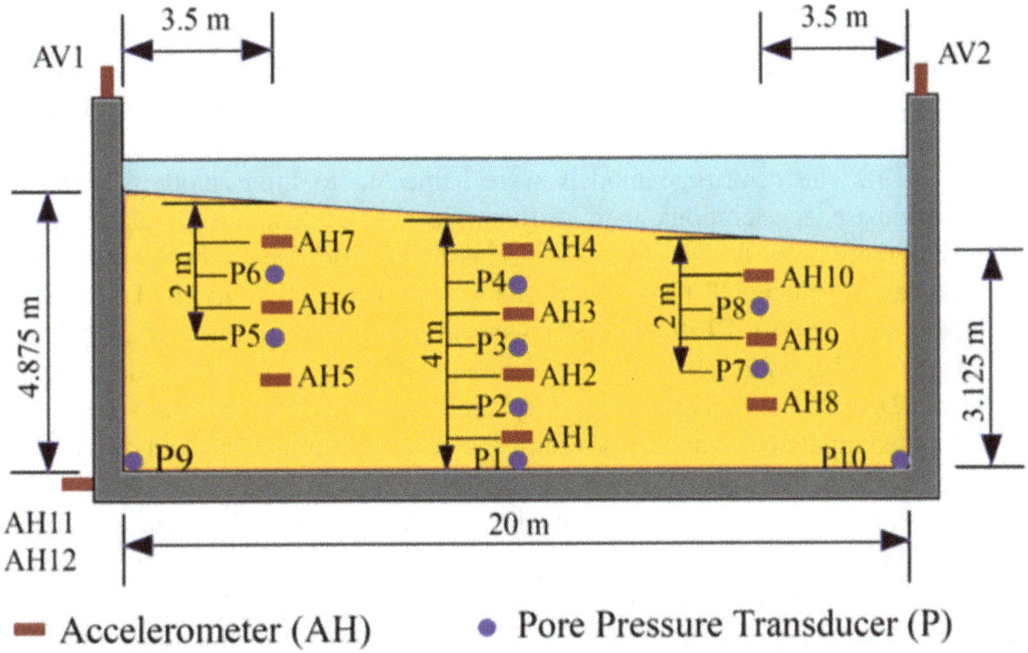


**Fig. 7.1** Schematic of the LEAP-2017 centrifuge model (dimensions are in prototype units)

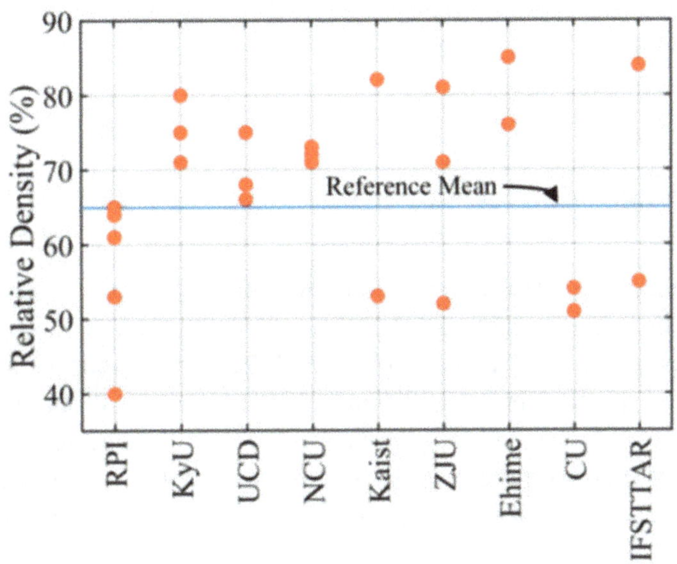


**Fig. 7.2** Relative density of the LEAP-2017 tests conducted at the nine centrifuge facilities

The deposits were saturated with a viscous fluid to achieve the same prototype hydraulic conductivity at the nine facilities. The models were instrumented with an extensive array of accelerometers (AH1–AH12, AV1, and AV2, with AH11 and AH12 measuring the horizontal input motion at the base of the model) and pore pressure transducers (P1 to P10), as shown in Fig. 7.1. Surface markers were used to measure the permanent lateral displacements (by surveying the location of the markers before and after shaking, as described in Kutter et al. 2018). A CPT (cone penetration test) was used during most of the centrifuge tests to characterize the deposit conditions before and after shaking. A comprehensive description of the model and experimental conditions is given by Kutter et al. (2018).

A total of 24 centrifuge test replicas of the sloping deposit were conducted at the 9 centrifuge facilities during LEAP-2017 (Carey et al. 2019; Escoffier and Audrain 2019; Hung and Liao 2019; Kim et al. 2019; Korre et al. 2019; Liu et al. 2019; Madabhushi et al. 2019; Okamura and Nurani Sjafruddin 2019; Vargas Tapia et al. 2019). The centrifuge models were subjected to input motions aimed at achieving base accelerations with different levels of closeness to a prescribed reference motion, as shown in Fig. 7.3. This figure also shows the input motions of two tests termed RPI0 and RPI4. RPI0 was conducted during LEAP-2015 (Kokkali et al. 2019). RPI0 and RPI1 were intended to be replica of each other. RPI4 was conducted in 2018 and was planned to be for a loose soil model (with an achieved $D_r = 40\%$). A qualitative assessment of the recorded motions reveals that the obtained input accelerations have different levels of similarities and differences. These differences are due to variability in equipment (e.g., shaker actuators) along with other unknown uncertainties and lead to dissimilarities in input amplitude, phase, and frequency contents. The recorded soil accelerations also showed a significant level of variability among the different centrifuge tests, as illustrated by the AH4 motions in Fig. 7.4. In addition to the dispersion in input motions, the response accelerations were also affected by the variability in properties and characteristics of the analyzed soil deposits (such as the relative density, as shown in Fig. 7.2), which, from a broad perspective, may lead to amplification, de-amplification, changes in response frequency and phase, and possibly other aspects of variations in soil accelerations.

## 7.3 Difference Metrics

A validation exercise of soil liquefaction computational tools involves (in addition to other requisites (Oberkampf and Roy 2010)) (1) an assessment of the similarities and differences in achieved input motions and recorded responses and (2) a sensitivity analysis to quantify how the different uncertainties in input and initial conditions affect soil response. Metrics are needed to quantify the level of consistency in input motions, compare the responses of the test replicas, and evaluate the level of agreement between experimental and numerical results. A number of metrics

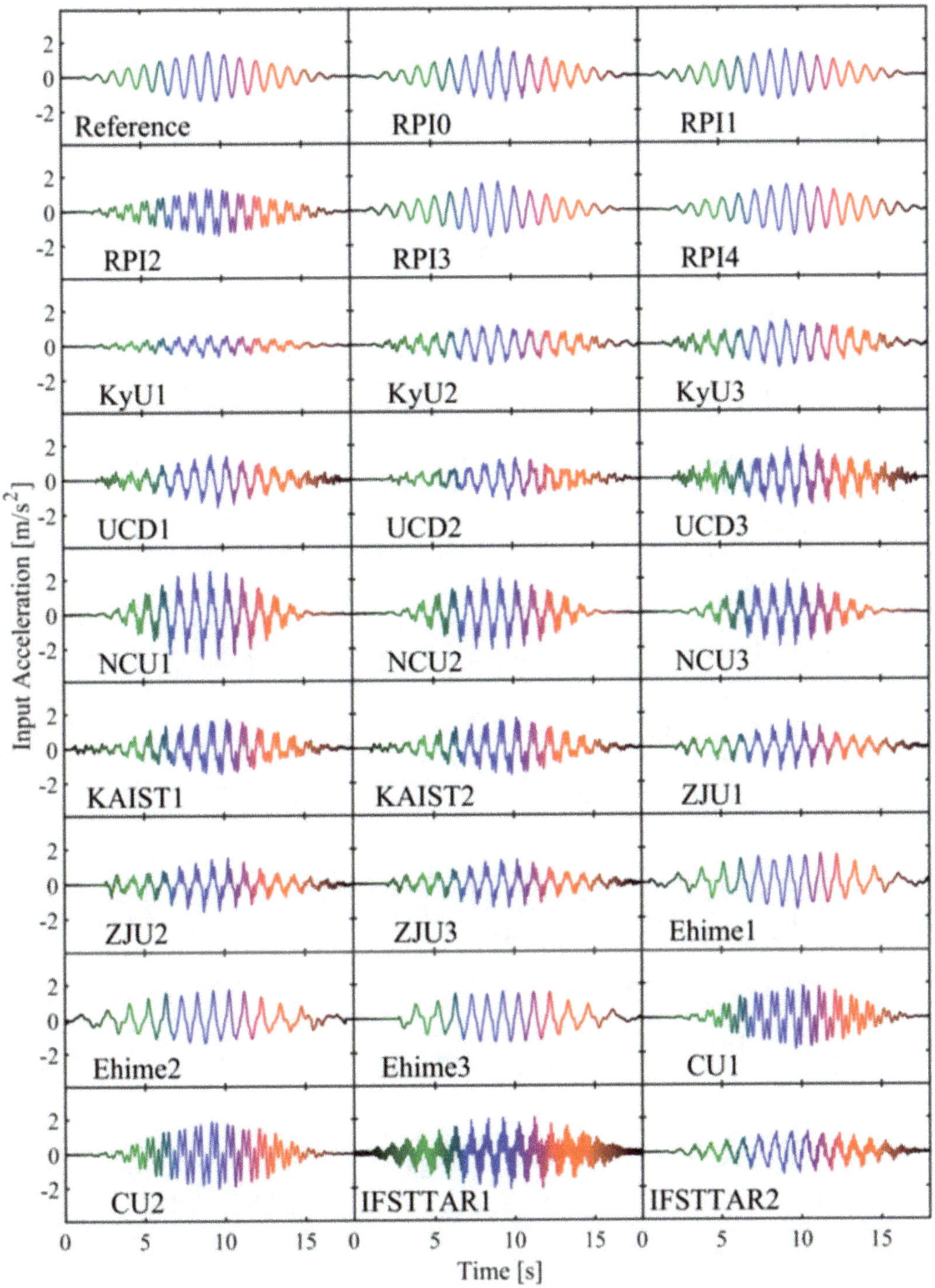


**Fig. 7.3** Reference and achieved input (AH11) accelerations of the analyzed 26 centrifuge tests

have been used by researchers to assess differences among dynamic time histories (e.g., accelerations), ranging from a simple vector norm to the Sprague and Geers metric (Geers 1984) which identifies magnitude and phase difference components, along with others. A brief overview of these metrics and some of the associated

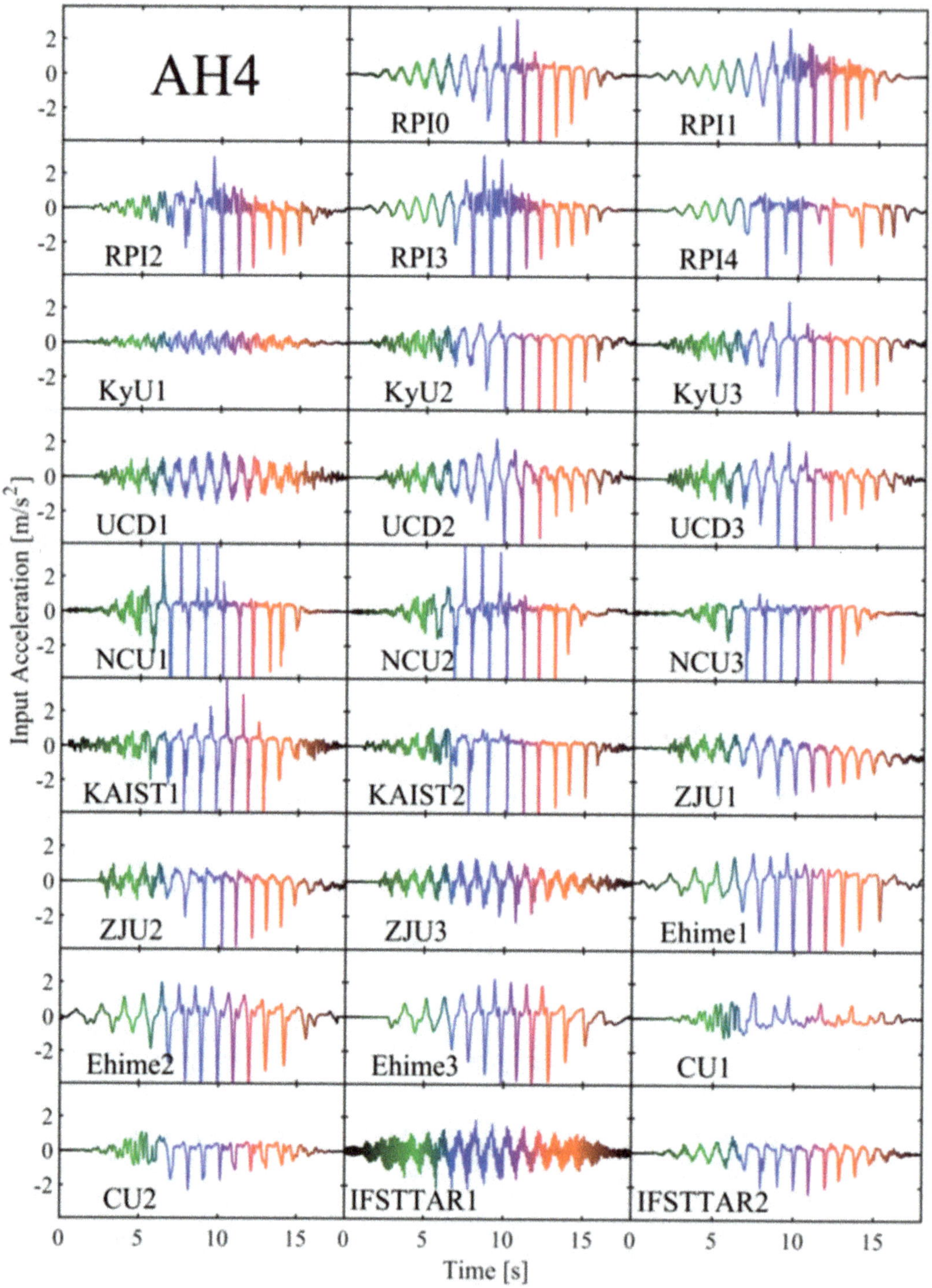


**Fig. 7.4** Recorded AH4 soil accelerations of the analyzed 26 centrifuge tests

characteristics are discussed by Zeghal et al. (2018). This article uses a new approach (Zeghal et al. 2018). The difference $d_{ij}$ between two corresponding acceleration time histories $a_i = a_i(t)$ and $a_j = a_j(t)$ of two different test replicas $i$ and $j$ is quantified using a normalized mean squared deviation (MSD):

$$d_{ij} = \frac{\int_0^W \left(a_i - a_j\right)^2 dt}{2\left(\int_0^W a_i^2 dt + \int_0^W a_j^2 dt\right)} \tag{7.1}$$

in which $t$ is time and $W$ is length of a time window of interest. This metric is normalized so that it varies between 0 and 1. A $d_{ij}$ metric approaching zero means that the two accelerations are essentially the same, whereas a metric of 1 is obtained, for instance, when two pure sinusoidal motions are 180 degrees out of phase with each other. The measure $d_{ij}$ is decomposed in terms of four specific fundamental components, namely, phase, frequency shift, amplitude at 1 Hz, and amplitude of frequency components higher than 2 Hz (referred to as 2 + Hz):

$$d_{ij} = d_{ij}^{\text{phase}} + d_{ij}^{\text{shape}} + d_{ij}^{F\text{shift}} \tag{7.2}$$

with:

$$d_{ij}^{\text{phase}} = \frac{\int_0^{+\infty} \left[2|A_i||A_j| - \left(A_i A_j^* + A_i^* A_j\right)\right] df}{2\left(\int_0^{+\infty} A_i^2 df + \int_0^{+\infty} A_j^2 df\right)} \tag{7.3}$$

$$d_{ij}^{\text{shape}} = \frac{DFW(|A_i|, |A_j|)}{2\left(\int_0^{+\infty} A_i^2 df + \int_0^{+\infty} A_j^2 df\right)} \tag{7.4}$$

$$d_{ij}^{F\text{shift}} = \frac{\int_0^{+\infty} \left(|A_i| - |A_j|\right)^2 df}{2\left(\int_0^{+\infty} A_i^2 df + \int_0^{+\infty} A_j^2 df\right)} - d_{ij}^{\text{shape}} \tag{7.5}$$

in which $f$ is frequency; $A_i$ and $A_j$ are the Fourier transforms of $a_i$ and $a_j$, respectively; and $A_i^*$ refers to the complex conjugate of $A_i$. The phase component $d_{ij}^{\text{phase}}$ reflects differences due to dissimilarities in acceleration phase angles. The shape component $d_{ij}^{\text{shape}}$ quantifies the difference associated with the geometrical shape (i.e., wave form and amplitude). DFW refers to a dynamic frequency warping (Goswami 2019), which is similar in concept to the dynamic time warping (DTW) used in speech recognition (Rabiner and Huang 1993). The use of DFW enables isolation of the magnitude differences associated with (slight) shifts in acceleration frequencies. For the LEAP-2017 input accelerations, the shape components were decomposed further:

$$d_{ij}^{\text{shape}} = d_{ij}^{\text{shape(1Hz)}} + d_{ij}^{\text{shape(2+Hz)}} \tag{7.6}$$

in which $d_{ij}^{\text{shape(1Hz)}}$ quantifies the shape (i.e., amplitude in this case) differences for the dominant 1 Hz component and $d_{ij}^{\text{shape(2+Hz)}}$ quantifies the difference related to the components at frequencies higher than 2 Hz (with the largest contribution often associated with the 3 Hz component). The frequency shift component $d_{ij}^{F\text{shift}}$ evaluates the difference stemming from variability in frequency of the acceleration components.

These metrics were verified using simple synthetic acceleration time histories with prescribed differences and were found to be effective (difference) identification and quantification tools (Goswami 2019). Also, the four difference metrics were used to evaluate equivalent (average) differences in amplitude at 1 Hz, amplitude for the 2 + Hz components, phase angle, and a shift in frequency at 1 Hz (referred to as $\Delta A_{ij}^{1\text{Hz}}$, $\Delta A_{ij}^{2+\text{Hz}}$, $\Delta \Phi_{ij}$, and $\Delta F_{ij}$, respectively) to characterize and quantify the specific factors responsible for the observed differences in accelerations. The details of this evaluation are presented in Goswami (2019). The relative values of the different metrics $d_{ij}^{\text{shape(1Hz)}}$, $d_{ij}^{\text{shape(2+Hz)}}$, $d_{ij}^{\text{phase}}$, and $d_{ij}^{F\text{shift}}$ and the differences $\Delta A_{ij}^{1\text{Hz}}$, $\Delta A_{ij}^{2+\text{Hz}}$, $\Delta \Phi_{ij}$, and $\Delta F_{ij}$ can be used as indicators to ascertain the difference that prevails.

### 7.3.1 *Input Motion Differences*

An analysis was conducted to assess the differences among the reference acceleration and the input motions that were recorded during the 26 centrifuge tests. First, the input motions were all cross-correlated to determine a consistent common time $t = 0$ for all the experiments. This eliminated all phase differences that are associated with a simple shift in the origin of time. The computed total differences $d_{ij}$ provided quantitative measures with numerical values varying from about 0.01 to 0.25, as exhibited in Fig. 7.5. In this figure, a 3D bar graph is used to display the whole set of difference metrics among the 26 recorded accelerations and the reference motion.

Overall, the input accelerations can be divided into three groups according to the difference metric values (Fig. 7.5): (1) Group 1 of accelerations that closely match the reference motion and also each other with $d_{ij} = 0.00$ to about 0.03, (2) Group 2 of accelerations that have an average match to the reference motion (and among the group) with $d_{ij}$ = about 0.03 to 0.08, and (3) Group 3 of accelerations that do not closely match the reference motion and also each other with $d_{ij}$ = about 0.08 to 0.25. Note that Group 3 includes the motion RPI3 (corresponding to the acceleration termed 5 in Fig. 7.5) which was generated so that it includes intentionally a significant 3 Hz component. The motions that are in a close match provide information that may be used, for instance, to assess reproducibility, while those with a loose match are advantageous in evaluating the sensitivities of the experiments to variation in input motions.

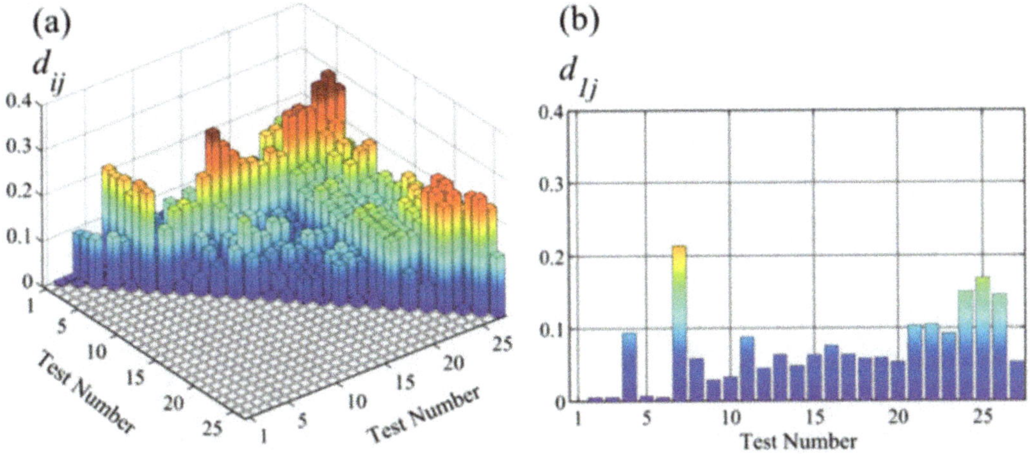


**Fig. 7.5** Input motion difference metrics: (**a**) among the reference motion (termed 1) and all analyzed centrifuge input accelerations (termed 2 to 26) and (**b**) between the reference motion and recorded accelerations (corresponding to the first (back) row of the left figure)

The total difference metrics were decomposed into 1 Hz amplitude, 2 + Hz amplitude, phase, and frequency shift component measures to assess the nature of the associated dissimilarities and reasons for these differences, as shown in Fig. 7.6. This figure also shows the quantitative values of the corresponding differences $\Delta A_{ij}^{1\text{Hz}}$, $\Delta A_{ij}^{2+\text{Hz}}$, $\Delta \Phi_{ij}$, and $\Delta F_{ij}$ among the accelerations. The decomposition shows that the difference metrics are overall comparable in values, with $d_{ij}^{\text{shape}(1\text{Hz})}$ being somewhat lower and $d_{ij}^{F\text{shift}}$ slightly higher than the other metrics. There is, however, a group of about five input accelerations that have relatively larger difference metrics (than the rest of the motions). The $d_{ij}^{\text{shape}(2+\text{Hz})}$ values clearly show that there are four input motions which have larger difference metrics with the rest of the accelerations and are consistent with the visual assessment provided by Fig. 7.3. Figure 7.6 also shows the quantitative values of the differences $\Delta A_{ij}^{1\text{Hz}}$, $\Delta A_{ij}^{2+\text{Hz}}$, $\Delta \Phi_{ij}$, and $\Delta F_{ij}$ among the accelerations. The differences in phase angle $\Delta \Phi_{ij}$ and frequency $\Delta F_{ij}$ are relatively minor from a practical perspective. The substantially low values of $\Delta \Phi_{ij}$ are explained by the use (in the conducted analyses) of a consistent common time $t = 0$ for all the experiments and the fact that all experiments properly achieved an input motion with a dominant 1 Hz component. The $\Delta A_{ij}^{1\text{Hz}}$ and $\Delta A_{ij}^{2+\text{Hz}}$ were more significant (especially for a set of about six input motions), even though the corresponding difference metrics were not substantially large. This is explained by the lower sensitivity of the total difference metric $d_{ij}$ to $\Delta A_{ij}^{1\text{Hz}}$ and $\Delta A_{ij}^{2+\text{Hz}}$ (compared to $\Delta \Phi_{ij}$ and $\Delta F_{ij}$). Overall, the $\Delta A_{ij}^{2+\text{Hz}}$ values were larger than those of $\Delta A_{ij}^{1\text{Hz}}$ and were as large as 0.6 m/s$^2$. A summary of the differences $\Delta A_{ij}^{1\text{Hz}}$ and $\Delta A_{ij}^{2+\text{Hz}}$ between the reference and input motions is presented in Fig. 7.7.

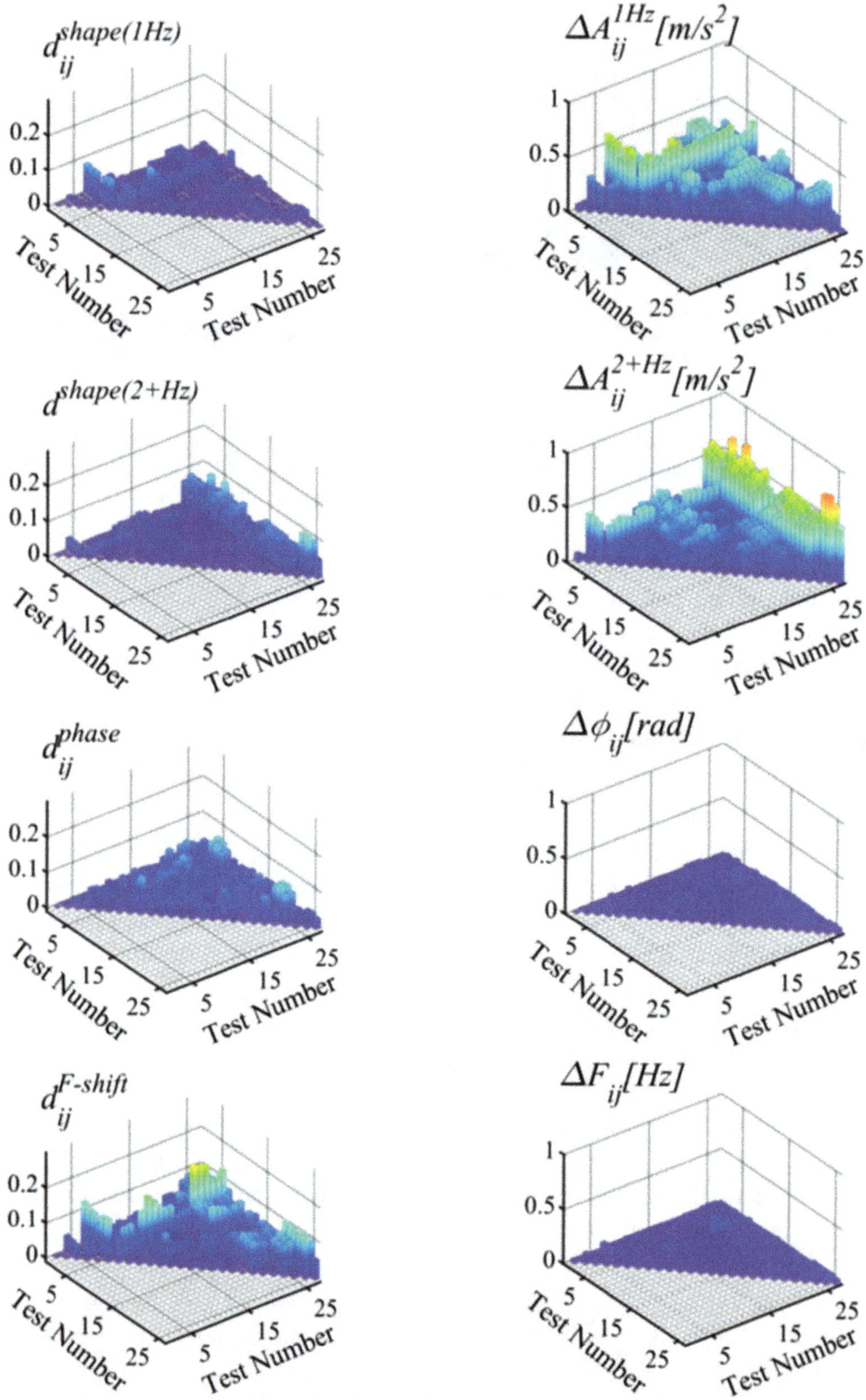


**Fig. 7.6** Difference metric components of the reference and input accelerations and corresponding $\Delta A_{ij}^{1\text{Hz}}$, $\Delta A_{ij}^{2+\text{Hz}}$, $\Delta \Phi_{ij}$, and $\Delta F_{ij}$

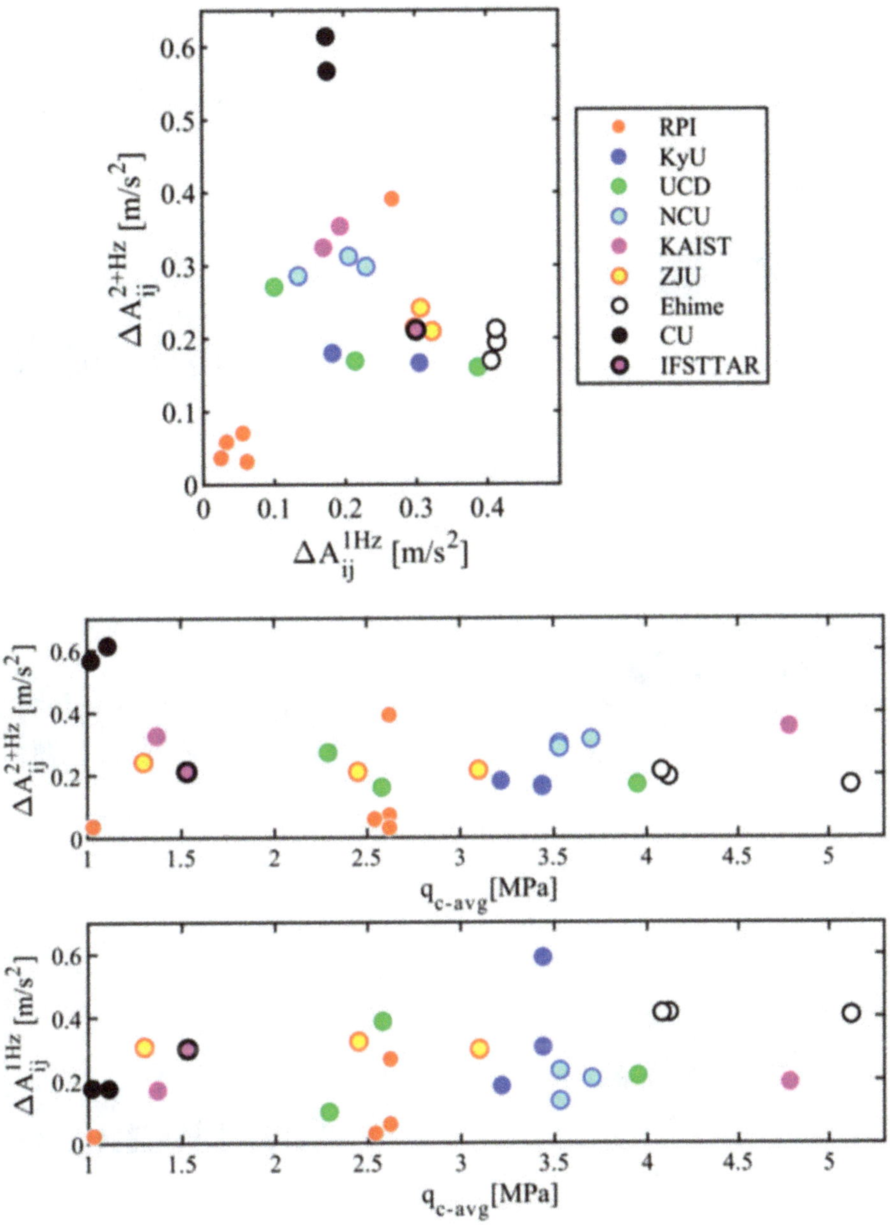


**Fig. 7.7** Summary of differences in input motion and initial conditions among the 26 replica tests (in terms $\Delta A_{ij}^{1Hz}$ and $\Delta A_{ij}^{2+Hz}$ of the base motions and cone penetration resistance $q_{c\text{-avg}}$)

### 7.3.2 *Response Motion Differences*

The total discrepancy measures and corresponding components were evaluated for the recorded soil responses at different depths. Herein, only the total metrics are presented and discussed (because of space limitations). The decomposition of these metrics and additional details are given in Goswami (2019). The values of the difference metric for the accelerations AH1 to AH4 (Fig. 7.8) increased from the base of the deposit to the free surface and were considerable at the shallow depth location AH4 (reaching values

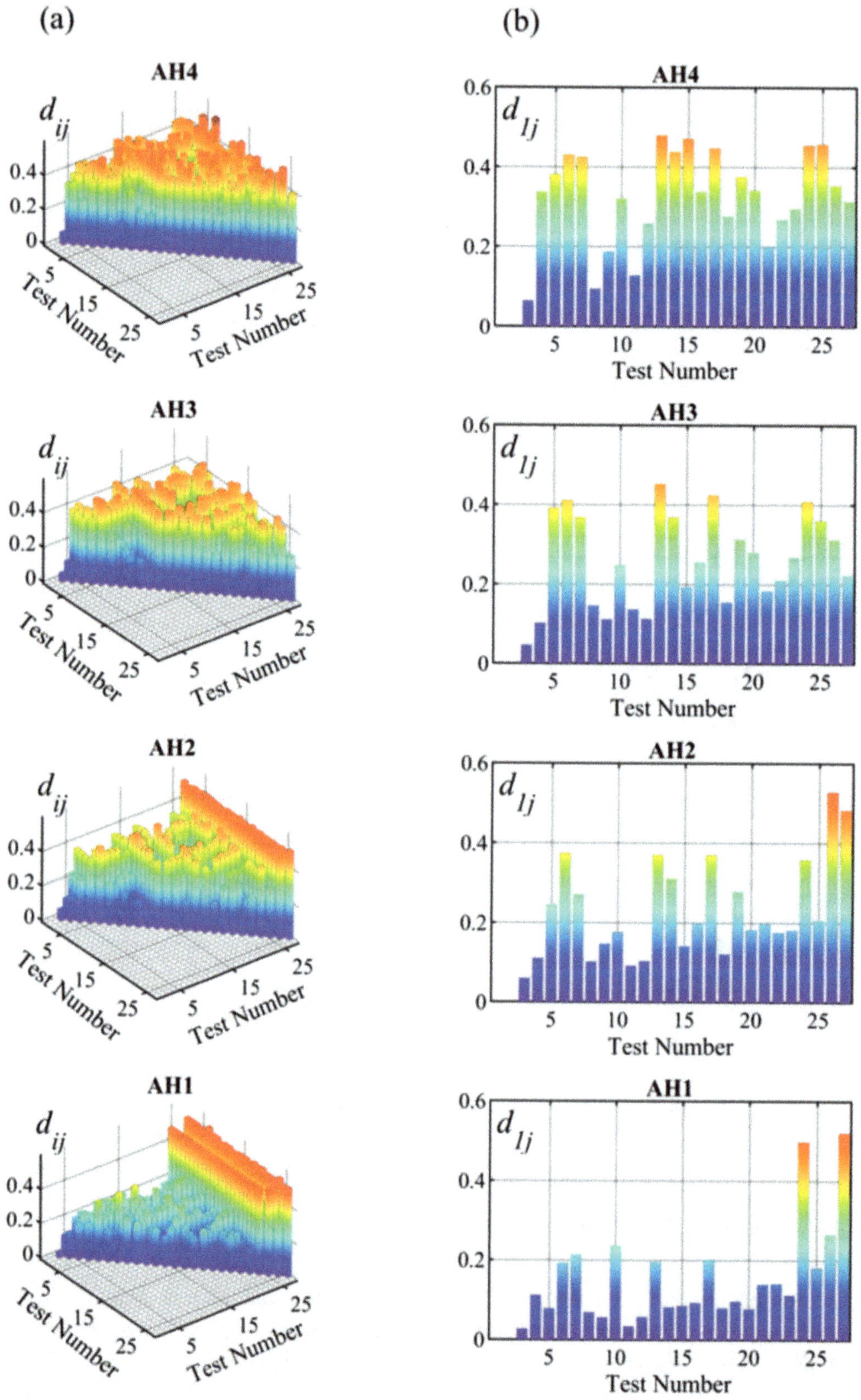


**Fig. 7.8** Response motion difference metrics for AH1, AH2, AH3, and AH4: (**a**) among the analyzed 26 centrifuge test accelerations and (**b**) between RPI0 and the other tests

of about 0.5). The difference metrics between the RPI0 accelerations and the other tests are shown in Fig. 7.8b to illustrate the range and level of likeliness and dissimilarities that these tests had with the response of one of the tests with an input motion close to the reference. These dissimilarities provide valuable information and are used below to assess the sensitivity and relative effects of the initial conditions on the difference metrics for the response acceleration at different depths.

## 7.4 Sensitivity Analysis

Evaluation of reliable estimates of the response (e.g., acceleration, displacement, etc.) sensitivity of a physical system, such as the analyzed LEAP sloping deposit, using a relatively limited set of test data represents a significant challenge. Diverse approaches have been used in a number of fields to evaluate local and global system sensitivities. A local sensitivity is evaluated based on a numerical differentiation concept and is usually obtained as the simple ratio of (observed or evaluated) changes in response quantities to a corresponding small variation in input parameters (at a specific value of these parameters). In contrast, a global sensitivity is based on a statistical framework and considers a range of input variations (in contrast to small variations). A number of review papers have been dedicated to these topics (e.g., Iooss and Lemaître 2015). In this study, the local approach is not adequate in view of the involved experimental uncertainties and limitations, while the common global approaches are hindered by the limited amount (from a statistical point of view) of experimental data available. The sensitivities were therefore estimated on the basis of a kriging analysis.

Kriging is a semi-parametric Gaussian process regression method that was originally developed in the field of geostatistics (Chilès and Delfiner 1999; McCullough et al. 2017). It provides an effective means that may be employed to estimate response quantities, and corresponding derivatives and integrals, over a domain of associated input parameters using only noisy observations or measurements for a limited irregularly spaced set of these parameters. The method involves an averaging process and provides an estimate of uncertainty (in terms of a standard deviation). Herein, kriging was used to assess the sensitivity of the difference metrics of the recorded AH1 to AH4 accelerations to variations in input motion and initial soil fabric condition associated with soil deposition and achieved grain packing.

### 7.4.1 *Acceleration Sensitivity*

The difference analysis above showed that $\Delta A_{ij}^{1\text{Hz}}$ and $\Delta A_{ij}^{2+\text{Hz}}$ are the two main input motion parameters that varied during the analyzed 26 LEAP centrifuge tests. The tested soil models also had variability in soil deposition and achieved grain packing (as documented by the mass densities reported by the different centrifuge facilities (Kutter et al. 2018, 2019)). These variations have a direct effect, for instance, on stiffness properties which in turn affect the deposit response. An average over depth of the CPT (cone penetration test) tip resistance is deemed herein to correlate better with the deposit initial fabric and packing conditions than relative or mass densities (mainly due to the high sensitivity of density computation to errors in the measurement of volume). The average of the measured CPT resistance at 1.5 m, 2.0 m, 2.5 m, and 3.0 m depths (hereafter referred to as $q_{\text{c-avg}}$) for the different centrifuge tests is presented in Fig. 7.7 along with the differences in input motion.

A kriging analysis was conducted to assess the sensitivity of the recorded accelerations to the parameters $\Delta A_{ij}^{1\text{Hz}}$, $\Delta A_{ij}^{2+\text{Hz}}$, and $q_{\text{c-avg}}$. The sensitivity analysis was performed using a subset of 17 tests of the conducted 26 centrifuge experiments. The tests were selected based on an investigation of the associated stress and strain time histories. This investigation showed a consistency among the stress-strain response of these tests and fundamental differences with the remaining nine tests, as shown in Fig. 7.9. The details of the stress-strain analysis, rational for the 17 test selection, and details of the kriging analysis are given in Goswami (2019). The following paragraphs focus on the conducted analysis results for brevity.

The sensitivity analysis is based on an estimate of the variation of the total difference measures $d_{ij}$ of the recorded accelerations at AH1 to AH4 as a function of $\Delta A_{ij}^{1\text{Hz}}$, $\Delta A_{ij}^{2+\text{Hz}}$, and $q_{\text{c-avg}}$. Specifically, the analysis provided a kriging hypersurface representing $d_{ij}$ (for each of AH1 to AH4) as a function of the variables $\Delta A_{ij}^{1\text{Hz}}$, $\Delta A_{ij}^{2+\text{Hz}}$, and $q_{\text{c-avg}}$ (over the domain associated with these variables, as shown in Fig. 7.7). The differences among the tests in input motions, $\Delta A_{ij}^{1\text{Hz}}$ and $\Delta A_{ij}^{2+\text{Hz}}$, and in response metric (at AH1 to AH4), $d_{ij}$, had to be computed with respect to a common reference, which was selected to be the RPI1 test. This test had an input acceleration substantially close to the reference and a relative density close to the reference mean value (Fig. 7.2). Other tests could also be used as a reference and would lead similar results to the ones presented below.

Three sets of surfaces and corresponding sensitivity functions are employed herein to visualize the obtained $d_{ij}$ kriging hypersurface results, as shown in Figs. 7.10, 7.11, 7.12, 7.13, 7.14, and 7.15. Thus, Fig. 7.10 shows the difference metric variations as a function of $\Delta A_{ij}^{1\text{Hz}}$ and $\Delta A_{ij}^{2+\text{Hz}}$ for a $q_{\text{c avg}} = 2.6$ MPa. This value corresponds to the mid-point of the domain of variations for $q_{\text{c-avg}}$ and corresponds to a $D_{\text{r}}$ of about 65%. The obtained results show that the AH1 accelerations have comparable sensitivities to variations in $\Delta A_{ij}^{1\text{Hz}}$and $\Delta A_{ij}^{2+\text{Hz}}$ (for $q_{\text{c-avg}} = 2.7$ MPa) and the associated discrepancy metric practically varies linearly as a function of these two parameters, as shown in Fig. 7.10. In contrast, the response at AH4 is about two times as sensitive to a $\Delta A_{ij}^{2+\text{Hz}}$as to $\Delta A_{ij}^{1\text{Hz}}$. The discrepancy metrics for AH1 to AH4 show a sensitivity that increased from the bottom of the deposit to the free surface. The associated sensitivity functions $\partial d_{ij}/\partial A_{ij}^{1\text{Hz}}$ and $\partial d_{ij}/\partial A_{ij}^{2+\text{Hz}}$ for $q_{\text{c-avg}} = 2.7$ MPa (Fig. 7.11) confirmed that AH1 has mostly constant sensitivity functions and AH4 has a sensitivity $\partial d_{ij}/\partial A_{ij}^{2+\text{Hz}}$ with large variations (especially as a function of $\Delta A_{ij}^{2+\text{Hz}}$). The sensitivities for AH2 and AH3

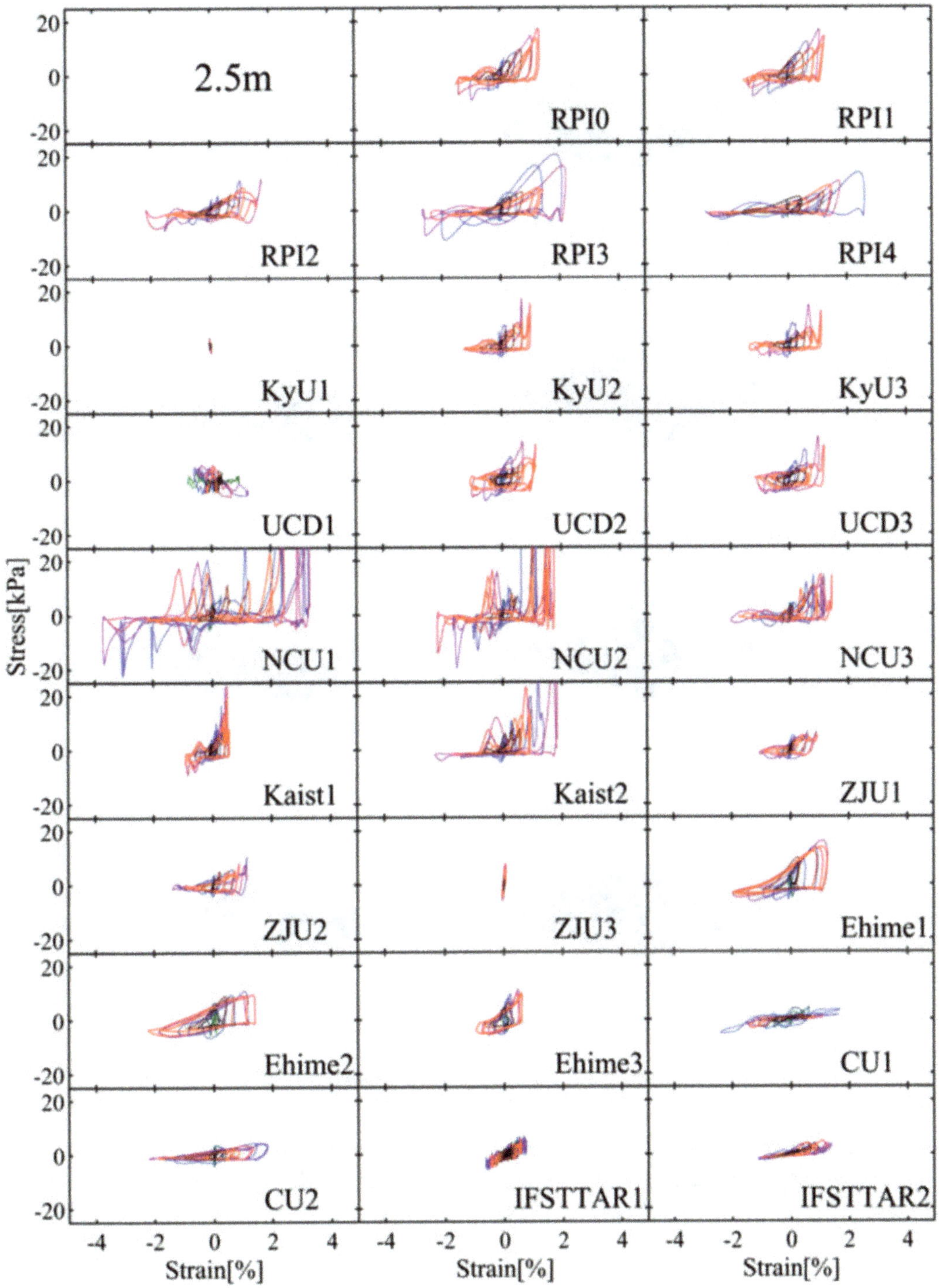


**Fig. 7.9** Stress-strain response of the analyzed centrifuge tests (at 2.5 m depth along the central accelerometer array)

had values that varied between those of AH1 and AH4. Note however that the estimated variations of the discrepancy metric (and corresponding sensitivities) are associated with larger standard deviation (i.e., lower level of confidence) at large $\Delta A_{ij}^{2+\mathrm{Hz}}$ values, especially for AH3 and AH4. This is explained by the sparsity of

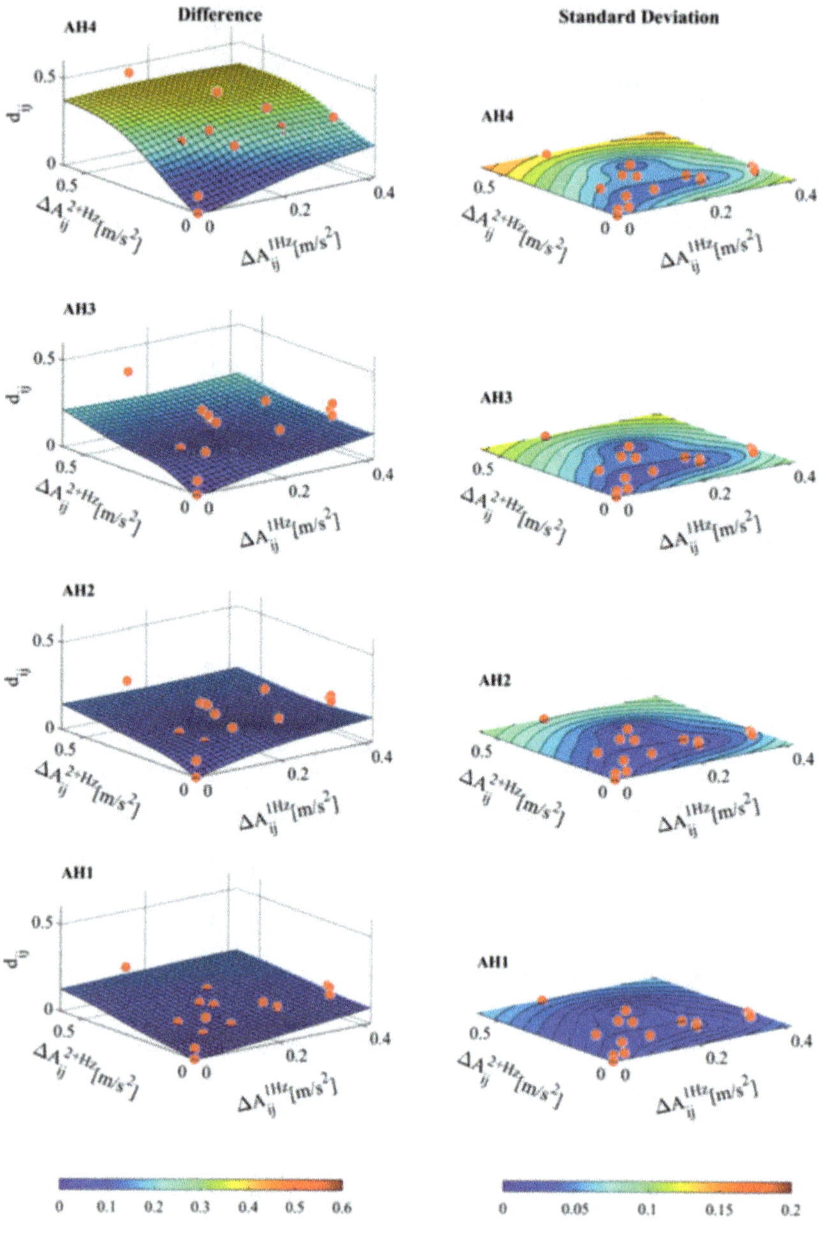


**Fig. 7.10** Variation of the difference metric $d_{ij}$ of the recorded accelerations (at AH1 to AH4) as a function of $\Delta A_{ij}^{1\text{Hz}}$ and $\Delta A_{ij}^{2+\text{Hz}}$ for a $q_{\text{c-avg}} = 2.7$ MPa (the red dots correspond to the analyzed 17 tests)

data for large values of $\Delta A_{ij}^{2+\text{Hz}}$ (only one experiment had a $\Delta A_{ij}^{2+\text{Hz}}$ larger than 0.04 m/s$^2$).

The difference metric surfaces as a function of $\Delta A_{ij}^{1\text{Hz}}$ and $q_{\text{c-avg}}$ for $\Delta A_{ij}^{2+\text{Hz}} = 0$, and $\Delta A_{ij}^{2+\text{Hz}}$ and $q_{\text{c-avg}}$ for $\Delta A_{ij}^{1\text{Hz}} = 0$, and the corresponding sensitivity functions (Figs. 7.12, 7.13, 7.14, and 7.15) were employed to explore the effects of the observed variation in CPT resistance. The obtained metric surfaces and sensitivity

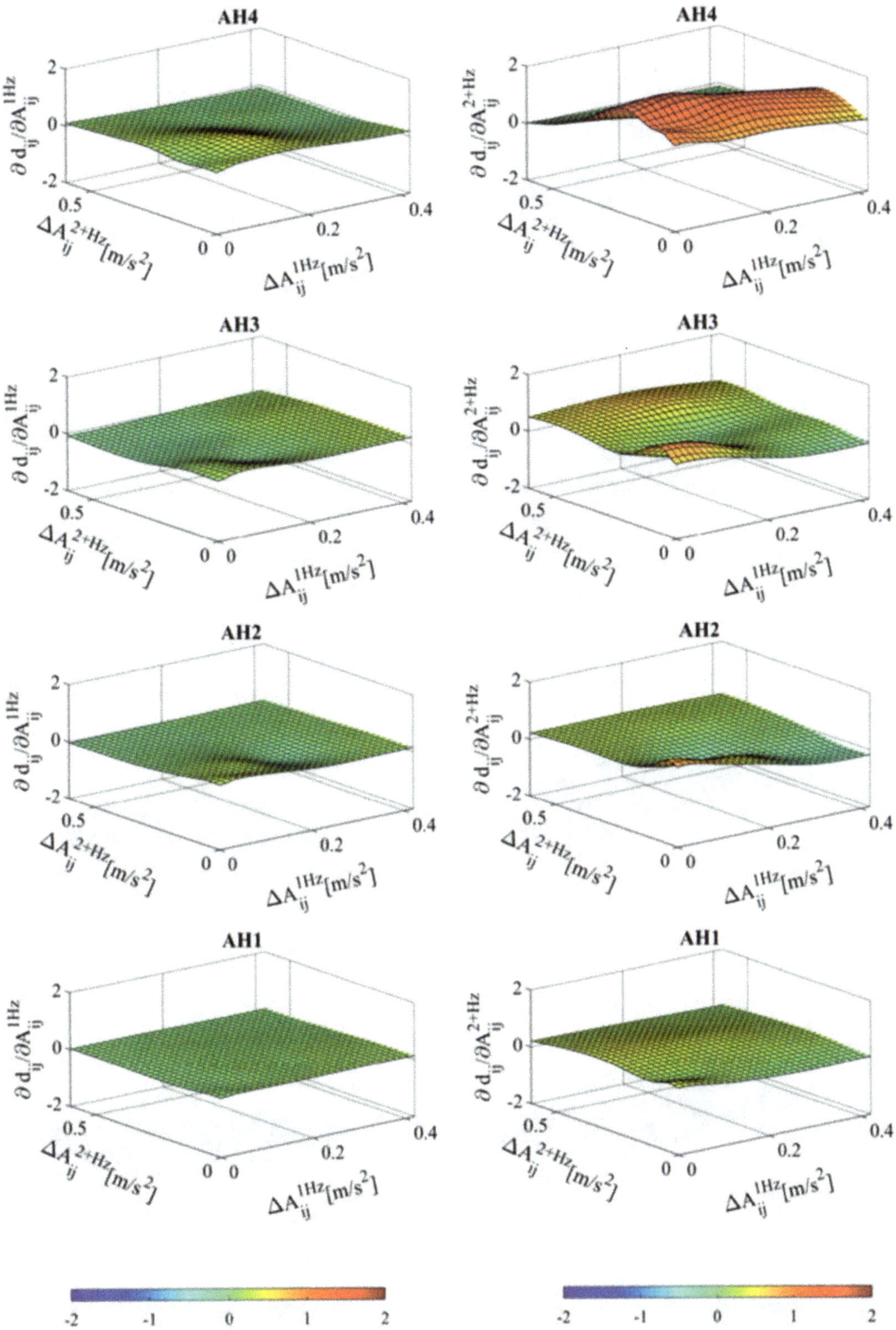

**Fig. 7.11** Sensitivity functions of the total difference metric of the recorded accelerations (at AH1 to AH4) with respect to variations in $\Delta A_{ij}^{1\text{Hz}}$ and $\Delta A_{ij}^{2+\text{Hz}}$ for $q_{\text{c-avg}} = 2.7$ MPa

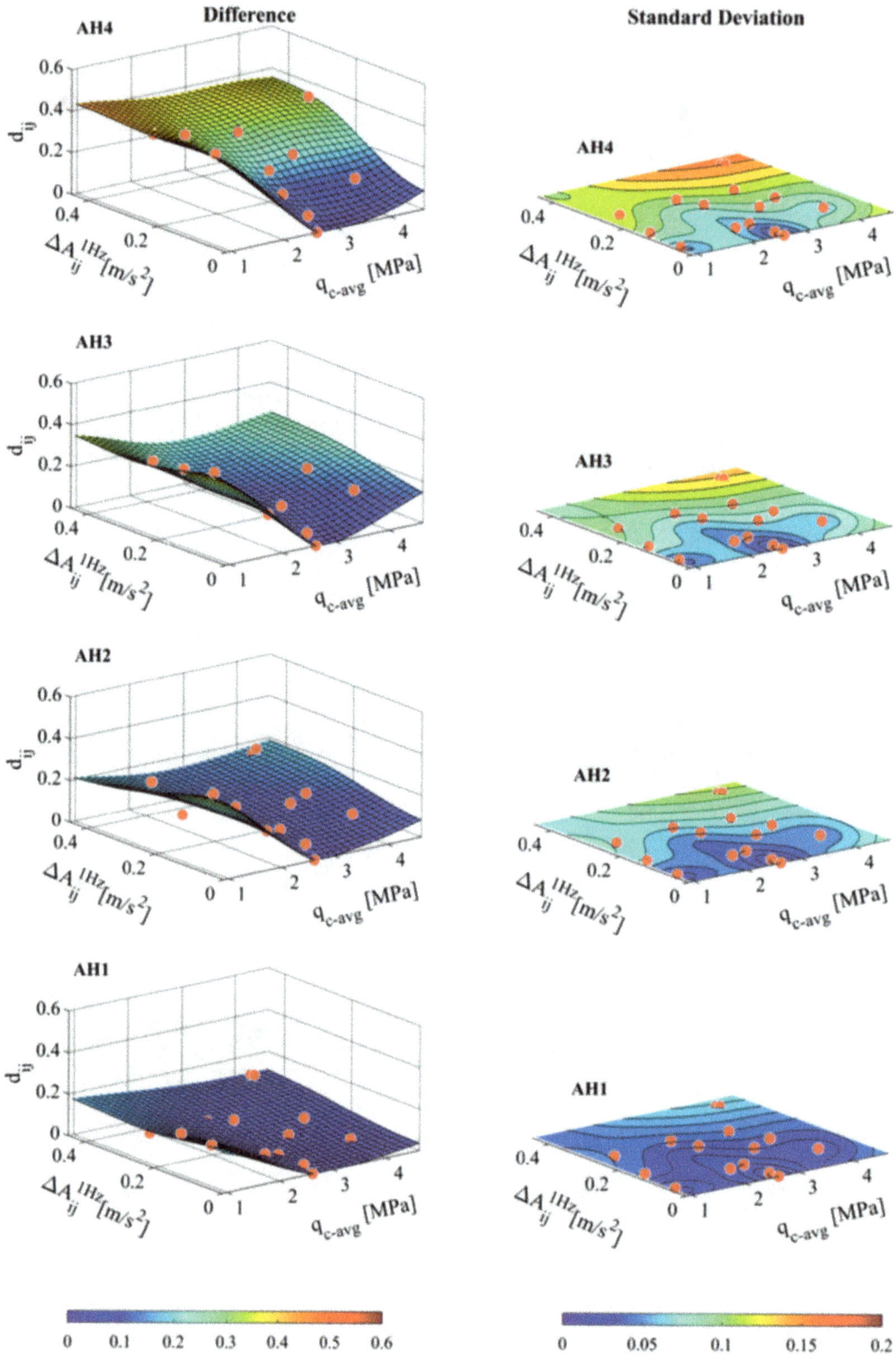


**Fig. 7.12** Variation of the total difference metric of the recorded accelerations (at AH1 to AH4) as a function of $\Delta A_{ij}^{1\text{Hz}}$ and $q_{\text{c-avg}}$ for $\Delta A_{ij}^{2+\text{Hz}} = 0$ (the red dots correspond to the analyzed 17 tests)

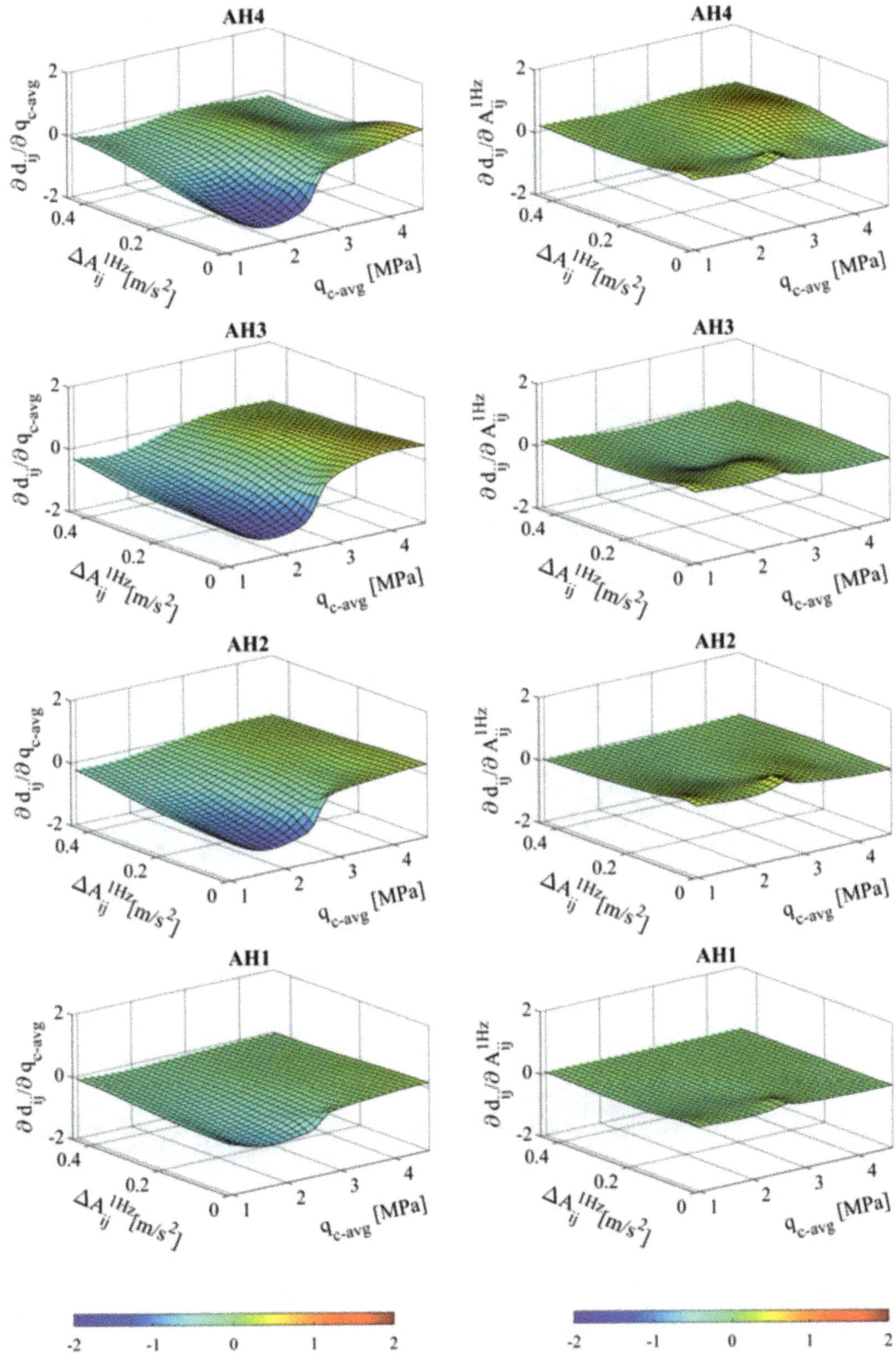


**Fig. 7.13** Sensitivity functions of the total difference metric of the recorded accelerations (at AH1 to AH4) with respect to variations in $\Delta A_{ij}^{1\mathrm{Hz}}$ and $q_{\text{c-avg}}$ for $\Delta A_{ij}^{2+\mathrm{Hz}} = 0$

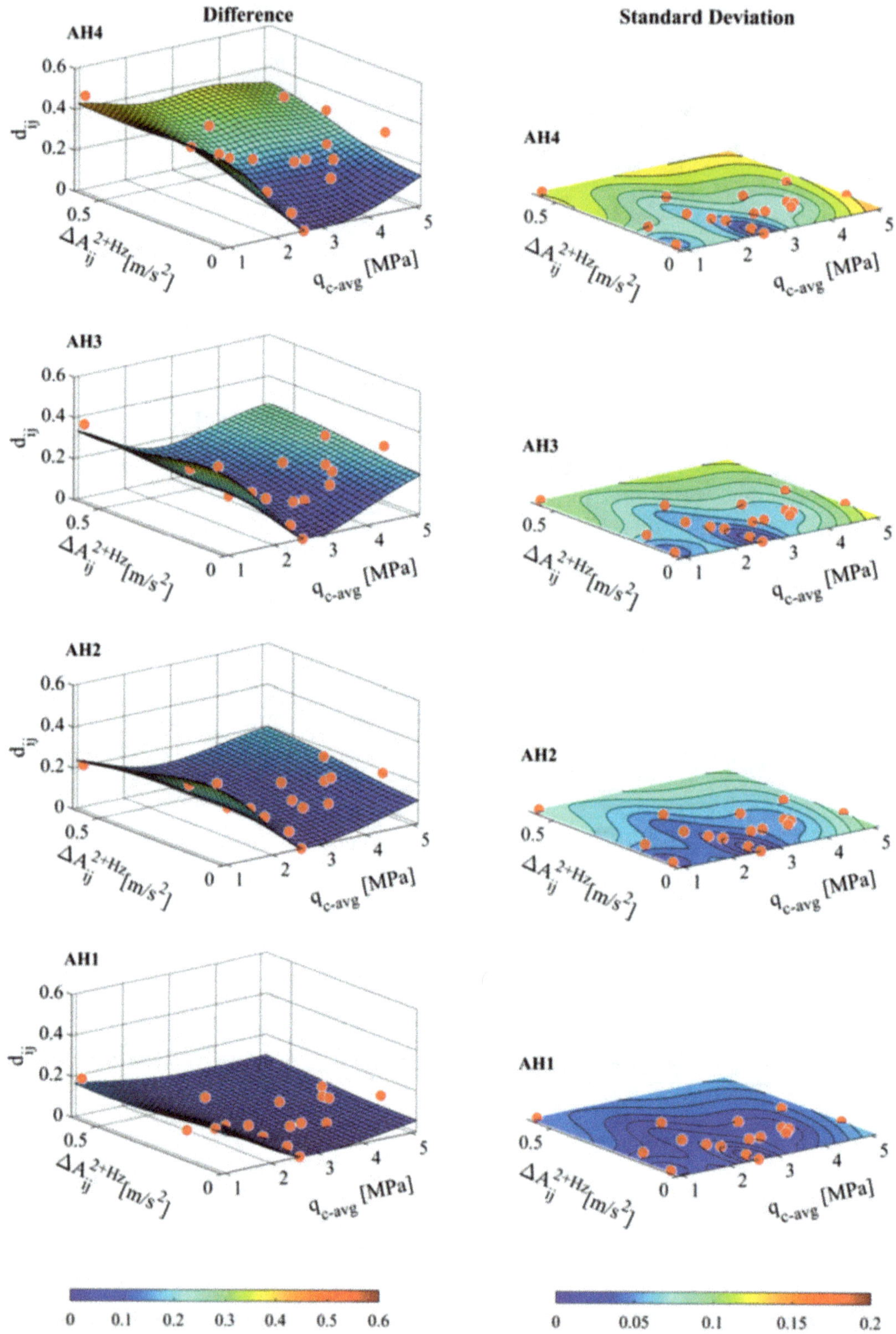


**Fig. 7.14** Variation of the total difference metric of the recorded (AH1 to AH4) accelerations as a function of $\Delta A_{ij}^{2+\text{Hz}}$ and $q_{\text{c-avg}}$ for $\Delta A_{ij}^{1\text{Hz}}=0$ and corresponding standard deviation (the red dots are the analyzed 17 tests)

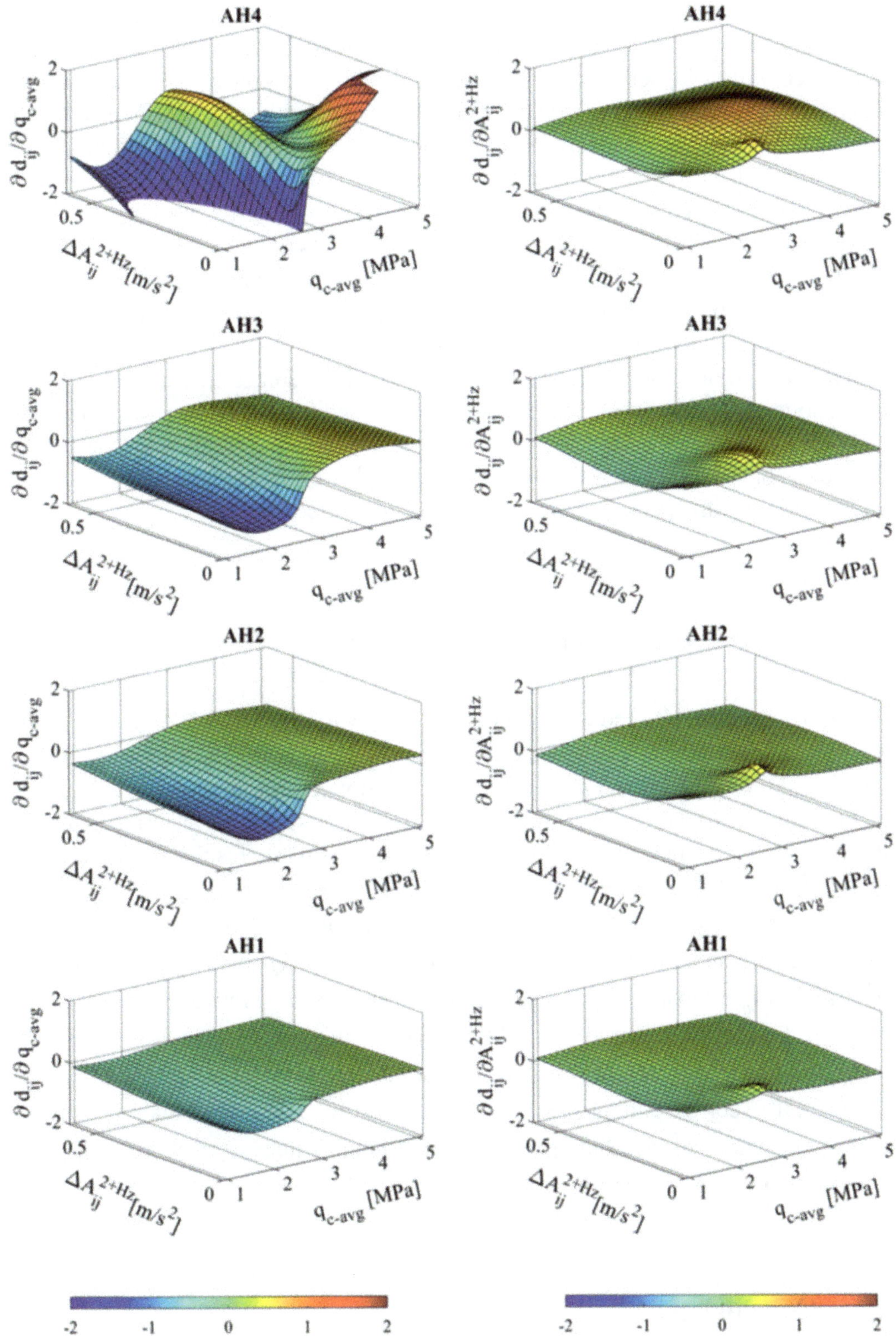


**Fig. 7.15** Sensitivity functions of the total difference metric of the recorded accelerations (AH1 to AH4) with respect to variations in $\Delta A_{ij}^{2+\mathrm{Hz}}$ and $q_{\text{c-avg}}$ for $\Delta A_{ij}^{1\mathrm{Hz}}=0$

functions show a response that is significantly more sensitive to a decrease in $q_{\text{c-avg}}$ than an increase. This is explained by the fact that lower values of $q_{\text{c-avg}}$ are associated with a looser more contractive soil with a response that contrasts substantially with that of the (dilative) reference deposit with a $D_r$ of about 65%. In contrast, larger $q_{\text{c-avg}}$ values are indicative of a denser soil that is only slightly more dilative and has only a somewhat different response. The sensitivity values increased from AH1 to AH4, and the sensitivities with respect to $q_{\text{c-avg}}$ were significantly larger than those associated with $\Delta A_{ij}^{\text{1Hz}}$ and $\Delta A_{ij}^{\text{2+Hz}}$. Overall, the obtained sensitivity functions (Figs. 7.11, 7.13, and 7.15) vary nonlinearly with variations in parameters. The level of nonlinearity increases from AH1 to AH4 and is more remarkable for the sensitivities that depends on $q_{\text{c-avg}}$. Figures 7.10, 7.12, and 7.14 also show the standard deviations corresponding to the estimated difference metric surfaces (and corresponding sensitivities). The deviations increased from AH1 near the bottom to AH4 close to the free surface of the deposit. Also, for any (acceleration) level, the deviations have the lowest values in the zones with well-distributed data points and largest values with sparse or no data points (as expected).

### 7.4.2 *Permanent Displacement Sensitivity*

A kriging analysis was used to assess the effects of variations in $\Delta A_{ij}^{\text{1Hz}}$, $\Delta A_{ij}^{\text{2+Hz}}$, and $q_{\text{c-avg}}$ on the permanent surface displacement (referred to as $D$). The analysis was performed for the selected 17 tests using the mean values of the measured displacement of the 2 central markers (Kutter et al. 2019). Three sets of surfaces are exhibited in Fig. 7.16 to visualize the obtained $D$ kriging hypersurface results as a function of (1) $\Delta A_{ij}^{\text{1Hz}}$ and $\Delta A_{ij}^{\text{2+Hz}}$ for a $q_{\text{c avg}} = 2.6$ MPa, (2) $\Delta A_{ij}^{\text{1Hz}}$ and $q_{\text{c-avg}}$ for $\Delta A_{ij}^{\text{2+Hz}}=0$, and (3) $\Delta A_{ij}^{\text{2+Hz}}$ and $q_{\text{c-avg}}$ for $\Delta A_{ij}^{\text{1Hz}} = 0$. The obtained results show that the measured permanent displacements are generally more sensitive to variations in $\Delta A_{ij}^{\text{1Hz}}$ and $\Delta A_{ij}^{\text{2+Hz}}$ than $q_{\text{c-avg}}$. In fact the displacements have rather a low sensitivity to $q_{\text{c-avg}}$ for $\Delta A_{ij}^{\text{1Hz}} \approx 0$ and $\Delta A_{ij}^{\text{2+Hz}} \approx 0$. The sensitivity to the CPT resistance increases when the variations in $q_{\text{c-avg}}$ are combined with variations in $\Delta A_{ij}^{\text{1Hz}}$ and $\Delta A_{ij}^{\text{2+Hz}}$. The corresponding standard deviations (Fig. 7.16) were again reasonable in the zone with well-distributed data points, and high values were observed in areas with limited or no data points.

## 7.5 Conclusions

This article presented an analysis of the differences and sensitivities among the acceleration time histories of 26 centrifuge LEAP (Liquefaction Experiments and Analysis Projects) test replicas of a saturated sloping deposit. A normalized mean squared deviation is used as difference metric to quantify the dissimilarities between

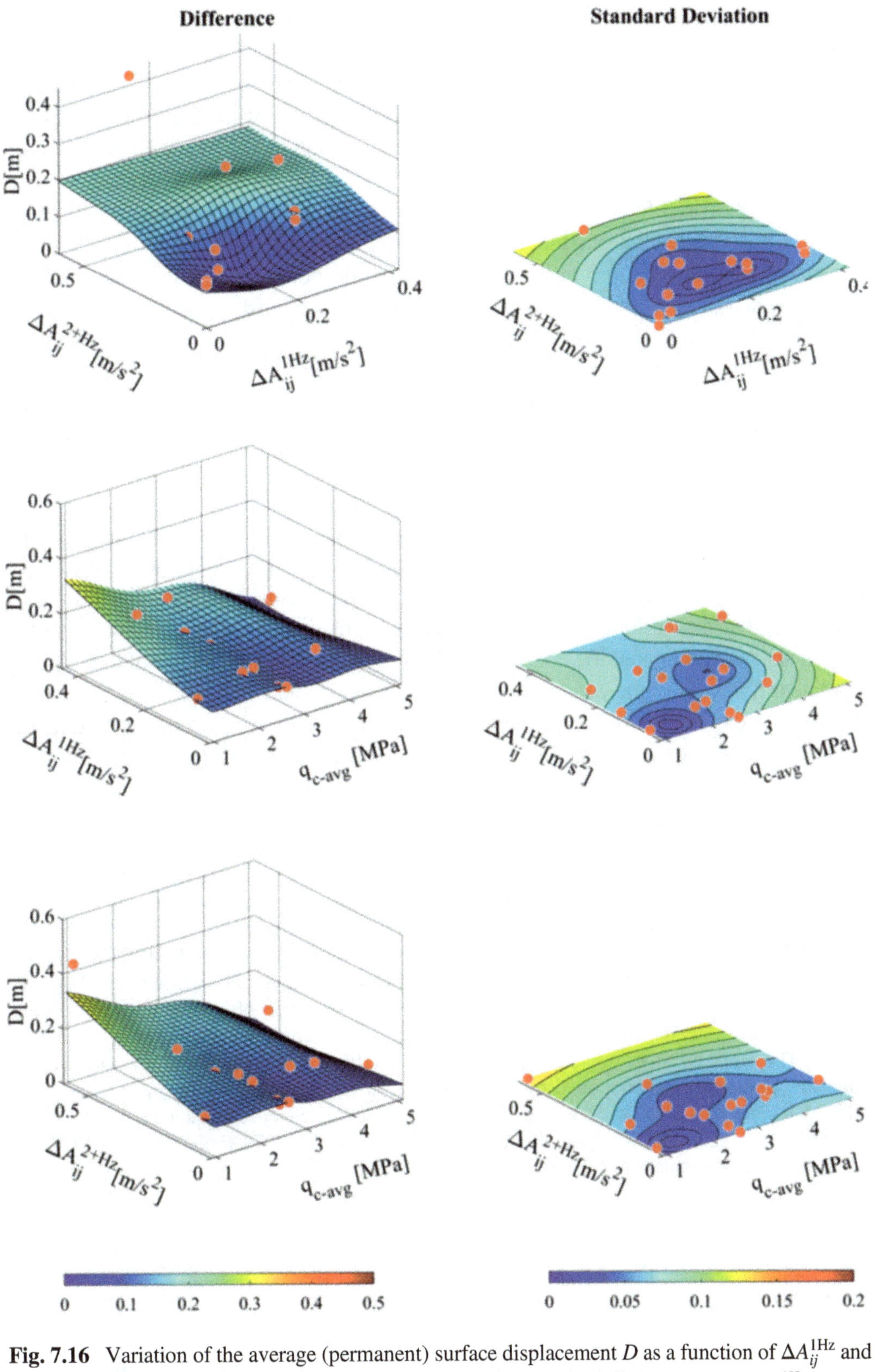


**Fig. 7.16** Variation of the average (permanent) surface displacement $D$ as a function of $\Delta A_{ij}^{1\text{Hz}}$ and $\Delta A_{ij}^{2+\text{Hz}}$ for a $q_c = 2.7$ MPa, $\Delta A_{ij}^{1\text{Hz}}$ and $q_{\text{c-avg}}$ for $\Delta A_{ij}^{2+\text{Hz}}$=0, and $\Delta A_{ij}^{2+\text{Hz}}$ and $q_{\text{c-avg}}$ for $\Delta A_{ij}^{1\text{Hz}}$=0 (the red dots are the analyzed 17 tests)

recorded acceleration time histories. This metric is uniquely decomposed in four terms associated with phase, frequency shift, amplitude at 1 Hz, and amplitude of frequencies higher than 2 Hz (2 + Hz) components. These metrics and measures were employed to assess and quantify the discrepancies of the input and response accelerations of the 26 different test replicas. The obtained difference metric values showed that the input accelerations can be divided into three broad classes: (1) Group 1 of accelerations that closely match the reference and also each other, (2) Group 2 of accelerations that have an average match to the reference and among the group, and (3) Group 3 of accelerations that did not closely match the reference and also each other with. This broad range of motions provides valuable information to assess both the repeatability of the tests and sensitivity of the recorded responses to variations in input motion. The differences among input motions were found to be associated mostly with variation in amplitude of the dominant component at 1 Hz and the components with frequencies higher than 2 Hz (2 + Hz).

A Gaussian process-based kriging was used to assess the sensitivity of the deposit response acceleration to differences in input motion amplitude at 1 Hz and 2 + Hz and average CPT (cone penetration test) resistance (used as a measure reflecting deposit fabric condition and initial grain packing). The conducted analyses showed that the analyzed deposit accelerations are relatively more sensitive to variations in CPT resistance than to the input motion and that this sensitivity is larger for a decrease in CPT resistance compared to an increase. The sensitivities were also found to be highly nonlinear functions of the variability in the input motion and CPT resistance. In contrast, the measured permanent displacements were generally more sensitive to differences in input motion amplitude at 1 Hz and 2 + Hz than the average CPT resistance.

**Acknowledgments** The experimental work of the LEAP project was supported by different institutions. In the USA, the work was funded by the US National Science Foundation Geotechnical Engineering program directed by Dr. Richard Fragaszy (NSF grants CMMI-1635524, CMMI-1635307, and CMMI-1635040 to Rensselaer Polytechnic Institute, George Washington University, and University of California Davis, respectively). The work at Ehime U. was supported by JSPS KAKENHI Grant Number 17H00846. The work at Kyoto U. was supported by JSPS KAKENHI Grant Numbers 26282103, 5420502, and 17H00846. The work at Kansai U. was supported by JSPS KAKENHI Grant Number 17H00846. The work at KAIST was part of a project titled "Development of performance-based seismic design," funded by the Ministry of Oceans and Fisheries, Korea. The work at NCU was supported by MOST: 106-2628-E-008-004-MY3. The work at Zhejiang University was supported by the National Natural Science Foundation of China, Nos. 51578501 and 51778573; Zhejiang Provincial Natural Science Foundation of China, LR15E080001; and National Basic Research Program of China (973 Project), 2014CB047005. These supports are gratefully acknowledged.

## References

Carey, T. J., Stone, N., Hajialilue Bonab, M., & Kutter, B. L. (2019). LEAP-UCD-2017 centrifuge test at University of California, Davis. In B. Kutter et al. (Eds.), *Model tests and numerical simulations of liquefaction and lateral spreading: LEAP-UCD-2017*. New York: Springer.

Chilès, J.-P., & Delfiner, P. (1999). *Geostatistics: Modeling spatial uncertainty*. New York: Wiley.

Escoffier, S., & Audrain, P. (2019). LEAP-UCD-2017 centrifuge test at IFSTTAR. In B. Kutter et al. (Eds.), *Model tests and numerical simulations of liquefaction and lateral spreading: LEAP-UCD-2017*. New York: Springer.

Geers, T. (1984). Objective error measure for the comparison of calculated and measured transient response histories. *Shock and Vibration Bulletin, 54*, 99–102.

Goswami, N. (2019). *Validation Framework for Assessment of numerical Predictions Using Leap Experiments*. Troy, NY: PhD Thesis, Rensselaer Polytechnic Institute.

Hung, W.-Y., & Liao, T.-W. (2019). LEAP-UCD-2017 centrifuge tests at NCU. In B. Kutter et al. (Eds.), *Model tests and numerical simulations of liquefaction and lateral spreading: LEAP-UCD-2017*. New York: Springer.

Iooss, B., & Lemaître, P. (2015). A review on global sensitivity analysis methods. In B. Iooss & P. Lemaitre (Eds.), *Uncertainty Management in Simulation-Optimization of Complex Systems* (pp. 101–122). Boston, MA: Springer.

Kim, S.-N., Ha, J.-G., Lee, M.-G., & Kim, D.-S. (2019). LEAP-UCD-2017 centrifuge test at KAIST. In B. Kutter et al. (Eds.), *Model tests and numerical simulations of liquefaction and lateral spreading: LEAP-UCD-2017*. New York: Springer.

Kokkali, P., Abdoun, T., & Zeghal, M. (2019). Physical modeling of soil liquefaction: Overview of LEAP production test 1 at Rensselaer Polytechnic Institute. *Soil Dynamics and Earthquake Engineering, 113*, 629–649.

Korre, E., Abdoun, T., & Zeghal, M. (2019). Verification of the repeatability of soil liquefaction centrifuge testing at Rensselaer. In B. Kutter et al. (Eds.), *Model tests and numerical simulations of liquefaction and lateral spreading: LEAP-UCD-2017*. New York: Springer.

Kutter, B. L., Carey, T. J., Zheng, B. L., Gavras, A., Stone, N., Zeghal, M., Abdoun, T., Korre, E., Manzari, M., Madabhushi, G. S., Haigh, S., Madabhushi, S. S., Okamura, M., Sjafuddin, A. N., Escoffier, S., Kim, D.-S., Kim, S.-N., Ha, J.-G., Tobita, T., Yatsugi, H., Ueda, K., Vargas, R. R., Hung, W.-Y., Liao, T.-W., Zhou, Y.-G., & Liu, K. (2018). Twenty-Four Centrifuge Tests to Quantify Sensitivity of Lateral Spreading to Dr and PGA. In S. J. Brandenberg & M. T. Manzari (Eds.), *Geotechnical Earthquake Engineering and Soil Dynamics V, GSP 293* (pp. 383–393). Alexandria, VA: ASCE. https://doi.org/10.1061/9780784481486.040.

Kutter, B. L., Carey, T., Stone, N., Zheng, B. L., Gavras, A., Manzari, M. T., Zeghal, M., Abdoun, T., Korre, E., Escoffier, S., Haigh, S., Madabhushi, G., Madabhushi, S. S. C., Hung, W.-Y., Liao, T.-W., Kim, D.-S., Kim, S.-N., Ha, J.-G., Kim, N. R., Okamura, M., Sjafuddin, A. N., Tobita, T., Ueda, K., Vargas, R., Zhou, Y.-G., & Liu, K. (2019). LEAP-UCD-2017 comparison of centrifuge test results. In B. Kutter et al. (Eds.), *Model tests and numerical simulations of liquefaction and lateral spreading: LEAP-UCD-2017*. New York: Springer.

Liu, K., Zhou, Y.-G., She, Y., Xia, P., Meng, D., Huang, J.-S., Yao, G., & Chen, Y.-M. (2019). Specifications and results of centrifuge model test at Zhejiang University for LEAP-UCD-2017. In B. Kutter et al. (Eds.), *Model tests and numerical simulations of liquefaction and lateral spreading: LEAP-UCD-2017*. New York: Springer.

Madabhushi, S. S. C., Dobrisan, A., Beber, R., Haigh, S. K., & Madabhushi, S. P. G. (2019). LEAP-UCD-2017 centrifuge tests at Cambridge. In B. Kutter et al. (Eds.), *Model tests and numerical simulations of liquefaction and lateral spreading: LEAP-UCD-2017*. New York: Springer.

Manzari, M. T., El Ghoraiby, M. A., Kutter, B. L., Zeghal, M., Wang, R., Chen, R., Zhang, J.-M., Osamu Ozutsumi, O., Fukutake, K., Kiriyama, T., Fasano, G., Chiaradonna, A., Bilotta, E., Montgomery, J., Ziotopoulou, K., Chen, L., Ghofrani, A., Arduino, P., Wada, T., Ueda, K., Mercado, V., Fuentes, W., Lascarro, C., Yang, M., Barrero, A. R., Taiebat, M., Tsiaousi, D., Ugalde, J., Thaleia Travasarou, T., Ichii, K., Uemura, K., Orai, N., Hyodo, M., Abdoun, T., Haigh, S., Madabhushi, S., Tobia, T., Hung, W.-Y., Kim, D. S., Okamura, M., Zhou, Y.-G., & Escoffier, S. (2018). Liquefaction experiment and analysis projects (LEAP): Summary of observations from the planning phase. *Soil Dynamics and Earthquake Engineering, 113*, 714–743.

McCullough, M., Jayakumar, P., Dasch, J., & Gorsich, D. (2017). The next generation NATO reference mobility model development. *Journal of Terramechanics, 73*, 49–60.

Oberkampf, W. L., & Roy, C. J. (2010). *Verification and validation in scientific computing*. New York, NY: Cambridge University Press.

Okamura, M., & Nurani Sjafruddin, A. (2019). LEAP-2017 centrifuge test at Ehime University. In B. Kutter et al. (Eds.), *Model tests and numerical simulations of liquefaction and lateral spreading: LEAP-UCD-2017*. New York: Springer.

Rabiner, L., & Huang, B. (1993). *Fundamentals of speech recognition*. Eaglewood Cliffs, NJ: PTR Prentice Hall.

Vargas Tapia, R. R., Tobita, T., Ueda, K., & Yatsugi, H. (2019). LEAP-UCD-2017 centrifuge test at Kyoto University. In B. Kutter et al. (Eds.), *Model tests and numerical simulations of liquefaction and lateral spreading: LEAP-UCD-2017*. New York: Springer.

Zeghal, M., Goswami, N., Manzari, M., & Kutter, B. (2018). *Discrepancy metrics and sensitivity analysis of dynamic soil response*. Austin, TX: Geotechnical Earthquakes and Soil Dynamics.

# Chapter 8
# LEAP-2017 Simulation Exercise: Overview of Guidelines for the Element Test Simulations

**Majid T. Manzari, Mohamed El Ghoraiby, Bruce L. Kutter, and Mourad Zeghal**

**Abstract** This paper summarizes the guidelines provided to the numerical simulation/prediction teams that participated in the LEAP-2017 prediction exercise. These guidelines are developed for the Phase 1 of the simulations that focused on the use of cyclic triaxial element tests for calibration of constitutive models that the participating teams used in their numerical simulations/predictions.

## 8.1 Introduction

LEAP-2017 is a collaboration within the framework of the Liquefaction Experiments and Analysis Projects to produce high-quality centrifuge data of a mild slope and validate constitutive models and numerical tools used in soil liquefaction analysis. This collaboration involved a simulation exercise which provided an opportunity to evaluate the capabilities and limitations of a large number of commonly used constitutive and numerical modeling techniques for liquefaction analysis. A number of simulation teams from the geotechnical engineering community across the world participated in modeling a selection of the performed centrifuge experiments. The simulation exercise consisted of constitutive model calibration, Type-B prediction, and Type-C prediction.

---

M. T. Manzari (✉)
Department of Civil and Environmental Engineering, George Washington University, Washington, DC, USA
e-mail: manzari@gwu.edu

M. El Ghoraiby
The George Washington University, Washington, DC, USA

B. L. Kutter
Department of Civil and Environmental Engineering, University of California, Davis, CA, USA

M. Zeghal
Department of Civil and Environmental Engineering, Rensselaer Polytechnic Institute, Troy, NY, USA



B. Kutter et al. (eds.), *Model Tests and Numerical Simulations of Liquefaction and Lateral Spreading*, https://doi.org/10.1007/978-3-030-22818-7_8

A series of laboratory experiments were first conducted on the designated soil for LEAP (Ottawa F65 sand) and made available to the simulation teams (hereafter referred to as predictors) for calibration of the constitutive models that were later used in Phase II and Phase III of the simulation exercise (i.e., Type-B and Type-C simulations). A Type-B simulation consists of analyzing the centrifuge experiments without any knowledge of the experiment results while the base excitation and achieved soil density are provided. For the Type-C prediction, the predictors were provided with the experimental results.

An outline of the essential elements of the calibration phase of the simulation exercise is presented in this paper. Each simulation team was requested to submit a detailed report discussing the process followed in the calibration of their selected constitutive model. The submitted report by each team was intended to provide a description of the constitutive model used by the team as well as the method used for calibration.

The submitted simulation results were requested to follow specific formats to facilitate future assessment and comparison efforts. The formatting of the data files is described in this report along with a template of the results. Matlab-based visualization tools were also provided to the numerical simulation teams for plotting the available cyclic triaxial experiments and the simulation results before submittal.

## 8.2 Soil Characterization and Element Tests

### *8.2.1 LEAP-2017 Tests*

The soil characterization and element tests on Ottawa F65 sand are discussed in El Ghoraiby et al. (2017, 2018). The soil characterization tests conducted included particle-size distribution, specific gravity, and permeability tests. The results of these tests are reported for five different batches of Ottawa F65 sand. The means and standard deviations of the measured results were also presented. A series of cyclic stress-controlled triaxial tests conducted by El Ghoraiby et al. (2017) were also made available to the prediction teams for three different soil densities. Details of sample preparation using "Constant Height Dry Pluviation" technique are presented in El Ghoraiby et al. (2017). Tables 8.1, 8.2, and 8.3 show a summary of the triaxial tests conducted for LEAP-2017 project.

Figure 8.1 shows the liquefaction strength curves for the three series of cyclic stress-controlled triaxial tests.

The element tests of LEAP-2017 project supplemented an earlier series of tests conducted by Vasko (2015) which focused on monotonic and cyclic strain-controlled tests. The results of the tests reported in Vasko (2015) were also made available to the numerical simulation teams via the NEEShub data repository and are now available on DesignSafe (Vasko et al. 2018).

The data resulted from the *cyclic stress-controlled tests* were saved as *tab delimited text files* in the folder named "CyclicStressControlledTests". The file

**Table 8.1** Summary of cyclic stress-controlled triaxial tests on specimens with $e_o = 0.515$

| $e_c = 0.515 - \rho_d = 1744$ kg/m$^3$ | | | | |
|---|---|---|---|---|
| Date | File name | $e_c$ | CSR | No. of cycles to 2.5% S.A. |
| 10/10/2016 | eo_0_515_po_100_csr_0_6.txt | 0.515 | 0.6 | 14 |
| 10/6/2016 | eo_0_515_po_100_csr_0_5.txt | 0.522 | 0.5 | 17 |
| 10/7/2016 | eo_0_515_po_100_csr_0_45.txt | 0.512 | 0.45 | 19 |
| 9/16/2016 | eo_0_515_po_100_csr_0_375.txt | 0.515 | 0.375 | 26 |
| 9/27/2016 | eo_0_515_po_100_csr_0_365.txt | 0.514 | 0.365 | 29 |
| 9/30/2016 | eo_0_515_po_100_csr_0_325.txt | 0.507 | 0.325 | 41 |
| 9/22/2016 | eo_0_515_po_100_csr_0_315.txt | 0.514 | 0.315 | 46 |
| 9/28/2016 | eo_0_515_po_100_csr_0_3.txt | 0.516 | 0.3 | 48 |
| 9/13/2016 | eo_0_515_po_100_csr_0_275.txt | 0.517 | 0.275 | 60 |
| 9/29/2016 | eo_0_515_po_100_csr_0_265.txt | 0.515 | 0.265 | 70 |
| 10/12/2016 | eo_0_515_po_100_csr_0_225.txt | 0.513 | 0.225 | 191 |

**Table 8.2** Summary of cyclic stress-controlled triaxial tests with $e_o = 0.542$

| $e_c = 0.542 - \rho_d = 1713$ kg/m$^3$ | | | | |
|---|---|---|---|---|
| Date | File name | $e_c$ | CSR | No. of cycles to 2.5% S.A. |
| 11/20/2016 | eo_0_542_po_100_csr_0_28.txt | 0.55 | 0.28 | 16 |
| 11/18/2016 | eo_0_542_po_100_csr_0_24.txt | 0.54 | 0.24 | 18 |
| 11/16/2016 | eo_0_542_po_100_csr_0_22.txt | 0.535 | 0.22 | 22 |
| 11/16/2016 | eo_0_542_po_100_csr_0_21.txt | 0.538 | 0.21 | 28 |
| 11/16/2016 | eo_0_542_po_100_csr_0_2.txt | 0.55 | 0.2 | 41 |
| 11/21/2016 | eo_0_542_po_100_csr_0_19.txt | 0.538 | 0.19 | 50 |

**Table 8.3** Summary of cyclic stress-controlled triaxial tests $e_o = 0.585$

| $e_c = 0.585 - \rho_d = 1666$ kg/m$^3$ | | | | |
|---|---|---|---|---|
| Date | File name | $e_c$ | CSR | No. of cycles to 2.5% S.A. |
| 11/10/2016 | eo_0_585_po_100_csr_0_2.txt | 0.581 | 0.2 | 9 |
| 11/1/2016 | eo_0_585_po_100_csr_0_17.txt | 0.584 | 0.17 | 15 |
| 11/2/2016 | eo_0_585_po_100_csr_0_16.txt | 0.587 | 0.16 | 17 |
| 11/4/2016 | eo_0_585_po_100_csr_0_14.txt | 0.575 | 0.14 | 33 |
| 11/7/2016 | eo_0_585_po_100_csr_0_12.txt | 0.598 | 0.12 | 59 |
| 11/14/2016 | eo_0_585_po_100_csr_0_1.txt | 0.583 | 0.1 | 188 |

names are listed in Tables 8.1, 8.2, and 8.3. Each file is composed of six columns. The first column is time in minutes. The second and third columns show the axial and volumetric strains, respectively, in %. The fourth, fifth, and sixth columns, respectively, show $\sigma'_3$, $\sigma_3$, and $\sigma_d$, all in kPa. This folder and its corresponding files is archived in DesignSafe.

To graph the test results reported in each file, a Matlab script (PlotExpResults.m) is provided in the same folder. The script allows the user to plot time histories of axial strain, deviatoric stress, and mean effective stress. PlotExpResults.m also provides plots of deviatoric stress vs. axial strain, effective stress path, as well as excess pore pressure ratio and mean effective stress vs. axial strain. Summery plots

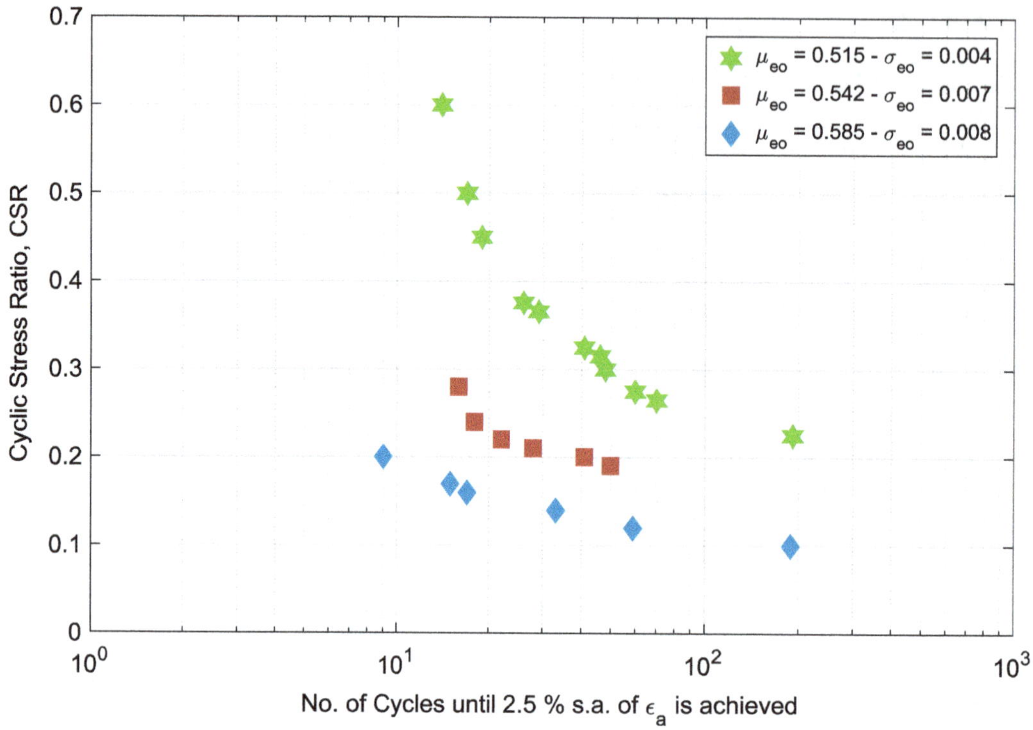


**Fig. 8.1** Liquefaction strength curves obtained from cyclic stress-controlled triaxial tests on Ottawa F65 sand (El Ghoraiby et al. 2017)

of the cyclic stress-controlled tests are presented in the appendices A–C of "LEAP-2017 soil characterization and element tests of Ottawa F65 sand" (El Ghoraiby et al. 2017).

### *8.2.2 Additional Available Element Tests on Ottawa Sand*

Additional experimental data on Ottawa F65 was also available and could be used by the simulation teams for their model calibration purpose, if needed. The simulation teams were made aware of the facts that the sample preparation techniques (which influence the fabrics of the prepared samples) used in these tests may be different from the cyclic stress-controlled triaxial tests performed for LEAP-2017.

The experimental data that was available during the planning phase of LEAP and for LEAP-2015 prediction exercise could be accessed at the following link (Vasko et al. 2018):

https://doi.org/10.17603/DS2TH7Q.

The data accessible via the above link consists of two sets of experiments performed by different researchers.

The first set of experiments consists of triaxial experiments performed by Andrew Vasko at the George Washington University (Vasko 2015). The soil samples in these experiments were prepared using dry pluviation with minor tapping on the mold to achieve the desired density. For additional information the reader is referred to Vasko (2015).

The second set of experiments is the direct simple shear experiments (monotonic and cyclic) performed by Ana Maria Parra Bastidas at the University of California, Davis. These experiments were performed using dry funnel deposition (for loose samples) and air pluviation (for dense samples) methods. For additional information the reader is referred to Bastidas (2016) and Bastidas et al. (2017).

If the simulation teams decided to rely on the additional experiments discussed above, they were asked to clearly state that in the calibration report. The specific additional tests used should be clearly identified in the report. Despite the dataset used for calibration, the numerical simulation teams were asked to simulate the cyclic triaxial tests presented in El Ghoraiby et al. (2017) using the final set of calibration parameters.

## 8.3 Model Calibration Report by Simulation Teams

Each simulation team was requested to submit a *calibration report* discussing the process that was followed in the calibration process. The calibration report was intended to cover the essential features of the constitutive model, the final model parameters, the calibration philosophy, and any assumptions used in the calibration process. It also presented a comparison of the liquefaction strength curves obtained from the model simulation against the results obtained from cyclic stress-controlled tests (Fig. 8.1).

### *8.3.1 Model Description*

Each simulation team was asked to briefly describe the constitutive model used in their element test simulation in the calibration report. If the model *in the exact form used in calibrations* was published elsewhere, a reference to the published article was included in the report. The reports were also to cover the main features of the model including the framework on which the model formulation was based (classical, bounding surface, two-surface or multi-surface plasticity, etc.), a description of the model internal variables and their functionality within the model, and any additional modifications to the original model formulation. Furthermore, the model description was to include the type of analysis platform on which the constitutive model was implemented (FEM, FDM, Mesh-free, etc.) and the integration scheme used in the implementation (implicit, explicit, and semi-implicit).

### 8.3.2 Model Parameters

The calibration reports had to list the values of all the model parameters obtained from the calibration process. The list of parameters included the values of any parameter that is kept at a default value. If more than one set of parameters were used for different soil densities, separate lists of model parameters for each case and the rationale for using more than one set of parameters were to be provided. If the model parameters were obtained by using the additional element test data (other than the cyclic triaxial stress-controlled tests shown in Fig. 8.1), the numerical simulation teams were asked to clearly state that in the calibration report.

### 8.3.3 Calibration Method

The numerical simulation teams were asked to discuss the approach used in the calibration of the constitutive model parameters and to explain how the element tests (cyclic triaxial and cyclic direct simple shear) were simulated. For example, the calibration reports were expected to include a discussion on how the internal variables were initialized for each test. The initial values assigned to the internal variables were also to be reported for each simulated test.

### 8.3.4 Liquefaction Strength Curves

Each calibration report was to include a plot comparing the simulated liquefaction strength curves with the results of cyclic stress-controlled triaxial tests reported by El Ghoraiby et al. (2017). The simulation results were also listed in a table which reported the number of cycles until 2.5% single amplitude axial strain is achieved for each test.

## 8.4 Simulation Results

In addition to the report on model calibration, the results of the simulation of the cyclic triaxial tests were asked to be reported in separate files with the formats discussed below.

### 8.4.1 Results Data Files

The results files were requested to be submitted as *tab delimited text* files. The files were designated to be placed in a folder called "calibration". The number of cycles until a 2.5% single amplitude axial strain is achieved was to be reported in a separate file for each density. The file names were to be labeled "eo_0_515_csr_nocyc.txt", "eo_0_542_csr_nocyc.txt", and "eo_0_585_csr_nocyc.txt" for $e_o$ of 0.515, 0.542, and 0.585, respectively. Each file was composed of a table of three columns. The first column was the initial void ratio of the soil before the shearing phase, the second column was the cyclic stress ratio, and the third column was the number of cycles until a 2.5% single amplitude axial strain is achieved.

The simulation results for each cyclic triaxial test were to be reported in a *tab delimited text* file with the same name as the simulation performed. The file was composed of a three-column table. The first column was the axial strain, $\varepsilon_a$; the second column was the deviatoric stress, $\sigma_d$; and the third column was the confining effective stress, $\sigma'_3$. The axial strain was reported in %, while the stresses were reported in kPa. Although simulations could have been run with variable time steps, the submitted results were asked to be reported with a *constant time step*.

The simulation results for cyclic direct simple shear tests were also reported in *tab delimited text* file with the same name as the simulation performed. The file was composed of a three-column table. The first column was the shear strain, $\varepsilon_a$, the second column was the shear stress, $\tau$, and the third column was the vertical effective stress, $\sigma'_v$. The shear strain was reported in %, while the stresses were reported in kPa. Although simulations could have been run with variable time steps, the submitted results were asked to be reported with a *constant time step*.

### 8.4.2 Matlab Scripts

A series of Matlab scripts were provided to the numerical simulations teams so that they can confirm the format consistency of the calibration results files. The first Matlab script was called "LiqStrengthCurve.m". This file was intended to load the *tab delimited text* files of the CSR vs. No. of Cycles and plot the liquefaction strength curve, where the simulation results were plotted against the experiment results. This file was provided to the simulation teams to ensure that the submitted simulation files conform to the specified format and facilitate future comparisons of the simulation with experimental data.

The second Matlab script was called "SimResultsPlot.m". The script loads the results file of the simulations of the cyclic stress-controlled triaxial tests. It plots the results in terms of a deviatoric stress vs. axial strain and deviatoric stress vs. mean effective stress. These scripts are provided in El Ghoraiby et al. (2018).

## 8.5 Concluding Remarks

The predictor teams were provided with the calibration data and requested to submit a detailed report by June 1, 2017. The *calibration report* and all the *simulation data files* were asked to be uploaded in the designated dropbox folder created for each simulation team. A summary of the calibration efforts and the results reported by each numerical simulation team are provided in Manzari et al. (2019).

## References


Bastidas, A. M. P. (2016). *Ottawa F-65 Sand Characterization*. PhD Dissertation, University of California, Davis.

Bastidas, A. M. P, Boulanger, R. W., Carey, T., & DeJong, J. (2017). *Ottawa F-65 Sand Data from Ana Maria Parra Bastidas*. https://datacenterhub.org/resources/ottawa_f_65. https://doi.org/10.17603/DS2MW2R

El Ghoraiby, M. A., Park, H., & Manzari, M. (2017). *LEAP 2017: Soil Characterization and Element Tests for Ottawa F65 Sand*. Washington, DC: The George Washington University.

El Ghoraiby, M. A., Park, H., & Manzari, M. (2018). *LEAP-2017 GWU Laboratory Tests*. DesignSafe-CI, Dataset. https://doi.org/10.17603/DS2210X.

Manzari, M. T., El Ghoraiby, M. A., Zeghal, M., Kutter, B. L., Arduino, P., Barrero, A. R., Bilotta, E., Chen, L., Chen, R., Chiaradonna, A., Elgamal, A. E., Fasano, G., Fukutake, K., Fuentes, W., Ghofrani, A., Ichii, K., Kiriyama, K., Lascarro, C., Mercado, V., Montgomery, J., Ozutsumi, O., Qiu, Z., Taiebat, M., Travasarou, T., Tsiaousi, D., Ueda, K., Ugalde, J., Wada, T., Wang, R., Yang, M., Zhang, J.-M., & Ziotopoulou, K. (2019). LEAP-2017 simulation exercise: Calibration of constitutive models and simulation of the element tests. In B. Kutter et al. (Eds.), *Model tests and numerical simulations of liquefaction and lateral spreading: LEAP-UCD-2017*. New York: Springer.

Vasko, A. (2015). *An Investigation into the Behavior of Ottawa Sand through Monotonic and Cyclic Shear Tests*. Masters Thesis, The George Washington University.

Vasko, A., El Ghoraiby, M., & Manzari, M. (2018). *LEAP-GWU-2015 Laboratory Tests*. DesignSafe-CI, Dataset. https://doi.org/10.17603/DS2TH7Q.

# Chapter 9
# LEAP-2017 Simulation Exercise: Calibration of Constitutive Models and Simulation of the Element Tests

**Majid T. Manzari, Mohamed El Ghoraiby, Mourad Zeghal, Bruce L. Kutter, Pedro Arduino, Andres R. Barrero, Emilio Bilotta, Long Chen, Renren Chen, Anna Chiaradonna, Ahmed Elgamal, Gianluca Fasano, Kiyoshi Fukutake, William Fuentes, Alborz Ghofrani, Koji Ichii, Takatoshi Kiriyama, Carlos Lascarro, Vicente Mercado, Jack Montgomery, Osamu Ozutsumi, Zhijian Qiu, Mahdi Taiebat, Thaleia Travasarou, Dimitra Tsiaousi, Kyohei Ueda, Jose Ugalde, Toma Wada, Rui Wang, Ming Yang, Jian-Min Zhang, and Katerina Ziotopoulou**

---

M. T. Manzari (✉)
Department of Civil and Environmental Engineering, George Washington University, Washington, DC, USA
e-mail: manzari@gwu.edu

M. El Ghoraiby
The George Washington University, Washington, DC, USA

M. Zeghal
Department of Civil and Environmental Engineering, Rensselaer Polytechnic Institute, Troy, NY, USA

B. L. Kutter
Department of Civil and Environmental Engineering, University of California, Davis, CA, USA

P. Arduino · L. Chen · A. Ghofrani
Department of Civil and Environmental Engineering, University of Washington, Seattle, WA, USA

A. R. Barrero · M. Taiebat · M. Yang
Department of Civil Engineering, University of British Columbia, Vancouver, BC, Canada

E. Bilotta · A. Chiaradonna · G. Fasano
Department of Civil, Architectural and Environmental Engineering, University of Napoli Federico II, Naples, Italy

R. Chen
Department of Hydraulic Engineering, Tsinghua University, Beijing, China

A. Elgamal · Z. Qiu
Department of Structural Engineering, University of California, San Diego, La Jolla, CA, USA

K. Fukutake · T. Kiriyama
Institute of Technology, Shimizu Corporation, Tokyo, Japan



B. Kutter et al. (eds.), *Model Tests and Numerical Simulations of Liquefaction and Lateral Spreading*, https://doi.org/10.1007/978-3-030-22818-7_9

**Abstract** This paper presents a summary of the element test simulations (calibration simulations) submitted by 11 numerical simulation (prediction) teams that participated in the LEAP-2017 prediction exercise. A significant number of monotonic and cyclic triaxial (Vasko, An investigation into the behavior of Ottawa sand through monotonic and cyclic shear tests. Masters Thesis, The George Washington University, 2015; Vasko et al., LEAP-GWU-2015 Laboratory Tests. DesignSafe-CI, Dataset, 2018; El Ghoraiby et al., LEAP 2017: Soil characterization and element tests for Ottawa F65 sand. The George Washington University, Washington, DC, 2017; El Ghoraiby et al., LEAP-2017 GWU Laboratory Tests. DesignSafe-CI, Dataset, 2018; El Ghoraiby et al., Physical and mechanical properties of Ottawa F65 Sand. In B. Kutter et al. (Eds.), *Model tests and numerical simulations of liquefaction and lateral spreading: LEAP-UCD-2017*. New York: Springer, 2019) and direct simple shear tests (Bastidas, Ottawa F-65 Sand Characterization. PhD Dissertation, University of California, Davis, 2016) are available for Ottawa F-65 sand. The focus of this element test simulation exercise is to assess the performance of the constitutive models used by participating team in simulating the results of undrained stress-controlled cyclic triaxial tests on Ottawa F-65 sand for three different void ratios (El Ghoraiby et al., LEAP 2017: Soil characterization and element tests for Ottawa F65 sand. The George Washington University, Washington, DC, 2017; El Ghoraiby et al., LEAP-2017 GWU Laboratory Tests. DesignSafe-CI, Dataset, 2018; El Ghoraiby et al., Physical and mechanical properties of Ottawa F65 Sand. In B. Kutter et al. (Eds.), *Model tests and numerical simulations of liquefaction and lateral spreading: LEAP-UCD-2017*. New York: Springer, 2019). The simulated stress paths, stress-

W. Fuentes
Universidad del Norte, Barranquilla, Colombia

K. Ichii
Faculty of Societal Safety Science, Kansai University (Formerly at Hiroshima University), Osaka, Japan

C. Lascarro · V. Mercado
University of British Columbia, Vancouver, BC, Canada

J. Montgomery
Department of Civil Engineering, Auburn University, Auburn, AL, USA

O. Ozutsumi
Meisosha Corporation, Tokyo, Japan

T. Travasarou · D. Tsiaousi · J. Ugalde
Fugro, Walnut Creek, CA, USA

K. Ueda
Disaster Prevention Research Institute, Kyoto University, Kyoto, Japan

T. Wada
Department of Civil and Earth Resources Engineering, Kyoto University, Kyoto, Japan

R. Wang · J.-M. Zhang
Department of Hydraulic Engineering, Tsinghua University, Beijing, China

K. Ziotopoulou
Department of Civil and Environmental Engineering, University of California, Davis, CA, USA

strain responses, and liquefaction strength curves show that majority of the models used in this exercise are able to provide a reasonably good match to liquefaction strength curves for the highest void ratio (0.585) but the differences between the simulations and experiments become larger for the lower void ratios (0.542 and 0.515).

## 9.1 Introduction

The LEAP-2017 project involved 11 numerical simulation teams from different academic institutions and geotechnical engineering firms across the world that participated in modeling of a selection of the performed centrifuge experiments. The simulation exercise consisted of constitutive model calibration, Type-B prediction, and Type-C prediction. This paper presents a summary of the results of the first phase (i.e., calibration) of this exercise. The main purpose of this phase was to provide the numerical simulation teams an opportunity to calibrate the constitutive models (that would be used in Type-B simulations) against the results of monotonic and cyclic triaxial tests conducted on Ottawa F-65 sand during the course of LEAP-2015 and LEAP-2017 projects. Moreover, the calibration phase would allow for a detailed documentation of the model performance against laboratory element tests before the same model is used in numerical simulations of LEAP-2017 centrifuge tests.

In preparation of the (calibration of constitutive models) phase, the following steps were followed:

- The George Washington University team performed a large number of laboratory tests (80 tests in total) on the selected soil (Ottawa F-65 sand) to obtain:
  - Basic properties of the soil (particle size distribution, specific gravity, hydraulic conductivity)
  - Stress-strain response and liquefaction strength of the soil using stress-controlled cyclic triaxial tests
- A detailed report on these experimental results (El Ghoraiby et al. 2017, 2018, 2019) was prepared and provided to the participating teams in the simulation exercise.
- Additional soil characterization and element tests data (monotonic and cyclic strain-controlled triaxial tests) produced during the LEAP (Vasko 2015; Vasko et al. 2018) were also provided to the simulation teams through "datacenterhub."
- In addition to the above data, the monotonic and cyclic direct simple shear tests performed by Bastidas (2016) were also made available through "datacenterhub" (Bastidas et al. 2017).
- The laboratory test data were provided on March 14, 2017, to 20 simulation teams from around the world (9 USA, 5 Japan, 1 China, 1 Germany, 1 Italy, 1 Canada, 1 Colombia, 1 New Zealand) that originally expressed interest in participating in the LEAP-2017 simulation exercise.
- The deadline for submission of the calibration simulations was set as June 1, 2017.

- Twelve teams submitted their calibration simulations by or near the deadline. The participating simulation teams submitted 17 different numerical simulations.

The element test results reported in El Ghoraiby et al. (2017, 2018, 2019) are the target of the calibration simulations. Each numerical simulation team submitted a detailed report discussing the process followed in the calibration of their selected constitutive model. The reports provided a description of the constitutive model used by the predictors as well as the method used for calibration. A brief summary of these reports is presented in the papers by each numerical simulation teams in the proceedings of LEAP-UCD-2017 workshop. A summary of the element tests simulations and results is presented in the following sections.

Here it is important to note that triaxial tests were chosen in the present study due to availability of equipment and based on the fact that the boundary conditions are more accurately defined in triaxial tests than in simple shear tests. Majority of the element tests were performed using an initial effective stress of about 100 kPa. This effective stress is larger than the vertical effective stress at mid-depth of the centrifuge model which was about 20 kPa in the middle of the deposit. The density of the sand specimens in the triaxial tests corresponded to relative densities of 71.5%, 87.5%, and 97.5%. The relative densities of the soil specimens in the centrifuge tests ranged from about 40% to 80%. In the following LEAP projects, a new series of direct simple shear and hollow cylinder torsional shear tests are planned so that constitutive models can be calibrated at conditions closer to that in the centrifuge tests.

## 9.2 The Numerical Simulation Teams

The numerical simulation teams that submitted their calibration reports and participated in the Type-B simulation exercise are listed in Table 9.1.

**Table 9.1** Numerical simulation teams

| No | Numerical simulation team | Constitutive model | Analysis platform |
|---|---|---|---|
| 1 | Tsinghua University | Tsinghua Constitutive Model | OpenSEES |
| 2 | Meisosha Corporation | Cocktail Glass Model | FLIP Rose |
| 3 | Shimizu Corporation | Bowl Model | HiPER |
| 4 | University of Napoli Federico II | Hypoplastic Model | Plaxis |
| 5 | UC Davis-Auburn University | PM4Sand Model | Flac-2D |
| 6 | University of Washington | Manzari-Dafalias Model/PM4Sand Model | OpenSEES |
| 7 | Kyoto University | Cocktail Glass Model | FLIP TULIP |
| 8 | Universidad del Norte | ISA-Hypoplasticity Model | ABAQUS |
| 9 | University of British Columbia | SANISand | FLAC-3D |
| 10 | University of California, San Diego | PDMY | OpenSEES |
| 11 | Fugro West | PM4Sand Model/UBCSAND | FLAC-2D |
| 12 | Hiroshima (Kansai) University | Cocktail Glass Model | FLIP Rose |

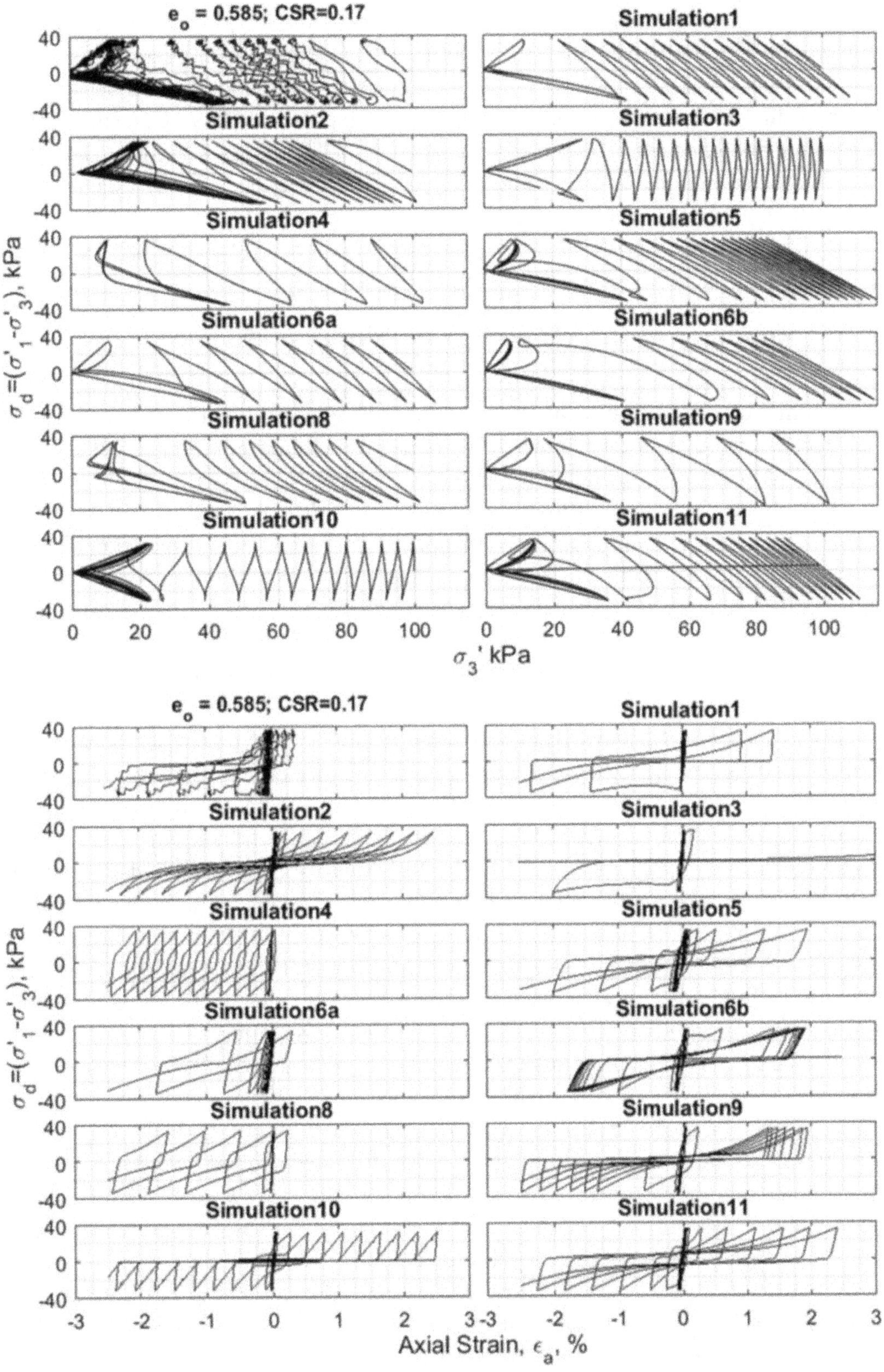


**Fig. 9.1** Comparison of the numerical simulations of a cyclic stress-controlled test on Ottawa F-65 sand for $\rho_d = 1666$ kg/m$^3$ ($e = 0.585$, $D_r$~71.5%), $p'_0 = 100$ kPa, CSR = 0.17

The constitutive model and the finite element/difference platform used by each numerical simulation team are also listed in the above table. More detailed information about each constitutive model and the numerical simulation techniques used by each team are provided in separate papers (Wang et al. 2019; Ozutsumi 2019; Fukutake and Kiriyama 2019; Fasano et al. 2019; Chen et al. 2019; Mercado et al. 2017; Qiu and Elgamal 2019; Wada and Ueda 2019; Yang et al. 2019; Tsiaousi et al. 2019; Ichii et al. 2019; Montgomery and Ziotopoulou 2019).

## 9.3 Summary of the Element Test Simulations

The numerical simulation teams were requested to present their simulations of the 23 cyclic triaxial tests on Ottawa F-65 sand (see El Ghoraiby et al. 2017, 2018, 2019). Majority of simulation teams simulated cyclic triaxial tests. The UCD-Auburn (Montgomery and Ziotopoulou 2019) and Fugro West teams (Tsiaousi et al. 2019) who worked with a plane strain constitutive model performed their single-element simulations under plane strain conditions and submitted a number of cyclic biaxial simulations for comparison with the experimental results. Hence, to be consistent in comparison of the numerical simulations, instead of the usual p′-q plot for the undrained stress path, the results are plotted in space of deviatoric stress ($\sigma_d = \sigma_1^{'} - \sigma_3^{'}$) versus minor principal effective stress ($\sigma_3^{'}$).

Figures 9.1 and 9.2 show detailed comparison of the simulated stress paths and stress-strain curves with two of the cyclic stress-controlled tests with a cyclic stress ratio, CSR = 0.17 and 0.12, performed on specimens with a density of 1666 kg/m$^3$ which, for Gs = 2.65, $(\rho_d)_{max} = 1757$ kg/m$^3$, and $(\rho_d)_{min} = 1491$ kg/m$^3$, corresponds to a void ratio of 0.585 and $D_r = 71.5\%$. The simulations are labelled simulations 1 to 11. The numbers refer to the order of the simulation teams in the list presented above. The numerical simulation team 6 submitted two different simulations with two different constitutive models but the same finite element platform which are labelled "a" and "b." It is also noted that a few simulation teams did not submit their simulations for all of the requested cyclic stress ratios. Figures 9.3 and 9.4 show similar comparisons of the numerical simulation with the results of cyclic stress-controlled triaxial tests on denser specimens $\rho_d = 1713$ kg/m$^3$ ($e = 0.542$, $D_r = 87.5\%$) for cyclic stress ratio (CSR) of 0.24 and 0.19, respectively. Figures 9.5 and 9.6 show the numerical simulations of the stress-controlled cyclic triaxial tests for a much denser specimen $\rho_d = 1744$ kg/m$^3$ ($e = 0.515$, $D_r = 97.5\%$) for cyclic stress ratio (CSR) of 0.365 and 0.225, respectively. Numerical simulation team 7 did not submit simulations for these cyclic stress ratios.

Figures 9.1, 9.2, 9.3, 9.4, 9.5, and 9.6 present selected examples of the simulations submitted by the numerical simulation teams. A review of Figs. 9.1, 9.2, 9.3, 9.4, 9.5, and 9.6 reveals the following trends:

1. The numerical simulation 1 showed stress-strain responses that are comparable to those observed in the experiments.
2. While the experimental results show ratcheting of the stress-strain curves toward the negative side (extension) of the axial strain, the stress-strain curves simulated

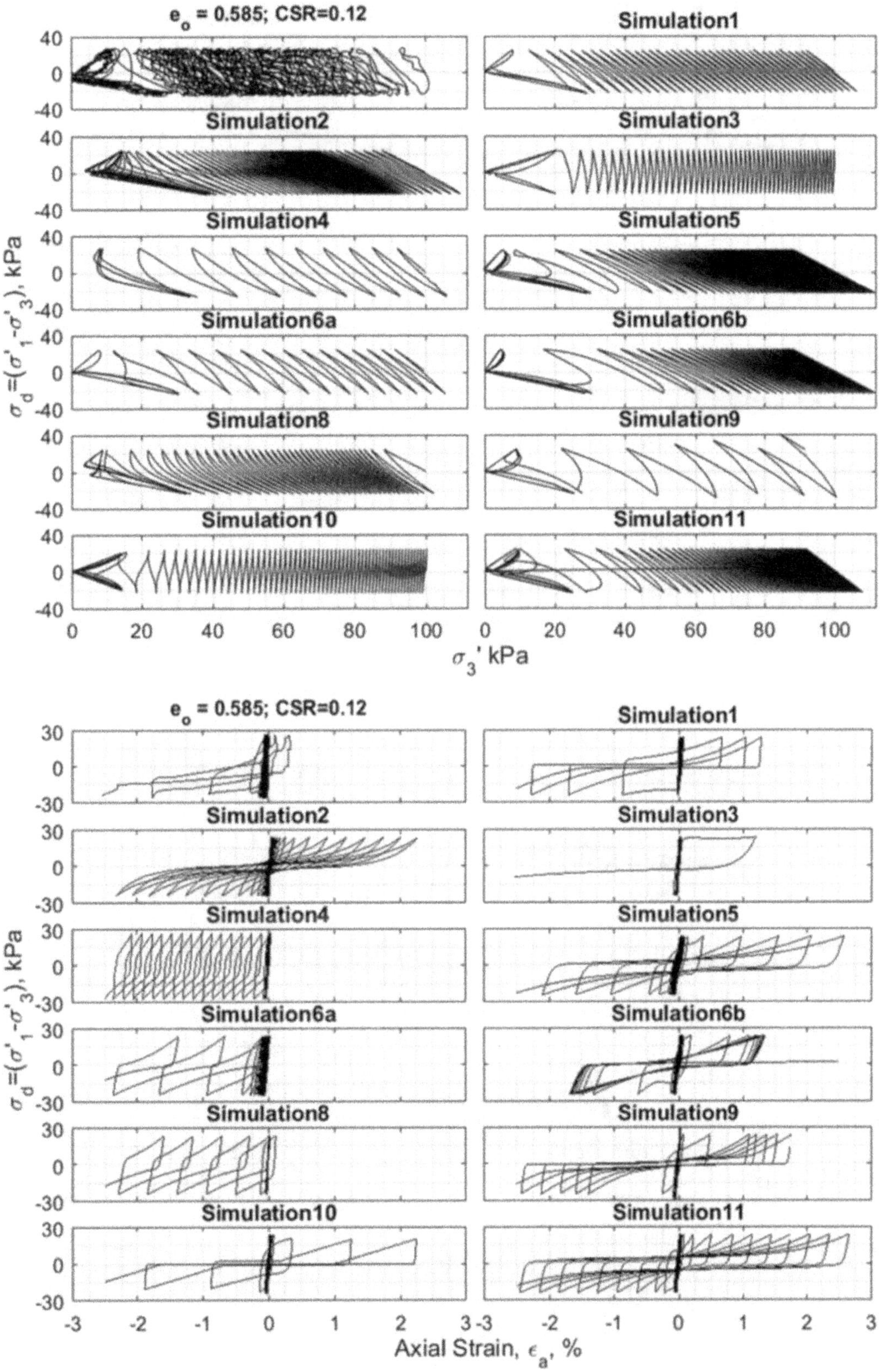


**Fig. 9.2** Comparison of the numerical simulations of a cyclic stress-controlled test on Ottawa F-65 sand for $\rho_d = 1666$ kg/m$^3$ ($e = 0.585$, $D_r$~71.5%), $p'_0 = 100$ kPa, CSR = 0.12

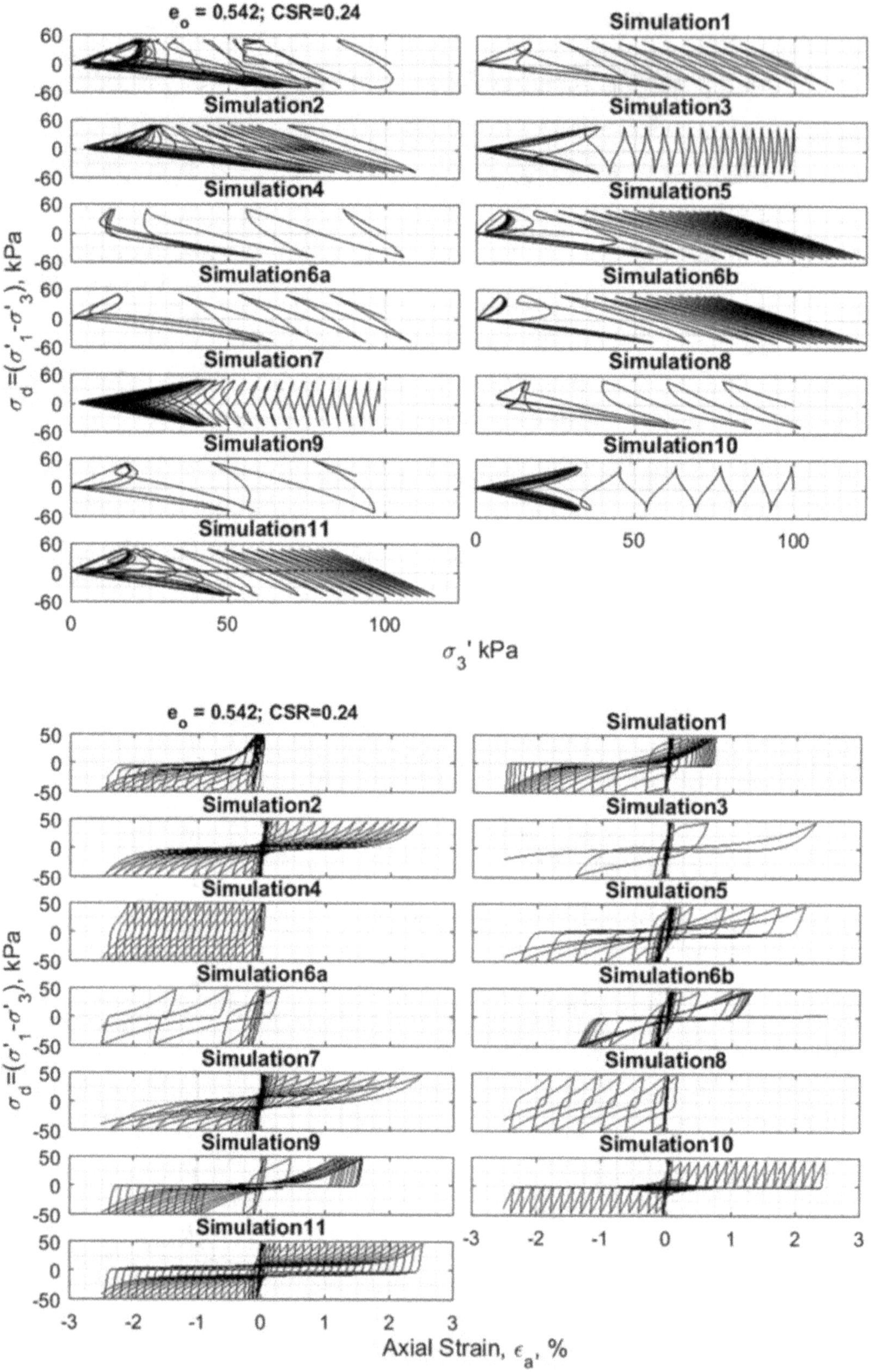


**Fig. 9.3** Comparison of the numerical simulations of a cyclic stress-controlled test on Ottawa F-65 sand for $\rho_d = 1713$ kg/m$^3$ ($e = 0.542$, $D_r$~87.5%), $p'_0 = 100$ kPa, CSR = 0.24

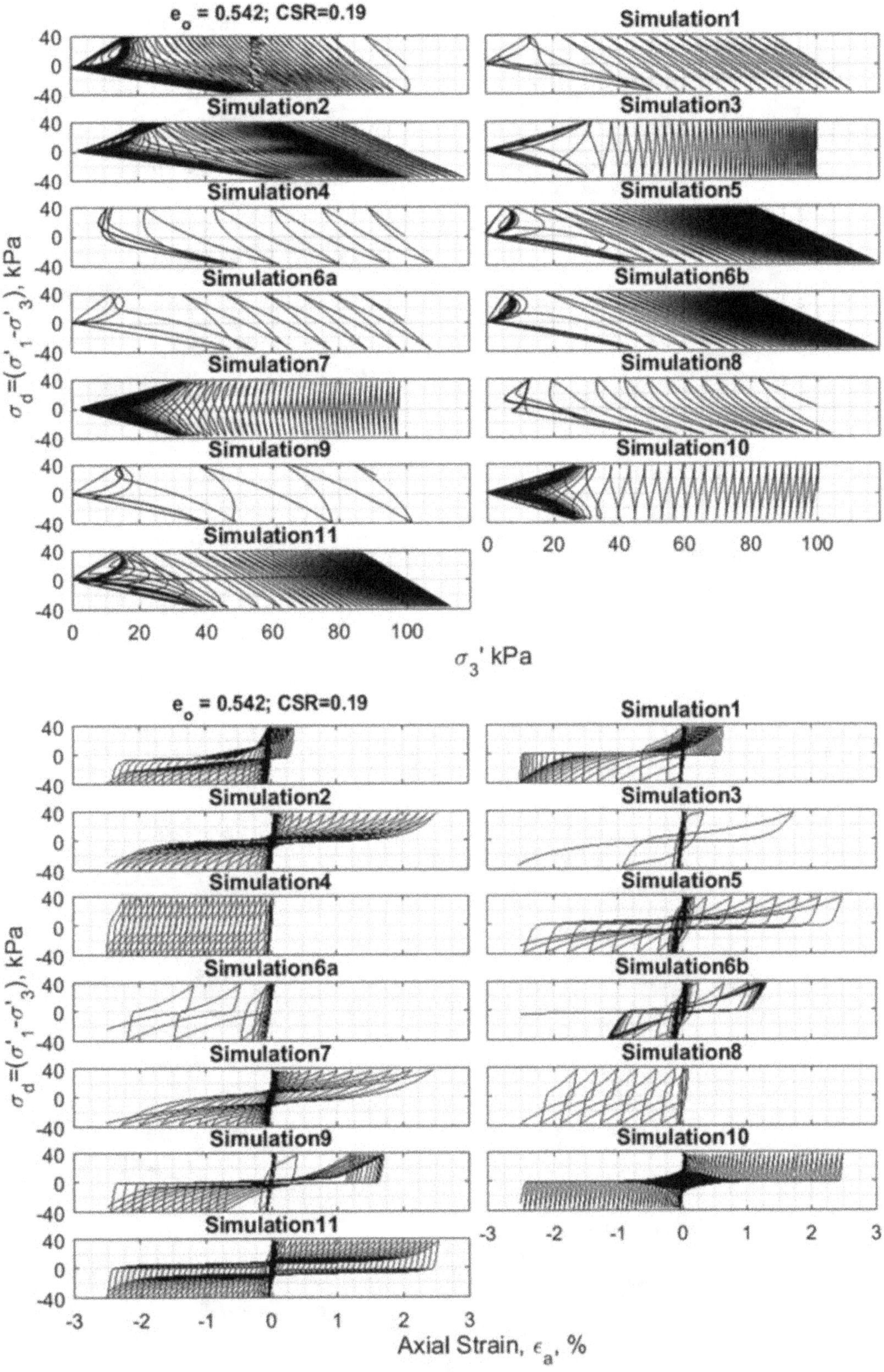


**Fig. 9.4** Comparison of the numerical simulations of a cyclic stress-controlled test on Ottawa F-65 sand for $\rho_d = 1713\ \mathrm{kg/m^3}$ ($e = 0.542$, $D_r$~87.5%), $p'_0 = 100$ kPa, CSR = 0.19

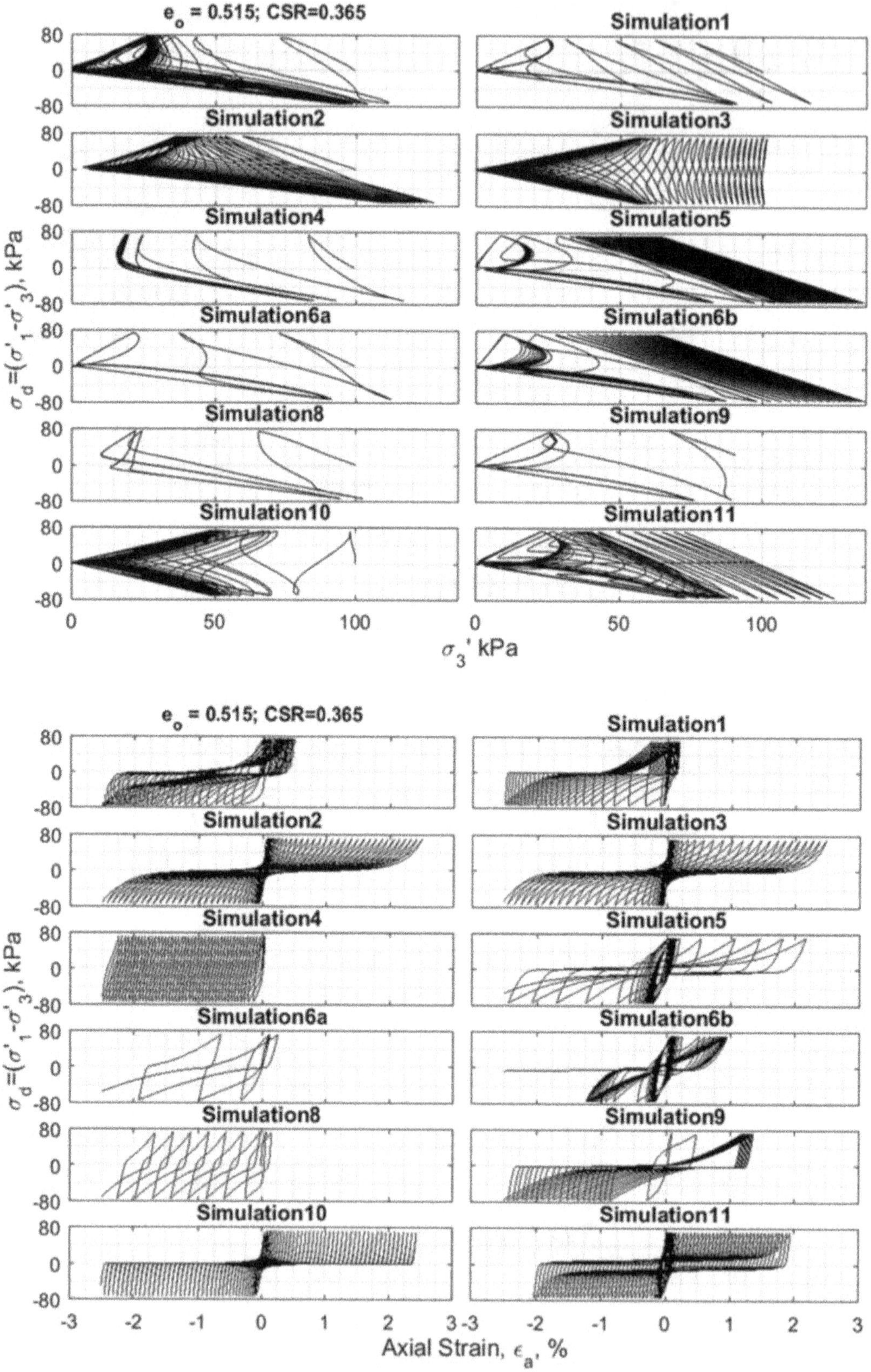


**Fig. 9.5** Comparison of the numerical simulations of a cyclic stress-controlled test on Ottawa F-65 sand for $\rho_d = 1744$ kg/m$^3$ ($e = 0.515$, $D_r = 97.5\%$), $p'_0 = 100$ kPa, CSR $= 0.365$

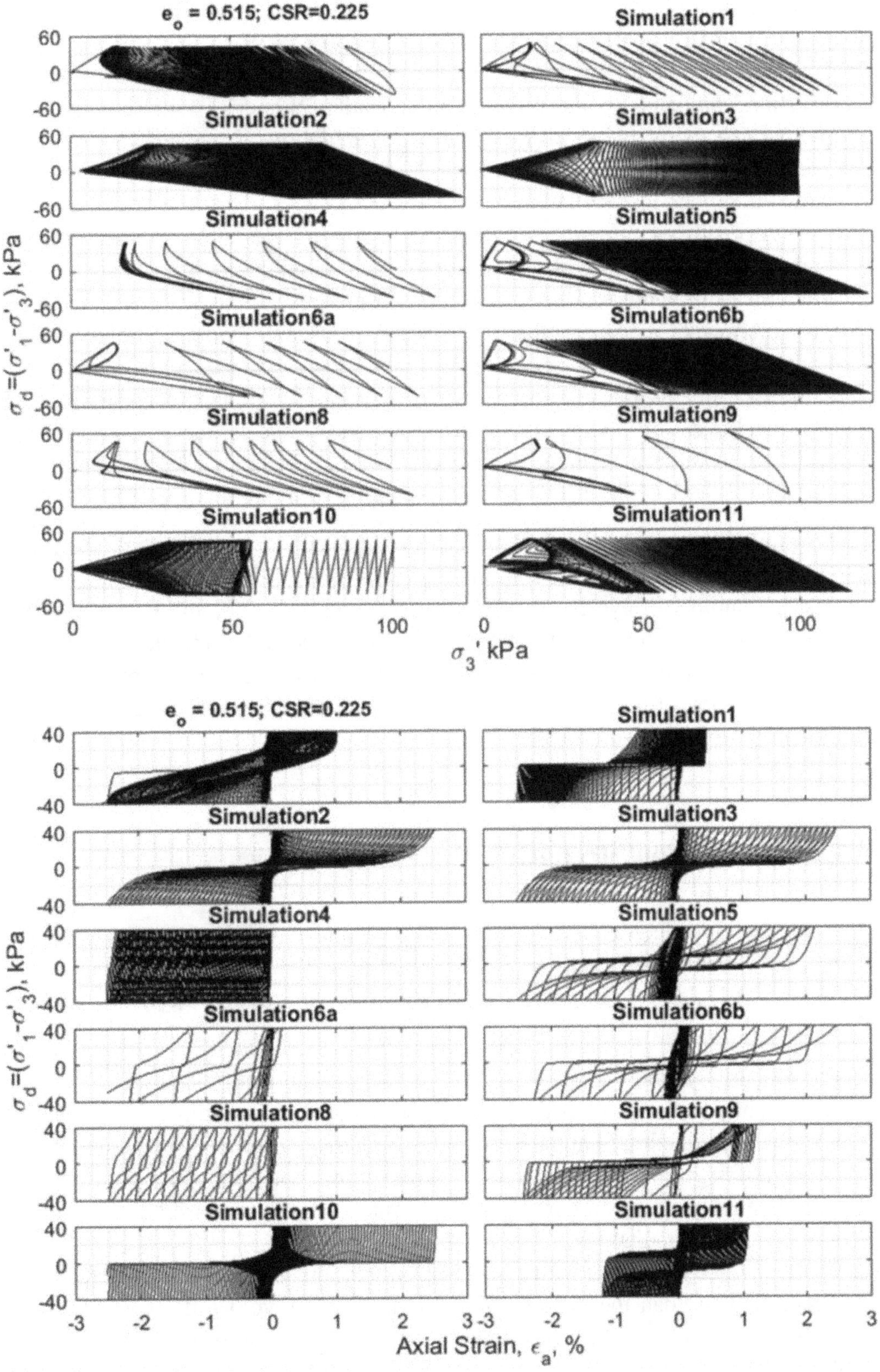


**Fig. 9.6** Comparison of the numerical simulations of a cyclic stress-controlled test on Ottawa F-65 sand for $\rho_d = 1744$ kg/m$^3$ ($e = 0.515$, $D_r = 97.5\%$), $p'_0 = 100$ kPa, CSR = 0.225

by the simulation teams 2, 3, 5, 6b, 7, 10, and 11 are more or less symmetric and do not show a visible bias in the ratcheting response.

3. The numerical simulations 6a show larger ratcheting in the triaxial extension direction compared to the experimental results. Numerical simulations 8 are qualitatively comparable to the numerical simulation of 6a, but show smaller ratcheting and seem to have stabilized at larger effective stresses than those shown in numerical simulation 6a.
4. The stress paths of numerical simulation 4 appear to have stabilized at an effective stresses larger than those observed in the cyclic triaxial experiments.
5. Numerical simulation 9 shows stress-strain responses similar to those observed in the experiments; however, they have reached a larger double amplitude axial strains at a lower number of cycles than those observed in the experiments.
6. In the experiments, the amplitude of the cyclic strains continues to grow with each additional cycle of loading. After some number of cycles, the amplitude of the cyclic strains stops growing in simulations 1, 4, 6a, 6b, and 8. The peak amplitude of cyclic strain is significantly underestimated by simulations 4 and 8.

To gain additional insights into the performance of each model, the experimentally observed and numerically simulated excess pore pressure ratios are plotted against the number of cycles for selected tests (Figs. 9.7, 9.8, and 9.9). All of the models predicted cyclic pore water pressures toward the end of the simulations. It is easy to compare the rate of pore pressure generation in the plots of pore pressure ratio as a function of number of cycles. The rate of pore pressure generation varied significantly from model to model.

## 9.4 Liquefaction Strength Curves

The liquefaction strength curves simulated by each team for the three void ratios of 0.585, 0.542, and 0.515 ($D_r$ = 71.5%, 87.5%, and 97.5%) are summarized in Figs. 9.10, 9.11, and 9.12, respectively. The following trends are observed from these curves:

1. For the largest void ratio ($e$ = 0.585), numerical simulations 6a and 9 showed significantly steeper curves than the experimental curves. Numerical simulation 9 showed a reduction in number of cycles for CSRs lower than 0.16 compared to larger CSRs. Numerical simulation 3, while showing a trend consistent with the experimentally observed curve, used different CSRs than those specified in the guidelines for simulations.
2. For the denser specimens with $e = 0.542$, simulations 6a, 8, and 9 predicted much steeper curves than the experimental curve. For this void ratio, the simulation 11 showed noticeably larger number of cycles than those observed in the experiments.
3. The quality of the match to the experimentally obtained curve for the densest case ($e$ = 0.515) appears to be significantly lower for a large number of models.

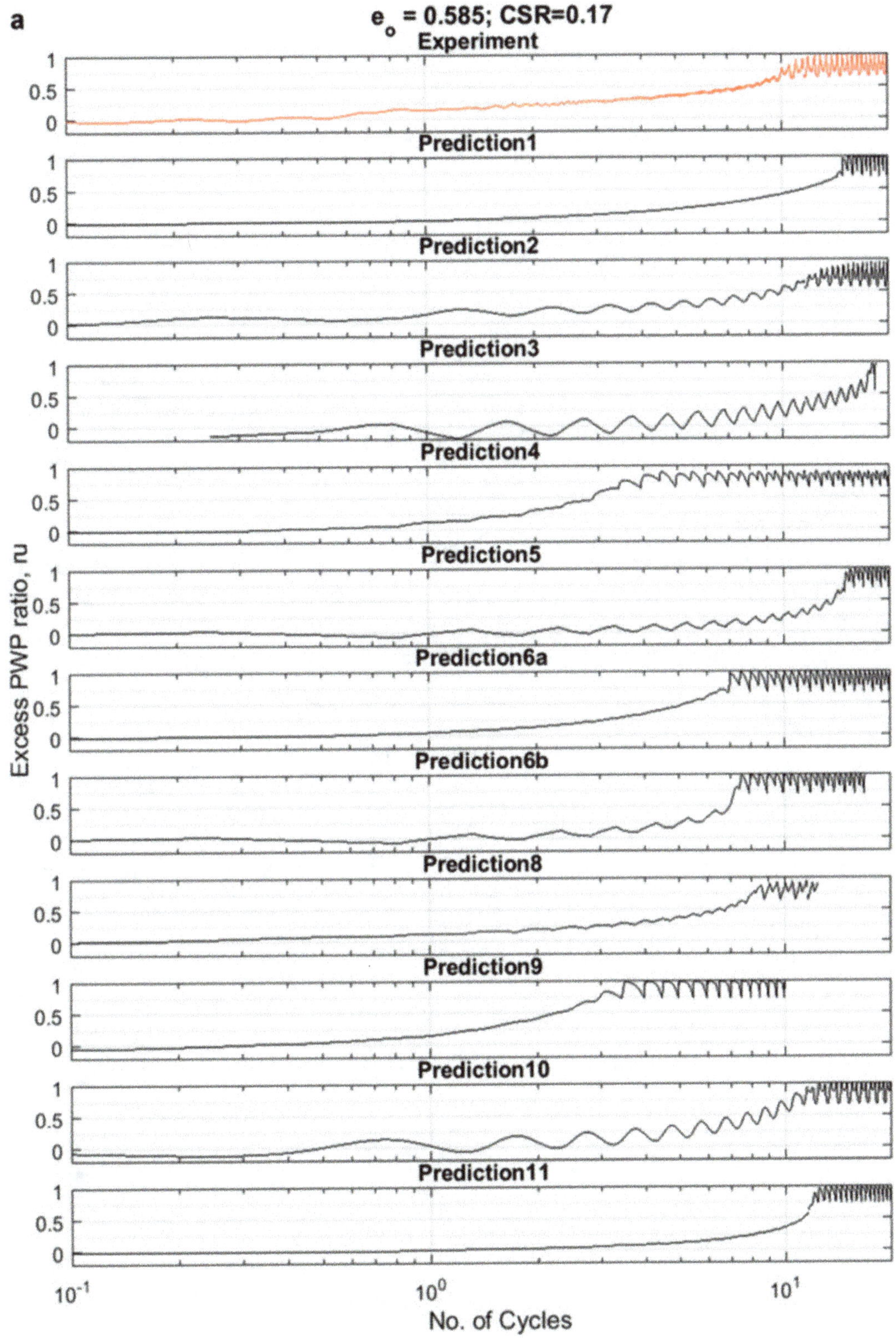


**Fig. 9.7** (**a**) Comparisons of the observed versus computed excess pore water pressure ratios with number of cycles for $\rho_d = 1666$ kg/m$^3$ ($e = 0.585$, $D_r$~71.5%) at CSR = 0.17. (**b**) Comparisons of the observed versus computed excess pore water pressure ratios with number of cycles for $\rho_d = 1666$ kg/m$^3$ ($e = 0.585$, $D_r$~71.5%) at CSR = 0.12

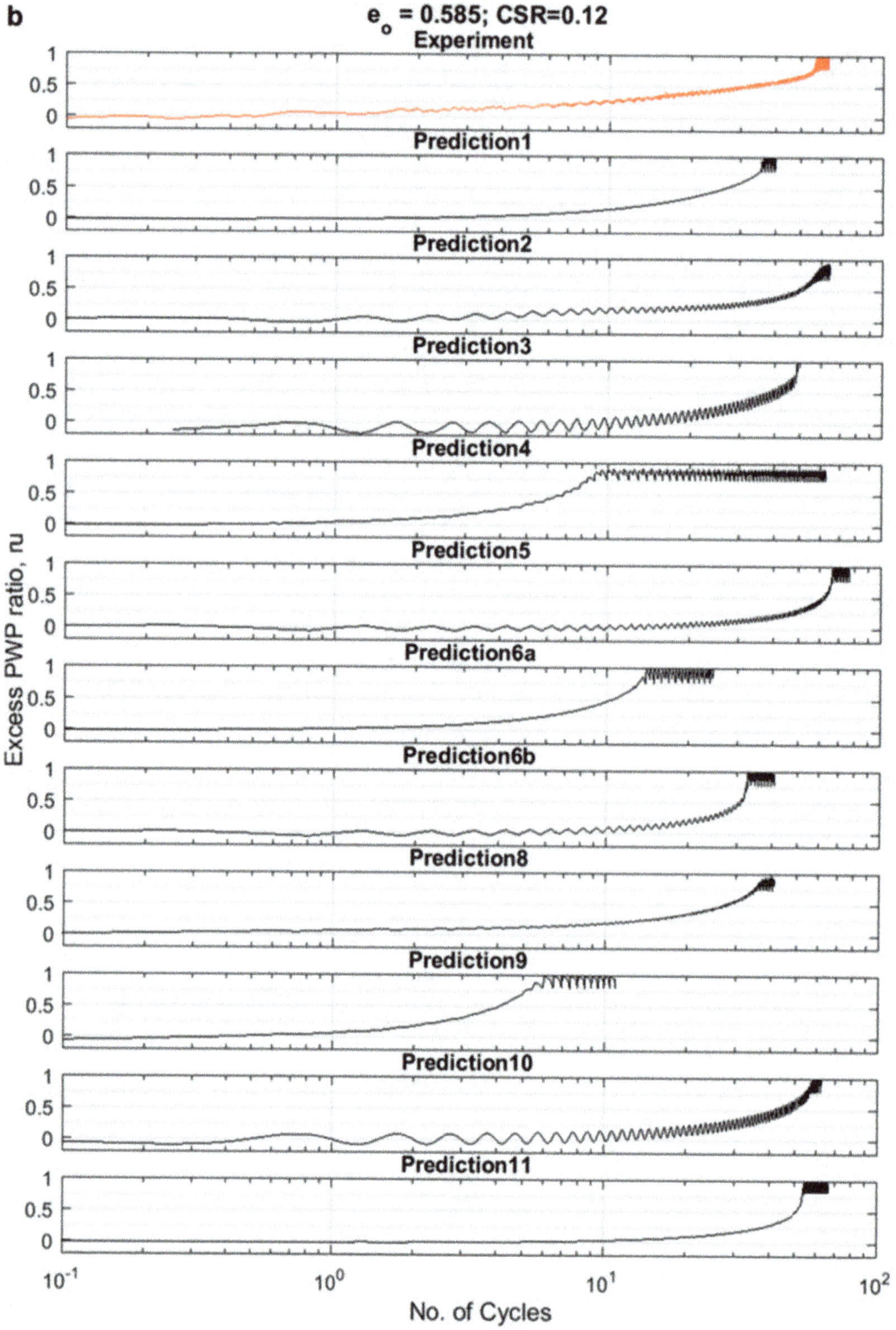


**Fig. 9.7** (continued)

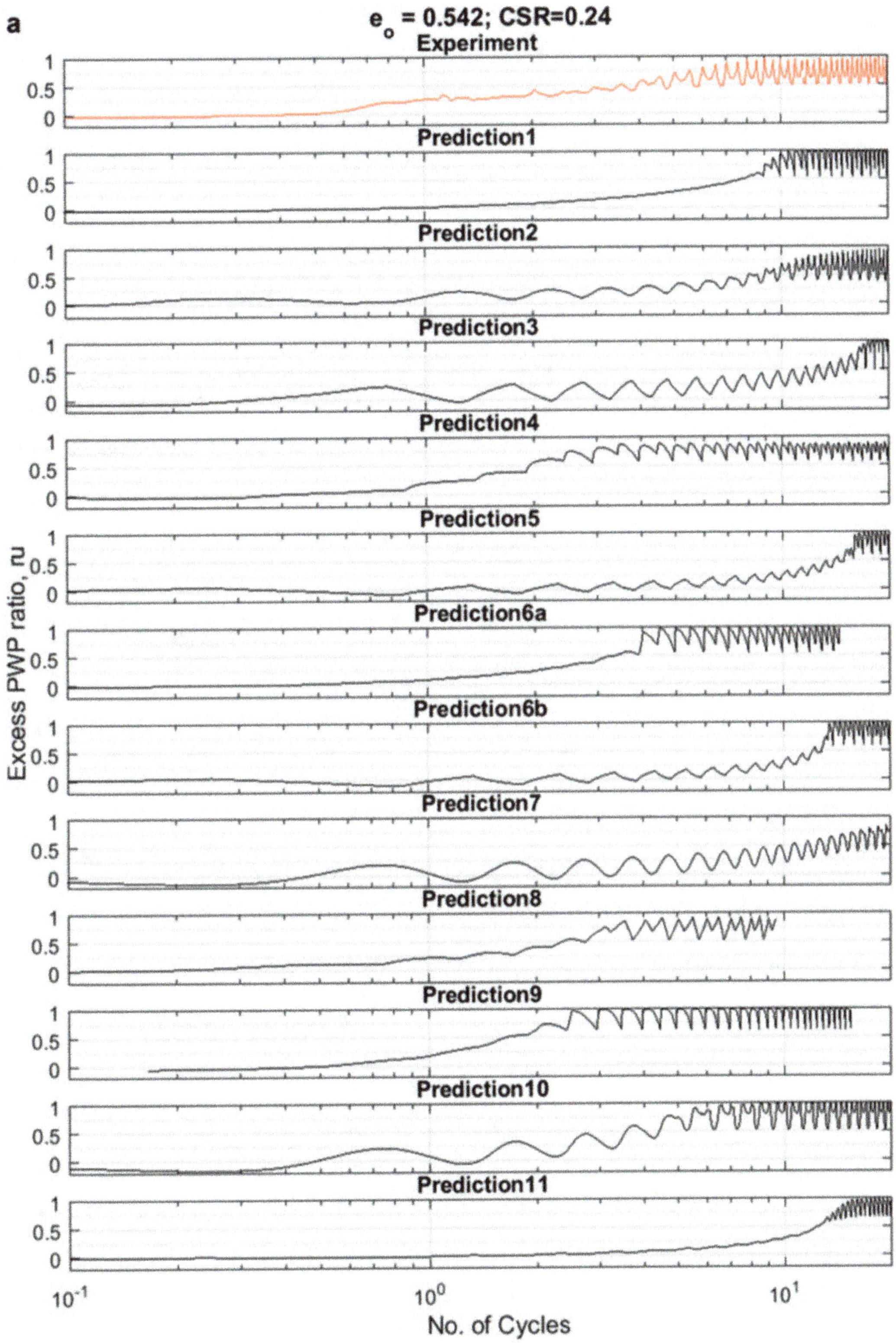


**Fig. 9.8** (**a**) Comparisons of the observed versus computed excess pore water pressure ratios with number of cycles for $\rho_d = 1713$ kg/m$^3$ ($e = 0.542$, $D_r$~87.5%) at CSR = 0.24. (**b**) Comparisons of the observed versus computed excess pore water pressure ratios with number of cycles for $\rho_d = 1713$ kg/m$^3$ ($e = 0.542$, $D_r$~87.5%) at CSR = 0.19

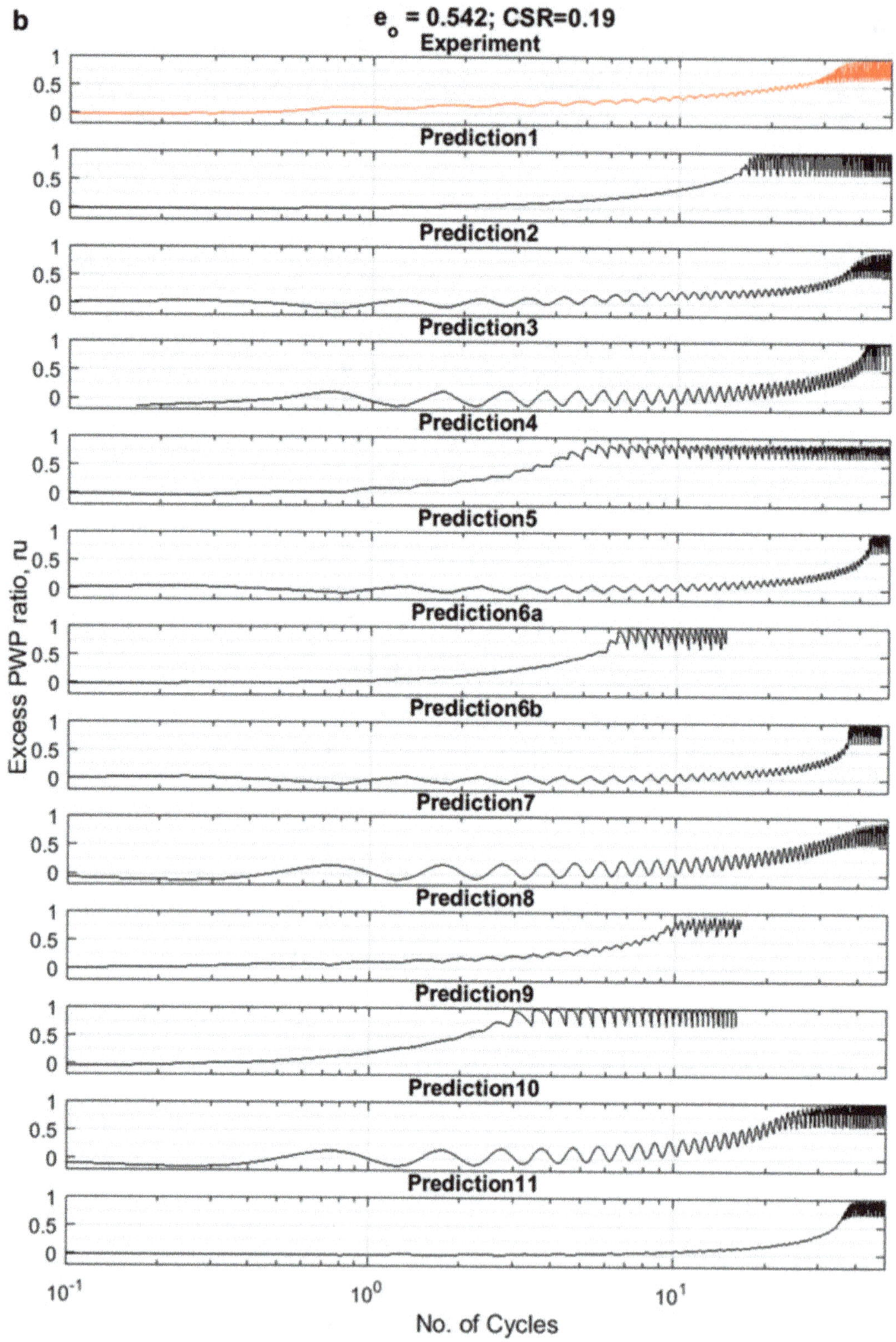


**Fig. 9.8** (continued)

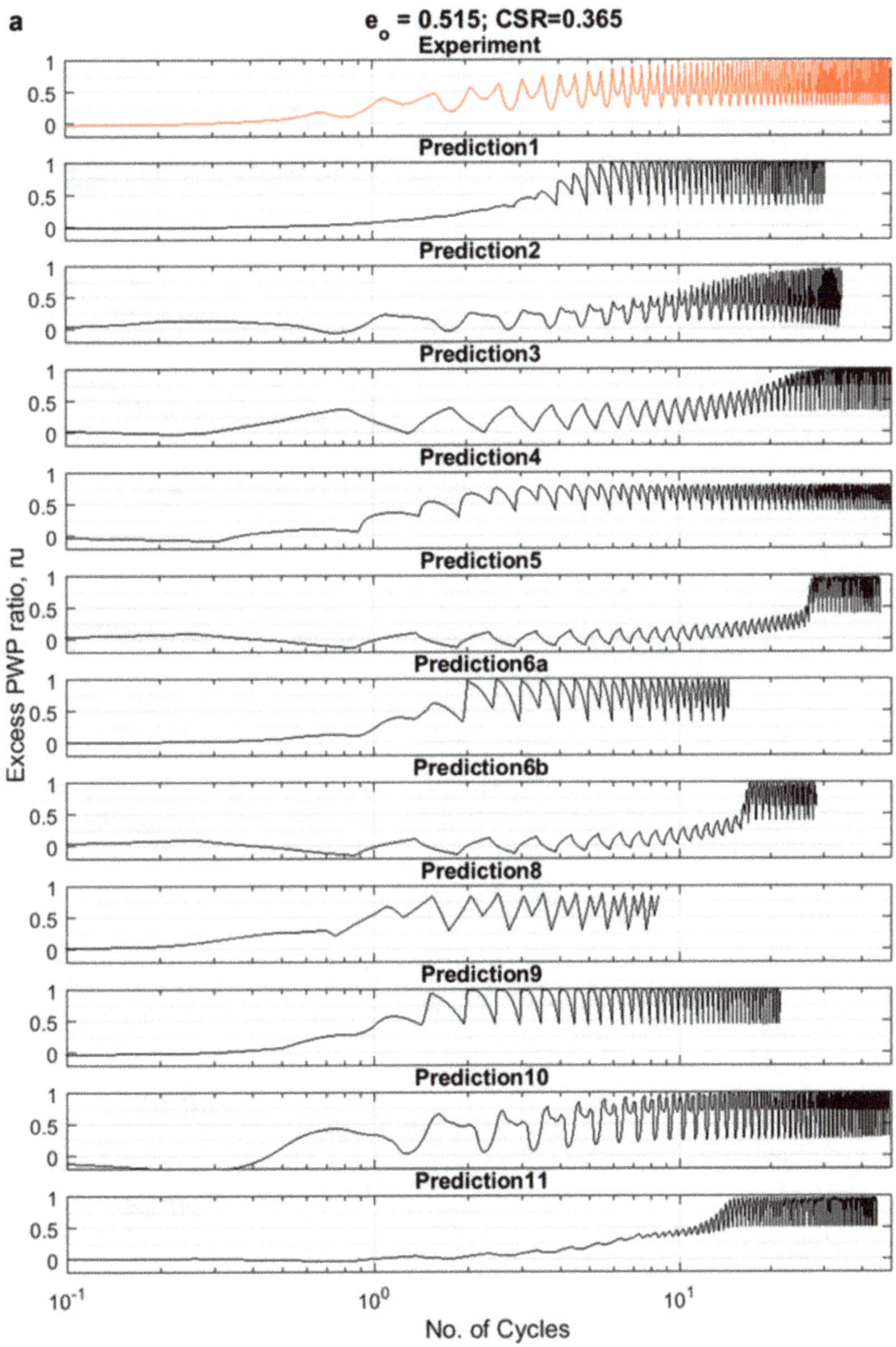


**Fig. 9.9** (**a**) Comparisons of the observed versus computed excess pore water pressure ratios with number of cycles. $\rho_d = 1744$ kg/m$^3$ ($e = 0.515$, $D_r$~97.5%) at CSR = 0.365. (**b**) Comparisons of the observed versus computed excess pore water pressure ratios with number of cycles. $\rho_d = 1744$ kg/m$^3$ ($e = 0.515$, $D_r$~97.5%) at CSR = 0.225

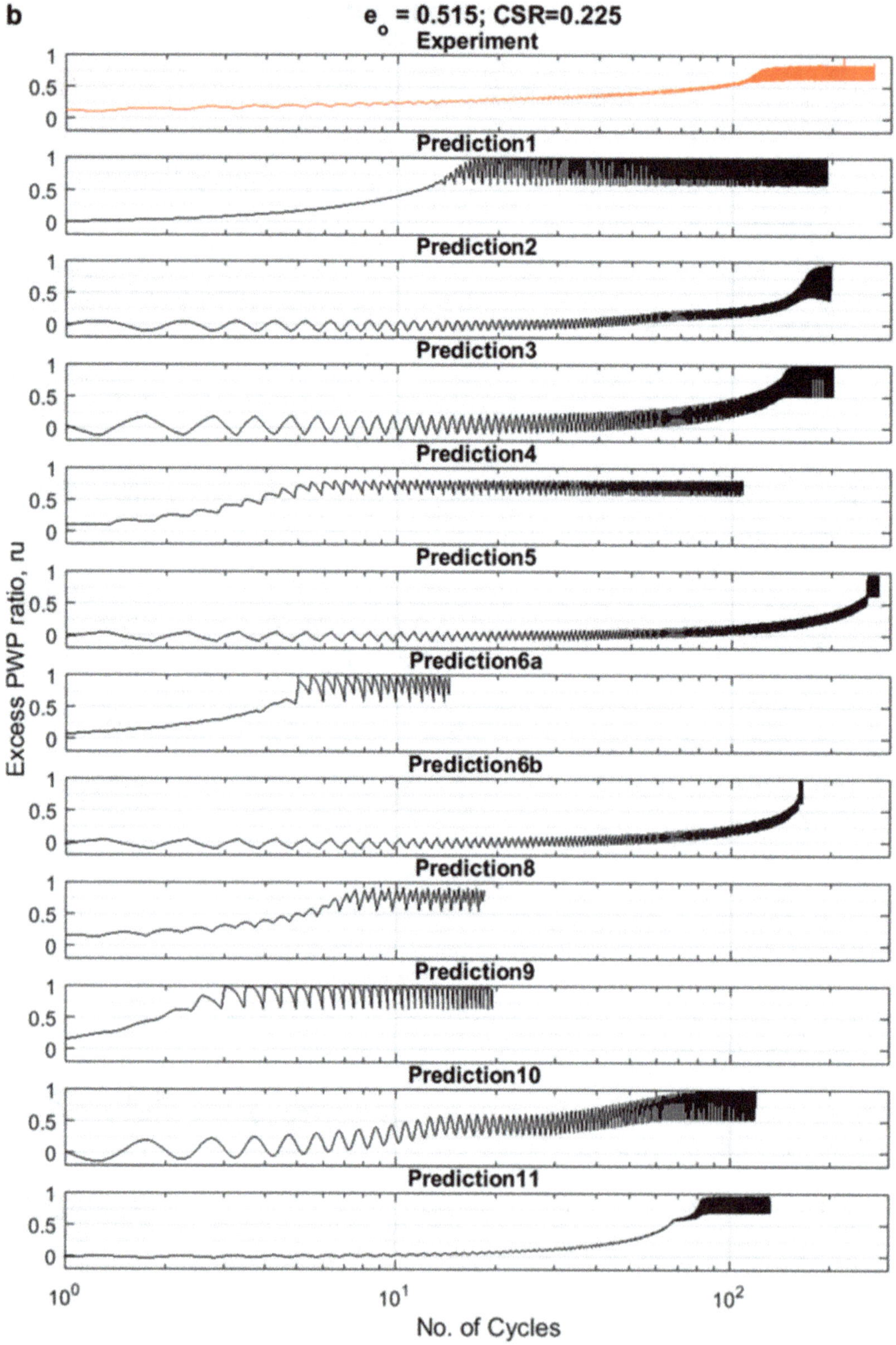


**Fig. 9.9** (continued)

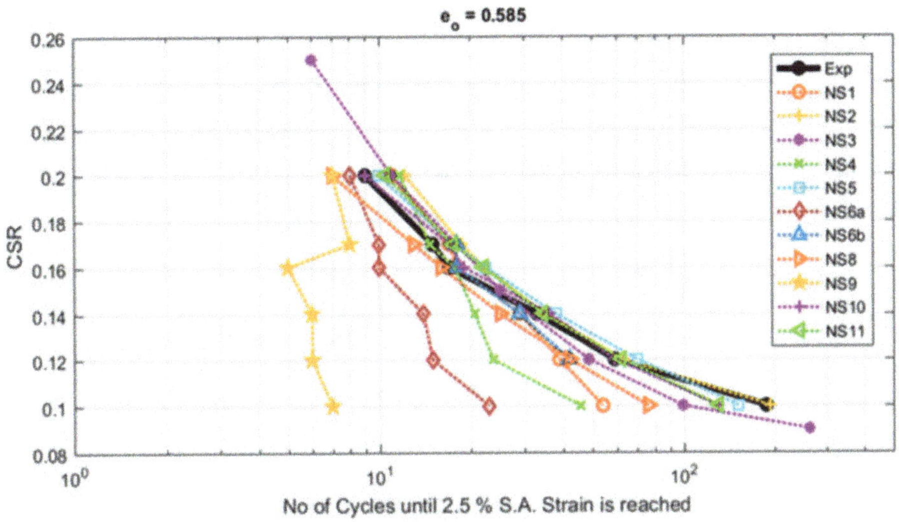


**Fig. 9.10** Comparison of the simulated liquefaction strength curves by different numerical simulations teams with the experimental results reported by El Ghoraiby et al. (2017, 2018, 2019) for $\rho_d = 1666$ kg/m$^3$ ($e = 0.585$, $D_r$~71.5%)

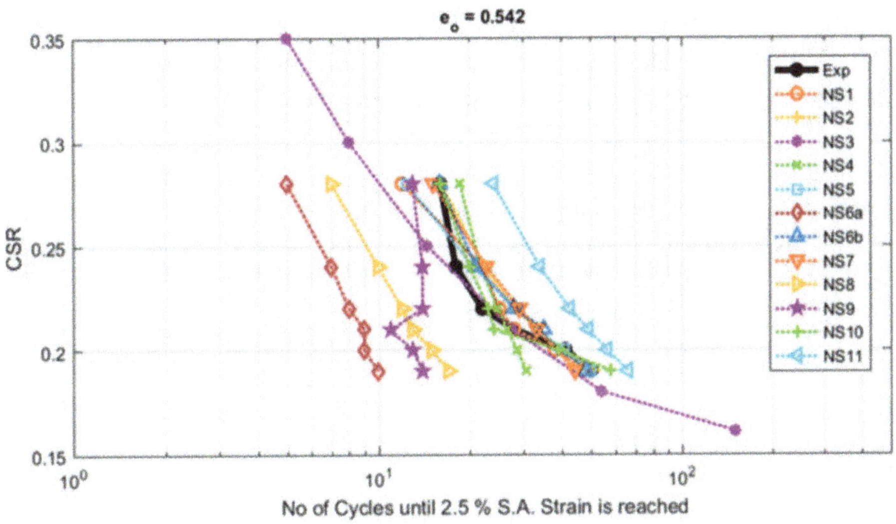


**Fig. 9.11** Comparison of the simulated liquefaction strength curves by different numerical simulations teams with the experimental results reported by El Ghoraiby et al. (2017, 2018, 2019) for $\rho_d = 1714$ kg/m$^3$ ($e = 0.542$, $D_r$~87.5%)

## 9.5 Conclusions

This article presented a summary of the numerical simulations of cyclic triaxial tests submitted by various numerical simulation teams as part of the calibration phase of LEAP-2017 project. These simulations show that a significant number of the constitutive models used in these simulations are able to capture the overall trends

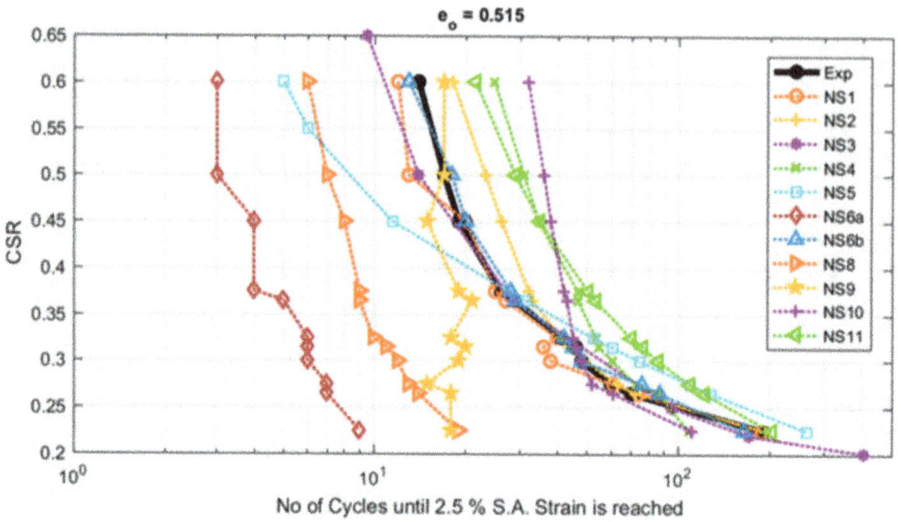


**Fig. 9.12** Comparison of the simulated liquefaction strength curves for $\rho_d$ = 1744 kg/m$^3$ ($e$ = 0.515, $D_r$~97.5%) by different numerical simulations teams with the experimental results reported by El Ghoraiby et al. (2017, 2018, 2019)

of the stress-strain response of Ottawa F-65 sand at medium dense condition ($e$ = 0.585, $D_r$~70%). However, capturing the liquefaction strength curves for all the three selected void ratios appeared to be quite challenging.

**Acknowledgments** The experimental work and numerical simulations on LEAP-UCD-2017 was supported by different funds depending mainly on the location of the work. The work by the US PIs (Manzari, Kutter, and Zeghal) is funded by National Science Foundation grants: CMMI 1635524, CMMI 1635307, and CMMI 1635040. Partial funding from Caltrans supported the efforts of co-authors Elgamal and Qiu.

## References

Bastidas, A. M. P. (2016). *Ottawa F-65 Sand Characterization*. PhD Dissertation, University of California, Davis.

Bastidas, A. M. P., Boulanger, R. W., Carey, T., & DeJong, J. (2017). *Ottawa F-65 Sand Data from Ana Maria Parra Bastidas*. https://datacenterhub.org/resources/ottawa_f_65.

Chen, L., Ghofrani, A., & Arduino, P. (2019). Prediction of LEAP-UCD-2017 Centrifuge test results using two advanced plasticity sand models. In B. Kutter et al. (Eds.), *Model tests and numerical simulations of liquefaction and lateral spreading: LEAP-UCD-2017*. New York: Springer.

El Ghoraiby, M. A., Park, H., & Manzari, M. T. (2017). *LEAP 2017: Soil characterization and element tests for Ottawa F65 sand*. Washington, DC: The George Washington University.

El Ghoraiby, M. A., Park, H., & Manzari, M. (2018). *LEAP-2017 GWU Laboratory Tests*. DesignSafe-CI, Dataset. https://doi.org/10.17603/DS2210X.

El Ghoraiby, M. A., Park, H., & Manzari, M. T. (2019). Physical and mechanical properties of Ottawa F65 Sand. In B. Kutter et al. (Eds.), *Model tests and numerical simulations of liquefaction and lateral spreading: LEAP-UCD-2017*. New York: Springer.

Fasano, G., Chiaradonna, A., & Bilotta, E. (2019). LEAP-UCD-2017 Centrifuge test simulation at UNINA. In B. Kutter et al. (Eds.), *Model tests and numerical simulations of liquefaction and lateral spreading: LEAP-UCD-2017*. New York: Springer.

Fukutake, K., & Kiriyama, T. (2019). LEAP-2017 Centrifuge test simulation using HiPER. In B. Kutter et al. (Eds.), *Model tests and numerical simulations of liquefaction and lateral spreading: LEAP-UCD-2017*. New York: Springer.

Ichii, K., Uemura, K., Orai, N., & Hyodo, J. (2019). Numerical simulation trial by cocktail glass model in FLIP ROSE for LEAP-UCD-2017. In B. Kutter et al. (Eds.), *Model tests and numerical simulations of liquefaction and lateral spreading: LEAP-UCD-2017*. New York: Springer.

Mercado, V., Fuentes, W. & Lascarro, C. (2017). *LEAP 2017 Simulation Exercise – Phase I: Model Calibration*. Calibration report, June 28, DesignSafe.

Montgomery, J., & Ziotopoulou, K. (2019). Numerical simulations of selected LEAP centrifuge. In B. Kutter et al. (Eds.), *Model tests and numerical simulations of liquefaction and lateral spreading: LEAP-UCD-2017*. New York: Springer.

Ozutsumi, O. (2019). LEAP-UCD-2017 numerical simulation at Meisosha Corp. In B. Kutter et al. (Eds.), *Model tests and numerical simulations of liquefaction and lateral spreading: LEAP-UCD-2017*. New York: Springer.

Qiu, Z., & Elgamal, A. (2019). Numerical simulations of LEAP dynamic centrifuge model tests for response of liquefiable sloping ground. In B. Kutter et al. (Eds.), *Model tests and numerical simulations of liquefaction and lateral spreading: LEAP-UCD-2017*. New York: Springer.

Tsiaousi, D., Ugalde, J., & Travasarou, T. (2019). LEAP-UCD-2017 simulation team Fugro. In B. Kutter et al. (Eds.), *Model tests and numerical simulations of liquefaction and lateral spreading: LEAP-UCD-2017*. New York: Springer.

Vasko, A. (2015). *An investigation into the behavior of Ottawa sand through monotonic and cyclic shear tests*. Masters Thesis, The George Washington University.

Vasko, A., El Ghoraiby, M., & Manzari, M., (2018). *LEAP-GWU-2015 Laboratory Tests*. DesignSafe-CI, Dataset, https://doi.org/10.17603/DS2TH7Q

Wada, T., & Ueda, K. (2019). LEAP-UCD-2017 type-B predictions through FLIP at Kyoto University. In B. Kutter et al. (Eds.), *Model tests and numerical simulations of liquefaction and lateral spreading: LEAP-UCD-2017*. New York: Springer.

Wang, R., Chen, R., & Zhang, J.-M. (2019). LEAP-UCD-2017 simulations at Tsinghua University. In B. Kutter et al. (Eds.), *Model tests and numerical simulations of liquefaction and lateral spreading: LEAP-UCD-2017*. New York: Springer.

Yang, M., Barrero, A. R., & Taiebat, M. (2019). Application of a SANISAND model for numerical simulations of the LEAP 2017 experiments. In B. Kutter et al. (Eds.), *Model tests and numerical simulations of liquefaction and lateral spreading: LEAP-UCD-2017*. New York: Springer.

# Chapter 10
# LEAP-2017: Comparison of the Type-B Numerical Simulations with Centrifuge Test Results

**Majid T. Manzari, Mohamed El Ghoraiby, Mourad Zeghal, Bruce L. Kutter, Pedro Arduino, Andres R. Barrero, Emilio Bilotta, Long Chen, Renren Chen, Anna Chiaradonna, Ahmed Elgamal, Gianluca Fasano, Kiyoshi Fukutake, William Fuentes, Alborz Ghofrani, Stuart K. Haigh, Wen-Yi Hung, Koji Ichii, Dong Soo Kim, Takatoshi Kiriyama, Carlos Lascarro, Gopal S. P. Madabhushi, Vicente Mercado, Jack Montgomery, Mitsu Okamura, Osamu Ozutsumi, Zhijian Qiu, Mahdi Taiebat, Tetsuo Tobita, Thaleia Travasarou, Dimitra Tsiaousi, Kyohei Ueda, Jose Ugalde, Toma Wada, Rui Wang, Ming Yang, Jian-Min Zhang, Yan-Guo Zhou, and Katerina Ziotopoulou**

M. T. Manzari (✉)
Department of Civil and Environmental Engineering, George Washington University, Washington, DC, USA
e-mail: manzari@gwu.edu

M. El Ghoraiby
The George Washington University, Washington, DC, USA

M. Zeghal
Department of Civil and Environmental Engineering, Rensselaer Polytechnic Institute, Troy, NY, USA

B. L. Kutter
Department of Civil and Environmental Engineering, University of California, Davis, CA, USA

P. Arduino · L. Chen · A. Ghofrani
Department of Civil and Environmental Engineering, University of Washington, Seattle, WA, USA

A. R. Barrero · M. Taiebat · M. Yang
Department of Civil Engineering, University of British Columbia, Vancouver, BC, Canada

E. Bilotta · A. Chiaradonna · G. Fasano
Department of Civil, Architectural and Environmental Engineering, University of Napoli Federico II, Naples, Italy

R. Chen · R. Wang · J.-M. Zhang
Department of Hydraulic Engineering, Tsinghua University, Beijing, China

A. Elgamal · Z. Qiu
Department of Structural Engineering, University of California, San Diego, La Jolla, CA, USA



B. Kutter et al. (eds.), *Model Tests and Numerical Simulations of Liquefaction and Lateral Spreading*, https://doi.org/10.1007/978-3-030-22818-7_10

**Abstract** This paper presents comparisons of 11 sets of Type-B numerical simulations with the results of a selected set of centrifuge tests conducted in the LEAP-2017 project. Time histories of accelerations, excess pore water pressures, and lateral displacement of the ground surface are compared to the results of nine centrifuge tests. A number of numerical simulations showed trends similar to those observed in the experiments. While achieving a close match to all measured responses (accelerations, pore pressures, and displacements) is quite challenging, the numerical simulations show promising capabilities that can be further improved with the availability of additional high-quality experimental results.

---

K. Fukutake · T. Kiriyama
Institute of Technology, Shimizu Corporation, Tokyo, Japan

W. Fuentes
Universidad del Norte, Barranquilla, Colombia

S. K. Haigh · G. S. P. Madabhushi
Department of Engineering, Cambridge University, Cambridge, UK

W.-Y. Hung
Department of Civil Engineering, National Central University of Taiwan, Taoyuan City, Taiwan

K. Ichii
Faculty of Societal Safety Science, Kansai University (formerly at Hiroshima University), Osaka, Japan

D. S. Kim
Department of Civil and Environmental Engineering, Korea Advanced Institute of Science and Technology, Daejeon, South Korea

C. Lascarro · V. Mercado
Department of Civil Engineering, University of British Columbia, Vancouver, BC, Canada

J. Montgomery
Department of Civil Engineering, Auburn University, Auburn, AL, USA

M. Okamura
Department of Civil Engineering, Ehime University, Matsuyama, Japan

O. Ozutsumi
Meisosha Corporation, Tokyo, Japan

T. Tobita
Department of Civil Engineering, Kansai University, Osaka, Japan

T. Travasarou · D. Tsiaousi · J. Ugalde
Fugro, Walnut Creek, LA, USA

K. Ueda
Disaster Prevention Research Institute, Kyoto University, Kyoto, Japan

T. Wada
Department of Civil and Earth Resources Engineering, Kyoto University, Kyoto, Japan

Y.-G. Zhou
Department of Civil Engineering, Zhejiang University, Hangzhou, China

K. Ziotopoulou
Department of Civil and Environmental Engineering, University of California, Davis, CA, USA

## 10.1 Introduction

LEAP-2017 project is a sequel to the LEAP-2015 project (Kutter et al. 2017; Manzari et al. 2017; Zeghal et al. 2017) that initiated an international collaboration for validation of constitutive and numerical modeling of soil liquefaction and its consequences using high-quality centrifuge tests. The goals of LEAP-2017 are to assess: (1) the repeatability and reproducibility of centrifuge tests, (2) the sensitivity of the experimental results to variation of testing parameters and conditions, and (3) the performance and validity of constitutive models and numerical modeling tools in predicting the observed response. To achieve these goals, the project consisted of a number of steps including a comprehensive numerical simulation exercise. For the LEAP-2017 project, this exercise consisted of three phases: (1) constitutive model calibration, (2) Type-B prediction, and (3) Type-C prediction and sensitivity analyses.

In the first phase, numerical simulation teams used a large number of laboratory tests (El Ghoraiby et al. 2018, 2019) to calibrate their constitutive models. Manzari et al. (2019) provide a summary of the results of the first phase. In the second phase of the project, twelve numerical simulation teams (hereafter referred to as predictors) from different academic institutions and the geotechnical engineering industry across the world participated in the numerical simulation of a selection of the performed centrifuge experiments. These numerical simulations are labeled as "Type-B" since the results of the experiments were unknown to the predictors at the time of their analyses, and only the test configuration and the achieved base excitations and densities were made available to them. As part of this exercise, each predictor team submitted a report discussing key details of the used numerical simulation techniques/platforms. A summary of these reports is also presented in the papers by each predictor team in the proceedings of LEAP-UCD-2017 workshop (Wang et al. 2019; Ozutsumi 2019; Fukutake and Kiriyama 2019; Fasano et al. 2019; Montgomery and Ziotopoulou 2019; Chen et al. 2019; Wada and Ueda 2019; Mercado et al. 2017; Yang et al. 2019; Tsiaousi et al. 2019; Ichii et al. 2019).

This paper presents a summary of key aspects of the Type-B numerical simulations and their comparisons with the experimental data obtained from centrifuge tests (Kutter et al. 2019).

## 10.2 LEAP-2017 Centrifuge Experiments

Similar to the LEAP-2015 project, the LEAP-2017 centrifuge experiments were designed to investigate the lateral spreading of a submerged mildly sloping liquefiable deposit. Figures 10.1 and 10.2 show the baseline schematic of these experiments (Kutter et al. 2017, 2019). The soil specimen is a sloping layer of Ottawa F65 sand with a height of 4 m (in prototype scale) at the center and a slope of

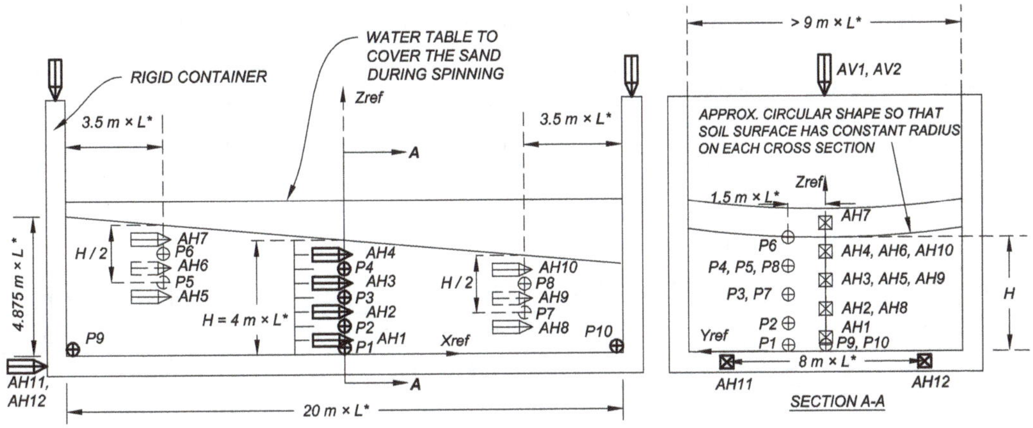


**Fig. 10.1** Baseline schematic for the LEAP-2017 experiment for shaking parallel to the axis of the centrifuge (Kutter et al. 2019)

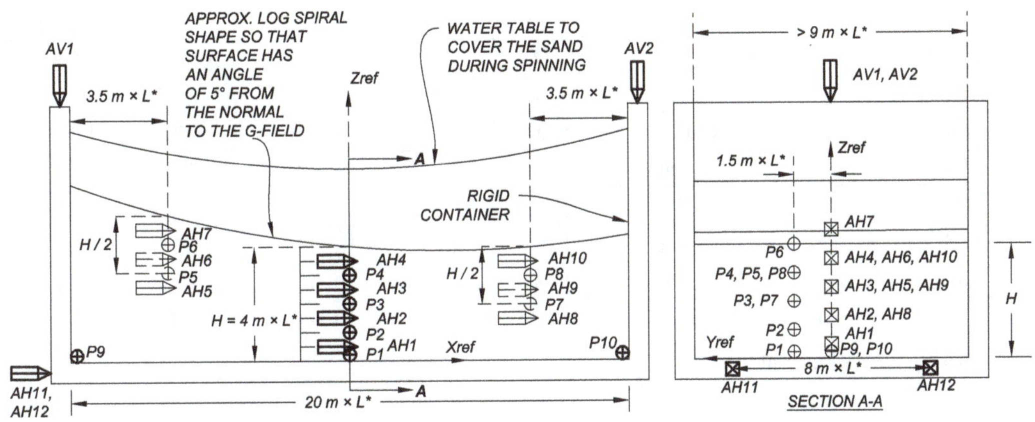


**Fig. 10.2** Baseline schematic for the LEAP-2017 experiment for shaking in the circumferential direction of the centrifuge (Kutter et al. 2019)

5 degrees. The prototype soil layer has a length of 20 m and a width of at least 9 m (in prototype scale).

The specimen is built in a container with rigid walls. Three arrays of accelerometers and pore pressure transducers are placed in the central section and at 3.5 m away from the side walls on the right and left of the model. In the transverse direction, the accelerometers are placed in the center while the pore pressure transducers are placed at a distance of 1.5 m away from the center. In the vertical direction, the sensors were 1.0 m apart. Tables 10.1a and 10.1b show the specified locations (distance from centerline and the depth) of the pore pressure sensors and accelerometers, respectively, at prototype scale. The actual sensor locations deviated from the specification; the deviation is documented in the papers by individual experimenters or in their data documented in the LEAP-UCD-2017 DesignSafe-CI.

Nine teams of centrifuge modelers from across the world performed 24 centrifuge tests covering a range of achieved densities and shaking intensities. Of these 24 tests, the numerical modelers were asked to simulate nine tests in the Type-B exercise.

**Table 10.1a** Specified positions of the pore pressure transducers of LEAP-2017 experiments

| Sensor | P1 | P2 | P3 | P4 | P5 | P6 | P7 | P8 |
|---|---|---|---|---|---|---|---|---|
| x-pos. (m) | 0.0 | 0.0 | 0.0 | 0.0 | −6.5 | −6.5 | 6.5 | 6.5 |
| Depth (m) | 4.0 | 3.0 | 2.0 | 1.0 | 2.0 | 1.0 | 2.0 | 1.0 |

**Table 10.1b** Specified positions of the accelerometers of LEAP-2017 experiments

| Sensor | AH1 | AH2 | AH3 | AH4 | AH5 | AH6 | AH7 | AH8 | AH9 | AH10 |
|---|---|---|---|---|---|---|---|---|---|---|
| x-pos. (m) | 0.0 | 0.0 | 0.0 | 0.0 | −6.5 | −6.5 | −6.5 | 6.5 | 6.5 | 6.5 |
| Depth (m) | 3.5 | 2.5 | 1.5 | 0.5 | 2.5 | 1.5 | 0.5 | 2.5 | 1.5 | 0.5 |

**Table 10.2** Summary of centrifuge experiments selected for LEAP-2017 Type-B simulations

| Centrifuge test | g* | Shaking direction | Radius of centrifuge (m) | μ*/g* (cSt) | Soil density (kg/m$^3$) | $D_r$ (%) |
|---|---|---|---|---|---|---|
| CU-2 | 40 | Tangential | 3.56 | 1.175 | 1605.8 | 46.3 |
| Ehime-2 | 40 | Axial | 1.184 | 1.0 | 1656.55 | 64.4 |
| KAIST-1 | 40 | Axial | 5.00 | 0.897 | 1701.2 | 79.4 |
| KAIST-2 | 40 | Axial | 5.00 | 0.936 | 1592.5 | 41.4 |
| KyU-3 | 44.4 | Tangential | 2.5 | 0.991 | 1637 | 57.6 |
| NCU-3 | 26 | Axial | 2.716 | – | 1652 | 62.8 |
| UCD-1 | 43 | Tangential | 1.094 | 1.0 | 1665 | 67.3 |
| UCD-3 | 43 | Tangential | 1.094 | 1.0 | 1658 | 64.9 |
| ZJU-2 | 30 | Axial | 4.315 | 1.0 | 1606 | 46.4 |

* Centrifugal acceleration

Table 10.2 shows a summary of the experiments selected for the LEAP-2017 Type-B simulations. The table lists the main characteristics of each experiment including: centrifugal acceleration scale g* with respect to gravity 1g, shaking direction relative to centrifugal axis, radius of centrifuge, viscosity of the pore fluid, and the reported achieved soil density.

## 10.3 Type-B Numerical Simulations

Following submission of the calibration simulations (phase 1), the recorded base motions along with the achieved geometry and density of the test specimens for the selected centrifuge tests were provided to all the predictors on July 28, 2017. The predictors were asked to submit their simulation of the nine centrifuge tests by September 30, 2017. The information provided for the majority of the selected tests included in-flight CPT measurements just before the first destructive base motion. Table 10.3 lists the numerical prediction teams (predictors) who participated in the Type-B simulation exercise.

**Table 10.3** Numerical simulation teams

| No | Numerical simulation team | Constitutive model | Analysis platform |
|---|---|---|---|
| 1 | Tsinghua University | Tsinghua constitutive model | OpenSees |
| 2 | Meisosha Corporation | Cocktail glass model | FLIP Rose |
| 3 | Shimizu Corporation | Bowl model | HiPER |
| 4 | University of Napoli Federico II | Hypoplastic model | Plaxis |
| 5 | UC Davis-Auburn University | PM4Sand model | FLAC-2D |
| 6 | University of Washington | Manzari-Dafalias model/PM4Sand model | OpenSees |
| 7 | Kyoto University | Cocktail glass model | FLIP TULIP |
| 8 | Universidad del Norte | ISA-hypoplasticity model | ABAQUS |
| 9 | University of British Columbia | SANISand | FLAC-3D |
| 10 | University of California, San Diego | PDMY | OpenSees |
| 11 | Fugro West | PM04 model/UBCSAND | FLAC-2D |
| 12 | Hiroshima (Kansai) University | Cocktail glass model | FLIP Rose |

The constitutive model and the finite element/difference platform used by each numerical simulation team are also shown in the following Table 10.3. It is noted that the fifth team in the above list (UC Davis-Auburn University Team) submitted four different predictions for each of the nine centrifuge tests. The basis of this selection was the variation between the relative densities achieved at the different facilities and between those and the target relative density. The differences between these simulations are discussed in details in by Montgomery and Ziotopoulou (2019). These simulations are labeled as 5a to 5d in this paper, with simulations 5a and 5b representing the team's "best estimate." The sixth team in the above list (University of Washington) used two different models (DM04 and PM4Sand) in their simulations which are identified as 6a and 6b in the following discussions. Similarly, the 11th numerical simulation team (Fugro-West team) used two different constitutive models (PM4Sand and UBC Sand) in their simulations of three of the nine centrifuge tests. These will be labeled, respectively, as 11a and 11b simulations in the following discussions.

The numerical simulation teams were requested to submit a Type-B simulation report discussing the steps followed in the prediction of the centrifuge experiments. The Type-B prediction report discussed the main features of the numerical analysis platform used in the simulation, the model geometry and the discretization details, the boundary conditions of the numerical model, the solution algorithm employed, and the assumptions used in the analyses. More detailed information about each constitutive model and the numerical simulation techniques used by each team are provided in separate papers and reports (Wang et al. 2019; Ozutsumi 2019; Fukutake and Kiriyama 2019; Fasano et al. 2019; Montgomery and Ziotopoulou 2019; Chen et al. 2019; Wada and Ueda 2019; Mercado et al. 2017; Yang et al. 2019; Tsiaousi et al. 2019; Ichii et al. 2019). The team from Hiroshima University submitted only one simulation for the Kyoto test which is described in details in Ichii et al. (2019) and will not be discussed herein.

## 10.4 Summary of Type-B Simulations Results

The results of the Type-B numerical simulations submitted by the 11 simulation teams for the selected LEAP-2017 centrifuge tests consist of 93 simulation sets, each containing time histories of predicted accelerations, excess pore water pressures, and displacements at selected locations within the centrifuge specimens. Due to space limitation, only a small subset of this data is presented herein. In the following sections, selected time histories of excess pore pressure, spectral accelerations, and lateral displacements are presented and discussed. While the data presented and analyzed here does not cover the entire dataset, the main objective is to provide representative samples of the performance of each simulation in comparison with the experimental data.

### *10.4.1 Excess Pore Water Pressure Time Histories*

Figures 10.3, 10.4, 10.5, 10.6, 10.7, 10.8, and 10.9 show detailed comparisons of the predicted time histories of excess pore water pressures computed at the central section of the centrifuge specimen (pore pressure sensors 1 to 4) with the results

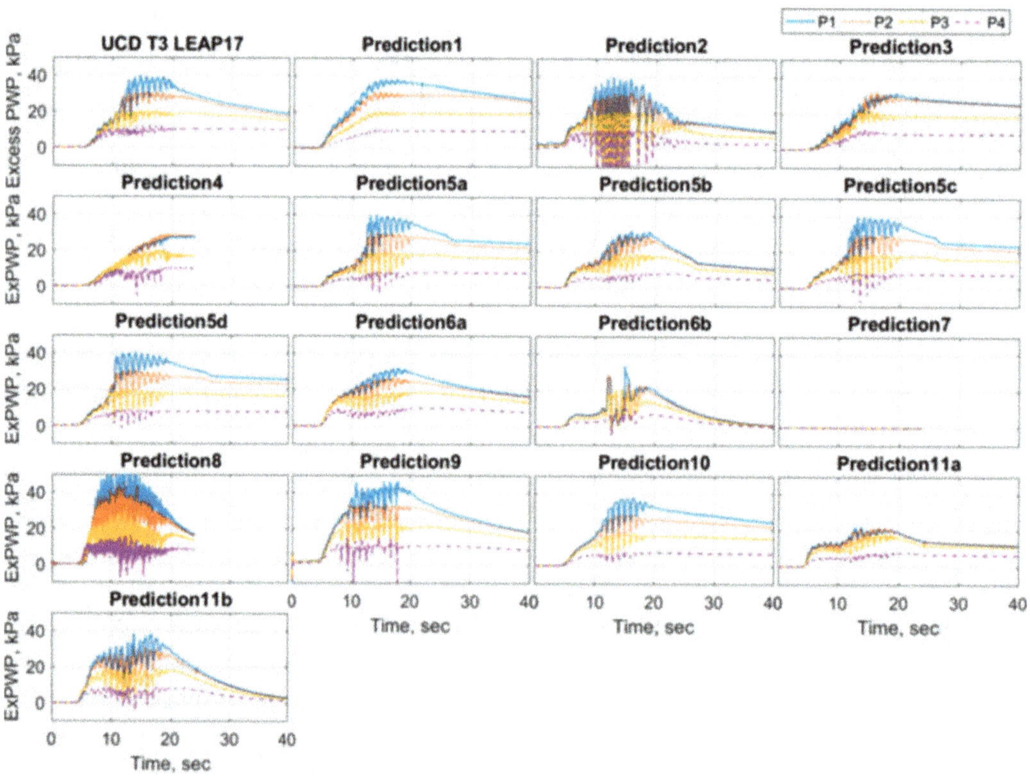


**Fig. 10.3** Comparison of the predicted time histories of excess pore water pressures computed at the central section of the soil specimen for UCD-3 test

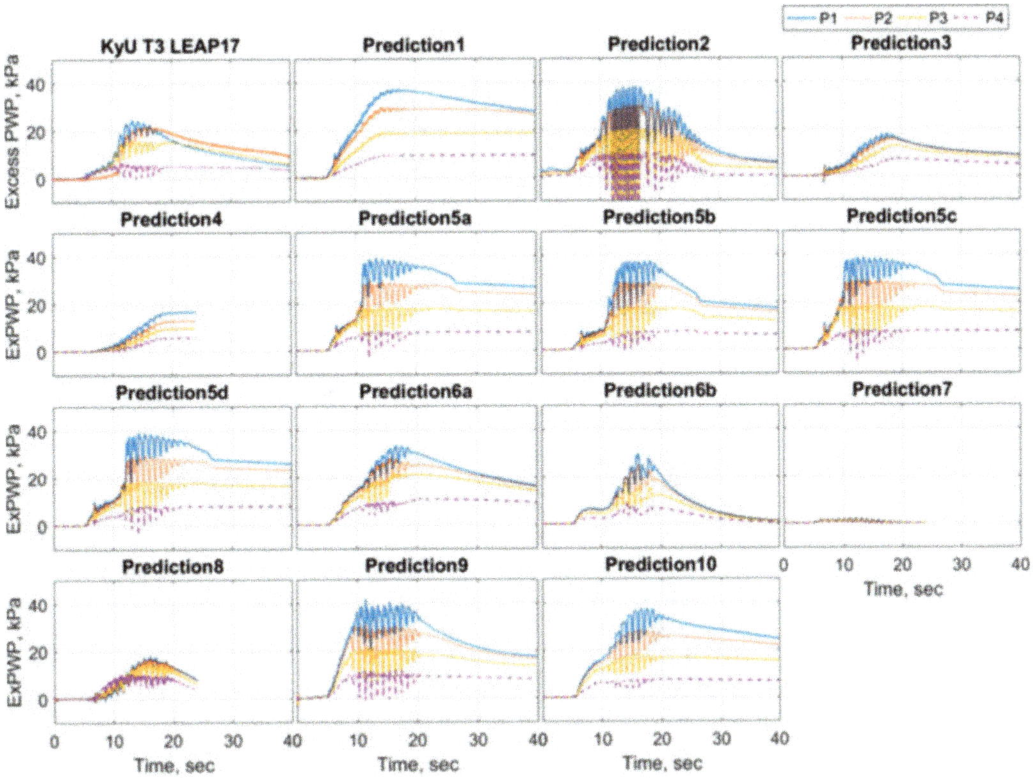


**Fig. 10.4** Comparison of the predicted time histories of excess pore water pressures computed at the central section of the soil specimen for KyU-3 test

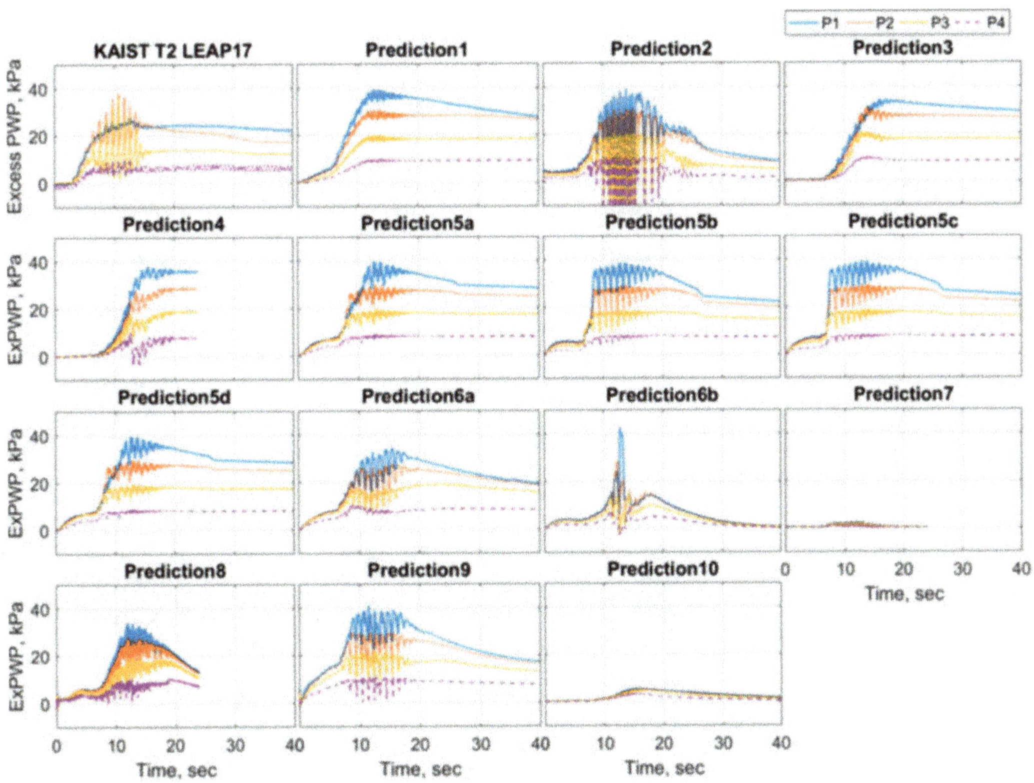


**Fig. 10.5** Comparison of the predicted time histories of excess pore water pressures computed at the central section of the soil specimen for KAIST-2 test

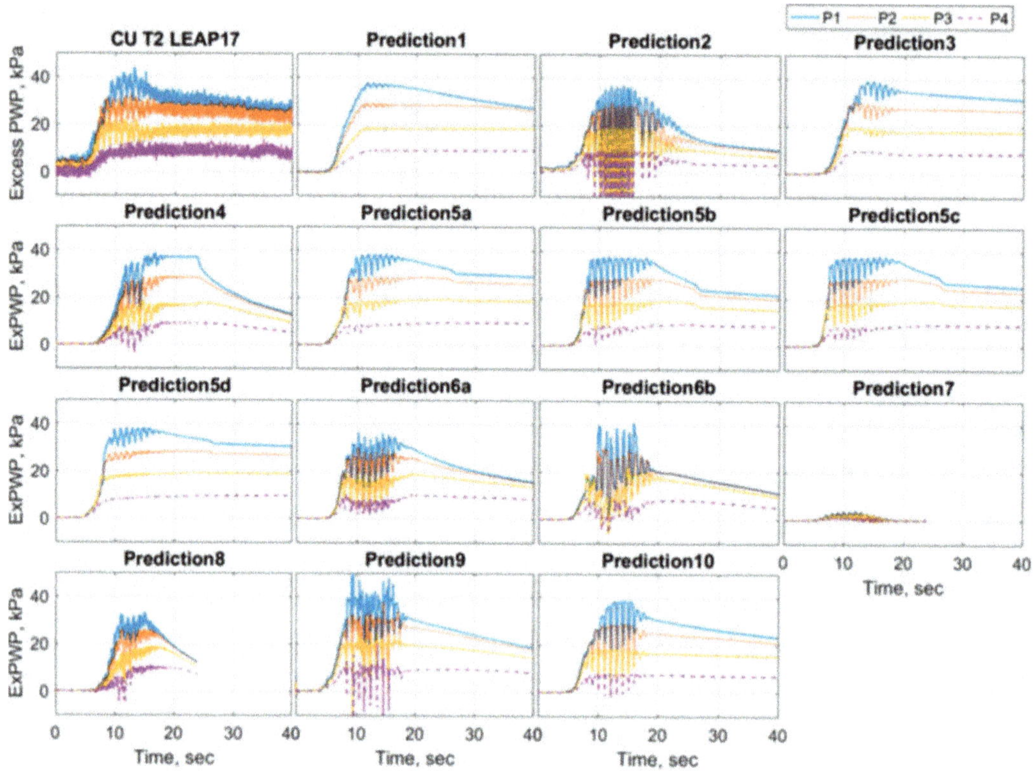


**Fig. 10.6** Comparison of the predicted time histories of excess pore water pressures computed at the central section of the soil specimen for CU-2 test

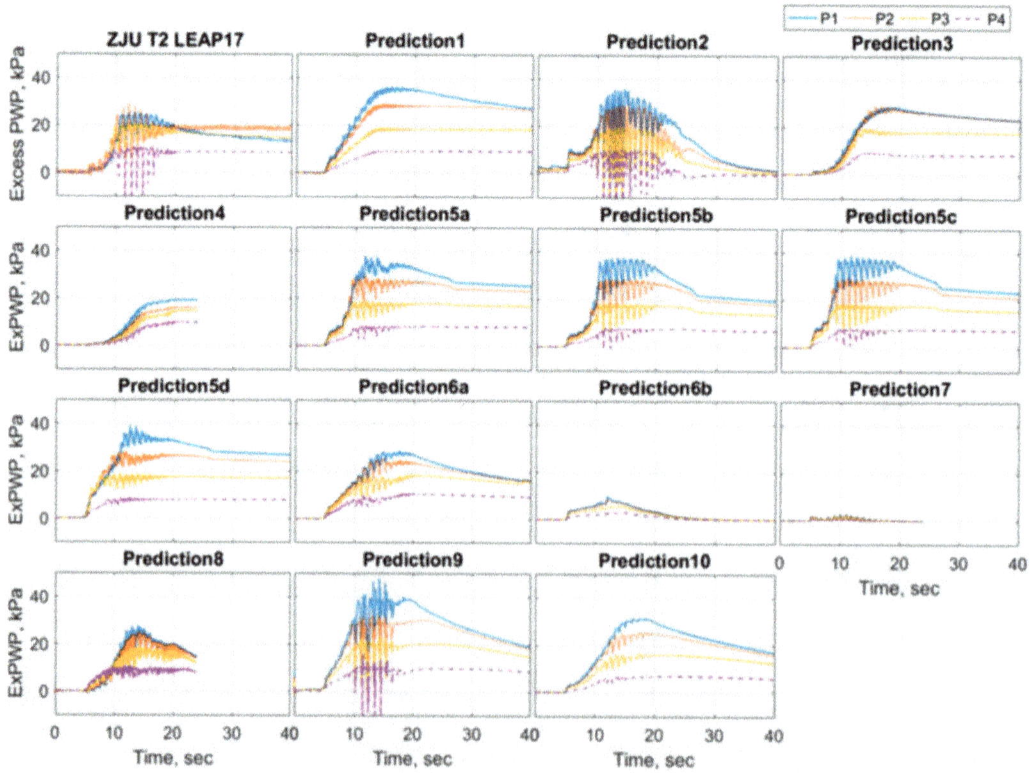


**Fig. 10.7** Comparison of the predicted time histories of excess pore water pressures computed at the central section of the soil specimen for ZJU-2 test

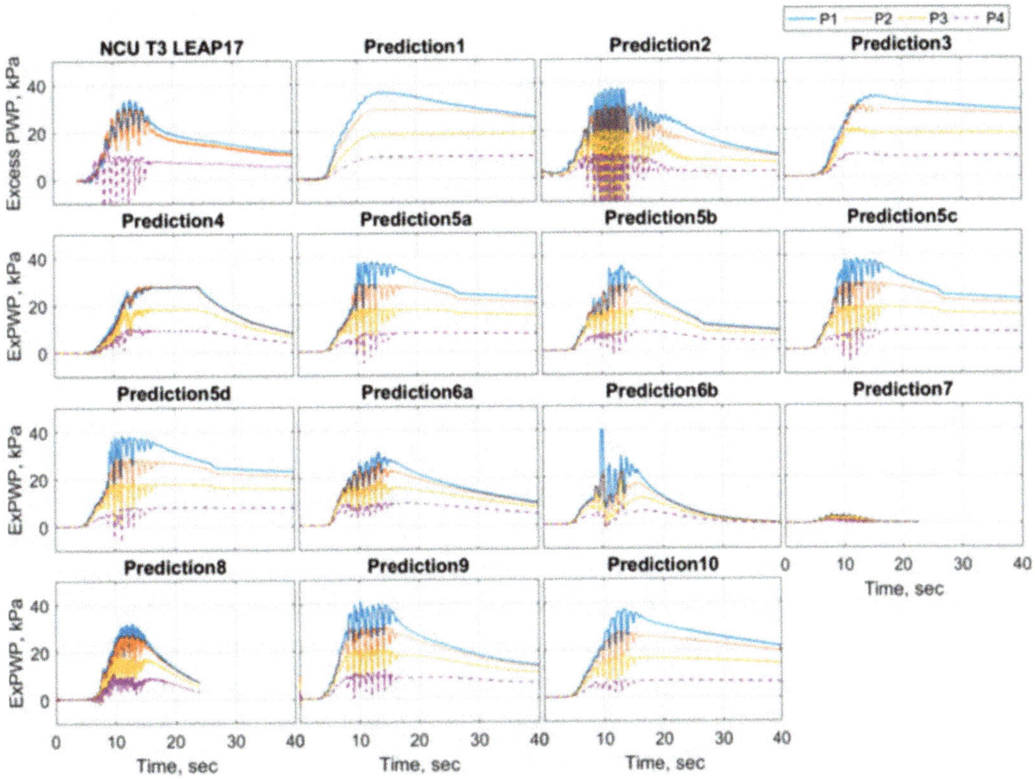


**Fig. 10.8** Comparison of the predicted time histories of excess pore water pressures computed at the central section of the soil specimen for NCU-3 (bottom) test

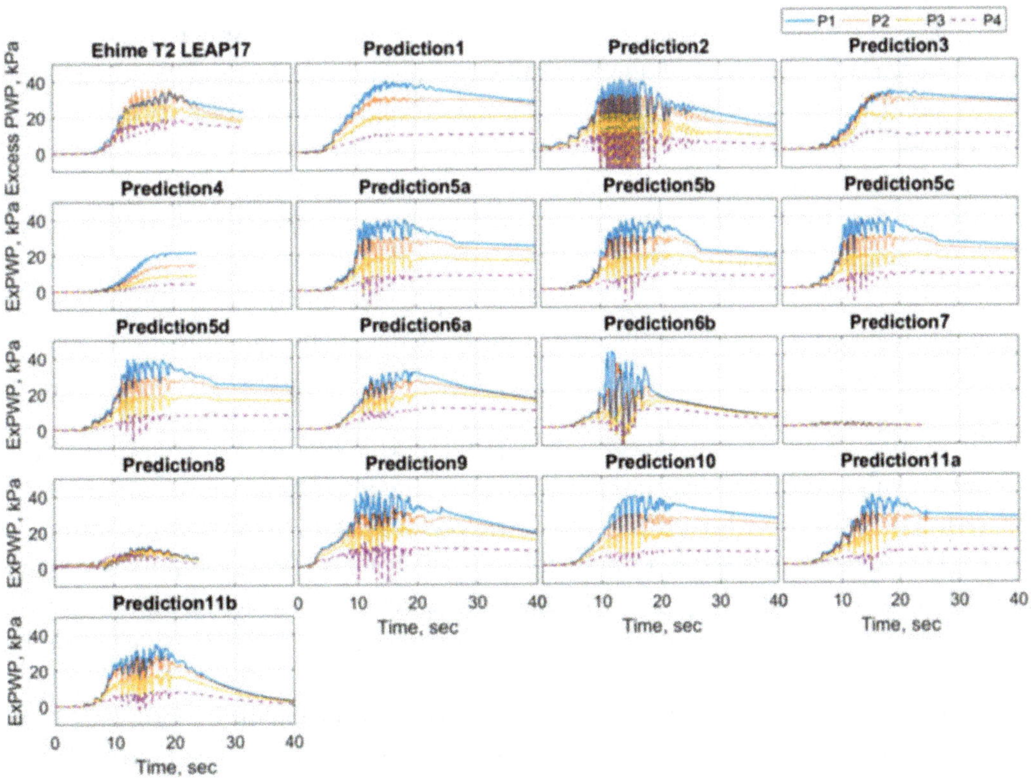


**Fig. 10.9** Comparison of the predicted time histories of excess pore water pressures computed at the central section of the soil specimen for Ehime-2 test

of a few selected centrifuge tests conducted at UC Davis (UCD T3, Carey et al. 2019), Kyoto University (KyU-3, Vargas Tapia et al. 2019), KAIST (KAIST-2, Kim et al. 2019), Cambridge University (CU-2, Madabhushi et al. 2019), Zhejiang University (ZJU-2, Liu et al. 2019), National Central University of Taiwan (NCU-3, Hung and Liao 2019), and Ehime (Ehime-2, Okamura and Nurani Sjafruddin 2019) as part of LEAP-2017 project (Kutter et al. 2019). Qualitative similarities with the experimental curves are noted for the majority of the predictions, except for Prediction 7 in which a calibration error, later noticed by the predictors, produced significantly underpredicted excess pore pressures.

A close examination of the pore pressures time histories illustrated in Figs. 10.3, 10.4, 10.5, 10.6, 10.7, 10.8, and 10.9 shows the following trends:

- Except for the simulation of the CU-2 test, Prediction 1 shows a larger maximum excess pore water pressure compared to the experimentally observed values with a relatively faster buildup and slower dissipation.
- Compared to the experimental results, Prediction 2 shows more dilative responses with larger oscillations when the excess pore pressures reach their maximum values. Moreover in a few cases (e.g., KyU-3), the maximum excess pore water pressures are overpredicted. The predicted pore water pressure time histories show faster dissipation than the experimentally observed time histories.
- In most cases, Prediction 3 shows trends that are reasonably close to the observed experimental responses, but in a few cases, the maximum excess pore pressures are slightly underpredicted.
- Prediction 4 only shows the computed responses up to the end of shaking and do not include the dissipation phase, except for the case of NCU-3. In few cases (e.g., UCD-3 test), the maximum excess pore water pressures are underpredicted.
- Predictions 5a, 5c, and 5d show similar buildup and dissipation trends and are different from Prediction 5b in the dissipation phase of the pore pressure time histories. The maximum pore pressures in these predictions are in reasonable agreement with the experimentally observed values in the majority of cases except for two cases where they are visibly underpredicted (UCD-3) or overpredicted (KyU-3) in Prediction 5b.
- Prediction 6a shows trends comparable to experimentally observed pore pressure time histories. However, the maximum excess pore pressures are visibly underpredicted in the case of UCD-3 and they are overpredicted in the cases of KyU-3.
- Prediction 6b shows some large spikes in the pore pressure time histories. These spikes might be related to numerical implementation of the constitutive model rather than the model performance. These predictions underpredict the observed excess pore water pressure time histories in the majority of the simulated tests.
- Prediction 7 significantly underpredicts the excess pore pressure time histories. This is likely due to a systematic error in application of the method used in the simulations.
- Prediction 8 shows a faster dissipation than that observed in the experiments in all the predicted time histories. The maximum excess pore pressures are reasonably

well predicted in a few cases and are underpredicted in the case of KyU-3, CU-2, and Ehime-2.

- Prediction 9 shows trends that are comparable to experimentally observed pore pressure time histories. However, the maximum excess pore pressures are overpredicted in the cases of KyU-3 and ZJU-2.
- Prediction 10 shows comparable trends with the experimental data. In most cases, a slower dissipation phase is predicted. The maximum excess pore pressures are overpredicted in the cases of KyU-3.
- Prediction 11a is in good agreement with the measured responses in the case of Ehime-2 and underpredicts the maximum excess pore pressures in the case of UCD-3. Predictions were not submitted for the other experiments.
- Prediction 11b shows comparable trends with the experimental data obtained in the case of UCD-3 and Ehime-2. Slightly faster dissipation rates are predicted compared to the experimentally observed rates. Predictions were not submitted for the other experiments.

### 10.4.2 Acceleration Time Histories and Spectral Accelerations

Figures 10.10, 10.11, 10.12, 10.13, 10.14, 10.15, and 10.16 show examples of acceleration time histories predicted by the different simulation teams and compare them with the time histories of accelerations obtained from the centrifuge test results. The following observations are noted:

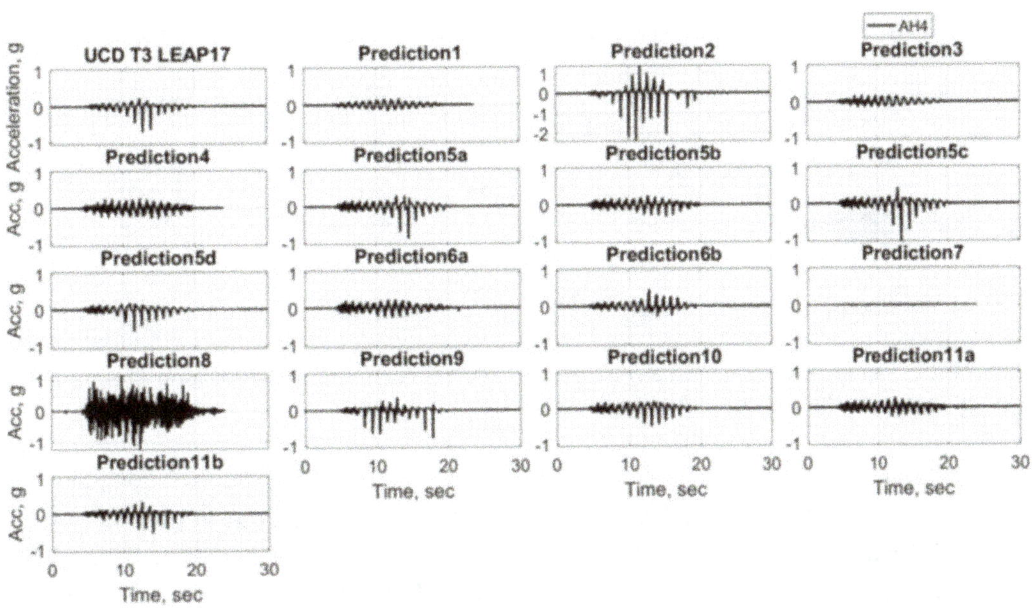


**Fig. 10.10** Comparison of the acceleration time histories computed at the location of AH4 in the central section of the soil specimen for UCD-3 test

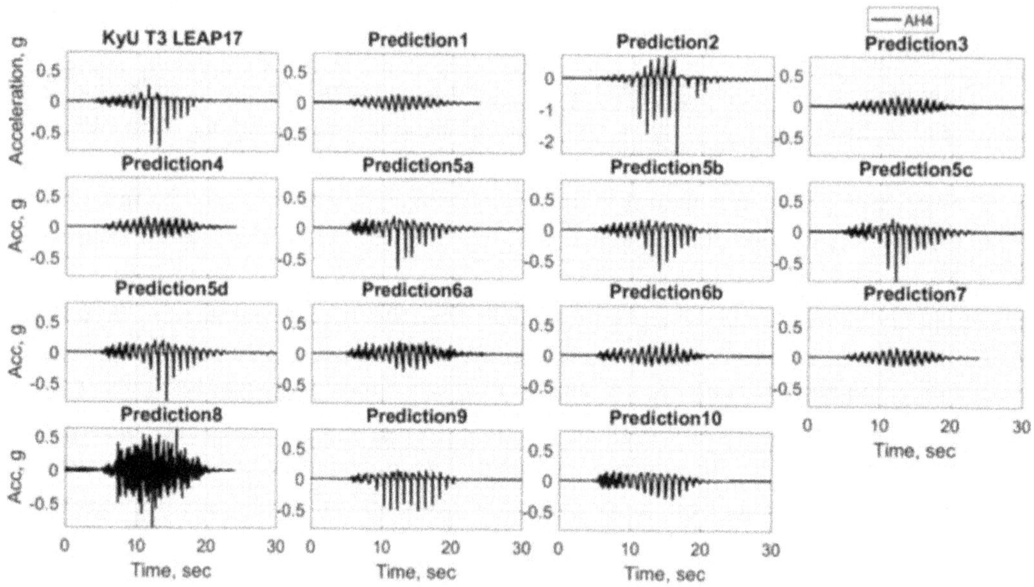


Fig. 10.11 Comparison of the acceleration time histories computed at the location of AH4 in the central section of the soil specimen for KyU-3 test

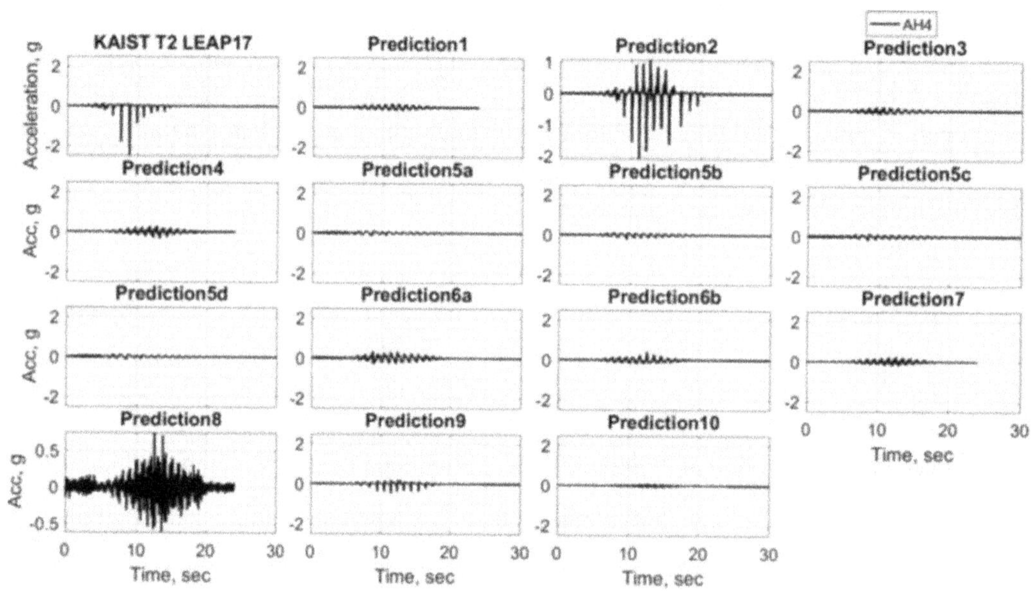


Fig. 10.12 Comparison of the acceleration time histories computed at the location of AH4 in the central section of the soil specimen for KAIST-2 test

- The experimental data of UCD-3 test show that the acceleration time history at AH4 (near the ground surface) is marked by dilation spikes (Fig. 10.10). This feature has been captured in a few predictions (Predictions 2, 5a, 5c, 5d, 6b, 9, 10, 11b) with varying degrees of success.

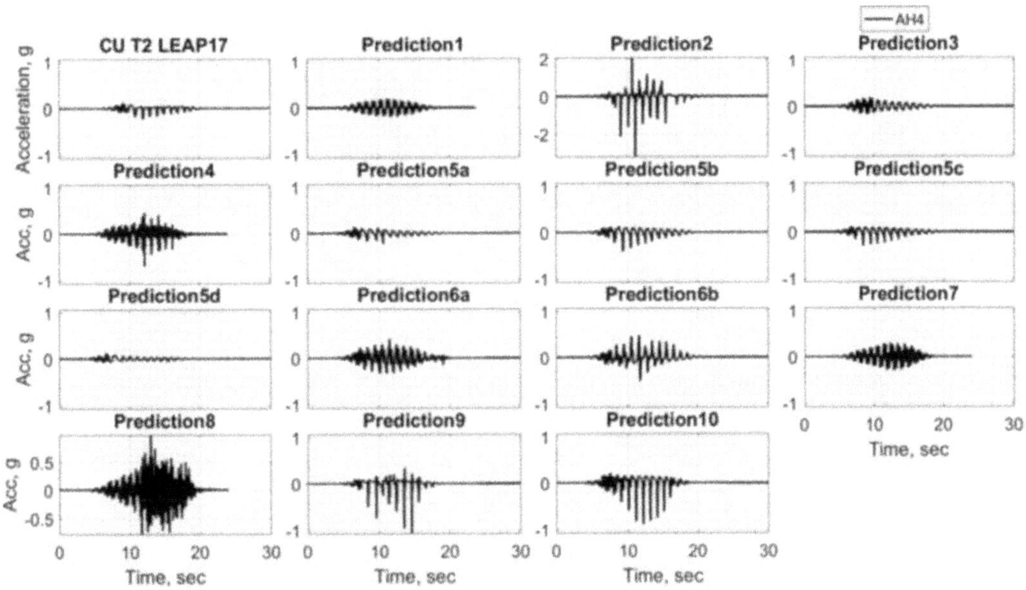


**Fig. 10.13** Comparison of the acceleration time histories computed at the location of AH4 in the central section of the soil specimen for CU-2 test

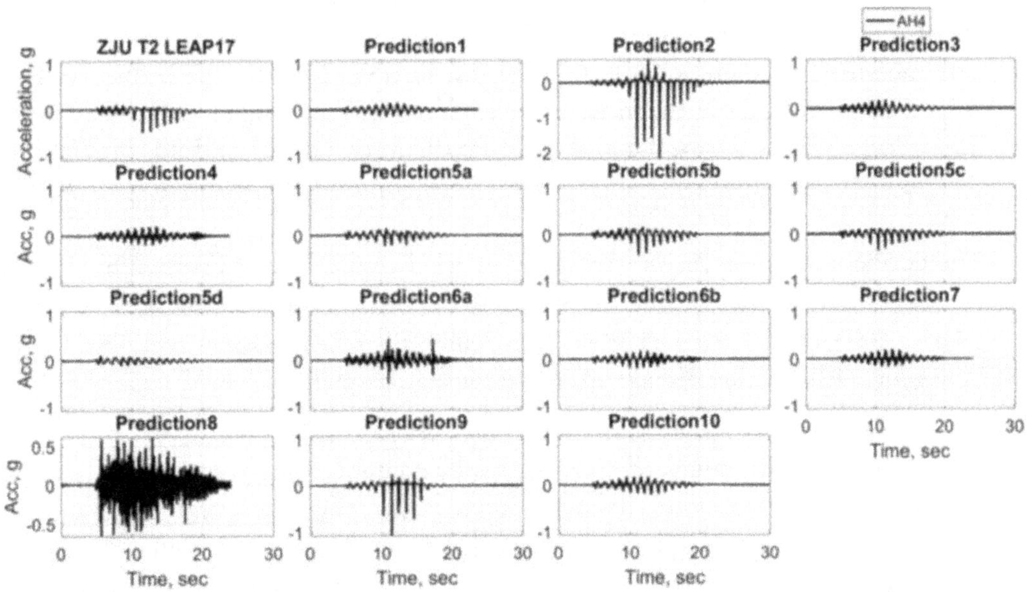


**Fig. 10.14** Comparison of the acceleration time histories computed at the location of AH4 in the central section of the soil specimen for ZJU-2 test

- Similar dilation spikes are also present in the AH4 acceleration time histories reported in KyU-3 test (Fig. 10.11). Again, a few predictions show similar dilation spikes (Predictions 2, 5a, 5b, 5c, 5d, 9, 10). Smaller dilation spikes appear in Predictions 6a and 6b as well.

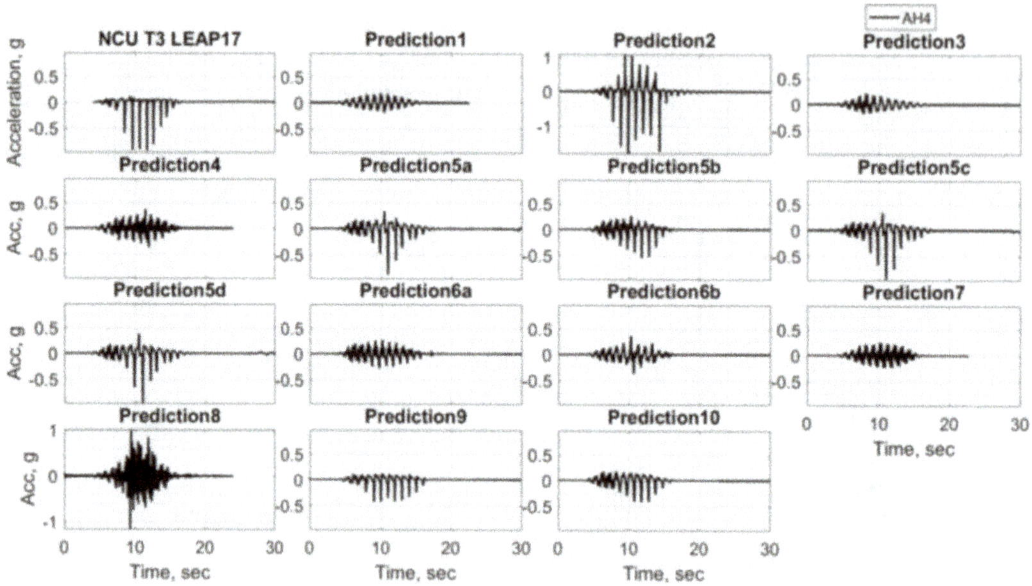


**Fig. 10.15** Comparison of the acceleration time histories computed at the location of AH4 in the central section of the soil specimen for NCU-3

- The dilation spikes observed in the CU-2 and ZJU-2 tests (Figs. 10.13 and 10.14) are also present in Predictions 2, 5a, 5b, 5c, 6a, 6b, 7, 9, and 10.
- The measured accelerations at AH4 in the KAIST-2 (Fig. 10.12) and NCU-3 (Fig. 10.15) tests show much stronger dilation spikes compared to other tests. These large spikes are well captured in Predictions 5 a, 5c, and 5d. Predictions 2, 9, and 10 also show large spikes in the acceleration time histories computed at AH4.
- The acceleration time history recorded at AH4 in the Ehime-2 test (Fig. 10.16) shows somewhat smaller dilation spikes (compared to those in the KAIST-2 and NCU-3 tests) in the positive direction. Again a few predictions (5a, 5b, 5c, 5d, 9, 10, 11a, 11b) show dilation spikes of similar magnitude.

The spectral accelerations shown in Figs. 10.17, 10.18, 10.19, 10.20, 10.21, 10.22, and 10.23 show details of high-frequency contents in each of the acceleration time histories discussed above. In these figures, Sa1, Sa2, Sa3, and Sa4 stand for spectral accelerations corresponding to accelerations A1, A2, A3, and A4, respectively. It is interesting to note that a number of predictions are able to capture the 1 Hz component of the accelerations at different depths quite well. Note also, that a few predictions have higher-frequency contents similar to the observed accelerations and a few other have damped out the high-frequency harmonics. These aspects of the numerical simulations are attributed less to the performance of the constitutive model and more to the numerical damping and other damping mechanisms (e.g., Rayleigh damping) introduced in the numerical simulations.

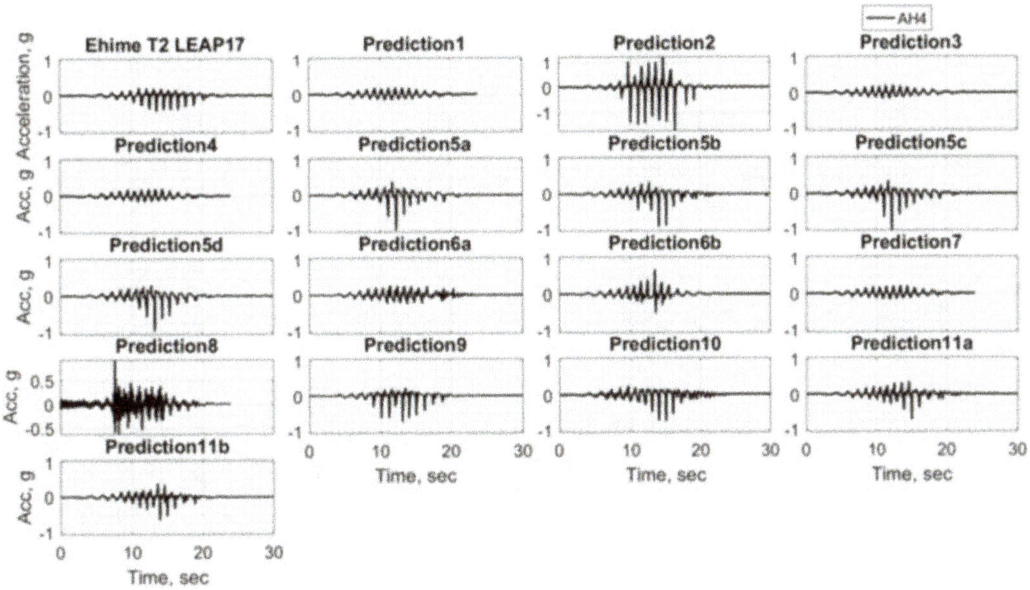


**Fig. 10.16** Comparison of the acceleration time histories computed at the location of AH4 in the central section of the soil specimen for Ehime-2 test

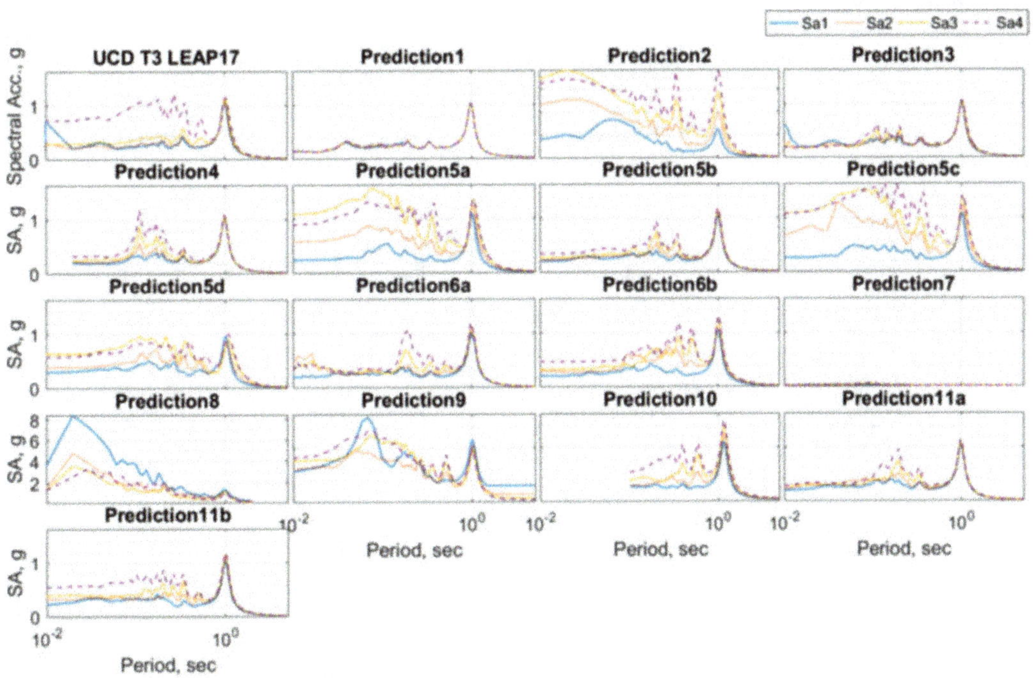


**Fig. 10.17** Comparison of the spectral acceleration computed at the central section of the soil specimen for UCD-3 test

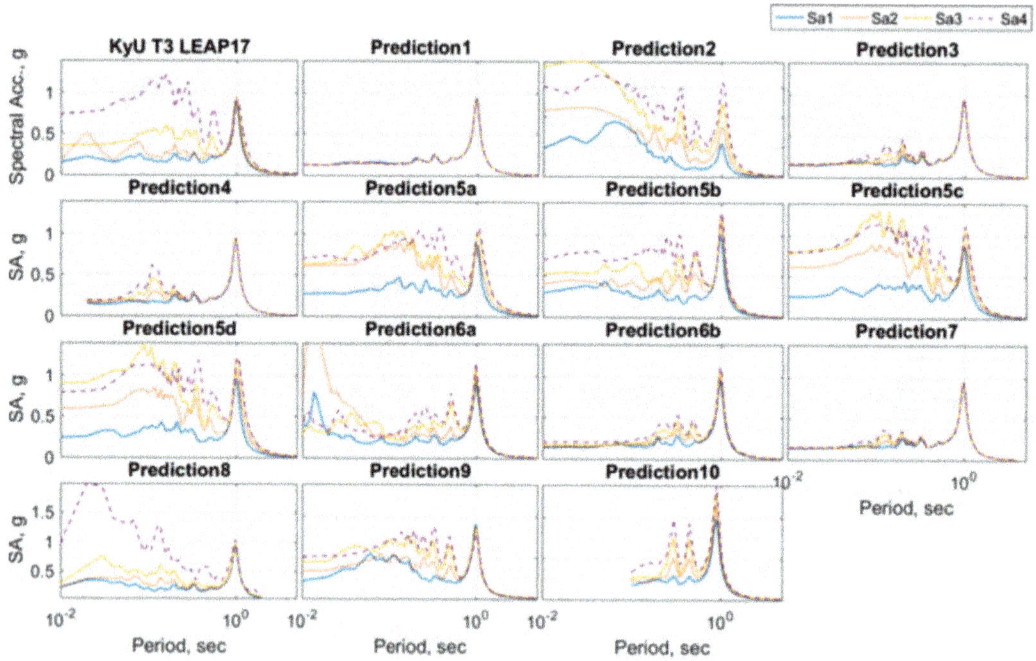


**Fig. 10.18** Comparison of the spectral acceleration computed at the central section of the soil specimen for KyU-3 test

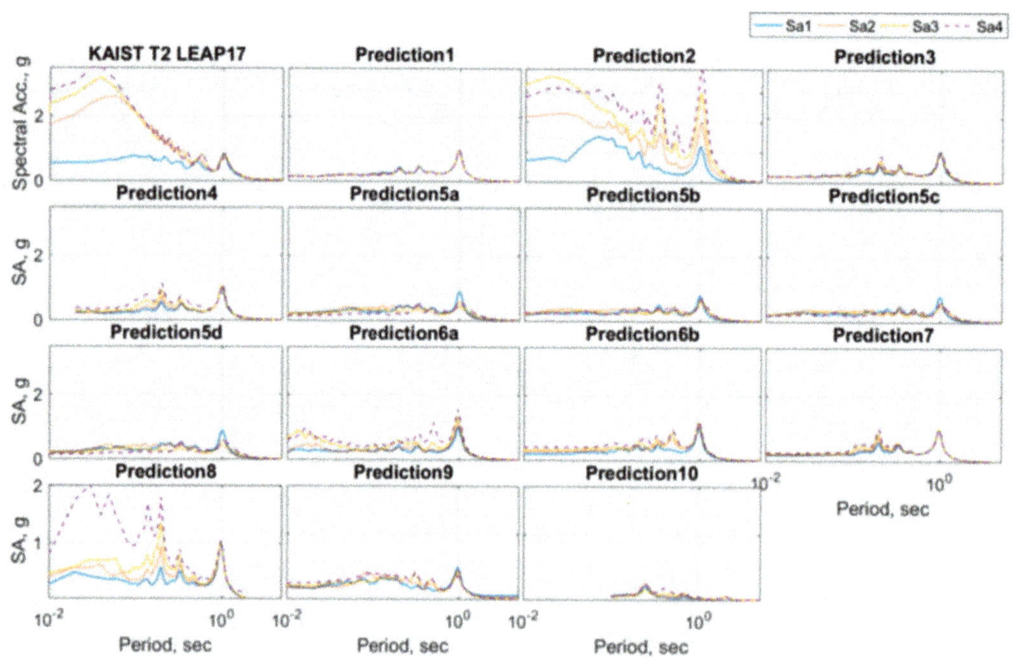


**Fig. 10.19** Comparison of the spectral acceleration computed at the central section of the soil specimen for KAIST-2 test

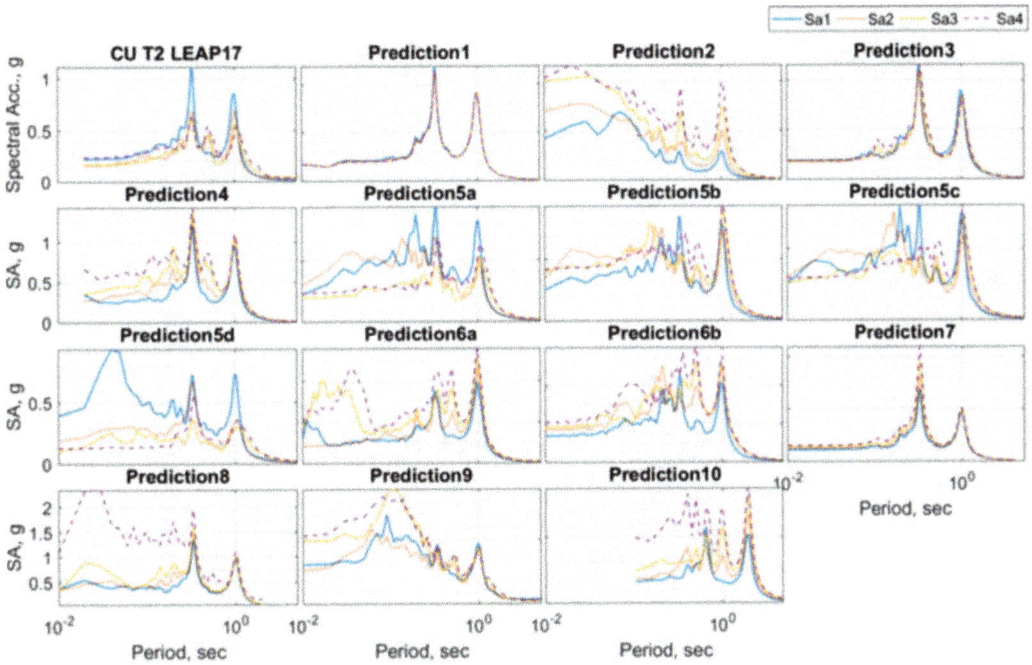


**Fig. 10.20** Comparison of the spectral acceleration computed at the central section of the soil specimen for CU-2 test

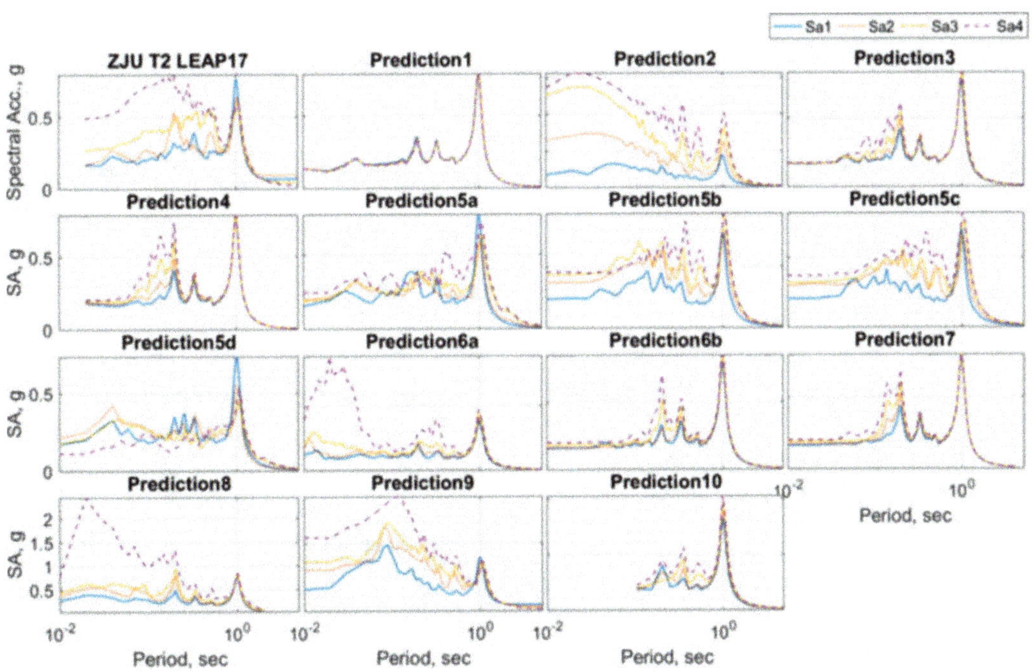


**Fig. 10.21** Comparison of the spectral acceleration computed at the central section of the soil specimen for ZJU-2 test

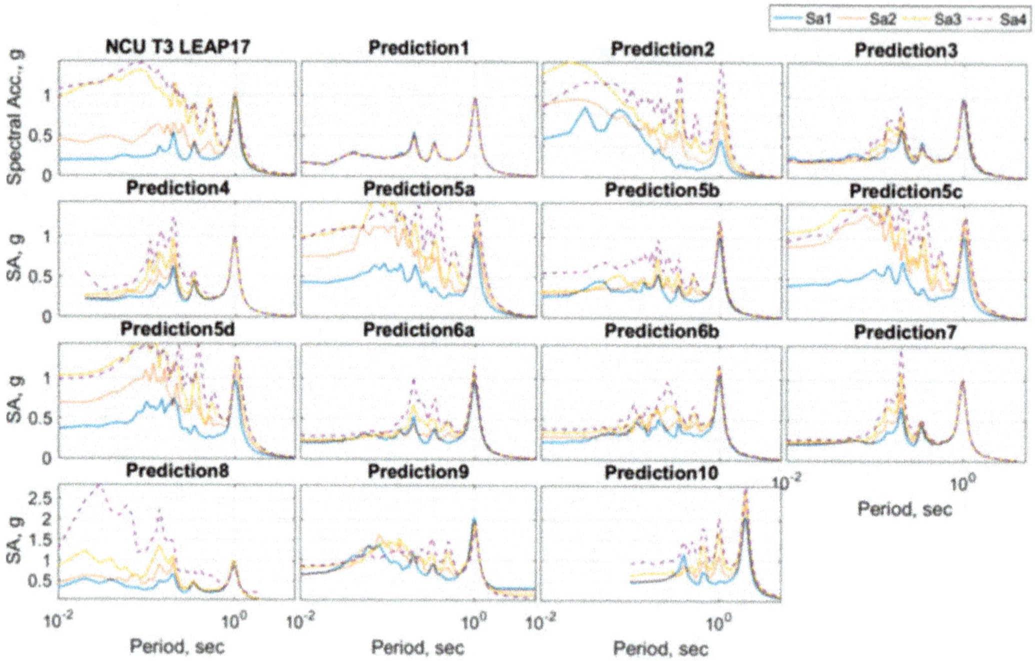


**Fig. 10.22** Comparison of the spectral acceleration computed at the central section of the soil specimen for NCU-3 test

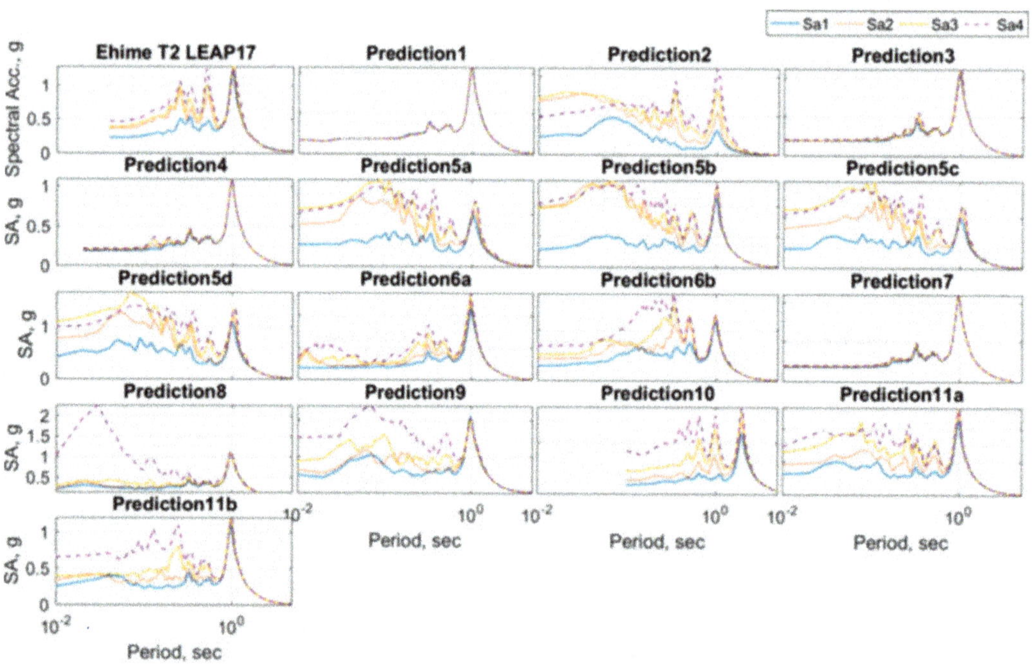


**Fig. 10.23** Comparison of the spectral acceleration computed at the central section of the soil specimen for Ehime-2 test

### 10.4.3 Lateral Displacements

The lateral displacements at the surface of the centrifuge specimens were measured by tracking the locations of surface markers using the photos taken by high-speed cameras as described in Kutter et al. (2019). In two (UCD-1 and UCD-3) out of the nine centrifuge tests selected for Type-B numerical simulations, the data obtained from high-speed cameras were used to construct a time history of surface displacement in the middle portion of the slope. In five other tests (CU-2, Ehime 2, KAIST-2, NCU-3, ZJU-3), the final locations of the surface markers measured after the end of the test and time histories of dynamic displacements computed from measured acceleration time histories were used to reconstruct surface displacement time histories (Kutter et al. 2019). In the case of the KAIST-1 and KyU-3 tests, surface marker data was not available or sufficiently consistent to reconstruct a time history of the surface displacements. In these cases, only dynamic components of the surface displacements were computed by double integration of the difference between the accelerations recorded at AH1 and AH4 (Kutter et al. 2019). Hence, it is important to note that in the following comparisons of the simulated time histories with the results of centrifuge tests except for UCD-1 and UCD-3 tests, the rest of the "experimental" results are computed surface displacement time histories obtained from the final locations of surface markers and double integration of the measured acceleration time histories.

Based on the results illustrated in Figs. 10.24, 10.25, 10.26, 10.27, 10.28, 10.29, 10.30, and 10.31, the following observations can be made:

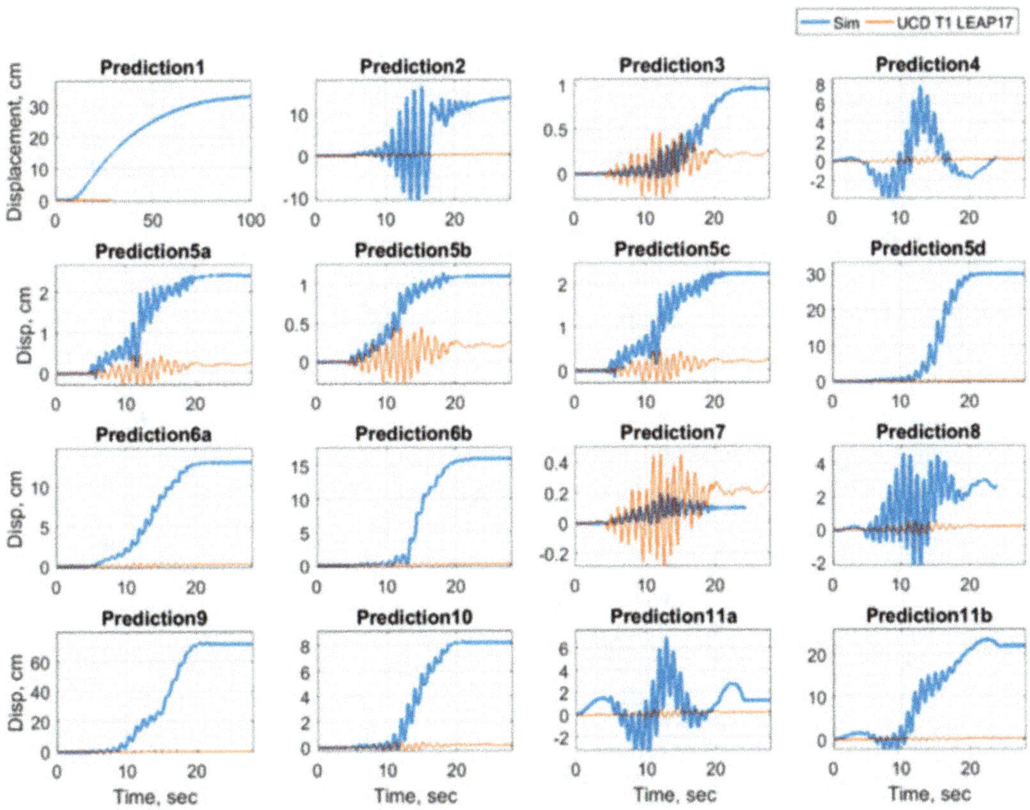


**Fig. 10.24** Comparisons of the numerical simulations of lateral displacement with the measured time history of surface displacement in the UCD-1 test

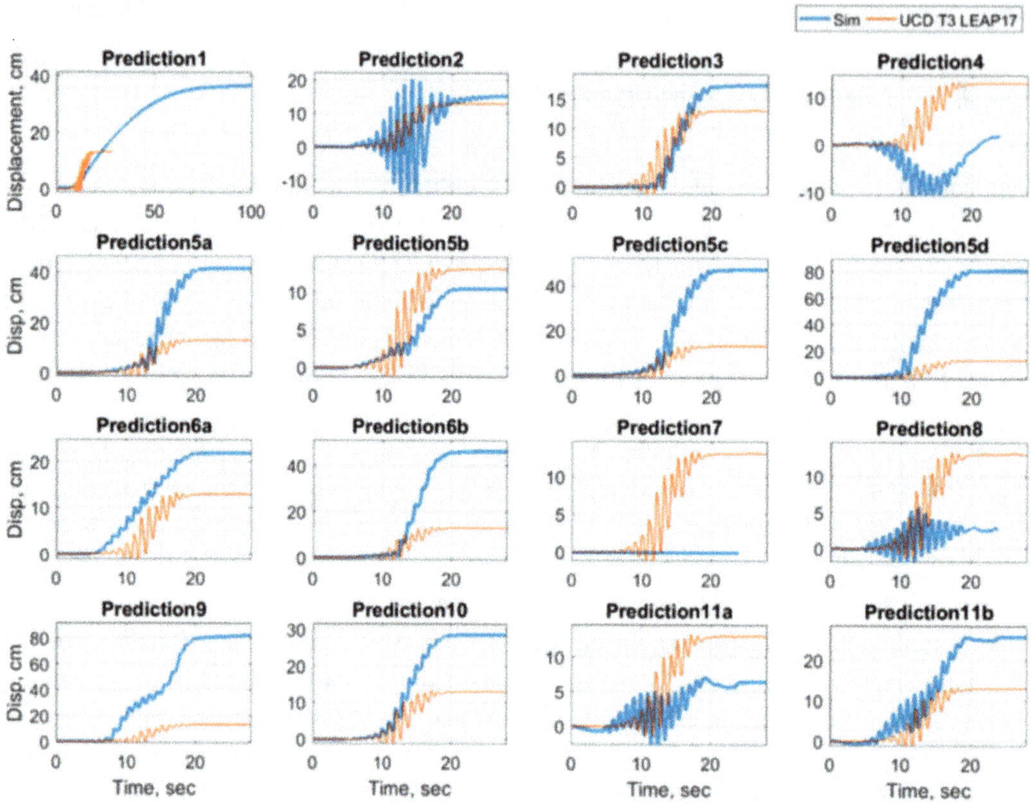


**Fig. 10.25** Comparisons of the numerical simulations of lateral displacement with the measured time history of surface displacement in the UCD-3 test

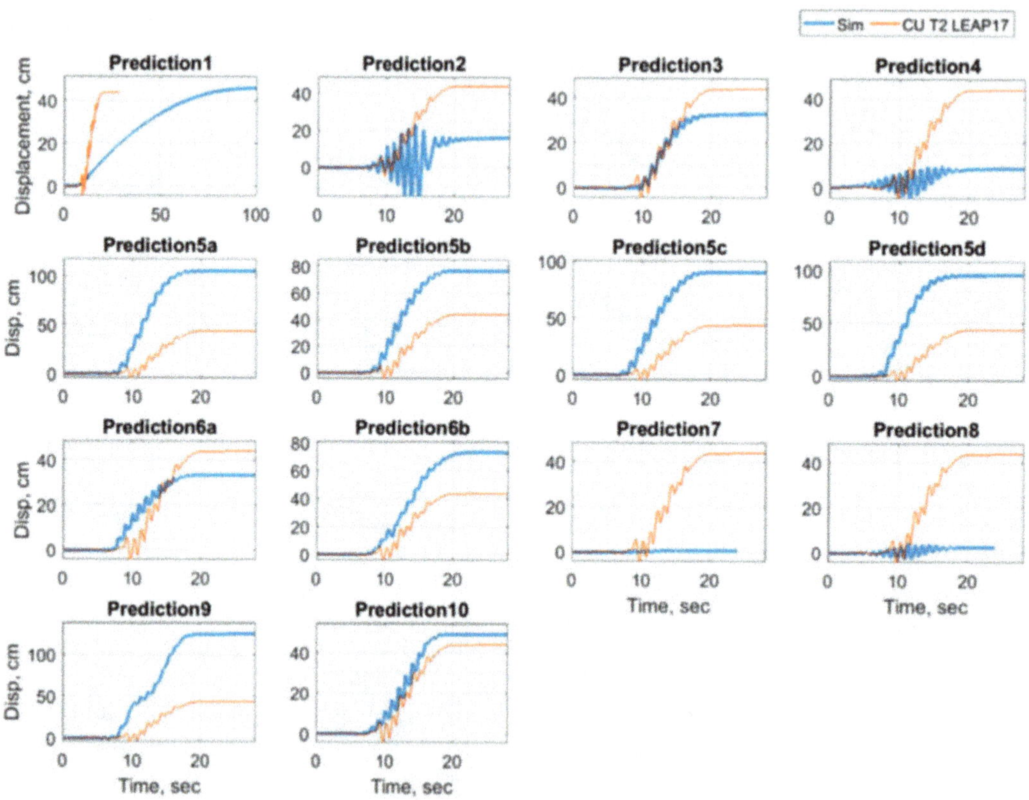


**Fig. 10.26** Comparisons of the numerical simulation of lateral displacement with the constructed time history of the surface lateral displacement in the CU-2 test

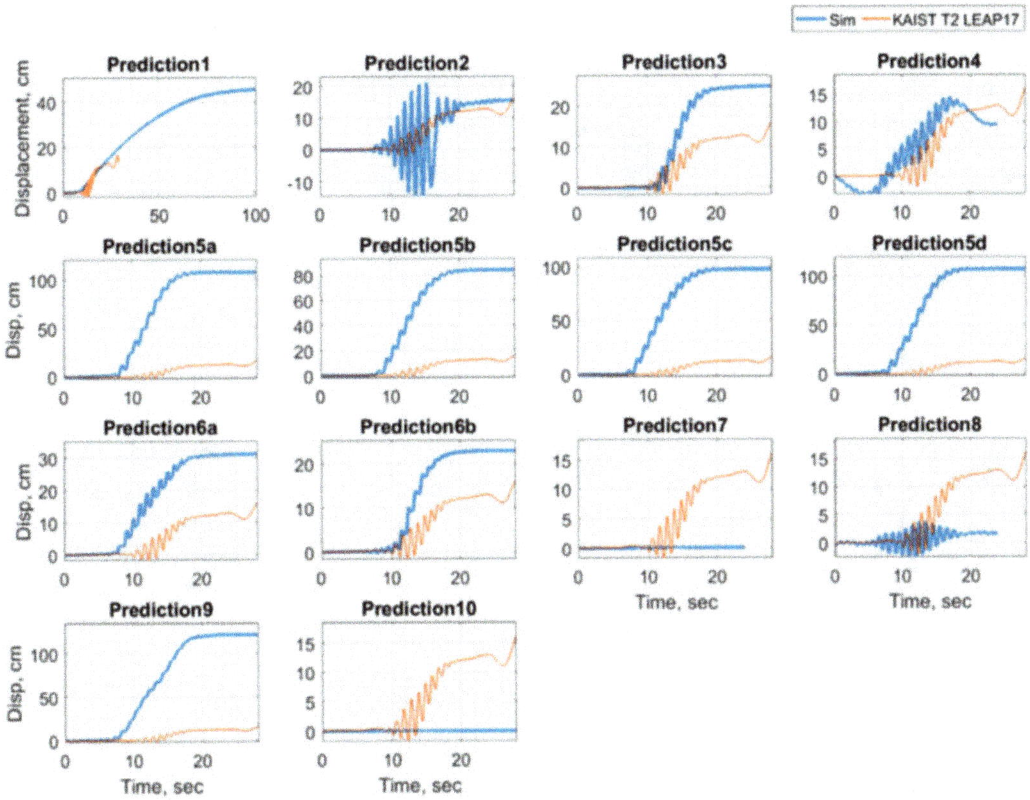


**Fig. 10.28** Comparisons of the numerical simulation of lateral displacement with the constructed time history of the surface lateral displacement in the KAIST-2 test

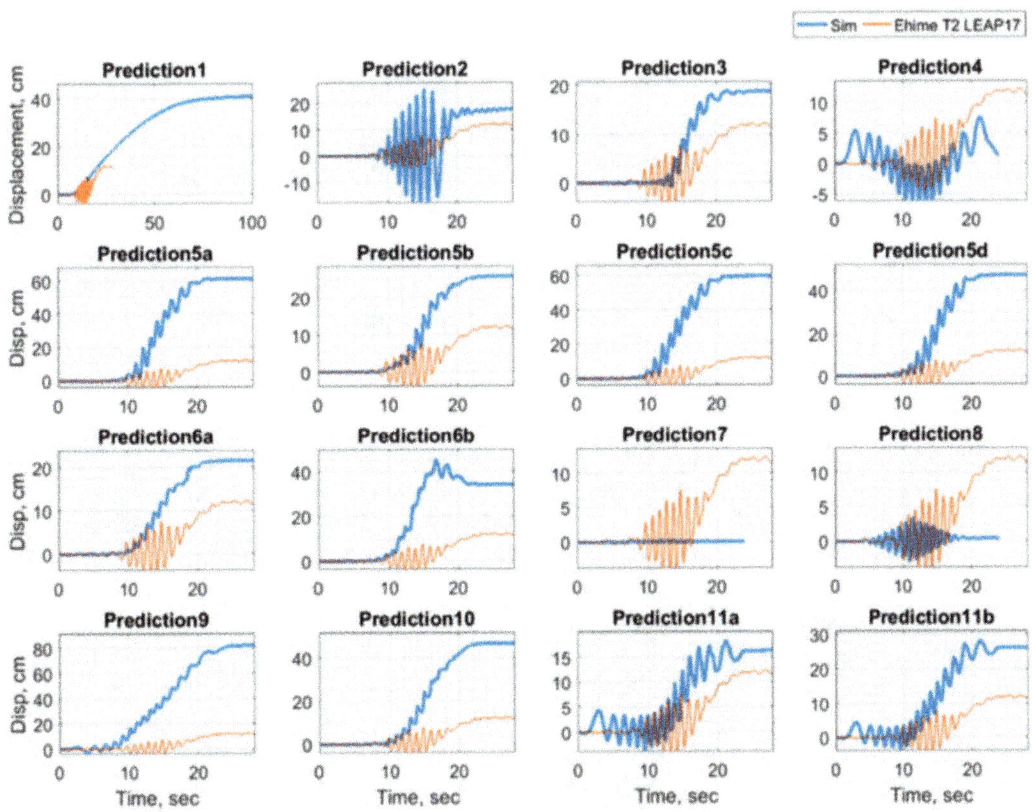


**Fig. 10.27** Comparisons of the numerical simulation of lateral displacement with the constructed time history of the surface lateral displacement in the Ehime-2 tests

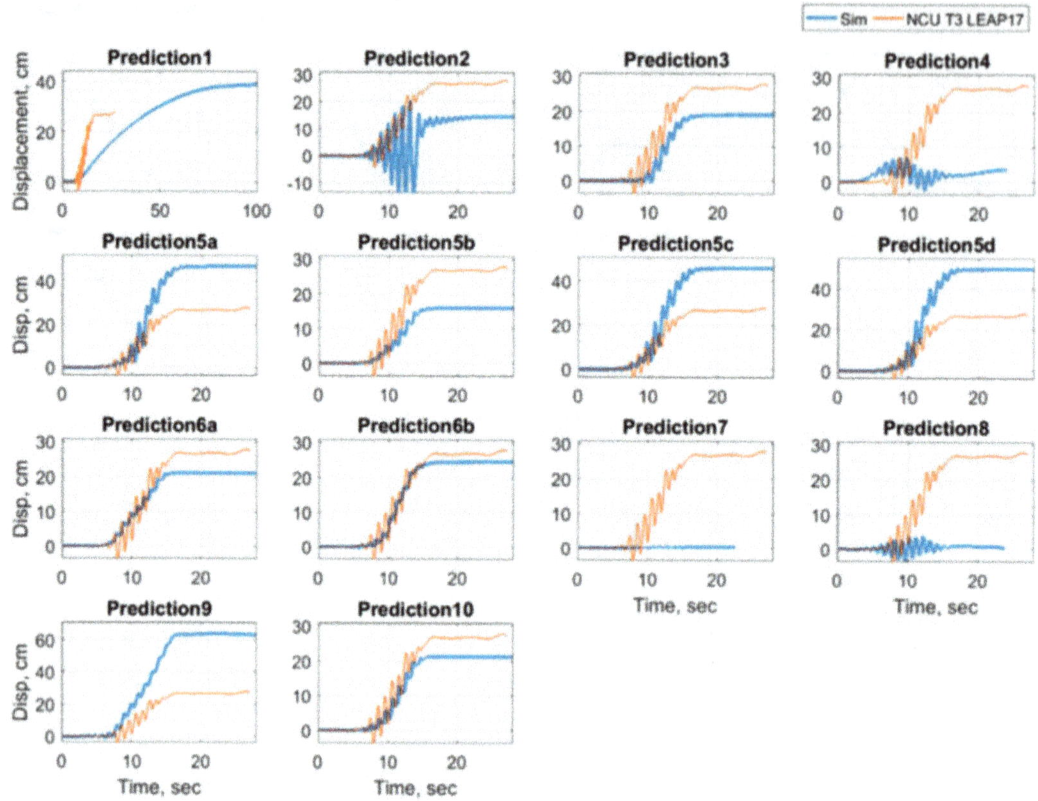


**Fig. 10.29** Comparisons of the numerical simulation of lateral displacement with the constructed time history of the surface lateral displacement in NCU-3 test

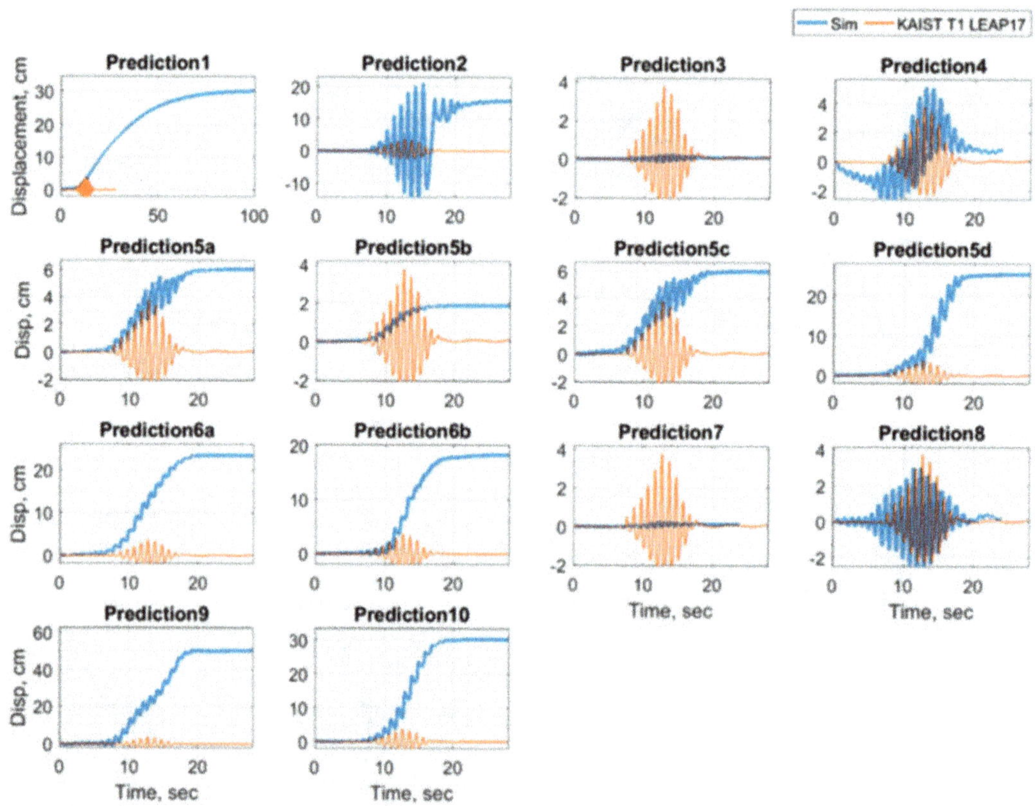


**Fig. 10.30** Comparisons of the numerical simulations of lateral displacement with the time history of dynamic components of lateral displacement in the KAIST-1 test

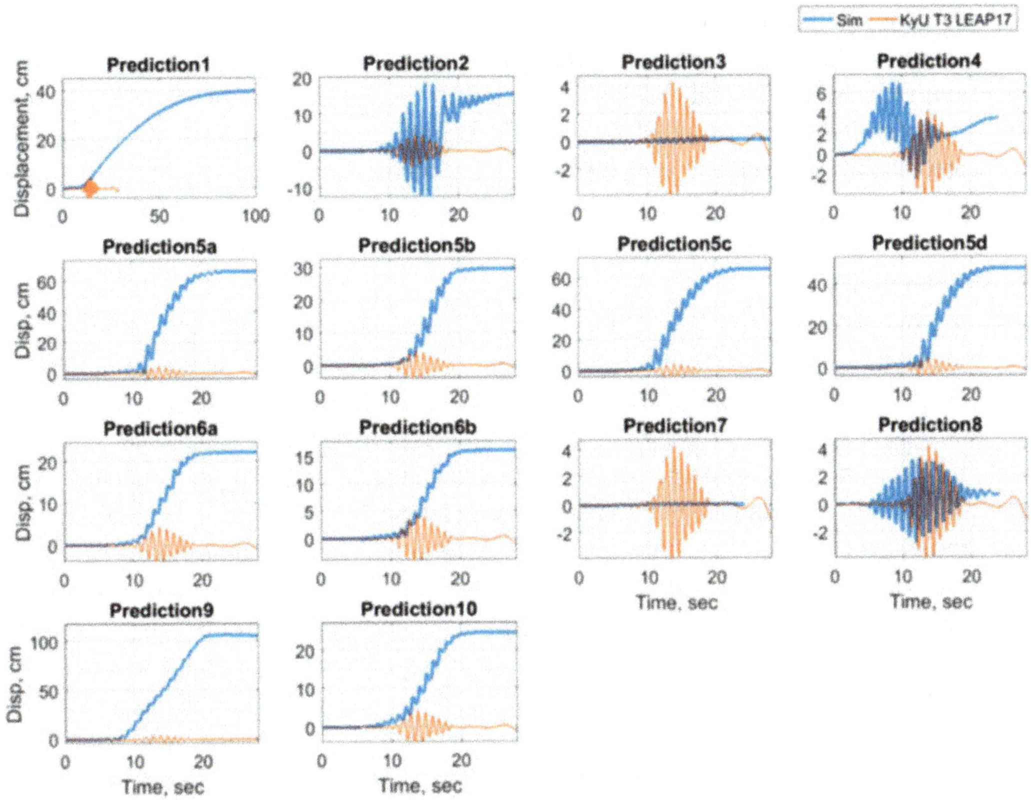


**Fig. 10.31** Comparisons of the numerical simulations of lateral displacement with the time history of dynamic components of lateral displacement in the KyU-3 tests

- The majority of Type-B simulations predicted much larger displacements than those observed in UCD-1 test. Liquefaction was not triggered in the experiment, but apparently it was triggered in many of the simulations. The final displacements (but not the cyclic components of displacement) predicted by Predictions 4, 7, and 11a are comparable with the very small displacement reported for the UCD-1 test.
- A number of predictions (i.e., 2, 3, 5b, 6a) showed maximum lateral displacements that favorably compare with the observed responses in the UCD-3 test. A large number of other simulations also showed comparable trends but overpredicted the lateral displacements. It is also noted that a number of simulations were able to predict the magnitude of the cyclic component of the lateral displacements quite well (4, 5a, 5c, 8, 11b). A few other predictions (1, 3, 6a, 10) underpredicted the cyclic component of the lateral displacement and showed a more or less smooth rise to the final displacement. This particular aspect of the simulation results is likely related to the cyclic stress-strain responses produced by the different constitutive models. Few models show small opening of the stress-strain loops and exhibit significant ratcheting, while a few others display increasingly larger stress-strain loops without much ratcheting (see Manzari et al. 2019).
- An interesting trend is observed in the simulations of the CU-2 test which produced the largest observed lateral displacement consistent with the lowest

value of cone penetration resistance measured in this test. Prediction 10 was closest to the "experimental" results. A number of numerical simulations overpredicted the lateral displacement in this test (5a, 5b, 5c, 6b, 9) while others significantly underpredicted (1, 2, 4, 7, 8) the lateral displacement. It must be noted, however, that the constitutive models used in the numerical simulations were calibrated for densities higher than 65% and the relatively low density achieved in CU-2 was outside of the range of densities the models were calibrated for.

- Significant underpredictions are observed in all the displacement simulations provided by prediction team 7. Based on the comments provided by the prediction team 7 in the LEAP-2017 workshop at UC Davis, these numerical simulations were based on a set of model parameters that were erroneously selected. This error has been corrected later and the corrected results are presented and discussed in the paper by Wada and Ueda (2019).
- Prediction 1 shows very little cyclic displacements (overly smooth curves) and a very long delayed rise of lateral displacements after the end of base motion. Interestingly, the ultimate displacement was approximately 40 cm for every simulation. This is not consistent with the observed experimental trends and the simulations provided by other predictors. The reason for this discrepancy has been discussed in Wang et al. (2019).
- Overall, the maximum lateral displacements predicted by prediction teams 2 and 3 appeared to be closer to the "measured" displacements. However, Prediction 2 displayed much larger cyclic displacements than those observed in the experiments and Prediction 3 tended to underpredict the cyclic components of displacement.

## 10.5 Overall Performance of Numerical Simulations

To further assess the overall performance of the predictions and the quality of their fit to the centrifuge test results, the following indicators are selected: (1) the maximum lateral displacement at the center of the soil surface, (2) the maximum excess pore water pressure ratio achieved at the depth of 1 m (P4), and (3) a scalar representing a measure of spectral acceleration (MSA). The root mean square error (RMSE) of these indicators was computed for each numerical team based on the nine centrifuge experiments as:

$$\mathrm{RMSE} = \sqrt{\frac{1}{N}\sum_{N}\left(P^{e} - P^{s}\right)^{2}} \tag{10.1}$$

where $P^e$ and $P^s$ are the values of an indicator for the experiments and simulation, respectively, and $N$ is the number of experiments which is equal to 9 for all the predictions except for the predictions 11a and 11b which were reported for only three centrifuge experiments.

The following measures are used to represent the spectral accelerations:

$$MSa_1 = \frac{(Sa)_{1\text{Hz}} + (Sa)_{3\text{Hz}} + (Sa)_{20\,\text{Hz}}}{3}; \quad MSa_2 = \int_{0.5}^{20} S_a\, df \tag{10.2}$$

$MSa_1$ is the average of spectral values at 1 Hz, 3 Hz, and 20 Hz that represents the peak of each acceleration time history, while $MSa_2$ is essentially the area under the spectral acceleration graph ($Sa$ versus $f$).

Figures 10.32 and 10.33 summarize the results of the computed RMSEs for the lateral displacement on the ground surface ($dx$), the maximum excess pore pressure ratio at the location of P4 ($r_u$-P4), and the spectral acceleration at the location of AH4 (MSa-AH4). Figure 10.32 presents the RMSEs computed based on $MSa_1$ and Figure 10.33 presents the RMSE computed based on $MSa_2$. These figures present a sample of the overall fit of the predictions to the experimental data. The RMSEs of $MSa_1$ and $MSa_2$ for Prediction 2 exceed the limits of the plot. This is mainly due to the presence of large high-frequency components in the acceleration time history as shown in Figs. 10.10, 10.11, 10.12, 10.13, 10.14, 10.15, and 10.16.

It is observed that Predictions 2, 3, 6a, and 11a show reasonably small RMSEs for lateral displacements, while Predictions 1, 2, 3, 5a, 5d, 6a, 8, and 9 show relatively small RMSEs for excess pore pressure ratios, $r_u$, at $P_4$. The spectral accelerations predicted by team 11 (11b and 11a) at AH4 show the lowest RMSEs.

The RMSEs are recomputed for the three centrifuge tests that were predicted by numerical simulation team 11 in order to assess the effect of number of centrifuge tests used. Figure 10.34 shows the results of these recalculated RMSEs for lateral displacements, pore pressure ratios at P4, and spectral accelerations ($MSa_2$) at AH4. While there are some changes (e.g., significant reduction in RMSE of lateral displacements predicted by teams 1 and 5b) in the relative performance of individual predictions, the overall trend observed in Fig. 10.33 remains the same. Hence, it was decided that it is not unreasonable to include the predictions made by team 11 in the comparisons with other predictions. In the following figures, the approach used in Fig. 10.33 will be used (i.e., $N = 9$ for all predictions, except for team 11 where $N = 3$).

Figures 10.35, 10.36, and 10.37 show similar comparisons as those shown in Figs. 10.32 and 10.33 for excess pore water pressure at locations of P3, P2, and P1 and for spectral accelerations AH3, AH2, and AH1, respectively. It is interesting to note that differences between the results of experiments and numerical simulations steadily increase as the point of interest moves from the bottom of the soil deposit to the ground surface.

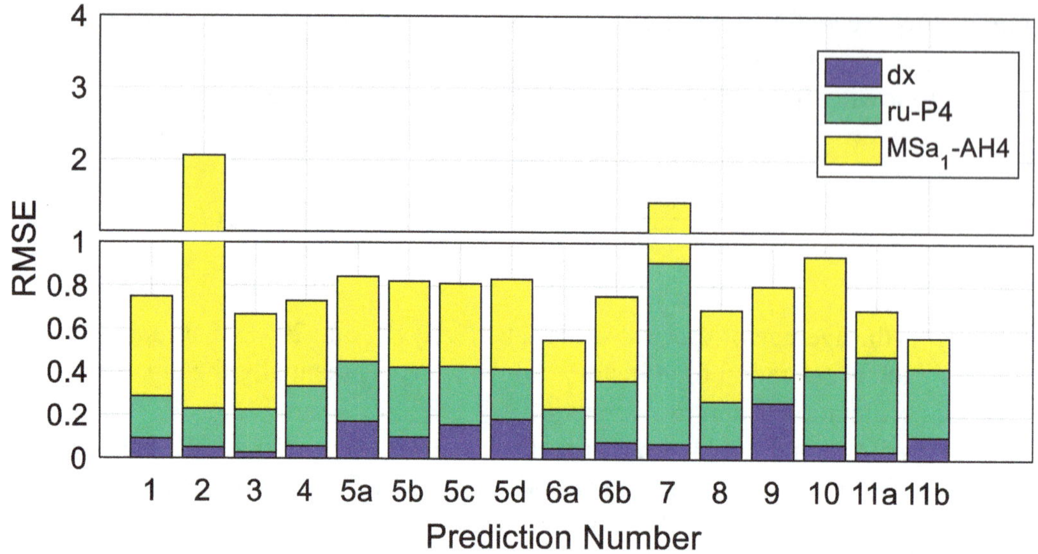


**Fig. 10.32** RMSE values for the Type-B simulations compared to the observed lateral displacements, pore pressure ratios at P4, and spectral accelerations ($MSa_1$) at AH4

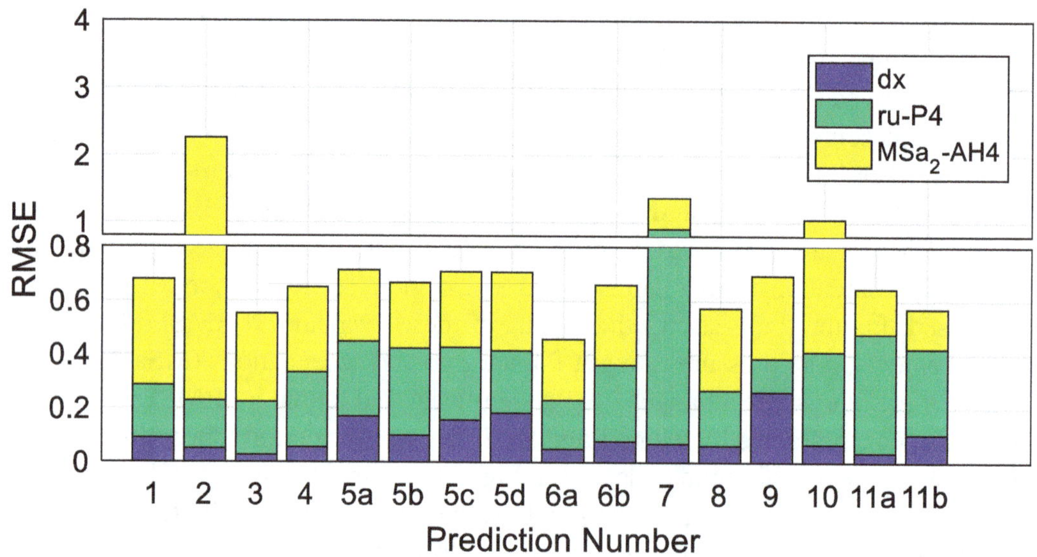


**Fig. 10.33** RMSE values for the Type-B simulations compared to the observed lateral displacements, pore pressure ratios at P4, and spectral accelerations ($MSa_2$) at AH4

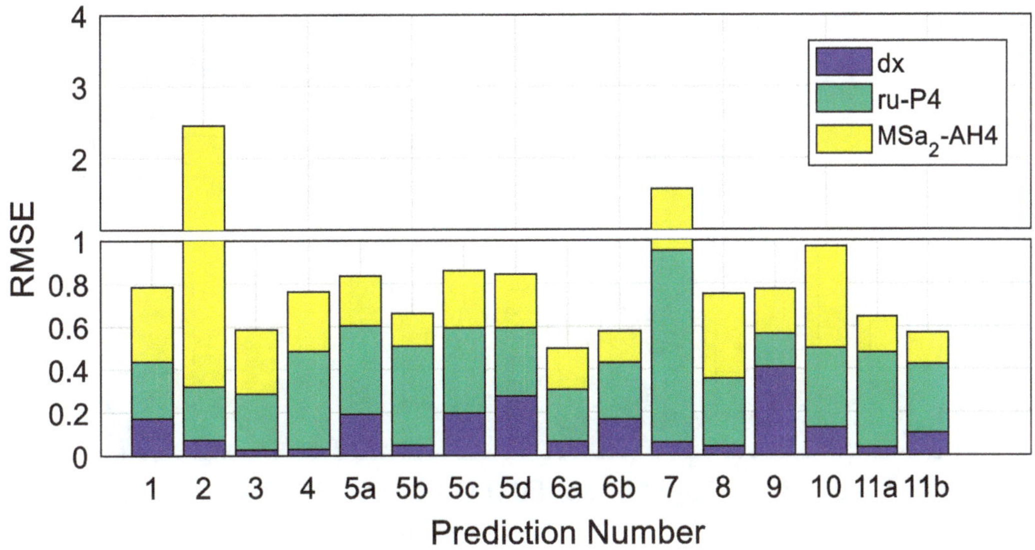


**Fig. 10.34** Recomputed RMSE values considering only the three centrifuge test that were predicted by the numerical simulation team 11

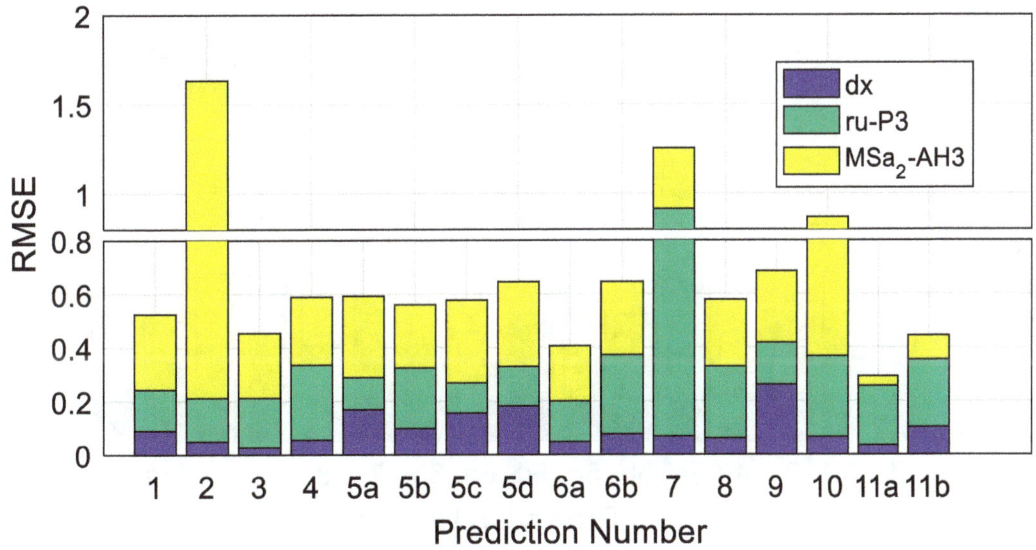


**Fig. 10.35** RMSE values for the Type-B simulations compared to the observed lateral displacements, pore pressure ratios at P3, and spectral accelerations ($MSa_2$) at AH3

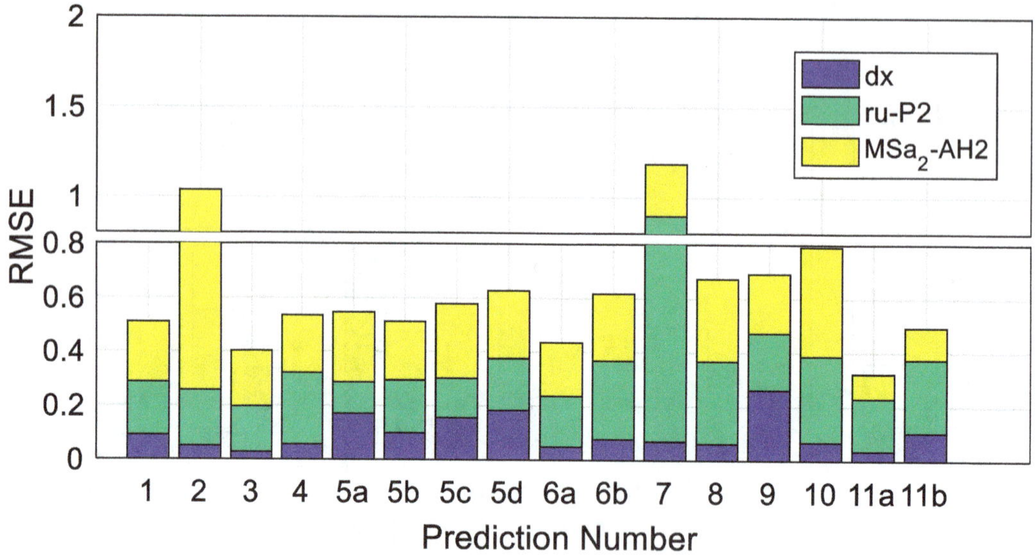


**Fig. 10.36** RMSE values for the Type-B simulations compared to the observed lateral displacements, pore pressure ratios at P2, and spectral accelerations ($MSa_2$) at AH2

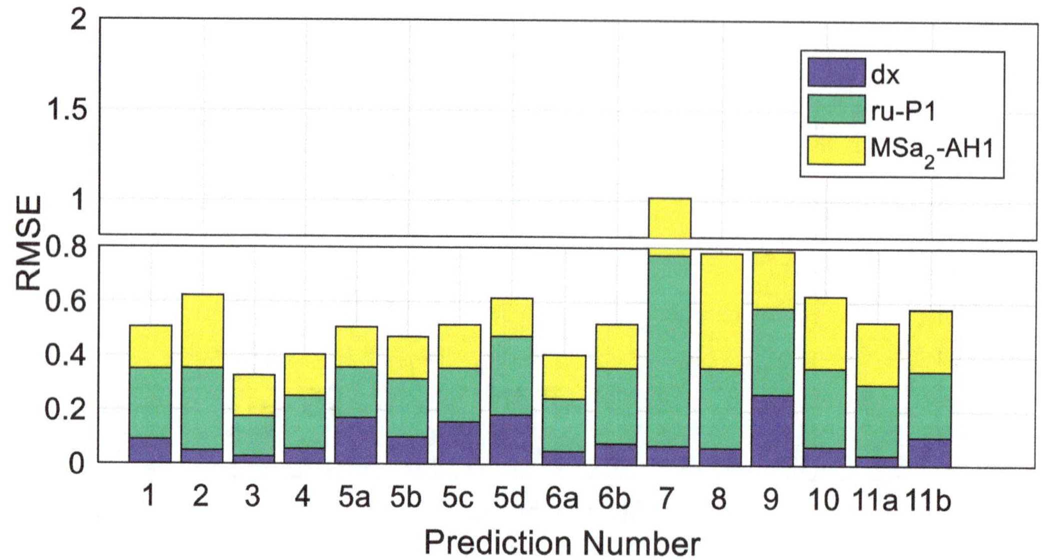


**Fig. 10.37** RMSE values for the Type-B simulations compared to the observed lateral displacements, pore pressure ratios at P1, and spectral accelerations ($MSa_2$) at AH1

## 10.6 Conclusions

Many of the Type-B simulations submitted by the participating prediction teams show trends that are comparable to the measured results obtained for the selected LEAP-2017 centrifuge tests. While the differences between the simulations and experiments are smaller for a few predictions, predicting all the key quantities (excess pore pressure, displacement, cyclic displacement, and acceleration) for the selected tests remains a challenging task that can be potentially addressed by reassessing the performance of the participating numerical simulation platforms in simulation of all the centrifuge data produced in LEAP-2017 project. Moreover, additional centrifuge tests with detailed measurements of displacement time histories can provide a more solid basis for future comparisons of the numerical simulations and experiments in geotechnical problems where permanent displacements are of significant interest.

**Acknowledgments** The experimental work and numerical simulations on LEAP-UCD-2017 were supported by different funds depending mainly on the location of the work. The work by the US PIs (Manzari, Kutter, and Zeghal) is funded by National Science Foundation grants: CMMI 1635524, CMMI 1635307, and CMMI 1635040. Partial funding from Caltrans supported the efforts of co-authors Elgamal and Qiu.

## References

Carey, T. J., Stone, N., Hajialilue Bonab, M., & Kutter, B. L. (2019). LEAP-UCD-2017 centrifuge test at University of California, Davis. In B. Kutter et al. (Eds.), *Model tests and numerical simulations of liquefaction and lateral spreading: LEAP-UCD-2017*. New York: Springer.

Chen, L., Ghofrani, A., & Arduino, P. (2019). Prediction of LEAP-UCD-2017 centrifuge test results using two advanced plasticity sand models. In B. Kutter et al. (Eds.), *Model tests and numerical simulations of liquefaction and lateral spreading: LEAP-UCD-2017*. New York: Springer.

El Ghoraiby, M. A., Park, H., & Manzari, M. (2018). *LEAP-2017 GWU laboratory tests*. DesignSafe-CI, Dataset. https://doi.org/10.17603/DS2210X.

El Ghoraiby, M. A., Park, H., & Manzari, M. T. (2019). Physical and mechanical properties of Ottawa F65 Sand. In B. Kutter et al. (Eds.), *Model tests and numerical simulations of liquefaction and lateral spreading: LEAP-UCD-2017*. New York: Springer.

Fasano, G., Chiaradonna, A., & Bilotta, E. (2019). LEAP-UCD-2017 centrifuge test simulation at UNINA. In B. Kutter et al. (Eds.), *Model tests and numerical simulations of liquefaction and lateral spreading: LEAP-UCD-2017*. New York: Springer.

Fukutake, K., & Kiriyama, T. (2019). LEAP-2017 centrifuge test simulation using HiPER. In B. Kutter et al. (Eds.), *Model tests and numerical simulations of liquefaction and lateral spreading: LEAP-UCD-2017*. New York: Springer.

Hung, W.-Y., & Liao, T.-W. (2019). LEAP-UCD-2017 centrifuge tests at NCU. In B. Kutter et al. (Eds.), *Model tests and numerical simulations of liquefaction and lateral spreading: LEAP-UCD-2017*. New York: Springer.

Ichii, J., Uemura, K., Orai, N., & Hyodo, J. (2019). Numerical simulation trial by cocktail glass model in FLIP ROSE for LEAP-UCD-2017. In B. Kutter et al. (Eds.), *Model tests and numerical simulations of liquefaction and lateral spreading: LEAP-UCD-2017*. New York: Springer.

Kim, S.-N., Ha, J.-G., Lee, M.-G., & Kim, D.-S. (2019). LEAP-UCD-2017 centrifuge test at KAIST. In B. Kutter et al. (Eds.), *Model tests and numerical simulations of liquefaction and lateral spreading: LEAP-UCD-2017*. New York: Springer.

Kutter, B., Carey, T., Hashimoto, T., Zeghal, M., Abdoun, T., Kokalli, P., Madabhushi, G., Haigh, S., Hung, W.-Y., Lee, C.-J., Iai, S., Tobita, T., Zhou, Y. G., Chen, Y., & Manzari, M. T. (2017). LEAP-GWU-2015 experiment specifications, results, and comparisons. *International Journal of Soil Dynamics and Earthquake Engineering, 113*, 616. https://doi.org/10.1016/j.soildyn.2017.05.018.

Kutter, B. L., Carey, T. J., Bonab, M. H., Stone, N., Manzari, M., Zeghal, M., Escoffier, S., Haigh, S. K., Madabhushi, G., Hung, W., Kim, D., Kim, N., Okamura, M., Tobita, T., Ueda, K., & Zhou, Y. (2019). LEAP-UCD 2017 version 1.01 model specifications. In B. Kutter et al. (Eds.), *Model tests and numerical simulations of liquefaction and lateral spreading: LEAP-UCD-2017*. New York: Springer.

Liu, K., Zhou, Y.-G., She, Y., Xia, P., Meng, D., Huang, J.-S., Yao, G., & Chen, Y.-M. (2019). Specifications and results of centrifuge model test at Zhejiang University for LEAP-UCD-2017. In B. Kutter et al. (Eds.), *Model tests and numerical simulations of liquefaction and lateral spreading: LEAP-UCD-2017*. New York: Springer.

Madabhushi, S. S. C., Dobrisan, A., Beber, R., Haigh, S. K., & Madabhushi, S. P. G. (2019). LEAP-UCD-2017 centrifuge tests at Cambridge. In B. Kutter et al. (Eds.), *Model tests and numerical simulations of liquefaction and lateral spreading: LEAP-UCD-2017*. New York: Springer.

Manzari, M., Ghoraiby, M. E., Kutter, B. L., Zeghal, M., Abdoun, T., Arduino, P., Armstrong, R. J., Beaty, M., Carey, T., Chen, Y.-M., Ghofrani, A., Gutierrez, D., Goswami, M., Haigh, S. K., Hung, W.-Y., Iai, S., Kokkali, P., Lee, C.-J., Madabhushi, S. P. G., Mejia, L., Sharp, M., Tobita, T., Ueda, K., Zhou, Y.-G., & Ziotopoulou, K. (2017). Liquefaction analysis and experiment projects (LEAP): Summary of observations from the planning phase. *International Journal of Soil Dynamics and Earthquake Engineering, 113*, 714. https://doi.org/10.1016/j.soildyn.2017.05.015.

Manzari, M. T., El Ghoraiby, M. A., Kutter, B. L., Zeghal, M., Wang, R., Chen, R., Zhang, J.-M., Osamu Ozutsumi, O., Fukutake, K., Kiriyama, T., Fasano, G., Chiaradonna, A., Bilotta, E., Montgomery, J., Ziotopoulou, K., Chen, L., Ghofrani, A., Arduino, P., Wada, T., Ueda, K., Mercado, V., Fuentes, W., Lascarro, C., Yang, M., Barrero, A. R., Taiebat, M., Tsiaousi, D., Ugalde, J., Thaleia Travasarou, T., Ichii, K., Uemura, K., Orai, N., Hyodo, M., Abdoun, T., Haigh, S., Madabhushi, S. P. G., Tobia, T., Hung, W.-Y., Kim, D. S., Okamura, M., Zhou, Y.-G., & Escoffier, S. (2019). LEAP-2017 simulation exercise: Calibration of constitutive models and simulation of the element test. In B. Kutter et al. (Eds.), *Model tests and numerical simulations of liquefaction and lateral spreading: LEAP-UCD-2017*. New York: Springer.

Mercado, V., Fuentes, W., & Lascarro, C. (2017). *LEAP 2017 simulation exercise – Phase I: Model calibration*. Calibration report, June 28, DesignSafe.

Montgomery, J., & Ziotopoulou, K. (2019). Numerical simulations of selected LEAP centrifuge. In B. Kutter et al. (Eds.), *Model tests and numerical simulations of liquefaction and lateral spreading: LEAP-UCD-2017*. New York: Springer.

Okamura, M., & Nurani Sjafruddin, A. (2019). LEAP-2017 centrifuge test at Ehime University. In B. Kutter et al. (Eds.), *Model tests and numerical simulations of liquefaction and lateral spreading: LEAP-UCD-2017*. New York: Springer.

Ozutsumi, O. (2019). LEAP-UCD-2017 numerical simulation at Meisosha Corp. In B. Kutter et al. (Eds.), *Model tests and numerical simulations of liquefaction and lateral spreading: LEAP-UCD-2017*. New York: Springer.

Tsiaousi, D., Ugalde, J., & Travasarou, T. (2019). LEAP-UCD-2017 simulation team Fugro. In B. Kutter et al. (Eds.), *Model tests and numerical simulations of liquefaction and lateral spreading: LEAP-UCD-2017*. New York: Springer.

Vargas Tapia, R. R., Tobita, T., Ueda, K., & Yatsugi, H. (2019). LEAP-UCD-2017 Centrifuge test at Kyoto University. In B. Kutter et al. (Eds.), *Model tests and numerical simulations of liquefaction and lateral spreading: LEAP-UCD-2017*. New York: Springer.

Wada, T., & Ueda, K. (2019). LEAP-UCD-2017 type-B predictions through FLIP at Kyoto University. In B. Kutter et al. (Eds.), *Model tests and numerical simulations of liquefaction and lateral spreading: LEAP-UCD-2017*. New York: Springer.

Wang, R., Chen, R., & Zhang, J.-M. (2019). LEAP-UCD-2017 simulations at Tsinghua University. In B. Kutter et al. (Eds.), *Model tests and numerical simulations of liquefaction and lateral spreading: LEAP-UCD-2017*. New York: Springer.

Yang, M., Barrero, A. R., & Taiebat, M. (2019). Application of a SANISAND model for numerical simulations of the LEAP 2017 experiments. In B. Kutter et al. (Eds.), *Model tests and numerical simulations of liquefaction and lateral spreading: LEAP-UCD-2017*. New York: Springer.

Zeghal, M., Goswami, N., Kutter, B., Manzari, M. T., Abdoun, T., Arduino, P., Armstrong, R., Beaty, M., Chen, Y.-M., Ghofrani, A., Haigh, S., Hung, Y.-W., Iai, S., Kokkali, P., Lee, C.-J., Madabhushi, G., Tobita, T., Ueda, K., Zhou, Y.-G., & Ziotopoulou, K. (2017). Stress-strain response of the LEAP-2015 centrifuge tests and numerical predictions. *International Journal of Soil Dynamics and Earthquake Engineering., 113*, 804. https://doi.org/10.1016/j.soildyn.2017.10.014.

# Chapter 11
# Numerical Sensitivity Study Compared to Trend of Experiments for LEAP-UCD-2017

**Bruce L. Kutter, Majid T. Manzari, Mourad Zeghal, Pedro Arduino, Andres R. Barrero, Trevor J. Carey, Long Chen, Ahmed Elgamal, Alborz Ghofrani, Jack Montgomery, Osamu Ozutsumi, Zhijian Qiu, Mahdi Taiebat, Tetsuo Tobita, Thaleia Travasarou, Dimitra Tsiaousi, Kyohei Ueda, Jose Ugalde, Ming Yang, Bao Li Zheng, and Katerina Ziotopoulou**

---

B. L. Kutter (✉)
Department of Civil and Environmental Engineering, University of California, Davis, CA, USA
e-mail: blkutter@ucdavis.edu

M. T. Manzari
Department of Civil and Environmental Engineering, George Washington University, Washington, DC, USA

M. Zeghal
Department of Civil and Environmental Engineering, Rensselaer Polytechnic Institute, Troy, NY, USA

P. Arduino · L. Chen · A. Ghofrani
Department of Civil and Environmental Engineering, University of Washington, Seattle, WA, USA

A. R. Barrero · M. Taiebat · M. Yang
Department of Civil Engineering, University of British Columbia, Vancouver, BC, Canada

T. J. Carey · B. L. Zheng · K. Ziotopoulou
Department of Civil and Environmental Engineering, University of California, Davis, CA, USA

A. Elgamal · Z. Qiu
Department of Structural Engineering, University of California, San Diego, La Jolla, CA, USA

J. Montgomery
Department of Civil Engineering, Auburn University, Auburn, AL, USA

O. Ozutsumi
Meisosha Corporation, Tokyo, Japan

T. Tobita
Department of Civil Engineering, Kansai University, Osaka, Japan

T. Travasarou · D. Tsiaousi · J. Ugalde
Fugro, Walnut Creek, CA, USA

K. Ueda
Disaster Prevention Research Institute, Kyoto University, Kyoto, Japan


B. Kutter et al. (eds.), *Model Tests and Numerical Simulations of Liquefaction and Lateral Spreading*, https://doi.org/10.1007/978-3-030-22818-7_11

**Abstract** This paper describes the numerical sensitivity study requested prior to the December 2017 LEAP workshop. Several but not all of the simulation teams participated in this sensitivity study. The results of the sensitivity study are used to begin to map out the simulation response surfaces that relate residual displacement to $PGA_{eff}$ and relative density. The simulation response surfaces are compared to the corresponding response surfaces determined by nonlinear regression of the centrifuge test data. The definition of the experimental response surface allows a means to objectively reduce the influence of outliers in the experiment dataset. The residuals between the experiments and the regression surface are used to quantify the uncertainty associated with experiment-experiment variability. Some metrics for assessing the comparison between simulations and experiments are explored; it is suggested that differences in the logarithm of displacement are more meaningful than arithmetic differences. As expected, some models predicted the average displacement well and some predicted triggering of liquefaction and the shape of the response function better than others. LEAP-UCD-2017 is not a final assessment of simulation procedures; instead, the results can be used to improve simulation specifications and calibration procedures and as a stimulus for more careful review of simulation results before they are submitted.

## 11.1 Description of the Requested Sensitivity Study

Manzari et al. (2019a, b) describe the calibration phase and the Type-B numerical simulation exercise for LEAP-UCD-2017. After receiving the Type-B simulations, the organizing team (Manzari, Zeghal, and Kutter) invited the simulation teams to predict the results of an additional centrifuge test performed at RPI and to conduct a study to illustrate the sensitivity of the simulations to relative density and ground motion intensity. Considering the time constraints, about half of the simulation teams listed in Table 11.1 were able to provide the requested sensitivity study. This paper analyzes the predicted residual displacements from the sensitivity study.

A unique feature of the LEAP-UCD-2017 exercise is the intention to quantify the uncertainty of centrifuge experimental results. To understand the significance of the uncertainty, it is necessary to quantify the sensitivity of results to variations in key input parameters. For the present study we focus on sensitivity to three specific parameters: (1) the intensity of the input motion, (2) the presence of high-frequency excitation superimposed on the 1 Hz ramped sine wave component of shaking, and (3) the relative density of the sand. The RPI1 centrifuge test was arbitrarily chosen as the base case for the sensitivity study.

The input parameters varied to generate seven unique simulations, NS-1 to NS-7, are listed in the first several rows of Table 11.2. A subset of three conditions (NS-1, NS-2, and NS-3) were planned to illustrate the sensitivity of simulations to relative density. A subset of three simulations (NS-1, NS-4, and NS-5) were included to illustrate the sensitivity of the simulations to input ground motion intensity. A pair of simulations (NS-1 and NS-6) have similar effective PGAs but different high-frequency contents (a 3 Hz component was superimposed on the 1 Hz ramped sine wave). Finally, a pair of simulations (NS-6 and NS-7) have the same proportions of the

**Table 11.1** Teams that provided multiple Type-B predictions

| Numerical simulation team | Constitutive model | Analysis platform | Simulation label | Sensitivity study?[a] |
|---|---|---|---|---|
| Wang et al. (2019) | Tsinghua constitutive model | OpenSees | 1-Os-Ts | Yes* |
| Ozutsumi (2019) | Cocktail glass model | FLIP rose | 2-Fr-co | Yes |
| Fukutake and Kiriyama (2019) | Bowl model | HiPER | 3-hi-Bo | No |
| Fasano et al. (2019) | Hypoplastic model | PLAXIS | 4-pa-Hy | Yes* |
| Montgomery and Ziotopoulou (2019) | (a) PM4Sand – Cal 1<br>(b) PM4Sand – Cal 2 - $G_o$ larger (default function of $D_r$)<br>(c) PM4Sand – all $D_r = 62\%$<br>(d) PM4Sand – even lower $D_r$ (obtained from CPT)<br>(e) PM4Sand – corrected $D_r$ | FLAC-2D | 5a-F2-pm<br>5b-F2-pm<br>5c-F2-pm<br>5d-F2-pm<br>5 s-F2-pm | Yes<br>No<br>No<br>No<br>Yes |
| Chen et al. (2019) | (a) DM04 model<br>(b) DM04 model<br>(c) PM4Sand | OpenSees (2D) OpenSees (1D) OpenSees (1D) | 6-Os-ma<br>6-Os1-ma<br>6-Os1-pm | Yes<br>No<br>Yes |
| Wada and Ueda (2019) | Cocktail glass model | FLIP | 7-Fl-co | No |
| Mercado et al. (2017) | ISA-hypoplasticity model | ABAQUS | 8-Ab-is | No |
| Yang et al. (2019) | SANISAND | FLAC-3D | 9-F3-Sa | Yes |
| Qiu and Elgamal (2019) | PDMY | OpenSees | 10-Os-PD | Yes |
| Tsiaousi et al. (2019) | (a) UBCSAND<br>(b) PM4Sand | FLAC-2D | 11a-F2-Ub<br>11b-F2-pm | Yes<br>Yes |

[a]The last column indicates if the team participated in the sensitivity study
Yes* indicates that they did participate and results are shown in their papers in this book, but not shown in this paper due to time limitations.

1 Hz and 3 Hz components but different amplitudes to allow a separate assessment of the sensitivity to shaking intensity for an input motion with significant high-frequency content. The simulation teams were specifically requested to provide the calculated lateral displacement of ground surface in the middle of the slope (Fig. 11.1).

## 11.2 Characterization of Displacements from Experiments

In this paper, the displacement results of the numerical simulations are compared to the two smoothed experimental regression surfaces presented by Kutter et al. (2018, 2019b). Two regression equations were used based on different underlying equations to obtain different regression surfaces. The first surface, shown in Fig. 11.2, was based on a six-parameter regression surface:

**Table 11.2** Table for preliminary reporting results of sensitivity study

| Simulation # | NS-1 | NS-2 | NS-3 | NS-4 | NS-5 | NS-6 | NS-7 |
|---|---|---|---|---|---|---|---|
| Dry density (kg/m$^3$) | 1651 | 1608 | 1683 | 1651 | 1651 | 1651 | 1651 |
| Soil | Ottawa F-65 | Ottawa F-65 | Ottawa F-65 | Ottawa F-65 | Ottawa F-65 | Ottawa F-65 | Ottawa F-65 |
| $D_r$ (assuming $\rho_{dmax} = 1765$ & $\rho_{dmin} = 1476$ kg/m$^3$).[a] | 65%[a] | 50%[a] | 75%[a] | 65%[a] | 65%[a] | 65%[a] | 65%[a] |
| Motion to be used for simulation provided in excel sheet | Achieved RPI-1, motion 1 | Achieved RPI-1, motion 1 | Achieved RPI-1, motion 1 | Achieved RPI-1, motion 1 scaled up | Achieved RPI-1, motion 1 scaled down | Achieved RPI-2 motion 1 | Achieved RPI-2 motion 1 scaled up |
| PGA (g) | 0.150 | 0.150 | 0.150 | 0.250 | 0.110 | 0.140 | 0.200 |
| 1 Hz component, $PGA_{1Hz}$ (g) | 0.135 | 0.135 | 0.135 | 0.225 | 0.099 | 0.110 | 0.160 |
| High-frequency component, $PGA_{HF}$ (g) | 0.021 | 0.021 | 0.021 | 0.036 | 0.015 | 0.080 | 0.110 |
| $PGA_{eff} = PGA_{1Hz} + 0.5$ $(PGA_{HF})$, (g) | 0.146 | 0.146 | 0.146 | 0.243 | 0.107 | 0.150 | 0.215 |

[a]At the time the request for the sensitivity study was sent to the simulation teams, the maximum and minimum dry densities were estimated to be 1765 and 1476 kg/m$^3$. Based upon the study by Carey et al. (2019b), the maximum and minimum dry densities were updated to 1757 and 1490.5, respectively. Using the updated index dry densities, the relative densities corresponding to 1651, 1608, and 1683 kg/m$^3$ would be 64%, 48%, and 75% – slightly smaller than the relative densities listed in the above table

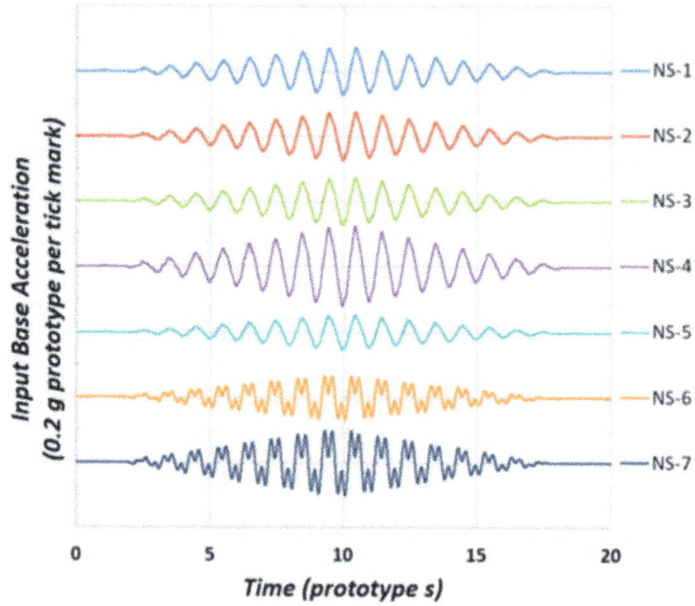


**Fig. 11.1** Input motions for numerical sensitivity study

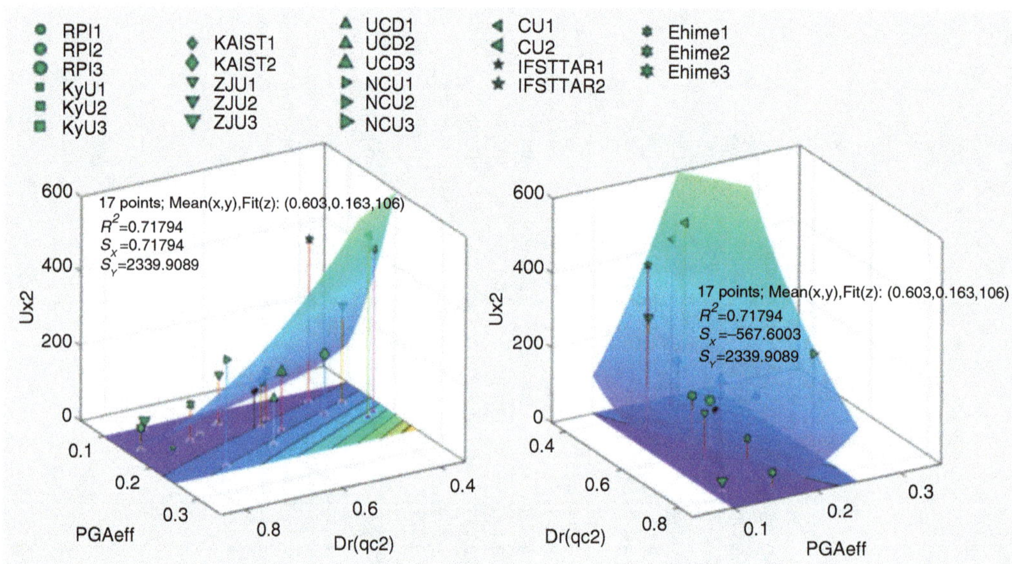


**Fig. 11.2** Two views of the regression surface for the six-parameter model fit to experimental data as described by Kutter et al. (2018)

$$\mathrm{Ux2} = \mathrm{b}_2 \left\langle \mathrm{b}_1 - \frac{(D_\mathrm{r} - 0.125)^{\mathrm{n3}} + 0.05}{1.3^{\frac{a_{\max}}{g}}} \right\rangle^{\mathrm{n1}} \left(\frac{a_{\max}}{g}\right)^{\mathrm{n2}} (1 - D_\mathrm{r})^{\mathrm{n4}} \qquad (11.1)$$

where $\mathrm{b}_1$, $\mathrm{b}_2$, n1, n2, n3, and n4 were used as regression parameters. For the experimental data, Ux2 is the average displacement from the two central surface markers.

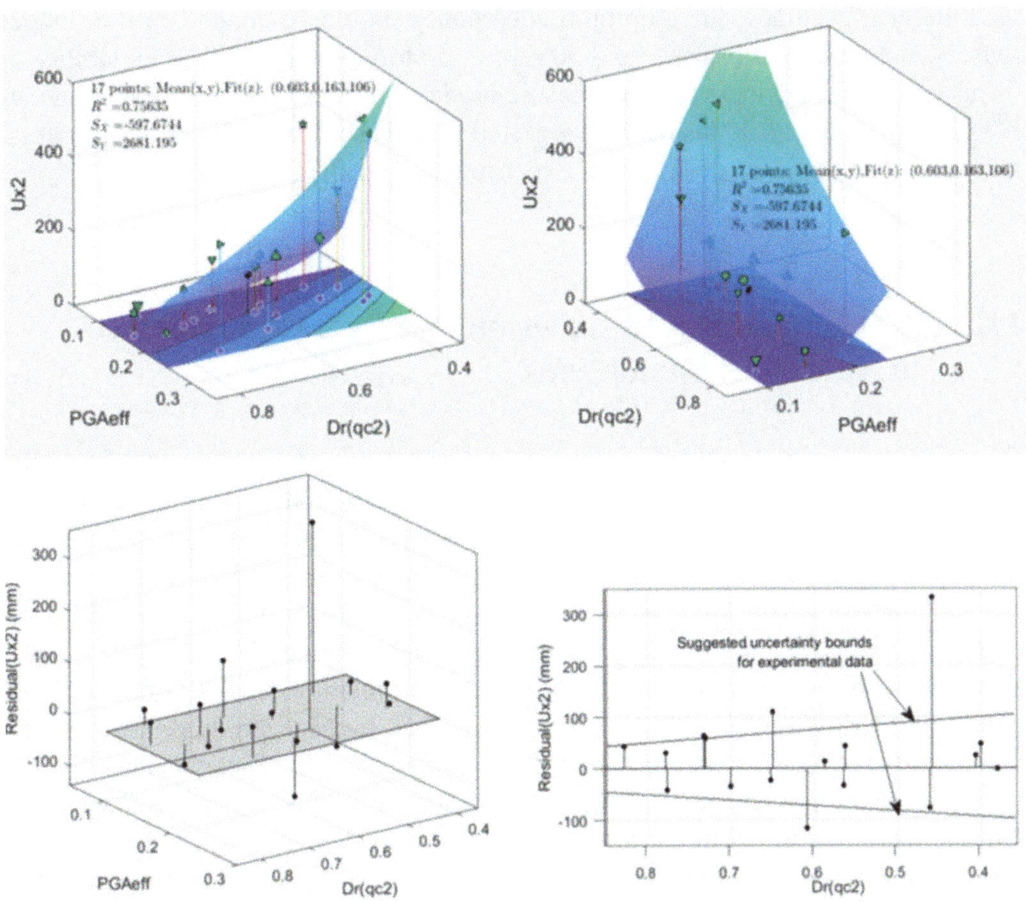


**Fig. 11.3** Two views of regression surface for four-parameter model to fit experimental data as described by Kutter et al. (2019b) (top). Two views of residuals of the data from the curve fit, with suggested uncertainty bounds representing experiment-experiment variability (bottom right). Displacements (Ux2) are given in units of mm and accelerations are given in ($g$)

Figure 11.3 shows a slightly different surface that is based only on the regression using the four parameters $b_1$, $b_2$, n1, and n3:

$$\mathrm{Ux2} = \mathrm{b}_2 \left\langle \mathrm{b}_1 - \frac{(D_\mathrm{r} - 0.125)^{\mathrm{n3}} + 0.05}{1.3^{\frac{a_{\max}}{g}}} \right\rangle^{\mathrm{n1}} \tag{11.2}$$

As explained by Kutter et al. (2019b), depending on the model, the method of estimating the parameters, and the exclusion of a small number of outliers, the correlation coefficients ($R^2$ values) varied between about 0.6 and 0.85, indicating that there is a meaningful relationship between the experimental data and the fitted surface. The bottom diagrams in Fig. 11.3 show the residuals between the experimental data and the regression surface. The bottom right diagram shows some lines that are suggested as being indicative of the uncertainty associated with experiment to experiment variability. The suggested uncertainty bounds were subjectively selected to be consistent with the data and these concepts: (1) based upon measurement accuracy, the minimum uncertainty should be about 1 mm in model scale, which corresponds to 20 and 50 mm in prototype scale; (2) the variability of displacement should increase with the amplitude of the displacement; and (3) about 2/3 of the data points lie inside the suggested bounds so that these bounds might be indicative of 1 standard deviation of experiment-experiment variability.

## 11.3 2D Comparisons of Experimental Regression Surfaces to Numerical Simulations

Figure 11.4 compares results from the numerical sensitivity study to sections of the regression surface. In Fig. 11.4a, three points from a section through the regression surfaces at $PGA_{eff} = 0.146$ g are connected by thick dashed lines; it is apparent that the 4-parameter and 6-parameter surfaces produced very similar results on this section. The thin dashed lines represent the uncertainty band associated with experiment-experiment variability for the 4-parameter model. Eight sets of simulations provided results for the sensitivity analysis for relative densities of 50%, 65%, and 75%. During the Type-B simulations described by Manzari et al. (2019b), the simulations by 5a-F2-Pm respected the prescribed densities but mistakenly used $\rho_{max}$ and $\rho_{min}$ reported by a different source, resulting in lower relative densities. During the sensitivity analysis, this team revised the relative density calculations to produce prediction 5s-F2-Pm. The thick dashed lines in Fig. 11.4b indicate sections through the regression surfaces at $D_r = 65\%$. The thinner dashed lines indicate the experimental uncertainty bands.

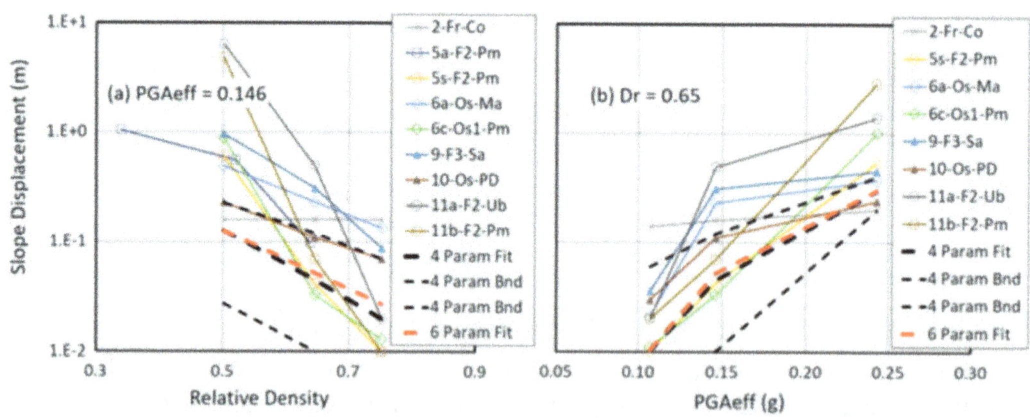


**Fig. 11.4** Comparison of experimental regression surfaces (dashed lines) to predictions for the numerical sensitivity study. (**a**) Simulations NS-1, NS-2, and NS-3 as a function of relative density with regression surface lines for $PGA_{eff} = 0.146$ g. (**b**) Simulations NS-1, NS-4, and NS-5 as a function of $PGA_{eff}$ with the regression surface lines for $D_r(qc2) = 0.65$. The plot also shows approximate uncertainty bounds for the experimental regression surfaces consistent with the suggested uncertainty bounds indicated in Fig. 11.3

Based on Fig. 11.4, one may conclude the following with respect to simulations:

1. On average, for the teams that participated in the sensitivity study, the NS-1 to NS-5 simulations overpredicted displacements. None of the NS-1 to NS-5 simulations fell below the uncertainty bounds of the experimental data. This observation is not applicable to the teams that did not participate in the sensitivity study.
2. Many of the simulations for the sensitivity study overpredicted displacements at the high ends of the displacement plots (low $D_r$ and high $PGA_{eff}$). Only two of the simulations fell within the uncertainty bounds for $D_r = 0.5$. Only three of the simulations fell within the uncertainty bounds for $PGA_{eff} = 0.288$.
3. On the low displacement ends of Fig. 11.4a, b, where the uncertainty bounds are wider on the log plot, most of the simulations fell within the uncertainty bounds. For $D_r = 0.75$, only three of the predictions were above the uncertainty bound of the regression surface. Five of the eight fell within the uncertainty bounds. For $PGA_{eff} = 0.1$, only one of the simulations fell above the uncertainty bounds for the data.
4. It is also interesting to see that a few of models (6a, 9, 11a) are somewhat insensitive to the PGA after it exceeds 0.15 g.
5. All of the simulation teams (except Team 5) used the relative density based on mass and volume measurements of the experiments instead of the method for determining the density from the cone penetration tests as recommended by Carey et al. (2019a, b). Simulation team 5 used different $\rho_{d\ max}$ and $\rho_{d\ min}$ data to determine relative density and thus their predictions (5a-F2-Pm) were for a looser state than the other simulation teams as is apparent in Fig. 11.4a. This discrepancy was corrected in a second set of simulations (5s-F2-Pm) which are more consistent with the other simulation teams and will be considered in later comparisons.

## 11.4 Error Measures and Ranking of Numerical Simulations

The results from the sensitivity study for simulations NS-1 to NS-5 are summarized in Table 11.3. The displacements from experiments are obtained from the regression surfaces at the specified $D_r$ and $PGA_{eff}$. The displacements from the simulations are for the top center surface of the sand slope.

Four error measures, M1-M4, were considered to evaluate the quality of the simulations. The first two are based on logarithmic differences of displacements and the second two are based on arithmetic differences in displacement:

$$M1 = \text{Median}\left|\ln\frac{\text{displacement from sim}}{\text{displacement from exp}}\right|$$

**Table 11.3** Comparison of displacement results of simulations for sensitivity study to displacements from the regression surfaces through the experimental data

| | | NS-1 | NS-2 | NS-3 | NS-4 | NS-5 |
|---|---|---|---|---|---|---|
| | PGAeff | 0.146 | 0.146 | 0.146 | 0.243 | 0.107 |
| | $D_r$ | 0.650 | 0.500 | 0.750 | 0.650 | 0.650 |
| Displacement from experiments | 4-parameter surface | 0.045 | 0.127 | 0.020 | 0.299 | 0.010 |
| | 6-parameter surface | 0.052 | 0.127 | 0.027 | 0.301 | 0.010 |
| Displacement from simulation (m) | 2-Fr-Co | 0.160 | 0.160 | 0.161 | 0.199 | 0.140 |
| | 5s-F2-Pm | 0.042 | 0.610 | 0.010 | 0.540 | 0.010 |
| | 6a-Os-Ma | 0.231 | 0.494 | 0.135 | 0.374 | 0.021 |
| | 6c-Os1-Pm | 0.033 | 0.889 | 0.013 | 1.004 | 0.011 |
| | 9-F3-Sa | 0.310 | 0.969 | 0.087 | 0.453 | 0.036 |
| | 10-Os-PD | 0.110 | 0.230 | 0.070 | 0.240 | 0.030 |
| | 11a-F2-Ub | 0.490 | 6.360 | 0.020 | 1.390 | 0.020 |
| | 11b-F2-Pm | 0.070 | 4.800 | 0.010 | 2.890 | 0.020 |

$$M2 = \text{RMS}\left(\ln \frac{\text{displacement from sim}}{\text{displacement from exp}}\right)$$

$$M3 = \text{Median}|(\text{displacement from sim}) - (\text{displacement from exp})|$$

$$M4 = \text{RMS}((\text{displacement from sim}) - (\text{displacement from exp}))$$

where (displacement from sim) is from the lower half of Table 11.3 for each simulation and (displacement from exp) is the average of the displacement from the 4-parameter and 6-parameter regression surfaces evaluated at the $PGA_{eff}$ and $D_r$ of NS-1 to NS-5. Tables 11.4 and 11.5 present the computed error measures for the various simulation teams. M1 is the median of the absolute values of the natural logarithm of the ratio of displacements listed in columns A–E (Table 11.4). Similarly, M2 is the root mean square (RMS) of the five values in columns A–E (Table 11.4). The values in parentheses next to M1 and M2 values are the rank orders of the error measures. The median error is less affected by a single simulation with a large error. For example, imagine four perfect simulations and one simulation that incorrectly predicted instability with huge displacements – the median error measures (M1 or M3) would be zero, while the RMS error (M2 or M4) could be large.

For displacements that span multiple orders of magnitude, and especially for data that is log-normally distributed, the statistics are best done by comparing the logarithms of displacements as is done with error measures M1 and M2. For error measures based on arithmetic differences (M3 and M4 in Table 11.5), a 10% error for a 1 m displacement prediction (0.1 m error) is treated as being equivalent to a

**Table 11.4** Logarithmic differences between the displacements from the simulations and the displacements from the experiment regression surfaces in columns A–E

| | | NS-1 | NS-2 | NS-3 | NS-4 | NS-5 | M1 = median of absolute value of columns A–E (R1) | M2 = RMS of columns A–E (R2) |
|---|---|---|---|---|---|---|---|---|
| | Sim ID | A | B | C | D | E | F | G |
| $\ln\left(\frac{\text{displacement from sim}}{\text{displacement from exp}}\right)$ | 2-Fr-Co | 1.194 | 0.228 | 1.927 | −0.410 | 2.624 | 1.19 (5) | 1.56 (6) |
| | 5s-F2-Pm | −0.144 | 1.567 | −0.852 | 0.588 | −0.015 | 0.59 (1) | 0.84 (2) |
| | 6a-Os-Ma | 1.562 | 1.355 | 1.748 | 0.220 | 0.745 | 1.36 (7) | 1.26 (4) |
| | 6c-Os1-Pm | −0.385 | 1.944 | −0.590 | 1.208 | 0.080 | 0.59 (2) | 1.07 (3) |
| | 9-F3-Sa | 1.855 | 2.030 | 1.317 | 0.411 | 1.263 | 1.32 (6) | 1.49 (5) |
| | 10-Os-PD | 0.819 | 0.592 | 1.094 | −0.223 | 1.084 | 0.82 (3) | 0.83 (1) |
| | 11a-F2-Ub | 2.313 | 3.911 | −0.159 | 1.533 | 0.678 | 1.53 (8) | 2.17 (8) |
| | 11b-F2-Pm | 0.367 | 3.630 | −0.852 | 2.265 | 0.678 | 0.85 (4) | 1.98 (7) |

Column F lists the median of the absolute values of columns A–E, and column G lists the RMS of columns A–E

**Table 11.5** Arithmetic differences between the displacements from the simulations and the displacements from the experiment regression surfaces in columns A–E

| | | NS-1 | NS-2 | NS-3 | NS-4 | NS-5 | M1 = median of absolute value of columns A–E ($R_1$) | M2 = RMS of columns A–E ($R_2$) |
|---|---|---|---|---|---|---|---|---|
| | Sim ID | A | B | C | D | E | F | G |
| displacement from sim − dsplacement from exp | 2-Fr-Co | 0.112 | 0.033 | 0.138 | −0.101 | 0.130 | 0.112 (6) | 0.244 (2) |
| | 5s-F2-Pm | −0.007 | 0.483 | −0.013 | 0.240 | 0.000 | 0.013 (1) | 0.539 (4) |
| | 6a-Os-Ma | 0.183 | 0.366 | 0.111 | 0.074 | 0.011 | 0.111 (5) | 0.431 (3) |
| | 6c-Os1-Pm | −0.016 | 0.762 | −0.010 | 0.704 | 0.001 | 0.016 (2) | 1.037 (6) |
| | 9-F3-Sa | 0.261 | 0.842 | 0.064 | 0.153 | 0.026 | 0.153 (7) | 0.897 (5) |
| | 10-Os-PD | 0.062 | 0.103 | 0.047 | −0.060 | 0.020 | 0.060 (4) | 0.143 (1) |
| | 11a-F2-Ub | 0.442 | 6.233 | −0.003 | 1.090 | 0.010 | 0.442 (8) | 6.343 (8) |
| | 11b-F2-Pm | 0.022 | 4.673 | −0.013 | 2.590 | 0.010 | 0.022 (3) | 5.343 (7) |

Column F lists the median of the absolute values of columns A–E, and column G lists the RMS of columns A–E

100% error for a 0.1 m displacement prediction. Arithmetic error measures such as M3 and M4 may be overly sensitive to small percentage errors in regions where displacements are large. It is interesting to note that in several cases the rank (in parentheses in Tables 11.4 and 11.5) changed by four places (out of eight possible) depending on the choice of error measure.

## 11.5 3-D Comparison of Simulations to Experimental Regression Surfaces

The 4-parameter and 6-parameter regression surfaces described above are compared to the Type-B predictions (Manzari et al. 2019b) along with the results of the numerical sensitivity study in Fig. 11.5. For all of the plots, the Ux value is the predicted horizontal component of the surface displacement in the middle of the soil profile.

The results from the sensitivity study (simulations NS-1 to NS-5) are shown in these figures as black circular data points connected by lines. The sensitivity study simulations NS-6 and NS-7 are indicated by the white circular data points. White points contain significant high-frequency content and black points contain little high-frequency content. Some of the simulations show a sensitivity to the high-frequency content that was not accounted for by the $PGA_{eff}$. In other words, the white circles of NS-6 and NS-7 plot above the black points for some simulations. For many of the simulations, the white points plot well above the black points suggesting that these simulations are more sensitive to high-frequency content than are the experiments.

The results of Type-B simulations for the nine model tests selected are also compared to the regression surfaces in Fig. 11.5. The shape of the data points indicates which experiment is being simulated, and the gray scale of the data points is indicative of the ratio of $PGA_{HF}/PGA_{1Hz}$ (the gray scale is consistent with that of the black and white points used for the sensitivity study). The density coordinates assumed by the simulation teams were used to plot their simulation data in Fig. 11.5; thus some errors associated with estimation of density are approximately accounted for in this data presentation.

Using this 3-D plot format allows for comparison of the trends and sensitivities of the simulations and experiments. Note that the vantage point chosen for the surface plots almost reduces the experimental regression surfaces to a curved line. Also note that the length of the stem below each data points indicates the predicted displacement and the intersection of the stem with the base plane indicates the density and PGA coordinates.

For most of the simulations, the results of the sensitivity study are consistent with the results of the Type-B simulations. Some of the predictions tend to be close to the experimental surface on average, despite having obviously different slopes and curvatures. Different metrics (as yet undetermined, but possibly similar to those

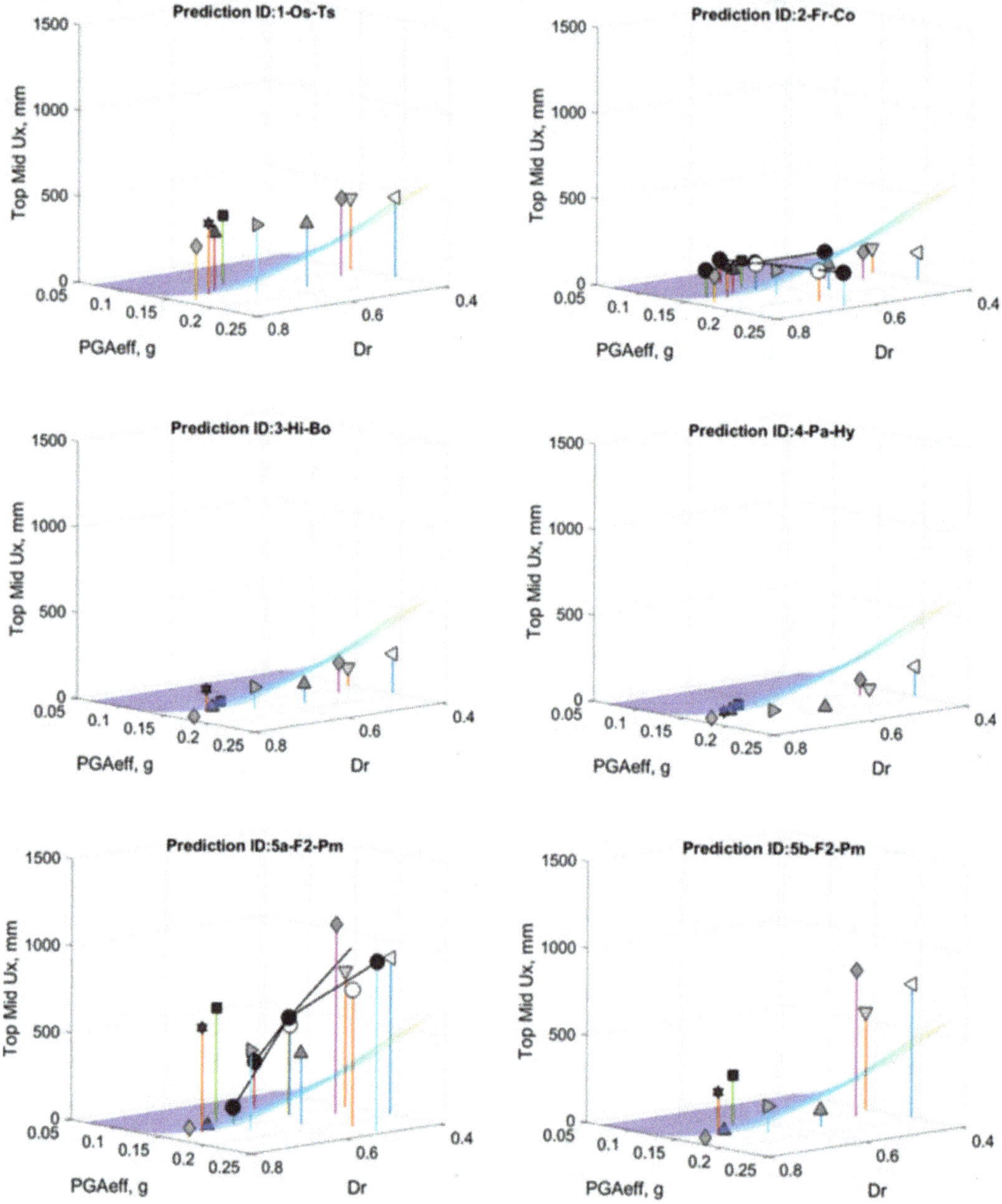


**Fig. 11.5** Comparison between regression to the experimentally determined response surfaces and the numerical predictions for the nine Type-B predictions and the seven predictions done for the sensitivity study. The black circles joined together are simulations of NS-1 to NS-5 of the sensitivity study. Comparison between regression to the experimentally determined response surfaces and the numerical predictions for the nine Type-B predictions and the seven predictions done for the sensitivity study. The black circles joined together are simulations of NS-1 to NS-5 of the sensitivity study. Comparison between regression to the experimentally determined response surfaces and the numerical predictions for the nine Type-B predictions and the seven predictions performed for the sensitivity study. The black circles joined together are simulations of NS-1 to NS-5 of the sensitivity study. The shape of the non-circular symbols indicates the experiment being simulated. The grey scale of the points indicates the intensity of the high frequency components (black = negligible high frequency content, white = very significant high frequency content)

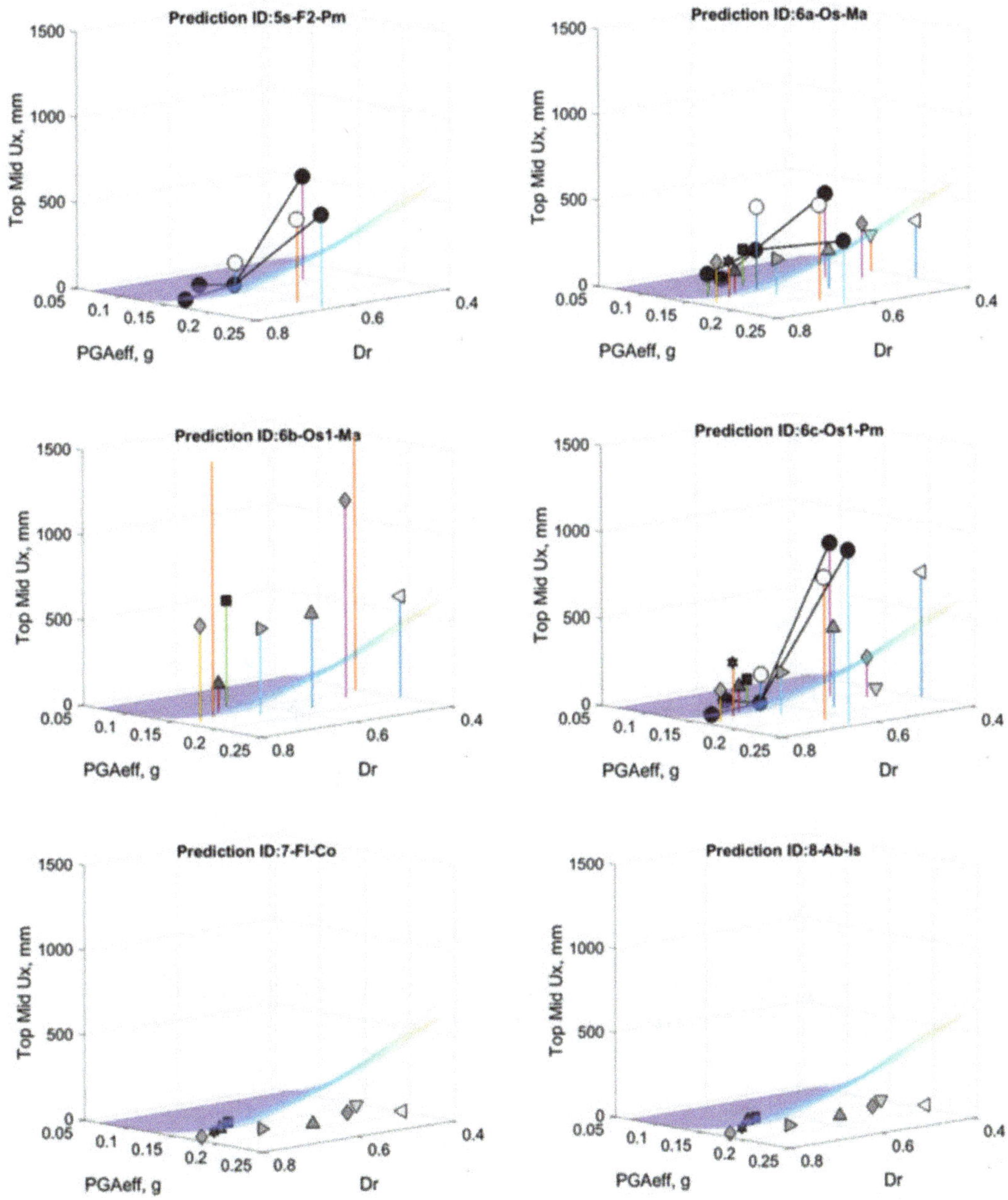


**Fig. 11.5** (continued)

explored by Goswami et al. (2019) to compare experimental data) should be used to quantitatively compare the shape and the amplitude of the trends.

An important feature of codes intended for simulation of liquefaction should be able to predict a triggering curve, where deformations increase for cases above the threshold cyclic stress ratio and deformations are small below the threshold cyclic stress ratio, and the threshold cyclic stress should increase as density increases. This behavior is not apparent for all of the simulations in Fig. 11.5. For some simulations that did predict a reasonable triggering threshold the simulated displacements

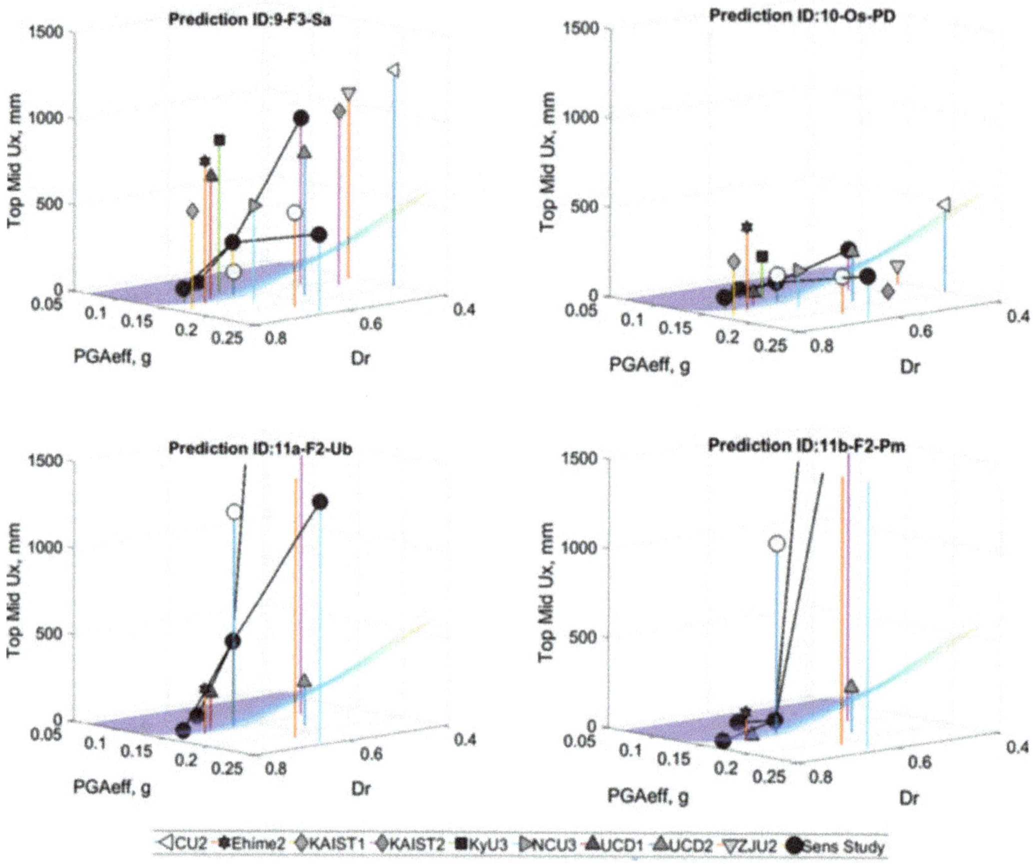


**Fig. 11.5** (continued)

increased too rapidly after passing the threshold for liquefaction, while the experimental displacements increase more gradually as PGA increases or density decreases.

## 11.6 Summary and Conclusions

The numerical sensitivity study was useful for LEAP because it allowed for the evaluation of the sensitivity of the numerical simulations to changes in the important input parameters for this problem. The evaluation of the quality of a simulation for a single point should account for the uncertainties associated with the experimental data point and the sensitivity of the simulation to these uncertainties. With enough data points from the sensitivity study and the centrifuge test matrix, we are able to map out response functions for the simulations and experiments and see how well they correspond to each other. In some cases, simulations may accurately model a subset of the experiments, even if the shape of the simulated response surface does not accurately mimic the shape of the experimental response surface.

It is not a goal of this paper to determine winners and losers, especially in light of difficulties with scalar metrics such as M1-M4 explored in this paper. For liquefaction-induced displacements, where the displacements vary by orders of magnitude, it is recommended that logarithmic differences (metrics M1 and M2) are better metrics than arithmetic differences (M3 and M4). Arithmetic error metrics may be unduly affected by small percentage errors in the large numbers and may not be appropriately sensitive to large percentage errors in the small-displacement portion of the function.

The results clearly show that most of the teams that participated in the sensitivity study overpredicted the experimentally observed displacements in the large displacement range, with the simulations falling well above the suggested uncertainty bounds due to experiment-experiment variability. Based upon the 3-D plots of Type-B predictions, some of the simulation teams (Teams 3 and 4) probably would not have overestimated displacements had they completed the sensitivity study.

A large number of experiments is needed to map out the experimental response function and a suitable regression analysis is required to provide an experimental response function to map against the numerical response function. The mapping of a response function by the parametric study in numerical simulations is one useful way to assess the quality of the comparisons between experiments and simulations, but many other aspects of the simulations should be considered. Numerical simulations should produce reasonable calibrations to element tests, and the time series data for pore pressure, displacement, and acceleration should also be considered as was done in many respects by Manzari et al. (2019a, b).

Some differences in the comparisons between the simulations and experiments can be attributed to simple discrepancies, such as miscalculation of relative density by the simulation team. The errors are understandable as different maximum and minimum index densities were reported to the simulation teams at different stages of LEAP. In future LEAP efforts using the same Ottawa Sand, the dry density should be used as the primary indicator of the state of the sand in all correspondence because this avoids confusion associated with estimation of the maximum and minimum dry density parameters and the small errors associated with conversion to void ratio using specific gravity reported by different sources. Whenever relative density is specified, the assumed maximum and minimum index dry densities should be explicitly stated; best estimates of these index densities, unfortunately, continue to change as more testing is done to determine the index densities.

At this time, LEAP-UCD-2017 should not be used to pass judgment on winners and losers of a contest. In many cases, the differences are more indicative of mistakes, imperfect instructions, and limited resources than validity of the numerical or constitutive behavior. The results of LEAP-UCD-2017 should be used constructively to improve simulation specifications, calibration procedures, and peer review practices that could minimize avoidable errors in future validation efforts.

**Acknowledgments** The experimental work on LEAP-UCD-2017 was supported by different funds depending mainly on the location of the work. The work by the US PIs (Manzari, Kutter, and Zeghal) is funded by National Science Foundation grants: CMMI 1635524, CMMI 1635307, and CMMI 1635040. Partial funding from Caltrans supported the efforts of co-authors Elgamal and Qiu.

## References

Carey, T. J., Gavras, A., & Kutter, B. L. (2019a). Comparison of LEAP-UCD-2017 CPT results. In B. Kutter et al. (Eds.), *Model tests and numerical simulations of liquefaction and lateral spreading: LEAP-UCD-2017*. New York: Springer.

Carey, T. J., Stone, N., & Kutter, B. L. (2019b). Grain size analysis and maximum and minimum dry density of Ottawa F-65 sand for LEAP-UCD-2017. In B. Kutter et al. (Eds.), *Model tests and numerical simulations of liquefaction and lateral spreading: LEAP-UCD-2017*. New York: Springer.

Chen, L., Ghofrani, A., & Arduino, P. (2019). Prediction of LEAP-UCD-2017 centrifuge test results using two advanced plasticity sand models. In B. Kutter et al. (Eds.), *Model tests and numerical simulations of liquefaction and lateral spreading: LEAP-UCD-2017*. New York: Springer.

Fasano, G., Chiaradonna, A., & Bilotta, E. (2019). LEAP-UCD-2017 centrifuge test simulation at UNINA. In B. Kutter et al. (Eds.), *Model tests and numerical simulations of liquefaction and lateral spreading: LEAP-UCD-2017*. New York: Springer.

Fukutake, K., & Kiriyama, T. (2019). LEAP-2017 centrifuge test simulation using HiPER. In B. Kutter et al. (Eds.), *Model tests and numerical simulations of liquefaction and lateral spreading: LEAP-UCD-2017*. New York: Springer.

Goswami, N., Zeghal, M., Kutter, B., Manzari, M., Abdoun, T., Carey, T., Escoffier, S., Ha, J.-G., Haigh, S., Hung, W.-Y., Kim, D.-S., Kim, S.-N., Kim, N. R., Korre, E., Liao, T.-W., Liu, K., Madabhushi, G., Madabhushi, S. S. C., Okamura, M., Sjafuddin, A. N., Tobita, T., Ueda, K., Vargas, R., & Zhou, Y.-G. (2019). Difference and sensitivity analyses of the LEAP-2017 Experiments. In B. Kutter et al. (Eds.), *Model tests and numerical simulations of liquefaction and lateral spreading: LEAP-UCD-2017*. New York: Springer.

Kutter, B. L., Carey, T. J., Zheng, B. L., Gavras, A., Stone, N., Zeghal, M., Abdoun, T., Korre, E., Manzari, M., Madabhushi, G. S., Haigh, S., Madabhushi, S. S., Okamura, M., Sjafuddin, A. N., Escoffier, S., Kim, D.-S., Kim, S.-N., Ha, J.-G., Tobita, T., Yatsugi, H., Ueda, K., Vargas, R. R., Hung, W.-Y., Liao, T.-W., Zhou, Y.-G., & Liu, K. (2018). Twenty-four centrifuge tests to quantify sensitivity of lateral spreading to $D_r$ and PGA. In S. J. Brandenberg & M. T. Manzari (Eds.), *Geotechnical earthquake engineering and soil dynamics V, GSP 293* (pp. 383–393). Reston, VA: ASCE. https://doi.org/10.1061/9780784481486.040.

Kutter, B. L., Carey, T. J., Bonab, M. H., Stone, N., Manzari, M., Zeghal, M., Escoffier, S., Haigh, S. K., Madabhushi, G., Hung, W., Kim, D., Kim, N., Okamura, M., Tobita, T., Ueda, K., & Zhou, Y. (2019a). LEAP-UCD-2017 V. 1.01 model specifications. In B. Kutter et al. (Eds.), *Model tests and numerical simulations of liquefaction and lateral spreading: LEAP-UCD-2017*. New York: Springer.

Kutter, B. L., Carey, T., Stone, N., Zheng, B. L., Gavras, A., Manzari, M. T., Zeghal, M., Abdoun, T., Korre, E., Escoffier, S., Haigh, S., Madabhushi, G., Madabhushi, S. S. C., Hung, W.-Y., Liao, T.-W., Kim, D.-S., Kim, S.-N., Ha, J.-G., Kim, N. R., Okamura, M., Sjafuddin, A. N., Tobita, T., Ueda, K., Vargas, R., Zhou, Y.-G., & Liu, K. (2019b). LEAP-UCD-2017 comparison of centrifuge test results. In B. Kutter et al. (Eds.), *Model tests and numerical simulations of liquefaction and lateral spreading: LEAP-UCD-2017*. New York: Springer.

Manzari, M. T., El Ghoraiby, M. A., Kutter, B. L., Zeghal, M., Wang, R., Chen, R., Zhang, J.-M., Osamu Ozutsumi, O., Fukutake, K., Kiriyama, T., Fasano, G., Chiaradonna, A., Bilotta, E., Montgomery, J., Ziotopoulou, K., Chen, L., Ghofrani, A., Arduino, P., Wada, T., Ueda, K., Mercado, V., Fuentes, W., Lascarro, C., Yang, M., Barrero, A. R., Taiebat, M., Tsiaousi, D.,

Ugalde, J., Thaleia Travasarou, T., Ichii, K., Uemura, K., Orai, N., Hyodo, M., Abdoun, T., Haigh, S., Madabhushi, M. S. P., Tobita, T., Hung, W.-Y., Kim, D. S., Okamura, M., Zhou, -Y.-G., & Escoffier, S. (2019a). LEAP-2017 Simulation exercise: Calibration of constitutive models and simulation of the element test. In B. Kutter et al. (Eds.), *Model tests and numerical simulations of liquefaction and lateral spreading: LEAP-UCD-2017*. New York: Springer.

Manzari, M. T., El Ghoraiby, M., Zeghal, M., Kutter, B. L., Arduino, P., Barrero, A. R., Bilotta, E., Chen, L., Chen, R., Chiaradonna, A., Elgamal, A., Fasano, G., Fukutake, K., Fuentes, W., Ghofrani, A., Haigh, S., Hung, W.-Y., Ichii, K., Kim, D. S., Kiriyama, T., Lascarro, C., Madabhushi, M. S. P., Mercado, V., Montgomery, J., Okamura, M., Ozutsumi, O., Qiu, Z., Taiebat, M., Tobita, T., Travasarou, T., Tsiaousi, D., Ueda, K., Ugalde, J., Wada, T., Wang, R., Yang, M., Zhang, J.-M., Zhou, Y.-G., & Ziotopoulou, K. (2019b). LEAP-2017: Comparison of the Type-B numerical simulations with centrifuge test results. In B. Kutter et al. (Eds.), *Model tests and numerical simulations of liquefaction and lateral spreading: LEAP-UCD-2017*. New York: Springer.

Mercado, V., Fuentes, W., & Lascarro, C. (2017). *LEAP 2017 simulation exercise – Phase I: Model calibration*. Calibration report, June 28, DesignSafe.

Montgomery, J., & Ziotopoulou, K. (2019). Numerical simulations of selected LEAP centrifuge experiments with PM4Sand in FLAC. In B. Kutter et al. (Eds.), *Model tests and numerical simulations of liquefaction and lateral spreading: LEAP-UCD-2017*. New York: Springer.

Ozutsumi, O. (2019). LEAP-UCD-2017 numerical simulation at Meisosha Corp. In B. Kutter et al. (Eds.), *Model tests and numerical simulations of liquefaction and lateral spreading: LEAP-UCD-2017*. New York: Springer.

Qiu, Z., & Elgamal, A. W. (2019). Numerical simulations of LEAP dynamic centrifuge model tests for response of liquefiable sloping ground. In B. Kutter et al. (Eds.), *Model tests and numerical simulations of liquefaction and lateral spreading: LEAP-UCD-2017*. New York: Springer.

Tsiaousi, D., Ugalde, J., & Travasarou, T. (2019). LEAP-UCD-2017 simulation team Fugro. In B. Kutter et al. (Eds.), *Model tests and numerical simulations of liquefaction and lateral spreading: LEAP-UCD-2017*. New York: Springer.

Wada, T., & Ueda, K. (2019). LEAP-UCD-2017 Type-B predictions through FLIP at Kyoto University. In B. Kutter et al. (Eds.), *Model tests and numerical simulations of liquefaction and lateral spreading: LEAP-UCD-2017*. New York: Springer.

Wang, R., Chen, R., & Zhang, J.-M. (2019). LEAP-UCD-2017 simulations at Tsinghua University. In B. Kutter et al. (Eds.), *Model tests and numerical simulations of liquefaction and lateral spreading: LEAP-UCD-2017*. New York: Springer.

Yang, M., Barrero, A. R., & Taiebat, M. (2019). Application of a SANISAND model for numerical simulations of the LEAP 2017 experiments. In B. Kutter et al. (Eds.), *Model tests and numerical simulations of liquefaction and lateral spreading: LEAP-UCD-2017*. New York: Springer.

# Part II
# Centrifuge Experiment Papers

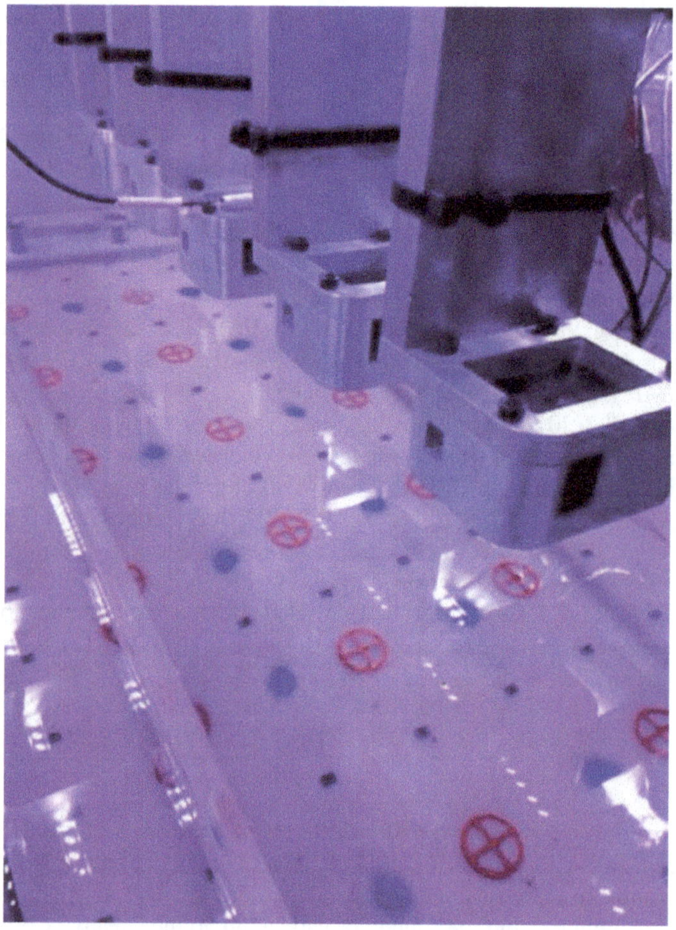

Centrifuge test at Zhejiang University

# Chapter 12
# LEAP-UCD-2017 Centrifuge Tests at Cambridge

**Srikanth S. C. Madabhushi, A. Dobrisan, R. Beber, Stuart K. Haigh, and Gopal S. P. Madabhushi**

**Abstract** As part of the LEAP project the seismic response of a liquefiable 5° slope was modelled at a number of centrifuges around the world. In this paper the two experiments conducted at Cambridge University are discussed. The model preparation is detailed with particular emphasis on the sand pouring, saturation and slope cutting process. The presence of the third harmonic in the input motion is shown and its significance discussed. The potential for wavelet denoising to filter random electrical noise from the pore pressure traces is illustrated. CPT strength profiles are highlighted and a possible softer layer in one of the tests is discussed. Whilst the specifications called for one dense and one loose test, the likelihood that both Cambridge tests were loosely poured is assessed. The PIV technique is used to obtain the displacements of the slope during the test. Finally, the correspondence between the PIV displacements and physical measurements of the marker movements is compared.

## 12.1 Introduction

The great advances in computation power of the last decade have greatly reduced the time and cost barriers of numerically studying highly coupled complex problems such as liquefaction. However, this necessitates high-quality experimental data on liquefaction problems to calibrate and validate the numerical assumptions against. Following in the steps of the VELACS project (Arulanandan and Scott 1993), the Liquefaction Experiments and Analyses Project (LEAP) strives to build a database of reliable centrifuge data from centres around the world to accompany the numerical research. The problem studied in this second phase of LEAP is that of a 5°

S. S. C. Madabhushi (✉) · A. Dobrisan · S. K. Haigh · G. S. P. Madabhushi
Department of Engineering, Cambridge University, Cambridge, UK

R. Beber
Department of Civil, Environmental and Mechanical Engineering, University of Trento, Trento, Italy



B. Kutter et al. (eds.), *Model Tests and Numerical Simulations of Liquefaction and Lateral Spreading*, https://doi.org/10.1007/978-3-030-22818-7_12

liquefiable slope subjected to 1 Hz destructive motions. This paper summarizes the two tests carried out at Cambridge within LEAP. The model preparation at Cambridge follows largely the same procedure originally detailed in the LEAP-GWU-2015 exercise (Madabhushi et al. 2017). In this paper emphasis will be placed on describing the salient differences of the model construction and results with respect to the other centres involved in the project.

## 12.2 Experiment Setup

Two tests were carried out at Cambridge as part of the LEAP 2017 exercise. At prototype scale, both tests modelled a 5° slope of uniform Ottawa F-65 sand, 4 m deep at the midpoint and 20 m in length. Figure 12.1 shows the general test setup, key dimensions and instruments at model scale. A centrifugal acceleration of 40 g at 1/3 the midpoint height was applied in both tests. To investigate the effect of sand density on the dynamic response of the slope, the two Cambridge University LEAP tests (CU01 and CU02) had different prescribed target densities: 1651 and 1599 kg/m$^3$, respectively. These corresponded to relative densities of 67% and 48%.

### *12.2.1 Sand Pouring*

An automated spot pluviator (Madabhushi et al. 2006) was used for sand pouring. The nozzle aperture (which controls the sand flow rate), drop height and to a less extent the pluviator translation characteristics determine the bulk density of the soil layers poured. This arrangement differs to the LEAP specification in terms of the sieve dimensions and whilst the same range of densities can be obtained the potential for differences in the soil fabric should be borne in mind.

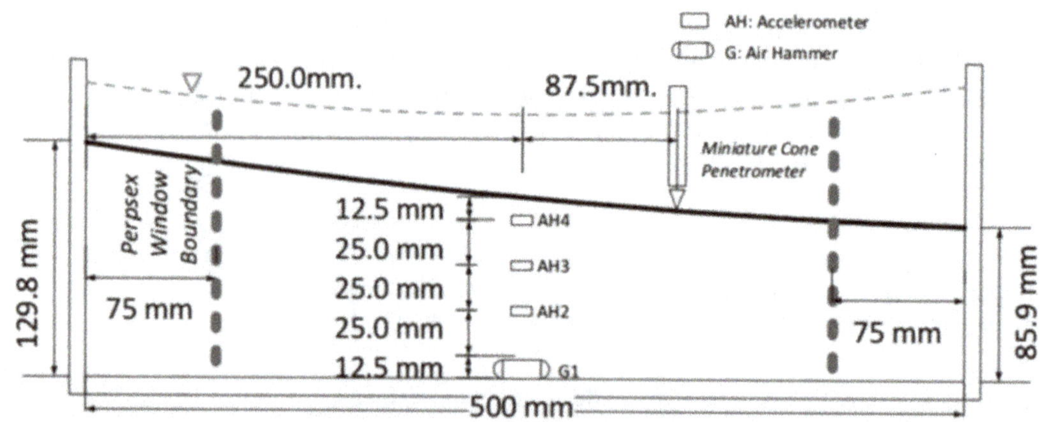


**Fig. 12.1** General model schematic—not all sensors shown

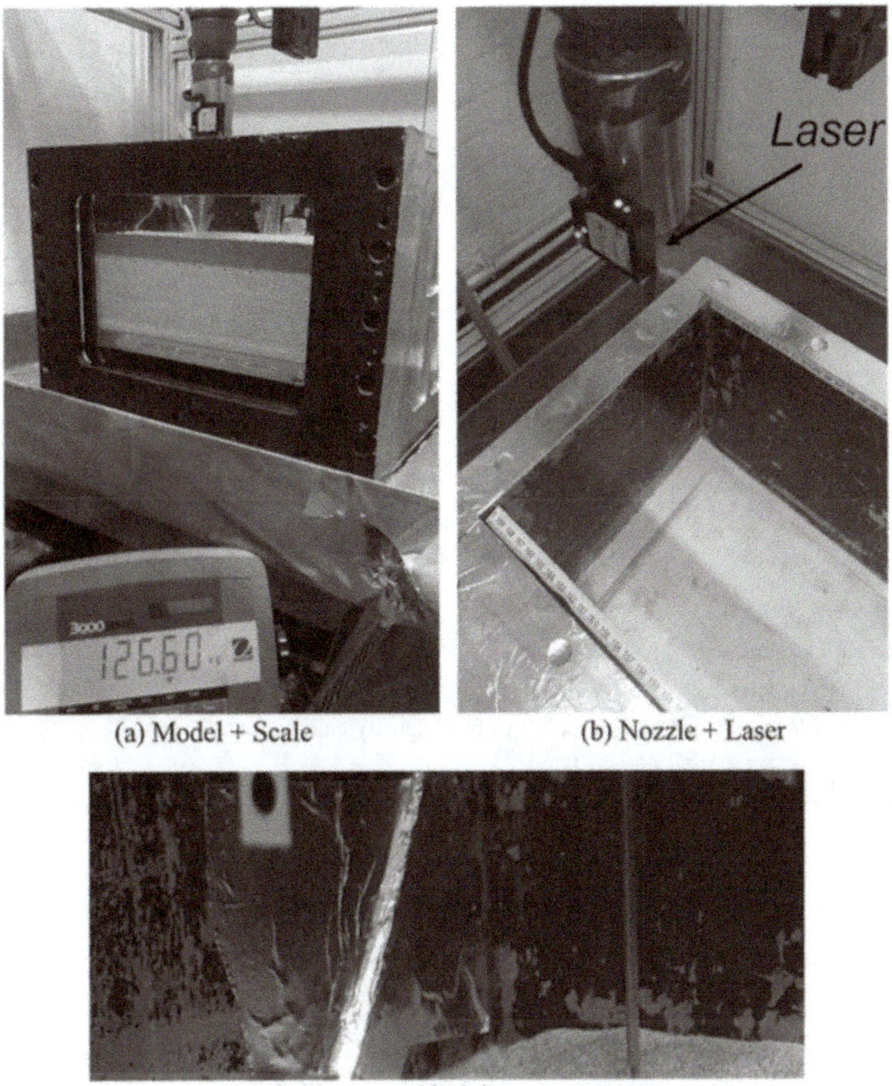


(a) Model + Scale (b) Nozzle + Laser

(c) Calliper sand height measurement

**Fig. 12.2** LEAP 2017 sand pouring. (**a**) Model + scale (**b**) nozzle + laser (**c**) calliper sand height measurement

The container was placed on a scale to monitor the sand mass during pluvation without disturbing the model. A laser unit was attached to the hopper nozzle to measure the sand height between the nozzle translations. The height of poured sand was also measured by callipers in conjunction with the laser measurement to increase the number of readings and to check consistency between the two methods of determining height. The pouring setup is presented in Fig. 12.2.

Four calliper measurements were taken for each layer of poured sand. Due to the operation mode of the pluviator, laser measurements could only be taken every fourth layer poured. Figure 12.3a highlights the general agreement between calliper and laser measurements.

Since the sand mass was recorded by the digital scale, the height measurements were used to estimate bulk density. As shown in Fig. 12.3b a relatively small scatter

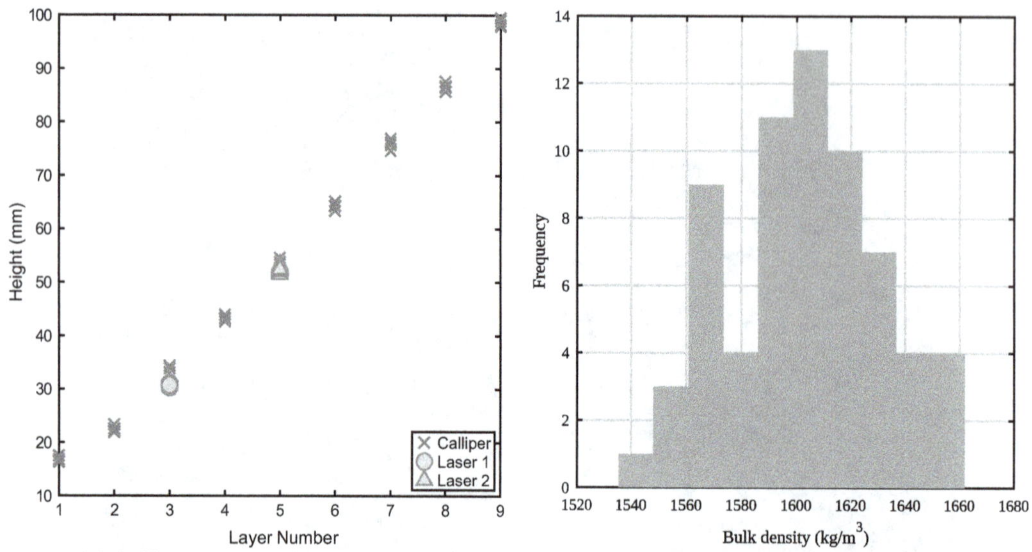


(a) Calliper and laser height measurements (b) Density histogram from height results

**Fig. 12.3** Comparison of laser and calliper results. Bulk density inference (CU02). (**a**) Calliper and laser height measurements. (**b**) Density histogram from height results

**Table 12.1** Mean and standard deviation of bulk density calculations

| Test | Target (kg/m3) | Mean Achieved (kg/m3) | Based on | Standard Deviation (kg/m3) |
|---|---|---|---|---|
| CU01 | 1652 | 1630 | Trimmed mean of 36 calliper measurements, interior 75 % | 43.1 |
| | | 1656 | Trimmed mean of 45 laser measurements, interior 75 % | 12.3 |
| CU02 | 1599 | 1610 | Trimmed mean of 35 calliper measurements, interior 60 % | 30.3 |
| | | 1606 | Trimmed mean of 30 calliper measurements, interior 75 % | 15 |

in height measurements translates to rather large uncertainties in density estimation. The same is accentuated in Table 12.1 which includes the standard deviation of the bulk density results.

During pouring instruments were positioned in the sand. An air hammer (Ghosh and Madabhushi 2002) was placed at the base of the model (Fig. 12.4a) in CU02. The impulses it generated, picked up by the piezoelectric accelerometers (Fig. 12.4b), can allow assessment of the shear wave velocity during the test. A list of the sensors employed and their positions can be found in Kutter et al. (2018).

### *12.2.2 Viscosity Measurement*

The viscosity of the methyl cellulose used to saturate the model was specified at 40 cSt to fulfil dynamic scaling laws (Schofield 1981). The viscosity was measured

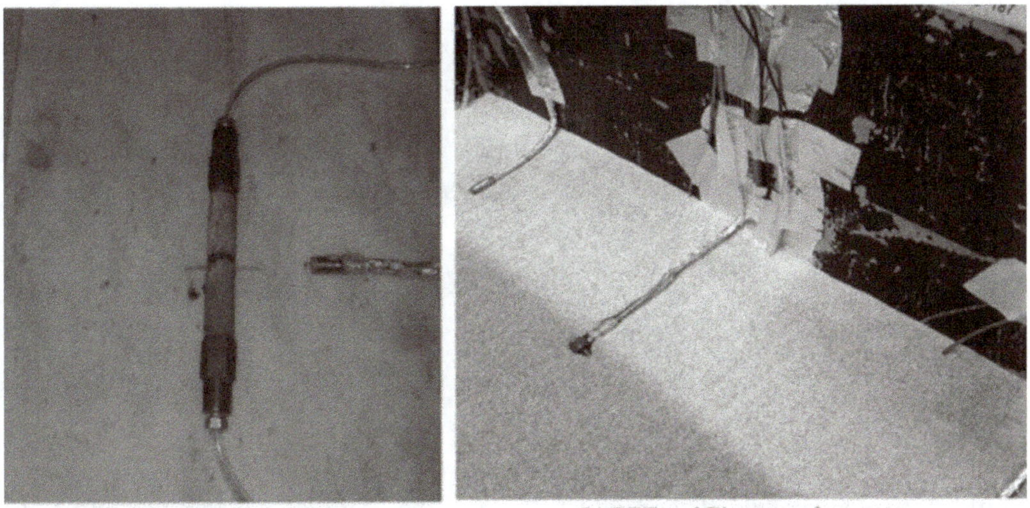

(a) Air-hammer at box base (b) PPT and Piezo accelerometer

**Fig. 12.4** Instruments in model container (CU02). (**a**) Air hammer at box base. (**b**) PPT and piezo accelerometer

**Fig. 12.5** Saturation setup

using a viscometer and values from before the test were 45 cSt for CU01 and 44 cSt for CU02. Owing to the temperature dependence of viscosity, the methyl cellulose was prepared with an assumed centrifuge temperature beforehand. A change of 1 °C in the centrifuge yields approximately a 5% difference in viscosity.

### *12.2.3 Saturation*

For both tests the models were saturated under vacuum using the CAM-SAT system (Stringer and Madabhushi 2009) that controls the rate of mass influx into the model base (Fig 12.5). A maximum rate of 0.5 kg/h was chosen to prevent fluidization of the soil. Prior to saturation the model was flushed with $CO_2$ gas in

**Fig. 12.6** Log-spiral guides mounted on top of model container

several cycles to improve the vacuum obtained. However, the degree of saturation could not be accurately determined following the method of Okamura and Soga (2006) due to the measured compliance in the saturation systems tubing. Cross comparison of the excess pore pressure generation between centres can be used to infer the degree of saturation in the Cambridge tests.

### *12.2.4 Slope Cutting*

After saturation the required slope was cut into a logarithmic spiral shape to accommodate the tangential shaking direction with respect to the package centrifugal motion. A cutting plate running along machined log-spiral guides was used to obtain the desired shape (Fig. 12.6). As discussed in Madabhushi et al. (2017), the fluid level was lowered during the cutting with the resulting capillary suction increasing the effective stress and aiding the shaping process.

In the Cambridge centrifuge the package sits horizontally so the cutting plate was designed to have a 1:40 slope to correct for the angle between the normal g field and the sand surface. The cut slope is shown in Fig. 12.7.

### *12.2.5 CPT*

An in-flight CPT mounted on the package (Fig. 12.8) was used to determine soil strength before and after the destructive motion. The CPT used was different to the UC Davis one; however additional centrifuge tests confirmed consistent results between the two devices (Carey et al. 2018).

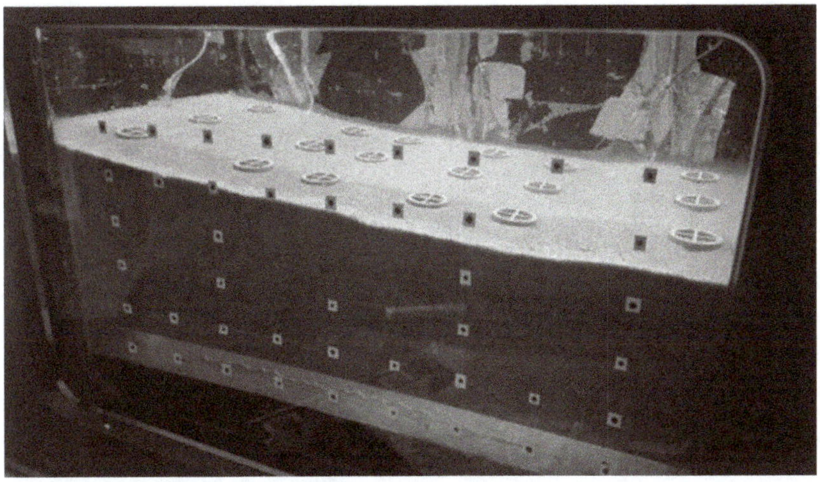

**Fig. 12.7** Cut slope with markers on top

**Fig. 12.8** Centrifuge package with CPT attached

## 12.3 Destructive Motions

The target prototype input motions for both CU01 and CU02 were ramped 1 Hz sines with PGA of 0.15 g. In an attempt to calibrate the desired input motion, dummy packages of the same mass as the actual LEAP models were loaded in an ESB container box (Brennan and Madabhushi 2002) and shaken. As shown in Fig. 12.9, the calibration runs exhibited little high-frequency noise, with most of the shake energy contained within the desired 1 Hz signal.

The recorded input acceleration during the two LEAP tests is shown in Fig. 12.10. As well as the PGA being closer to 0.2 g, the third harmonic at 3 Hz carried

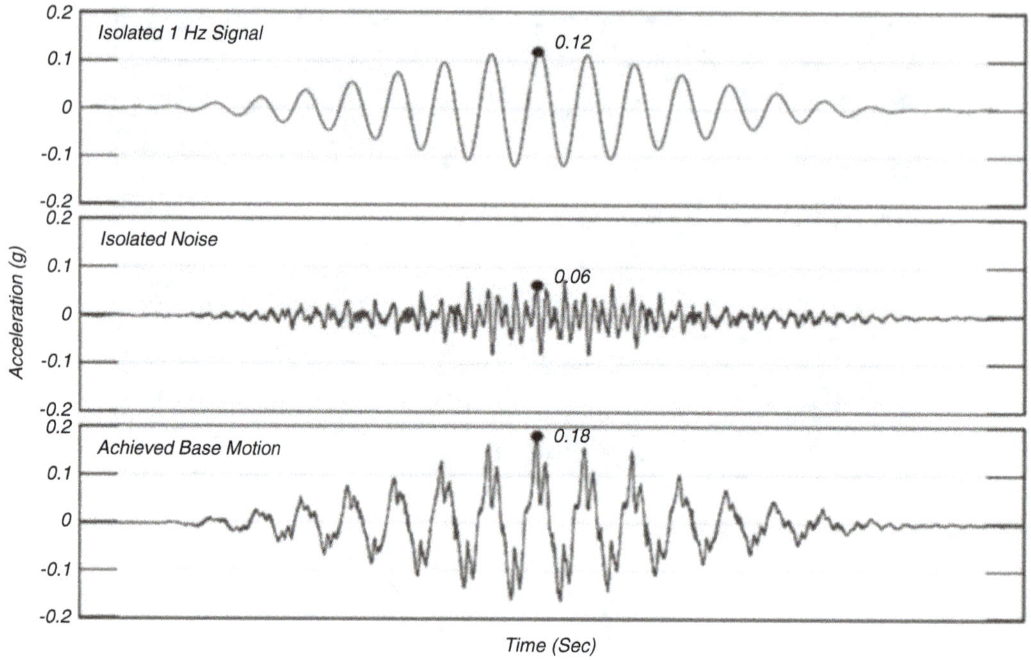


**Fig. 12.9** Input motion and isolated high-frequency component of calibration run

significant energy in both these runs. Although the mass was matched between the LEAP and dummy models, the LEAP tests were carried out in a window box. Potenitally, the different box stiffness might have triggered a resonant response of the shaker and model system at 3 Hz. A significant 3 Hz component was also observed previously by Madabhushi et al. (2017) in the LEAP 2015 exercise.

The impact of the third harmonic on the slope behaviour remains an area of research. Comparison between the centrifuge experiments at Cambridge and other centres shows the dilation spikes were less pronounced when compared with tests with a smaller third mode vibration (Fig. 12.11). The attenuated dilation could explain the larger slope displacements recorded during CU01 and CU02 when compared with other centrifuge centres (Fig. 12.12).

Accurate prediction of the dilation spikes during the slope shaking is numerically challenging. The numerical analyses in Madabhushi et al. (2017) could generally capture the behaviour of the Cambridge centrifuge tests in terms of the accelerations, excess pore pressures and slope displacements. However, the simulations showed little dependence on the third harmonic for the constitutive model and properties used. The potential sensitivity of both the experiments and numerical analyses to the third harmonic requires further comparison and investigation.

The increased noise on six of the eight pore pressure transducers was traced back to a malfunctioning amplifier card. The data presented in Beber et al (2018b) was filtered in the frequency domain using a low-pass filter. The potential for wavelet denoising to better recover the original signal is explored in Fig. 12.13. The relative magnitudes of the real signal and random electrical noise are clearly different and can be easily separated to allow reconstruction of a cleaner signal.

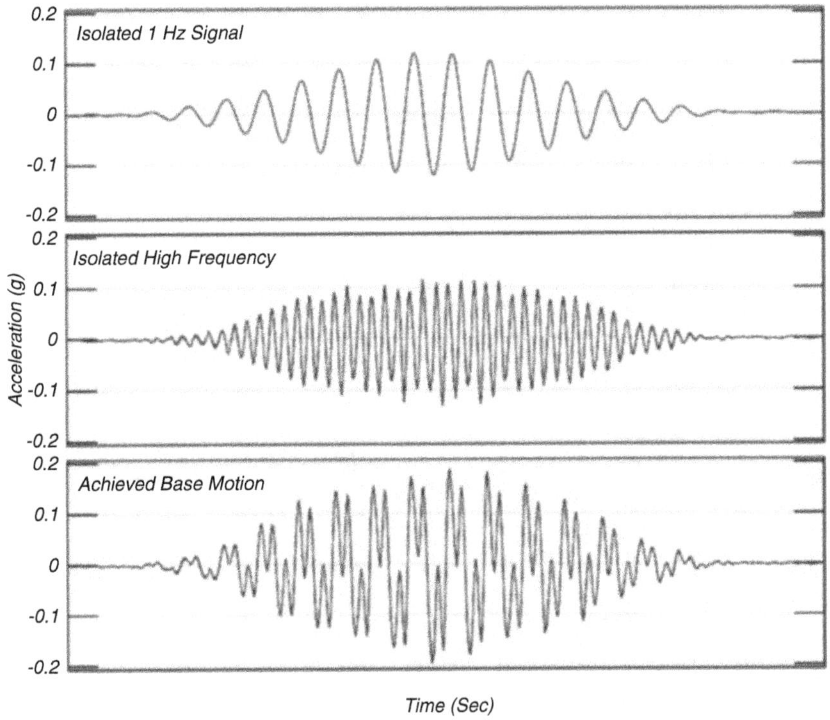


(a) CU01

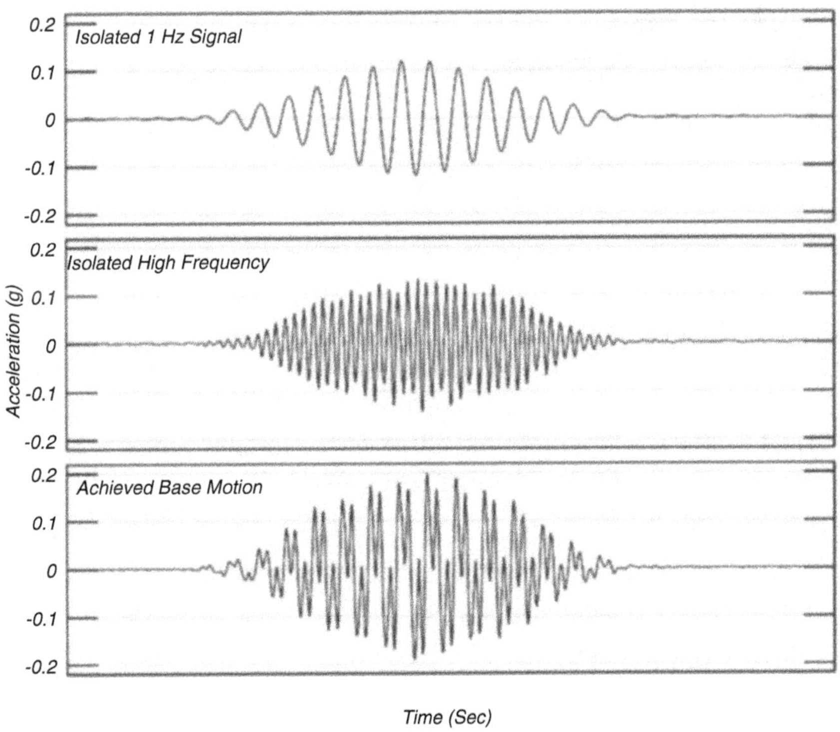


(b) CU02

**Fig. 12.10** Input motion and isolated high-frequency component. (**a**) CU01 (**b**) CU02

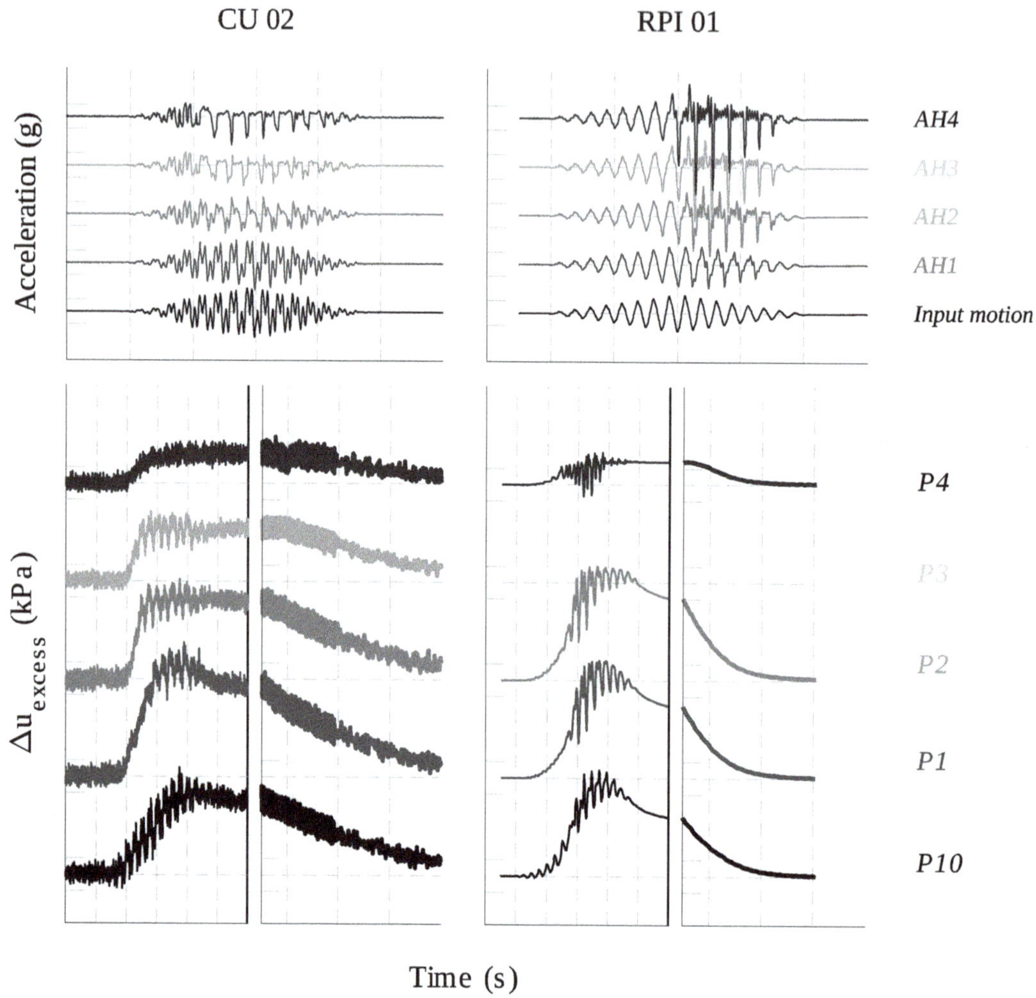


**Fig. 12.11** Presence of significant dilation spikes for input motions with reduced third harmonic

Figure 12.10 highlights the similarity between input accelerations for CU01 and CU02 and Fig. 12.14 shows the associated wavelet-denoised excess pore pressure generation during the two tests. CU01 was designed as a medium dense sand test (Rd 67%) and CU02 as a loose test (Rd 48%). However, the very similar excess pore pressure generation might suggest the difference in density between the two was not as large.

## 12.4 CPT Strength Profiles

The CPT strength traces with depth recorded in-flight before the destructive motion are shown in Fig. 12.15. The CU01 trace suggests a softer sand layer was encountered at 2 m depth. However, given the local nature of a CPT test and the overall consistency of the other instruments in CU01, the "bump" could have been a local effect rather than a general trend.

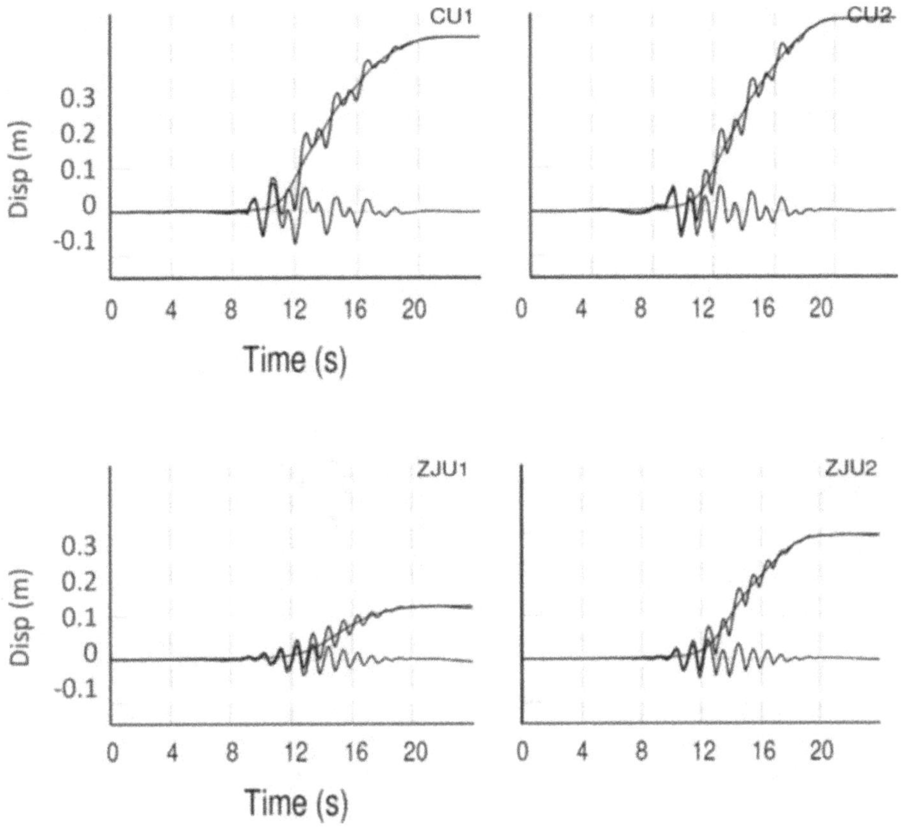


**Fig. 12.12** Slope displacement comparison to other test centres

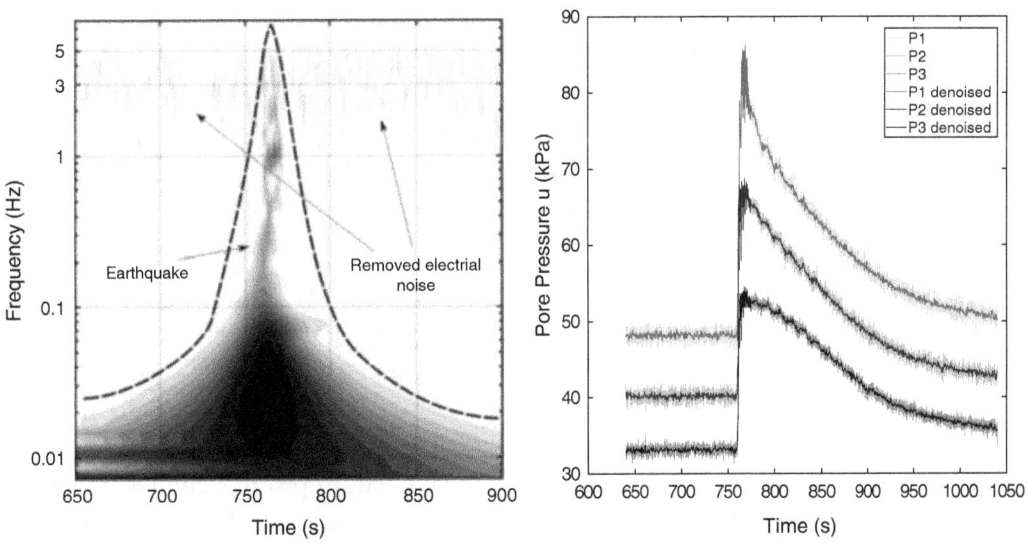


(a) Wavelet of original signal with noise visible

(b) Denoised PPTs (CU01)

**Fig. 12.13** Wavelet denoising. (**a**)Wavelet of original signal with noise visible. (**b**) Denoised PPTs (CU01)

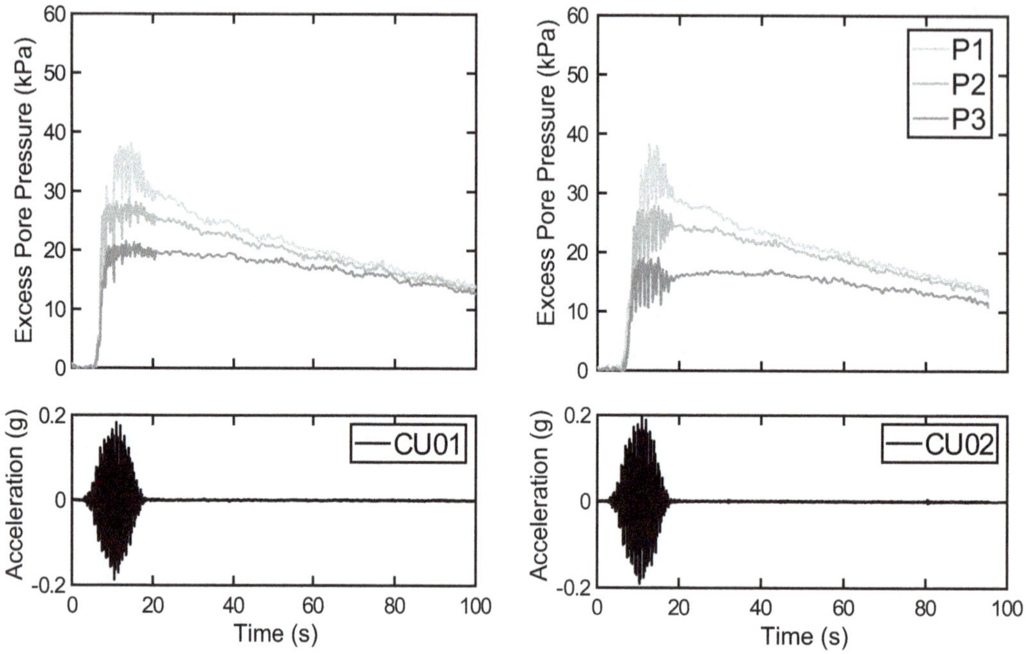


**Fig. 12.14** Excess pore pressure generation CU01 and CU02

**Fig. 12.15** CPT strength profiles

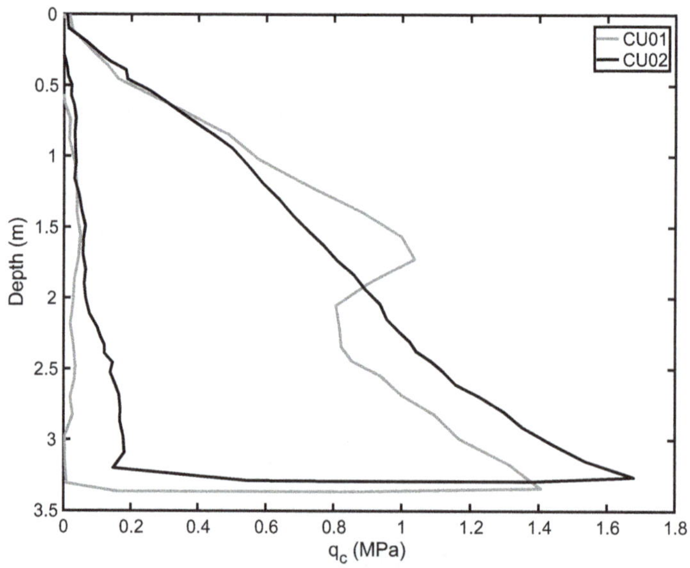


As shown by Beber et al. (2018a, b), the CPT strength profiles can be correlated with relative density through various empirical correlations. These indicate that CU01 and CU02 have a similarly loose average density.

Comparison of the CPT results between the LEAP 2017 tests reveals other medium-dense tests typically having cone resistances approximately double those recorded in CU01. Overall, caution is recommended if interpreting CU01 as a medium-dense sand test. Nevertheless, the similarity of the accelerations, excess pore pressure generations, marker displacements and average soil strength between CU01 and CU02 potentially recommend the first test as an additional dataset for a loose sand.

## 12.5 PIV

Images taken during the tests for PIV analysis are available at Beber et al. (2018a, b). These can be used for various analysis scenarios. Beber et al. (2018a, b) show from PIV that soil densification during swing up is small, but not necessarily negligible. Moreover, by tracking soil patches at the same location along the slope as the markers, the consistency between the PIV and marker measurements can be verified. The comparison is shown in Fig. 12.16a, which highlights a considerable difference between the two. As well as the limited measurement precision, the readings may be inaccurate if the markers did not remain fully coupled with the liquefying soil. Equally, the comparison necessitates that the PIV displacements near the Perspex are representative of the slope centreline. The container geometry is detailed by Cilingir and Madabhushi (2010) and may be considered rigid in the plane strain direction.

The dynamic component of the PIV displacement shown in Fig. 12.16b can be extracted. Likewise, the double integral of the accelerometer traces relative to the base motion can give the dynamic displacement of the slope centreline. The comparison between these two methods, shown in Fig. 12.16b, is favourable and suggests the dynamic displacement is fairly uniform across the slope width. This

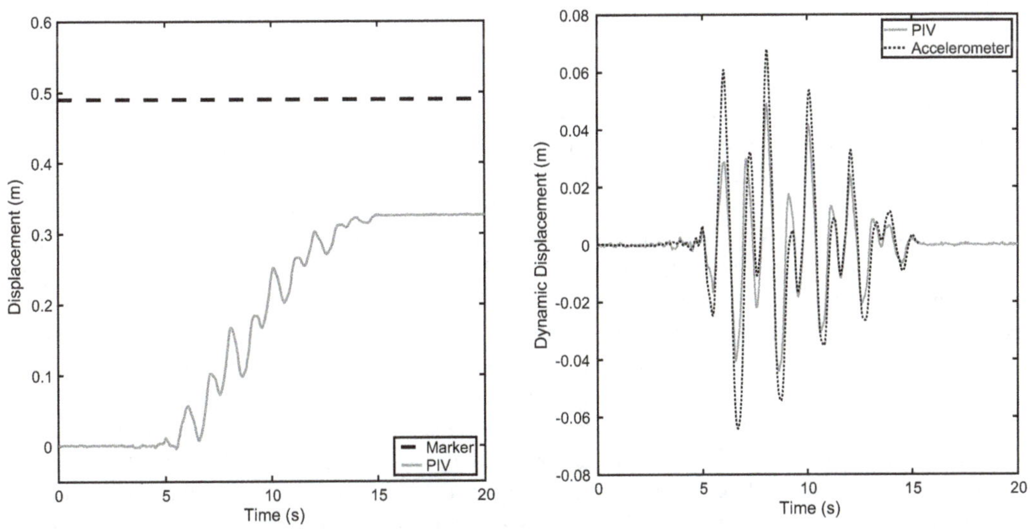


(a) Comparison with marker measurements (b) Dynamic component against accel. result

**Fig. 12.16** PIV displacements in CU01. (**a**) Comparison with marker measurements. (**b**) Dynamic component against accelerometer result increases the confidence in the PIV method to accurately determine the total displacement of the slope.

## 12.6 Conclusions

This paper summarized the methodology and results from the LEAP 2017 experiments carried out at Cambridge. The steps taken to measure the mass and height of sand during pouring are shown and the resulting density estimates and their uncertainty highlighted. Cross comparison between the two Cambridge tests, as well as placing them within the wider context of the LEAP database, suggests both CU01 and CU02 had similarly loose initial void ratios. During the tests, the potential for the container dynamics to influence the third harmonic in the input motion is raised, and the implication for the excess pore pressure generation and resulting slope displacement touched upon. The total displacement of the slope can be accurately determined from the PIV method, and the correspondence of the dynamic displacements between the front face and centreline of the slope was assessed. Overall, the importance of thorough examination of the internal consistency of centrifuge data is highlighted, and the value of a large database of results to better understand the relative sensitivity to experimental variations expounded.

**Acknowledgment** The authors wish to express their gratitude to all of the technicians at the Schofield centre for their help and support during the model preparation and testing.

## References

Arulanandan, K., & Scott, R. F. (1993). *Verification of numerical procedures for the analysis of soil liquefaction problems*. Rotterdam: A.A. Balkema.

Beber, R., Madabhushi, S. S. C., Dobrisan, A., Haigh, S. K., & Madabhushi, S. P. G. (2018a). LEAP GWU 2017: Investigating different methods for verifying the relative density of a centrifuge model. *International Conference in Physical Modelling in Geotechnics, London*.

Beber, R., Madabhushi, S. S. C., Dobrisan, A., Haigh, S. K., & Madabhushi, S. P. G. (2018b). CU1, CU2 – University of Cambridge Experiments. In *PRJ-1843: LEAP-UCD-2017*. https://www.designsafe-ci.org/data/browser/projects/7067616763427688936-242ac11d-0001-012/

Brennan, A. J., & Madabhushi, S. P. G. (2002). Design and performance of a new deep model container for dynamic centrifuge testing. In *International Conference on Physical Modelling in Geotechnics* (pp. 183–188). Rotterdam: Balkema.

Carey, T., Gavras, A., Kutter, B., Haigh, S. K., Madabhushi, S. P. G., Okamura, M., Kim, D. S., Ueda, K., Hung, W.-Y., Zhou, Y.-G., Liu, K., Chen, Y.-M., Zeghal, M., Abdoun, T., Escoffier, S., & Manzari, M. (2019). A new shared miniature cone penetrometer for centrifuge testing. In *Proceedings of 9th International Conference on Physical Modelling in Geotechnics, ICPMG 2018* (Vol. 1, pp. 293–229). London: CRC Press/Balkema.

Cilingir, U., & Madabhushi, S. P. G. (2010). Particle image velocimetry analysis in dynamic centrifuge tests. In *Proceedings of the 7th International Conference on Physical Modelling in Geotechnics (ICPMG)* (pp. 319–324). Zurich: CRC Press.

Ghosh, B., & Madabhushi, S. P. G. (2002). An efficient tool for measuring shear wave velocity in the centrifuge. In *Proceedings of the International Conference on Physical Modelling in Geotechnics* (pp. 119–124). Rotterdam: A.A. Balkema.

Madabhushi, S. P. G., Houghton, N. E., & Haigh, S. K. (2006). A new automatic sand pourer for model preparation at University of Cambridge. In *Proceedings of the 6th International Conference on Physical Modelling in Geotechnics* (pp. 217–222). London: Taylor & Francis.

Madabhushi, S. S. C., Haigh, S. K., & Madabhushi, S. P. G. (2017). LEAP-GWU-2015: Centrifuge and numerical modelling of slope liquefaction at the University of Cambridge. *Soil Dynamics and Earthquake Engineering*. https://doi.org/10.1016/j.soildyn.2016.11.009

Okamura, M., & Soga, Y. (2006). Effects of pore fluid compressibility on liquefaction resistance of partially saturated sand. *Soils and Foundations, 46*(5), 695–700.

Schofield, A. N. (1981). Dynamic and earthquake geotechnical centrifuge modelling. *International Conferences on Recent Advances in Geotechnical Earthquake Engineering and Soil Dynamics*.

Stringer, M. E., & Madabhushi, S. P. G. (2009). Novel computer-controlled saturation of dynamic centrifuge models using high viscosity fluids. *Geotechnical Testing Journal, 32*(6), 559–564.

# Chapter 13
# LEAP-UCD-2017 Centrifuge Test at University of California, Davis

**Trevor J. Carey, Nicholas Stone, Masoud Hajialilue Bonab, and Bruce L. Kutter**

**Abstract** Three centrifuge experiments were performed at the University of California, Davis, for LEAP-UCD-2017. LEAP is a collaborative effort to assess repeatability of centrifuge test results and to provide data for the validation of numerical models used to predict the effects of liquefaction. The model configuration used the same geometry as the LEAP-GWU-2015 exercise: a submerged slope of Ottawa F-65 sand inclined at 5 degrees in a rigid container. This paper focuses on presenting results from the two destructive ground motions from each of the three centrifuge models. The effect of each destructive ground motion is evaluated by accelerometer recordings, pore pressure response, and lateral deformation of the soil surface. New techniques were implemented for measuring liquefaction-induced lateral deformations using five GoPro cameras and GEO-PIV software. The methods for measuring the achieved density of the as-built model are also discussed.

## 13.1 Introduction

The current phase of LEAP, LEAP-UCD-2017, involved centrifuge experiments conducted at nine different research facilities, including the University of California, Davis (UCD). The experiment, similar to LEAP-GWU-2015 (Kutter et al. 2017), consisted of a submerged clean sand sloped at 5 degrees, subjected to a 1 Hz ramped sine wave ground motion inputted at the base of a rigid model container. Three experiments were performed on the 1 m radius Schaevitz centrifuge at the Center for Geotechnical Modeling. The 1 m centrifuge performs shaking in the circumferential direction of the centrifuge. Detailed specifications by Kutter et al. (2019a) were provided to facilitate replicability among the different centrifuge facilities.

Discussed herein are the specifications that pertained specifically to the UCD

T. J. Carey (✉) · N. Stone · B. L. Kutter
Department of Civil and Environmental Engineering, University of California, Davis, CA, USA
e-mail: tjcarey@ucdavis.edu

M. H. Bonab
Department of Civil Engineering, University of Tabriz, Tabriz, Iran



B. Kutter et al. (eds.), *Model Tests and Numerical Simulations of Liquefaction and Lateral Spreading*, https://doi.org/10.1007/978-3-030-22818-7_13

experiments and deviations from the specifications, both intended and unintended. The implementation of a new technique for measuring slope deformations during strong shaking is discussed and the process for measuring the achieved dry density of a constructed model and saturation protocol are detailed. Kutter et al. (2019b) provide a detailed comparison of all centrifuge experiments from the nine participating facilities, including the results presented herein.

## 13.2 UC Davis Test Specific Information

### 13.2.1 Description of the Model and Instrumentation

The container dimensions for the UCD experiments are 457.2 mm (L) × 279.4 mm (W) × 177.8 mm (H). The 457.2 mm length of the container was obtained by placing 25.4-mm-thick plastic plates on each end wall of the rigid container. The plastic plates were needed to ensure that the soil would remain completely submerged at 1 g without overtopping when the water surface curved during spinning. Figure 13.1 details the test geometry, sensor locations, and PVC blocks in model scale. Note that Carey et al. (2017) used a different model container and spacer configuration for LEAP-GWU-2015.

A flat soil surface in a centrifuge represents a hill in prototype. To model a flat surface in prototype, the model surface should be curved with the same radius as the centrifuge. A 5-degree slope from the normal of the radial g-field would have a varying radius along the slope of the model surface, which is described theoretically by a log-spiral. Carey et al. 2017 showed that rotating a circular arc by 5 degrees is a reasonable approximation for the log-spiral surface.

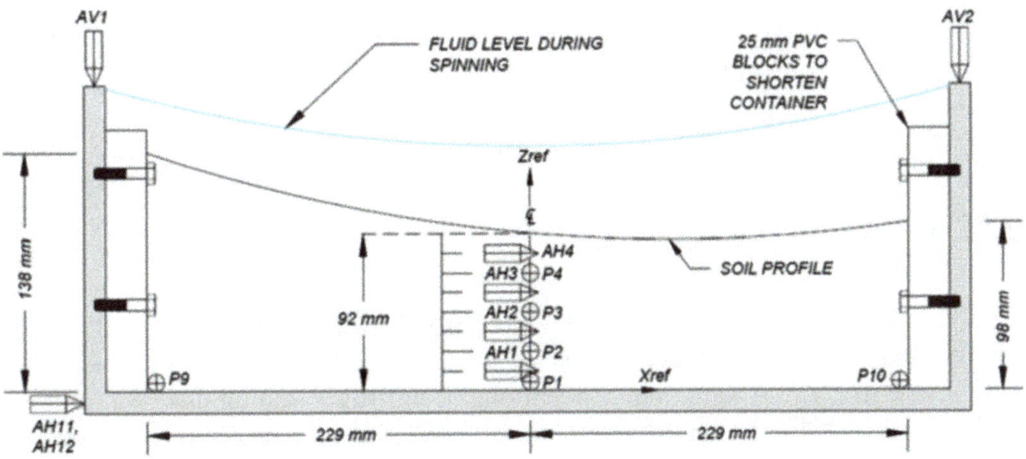


**Fig. 13.1** Model geometry and sensor layout (dimensions in model scale)

**Fig. 13.2** Wooden template and vacuum tool used to excavate sand to produce the log-spiral surface

Figure 13.2 shows the wooden template fabricated to approximate the log-spiral surface of the model. Sand was first pluviated to a depth above the desired final curved surface. Using a depth gauge, the amount of soil to remove was roughly calculated by measuring the depth of the model at the ends and midpoint. Excess soil was removed by vacuuming across and then down the slope. The final depths of the model were verified at each quarter point. With this method, the final surface depth could be constructed to ±1 mm of the specified height.

### *13.2.2 Sensors*

The number of sensors placed in the model was limited by the capacity of the data acquisition system; therefore, only the required pore pressure transducers (P1, P2, P3, P4, P9, and P10) and accelerometers (AH1, AH2, AH3, AH4, AH11, AH12, AV1, AV2) were included.

### *13.2.3 Scaling Laws*

The scaling laws for LEAP-UCD-2017 are provided by Kutter et al. (2019a). The length scale factor $L^*$ is defined as $L^* = L_{MODEL}/L_{PROTOTYPE} = (0.457\text{ m})/(20\text{ m}) = 1/43.75$. Using the conventional centrifuge scaling law for gravity, $g^* = 1/L^* = 43.75$. An angular velocity of 194 RPM was determined to produce $g^* = 43.75$ at the effective radius of 1.033 m (the radius to 1/3 of the depth of soil).

## 13.3 Test Results

UCD performed three experiments with a target dry density of 1651 kg/m$^3$. The measured dry densities of the three models ranged from 1648 to 1665 kg/m$^3$. Densities were calculated by the measurement of mass and volume. A more detailed explanation of the method used to measure the volume of each model is described later in this paper. Each specimen used the standard pluviator, which consisted of a No. 16 sieve with three slots (Kutter et al. 2019a) to place the soil. To provide the best likelihood to obtain the target density, the drop height from the standard pluviator to the model surface was adjusted before each experiment, based on the previous model's measured density. The calculated achieved density does, however, seem to be affected by unaccounted for factors such as humidity, temperature, personal habits and uncertain measurements of mass and volume.

### *13.3.1 Achieved Ground Motions*

Recognizing the achieved base motion will contain additional frequencies superimposed on the 1 Hz motion, Kutter et al. (2019a) introduced $PGA_{effective}$ as a ground motion intensity parameter, where $PGA_{effective} = PGA_{1Hz} + ½PGA_{highfreq}$. The PGA of the 1 Hz and high-frequency components are found using a filtering scheme described by Kutter et al. (2019a). Table 13.1 lists the sequence of shaking, and corresponding effective PGAs for each of the three models. Additional nondestructive motions were performed prior to and following each destructive motion, consisting of a series of square pulses with a PGA less than 0.04 g in an attempt to characterize the specimen shear-wave velocity. These motions were considered to be nondestructive due to their low magnitude and high-frequency content; however, these nondestructive motions have not been fully processed. The raw data from the nondestructive motions is archived in NHERI DesignSafe with the processed data of the destructive motions.

**Table 13.1** Ground motion PGAs for the three UCD experiments

| | Destructive motion 1 | | | | Destructive motion 2 | | | |
|---|---|---|---|---|---|---|---|---|
| | $PGA_{raw}$ (g) | $PGA_{eff}$ (g) | $PGA_{1Hz}$ (g) | $PGA_{HF}$ (g) | $PGA_{raw}$ (g) | $PGA_{eff}$ (g) | $PGA_{1Hz}$ (g) | $PGA_{HF}$ (g) |
| UCD1 | 0.165 | 0.149 | 0.119 | 0.060 | 0.378 | 0.296 | 0.214 | 0.163 |
| UCD2 | 0.339 | 0.210 | 0.149 | 0.122 | 0.248 | 0.220 | 0.179 | 0.083 |
| UCD3 | 0.192 | 0.183 | 0.134 | 0.099 | 0.180 | 0.172 | 0.132 | 0.081 |

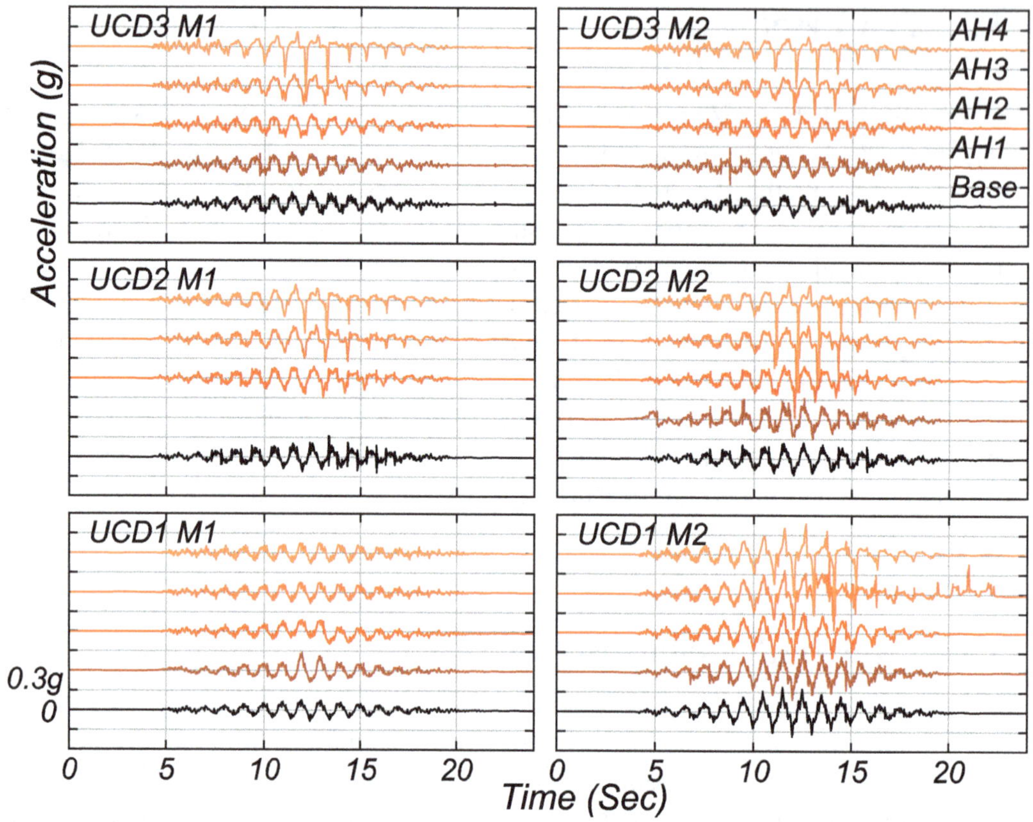


**Fig. 13.3** Horizontal acceleration time histories for UCD1, UCD2, and UCD3 (two destructive motions each)

## *13.3.2 Accelerometer Records During Destructive Motions*

The acceleration time histories of the input base motion and the four central horizontal accelerometers (AH1, AH2, AH3, and AH4) are reported in Fig. 13.3. The input base motion is taken as the average of the accelerations measured by AH11 and AH12.

- UCD1
  - Destructive motion 1: The surface accelerometers recorded an almost identical motion as the base input motion. Both motions appear to remain in phase with one another, implying the soil moved nearly as a rigid mass, with small shear strains, and no evidence of liquefaction.
  - Destructive motion 2: The second motion, M2, was the strongest of all the destructive motions. Initially, the central array sensors are in-phase with the input motion. As liquefaction occurs, beginning at the shallowest sensors first, dilation spikes, characterized as sharp, sudden accelerations, occur. The magnitudes of the spikes decrease with depth since the severity of liquefaction is decreasing with depth.

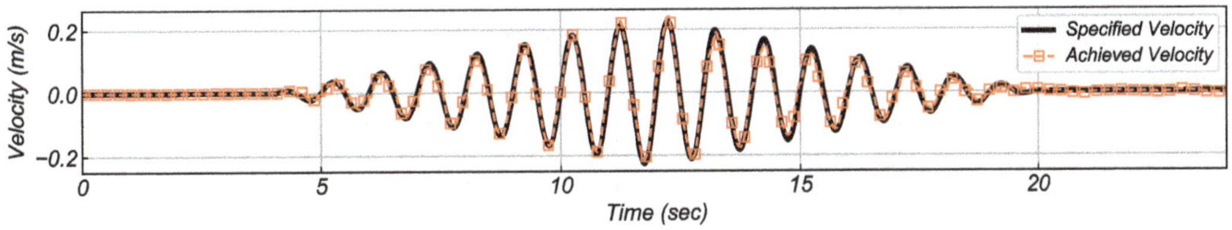


**Fig. 13.4** Specified versus achieved velocity for UCD2 destructive motion 1

- UCD2
  - Destructive motion 1: The input base motion had high-frequency components superimposed on the 1 Hz signal, especially from 12–16 seconds. It is believed that this behavior may have been caused by insufficient pressure in the supply accumulator for the hydraulic shaker. When the achieved velocity of the base motion (found by integrating the achieved base motion) is compared with the specified velocity in Fig. 13.4, good agreement is observed. This indicates the superimposed high-frequency accelerations had minimal effect on the base input motion velocity.
  - Destructive motion 2: The input motion contained a larger (0.18 g) 1 Hz component than M1 (0.15 g), but M2 had a smaller high-frequency component (0.083 g) compared to M1 (0.122 g) (see Table 13.1). The reduction of the high-frequency components was partly from increasing the hydraulic pressure to the accumulator. The dilation spikes for M2 have larger magnitudes than the spikes in M1, indicating more severe liquefaction in M2 than in M1. This also indicates that the severity of liquefaction dilation is more sensitive to the low-frequency components of the input motion.
- UCD3
  - Destructive motions 1 and 2: The input motions for M1 and M2 are nearly repeated in terms of magnitude, frequency content, and duration. Similar dilation spikes occur in both motions, but differences in the magnitudes can be observed. The spikes for AH4 are larger during M1 compared to M2, but the AH3 spikes are larger during M2. This could indicate localized zones of loose and dense material.

### *13.3.3 Excess Pore Pressures*

The excess pore pressures of the central array generated during strong shaking are reported in Fig. 13.5. P1 is located on the base of the container, at roughly a depth of 4 m, and P4 is at the surface, at 1 m depth. The initial vertical effective stress ($\sigma'$) prior to shaking is approximately 40, 30, 20, and 10 kPa for P1, P2, P3, and P4, respectively.

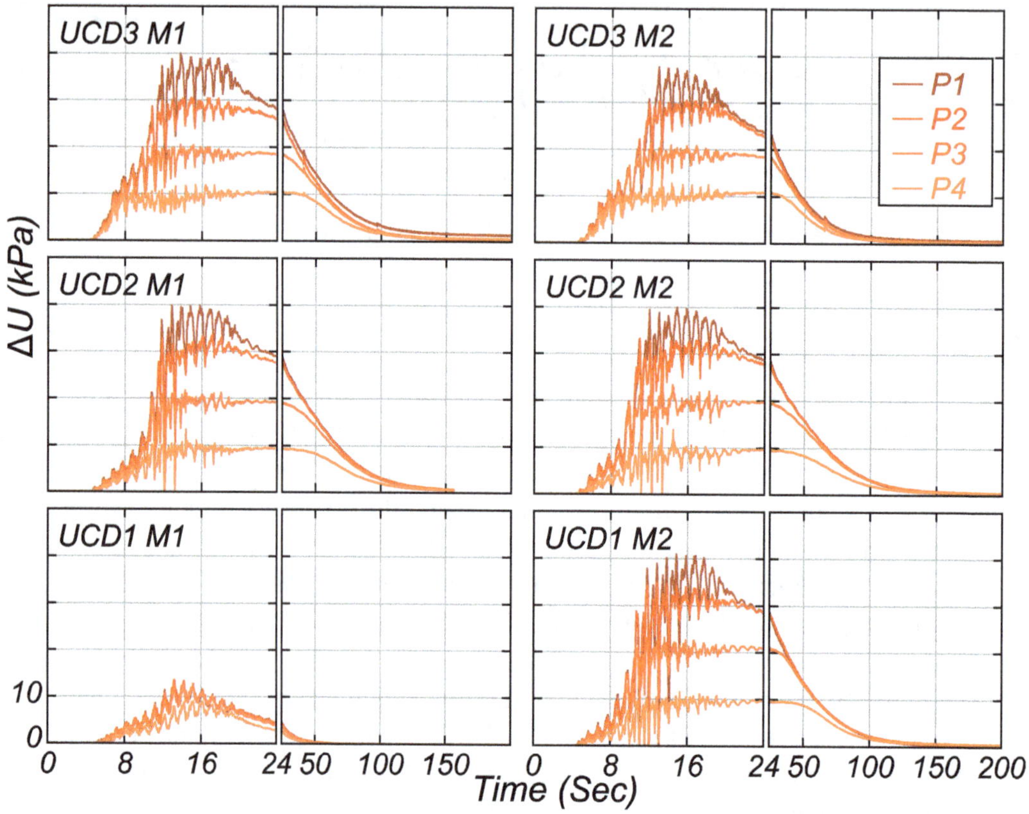


**Fig. 13.5** Central array excess pore pressures for UCD1, UCD2, and UCD3 (two destructive motions each)

- UCD1
  - Destructive motion 1: Similar to observations made of the acceleration response, the pore pressure response during M1 does not indicate that liquefaction occurred. The excess pore pressure ratios for P1 through P4 were 34%, 45%, 62%, and 90%. Once the magnitude of the base motion started to decrease at approximately 12 s, excess pore pressures immediately began to dissipate, indicating that at the depths of the PPTs the pore pressure never reached a constant state equal to their respective vertical effective stresses.
  - Destructive motion 2: The generation of excess pore pressure starts slow but as the magnitude of shaking increases each transducer reaches its initial vertical effective stress. Deliquefaction shockwaves, manifested as a sudden drop in porewater pressure due to the sudden dilation of a liquefied soil, are observed after the magnitude of the excess pore pressure is equal to the initial vertical effective stress (Kutter and Wilson 1999). Once shaking intensity decreases, pore pressures at P1, the deepest sensor, begin to dissipate first. Dissipation is followed then by the shallower sensors.

- UCD2
  - Destructive motions 1 and 2: The effect of the high-frequency component of the base motion is not easily seen in the pore pressure response for M1 and M2, indicating that the overall effect to the model performance is minimal. The stronger 1 Hz component in M2 is noticeable with the quicker accumulation of excess pore pressures for P1 and P2, compared with M1.
- UCD3
  - Destructive motions 1 and 2: Both M1 and M2 led to the fastest generation of excess pore pressure of all the tests conducted at UCD. At roughly 8 s the excess pore pressure ratio for P4 is already close to 100%; the other sensors indicated liquefaction later. Following shaking, the excess pore pressures were not maintained for the same duration when compared with the other experiments. The dissipation time for P1 and P2 after M2 was noticeably faster than after M1. This is possibly due to densification during M1.

Figure 13.6 shows the vertical component of acceleration for each destructive motion. The top acceleration trace is AV1 and the bottom is AV2. A fourth order band-pass filter with corner frequencies of 0.3–3 Hz was used to remove high-

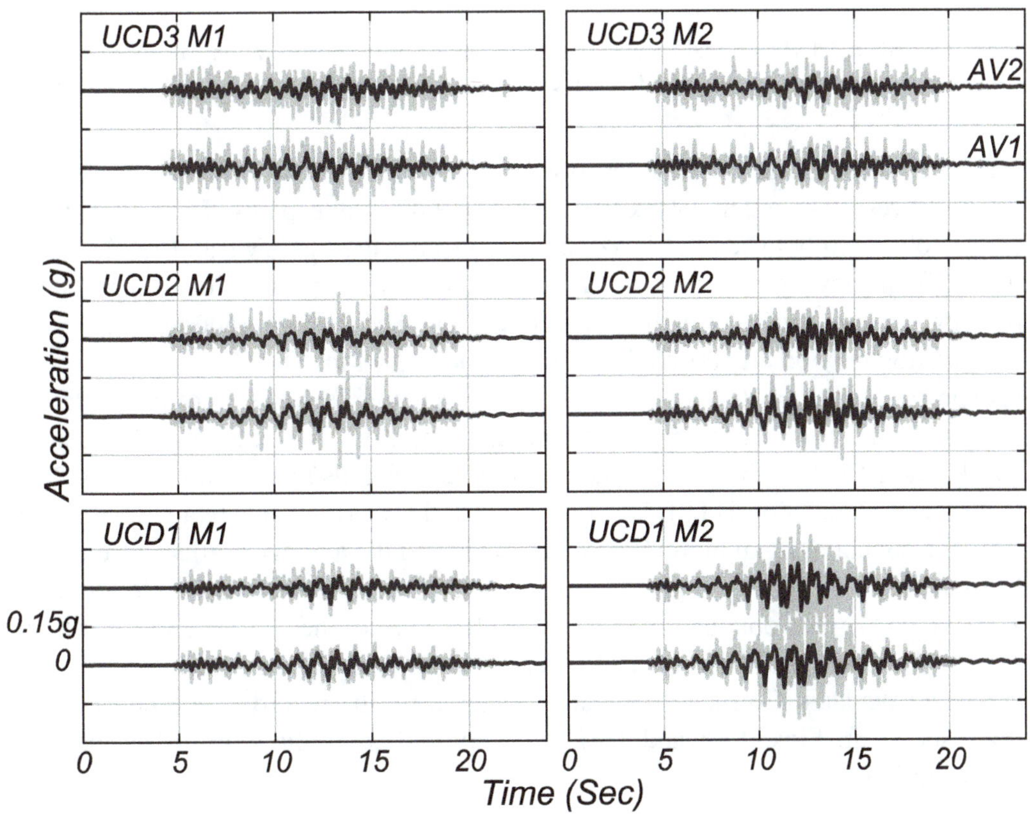


**Fig. 13.6** Filtered and non-filtered vertical acceleration time histories for UCD1, UCD2, and UCD3 (two destructive motions each)

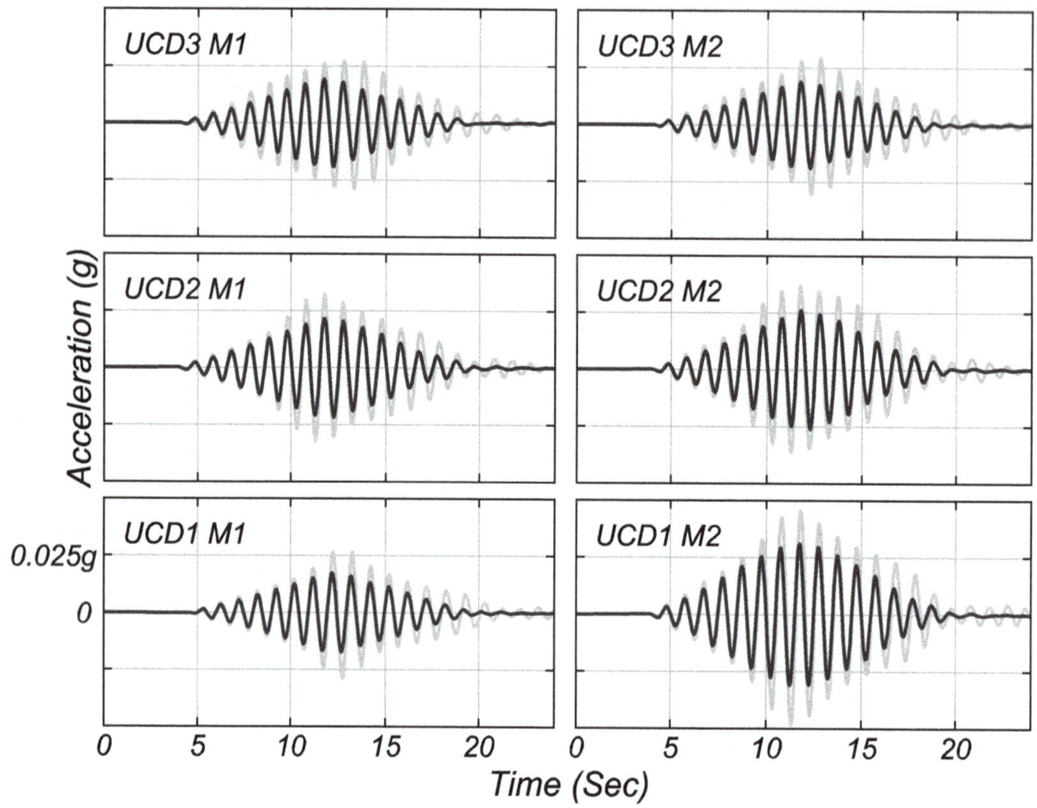


**Fig. 13.7** Filtered 1 Hz vertical acceleration (gray) and Coriolis acceleration calculated from the container input motion (black) for UCD1, UCD2, and UCD3 (two destructive motions each)

frequency noise, similar to that described by Kutter et al. 2017. The gray trace is the unfiltered response and the black trace shows the filtered response. Little to no phase shift is observed between the 1 Hz component measured at AV1 and AV2, but the higher-frequency components do appear to be out of phase. This suggests that rocking occurs at a 3 Hz or higher frequencies. The 1 Hz component of vertical acceleration may be the result of the Coriolis acceleration. The Coriolis acceleration is expressed as $\bar{a}_{cor} = 2\varpi v_{rel}$, where ω is rotational angular velocity of the centrifuge and $v_{rel}$ is the velocity of the container input motion. For example, the Coriolis acceleration for the UCD experiments ($\omega$ = 20.36 rads/s) with a 0.15 m/s 1 Hz component of velocity is 0.015 g. In Fig. 13.7 the Coriolis acceleration calculated from the container input motion shown as black and the average vertical acceleration from sensors AV1/AV2 in gray are compared. The 1 Hz vertical acceleration and 1 Hz velocity were obtained with a notch filter with corner frequencies of 0.5 and 1.5 Hz. The measured 1 Hz vertical acceleration in Fig. 13.7 is systematically larger than the calculated Coriolis acceleration. Therefore, the recorded vertical accelerations shown in Figs. 13.6 and 13.7 are amplified from the Coriolis acceleration shown in Fig. 13.7.

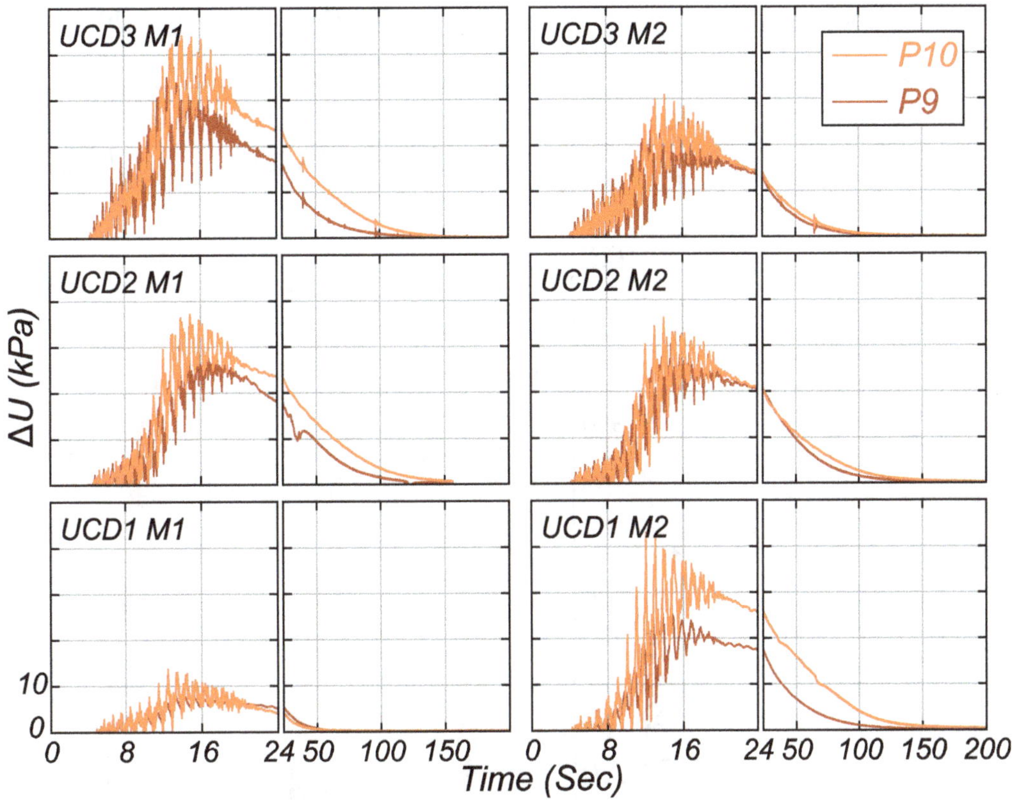


**Fig. 13.8** Bottom centerline pore pressure responses during UCD1, UCD2, and UCD3 (two destructive motions each)

Figure 13.8 shows the pore pressure transducer response of P9 and P10 for all destructive motions. Both P9 and P10 are located at the base of the container, but P9 has greater initial vertical effective stress since it is on the upslope side of the model. Suggested by Kutter et al. (2017), these two sensors are positioned to be sensitive to tilt of the model container that may occur because of friction generated when the bucket swings up. They also serve to confirm that the water level is constant; the return of the pore pressures to $\Delta u = 0$ in Fig. 13.8 indicates that the water table before and after shaking remained unchanged in the UCD experiments. Liquefaction at either of these sensors is difficult to achieve since no upward seepage of pore pressure can occur beneath the base of the container, and the rigid walls at the base of the container constrain against the development of shear strains. The oscillation of total stress along the ends of the container could be causing the dynamic oscillation in porewater pressure apparent in Fig. 13.8. The total stress at P9 and P10 is about 100 kPa; thus *a* $\pm$ 0.05 g vertical acceleration at the ends of the container (e.g., the low-frequency components in Fig. 13.6) could produce a cyclic total vertical stress of about $\pm$5 kPa, which is consistent with the low-frequency (~1 Hz) component of the pore pressures at P9 and P10.

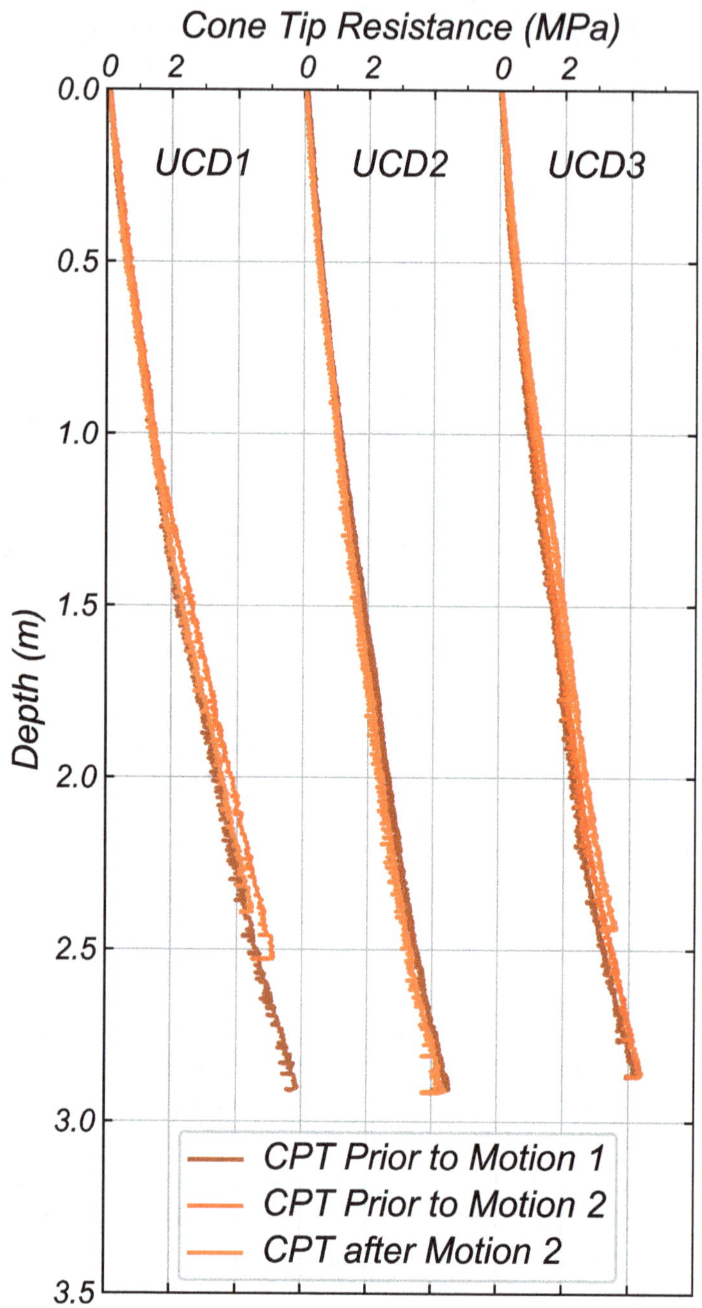


**Fig. 13.9** Cone penetration tests for UCD1, UCD2, and UCD3, before and after each destructive motion

### *13.3.4 Cone Penetration Tests*

Three CPT soundings were performed during each experiment using the CPT specifically designed for the LEAP project (Carey et al. 2018a). A sounding was performed prior to destructive motions 1 and 2 and after destructive motion 2. CPT soundings were used as an independent check of the density and state of the specimen prior to each destructive motion for numerical prediction of the experiment. Figure 13.9 shows the response obtained from each CPT. Overall, little change in cone tip resistance ($q_c$) was observed between destructive motions.

UCD3 shows an approximate 5% increase in $q_c$ at 2 m depth between M1 and M2. UCD2 shows very small (less than 2%) decrease in $q_c$ at 2 m depth between M1 and M2. $q_c$ increases by approximate 5% at 2 m depth after UCD1 M1, but an approximate 5% decrease in $q_c$ at 2 m depth after M2. As previously discussed, UCD1 M1 did not liquefy the model; therefore, the variation in the cone profiles may be attributed to a combination of effects of spatial variability, lateral stresses, and soil density.

### *13.3.5 Surface Marker Surveys*

Surface markers were placed in three longitudinal arrays, with each array consisting of 6 markers, for a total of 18 surface markers. Surface marker locations were recorded prior to and following each destructive motion to track lateral and vertical deformations. Figures 13.10, 13.11, and 13.12 show the displacement path of the central array of surface markers during the first destructive motion of UCD1, UCD2, and UCD3, respectively. The surface markers for UCD2 M1 and UCD3 M1 displaced with typical deformation tendencies of a sloping liquefiable deposit. All surface markers moved downslope; upslope markers tended to settle, while downslope markers tended to heave. Surface markers in UCD1 M1 underwent little deformation. Not seen in Figs. 13.10, 13.11, and 13.12, the outer arrays of surface markers followed similar deformation patterns as the central arrays.

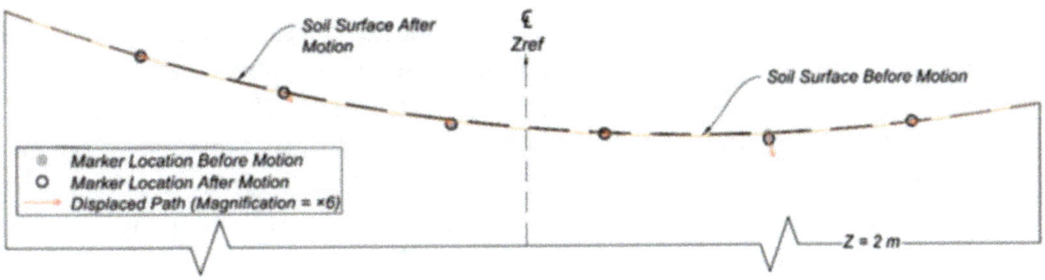


**Fig. 13.10** Surface marker displacement from hand measurements during UCD1 M1

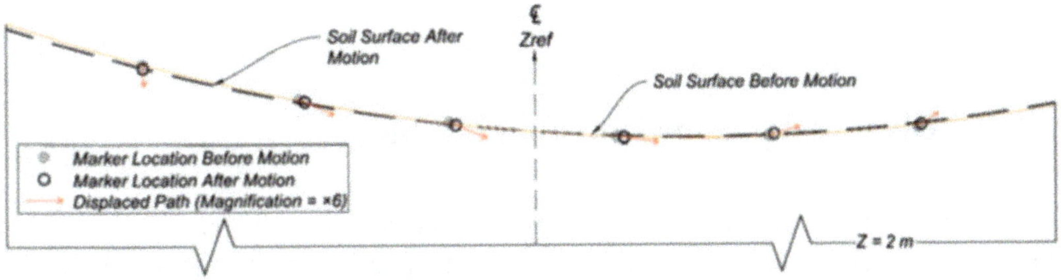


**Fig. 13.11** Surface marker displacement from hand measurements during UCD2 M1

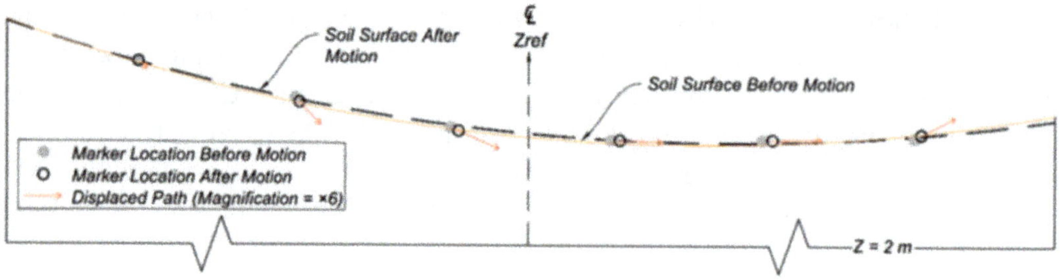


**Fig. 13.12** Surface marker displacement from hand measurements during UCD3 M1

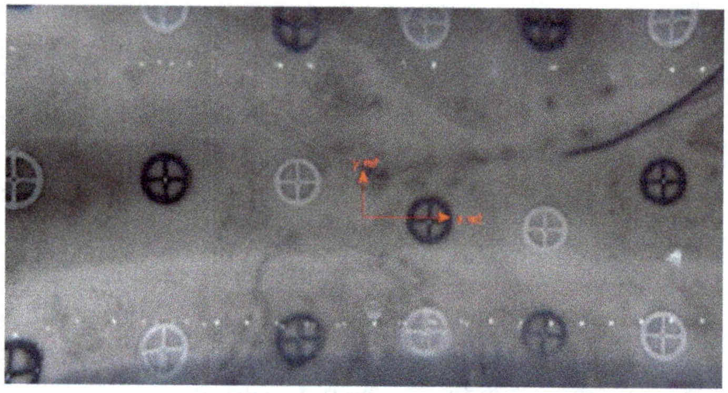

**Fig. 13.13** UCD2 surface marker array, with downslope center markers 4 and 5 offset to accommodate the wire from AH4 (image generated by superimposing images together from five cameras)

## 13.4 Nonconformities with Specifications

As discussed, each model contained three longitudinal arrays of six surface markers, for a total of 18 markers. In UCD2 and UCD3, the top accelerometer (AH4) cable was too close to the surface, making it impossible to embed the lower slope centerline surface markers without disturbing the sensor. In UCD2, the fourth and fifth surface markers from the upslope end of the model were offset from the centerline by 25 mm. In UCD3, the fourth, fifth, and sixth surface markers from the upslope end of the model were offset from the centerline by 25 mm. Figure 13.13 shows an example of the 25 mm offset from UCD2.

Several nondestructive motions were performed prior to and following each destructive motion as discussed earlier. For UCD1, additional destructive motions were applied after M2 for the purposes of calibrating shaker command motions for subsequent model tests; these motions are not analyzed in the present paper.

## 13.5 Advancements in Centrifuge Testing

LVDTs have been used previously to measure displacement of liquefied geosystems, but others (Fiegel and Kutter 1994) have noted the measurements may be unreliable during liquefaction. For LEAP-UCD-2017, a new procedure for tracking slope

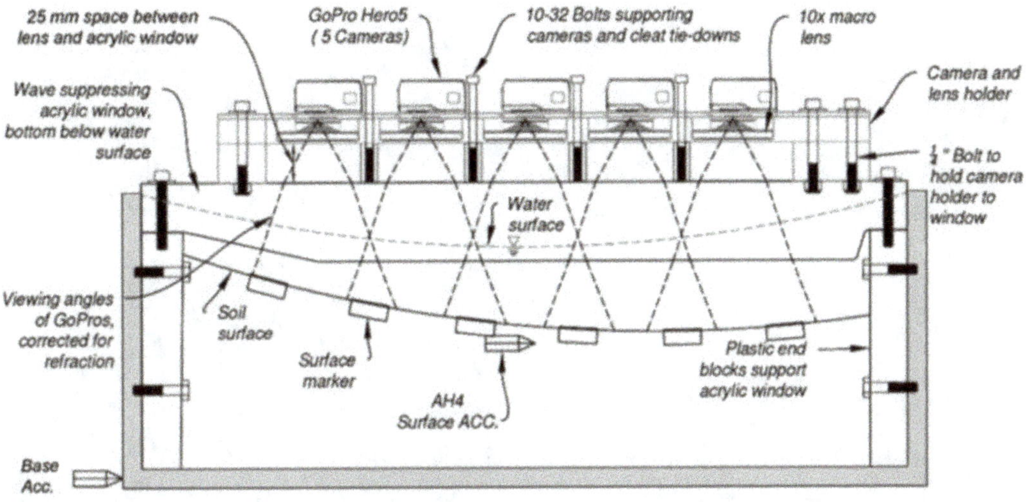


**Fig. 13.14** Profile view of centrifuge container with acrylic glass window and mounted GoPro cameras viewing through wide angle lenses

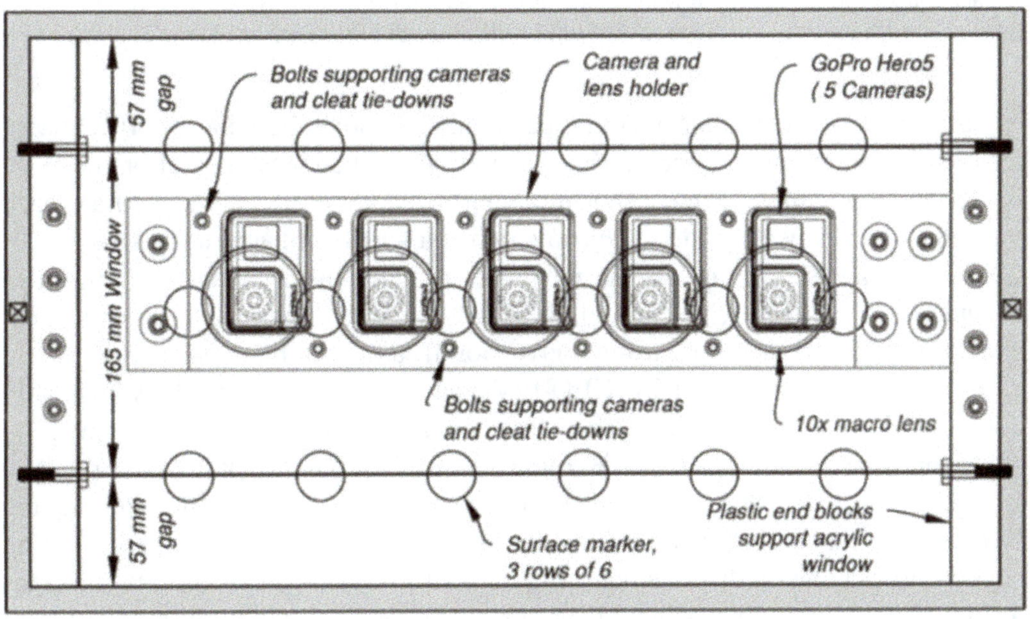


**Fig. 13.15** Top view of centrifuge container with acrylic glass window and mounted GoPro cameras viewing through wide angle lenses

displacements of a centrifuge model using a wave suppressing window, GoPro cameras, and GEO-PIV software (Carey et al. 2018b) was developed. The bottom of the wave suppressing window was located beneath the fluid surface to reduce distortion due to water surface waves, like a glass-bottomed boat as shown in Figs. 13.14 and 13.15. Note that the 57 mm gap between the window and the side walls of the container allowed for a free water surface and ensured that window vibrations could not cause significant water pressure oscillations. Five GoPro

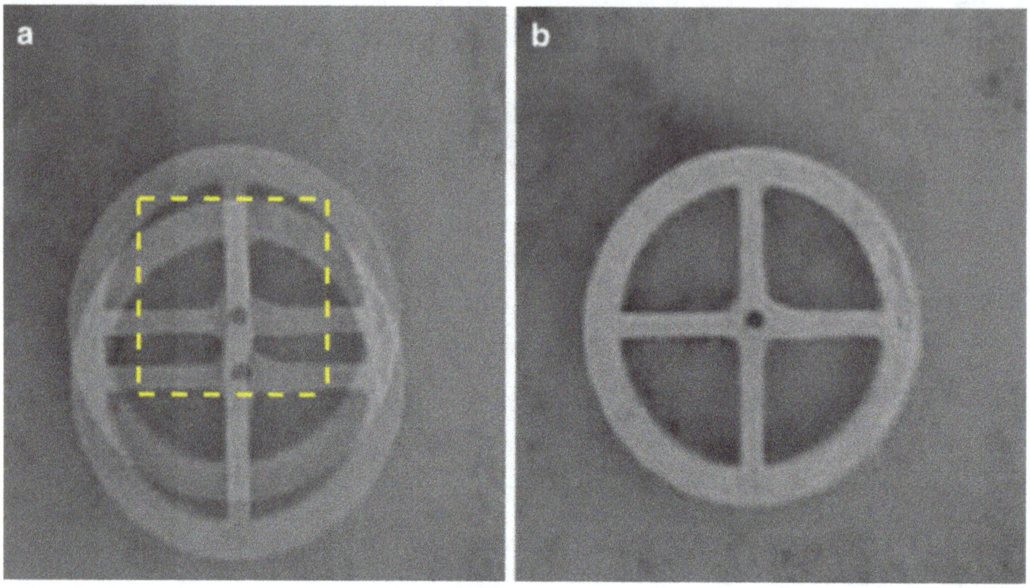


**Fig. 13.16** Validation of GEO-PIV using manual image processing in MATLAB, (**a**) location of a surface marker prior to and following shaking overlaid shown with a typical PIV patch. (**b**) the image recorded prior to shaking is shifted until a clear image of the surface marker is obtained

cameras were secured in a camera holder mounted above the window. 10× macro lenses are used to improve focus. Movement of the first and last marker on the slope is recorded by a single camera, but the interior markers (e.g., rows 2–5) are each viewed by two cameras. Videos are converted to a series of images using MATLAB. GEO-PIV was used to process the images to produce displacement time histories (Stanier et al. 2015). Carey et al. (2018b) describe how displacements were converted from pixels to mm using conversion factors specific to each camera.

Displacements calculated by GEO-PIV analyses were validated by manually counting the number of pixels the surface markers moved during shaking by superimposing images in MATLAB. Figure 13.16 shows an example of this process. In Fig. 13.16a, the location of a typical surface marker prior to and following shaking is overlaid. The dashed yellow box outlines the patch GEO-PIV tracks during shaking to calculate a displacement. Figure 13.16b overlays the same two images as Fig. 13.16a, but the image recorded prior to shaking is vertically shifted until there is a clear image of the surface marker. The pixel offset corresponds to the downslope displacement that marker experienced during shaking.

Figure 13.17 compares downslope displacements found by hand measurement of surface markers with calipers and rulers, GEO-PIV analysis, and manual pixel counting using MATLAB (Fig. 13.16). Surface markers 2–5 are viewed by two cameras; thus these marker locations have two manual pixel counting measurements. Overall, GEO-PIV was capable of accurately measuring the final displacement of surface markers, arguably better than the physical hand measurements. Table 13.2 provides the differences between GEO-PIV and manual pixel counting measurements were less than about 8%, with the exception of surface marker 1. The large displacement measurement error for surface marker 1 may be attributed to the fact

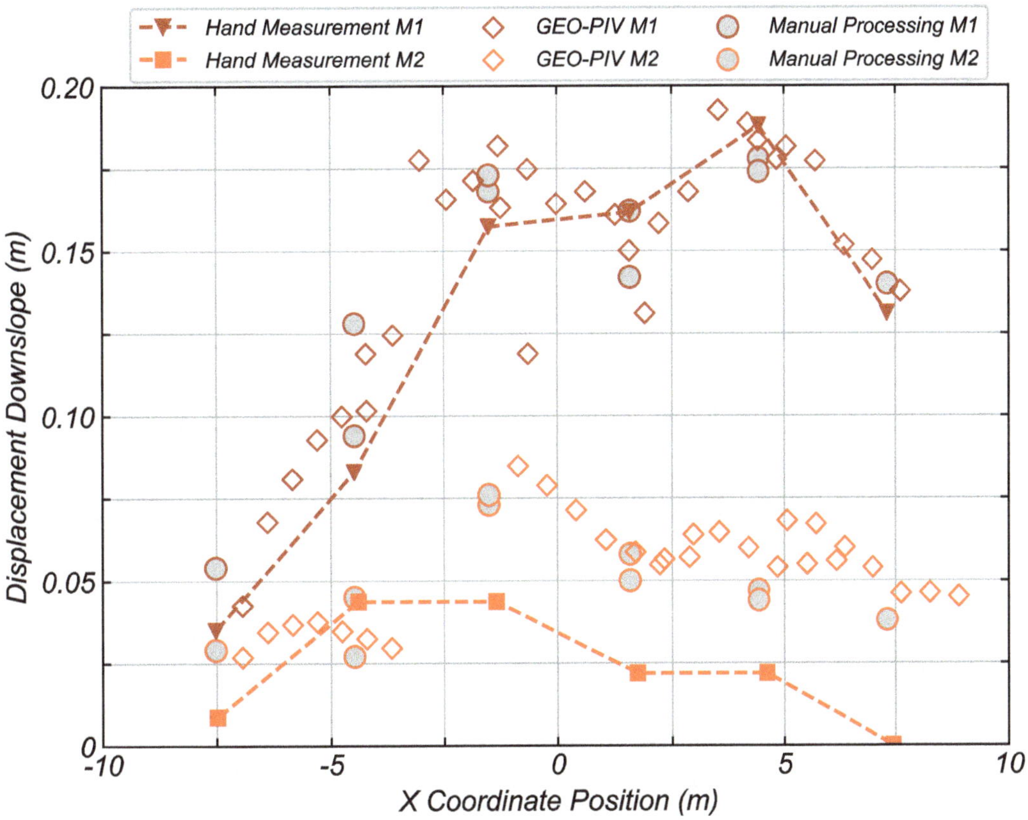


**Fig. 13.17** Comparison of surface marker displacement measurements using GEO-PIV, manual image processing in MATLAB, and hand measurements with rulers for centerline surface markers

**Table 13.2** Differences in surface marker displacement measurement between GEO-PIV and manual pixel counting

| | Upslope camera | Downslope camera |
|---|---|---|
| Surface marker 1 | 21.4% | NA |
| Surface marker 2 | 7.2% | 8.1% |
| Surface marker 3 | 1.0% | 2.0% |
| Surface marker 4 | 5.6% | 7.4% |
| Surface marker 5 | 5.4% | 3.0% |
| Surface marker 6 | NA | 1.6% |

that surface marker 1 lies only partially in the frame of camera 1 and the GEO-PIV patches do not completely overlap surface marker 1. An additional benefit of using GEO-PIV to determine downslope displacements is that displacements may be found at more locations because it can track the movement of patches of sand between surface markers.

Figure 13.18 compares displacement time histories generated by GEO-PIV to those found using the IPRV method and accelerometer data, a method to estimate displacement time histories developed by Kutter et al. (2017) and modified by Carey et al. (2018b) for UCD3 M1. The dynamic displacement of the IPRV method is the relative displacement between the base accelerometers and AH4, found by double

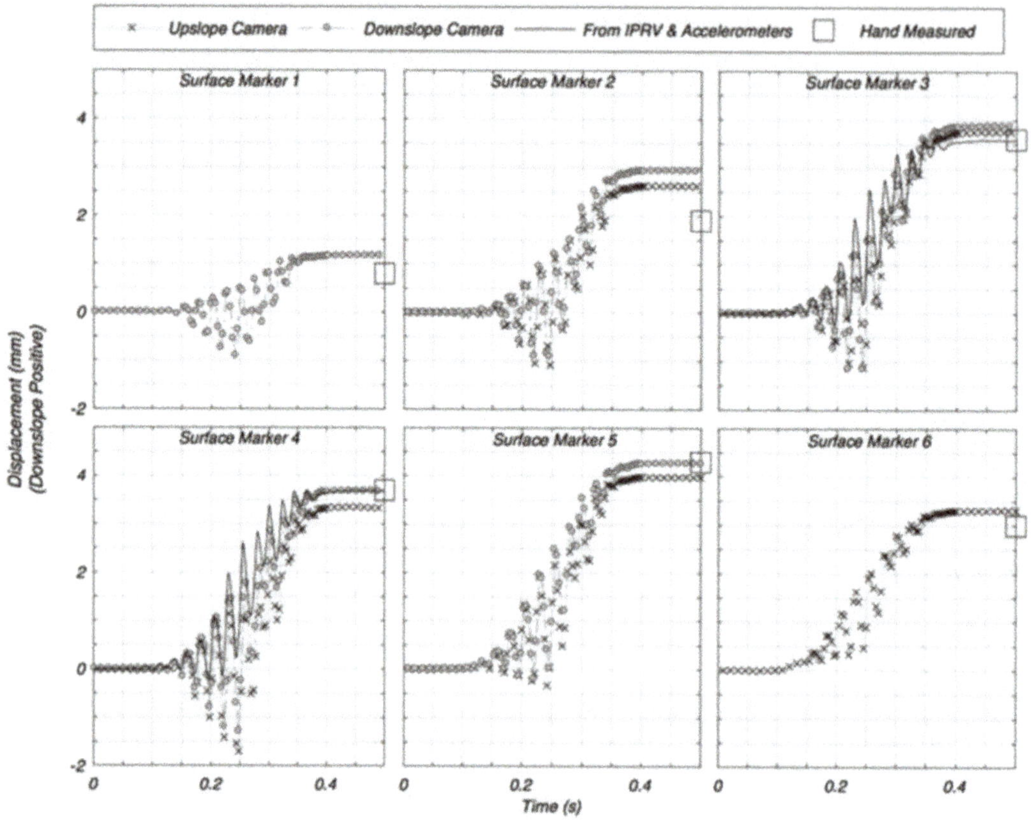


**Fig. 13.18** Displacement time series as measured using GEO-PIV and calculated from IPRV method and accelerometer data for UCD3 M1 from Carey et al. 2018b. (Note time and displacement are given in model scale units)

integration of the acceleration at those locations. The final displacement is determined by hand measurement of the surface marker, shown as a square at the end of each time history, and the ramp to the final displacement is the integrated positive relative velocity ramp scaled to match the hand measured marker locations. The IPRV method is only plotted for surface markers 3 and 4 since AH4 is located near those markers. Displacement time histories for markers 2 through 5 are provided for the up- and downslope cameras. Markers 1 and 6 are only recorded by one camera (see Fig. 13.14). Figure 13.18 shows (1) the up- and downslope cameras recorded similar displacement histories for a single marker; (2) the dynamic displacement from GEO-PIV matches well with the dynamic displacement from accelerometers, which are considered to be highly accurate; and (3) the IPRV is a reasonable method to approximate the ramp shape of displacement time histories.

## 13.6 Method of Measuring Density

Density was calculated by mass and volume measurements during model construction as an intermediate check to ensure the target density is achievable. At each intermediate lift, the surface was flattened by vacuuming excess sand and the depth

**Table 13.3** Achieved density of intermediate lifts and final model density

| UCD1 | | | UCD2 | | | UCD3 | | |
|---|---|---|---|---|---|---|---|---|
| Depth of lift (mm) | Avg. (kg/m$^3$) | Std Dev. (kg/m$^3$) | Depth of lift (mm) | Avg. (kg/m$^3$) | Std dev. (kg/m$^3$) | Depth of lift (mm) | Avg. (kg/m$^3$) | Std dev. (kg/m$^3$) |
| 52 | 1693 | 22 | 53 | 1655 | 19 | 57 | 1695 | 13.8 |
| 89 | 1693 | 25 | 115 | 1647 | 6 | 120 | 1659 | 5.96 |
| Curved surface (final density) | 1665 | 6 | Curved surface (final density) | 1648 | 3 | Curved surface (final density) | 1658 | 12.4 |

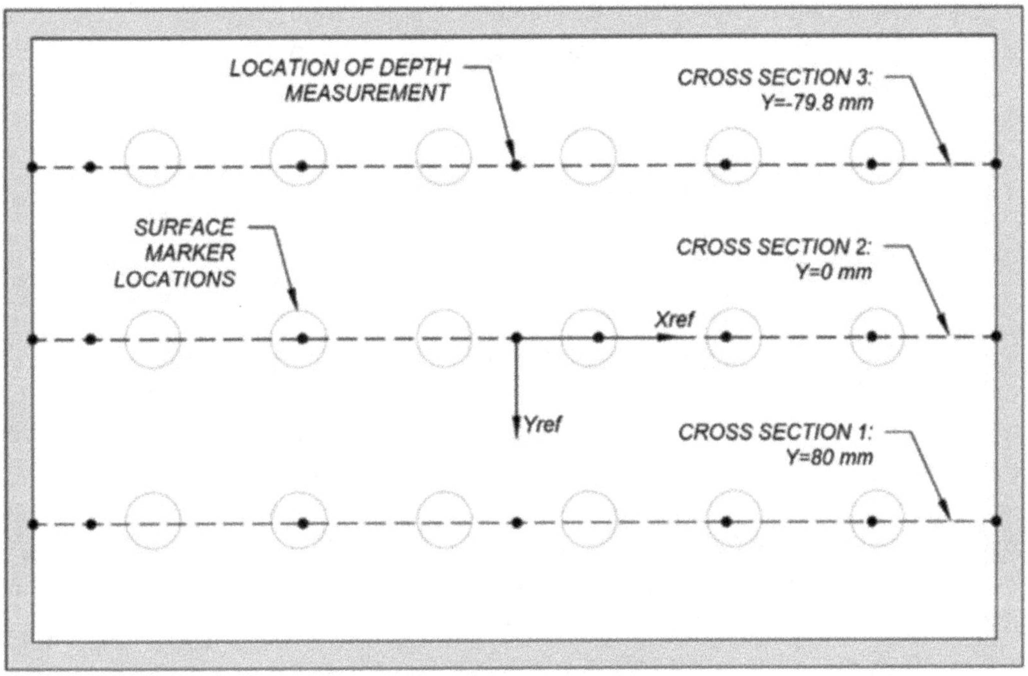


**Fig. 13.19** Location of lift height used for volume of model calculation

at 15 locations was measured. The measured depths were averaged, and the density was calculated using the mass of the lift. Reported in Table 13.3 are the depths of the layers measured and the average and standard deviations of the calculated density. The calculation of density for the shallower lifts is more sensitive to variations in lift height. The sensitivity was due to the relatively small mass of the lift and the uncertainty of density was dominated by uncertainty in the lift height.

The reported density of the model used volume and mass measurements of the final curved surface. Figure 13.19 shows the seven locations along the three longitudinal lines where the depth was measured. The three longitudinal lines of measurement roughly corresponded to specified longitudinal lines of the surface markers. Using AutoCAD, a curve in $x$–$z$ space was fit to the seven points on the

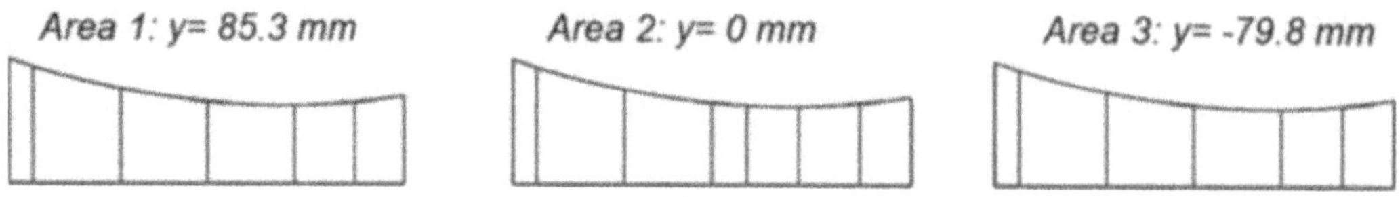


**Fig. 13.20** Three cross sections used to calculate the total volume of the completed model with curved surface

same longitudinal line. The MASSPROPS command in AutoCAD was then used to calculate the area of each region, as shown in Fig. 13.20. The volume was estimated by averaging the three areas and multiplying that average by the 279 mm width of the container.

The load cell used to measure the mass of the model was accurate to about 0.1 kg. The LEAP-UCD-2017 test template contains supplemental documentation of density calculation for each lift during construction of each model.

## 13.7 Pore Fluid Viscosity and Saturation

### *13.7.1 Pore Fluid Viscosity*

Methylcellulose was used for each of the UCD experiments to scale the viscosity of the pore fluid in accordance with the scaling law, $\mu = \mu_{\text{water}}/L^*$. Prior to centrifuge testing, several batches of viscous solution were mixed to determine the correct proportions of Dow F50 Food Grade methylcellulose powder and water by mass to achieve the desired viscosity of $\mu = 43.75$. The ratio of mass of methylcellulose to mass of water was 2.2%.

The methylcellulose was prepared in accordance with the chemical supplier recommendation for hot mixed solution. The procedure was as follows:

1. Warm roughly ¼ of the required deionized water to 90 °C. Add methylcellulose with a mass that is equal to 8.8% of mass of the deionized water.
2. Mix solution for 45 min.
3. Dilute the mixture with the same mass of water from step 1 at room temperature. Mix for additional 10 min. Following mixing, this stock solution should be at roughly double the required concentration.
4. Cool overnight.
5. Using an approximately 200 g sample of the stock solution, 208 g of room temperature deionized water was added and mixed. Viscosity of the solution was checked using a Cannon instrument size 2 Ubbelohde viscometer.
6. Step 5 was repeated if necessary, adjusting the amount of deionized water added, to determine the ratio of deionized water and stock solution to produce the desired viscosity.
7. Lastly, the entire batch of methylcellulose was mixed with the ratio of water and stock found in Step 6.

### 13.7.2 Model Saturation

Each of the UCD models followed the same procedure for saturation. Initially, the dry model and container were placed in a vacuum chamber. 97 kPa of vacuum was applied, then the vacuum was shut off, and the chamber was flooded with $CO_2$ up to until the vacuum was reduced to about 2 kPa. Then, the $CO_2$ flow was shut off and the 97 kPa vacuum was reapplied. This cycle was repeated two times and following the second cycle the vacuum was reapplied to 97 kPa. Following the third evacuation, the residual concentration of nitrogen and oxygen gas in the chamber should be $((101.4–97)/101.4)^3$ times its initial concentration in atmospheric air, and the partial pressure of $CO_2$ gas would be about 4% of an atmosphere. The saturation chamber and model were inclined so the toe of the model (left-hand side of Fig. 13.1) was the lowest point in the container. The methylcellulose solution was de-aired and dripped on a sponge on the top surface of the sand to maintain a pool of de-aired methylcellulose solution in the lowest edge of the container. As the wetting front progressed toward the top of the model, the size of the pool was allowed to grow. The top corner of the slope was the last location to saturate, allowing for any residual gas and air to escape from the model. Nine-seven kPa vacuum was maintained throughout the infiltration of methylcellulose. Once the model was completely saturated and the surface submerged, the vacuum was slowly released.

Once the model was saturated, the degree of saturation was then checked using the method described in the specification, modified from Okamura and Inoue (2012) and Carey et al. (2017). The chamber was opened, and a tethered float delineated with a 1 cm grid was placed on the surface of the methylcellulose. Laser pointers were positioned on the container walls pointing downward at the floating grid with a five-degree angle. The laser pointer and float configuration are shown in Fig. 13.21. The transparent cover was placed back over the vacuum chamber and the locations of the dots from the laser on the float were photographically recorded. A small, 10 kPa vacuum was applied and the location of the laser dots was recorded. Using Boyle's law (Carey et al. 2017) the degree of saturation for all three UCD experiments was calculated to be 99.99%. The tiny 0.01% bubble volume would easily dissolve in the water with time or with the increased pressure in the centrifuge.

## 13.8 Conclusions

This paper described the three experiments performed on the 1 m centrifuge at the University of California, Davis, as part of the LEAP-UCD-2017 exercise. The model performance and results for the two destructive motions were presented, and the nonconformities from the specifications were also discussed. The process to mix methylcellulose pore fluid was presented and the protocol to saturate and measure the degree of saturation of the model was also explained.

Motion 1 in UCD1 did not cause liquefaction, which is evident by the low excess pore pressures, and the nearly rigid acceleration response of the model. The second experiment had larger than expected high-frequency component superimposed on the

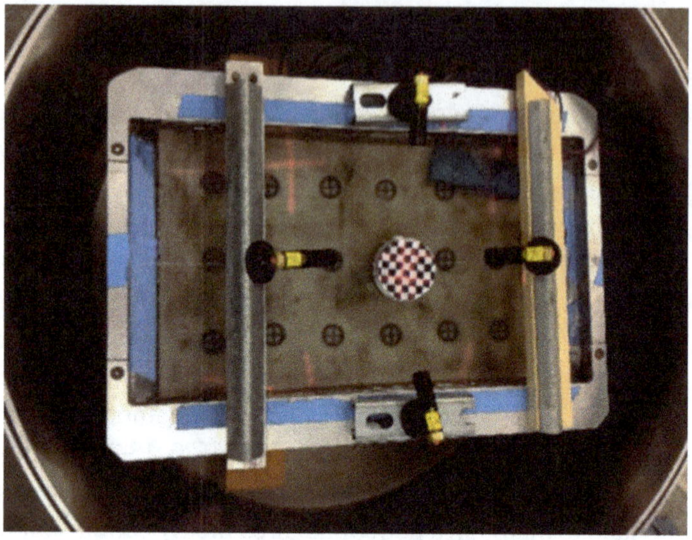

**Fig. 13.21** Lasers positioned so dots point at float on surface of methylcellulose

1 Hz specified input motion. The achieved velocity closely matched the specified velocity for both motions of the second test. The third experiment showed the best repetition of input motions and model performance, and can be used to determine if numerical simulations correctly model the evolution of soil properties from repeated shaking events. Other minor deviations from the specifications were discussed, such as alternate placement of surface markers, and the use of minor nondestructive ground motions before the destructive motions.

A new technique was presented that uses GoPro cameras mounted on a wave suppressing window acting like a glass-bottomed boat to record images of the deforming surface. GEO-PIV was used to process the images and create displacement time series. Results from the procedure were shown to be reliable when compared against hand measurements, sensor data, and the manual processing of images.

A protocol for measuring the achieved density of a specimen was described. The final density was calculated by measuring the depth of soil along three longitudinal transects and fitting a curve to those depths in AutoCAD. This method to find density, while still relying on point measurements, proved to be helpful for accurately determining the final density and the associated uncertainty of the density measurements.

**Acknowledgments** These experiments were supported by NSF CMMI Grant Number 1635307. The authors would also like to thank the staff at the Center for Geotechnical Modeling (CGM) for their assistance and technical insight throughout the series of tests.

## References

Carey, T., Gavras, A., Kutter, B., Haigh, S. K., Madabhushi, S. P. G., Okamura, M., Kim, D. S., Ueda, K., Hung, W.-Y., Zhou, Y.-G., Liu, K., Chen, Y.-M., Zeghal, M., Abdoun, T., Escoffier, S., & Manzari, M. (2018a). A new shared miniature cone penetrometer for centrifuge testing. In *Proceedings of 9th International Conference on Physical Modelling in Geotechnics, ICPMG 2018* (Vol. 1, pp. 293–229). London: CRC Press/Balkema.

Carey, T., Stone, N., Kutter, B., & Hajialilue-Bonab, M. (2018b). A new procedure for tracking displacements of submerged sloping ground in centrifuge testing. In *Proceedings of 9th International Conference on Physical Modelling in Geotechnics, ICPMG 2018* (Vol. 1, pp. 829–834). London: CRC Press/Balkema.

Carey, T. J., Hashimoto, T., Cimini, D., & Kutter, B. L. (2017). LEAP-GWU-2015 centrifuge test at UC Davis. *International Journal of Soil Dynamics and Earthquake Engineering, Elsevier, 113*, 663. https://doi.org/10.1016/j.soildyn.2017.01.030.

Fiegel, G. L., & Kutter, B. L. (1994). Liquefaction mechanism for layered soils. *Journal of Geotechnical Engineering, ASCE, 120*, 737–755. https://doi.org/10.1061/(asce)0733-9410(1994)120:4(737).

Kutter, B. L., Carey, T. J., Bonab, M. H., Stone, N., Manzari M., Zeghal M., Escoffier, S., Haigh, S., Madabhushi, G., Hung, W., Kim, D., Kim, N., Okamura, M., Tobita, T., Ueda, K., & Zhou, Y. (2019a). LEAP-UCD-2017 V. 1.01 model specifications. In B. Kutter et al. (Eds.), *Model tests and numerical simulations of liquefaction and lateral spreading: LEAP-UCD-2017*. New York: Springer.

Kutter, B. L., Carey, T., Stone, N., Zheng, B. L., Gavras, A., Manzari, M., Zeghal, M., Abdoun, T., Korre, E., Escoffier, S., Haigh, S., Madabhushi, G., Madabhushi, S. S. C, Hung, W.-Y., Liao, T.-W., Kim, D.-S., Kim, S.-N., Ha, J.-G., Kim, N. R., Okamura, M., Sjafuddin, A. N., Tobita, T., Ueda, K., Vargas, R., Zhou, Y.-G., & Liu, K. (2019b). LEAP-UCD-2017 comparison of centrifuge test results. In B. Kutter et al. (Eds.), *Model tests and numerical simulations of liquefaction and lateral spreading: LEAP-UCD-2017*. New York: Springer.

Kutter, B. L., Carey, T. J., Hashimoto, T., Zeghal, M., Abdoun, T., Kokkali, P., Madabhushi, S. P. G., Haigh, S. K., Burali d'Arezzo, F., Madabhushi, S. S. C., Hung, W.-Y., Lee, C.-J., Cheng, H.-C., Iai, S., Tobita, T., Ashino, T., Ren, J., Zhou, Y.-G., Chen, Y., Sun, Z.-B., & Manzari, M. T. (2017). LEAP-GWU-2015 experiment specifications, results, and comparisons. *International Journal of Soil Dynamics and Earthquake Engineering, Elsevier, 113*, 616. https://doi.org/10.1016/j.soildyn.2017.05.018.

Kutter, B. L., & Wilson, D. W. (1999). De-liquefaction shock waves. In *Proceedings of the seventh US–Japan workshop on earthquake resistant design of lifeline facilities and countermeasures against soil liquefaction, Seattle*. pp. 295–310.

Okamura, M., & Inoue, T. (2012). Preparation of fully saturated models for liquefaction study. *International Journal of Physical Modelling in Geotechnics, 12*, 39–46. https://doi.org/10.1680/ijpmg.2012.12.1.39.

Stanier, S. A., Blaber, J., Take, W. A., & White, D. J. (2015). Improved image-based deformation measurement for geotechnical applications. *Canadian Geotechnical Journal, 53*, 727–739. https://doi.org/10.1139/cgj-2015-0253.

# Chapter 14
# LEAP-2017 Centrifuge Test at Ehime University

Mitsu Okamura and Asri Nurani Sjafruddin

**Abstract** Three centrifuge tests were conducted at Ehime University for the LEAP-2017 exercise. The experiment consisted of a submerged clean sand, with target relative densities of either 65 or 80%, with a 5 degree slope in a rigid container. Models were prepared along with the specifications and each model was subjected to a 1 Hz ramped sine wave base motion twice. This paper provides overview of the models and some details of the effects on the pore pressure responses of relative density and shaking histories of the models. Typical residual deformation obtained by PIV analysis and liquefaction resistance estimated based on the model response and laboratory cyclic shear tests are also shown.

## 14.1 Introduction

This paper presents three centrifuge tests conducted at Ehime University (EU) in the LEAP-2017 project. Model configuration, preparation techniques, and sand used are in accordance with "LEAP UCD 2017 VERSION 1.01 MODEL SPECIFICATIONS" (Kutter et al. 2018) otherwise mentioned. In this paper, a brief description of facilities at Ehime University to perform the tests and general information of the models are described first, followed by typical responses of the models.

## 14.2 Centrifuge at Ehime University

The centrifuge at EU has symmetrical rotor arms with swing platforms at each end. The radius at the top of the platform from the centrifuge axis is 1.25 m when the platforms fully swing up. In this study, an electronic mechanical shaker was used to

M. Okamura (✉)
Department of Science and Engineering, Ehime University, Matsuyama, Japan
e-mail: okamura@cee.ehime-u.ac.jp

A. N. Sjafruddin
Department of Civil Engineering, Ehime University, Matsuyama, Japan



B. Kutter et al. (eds.), *Model Tests and Numerical Simulations of Liquefaction and Lateral Spreading*, https://doi.org/10.1007/978-3-030-22818-7_14

impart simulated input motions to the models. A 0.6 kW AC inverter motor drives the shaking table horizontally through a camshaft of an eccentricity 0.7 mm. Horizontal displacement amplitude was approximately 0.7 mm and acceleration was controlled by rotation rate of the driving motor in the range between 0 and 50 Hz.

Data acquisition system equipped on the rotation arm allows to measure 48 sensors. All the signals are converted on board and sent through wireless USB connection to the PC in the laboratory.

## 14.3 Centrifuge Model

### *14.3.1 Model Description*

A rigid model container was used with internal dimensions of 50 cm long, 12 cm wide, and 23 cm deep. Three models were prepared with target relative densities of either 65 or 80%, all saturated with 40 cSt viscous fluid and tested at 40 g with an exception that the second shaking for model 1 (EU1) was conducted at 20 g. Figure 14.1 shows a side view of the models. All the corresponding prototype dimensions are the same as those specified with an exception that the prototype width of 4.8 m was about a half that specified.

### *14.3.2 Sand*

Ottawa F-65 sand was used in all the experiments, which was shipped out from UC Davis on March 2017. Sieve tests on the sand confirmed that grain size distributions

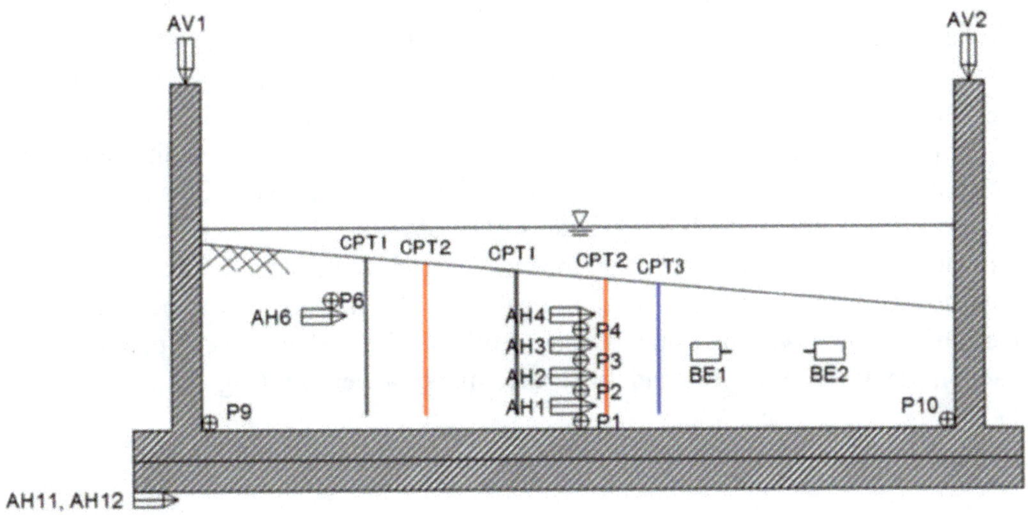


**Fig. 14.1** Schematic of models

**Table 14.1** Sieve test result

| | This study | Specifications |
|---|---|---|
| $D_{10}$ (mm) | 0.113 | 0.133 |
| $D_{30}$ (mm) | 0.171 | 0.173 |
| $D_{40}$ (mm) | 0.248 | 0.203 |
| $D_{60}$ (mm) | 0.265 | 0.215 |

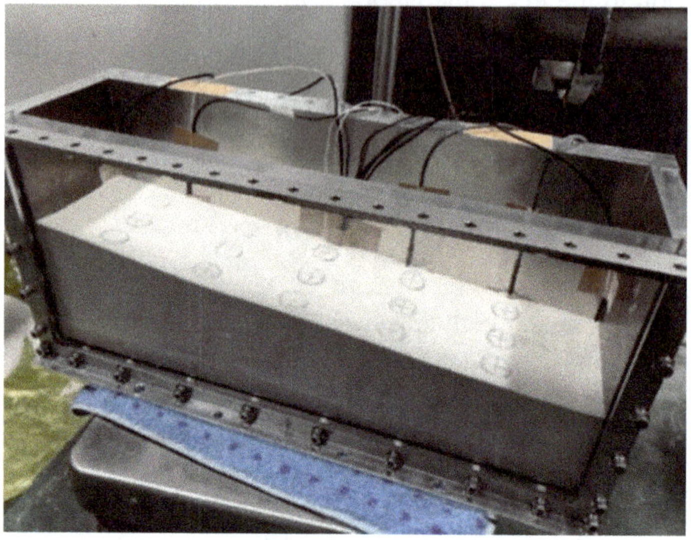

**Fig. 14.2** Model before saturation

of the sand used in EU were consistent with those shown in the specifications (Table 14.1).

## 14.3.3 Placement of Sand

Dry sand stored in an air-conditioned room, and thus humidity was kept low, was pluviated into the container through a screen with an opening size 1.0 mm, rather than 1.2 mm specified in the specification, because openings of the standard sieve in JIS (Japanese Industrial Standard) is slightly different from ASTM.

Three arrangements of the screen masked off were used to achieve different target relative densities. 12 mm slots spaced at 25 or 7 mm slots spaced at 25 mm were used for preparing sand bed with relative density of 65% or 80%, respectively. After the pluviation of the sand, the surface of the model was leveled using a vacuum device to measure the height of model surface to calculate dry density. The procedure of the measuring height was the same as that specified. The surface was sloped using the vacuum device to 5 degrees and 18 surface markers were set (Fig. 14.2).

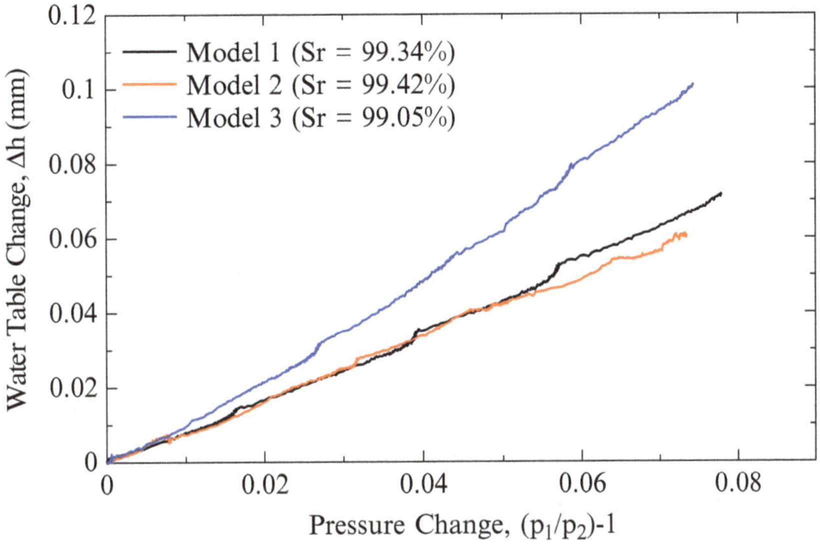


**Fig. 14.3** Relationship between pressure in the chamber and change in water level

### 14.3.4 *Saturation*

The model container was moved into a vacuum chamber and the air in the chamber was replaced by $CO_2$. This was achieved by introducing vacuum pressure of −95 kPa and flooding $CO_2$ gas. De-aired viscous fluid was dripped into the low end of the model slope while keeping the vacuum of −95 kPa constant in the chamber and a fluid supply tank (Okamura and Inoue 2012). The viscous fluid was a mixture of water and hydroxypropyl methylcellulose (type 60SH-50) termed Metolose from Shin-Etsu Chemical Company (Shin-etsu Chemical Co. Ltd. 1997). This Metolose solution was prepared by dissolving 1.8% Metolose by weight in water, so as to achieve a viscosity of 40 times that of water (40 cSt kinematic viscosity). The viscosity of the fluid in each model was measured at room temperature before and after the tests with a rotational viscometer. The measured viscosity was in a range between 39 and 44 cSt.

On completion of the saturation process, the vacuum in the chamber was released and the model was rested in the atmospheric pressure for a few hours. A small change in the pressure of approximately 10 kPa was applied to the chamber at a constant rate approximately 5 kPa/min, and water level was measured with a LED displacement transducer with the resolution of 10 μm. Figure 14.3 indicates relationship between change in water level, Δh, and change in pressure (p1/p2 -1) observed for the three models, where p2 and p1 are the atmospheric pressure (101.3 kPa) and the pressure in the chamber, respectively (in absolute pressure). In each test linear relationships were observed. The degree of saturation of the models was estimated from the slope of the relationship (Okamura and Inoue 2012) and summarized in Table 14.2. The degree of saturation was in a range between 99.1 and 99.4%. In the centrifuge, hydrostatic pressure of the model was enhanced and most

**Table 14.2** Relative density and degree of saturation of the models

| Test code | Target $D_r$ (%) | Before motion #1 | | Before motion #2 | After motion #2 |
|---|---|---|---|---|---|
| | | $D_r$ (%) | Degree of saturation, $S_r$ (%) | $D_r$ (%) | $D_r$ (%) |
| EU1 | 65 | 64 | 99.3 | 71 | 80 |
| EU2 | 65 | 67 | 99.4 | 77 | 79 |
| EU3 | 80 | 83 | 99.1 | 87 | 88 |

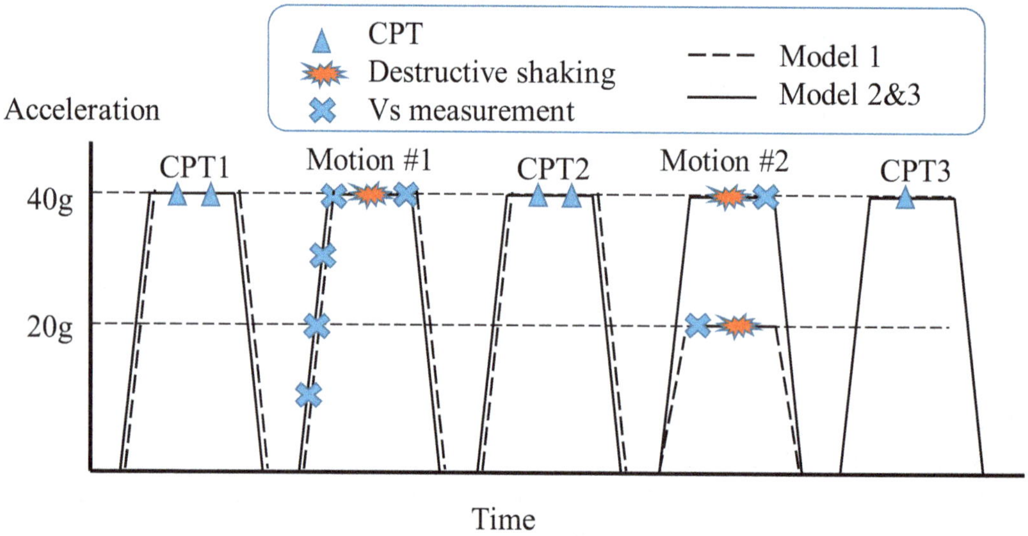


**Fig. 14.4** Centrifuge test sequence

of the remaining air and $CO_2$ bubbles in soil pore are considered to have dissolved (Kutter 2013), which increased further the degree of saturation.

### 14.3.5 Test Procedure

Three tests were conducted in this study. Models EU1 and EU2 had the same relative density of approximately 65%, subjected to the same input motion. EU2 is the duplicate of EU1 to confirm reproducibility of test. Model EU3 had higher relative density of 85% and subjected to the same input motion as EU1 and EU2 to study effects of initial relative density of the models.

All shaking tests and CPT were carried out at 40 g except for the second shaking of model 1. The centrifuge was spun up and down to mount and unmount the CPT device as indicated in Fig. 14.4. Locations of the surface markers were measured with a ruler and relative density of the model was estimated from the average settlement which is summarized in Table 14.2.

Cone penetration tests were performed before and after every destructive ground motion with a CPT device designed and fabricated at UCD. Two penetration tests were conducted at a time at intervals of 5 cm (12.5 times the cone diameter). The rate of cone penetration was 0.6 mm/s in model scale, which was slower than that specified.

Horizontally travelling shear wave velocity was measured with a pair of bender elements installed at mid-depth in the model layers before and after every destructive ground motion. The bender elements of 12 mm $\times$ 7 mm in area were fix to acrylic blocks and set in the sand at a distance of 50 mm (tip to tip).

Two destructive shaking tests, Motion #1 and #2, were conducted for all the models. Motion #1 was the tapered sine wave with a maximum prototype acceleration amplitude of 0.15 g. In the second destructive shaking tests, the same motion as the first one was imparted again to evaluate the evolution of the behavior of the model due to the previous shaking event.

All the data was recorded at a sampling rate of 2000 per second during shaking and after shaking for approximately 20 s until generated pore pressure completely dissipated. In the subsequent section in this paper, all the test results are in prototype scale otherwise mentioned. Note that data were not applied any numerical filtering.

## 14.4 Results

### *14.4.1 Shear Wave Velocity*

Shear wave velocities ($V_s$) measured with the bender elements at the mid-depth in EU1 and EU3 are indicated in Fig. 14.5. The measurement was performed before Motion #1, at 10 g, 20 g, 30 g, and 40 g, and velocities are plotted versus effective vertical stress at the depth of the bender element. $V_s$ increased with increasing effective vertical stress, with $V_s$ of EU3 being larger than EU1 because of the higher relative density.

### *14.4.2 Input Acceleration*

Accelerometers fixed to the base of the container measured input acceleration and a typical record is shown in Fig. 14.6a together with the specified motion. In order to duplicate the specified tapered sine wave by the mechanical shaker, rotation rate of camshaft changed with time accordingly. The shaking initiated at the time $t = 0$ and the rotation rate increased linearly with time until $t = 10$ s and decreased thereafter. Therefore, the acceleration frequency was very close to 1.0 Hz for $t = 9$–11 s and between 0.7 and 0.9 Hz for $t = 5$–9 s and $t = 11$–15 s. Fourier spectrum of the input acceleration (AH11) is shown in Fig. 14.6b.

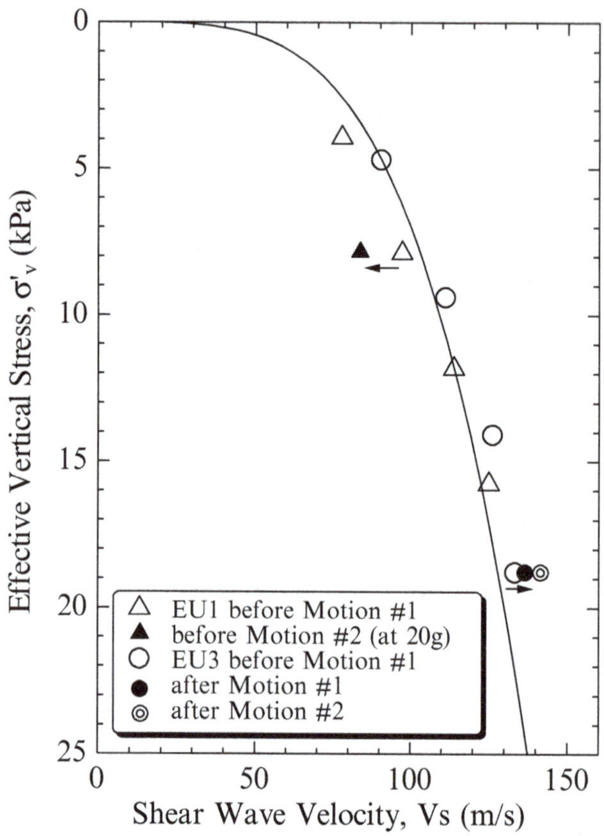


**Fig. 14.5** Shear wave velocity

## 14.4.3 Excess Pore Pressure Response

1. Excess pore pressure ratio
   Figure 14.7 shows excess pore pressure ratios (EPPR) measured on the centerline of the models for the Motion #1 shaking events. EPPR reached unity except for p1, and EPPR of p1 was higher than 0.8, indicating that the soils in all the models liquefied from the surface to the depth close to the bottom. Comparison of EU1 and EU2 shows that EPPRs are very similar to each other in terms of maximum EPPR and time duration for liquefaction condition lasted after shaking ended. For EU3, with higher relative density than EU1 and EU2, EPPR of p2, p3, and p4 also reached unity, which was the same as the two models, but duration for liquefaction condition lasted after shaking ended was clearly shorter.
2. Reproducibility of test
   Figure 14.8 compares excess pore pressure (EPP) responses for Motion #1 of EU1 and EU2. The target relative density of these models was the same and EU2 was replication test of EU1. The input accelerations were quite similar in these tests. EPP was similar at all the depths except for p1 installed on the base of the container, showing a good reproducibility of the tests.
3. Effect of initial relative density

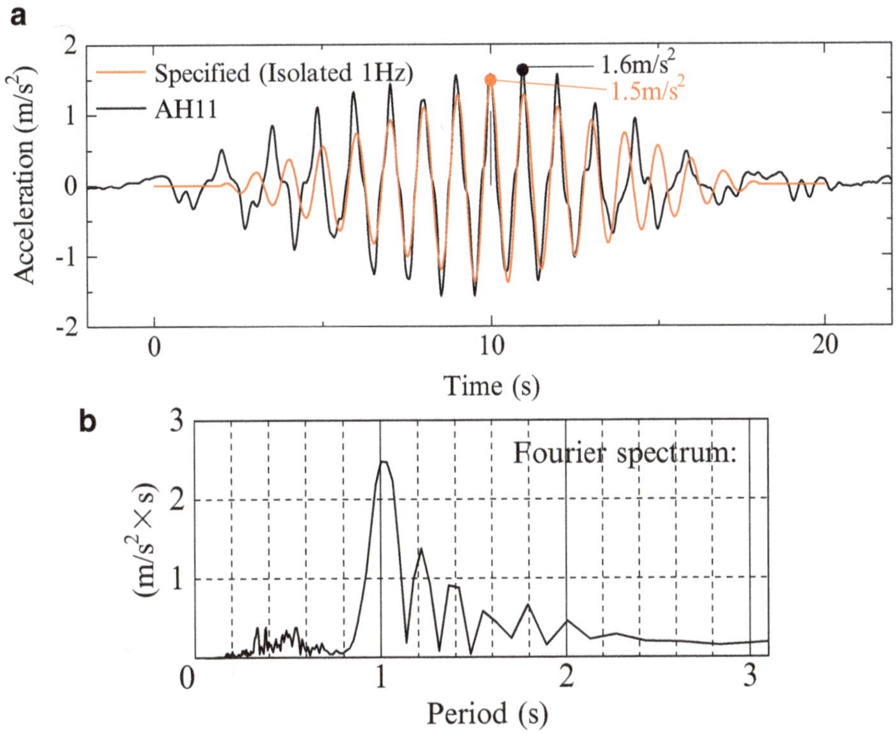


**Fig. 14.6** Input acceleration of Motion #1 for EU1. (**a**) Acceleration recorded on the shaking table. (**b**) Fourier spectrum of input acceleration

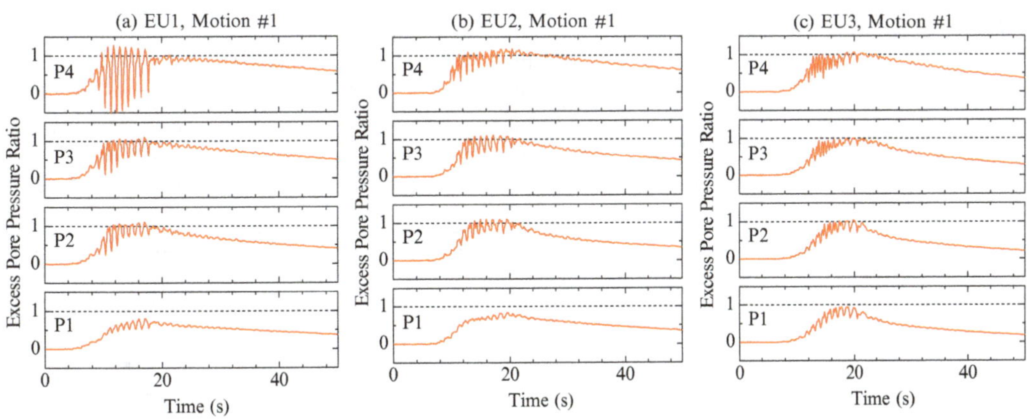


**Fig. 14.7** Time histories of excess pore pressure ratio

Figure 14.9 compares EPP response of models with different initial density, EU2 ($D_r = 65\%$) and EU3 ($D_r = 80\%$) for Motion #1. Maximum excess pore pressure attained at depths was quite similar for the two models; however, the rate of the pore pressure generation was clearly different. For all the depths, EPP of EU2 reached liquefaction condition 1–2 s earlier than EU3. Shear wave velocity and cone tip resistance ($q_t$) were consistent with this. $V_s$ and $q_t$ of the model before Motion #1 were higher for EU3 ($D_r = 85\%$) than EU1 and EU2 ($D_r = 65\%$).

4. Effect of shaking history

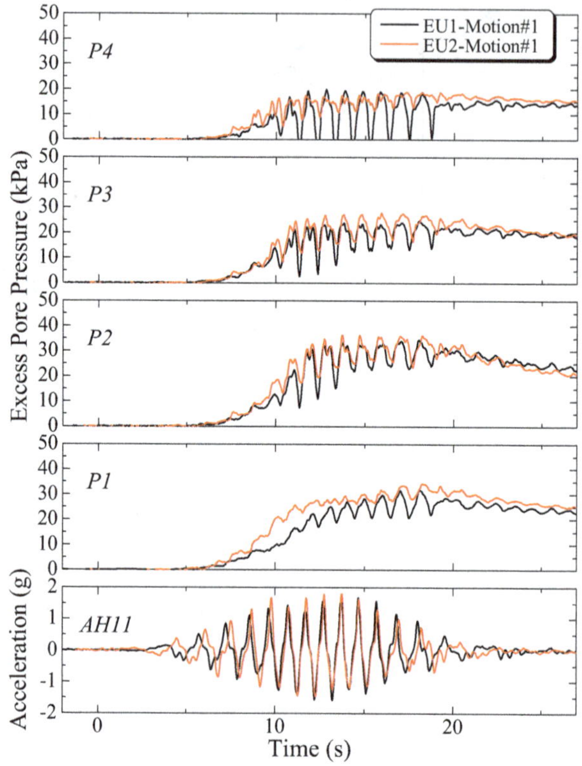


**Fig. 14.8** Excess pore pressure response for Motion #1 of EU1 and EU2

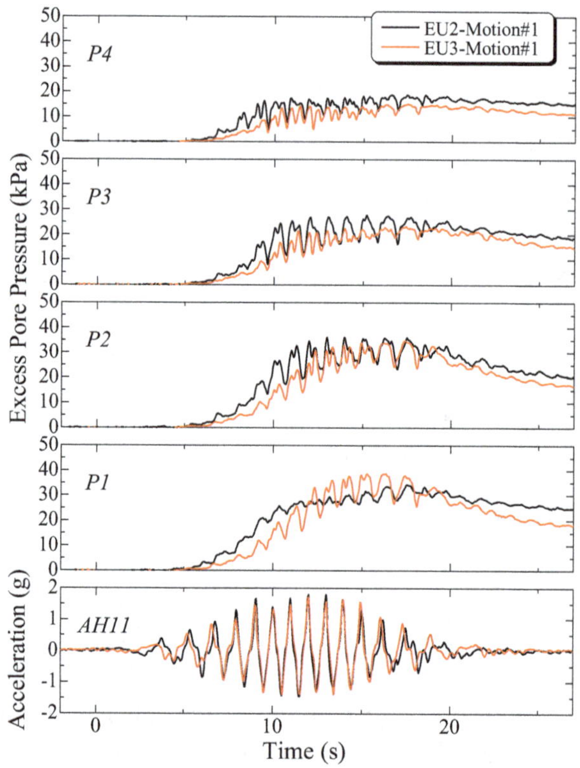


**Fig. 14.9** Excess pore pressure response for Motion #1 of EU2 and EU3

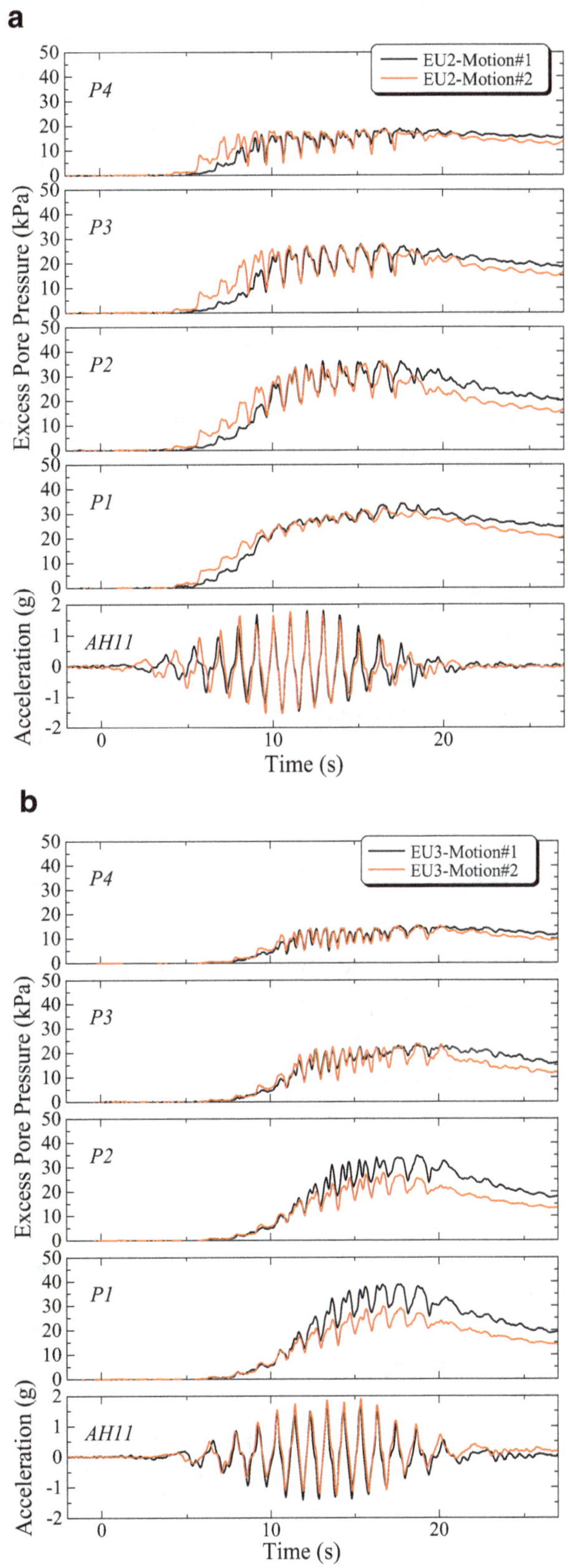


**Fig. 14.10** Excess pore pressure response to Motions #1 and #2. (**a**) Model 2. (**b**) Model 3

EU2 and EU3 were subjected to the same input acceleration as Motions #1 and #2. Comparisons of the model responses to Motions #1 and #2 reveal the effect of the first shaking event (Motion #1) on the responses to the second shaking. Figure 14.10a shows such direct comparisons for EU2. Relative density of EU2 before Motion #1 was 67% and 77% before second event. Despite the increased relative density due to Motion #1, EPP for Motion #2 generated faster than Motion #1. The sand after first shaking was more prone to generate excess pore pressures. It is also noticed that $q_c$ of EU2 after the first shaking event was very close to or slightly lower than that before first shaking event. Smaller $V_s$ was observed after Motion #1.

EPP response for EU3 is compared in the same manner in Fig. 14.10b. Relative density of EU3 before first shaking event was 83% and 87% before second event. At shallower depth recorded EPP at p3 and p4 before and after Motion #1 practically coincided. At p2, EPP for Motions #1 and #2 also coincided until input acceleration reached its maximum at $t = 12.5$ s. Thereafter, EPP for Motion #1 kept high until $t = 20$ s while EPP for Motion #2 started to decrease. The first shaking had no effect on EPP response of the sand with $D_r = 80\%$ or slightly enhanced the resistance to cyclic pore pressure generation. $V_s$ increased from 127 m/s before Motion #1 to 136 m/s before Motion #2 and further increased to 141 m/s after Motion #2 (see Fig. 14.5).

### 14.4.4 Liquefaction Triggering

In this study, an attempt was made to estimate liquefaction resistance of each model from model responses of Motion #1. Recorded acceleration was used to estimate cyclic shear stress ratio (CSR) at a depth of p3 assuming one-dimensional soil column condition. Because of the different number of cycles to liquefy in each test and nonuniform nature of CSR of each cycle as indicated in Fig. 14.11, the

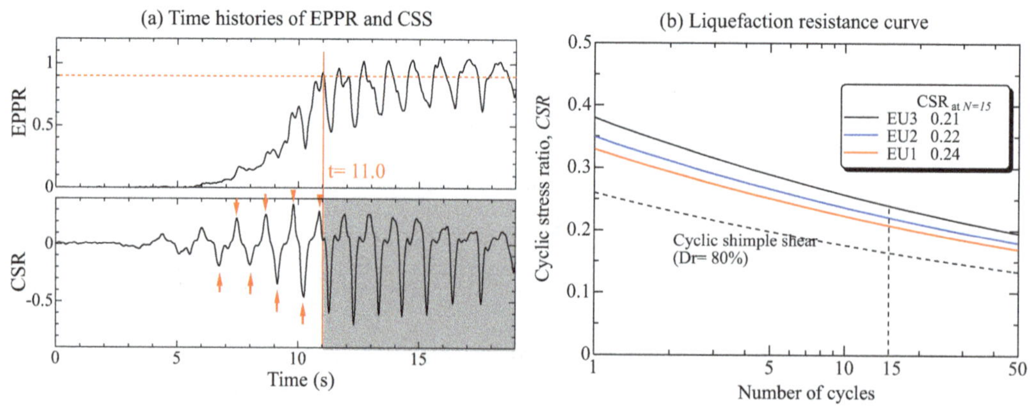


**Fig. 14.11** CSR time histories of EU2 for Motion #1 and liquefaction resistance curves

cumulative damage theory was used. Results of undrained cyclic simple shear test on the sand (Bastidas et al. 2017), of which test conditions were similar to those of the centrifuge tests including the sand deposition method and relative density (40 and 80%), indicated that the liquefaction resistance curves can be approximated as:

$$\text{CSR} = a \cdot N^{-0.15} \text{ for the sand with } D_r = 40\% \tag{14.1}$$

and

$$\text{CSR} = b \cdot N^{-0.17} \text{ for } D_r = 80\% \tag{14.2}$$

where CSR is cyclic stress ratio to reach the liquefaction condition in N uniform cycles and $a$ and $b$ are constants. Since test results for relative density of 65% were not given and the exponents in these equations are close to each other, Eq. (14.2) was used to determine the coefficients $b$ and thus cyclic resistance ratio of the centrifuge models.

Figure 14.11a shows typical time histories of excess pore pressure ratio and cyclic stress ratio at a depth 2 m from the surface. Based on the EPPR the soil is judged liquefied at $t = 11.0$ s (criteria is EPPR = 0.9). The peak values of CSR before this moment, shown by red triangles in Fig. 14.11a, were employed to calculate the cumulative damage.

Figure 14.11b shows relationship between CSR and number of cycles obtained from the simple shear test and those estimated for the centrifuge models for Motion #1. Liquefaction resistance ratio (CRR), which is defined in this study as the cyclic stress ratio to cause liquefaction condition in 15 uniform cycles, for EU1 and EU2 are depicted in Fig. 14.11b. CRR of EU1 and EU2 are similar and considerably lower than that for EU3, as expected. It is also noticed that CRR of the models are higher than the simple shear test result. A possible reason for this is existence of initial shear stress due to the sloping surface of the models. When the initial shear stress exists, the effective stress path can reach the failure line earlier which causes large acceleration spikes to the soil. The large acceleration spikes make the CSR significantly large. The other possible reasons are the use of the rigid container rather than a flexible laminar box and small vibrations which might have happened during model preparation.

### 14.4.5 Deformation of the Model

Side views of EU2 before and after Motion #1 in-flight are shown in Fig. 14.12. These pictures were taken with a USB3.0 camera with 5 megapixels at a frame rate of 30 Hz. Note that the width of the pictures is 12 m, while the model width is 20 m; therefore, areas near both side walls are not included. Particle image velocimetry analysis was conducted using the pictures and obtained deformation vectors are

**Fig. 14.12** Photos of EU2 before and after Motion #1

**Fig. 14.13** Deformation vectors caused by Motion #1 of EU2

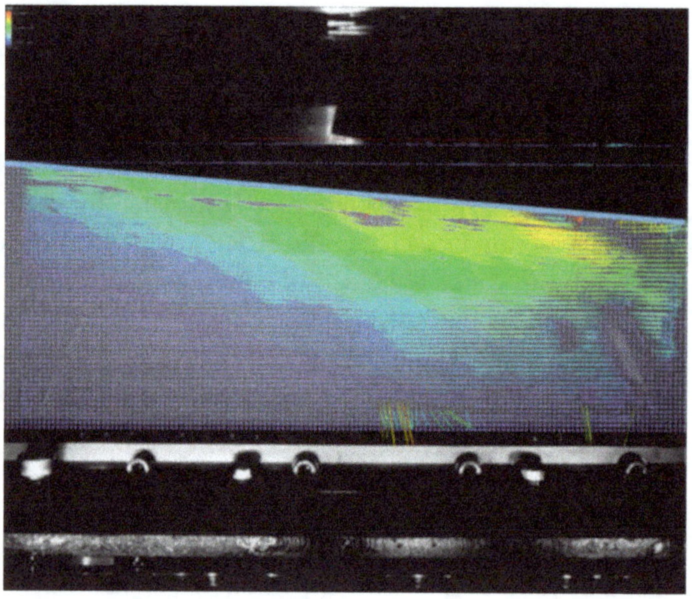

presented in Fig. 14.13. It can be seen that the movement of the soil was not uniform; soil at the upper right moved largely (yellow area), while displacement was limited in the lower left (dark blue area).

## 14.5 Conclusion

This paper describes three centrifuge tests conducted at Ehime University for LEAP-2017. The tested models were fully saturated uniform sand slope with relative density of either 65 or 80%. All the models were subjected the same ramped sine wave twice except for Motion #2 of EU2. The mechanical shaker was used to shake the models. The middle part of the input motion where the acceleration amplitude is largest had 1 Hz frequency, while the earlier and later part of the input motion had frequency lower than 1 Hz. Despite this nonconformity from the specification, input acceleration time histories of all shaking event for three models matched very closely. This makes it easy to study reproducibility of tests and effects of initial relative density, shaking history.

The same test was repeated in EU1 and EU2; the excess pore pressure responses of the two tests were quite similar, confirming intra-laboratory reproducibility. Destructive shaking and occurrence of liquefaction makes the medium dense sand (EU1 and EU2) more prone to generate excess pore pressure in the following shaking event. Shear wave velocity and cone tip resistance are also degraded by the destructive shaking. However, this is not the case for dense sand model (EU3) where pore pressure responses for Motions #1 and #2 practically coincided, and shear wave velocity increased from one shaking event to another.

An attempt was made to estimate liquefaction resistance of each model from model responses for Motion #1. Recorded acceleration was used to estimate cyclic shear stress ratio (CSR) assuming one-dimensional soil column condition, and the cumulative damage theory was used to account for the irregular nature of CSR time histories. The estimated liquefaction resistance of models are higher than those obtained from cyclic simple shear tests.

The distribution of the deformations in the model is analyzed with PIV. The displacement vectors are useful to obtain a broad view of entire deformation patterns.

**Acknowledgments** The experimental work was supported by JSPS KAKENHI Grant Number 26282103 (PI: Prof. S. Iai) and 17H00846 (PI: Prof. T. Tobita). The authors would like to thank a technical staff Mr. R. Tamaoka and students in the geotechnical group of EU T. Inoue, N. Sahashi, and S. Kubouchi for their assistance.

## References

Bastidas, A. M. P., Boulanger, R. W., Carey, T. C., & DeJong, J. (2017). *Ottawa F-65 Sand Data from Ana Maria Parra Bastidas*. Retrieved from https://datacenterhub.org/resources/ottawa_f_65, https://doi.org/10.1002/glia.23161

Kutter, B. L. (2013). Effects of capillary number, bond number, and gas solubility on water saturation of sand specimens. *Canadian Geotechnical Journal, 50*(2), 133–144. https://doi.org/10.1139/cgj-2011-0250.

Kutter, B. L., Carey, T. J., Stone, N., Bonab, M. H., Manzari, M., Zeghal, M., Escoffier, S., Haigh, S., Madabhushi, G., Hung, W.-Y., Kim, D.-S., Kim, N.-R., Okamura, M., Tobita, T., Ueda, K., & Zhou, Y.-G. (2019). LEAP-UCD-2017 V. 1.01 model specifications. In B. Kutter et al. (Eds.), *Model tests and numerical simulations of liquefaction and lateral spreading: LEAP-UCD-2017*. New York: Springer. https://doi.org/10.1016/j.neuron.2018.08.036.

Okamura, M., & Inoue, T. (2012). Preparation of fully saturated model for liquefaction study. *International Journal of Physical Modelling in Geotechnics, 12*(1), 39–46. https://doi.org/10.1680/ijpmg.2012.12.1.39.

Shin-etsu Chemical Co., Ltd. (1997). *Metolose Brochure*. Cellulose Department, DOI: https://doi.org/10.17226/5702.

# Chapter 15
# LEAP-UCD-2017 Centrifuge Test at IFSTTAR

**Sandra Escoffier and Philippe Audrain**

**Abstract** In the framework of the LEAP 2017 exercise, two dynamic centrifuge tests on a gentle slope of saturated Ottawa-F64 have been performed at the IFSTTAR centrifuge. These tests were conducted in parallel with other tests performed in nine other centrifuge centers. The objectives were to compare the experimental results, e.g., effect of the experimental procedure or of test parameters on the results, and to provide a database for numerical modeling. In this framework, all the tests were performed on the same prototype geometry and the first base shaking was a 1 Hz ramped sine with an effective amplitude of 0.15 g at the prototype scale. Among some of the tests performed in the various centrifuge facilities, several parameters were modified, such as the density and the second input motion, in order to study their impact on the slope behavior when it is subjected to base shaking. This paper details the procedure followed at the IFSTTAR center for the buildup of dense sand and medium loose sand models and the deviations from the specifications provided by the leaders of the LEAP-UCD-2017 program. The main deviations are, for both tests, the absence of the measurement of the saturation degree and, for the first, the characterization of the soil properties with CPT test and bender element measurements during the test. Some results of the tests performed are briefly presented such as the acceleration, pore pressure buildup, and the deformation of the slope surface.

## 15.1 Introduction

Actual researches in numerical modeling on liquefaction phenomena, for instance, advanced numerical techniques based on multiscale approach in large deformation (Callari et al. 2010), highlight the need of experimental database for the calibration and the validation processes. In this framework one of the objectives of the LEAP-UCD-2017 research program is to provide high-quality laboratory and centrifuge test

---

S. Escoffier (✉)
IFSTTAR, GERS, SV, Bouguenais, France
e-mail: sandra.escoffier@ifsttar.fr

P. Audrain
IFSTTAR, GERS, GMG, Bouguenais, France



B. Kutter et al. (eds.), *Model Tests and Numerical Simulations of Liquefaction and Lateral Spreading*, https://doi.org/10.1007/978-3-030-22818-7_15

data. A total of ten centrifuge teams were involved in this experimental research work.

Following the model specifications (Kutter et al. 2019), each team has performed a series of dynamic tests on a gentle slope of saturated Ottawa F-65 sand. The objectives of the specifications were to minimize the discrepancies between the experimental procedures followed in each centrifuge team in order to evaluate the quality of liquefaction centrifuge tests and the effects of procedure deviations on the obtained results through cross testing. In addition to this repeatability step, additional tests with different densities and with different second and eventually third base shaking were performed.

Two tests were performed at the IFSTTAR center with the same base shakings but with different densities. In this paper, the procedure and some of the results obtained on the dense sand model, S02, and on the medium loose sand model, S03, are presented.

## 15.2 As Built Model

### *15.2.1 Soil Material and Placement of the Sand by Pluviation*

The soil selected for the LEAP-UCD-2017 centrifuge experiments is Ottawa F-65, a clean sand with less than 0.5% of fines. Control quality checks on the grain characteristics (granulometry and maximum and minimum densities of the sand delivered) were performed following the model specifications established for this project. The maximum and minimum densities were measured based on the modified Lade et al. (1998) method and the modified ASTM D4254 method, respectively. In addition, the French NF P94–059 procedure was used for both. The average values for the maximum densities are 1737 and 1775 kg/m$^3$ for the NFP and the modified Lade et al. method, respectively. The average values for the minimum densities are 1509 and 1472 kg/m$^3$ for the NFP and the modified ASTM methods, respectively. From the LEAP-UCD-2017 Version 1.0 model specifications, the average values of all the tests that have been performed with different methods are 1475 and 1756 kg/m$^3$ for the minimum and the maximum densities, respectively. Due to the dispersion that has been observed, the values obtained by IFSTTAR are in the range of the previous ones (data are available in Kutter et al. 2019).

Figure 15.1 presents the granulometry curves that were obtained. The D10, D30, D50, and D60 are 0.139, 0.196, 0.226, and 0.247 mm, respectively. The D30, D50, and D60 values are somewhat higher than the one given by Kutter et al. (2017) due to the use of additional sieves for the sieves with an opening between 0.125 and 0.250 mm.

The Ottawa sand was dry pluviated through a sieve attached to the bottom of the hopper of the IFSTTAR dry pluviation system (Ternet 1999). The initial target densities for S02 and S03 tests at the IFSTTAR center were 1703 and 1651 kg/m$^3$, respectively. Due to the lack of data concerning medium dense sand, the target

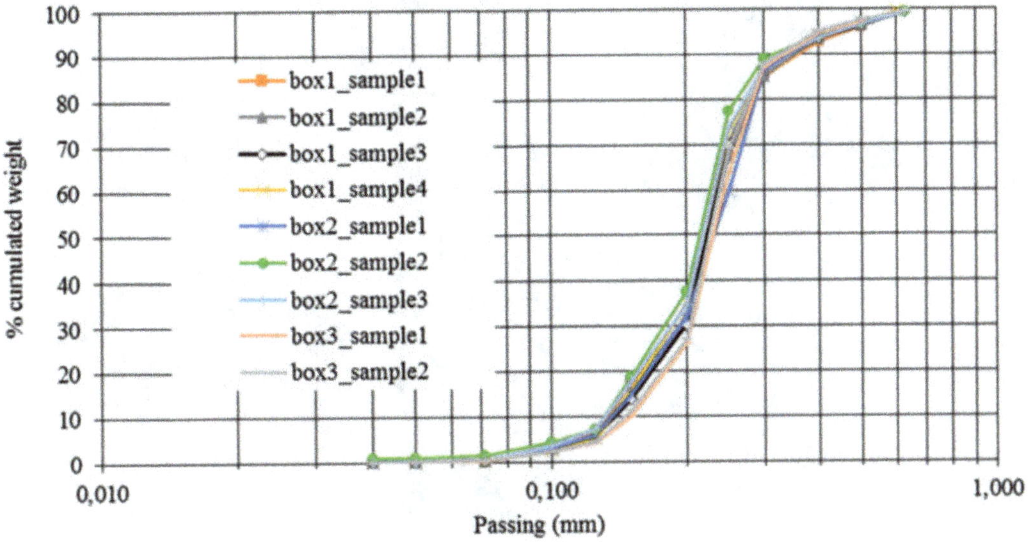


**Fig. 15.1** Granulometric curves of the OTTAWA sand obtained at the IFSTTAR center

density for the S03 test has been increased to 1599 kg/m$^3$. To obtain these request densities the specifications concerning the opening size of the sieve were not followed (e.g., for S02 a sieve with an opening of 1.2 mm and for the flow restriction the recommended design was three slots of 1.2 mm width with a center to center spacing of 26 mm).

First, due the standard sieves in France, the selected sieve for the calibration has an opening of 1.25 mm. In the case of the dense sand, following the recommended design for the slots, a first air pluviation test was performed but the sand flow stopped rapidly. As suggested in the specification document, the clogging effect could be due to the humidity of the room that was measured at 65%. Consequently, the geometry of the open part was modified. The width of the slots was increased up to 6 mm that corresponds to three-mesh width of the sieve and the center to center distance between the opening slots was 22 mm. Furthermore, due to the container width (i.e., internal width 200 mm) the slot opening was 240 mm long. Finally, a drop height of 50 cm and a horizontal velocity of 48 mm/s of the hopper were selected to obtain a density of 1696 $\pm$ 4.22 kg/m$^3$ (three measurements).

In the case of the medium loose sand, the minimum density that can be obtained with the pluviation system was 1624 $\pm$ 3.8 kg/m$^3$. It has been obtained with two slots of 2.9 cm width that correspond to 15 meshes and a center to center distance of 4.8 cm.

During the calibration procedure of the pluviation, the measurement of the density of the sand was performed using cylindrical density boxes (approximate volume 390 mm$^3$). Once the pluviation was made the top surface of the density box was leveled using a specific tool to avoid unintentional densification of the sand.

**Fig. 15.2** Rigid steel box especially built for the LEAP project at the IFSTTAR center and placement of the rigid box inside the fixed ESB container

## 15.2.2 Rigid Container Configuration and Sensor Layout

The specifications for the model are a gentle slope (5°) with a length of 20 m, a width of at least 9 m, and a top height of the slope of 4.875 m at the prototype scale. Not one container available at the IFSTTAR center enabled the buildup of such slope geometry. Consequently, a rigid steel container was especially built for the LEAP project (Fig. 15.2). The inner dimensions are 400 mm × 200 mm × 200 mm (L × W × H) which enable to build the gentle slope at the request prototype size for a centrifuge acceleration of 50 g.

Due to the shaker configuration, this rigid container was fixed through 12 screws into an ESB. In addition, the shearing movement at the extremity of the ESB container was blocked by four rods (one at each corner, Fig. 15.2). Before the first test, a series of drives was performed in order to obtain the time histories of the commands to the servo valves of the shaker (i.e., drives) that correspond to the request base shakings. This preliminary test that was performed on the two-container assembly highlighted the presence of non-negligible high frequencies not far from the predominant one (i.e., 50 Hz at the model scale). Consequently, a sufficient amount of sand was put in place between both containers to remove these high frequencies that should be due to resonance phenomena of the embarked system (Fig. 15.2).

The relative rigidity of cables of the pore pressure transducers (Druck PDCR81) induces some problems for their positioning during the sand pluviation. In order to facilitate their positioning and limit the length of cables in the sand mass, cable glands were inserted on one of the lateral side of the container (Fig. 15.2). This configuration enables the placement of the pore pressure sensors inside the container with the request length of cable before the starting of the pluviation. Sensors were taped up on the lateral side of the container and put in place once the request level of sand was pluviated.

Figure 15.3 details the model scale dimensions and the sensor layout.

The required accelerometers and pore pressure sensors (AH1 to AH4, AH12, AH11, AV1, AV2, P1 to P4, and P9 and P10) were all put in place during the pluviation. In order to facilitate the positioning of the sensors, the pluviation has also been calibrated to drop off a layer of approximately 1 cm thick each back and forth movement of the hopper, which is the model distance between each level of sensors of the central array (AH1 to AH4 and P1 to P4). In addition, two other accelerometers and pore pressure sensors (S02, P5 and P7; S03, P6 and P8) were installed to complete the instrumentation.

The location in the coordinate system defined in Fig. 15.3 was measured during placement and during excavation after the test. The difference between as-built and "after test" coordinates of the sensors embedded in soil could be partly due to errors, such as inaccurate measurement that was performed using a steel rule, inadvertent tugging on wires after placement or during excavation, and displacement of the sensors relative to the liquefiable soil. These location data are analyzed in the Sect. 15.4.2. of this paper and displacement vectors are provided.

A pair of bender element has also been installed. The pair of bender elements is the same as the fixed one presented by Brandenberg et al. (2006). However due to problem of amplification, the measurement of the shear wave velocity has not been performed in the model S02. In addition, as requested in the specifications no CPT test was performed during this test. These two main deviations from the model specifications do not enable the characterization of the soil properties for the S02 models. In the case of the S03 test performed on medium loose sand, both CPT tests and bender element measurements have been performed.

The shear wave velocity measurements were made just before and after each base shaking. For each soil state, 6 series of 36 measurements were made (2 series of measurement for the 3 selected input frequencies 5, 7.5, and 10 kHz). The use of several frequencies and the staking of data increase the reliability of the determined values. The CPT device used was developed and built at the University of California, Davis, and had an external diameter of 6 mm at the model scale. CPT measurements were made to characterize the initial state and the soil and the state after each base shaking. Due to the experimental configuration, the CPT was removed from the centrifuge before each shaking.

In the case of the IFSTTAR 1D shaker the direction of the solicitation is parallel to the axis of the centrifuge (Chazelas et al. 2008). In order to keep constant the radius between the surface of the soil in a transverse cross section and the center of rotation of the centrifuge, the surface should have a curved shape in the direction perpendicular to the base shaking. The distance between the axis of rotation of the centrifuge and the center of the soil surface is 5.063 m. Considering that the inner dimension of the container's width is 0.2 m, the difference in height between the midpoint and the corresponding point at the lateral side should be 1 mm. As this

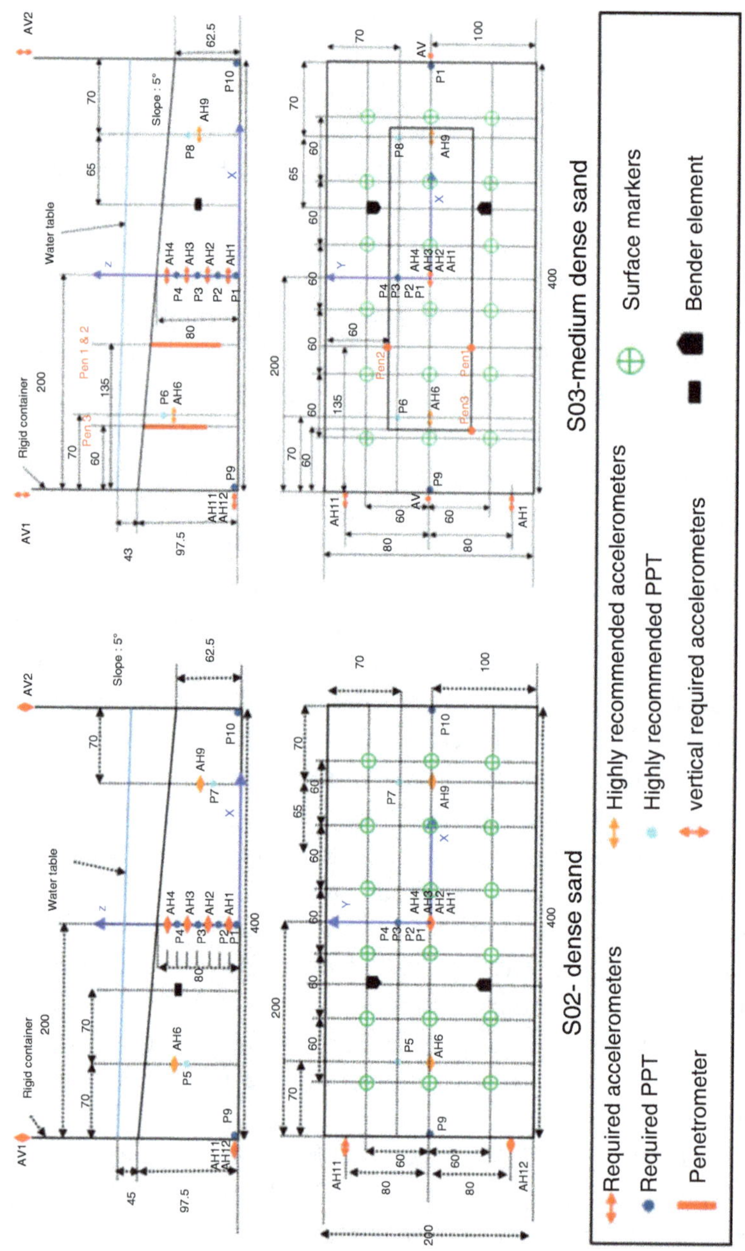


**Fig. 15.3** Model scale dimension and sensor layout of the test performed at IFSTTAR (dimensions are in mm at the model scale)

value is in the range of precision of the leveling of the surface, the soil surface was not curved in the Y direction.

In addition to the sensors, surface markers were placed in a grid pattern at the surface of the soil to measure the displacement during soil liquefaction. The diameter of the surface markers used was two times smaller than the recommended design (improved design with an external diameter of 13 mm at the model scale). The location of the markers in the X and Y directions was performed with a steel rule with an estimated precision of 1 mm and the Z location was performed with a laser sensor. The use of a laser sensor to measure the soil surface required a partial immersion of sensor (at least 10 mm) and consequently the water level was between 45 and 43 mm above the top of the slope (Fig. 15.3). The precision of the Z position is smaller than 0.5 mm as requested in the specifications. The surface markers have been put in place before the saturation process and their location has been measured at 1 g before the first spin up of the centrifuge and after each base shaking (Motion#1 and Motion#2) once the centrifuge was spun down.

### 15.2.3 Viscosity of Pore Fluid

The viscous fluid used for the test is a mixture of tap water, HMPC (Culminal MHPC50), and biocide that is added in order to avoid decrease of the viscosity with time. A series of viscosity measurements was performed on several mixtures with different amounts of HPMC (from 28 to 23.5 g per liter).

The viscosity was measured using a FungiLab smart series viscosimeter with a low viscosity adaptor combined with a thermostatic bath. This is a Couette-type viscosimeter. After the stabilization of the temperature, the kinematic viscosity was measured for different shear rate values (in a Couette-type viscosimeter the shear rate represents, under the hypothesis of a laminar flow and a Newtonian fluid, the gradient of the fluid velocity between the outer and the inner cylinder of the viscosimeter). The corresponding average kinematic viscosities that correspond to the centrifuge room temperature were 50.8 and 45.8 cst for the S02 and S03 tests, respectively. It should be noticed that the viscous fluid used for the test is rheofluidifying: the viscosity of the fluid decreases with the increase of the shear rate (Fig. 15.4).

### 15.2.4 Saturation Process

The saturation process was performed under vacuum at 1 g. Prior to the de-air fluid flow into the soil model, the rigid box was closed using an airtight lid. The container, the fluid tank, and the various connections were de-aired following the

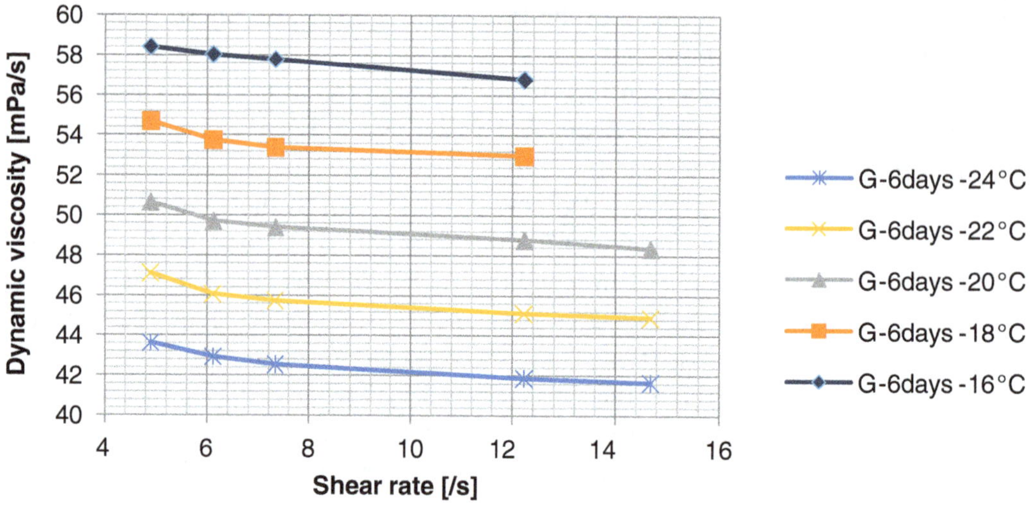


**Fig. 15.4** Evolution of dynamic viscosity for the viscous fluid used in S02 test (28 g/l) versus revolutions per minute and temperature

recommendations of Kutter (2013). Two cycles of vacuum/$CO_2$ were performed. Each cycle consists of, first, the application of a vacuum up to an absolute pressure of 100 to 94 mbar followed by flooding of $CO_2$ up to the atmospheric pressure. By this way, it is supposed that 99% of the air resting inside the container has been replaced by $CO_2$. After these two cycles, an absolute pressure of 94 mbar was applied and maintained during the fluid flow into the container. The fluid was dripped onto the surface near the slope tip (a sponge was put in place above the soil surface to prevent perturbations). The drip rate was not measured but once a pool was formed at the slop tip, the fluid flow was adjusted by hand to maintain the pool. The time request for saturation was around 10 h which corresponds roughly to a drip flow of 6.5 ml/min. The degree of saturation was not checked with the method proposed by Okamura and Inoue (2012) at the end of the saturation process.

One nonconformity took place during the saturation process in the case of the S02 test. Some air leakage was observed on the lateral side where the cable gland for pore pressure sensors P1 to P4 was located. However, the few air bubbles that appear concerned only a small localized zone of about 2–3 $cm^2$ at the surface.

## 15.3 Achieved Ground Motions

As previously mentioned the tests were performed at 50 g. The angular velocity (93.8 RPM) of the centrifuge was calculated to produce the 50 g centrifuge acceleration at a radius of 5.09 m (the radius to 1/3 depth, i.e., 26.7 mm at the model scale, below the soil surface at the central array of accelerometers).

### 15.3.1 Horizontal Component

Two base shakings were applied to the S02 and S03 models. Each one was a tapered sine, the first one obtained from a reference signal with a PGA of 0.1 g and the second one from a reference signal with a PGA of 0.3 g. The achieved base motions were analyzed by isolating the predominant frequency component (1 Hz at the prototype scale) from the noise. The filtering of the raw data was made by using a FIR filter with a passband frequency at the prototype scale of [0.76–1.24 Hz] with an attenuation of −40 dB. This filter type avoids introduction of phase lag between the frequency components of the signal. The time representations of the achieved base Motion#1 and Motion#2 for both tests (raw data, filtered data) are presented in Fig. 15.5.

The sampling frequency selected for the tests was 10,240 Hz except in the case of Motion#1 of the S03 test that was only 1024 Hz. The analysis of the frequency content of the base shaking for all the base shakings highlights three high-frequency components that are non-negligible especially in the case of Motion#1: 7 Hz, 9 Hz, and 16.6 Hz components. It should be noticed that due to the sampling frequency used for Motion#1 of the S03 test, only the 7 Hz and the 9 Hz components were recorded. The 16.6 Hz component corresponds to the dither associated to the servo valve of the hydraulic shaker and cannot be removed. However the velocity and the displacement induced by this high frequency are negligible due to the 16.6 Hz frequency value compared to the 1 Hz component.

For Motion#1, the reference signal has been selected to obtain a $PGA_{eff}$ of about 0.15 g following the definition given by Eq. 15.1.

$$PGA_{effective} = PGA_{1Hz} + 0.5 \cdot PGA_{hf} \tag{15.1}$$

In this equation the $PGA_{hf}$ corresponds to the peak value induced by the higher-frequency components that occur within one cycle of the peak of the 1 Hz component. For all the base shakings, except for Motion#2 of the S03 test, the maximum value of the 1 Hz component and the maximum value of the raw acceleration were not reached at the same time as illustrated in Fig. 15.6.

Consequently in order to determine the $PGA_{eff}$, the maximum value of the noise reached in the vicinity of the maximum value of the 1 Hz component has been considered. Table 15.1 gives the values obtained for the $PGA_{eff}$.

If the velocity is considered, the harmonic components and the frequency component induced by the dither are negligible for all the motions. An example is given in Fig. 15.7 for the case of Motion#1 of the S02 test. As the liquefaction phenomena are linked to the distortion and the velocity of the distortion, a criteria for the quantification of the base shaking based on the effective velocity could be relevant too. The velocity has been obtained by integrating the acceleration and then by applying a passband filter FIR 1 with a frequency passband at the model scale of [12.5–5070 Hz] (due to the sampling frequency error for Motion#1 of the S03 test, the passband frequency was reduced to [12.5–507 Hz]). As illustrated in Fig. 15.7

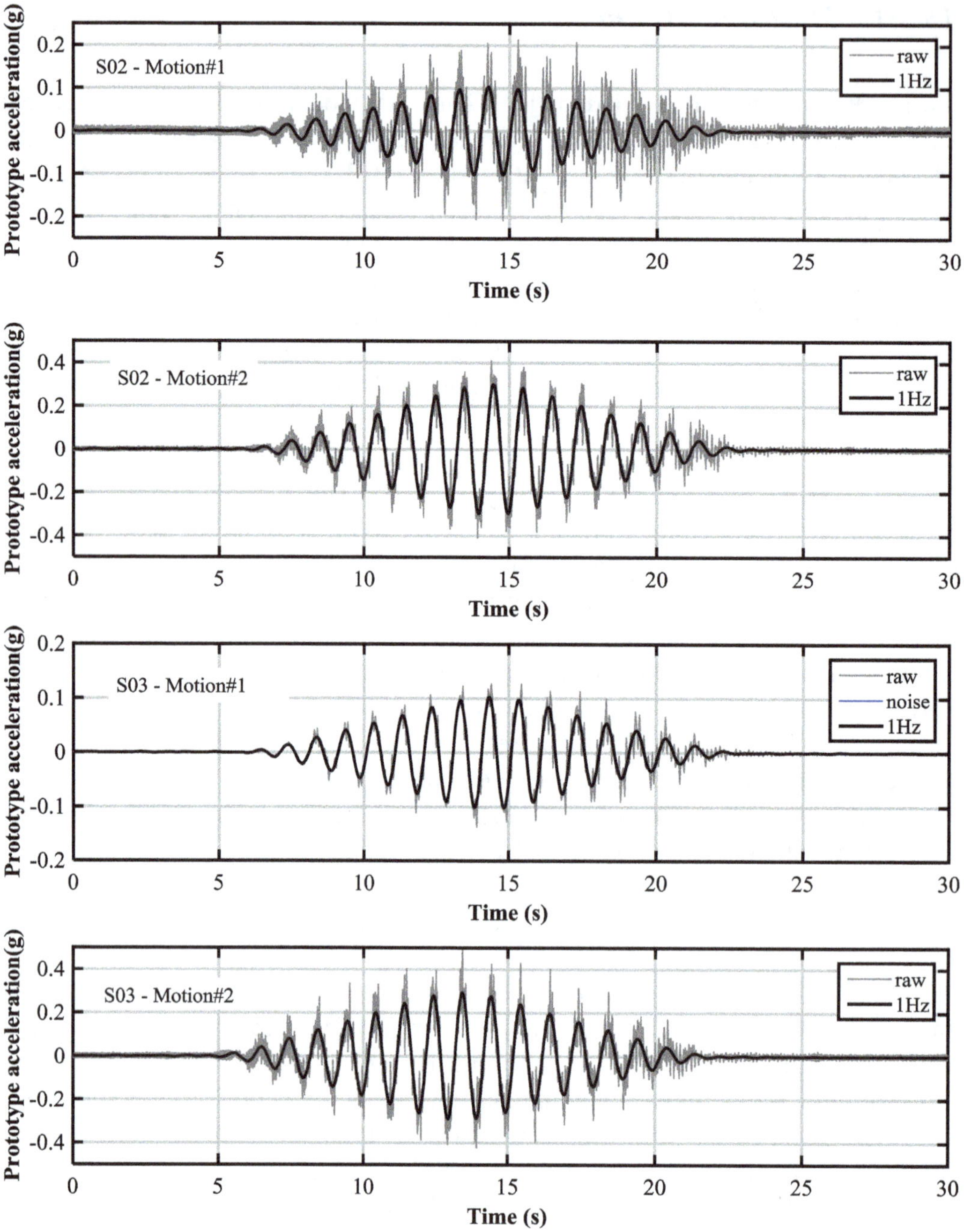


**Fig. 15.5** LEAP S02 and S03—time representations of the achieved base Motion#1 and Motion#2

the maximum value of the 1 Hz component and of the raw values of the velocity is reached at the same time. Considering the definition of the $PGV_{eff}$, the effective peak ground velocity, given by Eq. 15.2, the values obtained for each base shaking are also given in Table 15.1.

$$PGV_{effective} = PGV_{1Hz} + 0.5 \cdot PGV_{hf} \tag{15.2}$$

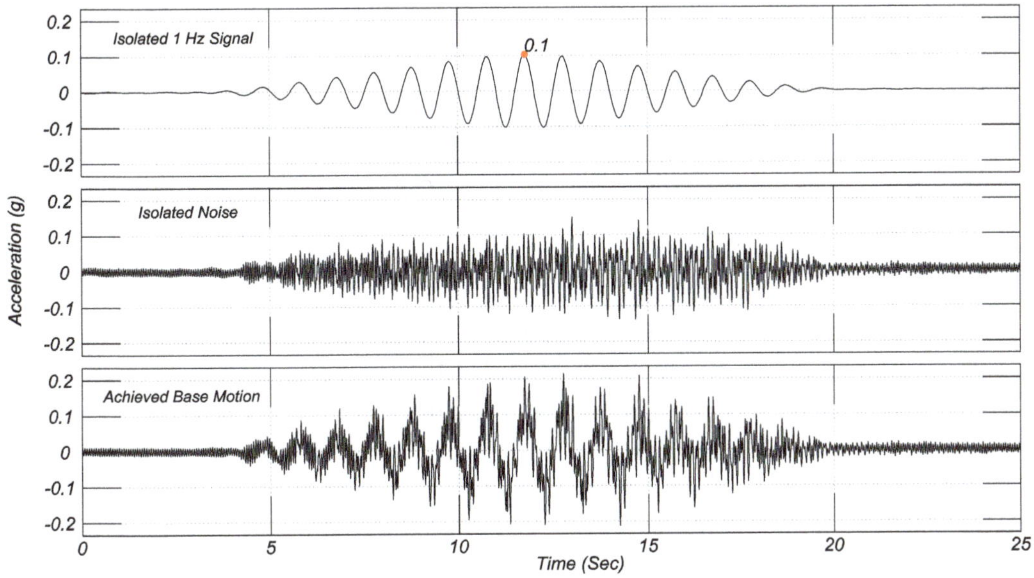


**Fig. 15.6** LEAP S02 Motion#1—achieved base acceleration time history (bottom). High-frequency noise isolated from achieved signal (middle). 1 Hz signal (top) (the red dot is the point at the time the maximum value of the 1 Hz component is reached)

### *15.3.2 Vertical Component*

The time representation of vertical accelerations and displacements measured at the top of the extremity of the rigid container (AV1 and AV2, Fig. 15.1) are represented in Fig. 15.8 for the test S03. The results obtained for the test S02 have a limited interest due to the absence of data for the vertical accelerometer AV1. The displacement is obtained by double integration process of the accelerometer. After each integration, the filter previously presented to obtain the velocity was applied.

From the frequency analysis of Motion#1 of test S02, a non-negligible part of the vertical acceleration is due to components with a frequency equal of higher than 7 Hz (up to around 50 Hz). As a consequence, as previously mentioned, due to an error in the selection of the sampling frequency (1024 Hz) for Motion#1 of test S03, the acceleration data are only relevant to frequency components up to about 10 Hz at the prototype scale and consequently the maximum vertical acceleration represents only 5% of the maximum horizontal acceleration against 71% in the case of Motion#1 of test S02 (the percentage 71% is based on the value of only one accelerometer).

For Motion#2 the maximum vertical acceleration represents 66.5 and 74%, respectively, of the maximum horizontal acceleration. Even if the level of vertical acceleration is non-negligible compared to that of the horizontal acceleration applied at the base of the container, due to the frequency content of the vertical acceleration, the vertical displacements are largely lower than the horizontal displacement. For Motion#1 of tests S02 and S03 the maximum dynamic vertical displacements are 4% and 3.4%, respectively, of the maximum dynamic horizontal displacement. In the case of Motion#2 the values are 8 and 3%, respectively.

**Table 15.1** Characteristics of the measured base motions for S02 and S03 tests

| Test | Base shaking | $PGA_{eff}$ (g) | PGA 1 Hz component (g) | % Due to high frequency[a] | $PGV_{eff}$ (m/s) | PGV 1 Hz component (m/s) | % Due to high frequencies[a] | Frequency of the first harmonic |
|---|---|---|---|---|---|---|---|---|
| S02 | Motion#1 | 0.153 | 0.102 | 34% | 0.166 | 0.160 | 4% | 7 Hz |
| | Motion#2 | 0.323 | 0.299 | 8% | 0.482 | 0.468 | 3% | 7 Hz |
| S03 | Motion#1 | 0.113 | 0.1 | 12% | 0.165 | 0.16 | 3% | 7 Hz |
| | Motion#2 | 0.37 | 0.28 | 24% | 0.445 | 0.03 | 3% | 7 Hz |

[a]This is, considering the equation, the percentage of the $PGA_{eff}$ and $PGV_{eff}$ due to high-frequency components

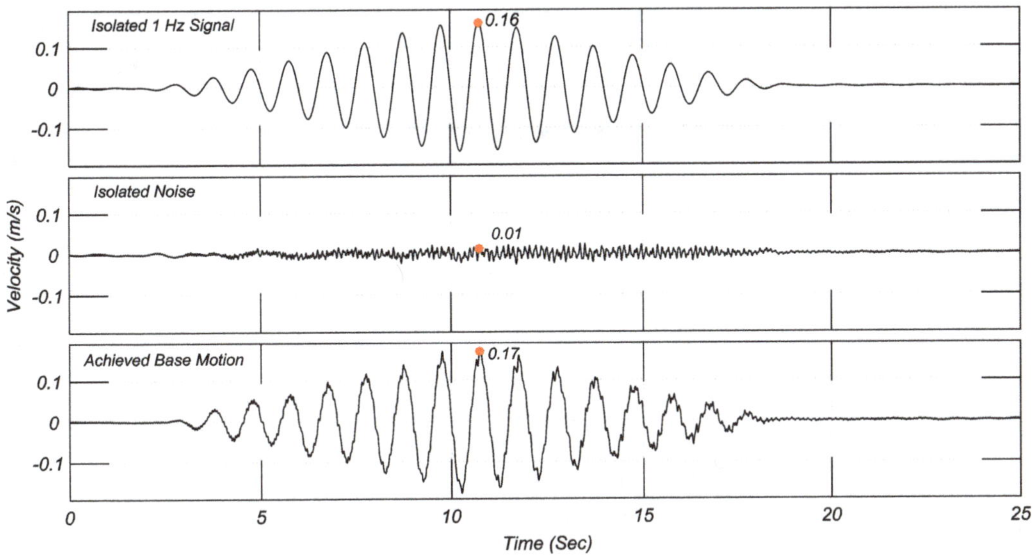


**Fig. 15.7** LEAP S02 Motion#1 achieved base velocity time history (bottom). High-frequency noise isolated from achieved signal (middle). 1 Hz signal (top)

The time representation of the dynamic vertical displacement for Motion#2 of S03 test illustrates an opposition phase between AV1 and AV2 that characterized a rotation of the container. In the case of Motion#1 both vertical accelerometers are in phase opposition but the displacement amplitude varies differently with time. In both cases the rotation is limited (it is estimated around $3.10^{-4\circ}$ and $8.10^{-3\circ}$ for Motion#1 and Motion#2, respectively).

## 15.4 Results

In the following all the data are given at the prototype scale otherwise mentioned.

### *15.4.1 Pore Pressure and Acceleration Responses*

As previously mentioned, due to the use of a laser sensor to record vertical displacement of the markers between each base shaking, a minimum value for height of the water table above the soil surface was necessary. The height of the water table for the S02 and S03 tests was 45 and 43 mm, respectively, at the model scale (Fig. 15.1). For both containers no system of wave limitation was used and the level of the water during the shaking was not recorded. Figure 15.9 illustrates the variation of the pore pressure during the shaking at the bottom of the container near each extremity (P9 and P10). In addition, the pore pressure at 70 mm (3.5 m at the prototype

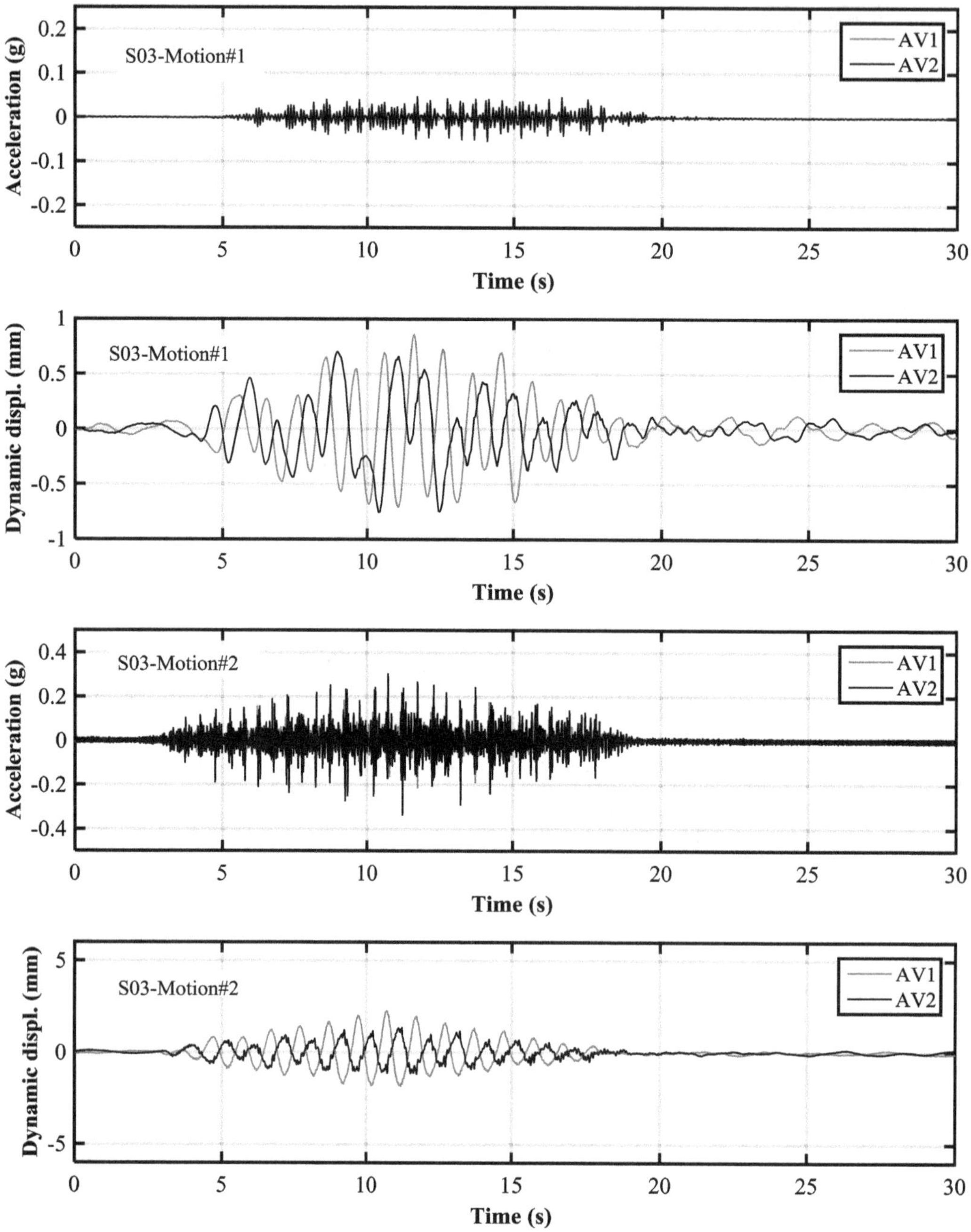


**Fig. 15.8** LEAP S03 Motion#1 and Motion#2 vertical acceleration and dynamic displacement measured at the top of both extremities of the rigid container

scale) of each extremity of the container and at about 2 m depth below the soil surface is also presented.

In this example, considering the initial position of the pore pressure sensors measured during the buildup of the model S03 and the level of the water table above the soil surface, the maximum pore pressure buildup, considering that there were no waves, was around 29.4, 47.1, 5.8, and 13.2 kPa for the P10, P0, P8, and P6 pore pressure sensors, respectively. These values are lower than the maximum pore

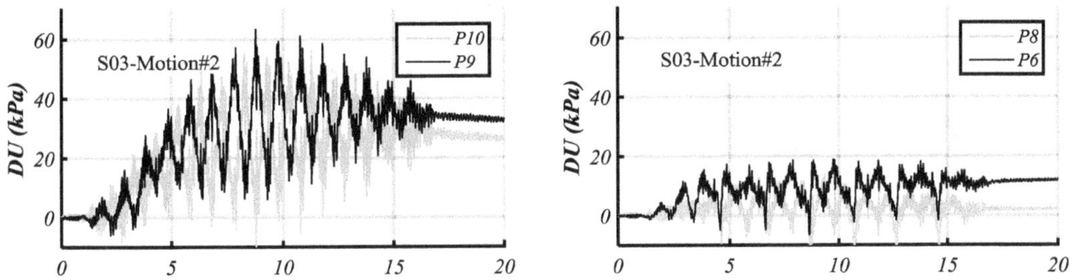


**Fig. 15.9** S03 Motion#2 pore pressure response near and at the extremity of the container

pressure buildup measured during Motion#2 of the S03 test (i.e., 56.6, 63.6, 13.6, and 19 kPa for the P10, P9, P8, and P6 sensors, respectively). In addition, the variation of the pore pressure measured by sensors P9 and P10 is phase opposite such as for the sensors P8 and P6. Even if further analysis is requested, the amplitude of the pore pressure and the phase opposition suggest that one part of the pore pressure fluctuations recorded by these four sensors is due to the waves. This analysis suggests that a wave reduction system should be built for future tests to avoid non-negligible effect of waves near the extremities of a rigid container.

Figure 15.10 illustrates the pore pressure variation, for both tests, of the central array of pore pressure sensors for each base shaking. Referring to the as-built sensor depths, the initial effective pore pressure for the P3 and P4 sensors of the S02 test is about 23 and 11 kPa, respectively. In the case of the S03 test, the initial effective vertical stress for the P2, P3, and P4 sensors is about 28, 16, and 13 kPa, respectively.

In the case of the central array of pore pressure sensors, the shape of the pore pressure buildup such as the maximum value suggests that the effect of the wave is limited in the middle of the container.

In the case of the S02 test (dense sand, $\rho_{\mathrm{d}} = 1696$ kg/m$^3$) Motion#1, the pore pressure ratio (i.e., ratio of the pore pressure buildup to the initial vertical effective stress) rapidly reached about 100% at shallow depth (P4). A pore pressure ratio of around 100% was also reached at lower depth (P3) 4 s later. Compared to the pore pressure evolution near the surface (P4), in the case of the P3 sensor, the pore pressure buildup remains at the same value a shorter duration after the end of the base shaking. For both levels the dissipation of the pressure buildup took about 90 s at the prototype scale. In addition, spikes appeared once the pore pressure ratio of 100% was reached. These sharp peaks of pore pressure are more pronounced at 2 m depth. These rapid decreases of pore pressure buildup can be due to dilatancy phenomena in dense sand when the saturated soil reaches the phase transformation line. This result is in accordance with the measured accelerations presented in Fig. 15.11. As for the pore pressure buildup, acceleration spikes are more pronounced in the case of the AH4 than in the case of the AH3: it should be supposed that the dilatancy phenomena should have occurred up to the depth of 3 meters.

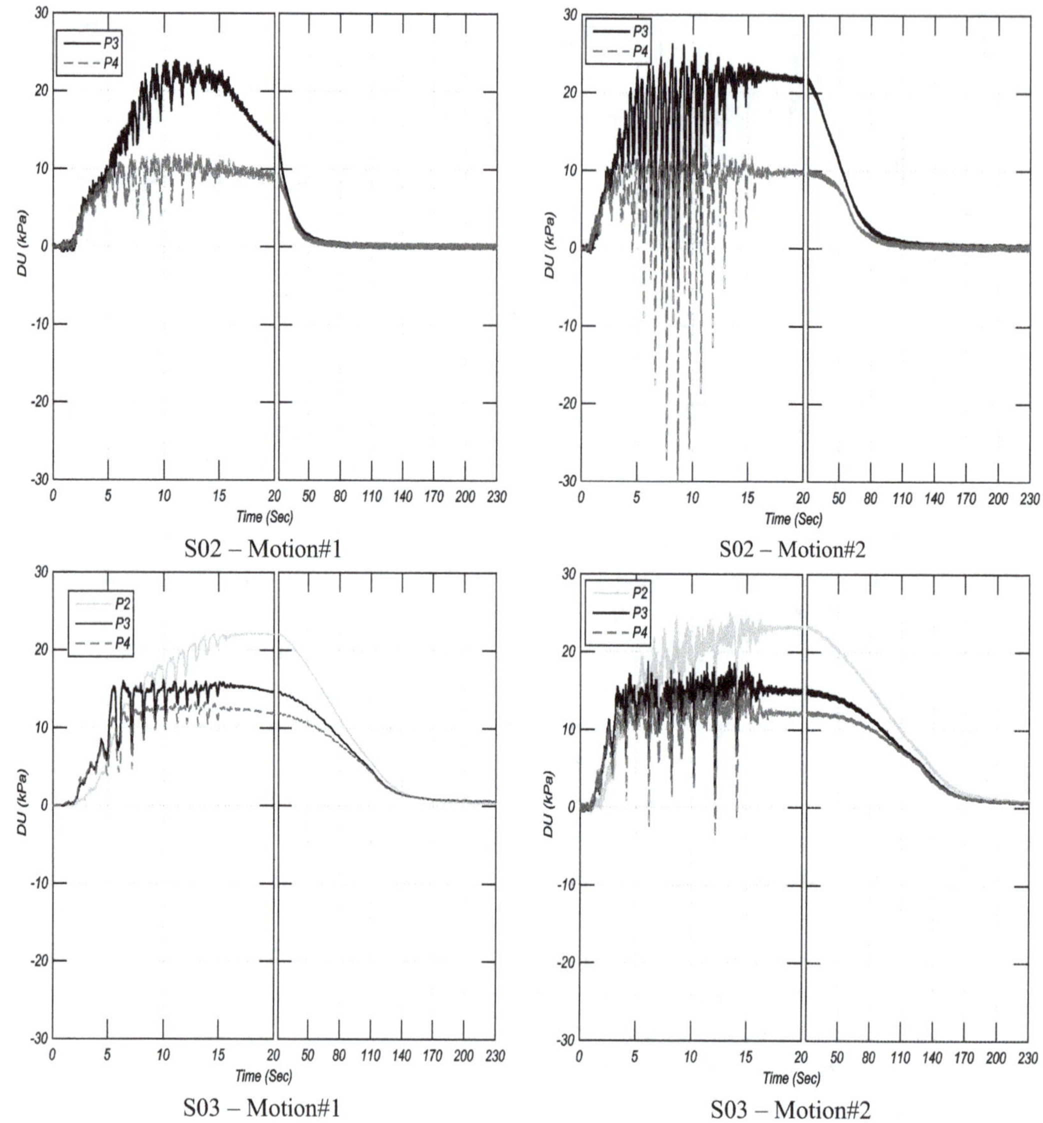


**Fig. 15.10** S02 and S03—Motion#1 and Motion#2 pore pressure variations in the central array

In the case of Motion#2 the ratio of the excess of pore pressure over the initial vertical effective stress reached 1 for all the reliable recorded pore pressure (Fig. 15.10). Compared to the case of the P4 sensor for Motion#1, sharper peaks of the pore pressure appeared for both levels of measurement. The variation of the pore pressure reached −4.9 and −30.9 kPa for the P3 ($z = 2.2$ m) and the P4 ($z = 1.05$ m) sensors, respectively. Taking into account the initial vertical effective stress at the location of the pore pressure sensor P4 (i.e., 10.9 kPa), a negative vertical effective stress is reached on very short periods at shallow depth. Due to the main objective of this descriptive paper, a deeper analysis of this observed phenomena will be provided in another paper.

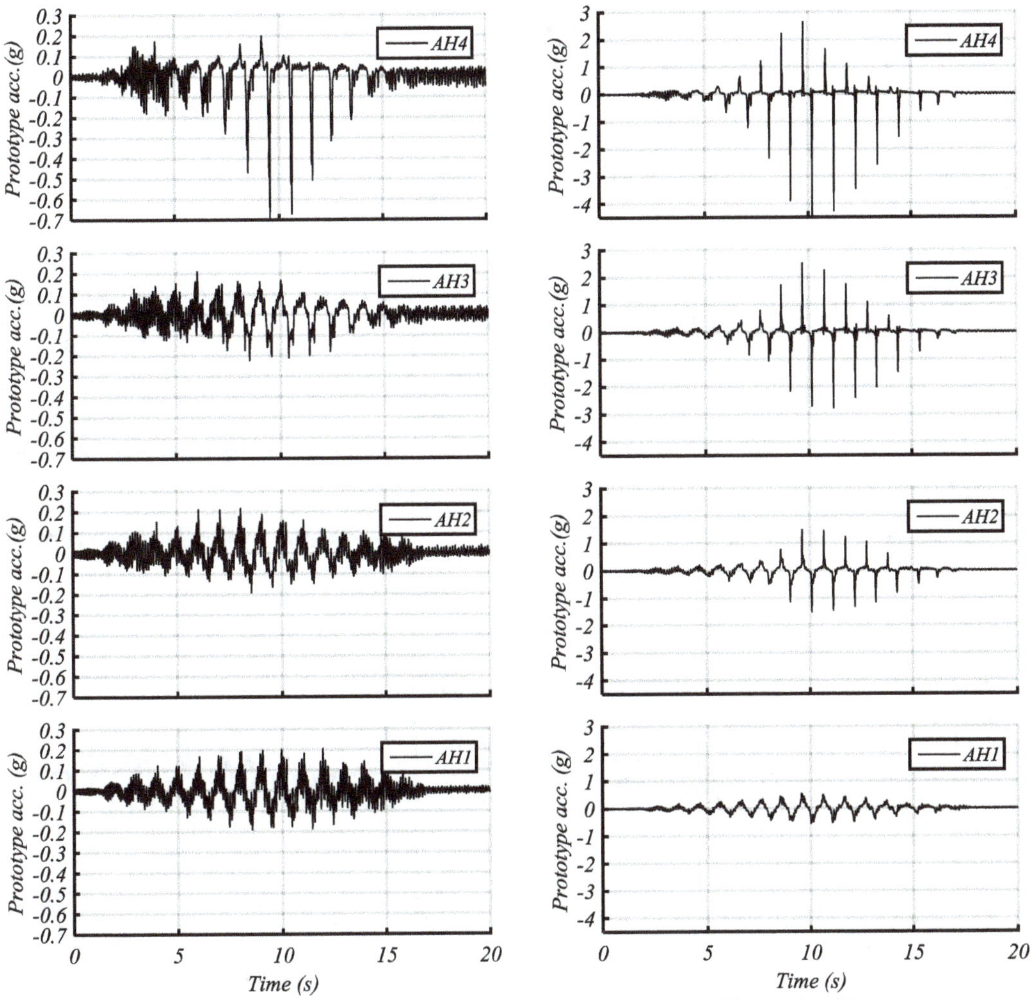


(a) Motion#1 (b) Motion#2

**Fig. 15.11** S02 test, acceleration response of the central array for (**a**) Motion#1 and (**b**) Motion#2

The peaks that appeared in the pore pressure recordings (Fig. 15.11) are in accordance with the sharp peaks of horizontal acceleration that appear for all the accelerations measured except for the AH1 that is the acceleration at 3.5 m depth. It should be supposed that the phenomena of dilatancy occurred up to 2.5–3.5 m depth.

In the case of test S03 (medium loose sand, $\rho_d = 1624$ kg/m$^3$) the pore pressure ratio for Motion#1 reached 88%, 100%, and 100% for P2, P3, and P4, respectively. The pressure buildup is more rapid than for the dense sand. Furthermore, the pressure decreases more slowly than for the dense sand.

Taking into account the difference of sampling frequency between both motions, no noticeable difference appears in the evolution of the pore pressure buildup in the case of Motion#2. In both cases, there are pore pressure spikes that coincide with acceleration spikes (Fig. 15.12).

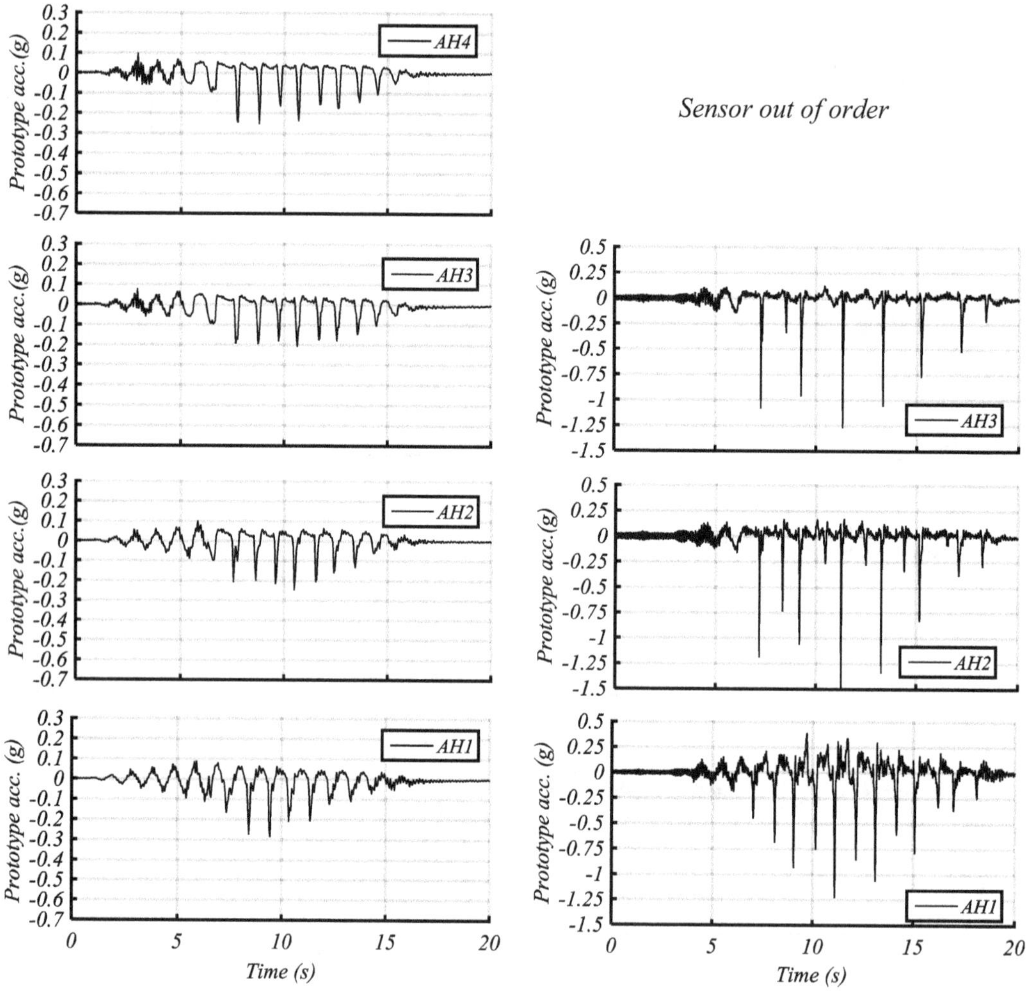


(a) Motion#1 (b) Motion#2

**Fig. 15.12** S04 acceleration response of the central array for: (**a**) Motion#1 and (**b**) Motion#2

Contrary to the dense sand, acceleration spikes appear at all the levels whatever the base shaking applied. However, the amplitude of the spike is generally larger and in the two directions (positive and negative) when they appear in dense sand. Furthermore, in the case of the medium loose sand subjected to large base shaking (S03 Motion#2), the number of spikes decreases with the decreasing of the depth.

### 15.4.2 *Surface Maker Response*

A cross view and a top view of the initial position and the vector of the total displacement of the surface markers and the embedded sensors are illustrated in

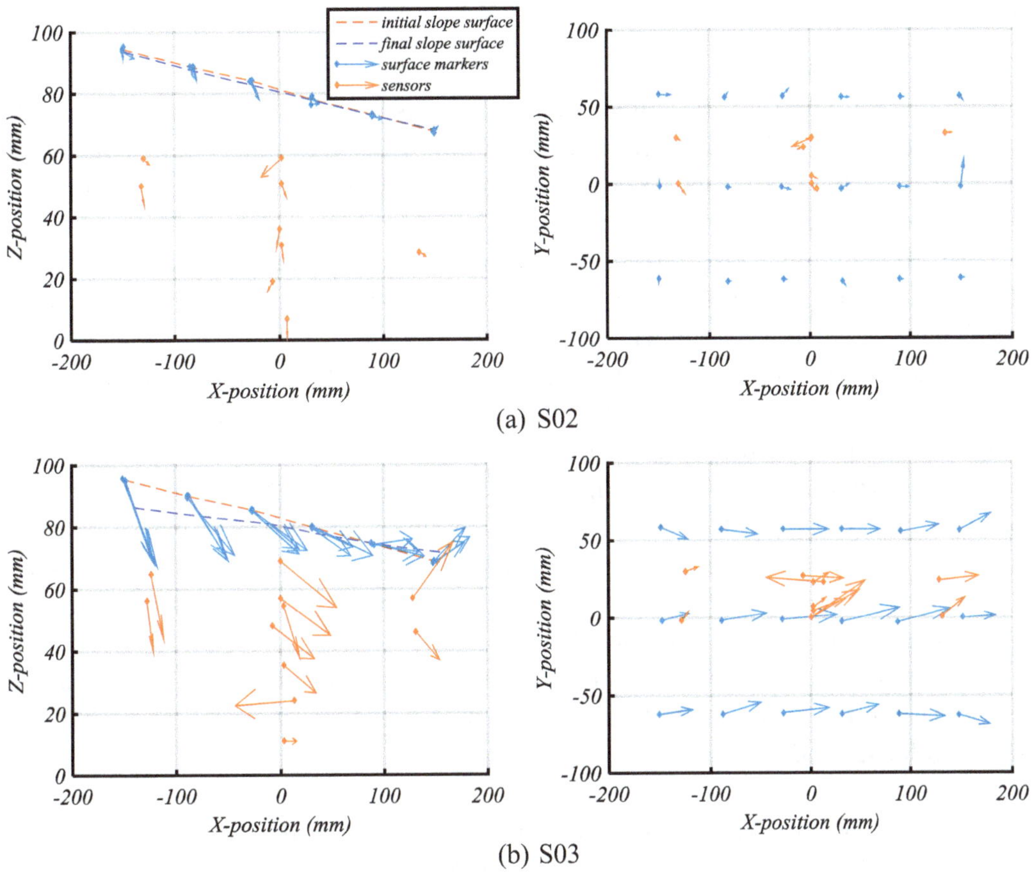


**Fig. 15.13** S02 and S03 tests: top and cross views of the surface markers and sensor displacement induced by the two successive base shakings

Fig. 15.13. The initial position is the one that corresponds to the first location measurement before the first spin up of the centrifuge for the surface markers and during the pluviation process for the embedded sensors. The final location measurement corresponds to the location measured after the second base shaking once the centrifuge was spun down for the surface markers and during the dismantle of the container for the embedded sensors. As the displacement has remained limited in test S02, the length of the displacement vector was magnified by 3 for both tests.

Based on the surface markers, in the case of the S02 test, the settlement does not exceed 10 cm in the upper part and -2 cm near the toe of the slope. In addition, the maximum average horizontal displacement is about 10 cm and was reached in the middle of the slope. For the medium loose sand, the displacements are largely higher with a maximum settlement of 47 cm in the upper part of the slope and $-17.5$ cm near the toe of the slope. The maximum horizontal displacement reaches 97 cm and is also located in the middle of the slope.

## 15.5 Conclusion

This paper summarized the buildup and some results of the two centrifuge tests performed at IFSTTAR in the framework of the LEAP-UCD-2017 series of tests.

Two centrifuge tests were performed by IFSTTAR; the tests were done on a dense and a medium loose Ottawa-F65 sand. The main nonconformities of the experiment buildup were the absence of the measurement of saturation degree for both tests and the measurement of the soil characteristics through CPT tests and bender element measurements.

Despite all care taken during the determination of the drives, non-negligible high-frequency components were recorded for Motion#1. Motion#2 with higher amplitude was less perturbed by high-frequency components.

For the two motions the recording of pore pressure sensors highlighted sharp pore pressure peak that illustrated dilatancy phenomena. The phenomena were more pronounced in dense sand and with increasing the $PGA_{eff}$. The dense Ottawa sand and in a more limited way the medium loose sand should have reached under the test conditions the phase transformation line.

The density has a large effect on the observed sensors and marker displacement; in the case of the dense sand the displacement is limited contrary to the case of the medium loose sand.

**Acknowledgments** This experiment has been made in the framework of the LEAP-UCD-2017 series of experiments. The authors would like to thank all the participating centrifuge teams that share their knowledge.

## References

Brandenberg, S. J., Choi, S., Kutter, B. L., Wilson, D. W., & Santamarina, J. C. (2006). A bender element system for measuring shear wave velocities in centrifuge models. *Proceedings, 6th International Conference on Physical Modeling in Geotechnics, 1*, 165–170. https://nees.org/data/get/NEES-2006-0149/Documentation/References/Brandenberg_2006.pdf.

Callari, C., Armero, F., & Abati, A. (2010). Strong discontinuities in partially saturated poroplastic solids. *Computer Methods in Applied Mechanics and Engineering*, 1513–1535.

Chazelas, J.-L., Escoffier, S., Garnier, J., Thorel, L., & Rault, G. (2008). Original technologies for proven performances for the new LCPC earthquake simulator. *Bulletin of Earthquake Engineering, 6*(4), 723–728.

Kutter, B. L. (2013). Effects of capillary number, bond number, and gas solubility on water saturation of and specimens. *Canadian Geotechnical Journal, 50*(2), 133–144.

Kutter, B. L., Carey, T. J., Hashimoto, T., Zeghal, M., Abdoun,T., Kokkali, P., Madabhushi, G., Haigh, S. K., Burali d'Arezzo, F., Madabhushi, S., Hung, W.-Y., Lee, C.-J., Cheng, H. C., Iai, S., Tobita, T., Ashino, T., Ren, J., Zhou, Y.-G., Chen, Y.-M.,Sun, Z.-B. & Manzari, M. T. (2019). LEAP-GWU-2015 experiment specifications, results, and comparisons. Soil Dynamics and Earthquake Engineering, Vol. 113, pp 616–628.

Kutter, B. L., Carey, T. J., Stone, N., Bonab, M. H., Manzari, M., Zeghal, M., Escoffier, S., Haigh, S., Madabhushi, G., Hung, W.-Y., Kim, D.-S., Kim, N.-R., Okamura, M., Tobita, T., Ueda, K., & Zhou, Y.-G. (2019). LEAP-UCD-2017 V. 1.01 model specifications. In B. Kutter et al. (Eds.), *Model tests and numerical simulations of liquefaction and lateral spreading: LEAP-UCD-2017*. New York: Springer.

Lade, P. V., Liggio, C. D., Jr., & Yamamuro, J. A. (1998). Effect of non-plastic fines on minimum and maximum void ratios of sands. *Geotechnical Testing Journal., 21*(4), 336–347.

Okamura, M., & Inoue, T. (2012). Preparation of fully saturated model for liquefaction study. *International Journal of Physical Modelling in Geotechnics, 12*(1), 39–46.

Ternet O. (1999). *Reconstitution and characterization of the sand samples*. Application to the tests out of the centrifuge and calibration chamber. PhD thesis, Univ. de Caen, 184p, in French.

# Chapter 16
# LEAP-UCD-2017 Centrifuge Test at KAIST

**Seong-Nam Kim, Jeong-Gon Ha, Moon-Gyo Lee, and Dong-Soo Kim**

**Abstract** Since the earthquakes of Niigata (Japan, 1964) and Alaska (USA, 1964), the dangers of liquefaction have been highlighted and research into liquefaction has been actively performed. Particularly, as part of the provision and verification of liquefaction data through physical modeling by using centrifuge and numerical prediction, the Liquefaction Experiments Analysis Project (LEAP) was launched. The purpose of the recent LEAP-UCD-2017, in which nine facilities participated, was to evaluate the uncertainty and repeatability of response in a previous study for LEAP-GWU-2015. The ground models were prepared in a rigid box with a 5° sloping model with relative densities of 85 and 50% by using Ottawa sand. The models were subjected to nondestructive and destructive motions based on a ground motion consisting of a tapered 1 Hz sine wave. This paper describes not only the experimental procedure in detail but also the difference in the sensor response of the ground model corresponding to the relative density of 85 and 50% during liquefaction. Moreover, it provides data for permanent horizontal–vertical displacement through liquefaction and for validation of the numerical model.

S.-N. Kim
Water Management Department, Korea Advanced Institute of Science and Technology, Daejeon, South Korea

J.-G. Ha
Korea Advanced Institute of Science and Technology, Daejeon, South Korea

M.-G. Lee
Earthquake Research Center, Korea Institute of Geoscience and Mineral Resources, Daejeon, Republic of Korea

D.-S. Kim (✉)
Department of Civil and Environmental Engineering, Korea Advanced Institute of Science and Technology, Daejeon, South Korea
e-mail: dongsookim@kaist.ac.kr



B. Kutter et al. (eds.), *Model Tests and Numerical Simulations of Liquefaction and Lateral Spreading*, https://doi.org/10.1007/978-3-030-22818-7_16

## 16.1 Introduction

Liquefaction is a phenomenon in which the strength and stiffness of soil is reduced because of the occurrence of earthquakes or other loading. Damage and ground failure due to liquefaction remain a major concern to the geotechnical engineers. Various methods, such as field investigation and laboratory tests, have been conducted to evaluate the phenomenon and consequences of liquefaction. Simultaneously, studies involving various numerical analyses have produced various insights on liquefaction.

Constitutive models and numerical analysis techniques that simulate complex liquefaction phenomena must be validated using well-defined experimental results (Ueda and Iai 2018). Under such a demand, a collaborative study between numerical modelers and centrifuge experimenters was conducted 20 years ago, termed VELACS (Arulanandan and Scott 1993–1994). The Liquefaction Experiments and Analysis Project (LEAP) is an ongoing project to reduce the inconsistency in the experimental results and to provide high-quality experimental results to numerical modelers. In LEAP-GWU-2015, various institutions conducted the centrifuge tests on liquefaction of sloping ground in a rigid box (Manzari et al. 2015).

Following LEAP-GWU-2015, the nine facilities participated in LEAP-2017 which includes RPI, Kyoto University, KAIST, Zhejiang University, UC Davis, Cambridge University, National Central University, IFSTTAR, and Ehime University. The purpose of LEAP-2017 was to characterize the sensitivity of the liquefaction phenomena according to the relative density of the sloping ground and input motion intensity. Through centrifuge model tests, the displacement and deformation of the sloping ground were mainly evaluated, and accelerations and pore water pressure records were compared.

Researchers from KAIST developed a centrifuge facility with 5 m radius and 240 g-ton capacity in 2009, and participated in LEAP 2017. By considering the various conditions of LEAP-2017, KAIST researchers selected two cases of different relative densities (dense and loose) with the same input motion intensity of 0.15 g. This paper provides details of the centrifuge model tests conducted by KAIST for LEAP 2017, including facility and equipment, test procedure, and results.

## 16.2 Centrifuge Facility and Earthquake Simulator at KAIST

A dynamic geotechnical centrifuge facility at KAIST was utilized to perform tests for LEAP 2017. An electrohydraulic earthquake simulator was mounted on the centrifuge with a platform radius of 5 m and a maximum capacity of 240 g-tons. The main body of the centrifuge has the unique feature of an automatic balancing system and includes parts, such as fluid rotary joints, slip rings, and a fiber optic

(a) (b)

**Fig. 16.1** Dynamic geotechnical centrifuge facility at KAIST centrifuge testing center: (**a**) centrifuge main body, (**b**) earthquake simulator

rotary, for general testing purposes (Kim et al. 2013a). The centrifuge test of KAIST was conducted with a gravitational acceleration of 40 g.

Target excitations were applied using an earthquake simulator that adopted a dynamic self-balancing technique to eliminate a large portion of the undesired reaction forces and vibrations transmitted to the main body. The earthquake simulator was designed to operate at up to 100 g centrifugal acceleration, and the base shaking acceleration can be exerted to a maximum value of 20 g at 40 g of centrifugal acceleration with a maximum payload of 700 kg. The dimensions of the slip table were 670 mm (length) × 670 mm (width) (Kim et al. 2013b) (Fig. 16.1).

## 16.3 Physical Modeling

### *16.3.1 Soil Material and Density*

Ottawa F-65 sand was used as the standard sand for LEAP-2017. The grain size characteristics and property of the soil are as follows: $Gs = 2.665$, $D_{10} = 0.13$ mm, $D_{30} = 0.17$ mm, $D_{50} = 0.20$ mm and $D_{60} = 0.21$ mm (Kutter et al. 2018). Based on the preliminary test results of the participating facilities of LEAP-2017, the minimum and maximum densities of sand were determined as $\rho_{dmax}= 1752$ kg/m$^3$ and $\rho_{dmin}= 1470$ kg/m$^3$, respectively. The target soil densities of the dense and loose centrifuge test conditions in KAIST were specified as 1703 and 1599 kg/m$^3$ and the relative densities were equivalent to 81% and 50%, respectively.

The sand model was constructed by dry pluviation through a screen with an opening size of approximately 1.20 mm; the screen was partially blocked to limit the flow. The density of the soil model was determined by the size of the opening slot and the drop height, and a calibration test is necessary when the geometry is

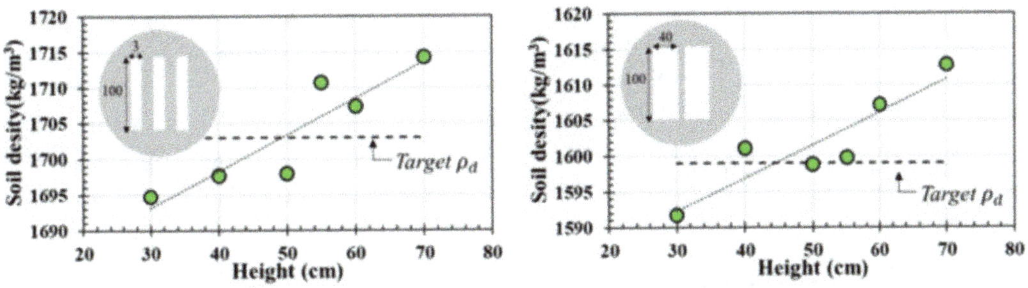


**(a)** Dense case (target $\rho_d$ = 1715 kg/m3) **(b)** Loose case (target $\rho_d$ = 1608 kg/m3)

**Fig. 16.2** Density versus drop height relationship for two sieve designs. (**a**) Dense case with small opening slots. (**b**) Loose case with large opening slots

changed. Figure 16.2 shows the geometry of the opening slots and the calibration test results by trial and error adjustments of the pluviation drop height for constructing the dense and loose soil models for the KAIST experiment.

The sizes of the opening slot were 3 mm × 100 mm for the dense soil and 40 mm × 100 mm for the loose soil. The drop heights in the calibration test were varied from 30 to 70 cm. From the calibration test results in shown Fig. 16.2, the final drop height was set to 55 cm in both cases.

The measured dry unit weights of soil grounds constructed in the rigid box were 1701.2 and 1592.5 kg/m$^3$ for dense and loose conditions, respectively. After the sand was pulviated to the target height, a 5° inclined guide was installed on the top of the box. The manufactured scraper was connected directly to the inclined guide and was carefully scraped according to the slope. The soil generated by the scraping was removed as carefully as possible by using a vacuum cleaner so that the possible ground was not disturbed, and a 5° ground model was achieved.

### *16.3.2 Viscous Fluid*

The viscous fluid was a mixture of water and methylcellulose F-65, and the target viscosity was set to 40 cSt as dictated by the scaling law defined by Garnier et al. (2007). The temperature and concentration are the influencing factors of the fluid viscosity; thus, the calibration tests were performed before model construction, as shown in Fig. 16.3. The viscosity was measured using a falling ball viscometer, which applies the Newton's law of motion under force balance when the falling spherical ball reaches its terminal velocity. The equation for calculating viscosity is expressed in Eq. (16.1) (Viswanath et al. 2007):

$$\mu = K(\rho_f - \rho)t, \tag{16.1}$$

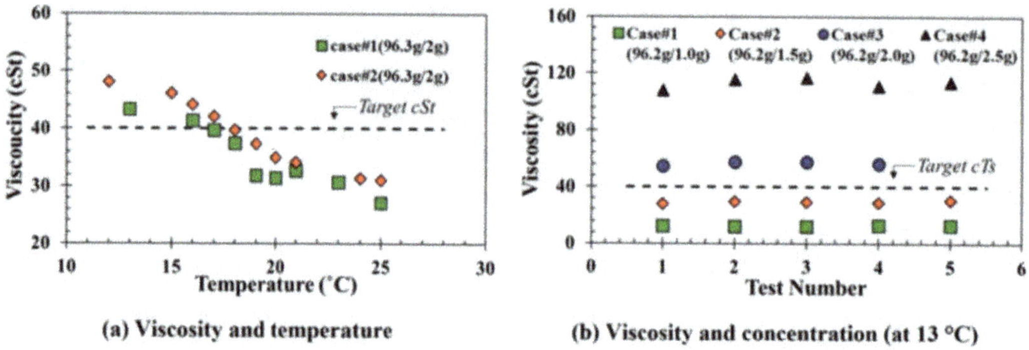


**Fig. 16.3** Calibration tests for viscosity according to temperature and concentration. (**a**) Viscosity and temperature. (**b**) Viscosity and concentration (at 13 °C)

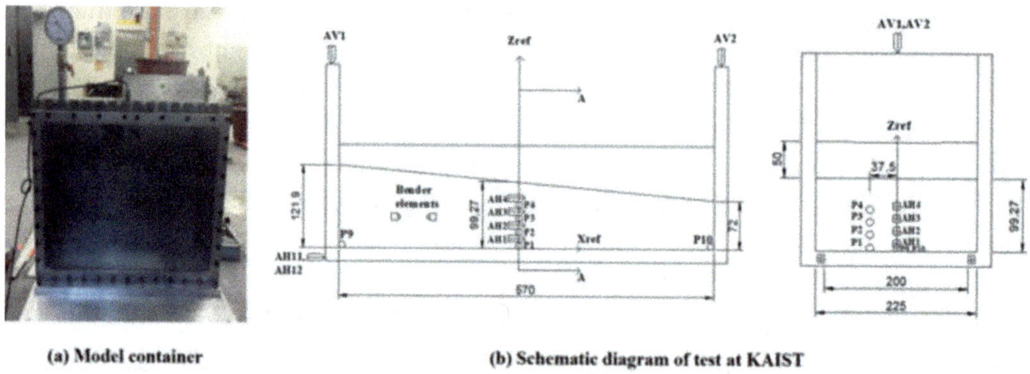


**Fig. 16.4** Image of used rigid model box and the schematic of centrifuge test model and instruments. (**a**) Model container. (**b**) Schematic diagram of test at KAIST

where $K$ is the viscometer constant, $\rho_f$ is the ball density, $\rho$ is liquid density, and $t$ is the time of descent (in minutes).

At a constant concentration, the viscosity increases with the decrease in temperature (Fig. 16.3a), and the concentration increases at a constant temperature (Fig. 16.3b). The viscous fluid of the centrifuge test was prepared based on the calibration result, and the final viscosities of the dense and loose test cases were 37.45 and 35.87 cSt, respectively. These viscosities were slightly lower than the target viscosity of 40 cSt because of the sensitivity to temperature.

### *16.3.3 Model Description and Instrumentations*

The condition for model construction in LEAP-2017 was the construction of a 5° sloping sand ground in a rigid box. A new rigid box with a transparent window was manufactured (Fig. 16.4a) with internal dimensions of 570 mm × 225 mm × 450 mm (length × width × depth). The model constructed through dry pluviation is described in Fig. 16.4b, with the following dimensions in the prototype scale: 22.8 m × 4 m × 9 m (length × depth at midpoint × width). The length of the slope (22.8 m) was about 15% greater than the specified length (20 m). In the KAIST centrifuge facility, the 5° inclination along the length of the model was not curved because the shaking plane was perpendicular to the plane of rotation of the centrifuge. The specifications suggested that the surface should be curved in the end view (right side of Fig. 16.4b); however, the KAIST models were not curved because the error due to lack of curvature was thought to be negligible for this small model in a large centrifuge. A 4 m radius curve across the 225 width of the container would only result in about 1.5 mm elevation difference between the middle and the side of the container (75 mm in prototype scale).

The responses of the soil model during shaking were monitored using eight accelerometers along the direction of shaking (AH1–AH4 in the soil mass and AH11–AH12 on the rigid container), two vertical accelerometers (AV1 and AV2), and six pore pressure transducers (P1–P6, P9–P10). These instruments were required in LEAP-2017, and their layout is shown in Fig. 16.4. To check the variation of shear wave velocity, one pair of bender elements was installed at the same depth as that of accelerometer AH3. Table 16.1 lists the details of the instruments used.

The required 18 surface markers were installed on the ground uniformly with a spacing of 2 m × 2 m. The markers, with a diameter of 26 mm, were manufactured using PVC material and were designed to be anchored to the soil and provide minimal restriction to pore pressure drainage. A high-speed camera was mounted at the centrifuge arm to measure the plan view lateral displacements of the surface markers during shaking. The high-speed camera at KAIST is a Phantom v5.1 HI-G, which can record videos at 1200 frames per second at a resolution of 1024 × 1024 pixels. The self-balanced system of the shaking table and the hinges connecting the basket to the centrifuge arm may isolate the camera from vibrations.

### 16.3.4 Saturation and Container Modifications

Figure 16.5 shows the schematic of the saturation system at KAIST. Before saturating, the box was confirmed to be completely sealed from external air. The procedure for the saturation process is as follows: vacuum pressure (>94 kPa) was applied and

**Table 16.1** Detailed information of instruments

| Instrument | Type | Name | Description |
|---|---|---|---|
| Accelerometer | 353B17 | AH1–AH4 | Soil horizontal acc. |
| | | AH11–AH12 | Box horizontal acc. |
| | | AV1–AV2 | Box vertical acc. |
| Pore pressure transducer | PDCR 81 | P1–P4 | Pore water pressure in soil |
| | EPB-PW-3.5BS-V5/ L5M | P9–P10 | Pore water pressure at boundary |

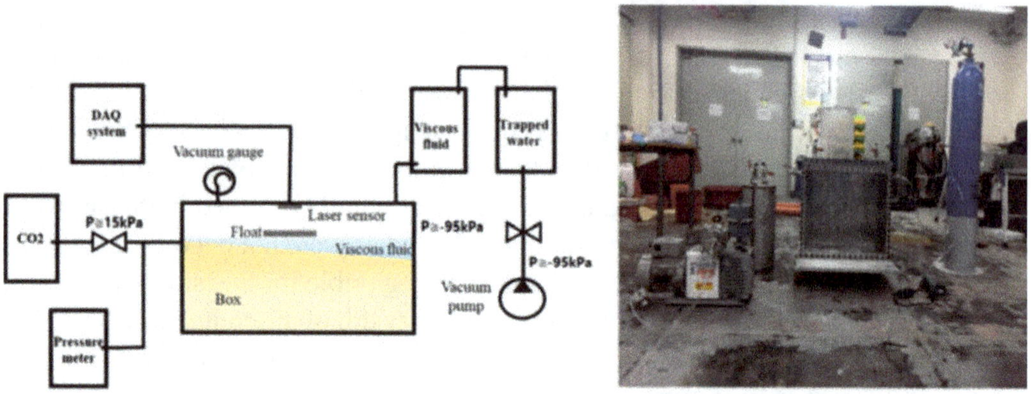


(a) Schematic of saturation system (b) Actual saturation system

**Fig. 16.5** Saturation system. (**a**) Schematic of saturation system (**b**) Actual saturation system

low pressure $CO_2$ (<15 kPa) was flooded in the box repeatedly. This process was performed five times for 20 min each. In addition, a strong vacuum pressure was applied to eliminate the trapped air in the viscous fluid container. While maintaining vacuum pressure in the rigid box and viscous fluid container, the viscous fluid slowly dripped into the ground model (Madabhushi et al. 2018). The dripped point was in the downward direction of the slope, and to minimize the impact of gravity, sponge was installed on the ground surface. After filling the fluid up to 5 cm higher than the soil, Okamura's method was used to measure the degree of saturation (Okamura and Inoue 2012). After the rigid box was released to the atmospheric pressure, a low vacuum pressure (<15 kPa) was applied, and a change in the level of viscous fluid was observed. At this time, the volume change of the viscous fluid was the same as the volume change of the air trapped in the ground, and the degree of saturation measurement was calculated based on this result, indicating more than 99.2% and 99.4% in both models, respectively.

### *16.3.5 Sequence of the Centrifuge Test*

Table 16.2 summarizes a typical sequence of the centrifuge tests. In each of the centrifuge tests of dense and loose ground model, four seismic excitations of two nondestructive and two destructive motions were applied. The frequency wavelets covering a wide range were used to represent the nondestructive motion to identify ground system before and after liquefaction. A tapered sine wave with a frequency of 1 Hz was designated as the destructive wave. In the KAIST tests, the target intensity was determined as 0.15 g for both the first (Motion #2) and second (Motion #3) destructive waves. By applying the destructive motions of the same intensity, the first and post liquefaction phenomena can be compared.

Cone penetration tests were conducted before and after the first destructive motion (Motion #2) to evaluate the soil condition. The cone was connected with

**Table 16.2** Experimental procedure of dense and loose ground model

| Event | g-level | Event description |
|---|---|---|
| Event #1 (CPT) | 40 g | CPT (before motion) |
| Event #2 (seismic excitations) | 40 g | Motion #1 and motion #2 (target: 0.15 g) |
| Event #3 (CPT) | 40 g | CPT(after motion) |
| Event #4 (seismic excitations) | 40 g | Motion #3 (target: 0.15 g) and motion #4 |

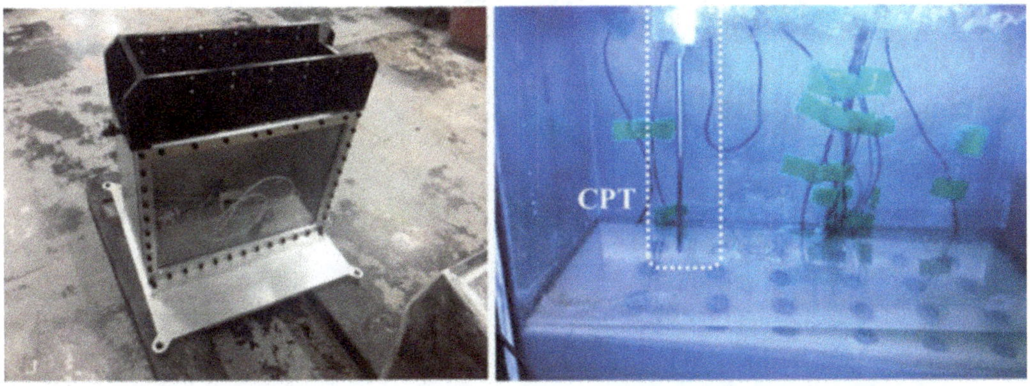


**(a) Installation rack** **(b) Cone penetration test**

**Fig. 16.6** (**a**) Rack installation onto the box. (**b**) Cone penetration test

the load cell and designed so that the strain gauge was not attached to the cone tip. The length of the cone was 200 mm, and the diameter of the cone tip was 6 mm. The penetration velocity was slow at 5 mm/s (model scale), and the penetration depth was above 75 mm (model scale). The CPTs were conducted at locations selected to avoid the sensor installed in the ground model and the marker on the ground surface. As shown in Fig. 16.6, because it was necessary to install the guide rack and loading actuator above the box for the CPT, the rotation of the centrifuge was stopped and restarted. When the centrifuge spinning was stopped, the position of the surface marker was also investigated.

## 16.4 Centrifuge Test Results

### *16.4.1 Achieved Input Motion*

In the previous study, the shakers of most facilities had generated high-frequency motions superimposed on the smooth ramped sine wave motion. Therefore, the $PGA_{eff}$ concept was introduced in the current project to attempt to compare the results among the facilities that generated different amounts of high-frequency content in the input motion:

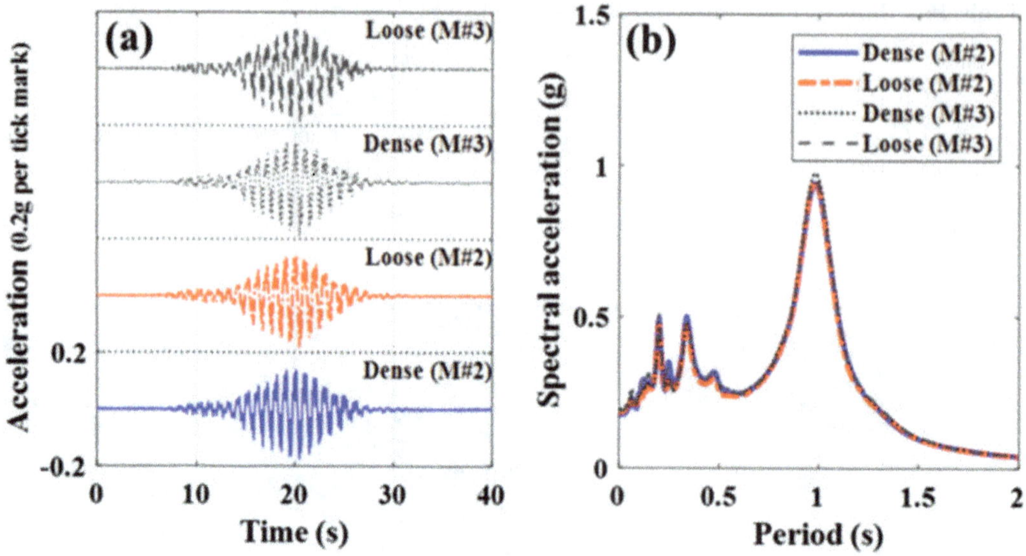


**Fig. 16.7** Main input motion (Motions #2 and #3) as recorded using AH11 and AH12: (**a**) time history, (**b**) response spectrum

$$PGA_{eff} = PGA_{1Hz} + 0.5 \times PGA_{hf}, \tag{16.2}$$

where $PGA_{1Hz}$ is a component in the frequency band of 1 Hz, and $PGA_{hf}$ is the peak acceleration of the high-frequency components of the input motion (Kutter et al. 2018).

The main input motions (Motions #2 and #3) that were achieved for dense and loose models are shown in Fig. 16.7a. These motions were respectively measured by AH11 and AH12 accelerometers attached to the bottom side of the model box. The acceleration response spectra for the input motions presented in Fig. 16.7b show that the input motions applied to a series of tests contain some high-frequency components. However, accurately controlling the high-frequency components of the base acceleration by using the hydraulic shaker on the end of a spinning centrifuge is difficult. Therefore, compared to the deviation from the input motions of other facilities participating in LEAP-2017, the input motions applied to KAIST tests are reasonable.

Table 16.3 lists the details of main motions. The $PGA_{eff}$ of the input motions applied in the tests were slightly higher than the target $PGA_{eff}$ (i.e. 0.15 g).

### *16.4.2 Investigation of Soil Model*

In the dense and loose sand models, CPT and the bender elements test (BET) were performed before and after Motion #2. The CPT results and shear wave velocity ($V_s$) profiles, which were deduced from the CPT results, are shown in Fig. 16.8. Here, the $V_s$ values were derived from the method suggested from Eq. (16.3) by Kim et al.

**Table 16.3** Details of main input motions applied to tests

| | | Target $PGA_{eff}$ (g)[a] | $PGA_{raw}$ (g)[b] | $PGA_{eff}$ (g)[c] | $PGA_{hf}$ (g)[d] | $PGA_{1Hz}$ (g)[e] | PGV (m/s)[f] | $CAV_s$ (m/s)[g] |
|---|---|---|---|---|---|---|---|---|
| Dense | Motion #2 | 0.15 | 0.178 | 0.168 | 0.098 | 0.119 | 0.209 | 7.18 |
| | Motion #3 | 0.15 | 0.192 | 0.182 | 0.096 | 0.124 | 0.216 | 7.49 |
| Loose | Motion #2 | 0.15 | 0.185 | 0.166 | 0.092 | 0.120 | 0.210 | 7.30 |
| | Motion #3 | 0.15 | 0.192 | 0.174 | 0.107 | 0.121 | 0.210 | 7.25 |

[a]Targeted PGA of the 1 Hz component plus half the PGA of the higher frequency component
[b]Maximum horizontal acceleration recorded at the container base
[c]PGA of the 1 Hz component plus half the PGA of the higher frequency component
[d]PGA of the higher frequency component
[e]PGA of the 1 Hz component
[f]Peak velocity of the base input motion obtained by integration and base line correction
[g]Cumulative absolute velocity of the base input motions with a threshold of 0.005 g (prototype)

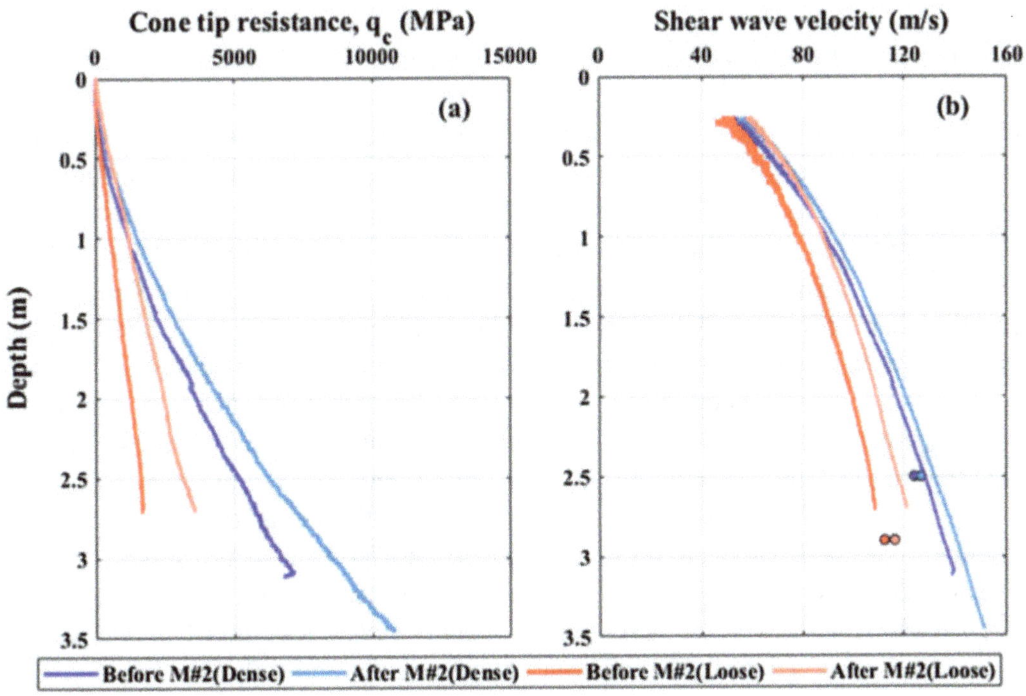


**Fig. 16.8** $q_c$ and $V_s$ profiles before and after Motion #2 deduced from CPT and BET results

(2017). Moreover, the $V_s$ values obtained at a measurement depth through BET are also plotted in Fig. 16.8 for comparison, and are confirmed to be similar to those derived from the CPT results at the same depth.

The $q_c$ and $V_s$ values increased after Motion #2, indicating that the soil stiffness was increased by Motion #2 owing to some densification. To quantitatively present the change in the soil stiffness by Motion #2, the relative densities for each soil depth derived from the $q_c$ profile (Kim et al. 2017) are summarized in Table 16.4. In addition, Table 16.5 describes the site periods of the soil model calculated using Eq. (16.4), in which the $V_s$ used is converted from the CPT results:

$$V_{s1} = 125.8(q_{c1})^{0.15}\ \text{(dry sand)}, V_{s1} = 108.2(q_{c1})^{0.15}\ \text{(Saturated sand)} \quad (16.3)$$

$$T = 4\sum_{i=1}^{n}\left(\frac{D_i}{V_{si}}\right) \quad (16.4)$$

where $T$ is the site period of the soil model (s), $D_i$ is the thickness of $i$th layer (m), and $V_{si}$ is the shear wave velocity of the $i$th layer (m/s).

### 16.4.3 Excess Pore Water Pressure

**Comparison of Central Array Pore Pressure Ratio with Respect to Depth**

Figure 16.9 shows the response of pore pressure sensors installed in the central array at each depth. In the prototype scale, P1–P4 are excess pore water pressure response records at depths of 4, 3, 2, and 1 m, respectively. The pore pressure ratio ($R_u = \Delta u/\sigma'_{vo}$) is defined as ratio of the excess pore pressure to the initial vertical effective stress. This is generally used in liquefaction evaluation, and the liquefaction can be assumed to occur when $R_u = 1$. The black and red lines indicate the pore pressure responses in dense and loose sand, respectively. The responses of all the sensors included the generation of excess pore water pressure, generation of cycle components by dilatancy of soil during the shaking, and dissipation process after the shaking. In the dense soil condition, for the sensor closer to the ground surface, the pore pressure ratio increases and a few negative spikes occured in P4 because of dilatancy. These spikes can occur after the excess pore pressure approaches the initial effective stress and is termed as the de-liquefaction shock waves (Kutter and Wilson 1999). However, the maximum $R_u$ value of P4 is less than 1.0, and a spike was clearly observed during the shaking, indicating that the soil behavior at a depth of 1 m is similar to that during liquefaction. In the loose soil condition, the pore pressure ratios of P2, P3, and P4 show the dilatancy spikes with the $R_u$ value exceeding 1; thus, the liquefaction can be said to occur from the depth of 3 m based on the $R_u$ value. However, the spikes in the positive direction occurred contrary to the negative pore pressure spikes, which was observed in the result of LEAP-GWU-2015.

Generally, it is impossible for the Ru value to exceed 1 and for the de-liquefaction shock wave to occur in the positive direction. The cause of this is presumed to be

**Table 16.4** Relative densities and shear wave velocities for each soil depth derived from the CPT results

| | | $\rho$ (kg/m$^3$) | $D_r$ | $q_{c.1.5m}$ (MPa) | $D_r\left(q_{c.1.5m}\right)$ | $q_{c.2.0m}$ (MPa) | $D_r\left(q_{c.2.0m}\right)$ | $q_{c.2.5m}$ (MPa) | $D_r\left(q_{c.2.5m}\right)$ |
|---|---|---|---|---|---|---|---|---|---|
| Before motion #2 | Dense | 1701.2 | 87 | 2.51 | 75 | 3.88 | 77 | 5.50 | 78 |
| | Loose | 1592.5 | 54 | 1.00 | 41 | 1.40 | 46 | 1.71 | 45 |
| After motion #2 | Dense | – | – | 2.92 | 76 | 4.63 | 76 | 6.44 | 78 |
| | Loose | – | – | 1.77 | 59 | 2.45 | 58 | 3.17 | 57 |

**Table 16.5** Site periods of soil model using shear wave velocity converted from the CPT results

| | Before motion #2 dense | After motion #2 dense | Before motion #2 loose | After motion #2 loose |
|---|---|---|---|---|
| Site period (s) | 0.144 | 0.142 | 0.159 | 0.156 |

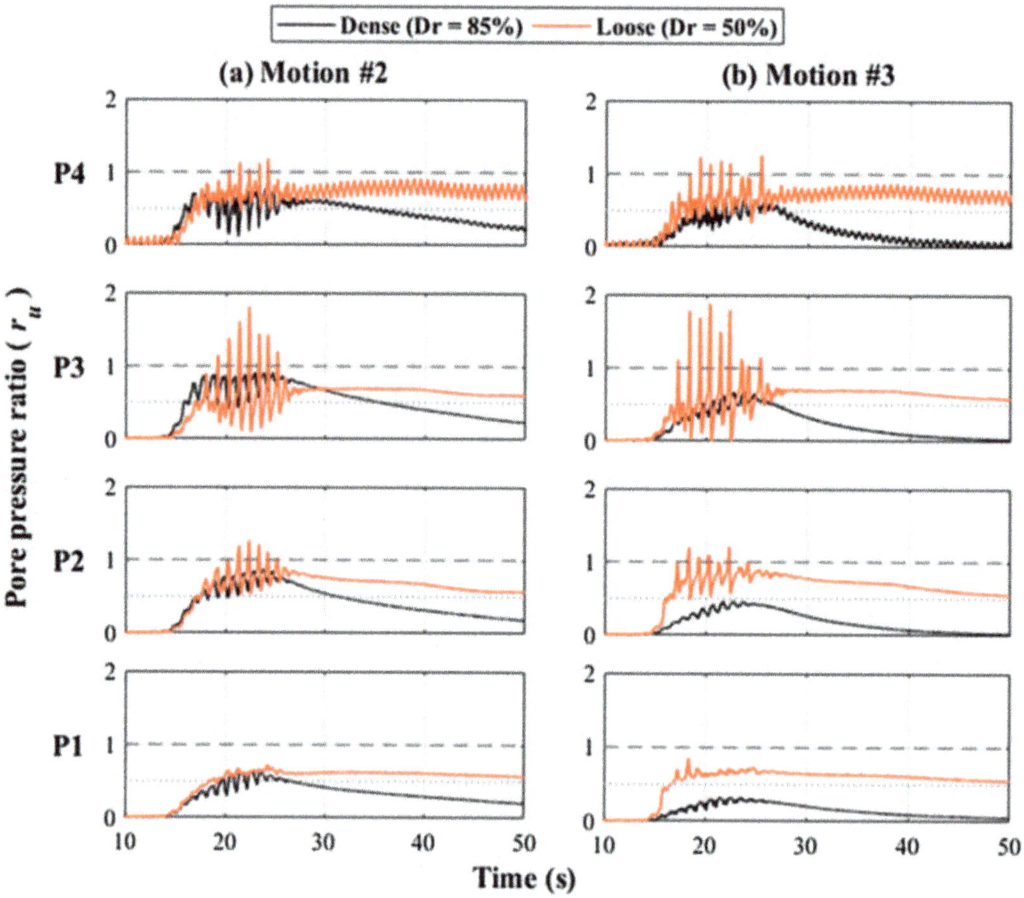


**Fig. 16.9** Response of pore water pressure is divided by effective stress, and represented according to the time histories at specific depth by pore pressure ratio (Black and red lines represent the responses of the dense and loose models, respectively): (**a**) Motion #2, (**b**) Motion #3

local dynamic compressive stress around sensors due to sensor installation and hand pluviation. Also sensors wire stiffness may be somewhat affecting the response.

Sensor P2 dissipated immediately after the earthquake at approximately 30 s. Therefore, the presence or absence of liquefaction based on the depth should be examined from other aspects also. Unlike P2, P3, and P4, P1 is installed at the bottom of the box and is not surrounded by sand in all directions, thus it has a relatively low cyclic response.

The pore pressure ratios in the dense soil condition during Motion #3 are described in Fig. 16.9b, which shows not only the overall low pore pressure ratio but also faster dissipation rate than that of Motion #2. In case of the loose soil condition during Motion #3, similar responses were generated as those in Motion #2; however, the spike shapes were slightly changed asymmetrically. This may be because the void ratio of soil decreased in both cases, and because the geometry of the sloping ground was changed by the movement from the previous shaking in Motion #2.

## Dissipation Time of Pore Pressure

Figure 16.10 shows the variations in excess pore water pressures of the central array with respect to time according to the soil relative density and main input motion. Figure 16.10a-1 and b-1 show response graphs in the dense and loose models when Motion #2 is imposed on the ground model. Similarly, Fig. 16.10a-2 and b-2 show response graphs when Motion #3 is imposed on the ground model.

In Fig. 16.10a-1 and b-1 displaying Motion #2, the excess pore water pressure, which was measured at a deeper depth, appeared higher and was dissipated at the end of the motion (28 s). The dissipation time in the loose model was longer than that

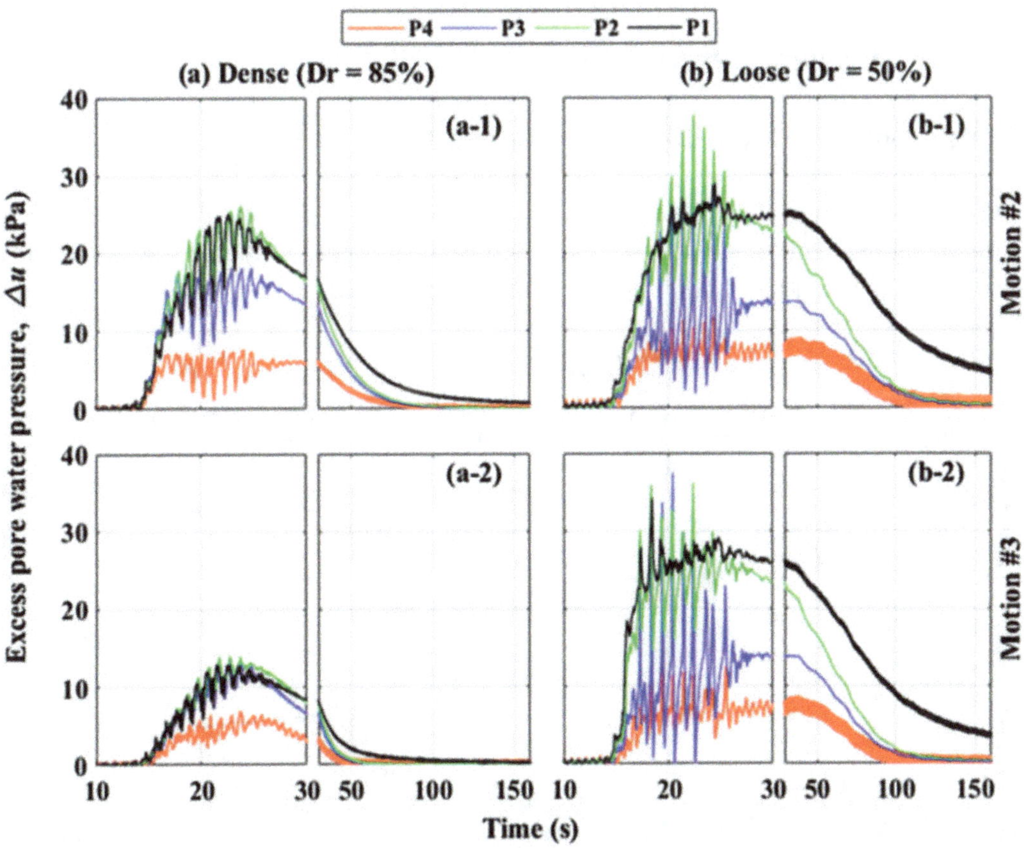


**Fig. 16.10** Excess pore water pressure data with respect to the depth of main input motion (Motions #2 and #3). P1, P2, P3, and P4 are the pore water pressure responses at depths of 4, 3, 2, and 1 m, respectively. (**a-1**) Motion #2 of dense model, (**a-2**) Motion #3 of dense model, (**b-1**) Motion #2 of loose model, and (**b-2**) Motion #3 of dense model

in the dense model. Almost all the excess pore water pressure was dissipated at 150 s in the dense model; however, the dissipation in the loose model took more than 150 s. In Fig. 16.10a-2 and b-2, showing the imposition of Motion #3, the dissipation rate patterns accelerated because of the change of void ratio and permeability following liquefaction.

#### Effect of Radial Gravity Field on P9 and P10

Figure 16.11 shows the results of time histories P9 and P10 installed in the side boundary of the rigid box (Hung et al. 2018). The figure shows the characterization of the radial gravity field generated by the change in the g-level while conducting the spinning centrifuge test. P9 and P10 are located at a position where the water level is relatively lower than P1 during spinning. As shown in Fig. 16.11, the pore pressure responses of P9 and P10 are smaller than that of P1. Similar with LEAP-GWU-2015 results, all facilities in LEAP-2017 showed this trend.

### *16.4.4 Acceleration Responses*

#### Comparison of Central Array Accelerations Based on Depth

Figure 16.12 shows the acceleration time histories when Motion #2 is applied to the ground model. The black and red lines represent the acceleration responses in the dense and loose models, respectively, and the depths from AH1 (4 m) to AH4 (1 m

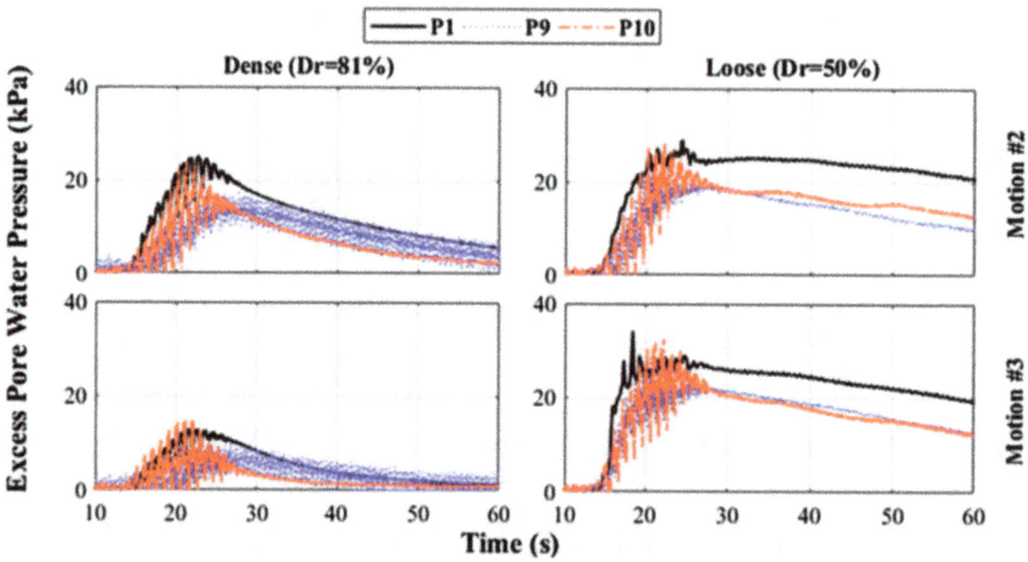


**Fig. 16.11** Comparison of excess pore water pressure of P1, P9, and P10. Verification of water level change due to radial gravity field during rotation of centrifuge

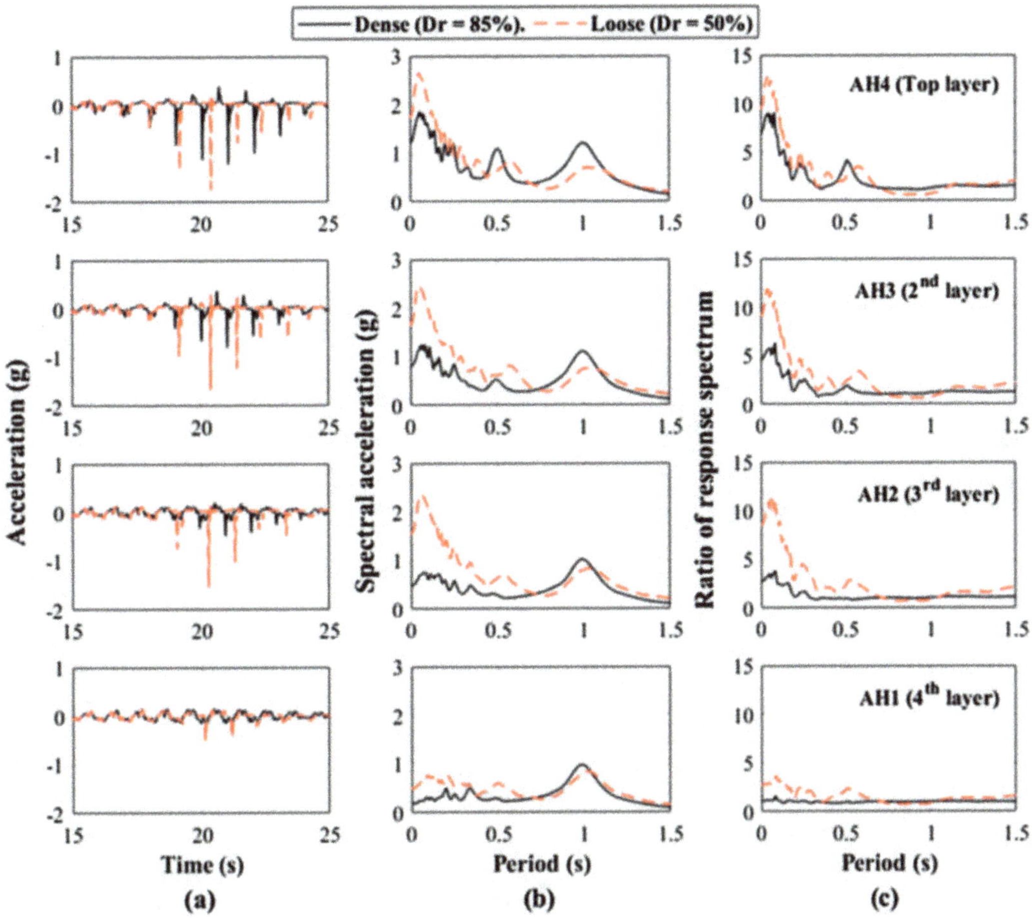


**Fig. 16.12** Acceleration response recorded with respect to depth when Motion #2 of target PGA of 0.15 g was applied to the ground model. The black and red lines indicate the responses of acceleration for the dense and loose models, respectively. Graphs of (**a**) acceleration, (**b**) response spectrum, and (**c**) ratio response spectrum with respect to time

deep) are continuously increased by 1 m. Figure 16.12a on the left shows the occurrence of spike due to soil dilatancy, caused by a de-liquefaction shock wave. The sharp spike in the dense model began to appear gradually at a depth of 3 m (AH2) but occurred from the depth of 4 m (AH1) in the loose model. In addition, the closer to the ground surface, the more the spike is amplified. This response is presented asymmetrically owing to the sloping ground model.

Figure 16.12b shows the response spectrum with respect to depth based on the measured acceleration record. The spectral acceleration tend to increase at approximately 0.1, 0.5, and 1 s. In the loose model, at approximately 0.1 s, the spectral acceleration response was larger than that in the dense model. This is because the de-liquefaction shock wave, which is a high-frequency component, was larger in the loose model. Moreover, for the loose model, the peak in the response spectrum near 1 s period was observed to increase above 1 s.

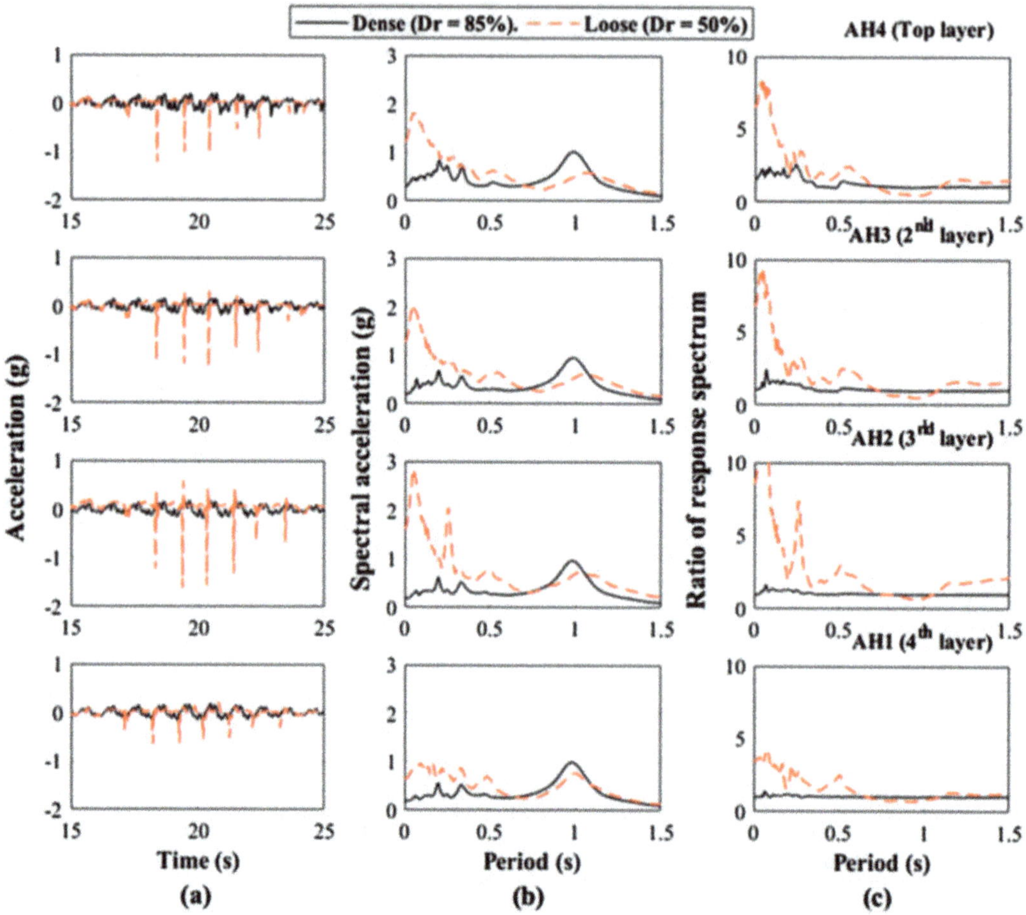


**Fig. 16.13** Acceleration response recorded with respect to depth when Motion #3 of target PGA of 0.15 g was applied to the ground model. Black and red lines indicate the responses of acceleration in the dense and loose models, respectively. Graphs of (**a**) acceleration, (**b**) response spectrum, and (**c**) ratio response spectrum with respect to time

The ratio of response spectrum in Fig. 16.12c is defined as the ratio of the response spectrum of the ground to the input motion measured by AH11 and AH12. This ratio is considerably similar to the trend of the response spectrum in Fig. 16.12b, except for the response near 1 s.

Figure 16.13 shows the acceleration response when Motion #3 is imposed on the model. In Fig. 16.13a, the spike due to soil dilantacy did not appear in the dense model. This implies that the relative density increases during the ground consolidation after application of Motion #2. Table 16.4 presents the quantitative values. In the loose model, the spike still occurred at all depths but was relatively smaller, and the occurrence phase became faster compared with the acceleration response of Motion #2. Figure 16.13b are response spectra, showing the same tendency as in Motion #2. Compared with Motion #2, the peak point at approximately 0.1 s for Motion #3 was relatively small in both models. This is also presumed to increase the relative density of the ground due to the soil consolidation. Therefore, the ratio response spectrum Fig. 16.13c was also smaller when compared with that in Motion #2 near 0.1 s.

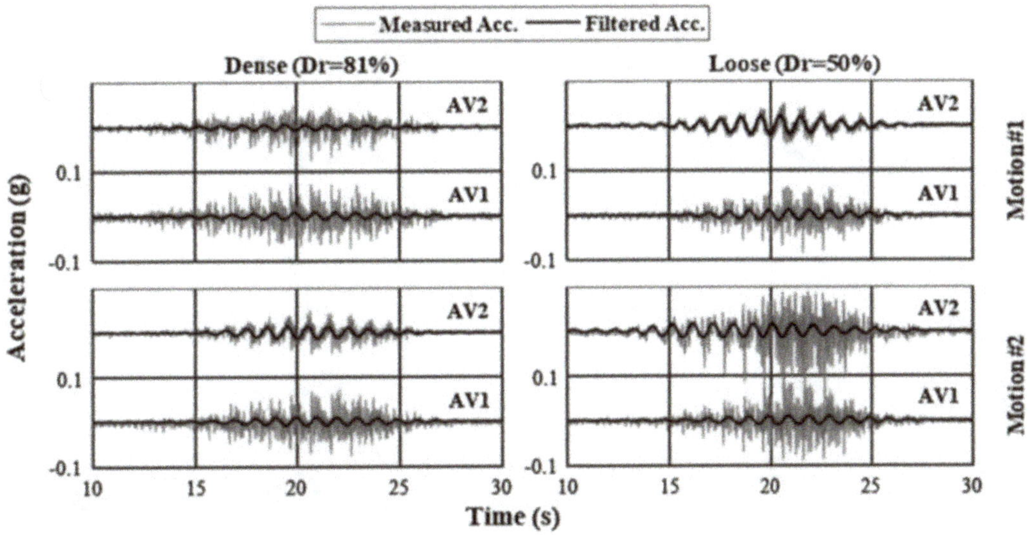


**Fig. 16.14** Vertical acceleration record measured using AV1 and AV2 for verifying the rocking effect. The black lines represent raw data, while the gray lines show the 0.3–3 Hz band pass filtered data

### Vertical Acceleration of the Rigid Box from AV1 and AV2

Figure 16.14 shows the rocking behavior measured using AV1 and AV2 accelerometers installed at the top and outside of the box. The gray lines indicate the unfiltered motions and the black lines represent the filtered band pass to show the components of motion between 0.3 and 3 Hz. Due to the rocking effect, an overturning moment can be induced by the inertial force between the ground model and box, and can be amplified by the resonance frequency.

The vertical acceleration has a peak, which is far lower than that in the base input motion, and AV1 (the upslope side of the model) and AV2 (the downslope side of the model) are asymmetrical in the loose model. The rocking accelerations are expected to be of the opposite sign (180° out of phase) for AV1 and AV2 (Kutter et al. 2018). Based on the measured acceleration, the rocking effect was seen because the phase difference of the two vertical accelerometers was contradictory in Motion #2. The results of Motion #3 show the rocking effect due to the asymmetric acceleration records AV1 and AV2, and future research is needed to determine how this rocking effect will affect the liquefaction behavior.

## *16.4.5 Comparison Between Acceleration and Pore Water Ratio*

Figure 16.15 shows the superimposed acceleration and pore pressure ratio time histories. This figure shows that the spike of the acceleration record, which is the

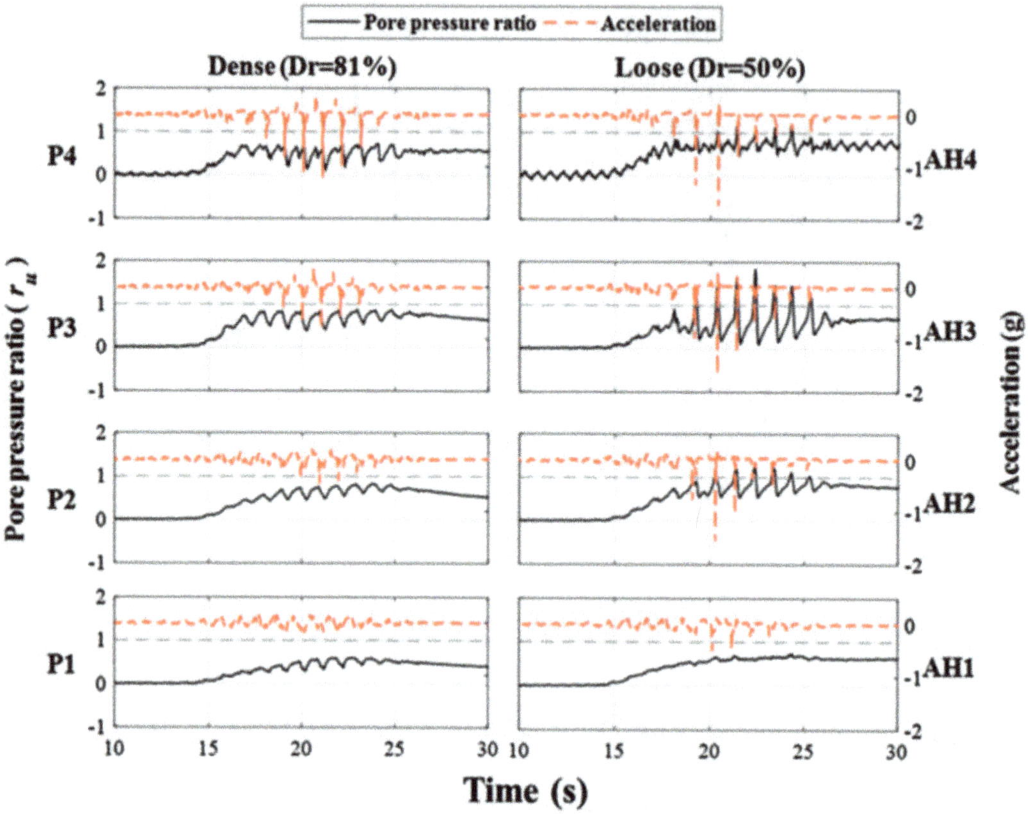


**Fig. 16.15** Acceleration and pore water ratio records superimposed by depth

de-liquefaction shock wave, coincides with the pulse where the excess pore water pressure is temporarily reduced. Such a phenomenon can be seen at the depth of AH4, with P4 installed on the ground surface of the dense model. In the loose model, the de-liquefaction shock wave appeared from P2 at the AH2 depth.

De-liquefaction is described as the temporary solidification of soil caused by dilatancy at large shear strains, and re-liquefaction is described as the return to a state with effective stress of zero due to the unloading of the shear stress (Kutter and Wilson 1999). That is, the spike pulses in the acceleration record coincide with pulses of negative pore pressure; this is called the de-liquefaction shock wave.

As shown in Fig. 16.15, the point of occurrence of the de-liquefaction shock wave in the loose model differs for each depth. For the depth of AH4, P4 is closest to the ground surface and the de-liquefaction shock wave occurred in approximately 17 s. In contrast, it occurred at approximately 18 s at the AH3 depth for P3. Moreover, it was generated at 24 s in AH2 for P2. The deeper it is from the ground surface, the more the generation time of the spike is delayed.

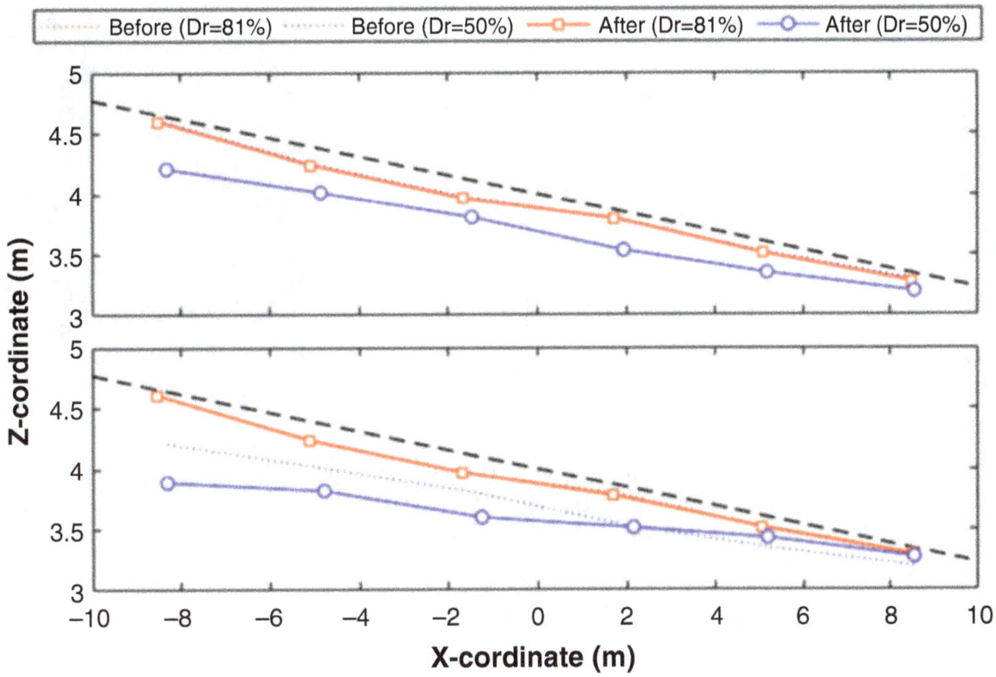


**Fig. 16.16** Measurement is executed at 1 g-level after completely stopping the centrifuge operation, and vertical displacement was measured directly from the front of the box (Dashed, red, and blue lines represent displacement of original ground model, dense model, and loose model, respectively)

## 16.4.6 *Displacement of Surface Marker*

### Comparison of Vertical Displacement

Figure 16.16 shows the measured vertical displacements of loose and dense ground models after Motions #2 and #3. The vertical displacement measurement was performed at 1 g after complete stopping of the centrifuge, and the displacement was directly measured from the front of the box according to the markers position. The dashed line represents the original ground model before motion was applied, and the geometries of both models were the same. The red and blue lines show the change in the vertical displacement of the dense and loose models, respectively, after the shakings.

When Motion # 2 was imposed on the model, vertical displacements of 9 and 223 mm were generated in the dense and loose models, respectively, at the prototype scale. In the dense model, almost no vertical displacement occurred compared to that in the loose model. As shown in Fig. 16.16, the vertical settlements in Motion #3 were smaller in both models compared to those in Motion #2. The dense model had showed a slight change in displacement, while the loose model produced an overall 97 mm (prototype) displacement. This implies that the relative density of the ground increased after Motion #2. In the loose model, the vertical settlement at the lower slope near the $x$-coordinates of 5–8 m was negative, because it is assumed that the sand of the upslope part drifted down, and the ground level was raised.

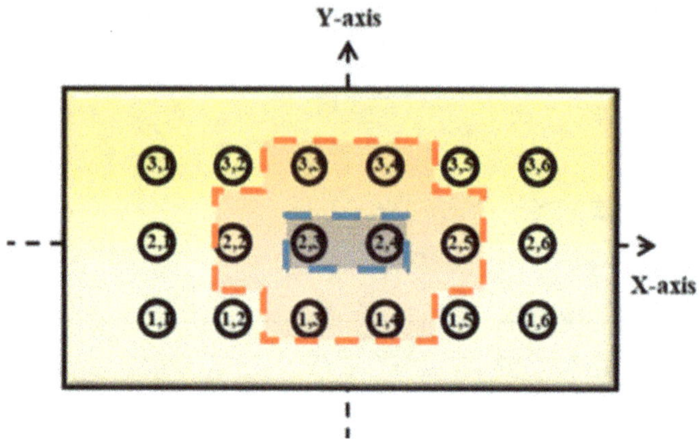


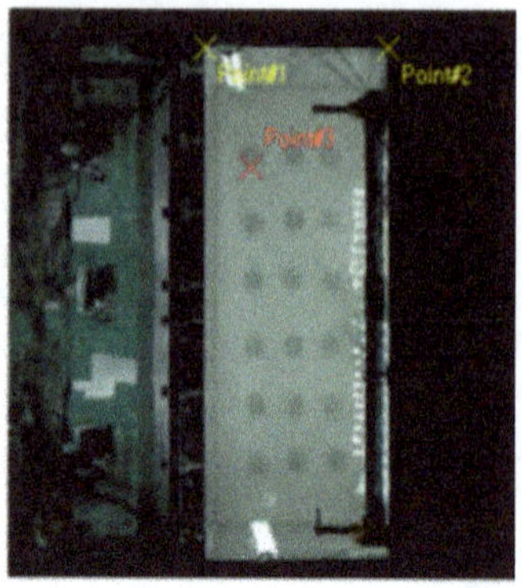


**Fig. 16.17** Surface markers coordinate system (the markers included in the red line have a relatively large permanent horizontal displacement) and TEMA software sets the points (points #1 and #2 are the reference points and point #3 is the tracking point)

## Comparison of Horizontal Displacement

The TEMA software was used to track targets and measure horizontal displacement of the makers based on the saved video. Before using this tool, the box width was fixed with two reference points, and the actual length was set by entering it. Actually, the box width was 225 mm and was scaled according to this length; the displacement was calculated by tracking the displacement of the marker. Figure 16.17 shows the reference points (points #1 and #2) and marker tracking point (point #3). After measuring the reference points of the width of the box to the scaling points, a permanent horizontal displacement, which was tracked by a marker point, was detected in real time. To achieve high-quality displacement results, a lighting system was built by installing LEDs on the rigid box.

Figure 16.18 shows the typical time histories of horizontal displacements obtained using the TEMA software. Permanent displacements during shaking are shown with respect to the time histories. The measured time was longer than the displayed time but was terminated at 40 s because of the difficulty in accurately measuring displacement due to water wave. It is recommended to install the thick cover plate presented in the LEAP guideline to minimize the water wave.

The changes in horizontal displacement before and after Motions #2 and #3 for the dense and loose models are shown in Fig. 16.19 through contour lines. Figure 16.17 shows a total of 18 surface markers installed on the ground surface, and the red part indicates the center markers where the displacement occurred relatively more than at other points. The contour line was presented by tracking the horizontal displacement for each marker during shaking. Notably, there was hardly any occurrence of displacement in the dense model. In contrast, in the loose model, the markers moved on average in the downslope direction by 134 mm (prototype). Especially, the horizontal displacement of the markers installed in the middle slope was 169 mm (prototype), which is higher than the average. A special point is that the displacement of the first horizontal marker line tends to occur significantly.

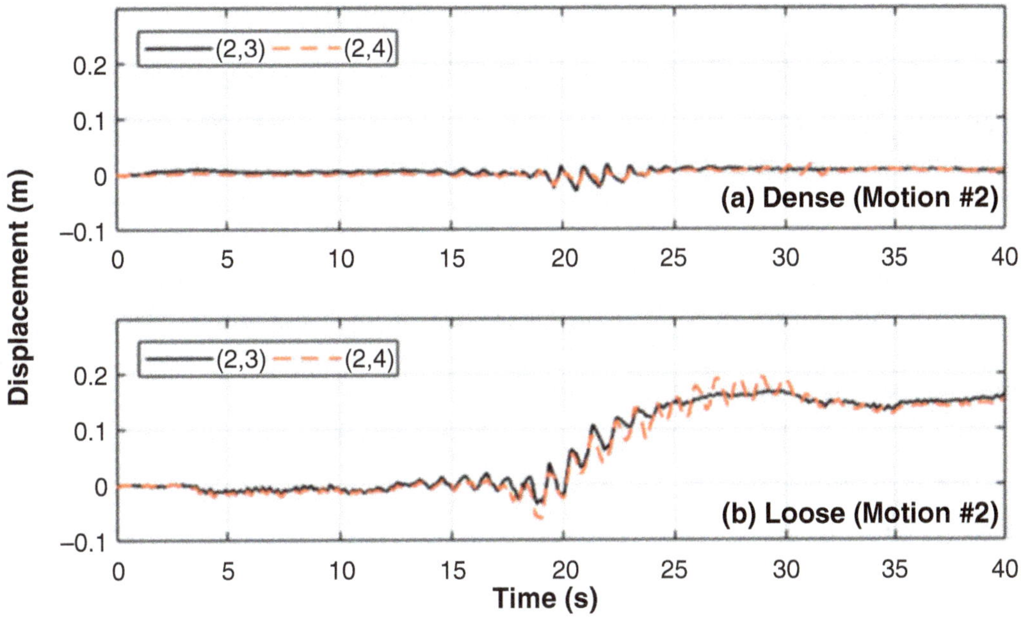


**Fig. 16.18** Representative results tracking the permanent horizontal displacement of markers by using TEMA software. Permanent horizontal displacement of the dense model hardly occurs, and approximately 200 mm occurs in the loose model

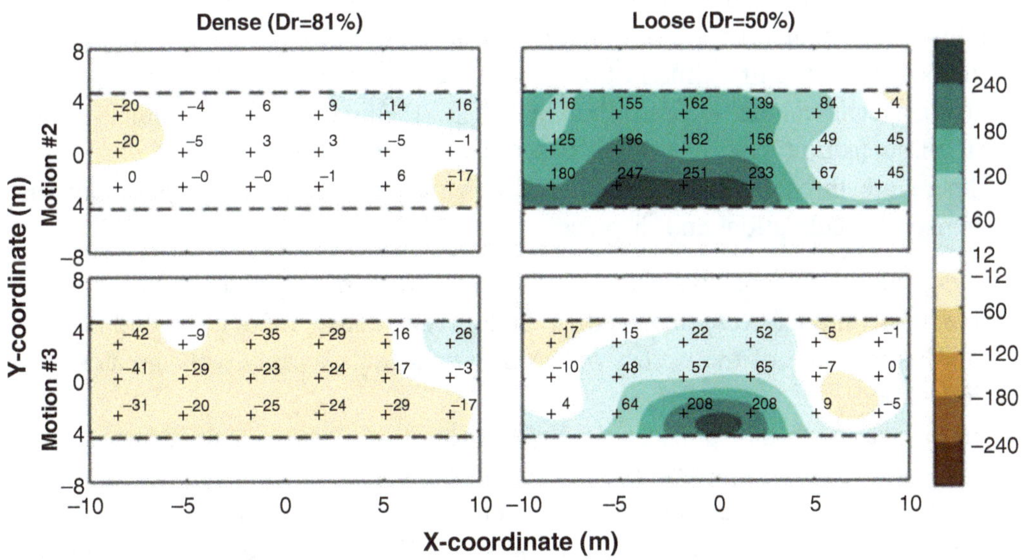


**Fig. 16.19** Contour line is presented using calculation of the horizontal displacement for each marker by using TEMA software

In this part, as the front of the box, the contact between the front part of the acrylic material and the soil particles makes the friction force smaller than the other parts. Therefore, there is possibility that a larger horizontal displacement may occur compared with other parts. The horizontal displacement of the markers placed at the other edge of the box was suspected to be smaller than other markers owing to the boundary effect.

The amount of displacement after Motion #3 is 50% or smaller than the displacement after Motion #2. Since a considerable amount of time passed after Motion #2, the ground stiffness is suspected to increase due to the rearrangement of particles. As shown in Table 16.4, relative density was measured through the CPT after Motion #2, and it is found that the relative density increased. Therefore, even with a similar PGA in Motion #3, the horizontal displacement occurred in both models was slight. However, the marker horizontal displacement of the dense model seemed to be moved toward the upslope direction. This reflects the difficulty of the accurate measurement by water wave (Kokkali et al. 2018).

## 16.5 Conclusion

Facilities in each country participated in LEAP-2017, and the same sand, viscous fluid, type of input motion, and box conditions were used to ensure data reliability. However, the relative density of the soil and the input motion intensity were set as variations and the liquefaction behavior by these two variations was evaluated. KAIST evaluated the liquefaction behavior at two different densities: a dense model ($D_r$ = 85%) and loose model ($D_r$ = 50%). Two destructive motions with amplitude of about 0.15 g (Motions #2 and #3) were applied to both models. By participating for the first time in this project, KAIST performed physical modeling by using a technical method described in the LEAP guidelines. The technical method for the liquefaction experiment was described in detail in Sect. 16.2.

There are three important results of this project: response of pore pressure, response of acceleration, and displacement.

1. De-liquefaction shock wave occurred in pore pressure response, and the spikes in the loose model were larger than that in the dense model. In addition, the dissipation time of the excess pore water pressure was relatively greater in the loose model.
2. In the acceleration response, a de-liquefaction shock wave occurred largely in the loose model. The spectral accelerations of the ground motions show the frequency content of the input motion and the de-liquefaction shock waves.
3. As a result of the displacement, vertical and horizontal displacements occurred in the loose model. In particular, the displacement of the markers near the center of the ground was relatively larger than the displacement of the markers located at the edge. This phenomenon was caused by the boundary conditions imposed by the rigid model container.

Both Motions #2 and #3 have the same input motion component and PGA. The comparison of the response to these motions showed that the responses of acceleration, pore water pressure, and displacement indicated less nonlinearity during Motion #3 than Motion #2. Densification or strengthening of the models is also consistent with the increase in cone penetration resistance due to applying Motion #2 before Motion #3.

As an organization that participated in LEAP project for the first time, there are some suggestions for the betterment of the project; this includes technical limitation and difficulties of KAIST while experimenting.

1. In the course of the creating of the ground model, there was a great difficulty in installing sensors. The line of the sensor is not flexible, and it is difficult to place the sensor at the accurate position. This may lead to ground disturbances and cause the relative density per layer to be asymmetric. Therefore, it is important to think about how you can safely and accurately position the sensor installation.
2. There was an insufficient quantity of light, an insufficient contrast of markers, and an error disturbing accurate measurement with respect to water wave in measuring displacement. If these problems are solved, better quality data could be extracted.

**Acknowledgments** This research was part of the project titled "Development of performance-based seismic design," funded by the Ministry of Oceans and Fisheries, Korea. The authors also gratefully acknowledge the KREONET service provided by Korea Institute of Science and Technology Information.

## References

Arulanandan, K., & Scott, R. F. (1993). *Verification of numerical procedures for the analysis of soil liquefaction problems*. Brookfield: A.A. Balkema.

Garnier, J., et al. (2007). Catalogue of scaling laws and similitude questions in geotechnical centrifuge modelling. *International Journal of Physical Modelling in Geotechnics, 7*(3), 1.

Hung, W.-Y., Lee, C.-J., & Hu, L.-M. (2018). Study of the effects of container boundary and slope on soil liquefaction by centrifuge modeling. *International Journal of Soil Dynamics and Earthquake Engineering, 113*, 682–697. https://doi.org/10.1016/j.soildyn.2018.02.012.

Kim, D.-S., et al. (2013a). A newly developed state-of-the-art geotechnical centrifuge in Korea. *KSCE Journal of Civil Engineering, 17*(1), 77–84.

Kim, D.-S., et al. (2013b). Self-balanced earthquake simulator on centrifuge and dynamic performance verification. *KSCE Journal of Civil Engineering, 17*(4), 651–661.

Kim, J.-H., Choo, Y. W., & Kim, D.-S. (2017). Correlation between the shear-wave velocity and tip resistance of quartz sand in a centrifuge. *Journal of Geotechnical and Geoenvironmental Engineering, 143*(11), 04017083.

Kokkali, P., Abdoun, T., & Zeghal, M. (2018). Physical modeling of soil liquefaction: Overview of LEAP production test 1 at Rensselaer Polytechnic Institute. *Soil Dynamics and Earthquake Engineering, 113*, 629–649.

Kutter, B., Carey, T., Hashimoto, T., Zeghal, M., Abdoun, T., Kokalli, P., Madabhushi, G., Haigh, S., Hung, W.-Y., Lee, C.-J., Iai, S., Tobita, T., Zhou, Y. G., Chen, Y., & Manzari, M. T. (2018). LEAP-GWU-2015 experiment specifications, results, and comparisons. *International Journal of Soil Dynamics and Earthquake Engineering, 113*, 616–628. https://doi.org/10.1016/j.soildyn.2017.05.018.

Kutter, B. L., & Wilson, D. W. (1999). De-liquefaction shock waves. In: *Proceedings of the seventh US–Japan workshop on earthquake resistant design of lifeline facilities and countermeasures against soil liquefaction* (p. 295–310). Seattle.

Madabhushi, S. S. C., Haigh, S. K., & Madabhushi, G. S. P. (2018). LEAP-GWU-2015: Centrifuge and numerical modelling of slope liquefaction at the University of Cambridge. *International Journal of Soil Dynamics and Earthquake Engineering, 113*, 671–681. https://doi.org/10.1016/j.soildyn.2016.11.009.

Manzari, M. T., Kutter, B. L., Zeghal, M., Iai, S., Tobita, T., Madabhushi, S. P. G., Haigh, S. K., Mejia, L., Gutierrez, D. A., & Armstrong, R. J. (2015). LEAP projects: concept and challenges. In *Proceedings of the fourth international conference on geotechnical engineering for disaster mitigation and rehabilitation*. (4th GEDMAR), 2014 Sept 16–18. Kyoto: Taylor & Francis.

Okamura, M., & Inoue, T. (2012). Preparation of fully saturated models for liquefaction study. *International Journal of Physical Modelling in Geotechnics, 12*(1), 39–46.

Ueda, K., & Iai, S. (2018). Numerical predictions for centrifuge model tests of a liquefiable sloping ground using a strain space multiple mechanism model based on the finite strain theory. *International Journal of Soil Dynamics and Earthquake Engineering, 113*, 771–792. https://doi.org/10.1016/j.soildyn.2016.11.015.

Viswanath, D. S., et al. (2007). Introduction. In *Viscosity of liquids* (pp. 72–80). Amsterdam: Springer.

# Chapter 17
# LEAP-UCD-2017 Centrifuge Test at Kyoto University

**Ruben R. Vargas, Tetsuo Tobita, Kyohei Ueda, and Hikaru Yatsugi**

**Abstract** As part of the LEAP-UCD-2017 exercise, 24 centrifuge tests were conducted at 9 centrifuge facilities around the world; among them, 3 tests were carried out in the facilities of the Disaster Prevention Research Institute at Kyoto University. The main objective of the tests is to characterize the median response and uncertainty of the dynamic response of a uniform-density sloping deposit of clean sand, subjected to a ramped sinusoidal wave; this chapter introduces specific details of the model preparation, testing procedures, and achieved response. Additionally, a brief discussion related to the uncertainties in the density estimation is presented.

## 17.1 Introduction

During the past 40 years, efforts and developments in computational modeling of geo-materials have contributed significantly to increase the accuracy of prediction of the dynamic response of soil systems. Due the catastrophic consequences (as result of seismic events), special emphasis has been pointed to the liquefaction-induced ground failures. However, despite the efforts, results of numerical simulations have a certain degree of discrepancy with results obtained in physical modeling. Therefore, it is important to highlight that exercises of verification and validation (V&V) of numerical simulations are still needed to enhance the reliability of numerical models for liquefaction prediction.

Previous V&V works such as "VELACS Project" (1994) and "LEAP-GWU-2015" have attempted to verify numerical simulations in comparison to centrifuge experiments, but more efforts are needed prior to reaching definitive conclusions.

R. R. Vargas (✉) · K. Ueda
Disaster Prevention Research Institute, Kyoto University, Kyoto, Japan
e-mail: vargas.rodrigo.35m@st.kyoto-u.ac.jp

T. Tobita
Department of Civil Engineering, Kansai University, Osaka, Japan

H. Yatsugi
Faculty of Environmental and Urban Engineering, Kansai University, Osaka, Japan



B. Kutter et al. (eds.), *Model Tests and Numerical Simulations of Liquefaction and Lateral Spreading*, https://doi.org/10.1007/978-3-030-22818-7_17

LEAP (Liquefaction Experiments and Analysis Projects) is a joint project that pursues the verification, validation, and uncertainty quantification of numerical liquefaction models, based on centrifuge experiments. This project will have several exercises, in the sense that each one would be focused on different aspects of soil liquefaction, such as dams, ports, lateral spreading, etc.

"LEAP-UCD-2017" is one of the LEAP's exercises, whose main objective is to perform a sufficient number of experiments to characterize the median response, and the uncertainty of a specific sloping deposit of sand. Centrifuge experiments in nine facilities around the world were developed for this exercise; this paper will make a review of the experiments performed in the installations of the Disaster Prevention Research Institute (DPRI) at Kyoto University. Three experiments were performed; two of which (KyU2 and KyU3) are presented in this paper with emphasis on model preparation, testing process, and special features that may cause uncertainty or variations in comparison to experiments developed in other facilities.

## 17.2 Test Specifications and Model Preparation

### *17.2.1 Description of the Model and Instrumentation*

Ottawa F-65 was chosen as standard sand for "LEAP-UCD-2017"; this sand was provided to the facilities by UC Davis (see details in Kutter et al. 2019). To avoid small errors due to uncertainty in specific gravity (Gs) and the maximum and minimum densities, the sand placement was prescribed based on target density (in replacement of relative density).

A uniform-density, 5-degree slope model inside a rigid container was specified for this exercise (see Fig. 17.1); a scaling ratio of 44.44 has been employed for the tests at Kyoto, which leads to prototype scale's dimensions to be 20 m in length, 4 m in height at midpoint, and 6.66 m of width.

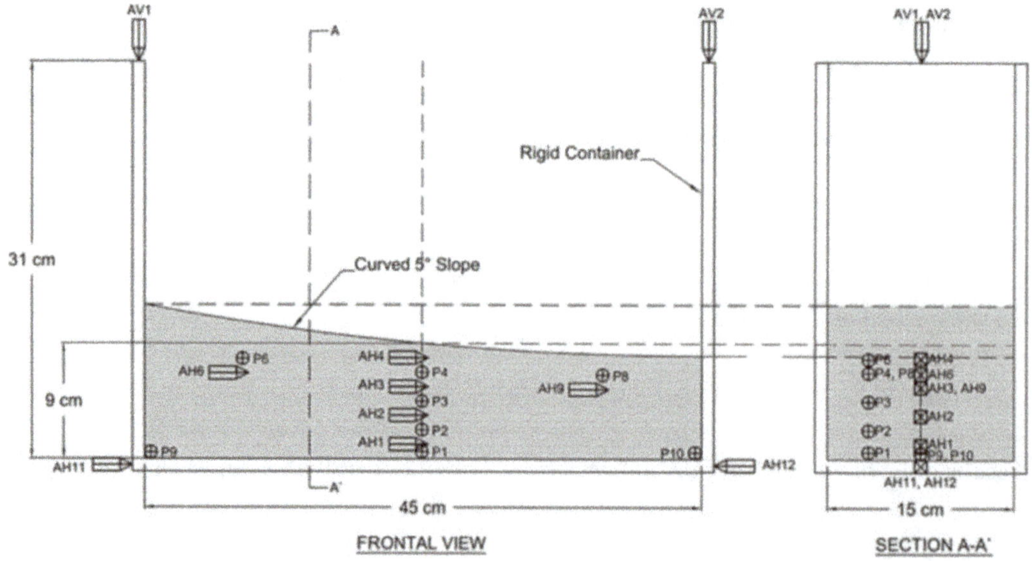


**Fig. 17.1** Dimensions of the tested model

Regarding instrumentation of the model, nine horizontal accelerometers (AH1, AH2, AH3, AH4, AH6, AH9, AH11, and AH12), two vertical accelerometers (AV1 and AV2), and eight pore pressure transducers (P1, P2, P3, P4, P6, P8, P9, and P10) were placed.

## 17.2.2 Air Pluviation and Surface Curving

The specified technique for sand placement was air pluviation. A U.S. Standard Sieve No. 16 with a three-slot arrangement (each one of 100 mm x 10.3 mm) was specified as the pluviation tool (see Fig. 17.2).

During the sand placement (after height calibration), density was measured at three different stages in each model, using a rigid steel ruler (with a wood support). Achieved densities of 1659 and 1637 kg/m$^3$ were obtained for tests KyU2 and KyU3, respectively. A brief analysis about sensitivity and density measurement is presented in Sect. 17.4.

The centrifuge facility at Kyoto University has an effective radius of 2.50 m, and is equipped with a shaking table in the circumferential direction. In order to avoid errors due to variations in the radial gravity field (as stated by Tobita et al. 2018), the surface was curved according to the geometry of the facility; therefore, all points in the surface will have the same slope relative to the gravity field, and would represent ground with a constant, 5-degree slope angle in the prototype scale. In order to achieve the curved surface: (a) a flat specimen was obtained by air pluviation, (b) an aluminum guide was fabricated with the shape of the target curved surface, (c) a fixed length aluminum pipe is conditioned to a vacuum apparatus, and (d) following the guide, vacuum is applied to reach the curved surface (see Fig. 17.3).

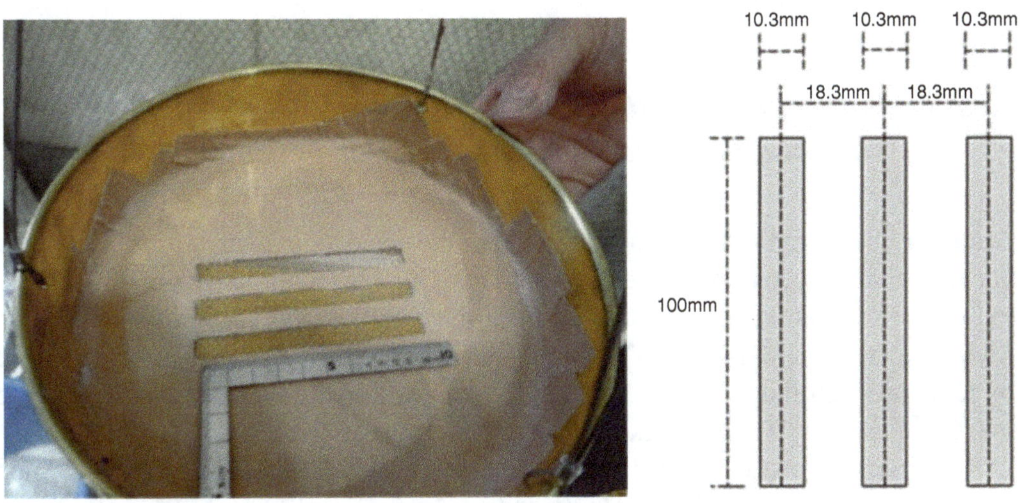


(a) (b)

**Fig. 17.2** (**a**) Tool employed for air pluviation procedure. (**b**) Geometry of the openings for the pluviation

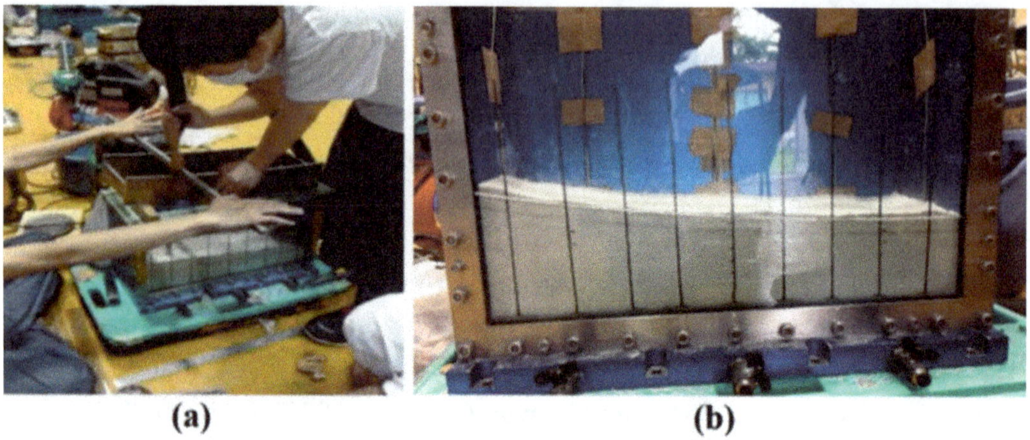


**Fig. 17.3** (**a**) Vacuum process to obtain the curved surface. (**b**) Achieved curved surface

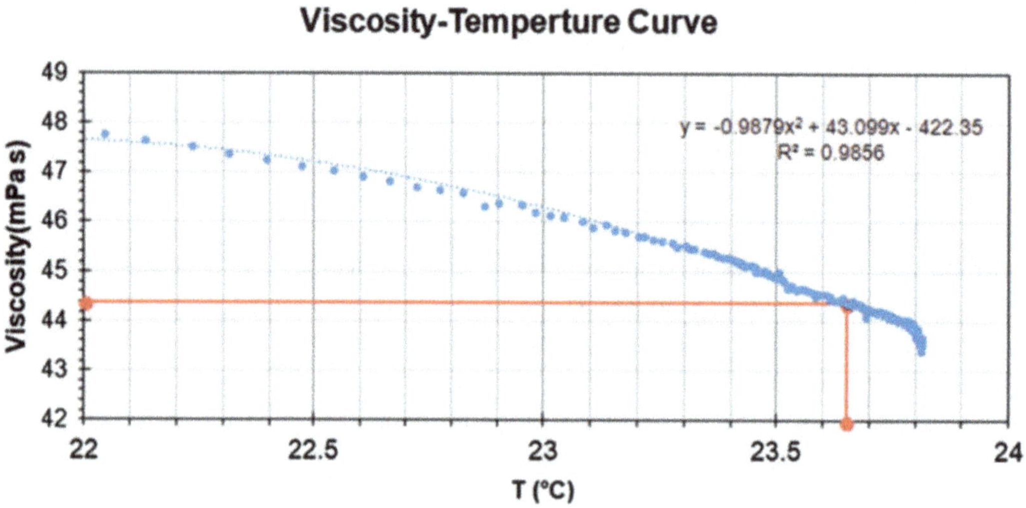


**Fig. 17.4** Viscosity–temperature curve

## *17.2.3 Saturation, Input Motions, Measurement of Surface Deformations and Cone Penetration Testing (CPT)*

After air pluviation and sensor placement, the rigid container was placed in a vacuum chamber, and one series of vacuum –$CO_2$– vacuum was applied in order to facilitate dissolution of gas bubbles, prior to saturation with a de-aired solution of Methylcellulose (SM-100, Shin'etsu Chemical Co.), following the method proposed by Okamura and Inoue (2012). Figure 17.4 shows the viscosity–temperature curve measured using a "Tuning-Fork Vibration Viscometer" (Izumo 2006).

The specified procedure includes a shake sequence of six input motions, which consists of ramped sinusoidal 1 Hz wave (as seen in Fig. 17.5a). Motions 2, 3, and

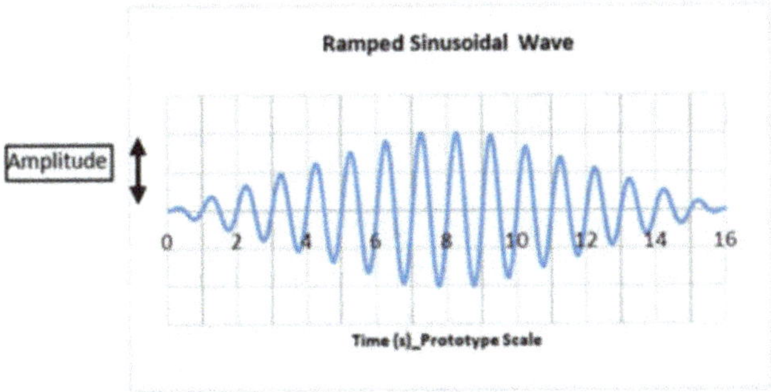


(a)

| Motion | Amplitude (g) |
|---|---|
| Motion 1 | 0.015 |
| Motion 2 | 0.15 |
| Motion 3 | 0.25 |
| Motion 4 | 0.015 |
| Motion 5 | 0.25 |
| Motion 6 | 0.015 |

(b)

**Fig. 17.5** (**a**) Specified ramped sine wave. (**b**) Amplitude of motions

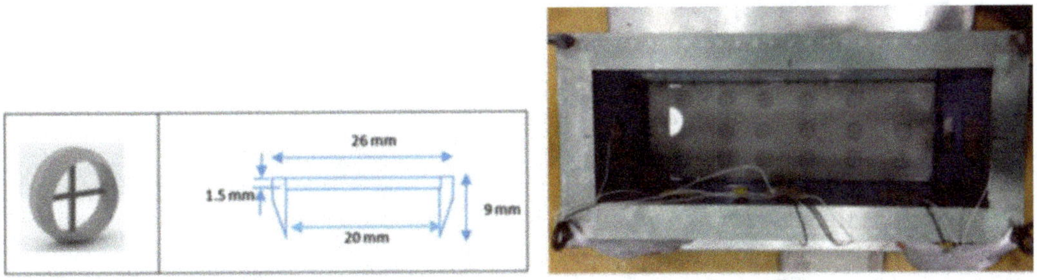


(a) (b)

**Fig. 17.6** (**a**) Geometry of surface markers. (**b**) Placement of surface markers

5 were considered as "destructive" and motions 1, 4, and 6 as "nondestructive" (see Fig. 17.5b).

Due to the presence of high-frequency vibrations in the achieved motions, and taking into account that higher frequency components have some but relatively small effect on the behavior of the model, this project (as a first approximation) used the effective PGA, defined as:

$$\mathrm{PGA}_{\mathrm{effective}} = \mathrm{PGA}_{1\mathrm{Hz}} + 0.5 \times \mathrm{PGA}_{\mathrm{hf}}$$

where: "$\mathrm{PGA}_{1\mathrm{Hz}}$" represents the PGA of the 1 Hz component of the achieved motion, and "$\mathrm{PGA}_{\mathrm{hf}}$" represents the higher frequency components of the ground motion.

To measure the surface displacements before and after each destructive motion, 18 surface markers provided by UC Davis were placed in the soil surface (see Fig. 17.6).

Additionally, three cone penetration tests in each experiment were performed (before each destructive motion) with a new 6 mm Mini-CPT provided by UC Davis; according to the specifications, the penetration rate was fixed as 2.12 mm/s.

## 17.3 Test Results

### 17.3.1 Achieved Ground Motions

As stated in the previous section, six input motions were applied to each model. For each destructive motion, $PGA_{1Hz}$ values were calculated using a notched band pass filter with corner frequencies between 0.9 and 1.1 times the predominant frequency (i.e., 1 Hz). An example of the signal filtering process and the estimated values of $PGA_{effective}$ for destructive motions are shown in Fig. 17.7a.

As shown in Fig. 17.7b, in order to get experimental data to verify the concept of "effective PGA" (as defined in Sect. 17.2.3), a calibration for the motions was

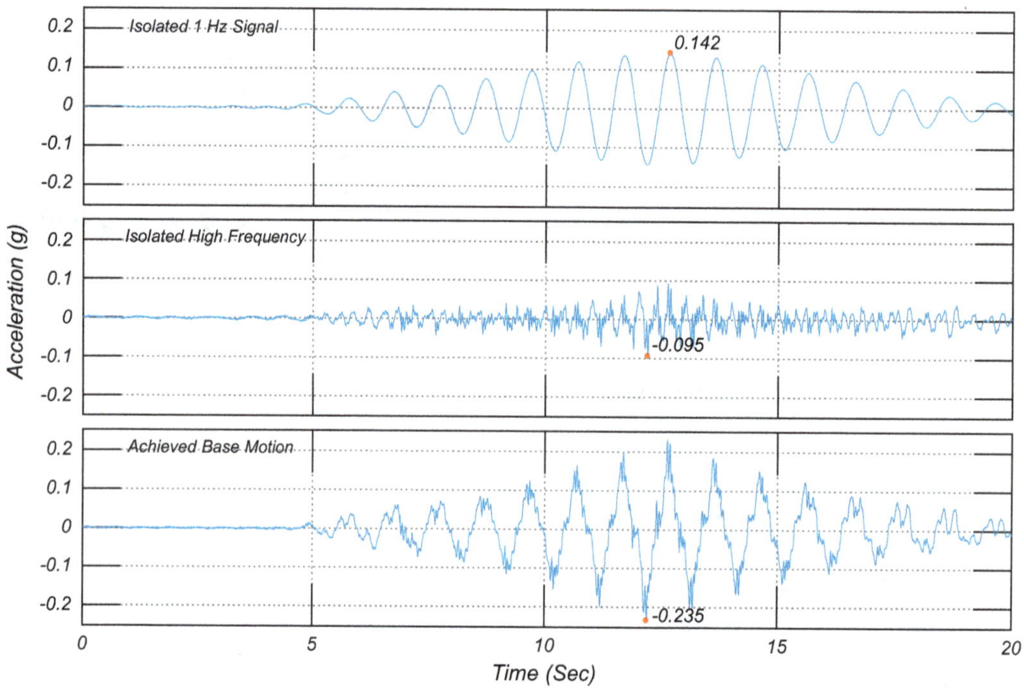


(a)

| Motion | | Target PGA (g) | Achieved PGA (g) | PGA of 1 Hz component (g) | Achieved PGA effective (g) |
|---|---|---|---|---|---|
| KyU2 | Motion 2 | 0.15 | 0.12 | 0.10 | 0.11 |
| | Motion 3 | 0.25 | 0.33 | 0.20 | 0.27 |
| | Motion 5 | 0.25 | 0.33 | 0.21 | 0.27 |
| KyU3 | Motion 2 | 0.15 | 0.14 | 0.09 | 0.12 |
| | Motion 3 | 0.25 | 0.24 | 0.14 | 0.19 |
| | Motion 5 | 0.25 | 0.23 | 0.14 | 0.19 |

(b)

**Fig. 17.7** (**a**) Filtering of input motion 3—KyU3. (**b**) Achieved input PGA developed in the sense that, for the KyU2 Model the values of "achieved PGA effective" would be close to the target values, and in the KyU3 Model, the values of "achieved PGA" would be close to the target values.

In Fig. 17.8, the whole input motions (without any filtering process) are reported.

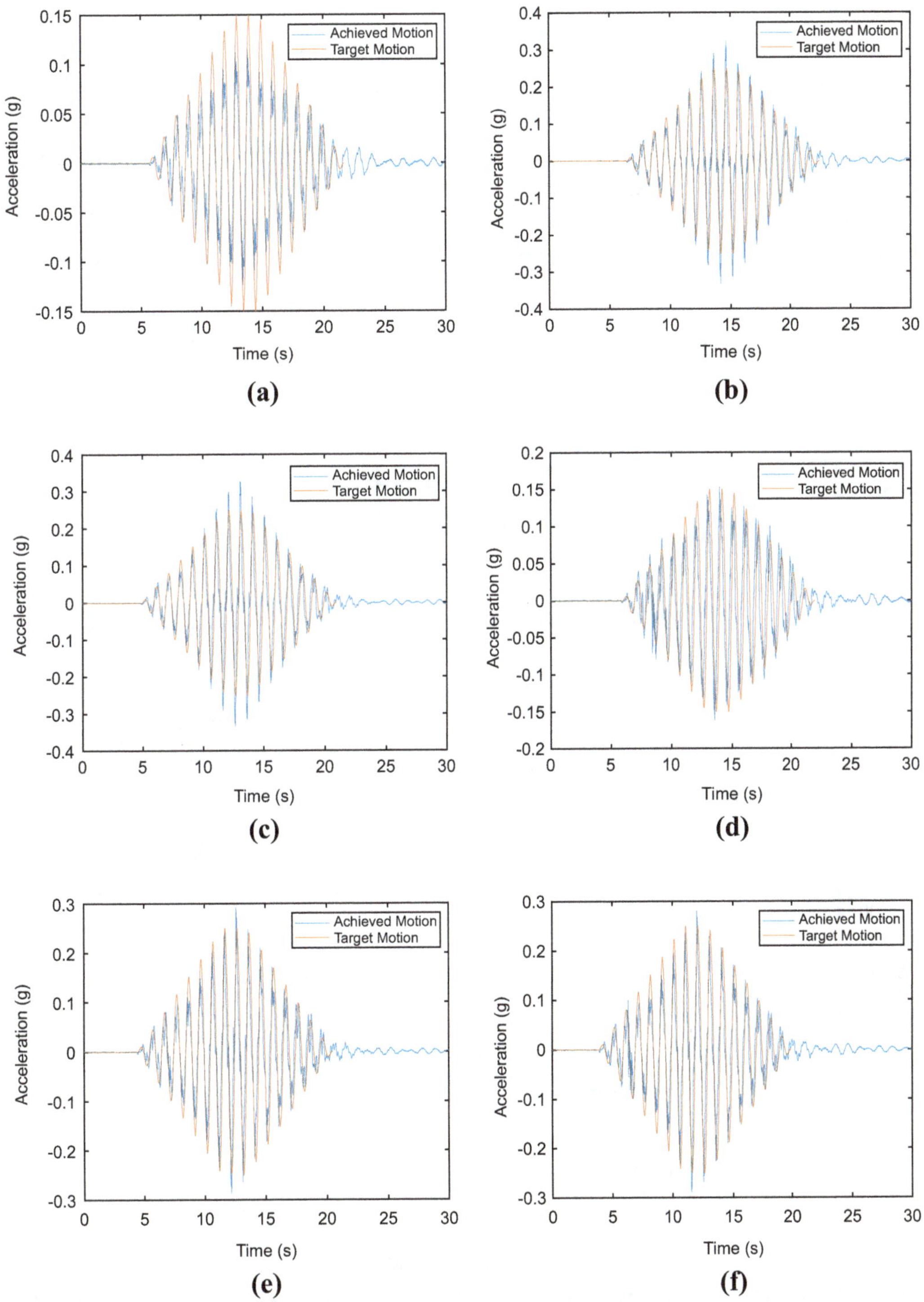


**Fig. 17.8** (**a**) Input Motion-Motion 2-Model KyU2; (**b**) Input Motion-Motion 3-Model KyU2; (**c**) Input Motion-Motion 5-Model KyU2; (**d**) Input Motion-Motion 2-Model KyU3; (**e**) Input Motion-Motion 3-Model KyU3; (**f**) Input Motion-Motion 5-Model KyU3

### 17.3.2 *Excess Pore Pressure During Motions*

As specified, pore pressure transducers (PPTs) were employed to measure the increase of excess pore water pressure (Δu); in Fig. 17.9, the correspondent values of Δu and the initial values of effective vertical stress are presented for Model KyU2.

In this paper, the excess pore pressure ratio is defined as: $r_u = \Delta u / \sigma'_v$. Regarding Model KyU2, full liquefaction (i.e. $r_u \approx 1$) for Motion 2 (Fig. 17.9a) was only achieved in shallow parts (sensors P6, P4, and P8). Values of $r_u$ between 0.75 and 0.9 were recorded for sensors P1, P2, P3, and P10; however, for sensor P9 a maximum value of $r_u$ equal to 0.35 was observed. Records of sensors in Motion 3 (Fig. 17.9b) indicate that values of $r_u \approx 1$ were achieved also in sensors P4, P6, P8, and P10 (corresponding to shallow areas and the base of the downslope side), and values between 0.85 and 0.95 were reported in sensors P1, P2, and P3; for sensor P9,

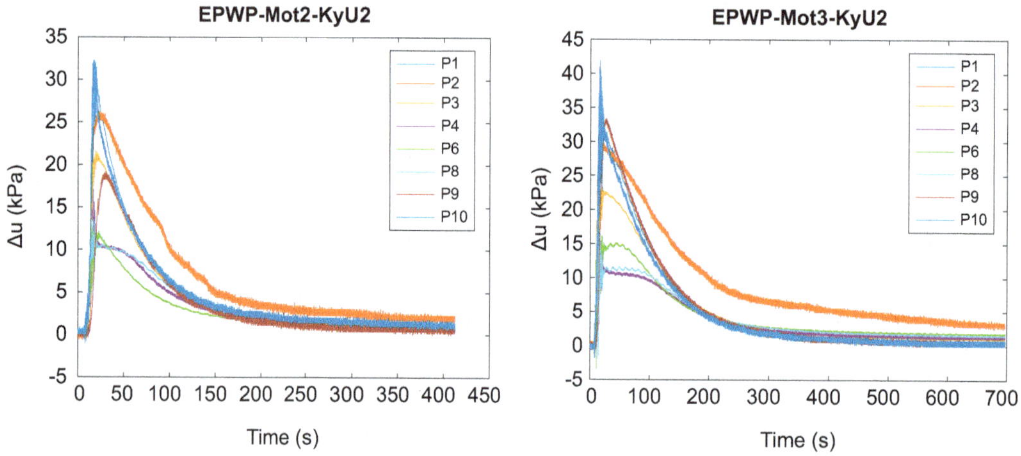


(a) (b)

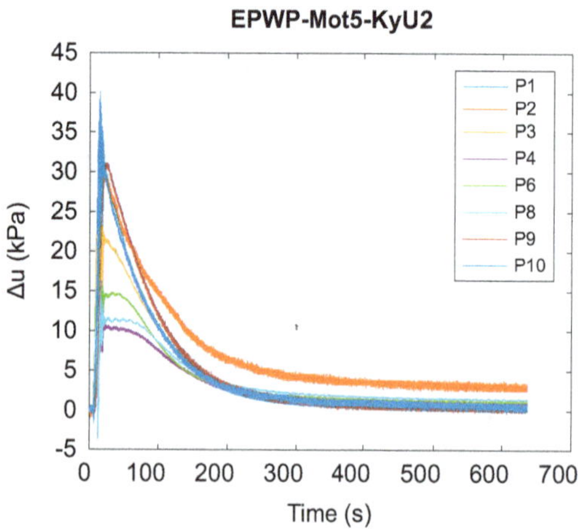


| Sensor | Depth (m) | σ'₁ (KPa) |
|---|---|---|
| P1 | 3.96 | 40.25 |
| P2 | 3.38 | 34.38 |
| P3 | 2.49 | 25.30 |
| P4 | 1.06 | 10.77 |
| P6 | 1.29 | 13.12 |
| P8 | 1.13 | 11.46 |
| P9 | 5.31 | 53.96 |
| P10 | 3.59 | 36.54 |

(c) (d)

**Fig. 17.9** (**a**) Excess pore water pressure motion 2—Model KyU2, (**b**) Excess pore water pressure motion 3—Model KyU2, (**c**) Excess pore water pressure motion 5—Model KyU2, (**d**) Initial stress state for sensors a value of 0.60 was found. In case of Motion 5 (Fig. 17.9c), the behavior remained the same as the one described in Motion 3. It is important to mention that despite the increase in the values of $q_c$ (obtained through the CPT test before Motion 3, and before Motion 5 (Fig. 17.17)), no increase in liquefaction resistance has been reported (with exception of sensors P1 and P9, where a small increase was recorded).

Regarding Model KyU3, it is important to mention that sensor P9 did not function during the test, so, its values will not be reported. For Motion 2 (Fig. 17.10a), full liquefaction (i.e. $r_u \approx 1$) was only achieved in the shallow part of the downslope (i.e. sensor P8); values of $r_u$ under 0.70 were recorded in the bottom of the model (sensors P1 and P10), and values from 0.75 to 0.90 were obtained in sensors P2, P3, P4, and P6. Records of sensors in Motion 3 (Fig. 17.10b) indicate that values of $r_u \approx 1$ were achieved almost in all sensors, except in sensors P2 and P4, where a value of around 0.9 was found. In case of Motion 5 (Fig. 17.10c), the behavior of sensors P3, P4, P6, and P8 remained the same as the one described in Motion

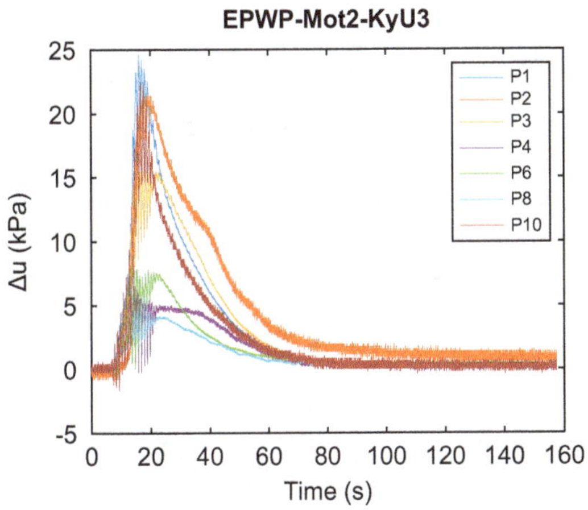


(a)

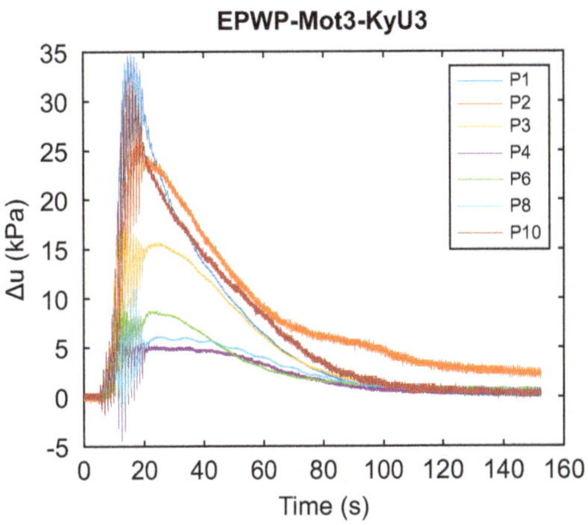


(b)

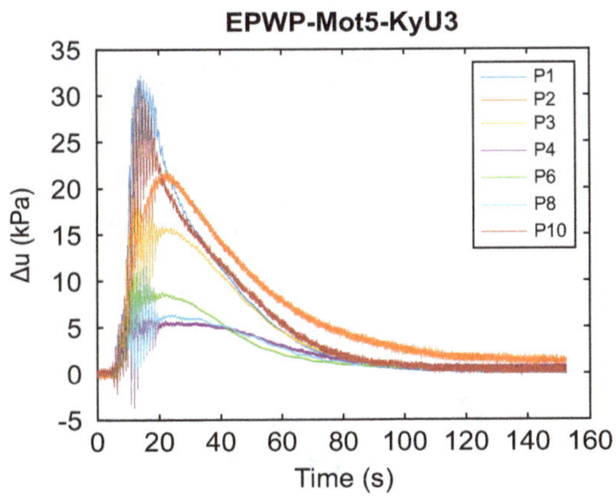


(c)

| Sensor | Depth (m) | $\sigma'_1$ (KPa) |
|---|---|---|
| P1 | 3.60 | 36.09 |
| P2 | 2.85 | 28.56 |
| P3 | 1.85 | 18.53 |
| P4 | 0.85 | 8.49 |
| P6 | 0.99 | 9.97 |
| P9 | 4.86 | 48.78 |
| P10 | 3.19 | 32.01 |

(d)

**Fig. 17.10** (**a**) Excess pore water pressure motion 2—Model KyU3. (**b**) Excess pore water pressure motion 3—Model KyU3. (**c**) Excess pore water pressure motion 5—Model KyU3. (**d**) Initial stress state for sensors

3. However, for sensors P1 and P2, a reduction in the $r_u$ value of 7% and 15% respectively was found; in this model, no significant difference among values of qc (obtained through the CPT test before Motion 3, and before Motion 5) were reported (Fig. 17.17).

### 17.3.3 Ground Motion Accelerations

Response time histories for all accelerometers placed inside the deposit are reported in Figs. 17.11, 17.12, 17.13, 17.14, 17.15, and 17.16. As expected for the results of excess pore water pressure (EPWP), for both models, in Motion 2 (i.e. Fig. 17.11 and 17.14) neither significant amplification nor distortion has been recorded for the

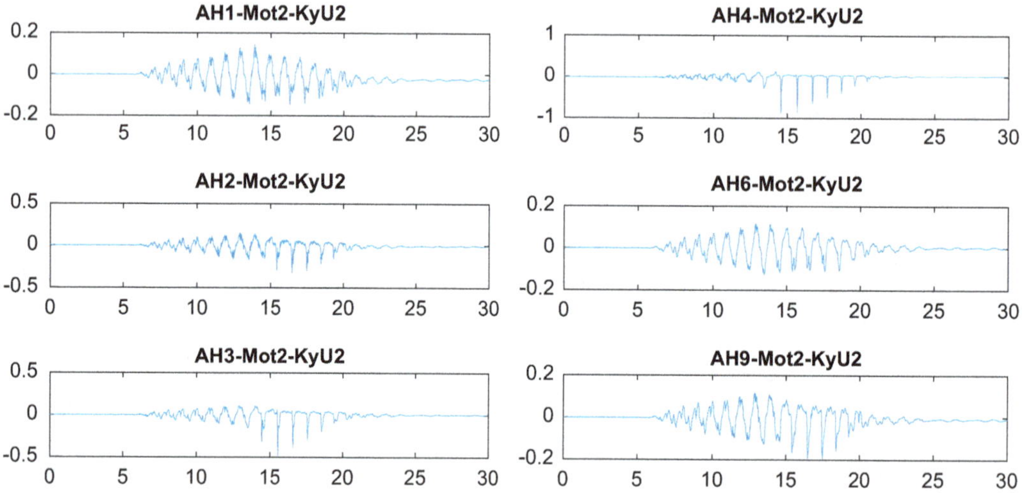


**Fig. 17.11** Ground motion accelerations (g) versus time (s) for Model KyU2—Motion 2

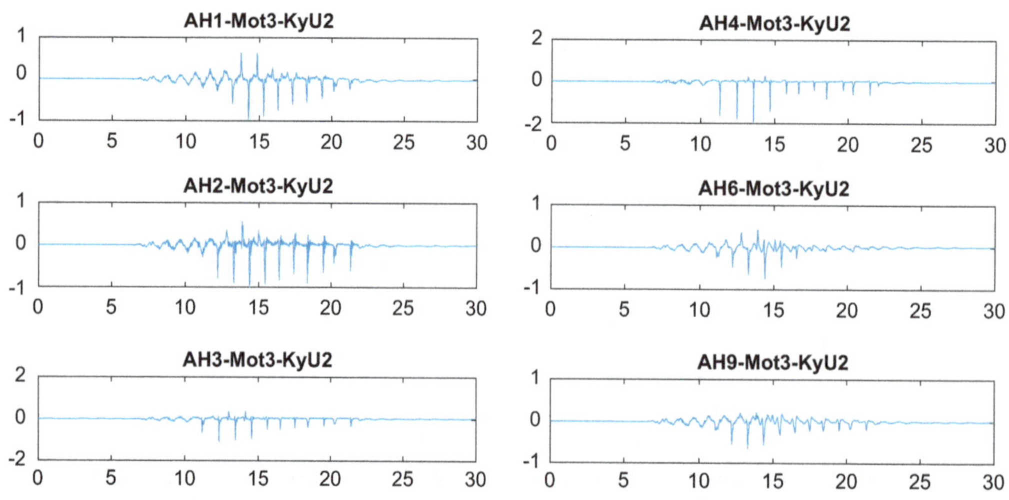


**Fig. 17.12** Ground motion accelerations (g) versus time (s) for Model KyU2—Motion 3

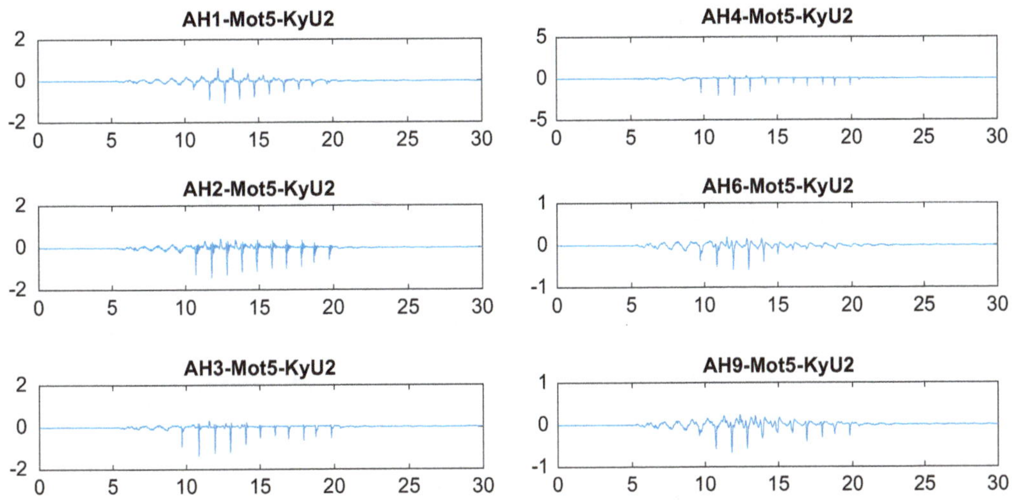


**Fig. 17.13** Ground motion accelerations (g) versus time (s) for Model KyU2—Motion 5

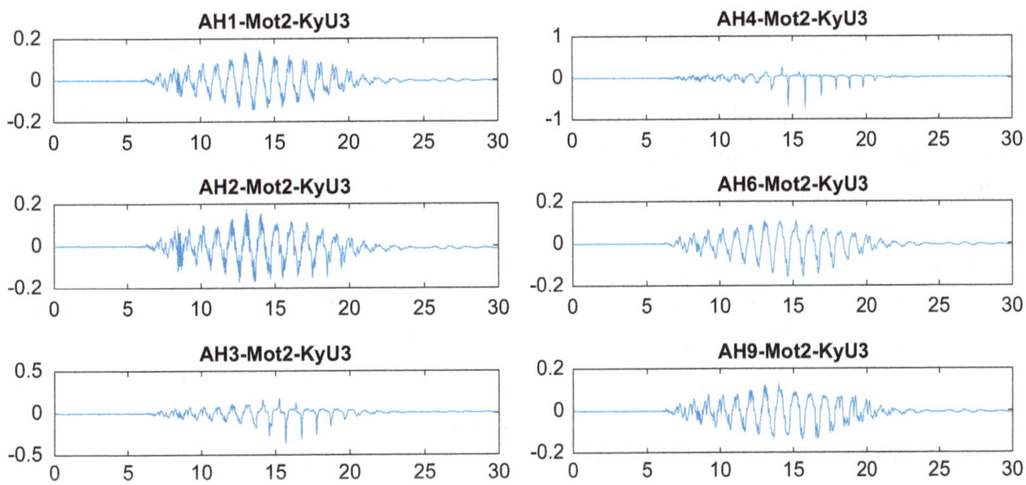


**Fig. 17.14** ground motion accelerations (g) versus time (s) for Model KyU3—Motion 2

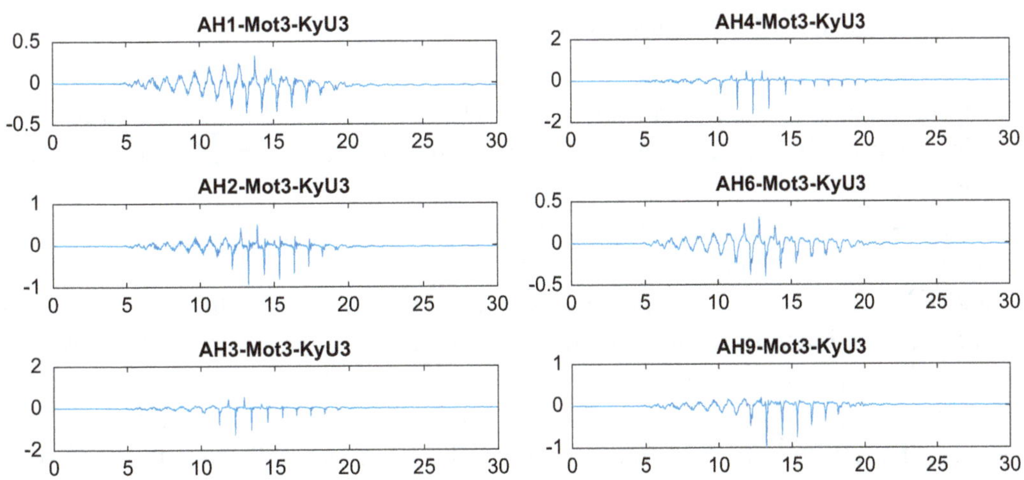


**Fig. 17.15** Ground motion accelerations (g) versus time (s) for Model KyU3—Motion 3

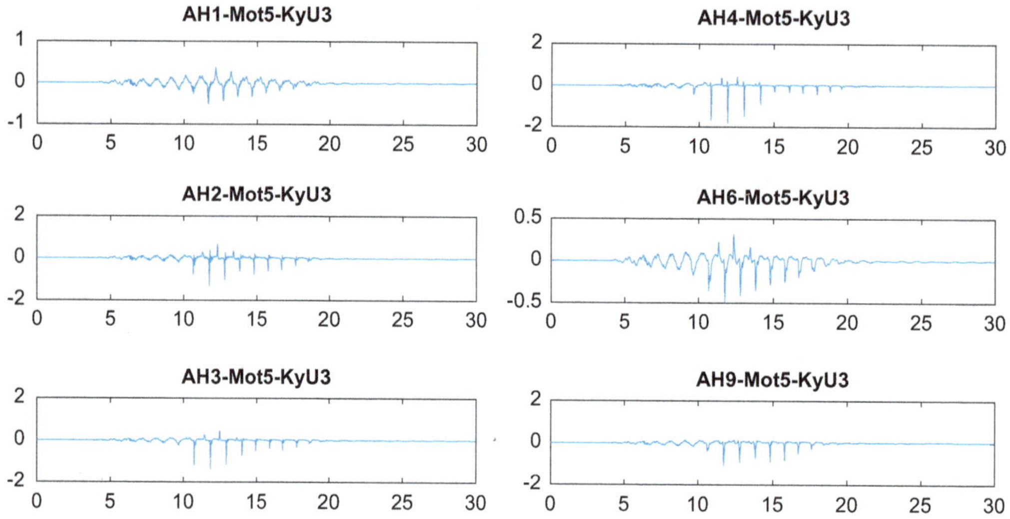


**Fig. 17.16** Ground motion accelerations (g) versus time (s) for Model KyU3—Motion 5

bottom accelerometers (i.e. AH1, AH2). However, for accelerometers AH3, AH4, AH6, and AH9, initially, the recorded acceleration remained similar to the input motion before the development of significant EPWP (at around 13 s); after that, the motion significantly changed and developed sharp spikes (which are characteristic of a dilative behavior of liquefied sand). It is important to remark that the distortion in the acceleration starts first (and also becomes larger) with a decrease in effective vertical stress, because the EPWP generation developed faster in those zones.

For Motions 3 and 5 (Figs. 17.11, 17.12, 17.15 and 17.16), the input wave is similar and greater than in Motion 2 and the recorded response of all accelerometers is similar to the response of accelerometers AH3, AH4, AH6, and AH9 in Motion 2. However, the start of distortion in the response waves starts sooner (due to the increase in the PGA), at around 10 s for the shallow accelerometers.

### *17.3.4 Cone Penetration Tests*

As specified, cone penetration tests (CPTs) were performed before each destructive motion; hence, CPT1, CPT2, and CPT3 were performed before Motions 2, 3, and 5, respectively.

Figures 17.17a and 17.17b shows the penetration test results for each model; in Model KyU2 penetration tests, no significant differences were obtained among CPT 1 and CPT 2; for CPT3 an increase between 20 and 30% in the value of qc is reported, so an increase in the liquefaction resistance would be reasonable; however, as explained before, little difference was found in the values of EPWP, between Motions 3 and 5. Yamada et al. (2010) found that in the case of re-liquefaction process (i.e. multiple liquefaction process in the same ground), anisotropy of soils becomes a key factor, rather than the relative density. This fact may explain the

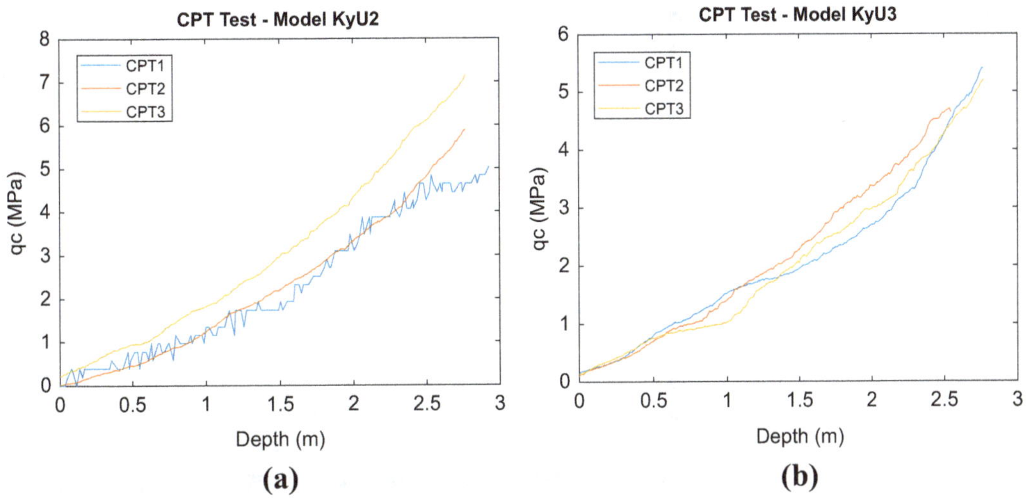


**Fig. 17.17** (**a**) CPT test results for model KyU2 (**b**) CPT test results for model KyU3

similar behavior in the development of EPWP in Motions 3 and 5 for Model KyU2, even when the $q_c$ value became larger prior Motion 5.

For Model KyU3, no significant differences were found among the three CPT realizations; also, as the previous case, no significant difference was found in the values of EPWP, between Motion 3 and 5.

It is also worth to mention that due to the characteristics of the CPT apparatus, it was not possible to perform shaking while the apparatus was mounted; consequently, each time the CPT tests were performed, the centrifuge had to: (1) be stopped to assemble the CPT, (2) increase the g-level at 44.4 g to perform the test, (3) stop again the facility to disassemble the CPT, and (4) increase the g-level for the shaking process. Effects of this procedure on soil density should be quantified in the future.

## 17.4 Discussion of Special Features

### *17.4.1 Sensitivity of Density Measurement by Volumetric Methods*

As indicated in Sect. 17.2.2, density was measured at three different stages in each model (as shown in Fig. 17.18), and variability among each layer was found. It is important to remark that models were prepared as a "flat ground" by air pluviation, and density measurements were performed during this process; the curved surface was obtained thereafter by applying vacuum (as stated in Sect. 17.2.2).

Taking into account that the dimensions of the rigid box that were used for both models are L = 45 cm, W = 15 cm, and H = 31 cm, Fig. 17.19 shows a very simple sensitivity analysis that was performed to estimate the limits of the Dr values that

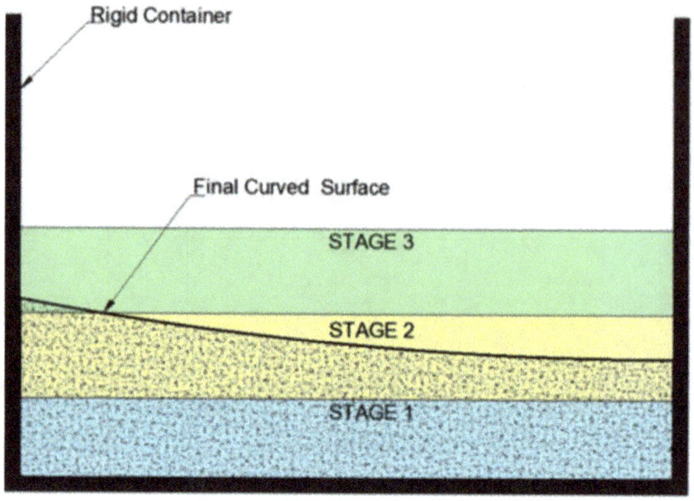


**Fig. 17.18** Stages of air pluviation procedure and density measurement

could be achieved if a measurement error would be committed. According to Fig. 17.18, the achieved density of sand placed in Stages 1 and 2 covers more than 99% of the final curved surface, so stage 3 is not presented.

For the sensitivity analysis due to the conditions of the experiment, weight of sand, and depth and width dimensions of the container were taken as constant values, and the measurement of the height of pluviated sand was taken as a variable. It is important to mention that, for each density measurement, height of sand was measured at 15 uniformly distributed points over the surface of ground; with this information, dry density was calculated as the ratio of weight to volume.

As seen in Fig. 17.19 for Stage 1, an error in the measurement of the height of sand of just 1 mm would lead to a constant deviation in the Dr value of 10% (Fig. 17.19a), which for some analysis may be considered as an important deviation. For the same 1 mm error in the height of sand, if we are interested in the measurement of density for only Stage 2, the deviation in the Dr value would remain the same; however, if we want to estimate the overall average density, the deviation associated with 1 mm height error would be reduced to 5% (Fig. 17.19b).

With regard to the achieved measurements of the sand's height, values ranging from 10% (for Stage 1 measurements) to 3% (for Stage 2 measurements) correspondent to the "coefficient of variability" (i.e. standard deviation normalized by the mean value) were obtained; it seems that this variability was mainly caused by the "edge effect," which led to obtain lower surface values near the container's edges during the air pluviation procedure.

Figure 17.20 shows the results for an air pluviation calibration exercise (20 trials were performed and each point corresponds to the average of 3 or 4 trials) realized prior the development of models. A strong correlation was not obtained ($r^2 \approx 0.68$), which lead us to think that homogeneity may not be ensured in the whole process.

The causes of the nonachievement of strong correlation were not still fully determined; but one important reason may be that the container selected for air pluviation (as shown in Fig. 17.2) produced a "high" volume flow of sand during the

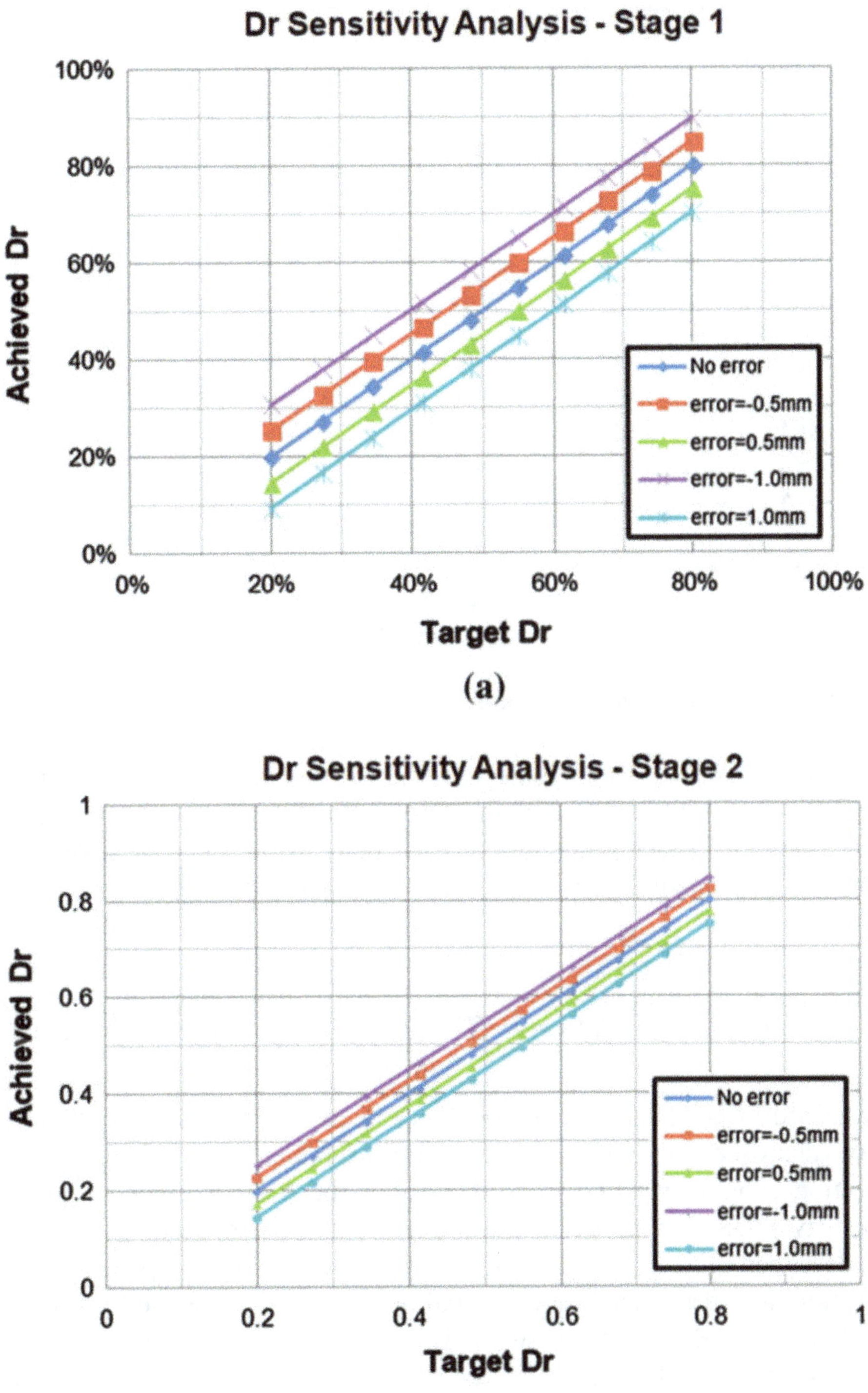


(b)

**Fig. 17.19** (**a**) Sensitivity analysis for density measurement—stage 1. (**b**) Sensitivity analysis for density measurement—stage 2

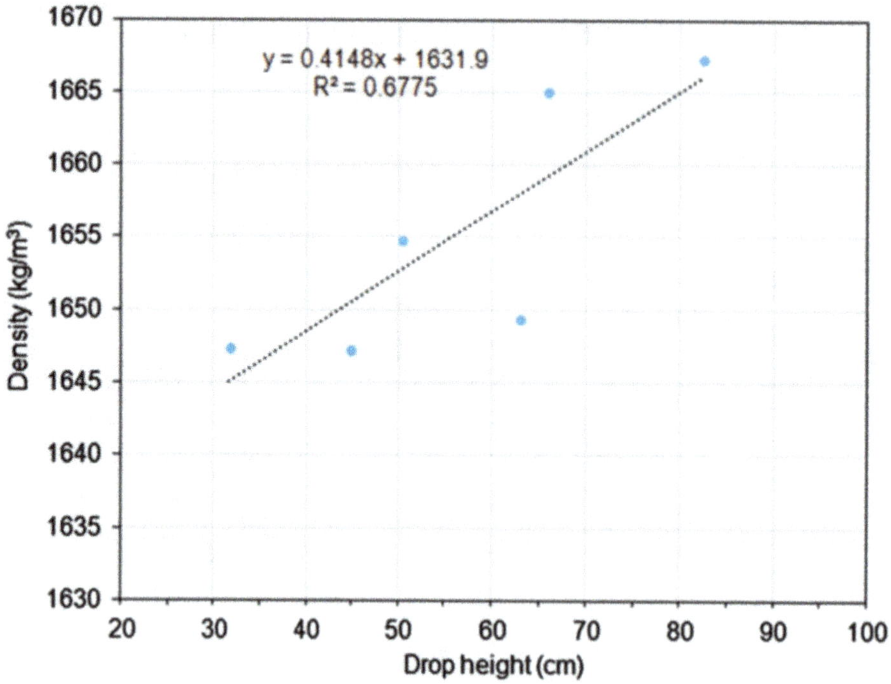


**Fig. 17.20** Drop height calibration for the air pluviation process

process. T his caused that in some intervals of time, an horizontal and uniform surface was not kept (increasing the "edge effect"). Therefore, decreasing the volume flow (i.e. reducing the area of each hole, or keeping one hole instead of three) could be a possibility to improve the correlation degree and homogeneity.

### *17.4.2 Correlation Between CPT and Density Measurements*

As described in previous sections for the characteristics of this project, measured density values seem to be very sensitive to measurements of volume. Therefore, in this section a comparison of density obtained by the "mass–volume measurements" and density obtained from "correlations from CPT" is performed.

As described in Sect. 17.3.4, three CPT tests were performed per each model; in this section, in order to analyze the initial density, results obtained just in the first CPT of each model at depths between 1.5 and 2.5 m are discussed. An empirical correlation (provided by UC Davis) between dry density and $q_c$ is employed for comparative purposes; this correlation has the form: $\rho_{dry} = a \times q_c + b$, where $\rho_{dry}$ is the dry density obtained by the correlation, $q_c$ is the tip resistance measured at CPT test, and "a" and "b" are depth dependent constants determined by linear regression between qc and density at a particular depth of measurement of $q_c$. The $a$ and $b$ values were provided at three depths (1.5, 2, and 2.5 m) in a spreadsheet described by Kutter et al. (2019). A quadratic curve fit was used to interpolate $a$ and $b$ values at intermediate depths; using this interpolated values, it was then possible to obtain a

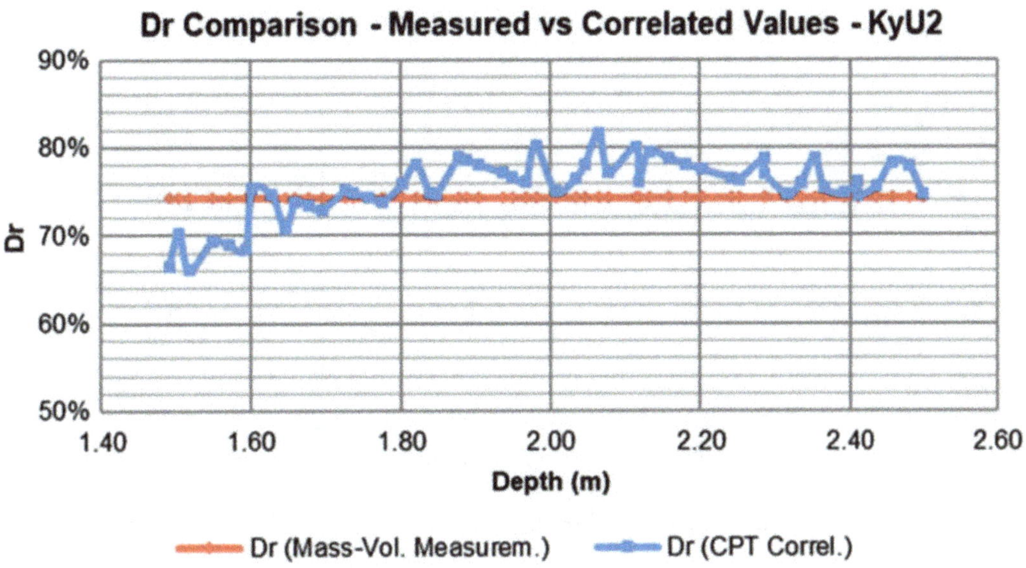


**Fig. 17.21** Comparison between the achieved Dr values obtained by "correlations from CPT" and "mass–volume measurements" for KyU2 Model

continuous estimate of the "correlated" density. Figures 17.21 and 17.22 show a comparison between the achieved Dr values obtained by "correlations from CPT" and "mass–volume measurements" for the KyU2 and KyU3 Models, respectively.

Figure 17.21 shows that on average, there is no significant difference in the Dr values obtained by the two different methods; however, Fig. 17.22 shows a difference of 4% on average. A more formal definition of the weighted average deviation between the correlated relative density and the measured relative density was determined using the $\Delta_{D_r}$ value, which was estimated as follows.

$$\Delta_{\mathrm{Dr}} = \frac{\sum_{i=1}^{n}\left[(d_i - d_{i-1})\cdot\left(\frac{|\mathrm{Dr}_m - \mathrm{Dr}_{c_i}| + |\mathrm{Dr}_m - \mathrm{Dr}_{c_{i-1}}|}{2}\right)\right]}{d_n - d_0}$$

where:

- $\Delta_{\mathrm{Dr}}$: Weighted average of difference between "measured Dr" and "correlated Dr"
- $d_i$: Depth of CPT on the $i$th measurement
- $\mathrm{Dr}_m$: Measured relative density
- $\mathrm{Dr}_{c_i}$: Estimated relative density on the $i$th measurement (based on $q_c$ values)
- $n$: Number of CPT measurements in each test

The weighted average $\Delta_{\mathrm{Dr}}$ values were tabulated for all of the LEAP-UCD-2017 centrifuge tests for which CPT data was available. The distribution of the average deviation is summarized in Fig. 17.23. It can be seen that in most cases the $\Delta_{\mathrm{Dr}}$ was less than 5%, but in 26% of the cases the deviation was more than 5%.

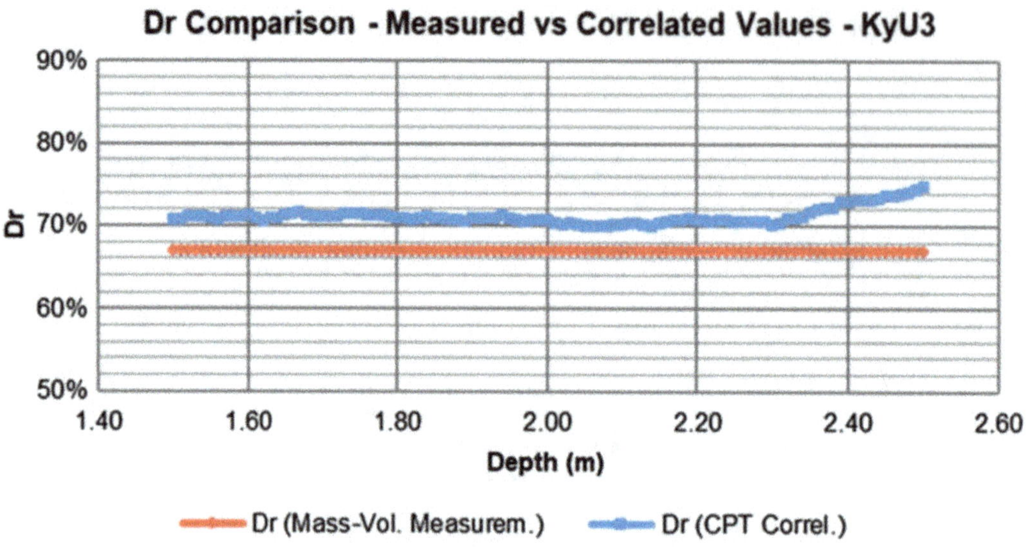


**Fig. 17.22** Comparison between the achieved Dr values obtained by "correlations from CPT", and "mass–volume measurements" for KyU3 Model

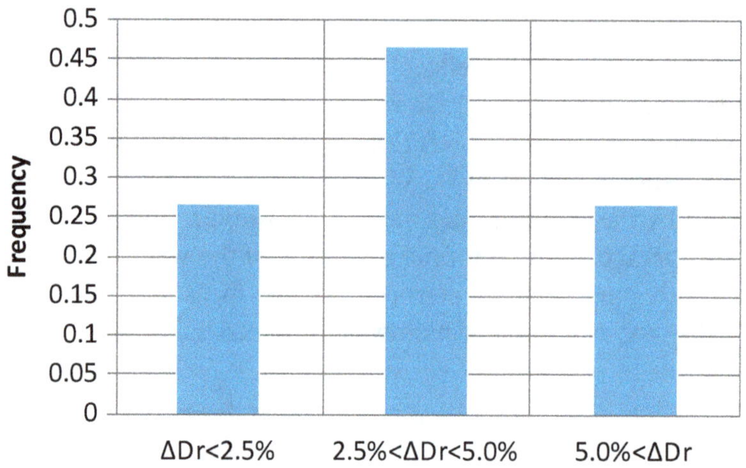


**Fig. 17.23** Distribution of $\Delta_{Dr}$ value among facilities

## 17.5 Conclusions

LEAP (Liquefaction Experiments and Analysis Projects) is a joint exercise that pursues the verification, validation, and uncertainty quantification of numerical liquefaction models. "LEAP-UCD-2017" is one of the LEAP's exercises, whose main objective is to perform a sufficient number of experiments to characterize the median response, and the uncertainty of response of a specific sloping deposit of sand.

A review of the KyU2 and KyU3 Models performed in the facilities of the Disaster Prevention Research Institute at Kyoto University for "LEAP-UCD-2017" exercise was presented.

- A brief review of the specifications for the realizations was explained, along with the characteristics and differences performed to obtain the "as-built" model.
- For both models, results of excess of pore water pressure in Motion 2 (PGA 0.15) showed values of $r_u \approx 1$ in shallow sensors; however, for Motions 3 and 5 (PGA 0.25), values of $r_u \approx 1$ were obtained almost in all of the sensors.
- For Model KyU2, despite the increase in the values of $q_c$ (obtained through the CPT test before Motion 3, and before Motion 5), no increase in liquefaction resistance was observed; these values are in accordance with the values obtained by Yamada et al. (2010), who found that in the case of re-liquefaction process, anisotropy of soils becomes a key factor, rather than the relative density.
- Relating to ground motion accelerations, as expected the recorded acceleration remained similar as the input motion before the development of significant EPWP; thereafter, significant distortion was recorded, presenting sharp spikes which are characteristic from dilatant behavior of liquefied soil.

Due to its importance in verification & validation tests, a discussion about density measurement was presented in Sect. 17.4.

- Volumetric measurement of density becomes very sensitive (even for small errors); therefore, special attention should be focused on this step in future experiments.
- Specified container for "air pluviation" may be checked for further experiments; size of the apertures might be reviewed to reduce the volume flow of sand, in order decrease the dependence on the operator's ability to keep horizontal and uniform surface during air pluviation.
- Comparison of density values obtained by correlations with CPT tests and by mass and volume measurements shows a discrepancy among research facilities. To reduce the discrepancy in future LEAP exercises, it is recommended that increased attention be devoted to improving the accuracy of density measurements by both methods.

## References

Izumo, N. (2006). Physical quantity measured by a vibration viscometer. In *Proceedings of 23th Sensing Forum, Tsukuba*. pp. 149–152.

Kutter, B. L., Carey, T. J., Stone, N., Bonab, M. H., Manzari, M., Zeghal, M., Escoffier, S., Haigh, S., Madabhushi, G., Hung, W.-Y., Kim, D.-S., Kim, N.-R., Okamura, M., Tobita, T., Ueda, K., & Zhou, Y.-G. (2019). LEAP-UCD-2017 V. 1.01 model specifications. In B. Kutter et al. (Eds.), *Model tests and numerical simulations of liquefaction and lateral spreading: LEAP-UCD-2017*. New York: Springer.

Okamura, M., & Inoue, T. (2012). Preparation of fully saturated models for liquefaction study. *International Journal of Physical Modeling in Geotechnics, 12*(1), 39–46. https://doi.org/10.1680/ijpmg.2012.12.1.39.

Tobita, T., Ashino, T., Ren, J., & Iai, S. (2018). Kyoto University LEAP-GWU-2015 tests and the importance of curving the ground surface in centrifuge modelling. *International Journal of Soil Dynamics and Earthquake Engineering, 113*, 650–662. https://doi.org/10.1016/j.soildyn.2017.10.012.

Yamada, S., Takamori, T., & Sato, K. (2010). Effects on reliquefaction resistance produced bu changes in anisotropy during liquefaction. *Soils and Foundations, 50*(1), 9–25.

# Chapter 18
# LEAP-UCD-2017 Centrifuge Tests at NCU

**Wen-Yi Hung and Ting-Wei Liao**

**Abstract** Liquefaction Experiments and Analysis Projects (LEAP) aim to use simple centrifuge modeling tests to validate and calibrate the numerical modeling results. In LEAP-UCD-2017 project, the design and specification of the model are quite uncomplicated than that of the earlier stage of LEAP. The model stored in a rigid container was construted by medium dense sand with 5 degrees of slope and subjected to a 1-Hz sinusoidal wave of base motion. Models were built, tested and cross-validated by many different research institutes. This paper describes in detail the result of experiments carried at National Central University (NCU), Taiwan (R.O.C.) This paper describes in details the unique part of the experiments at NCU.

## 18.1 Introduction

Since the 1960s, soil liquefaction is an important issue all over the world. In order to gain further insight into this issue, physical modeling and numerical modeling are widely used to simulate and investigate the behaviors of soil liquefaction. In fact, numerical simulations cannot prove its preciseness and applicability without good parameter correction and verification from the results of physical modeling. Therefore, many projects have been proposed to evaluate the effectively between numerical simulations and physical modeling tests, and early period works such as VELACS project (Arulanandan and Scott 1993–1994) tried to verify the accuracy of various analytical procedures, but the results were not conclusive at the end. In recent years, the experimental techniques used in centrifuge modeling and the instrumentation employed have improved significantly. Hence, a new international cooperation project called Liquefaction Experiments and Analysis Projects (LEAP) is carried out (Manzari et al. 2015). Several centrifuge facilities from the USA, UK, China, Japan, Korea, and Taiwan are participating in the LEAP.

---

W.-Y. Hung (✉) · T.-W. Liao
Department of Civil Engineering, National Central University, Taoyuan, Taiwan
e-mail: wyhung@ncu.edu.tw



B. Kutter et al. (eds.), *Model Tests and Numerical Simulations of Liquefaction and Lateral Spreading*, https://doi.org/10.1007/978-3-030-22818-7_18

For the LEAP-UCD-2017, NCU conducted three centrifuge modeling tests by the geotechnical centrifuge at the NCU centrifuge laboratory. The one-dimensional shaking table on the centrifuge at NCU is capable of offering different kinds of input motions used in the LEAP exercise.

## 18.2 Test Equipment and Materials

### *18.2.1 Shaking Table on Geotechnical Centrifuge and Rigid Container*

The geotechnical centrifuge in NCU has nominal radius of 3 m and a one-dimensional servo-hydraulic-controlled shaker which is assembled on its swing basket, as shown in Fig. 18.1a. The shaker has a capacity to hold a maximum mass of 400 kg at a maximum acceleration field of 80 g corresponding to maximum nominal force of 53.4 kN. The maximum table displacement is ±6.4 mm with a maximum input frequency of 250 Hz. The payload mounting area of the shaker is 1000 mm (length) × 550 mm (wide).

The container used for LEAP is a rigid box composed by aluminum alloy plates as shown in Fig. 18.1b. The inner dimensions are 767 mm (length) × 355 mm (width) × 400 mm (height), and the net weight of empty box is 106.8 kg.

### *18.2.2 Sand Material*

Ottawa sand (denoted as F-65) is provided by UC Davis and used to prepare and construct the gentle slope models in the LEAP. This kind of sand is a uniform clean sand and can be classified as SP (poorly graded sand) in the Unified Soil Classification System (USCS). The distribution curve of particle size of the Ottawa sand is shown in Fig. 18.2. It can be recognized that finer content is insignificant as the passing amount through No. 200 sieve is approximately 0.2% of total sand weight,

(a) (b)

**Fig. 18.1** (**a**) Geotechnical centrifuge at NCU (**b**) Rigid container

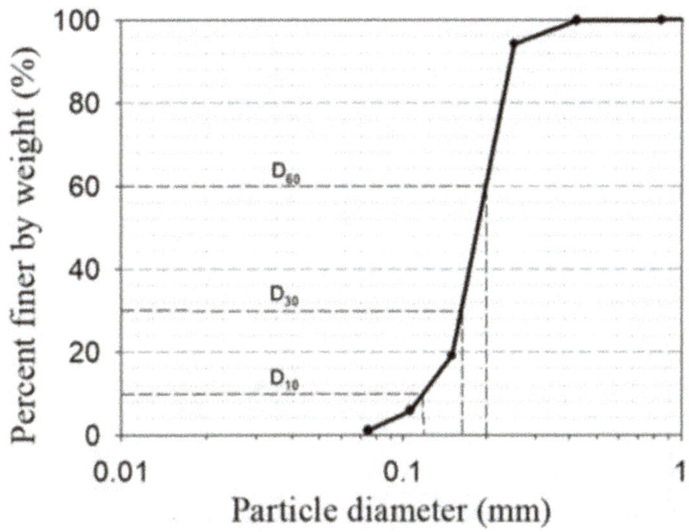


**Fig. 18.2** Particle size distribution of Ottawa sand (F-65)

the residual weight passing through No. 100 sieve is around 92.0%, and passing amount through No. 140 sieve is about 6.0% of total weight. Based on those observations, it demonstrated that the sand particle is very small and uniform. Some other physical parameters include specific gravity ($G_s$) of 2.66 and the mean diameter ($D_{50}$) of 0.203 mm.

The maximum and minimum dry unit weights under modified ASTM (American Society for Testing and Materials) method are 16.80 and 14.1 kN/m$^3$ measured by NCU, respectively, but they are 17.24 and 14.62 kN/m$^3$ measured by UCD (Carey et al. 2019). The target unit weight of slope model in LEAP is 16.20 kN/m$^3$ which corresponding with 80% of relative density according to NCU max-min dry unit weight result and 64% following UCD's results. Although there is a significant different in relative density, it is no doubt that the centrifuge models in different facilities are similar due to the use of identical material and unit weight.

## 18.3 Tests Arrangement

### *18.3.1 Arrangement of the Centrifuge Model*

As Fig. 18.3a, b shows that the longitudinal and cross sections of the specific arrangement follow the patterns of the LEAP-2017 project (Kutter et al. 2017), the model is constructed as a 5 degree-inclined saturated sandy slope with 4 m deep in prototype scale (153.8 mm in model type) at the middle of slope and 20 m long in prototype scale (767.0 mm in model type).The target unit weight is 16.20 kN/m$^3$, which is corresponding to a relative density of 65% as a medium dense sand. The curvature of ground surface relates to effective radius as 2.714 m, which is the distance from ground surface to rotation center of centrifuge, as shown in Fig. 18.3b.

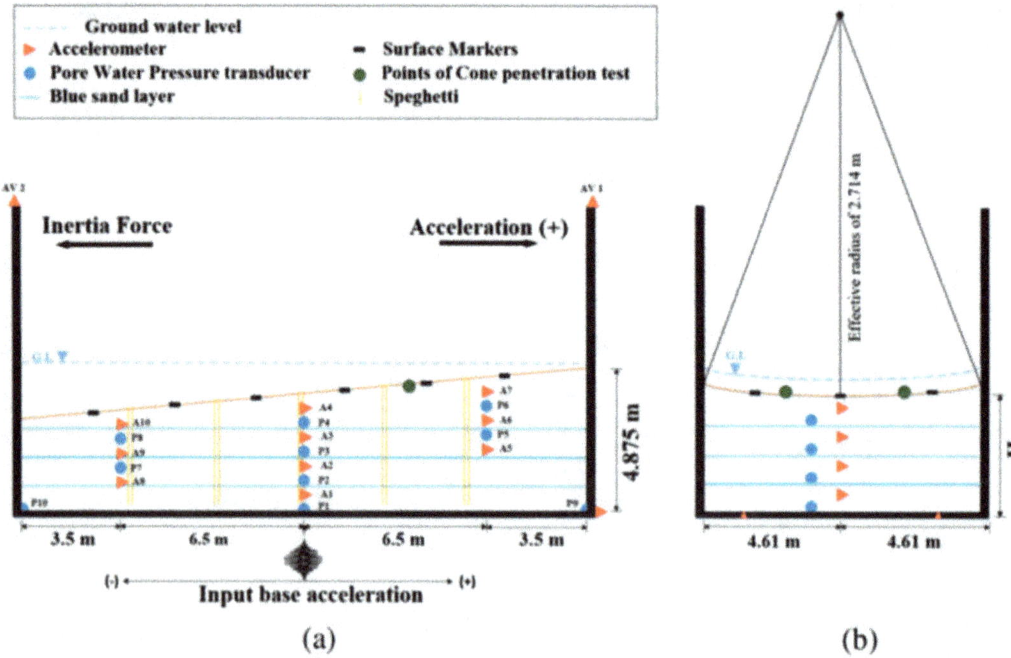


**Fig. 18.3** The arrangement of the centrifuge model (**a**) Longitudinal section (**b**) Cross section

There are totally ten accelerometers and eight pore water pressure transducers embedded into the strata divided into three arrays along depth from upslope to downslope, and the interval between each two sensors is constant. Besides, one accelerometer was fixed on the shaker, and two accelerometers are vertically assembled on the top of the rigid container to measure the rocking of container during shaking. Two of pore water pressure transducers were installed at the bottom corner of container. There are a total of 18 polyvinyl chloride (PVC) circular surface makers with outer diameter of 26 mm and inner diameter of 23 mm, and they are put as three arrays at the certain location on the ground surface as indicated in Fig. 18.3a. Three thin blue sand layers are placed at the same elevations of pore pressure transducers P4, P3, and P2 with depth at middle slope of 38 mm, 77 mm, and 115 mm, respectively. Several spaghetti noodles were inserted vertically into the soil deposit.

In this study, the positive acceleration defines that the shaking table moves toward the right side (positive *x*-axis direction), and the slope dip direction is toward the left side (negative *x*-axis direction). Thus, the inertial force caused by positive acceleration is toward negative *x*-axis direction.

### *18.3.2 Input Base Motions*

Three centrifuge modeling tests labeled as NCU 1, NCU 2, and NCU 3 were conducted with different input main-shaking events. Every test has two main-shaking events to investigate the effects of input motion waveforms on inclined saturated slope. Tables 18.1, 18.2, and 18.3 show the fundamental characteristics of

**Table 18.1** Fundamental characteristics of input motions for model NCU 1

| | Shape of wave | Frequency (Hz) | Number of cycles | Isolated 1 Hz signal | Isolated noise | Achieved base motion | PGA effective |
|---|---|---|---|---|---|---|---|
| Pre-shaking | SW | 3 | 1 | – | – | 0.09 | – |
| Motion 1 | TSW | 1 | 16 | 0.18 | 0.09 | 0.27 | 0.22 |
| Post-shaking | SW | 3 | 1 | – | – | 0.09 | – |
| Post-shaking | NTSW | 3 | 1 | – | – | 0.09 | – |
| Stop testing temporarily | | | | | | | |
| Pre-shaking | SW | 3 | 1 | – | – | 0.09 | – |
| Motion 2 | SW | 1 | 16 | 0.18 | 0.10 | 0.27 | 0.23 |
| Post-shaking | SW | 3 | 1 | – | – | 0.09 | – |
| Post-shaking | NTSW | 3 | 1 | – | – | 0.09 | – |

**Table 18.2** Fundamental characteristics of input motions for model NCU 2

| | Shape of wave | Frequency (Hz) | Number of cycles | Isolated 1 Hz signal | Isolated noise | Achieved base motion | PGA effective |
|---|---|---|---|---|---|---|---|
| Pre-shaking | SW | 3 | 1 | – | – | 0.09 | – |
| Motion 1 | TSW | 1 | 16 | 0.15 | 0.07 | 0.22 | 0.19 |
| Post-shaking | SW | 3 | 1 | – | – | 0.09 | – |
| Post-shaking | NTSW | 3 | 1 | – | – | 0.09 | – |
| Stop testing temporarily | | | | | | | |
| Pre-shaking | SW | 3 | 1 | – | – | 0.09 | – |
| Motion 2 | NTSW | 1 | 16 | 0.15 | 0.09 | 0.25 | 0.20 |
| Post-shaking | SW | 3 | 1 | – | – | 0.09 | – |
| Post-shaking | NTSW | 3 | 1 | – | – | 0.09 | – |

each motion for three tests. Before every main-shaking, a small 1-cycle sine wave of pre-shaking is applied to detect the shear wave velocity along depth and to identify the initial condition before main-shaking.

There are two main shaking events (Motion-1 and Motion-2) act on each model, all of them have the same frequency of 1 Hz and 16 number of cycles but different in

**Table 18.3** Fundamental characteristics of input motions for model NCU 3

| | Shape of wave | Frequency (Hz) | Number of cycles | Isolated 1 Hz signal | Isolated noise | Achieved base motion | PGA effective |
|---|---|---|---|---|---|---|---|
| Pre-shaking | SW | 3 | 1 | – | – | 0.10 | – |
| Motion 1 | TSW | 1 | 16 | 0.12 | 0.06 | 0.18 | 0.15 |
| Post-shaking | SW | 3 | 1 | – | – | 0.10 | – |
| Post-shaking | NTSW | 3 | 1 | – | – | 0.10 | – |
| Stop testing temporarily | | | | | | | |
| Pre-shaking | SW | 3 | 1 | – | – | 0.09 | – |
| Motion 2 | SW | 1 | 16 | 0.12 | 0.07 | 0.19 | 0.15 |
| Post-shaking | SW | 3 | 1 | – | – | 0.10 | – |
| Post-shaking | NTSW | 3 | 1 | – | – | 0.10 | – |

*SW* sine wave, *TSW* tapered sine wave, *NTSW* negative tapered sine wave

amplitude and waveform. Motion 1 is a tapered sine wave that applied in three models and returned three different peak base accelerations (PBA) which are 0.27 g, 0.22 g and 0.18 g, respectively. In NCU-1 and NCU-3 model, Motion 2 is a common sine wave and also yield two values of PBA which are 0.27 g and 0.19 g. Lastly, Motion 2 in NCU-2 is a negavtive tapered sine wave and provide 0.25 g PBA.

After every main-shaking, two small 1-cycle sine waves of post-shakings are applied to detect the shear wave velocity change along depth and to identify the soil properties again after main-shaking. For the two post-shaking, one is sine wave (SW) and another is negative sine wave (NTSW).

## 18.4 Test Procedure

### *18.4.1 Model Pluviation*

The models were prepared by air pluviation using No. 16 sieve as shown in Fig. 18.4. The flow rate and drop height must be determined to achieve the specific dry unit weight. Therefore, the flow rate of sand mass is controlled at approximately 2.5 kg/minute for passing the opening size of 1.18 mm with three slender slots. The travel path during pluviation is shown in Fig. 18.5. A constant drop height of 0.5 m was used due to the relationship between drop height and dry unit weight under a specific flow rate as shown in Fig. 18.6. The elevation of the sieve above the container during pluviation should be frequently adjusted to maintain a constant

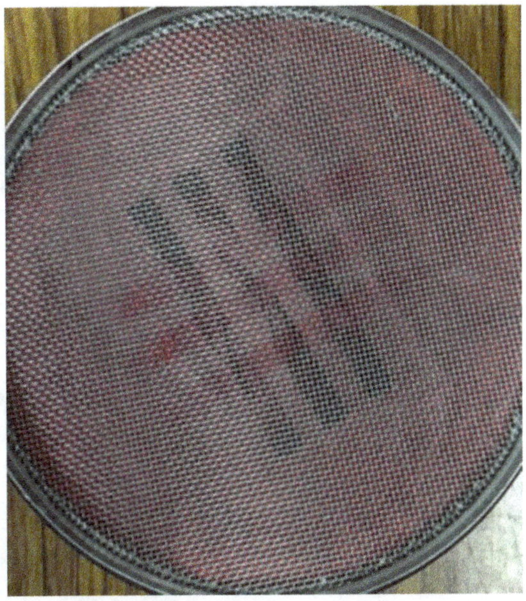

**Fig. 18.4** The pluviation equipment

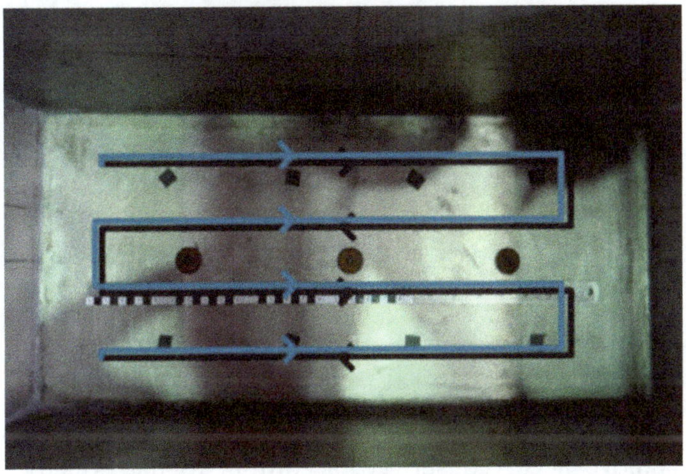

**Fig. 18.5** The travel path during deposition

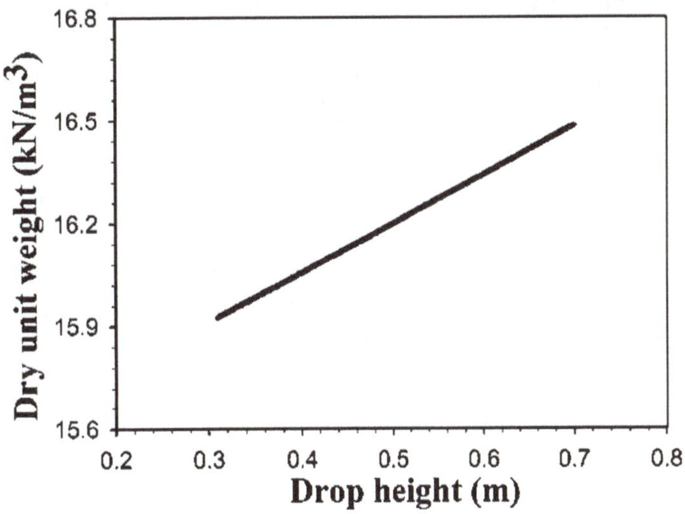


**Fig. 18.6** The relationship between dry unit weight and drop height

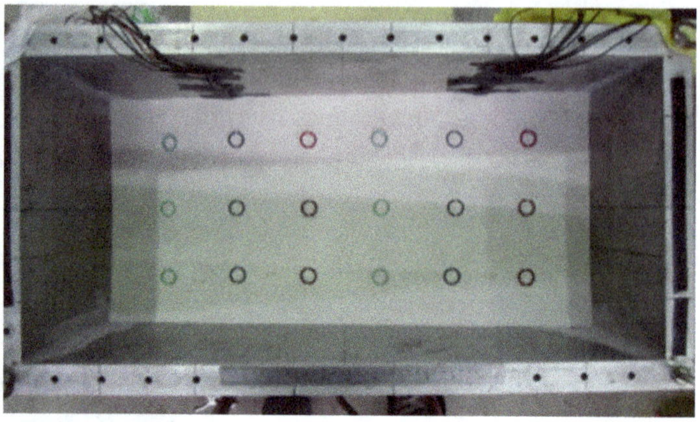

**Fig. 18.7** The top view of the dry sample

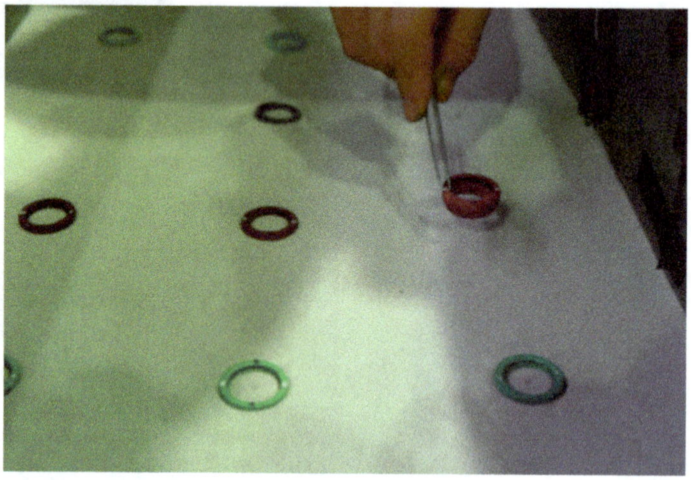

**Fig. 18.8** Embedment of surface markers

drop height from the sieve to the sand surface. The ground surface was required to achieve a curved surface, according to the effective radius of the geotechnical centrifuge of 2.714 m.

When pluviated to specific depths, the accelerometers and pore water pressure transducers were embedded at the specific location to measure the time histories of acceleration and pore water pressure during test, and also colored sand was pluviated as a thin color sand layer to observe the circumstances of settlement and lateral spreading. After completing the slope, the surface makers were put on the specific location to record dynamic displacement by cameras during test, and the spaghetti noodles were inserted into strata to observe the circumstances of settlement and lateral spreading. Figure 18.7 shows the top view of the dry model after pluviation was completed. Figures 18.8 and 18.9 show the method used to insert the surface markers and spaghettis, respectively.

Fig. 18.9 Insertion of noodles

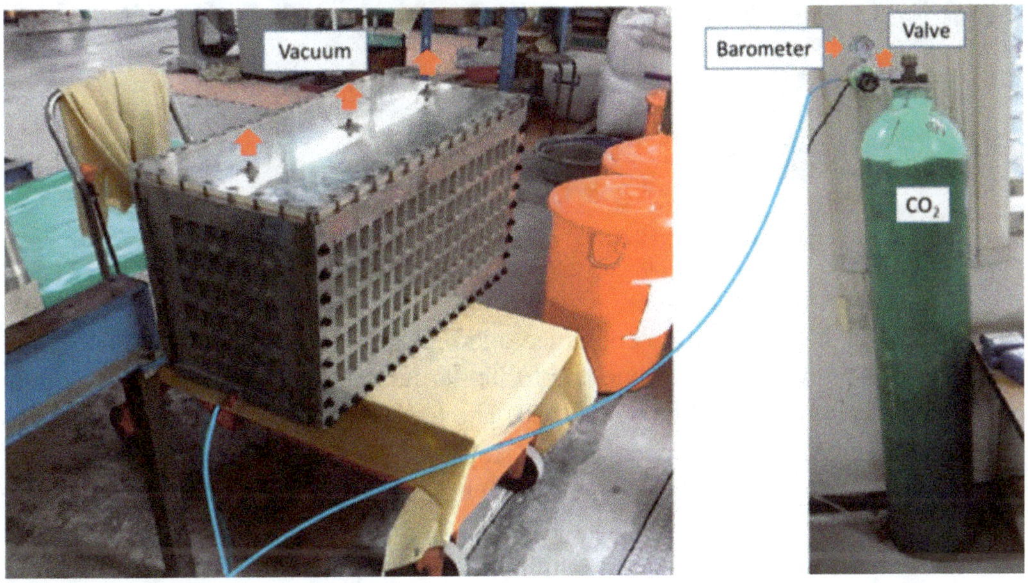


Fig. 18.10 $CO_2$ system

### 18.4.2 Model Saturation

Before fluid saturation, a cap was fixed on the top of aluminum container, and the air is vacuumed out from the channels of cap. Then the pure $CO_2$ floods the model from the bottom of container to fill the voids as shown in Fig. 18.10. The duration is 1.5 h with pressure of 0.25 kg/cm$^2$ monitored from the barometer. There is not enough space to assemble the frames for laser displacement transducer (LDT) during the following saturation process, and then the aluminum container was placed into a large container and the cap was removed. Frames and LDT are installed on the top of aluminum container. A thin polystyrene plate is placed on the slope surface to measure the water level height by LDT for validating the degree of saturation (Okamura and Inoue 2012). Another bigger acrylic cap and rubber o-ring were

**Fig. 18.11** Saturation system

used to seal the large saturation container, and Fig. 18.11 shows the whole saturation system. A vacuum pump is connected to the large container until a stable vacuum pressure of 68 cmHg (91 kPa) is obtained. The methylcellulose liquid with 26 times the viscosity of water is dropped on the slope model to saturate with flow rate of about 2.0 kg/h. A total liquid weight is 27.0 kg after completely soaking the whole model.

Based on Okamura's method, the higher degree of saturation of soil leads to the smaller change in water level at different vacuum pressure according to Eq. 18.1:

$$\Delta h = (1 - S_r)\frac{e}{e+1}\frac{V}{A}\left(\frac{p_1}{p_2} - 1\right) \tag{18.1}$$

where $p_1$ and $p_2$ are the chamber pressure (from $p_1$ to $p_2$), $\Delta h$ is the change of water level, $S_r$ is the degree of saturation, $e$ is the void ratio, $V$ is the volume of model, and $A$ is the horizontal section area of model. In Okamura and Inoue's study, the vacuum pressure varies from 15 kPa to 0 kPa(atmospheric pressure). In this study, the degree of saturation is 99.83% if the chamber vacuum pressure changes from 27 kPa to 0 kPa. But it reduces to about 99.4% when the chamber releases higher vacuum pressure. Since the saturation increases with pressure, the degree of saturation of the centrifuge model under positive pore pressure is expected to be greater than 99.8%.

### 18.4.3 Centrifuge Modeling Procedure

The completed model is then put on the platform of NCU geotechnical centrifuge. Figures 18.12 and 18.13 show the arrangement of model including cone penetration test (CPT) and camera systems. There are five cameras assembled at the top acrylic cap of container, and two sets of CPT cones with two cylinders were installed on the platform. The centrifuge modeling test is shown in Fig. 18.14. Firstly, the model is spun up to 26 g to simulate the same prototype as specified by Kutter et al. (2019—specifications paper, this proceedings). During testing, a non-destructive detect technique, defined as pre-shaking, is triggered before and after main-shaking to measure the shear wave velocity and to estimate the natural frequency of soil strata. It is a 3 Hz, 1 cycle, around 0.066 g (prototype) peak base acceleration sine pulse. Also, before and after the main-shaking, the cone penetration test (CPT) is done by two sets of CPT systems. The resistance is measured by load cell, and displacement

**Fig. 18.12** The arrangement of centrifuge modeling test

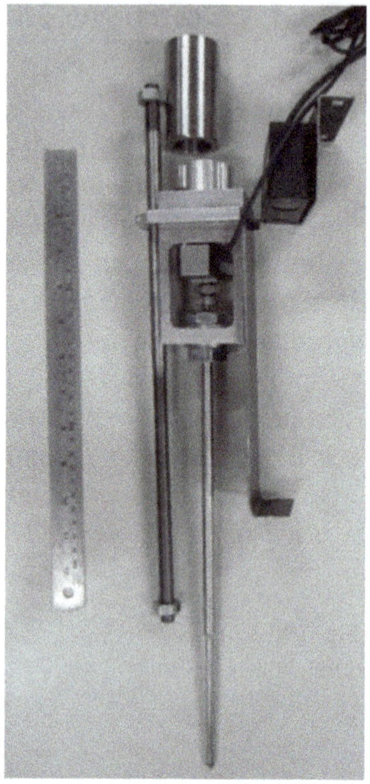

**Fig. 18.13** Equipment of cone penetration test

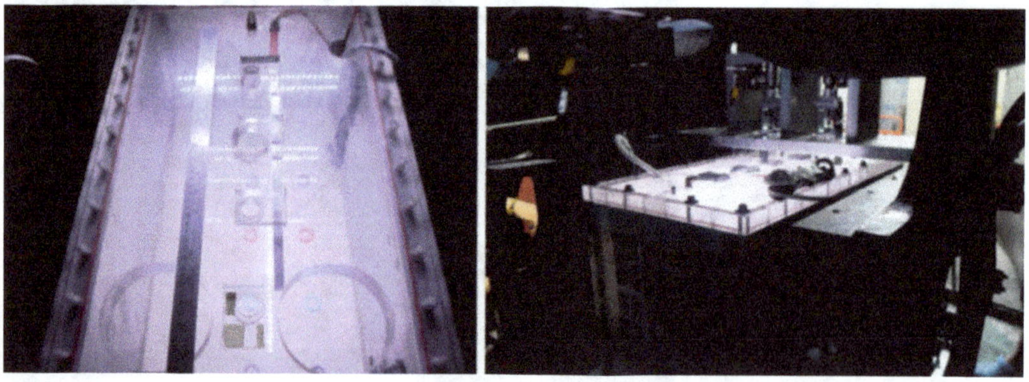

**Fig. 18.14** The arrangement of cameras installed at the top of the container

is measured by laser displacement transducer. After finishing the first main-shaking test, the centrifuge is stopped to measure the elevation and the $x$ and $y$ positions on the ground surface, and one set of main-shaking test is completed. The second set of main-shaking centrifuge modeling is then continued with for the same model. Finally, the tested model is cut to detect the deformation of blue thin sand layers and spaghettis. Figure 18.15 shows the process of LEAP tests in NCU.

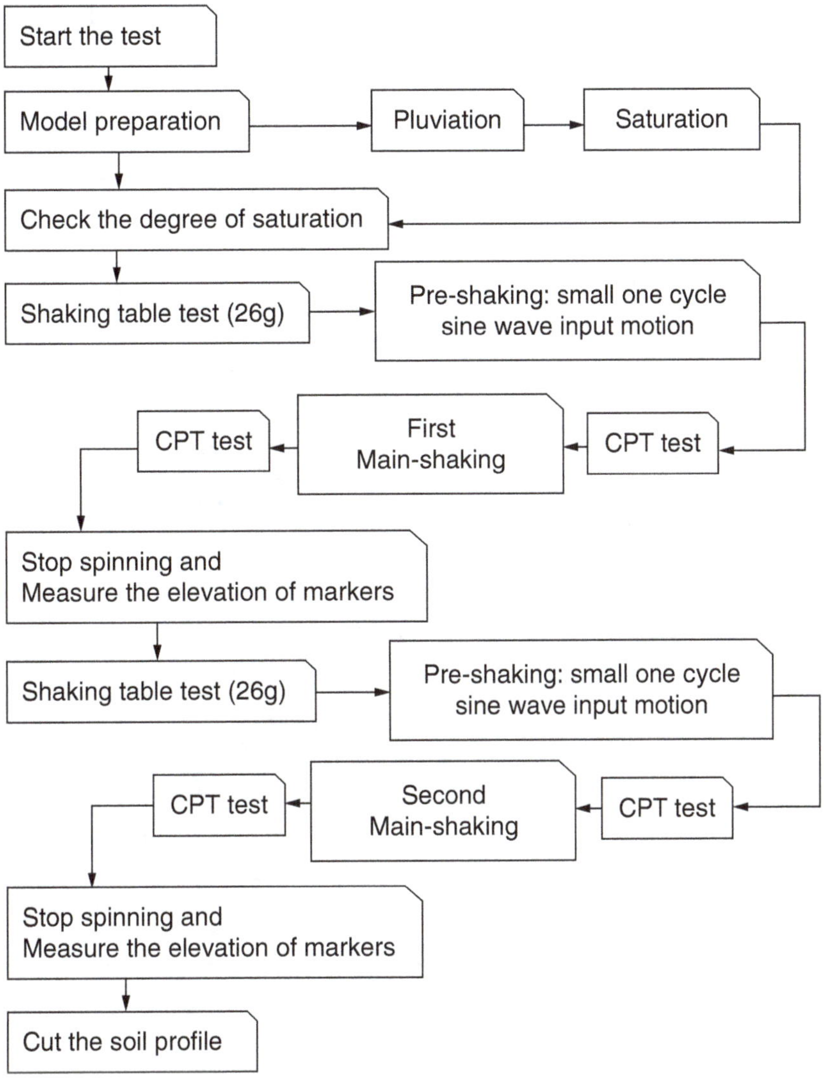


**Fig. 18.15** The process of LEAP tests in National Central University

## 18.5 Results

### *18.5.1 Acceleration Response and Pore Pressure Response*

Figure 18.16 illustrates the acceleration time histories of models NCU 1-M1 (PBA 0.27 g), NCU 2-M1 (PBA 0.22 g), and NCU 3-M1 (PBA 0.18 g) by red, blue, and black lines, respectively. Figure 18.17 shows the acceleration time histories of models NCU 1-M1 (PBA 0.27 g), NCU 2-M1 (PBA 0.22 g), and NCU 3-M1 (PBA 0.18 g) at middle slope array. On the basis of these two figures, all of them have the same phenomenon, and the shape of wave becomes incomplete as the depth under the ground surface decreases. The average of peak amplitude in each cycle of acceleration during the shaking section would attenuate due to the soil strata

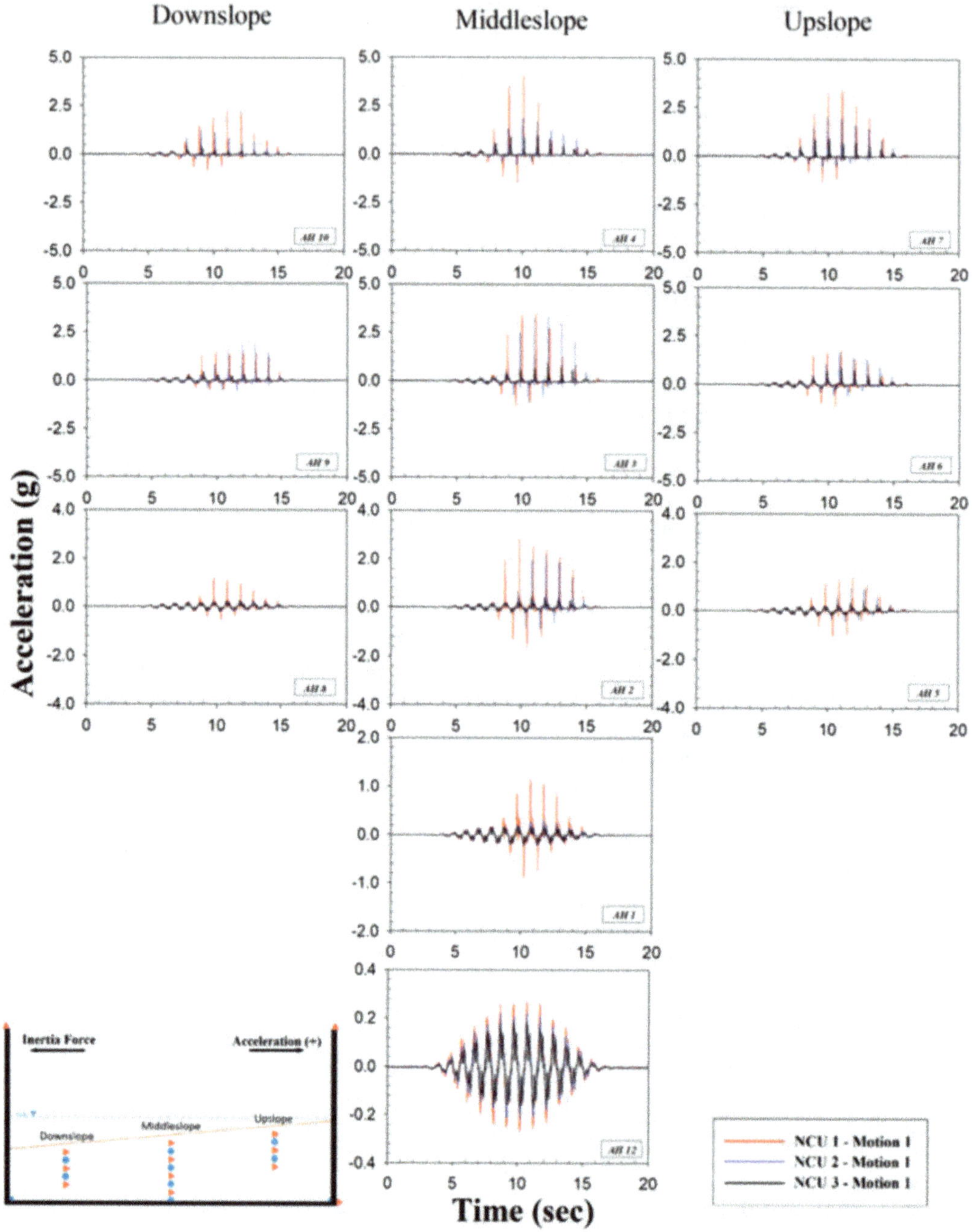

Fig. 18.16 Time histories of acceleration for models NCU 1-M1, NCU 2-M1, and NCU 3-M1

softening. In NCU 1-M1 at the depth of 3.5 m, the strata softened until to the shallowest of the accelerometers. The similarity can be observed at model NCU 2-M1 and NCU 3-M1 at the depths of 3.5 m and 2.5 m, respectively.

When soil strata cannot propagate the shear wave during liquefaction, it is explained due to the effective stress approaching zero; simultaneously, the acceleration amplitude would attenuate and even approach to zero without the spikes. In NCU 1-M1 at the depth of 3.5 m, the shape of wave is observed sufficiently, and the

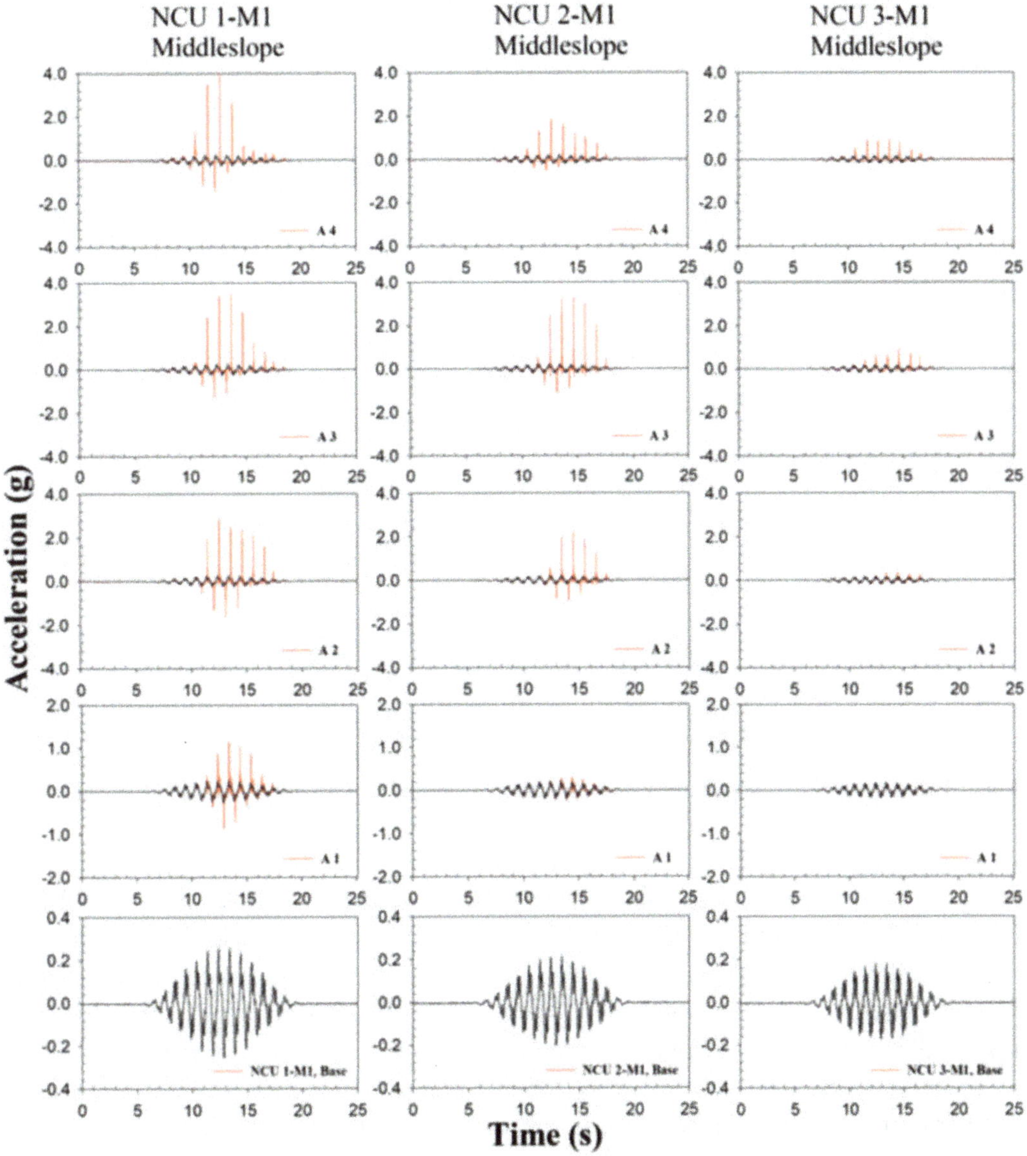


**Fig. 18.17** Time histories of acceleration for models NCU 1-M1, NCU 2-M1, and NCU 3-M1 at middle slope

attenuation of acceleration is observed at the 11th cycle, 10th cycle, and 9th cycle on depths of 2.5 m, 1.5 m, and 0.5 m, respectively, which means the soil liquefied at that time. In NCU 2-M1 at the depth of 3.5 m, a complete shape of wave is observed, and the attenuation of acceleration is catched at the 11th cycle, 9th cycle, and 7th cycle at depths of 2.5 m, 1.5 m, and 0.5 m respectively. In NCU 3-M1 at the depth of 1.5 m, the shape of wave is still complete, and the attenuation of acceleration is observed at the ninth cycle at the depth of 0.5 m. In summary, different amplitude of PBA would affect the required duration to reach liquefaction as also the depth of liquefaction.

Figure 18.18 illustrates the pore water pressure ratio time histories of models NCU 1-M1 (PBA 0.27 g), NCU 2-M1 (PBA 0.22 g) and NCU 3-M1 (PBA 0.18 g) by red, blue and black lines, respectively. The excess pore water pressure ratio

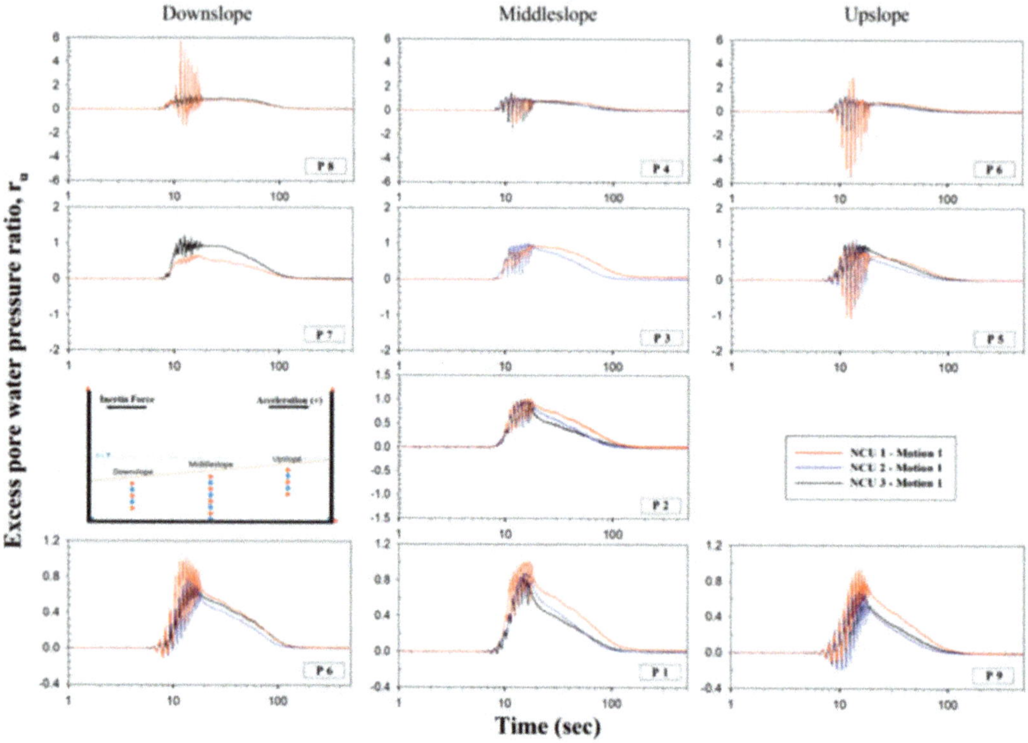


**Fig. 18.18** Time histories of pore water pressure for models NCU 1-M1, NCU 2-M1, and NCU 3-M1

achieved 1 at 1 meter depth in each test means that the soil strata is completely liquefied. Liquefaction happened at 2 m depth only in case NCU 1-M1. The pore water pressure in case NCU 1-M1 took the longest time to complete its dissipation than in cases NCU 2-M1 and NCU 3-M1. The waveforms of P9, P1, and P10 installed at the bottom of container are very similar to shaking histories.

In this paper, the coordinate system is indicated in Fig. 18.3. The positive acceleration means the shaking table moves toward upslope (positive $x$-axis direction), and the slope dip direction is toward downslope (negative $x$-axis direction). It is a different sign convention than other facilities. For the following figures, the response of upslope is also put at the right side and that of downslope is at the left side. As shown in Fig. 18.19, there are prominent amplitudes of spikes appearing in the phase where the positive accelerations exist at any depths of the upper slope array and the center slope array, especially in the cases NCU 1-M1 and NCU 2-M1, which is applied for large base acceleration. Because the direction of slope dip is toward the direction of negative base acceleration, it is opposite to the direction of the first half cycle of base acceleration. To demonstrate prominent amplitudes of spikes which appear in the phase of the positive acceleration, the shear strain induced by inertial force from the first half of time history of base acceleration coincides with the direction of slope dip, soil strata would dilate by the shear strain that conducted excess pore water pressure exciting by shaking decrees immediately, and effective stress of soil also increases at once too, so that it can propagate shear wave simultaneously. By contrast, the prominent amplitudes of spikes occur in the

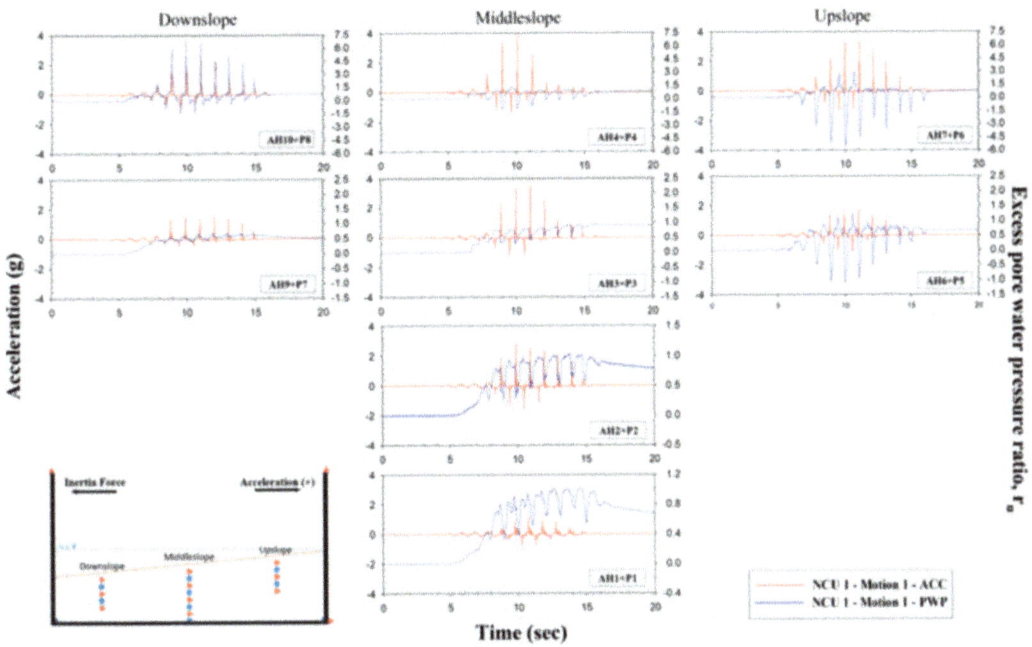


Fig. 18.19 The acceleration and excess pore water pressure histories of model NCU-1

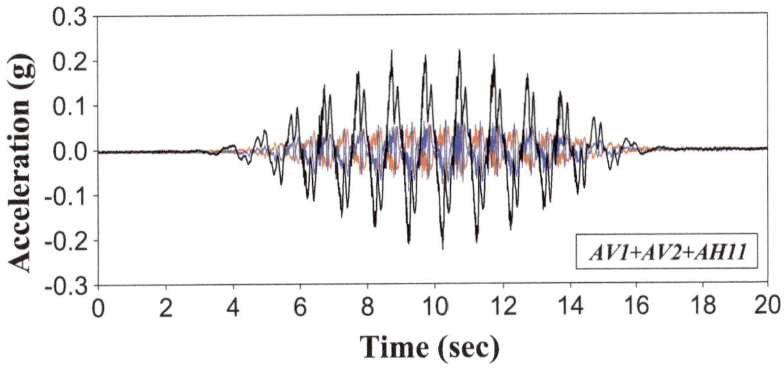


Fig. 18.20 Time histories of acceleration for AV1 and AV2 and AH11 in NCU 2-M1

phase of the positive acceleration histories at the lower slope; however, these spikes accompany with peak value of pore water pressure at the same time. The excess pore water pressure is higher several times than effective stress, especially the pore water pressure transducers are located at the lower part of the slope. This issue is discussed later in this chapter.

Figure 18.20 demonstrates the acceleration time histories of AV1, AV2, and AH11, by red, blue, and black lines in case NCU 2-M1, correspondingly. AV1 and AV2 have the same value of peak ground motion of about 0.05 g. The waveform of AH11 and AV2 is presented at the same phase of wave; however, the phase of waveform of AV1 is opposite to AH11, suggesting that the vertical accelerations are caused by rocking of the container.

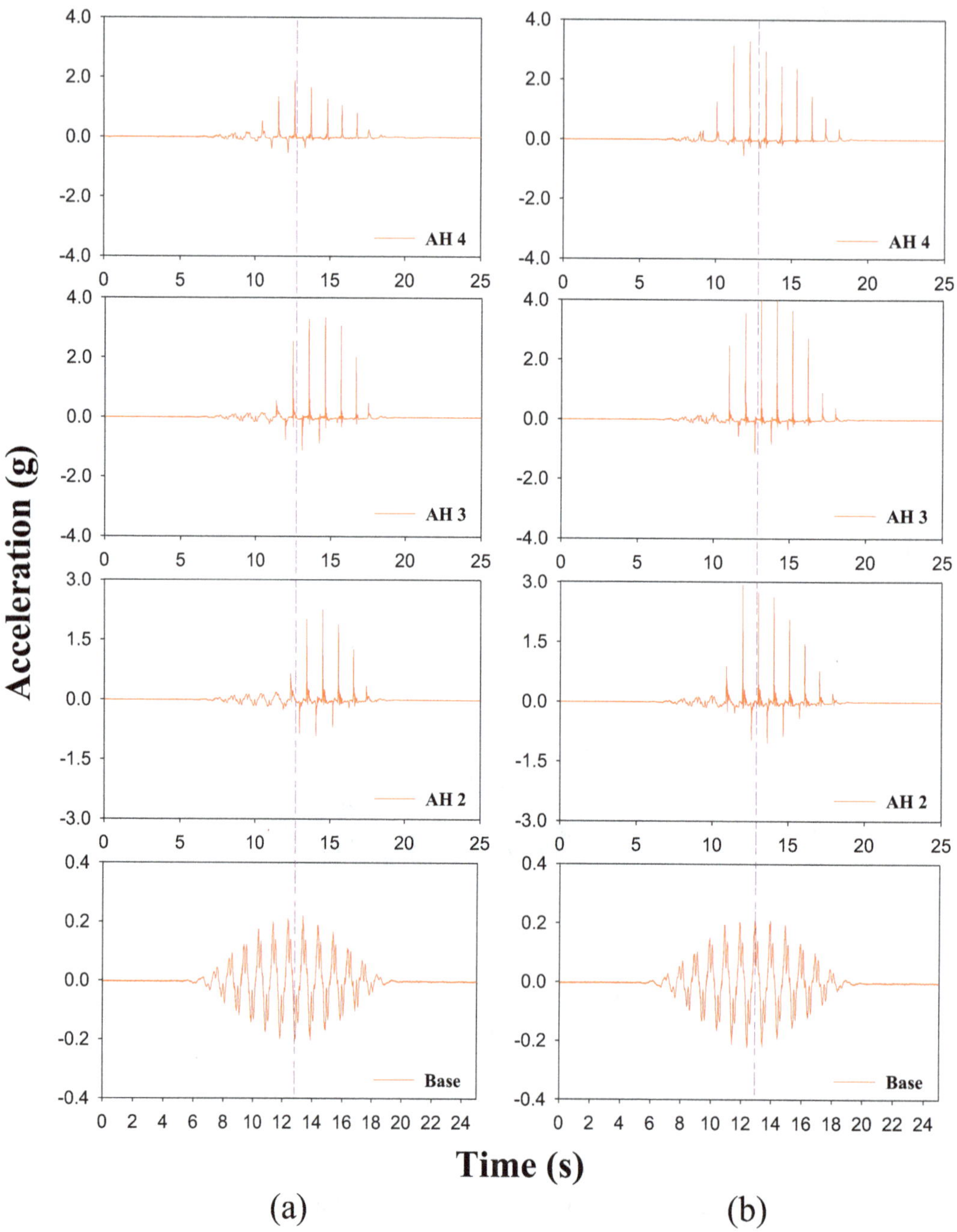


**Fig. 18.21** Time histories of acceleration at the middle slope. (**a**) NCU 2-M1 (**b**) NCU 2-M2

Figure 18.21 indicates the acceleration time histories at the middle slope in cases NCU 2-M1 and NCU 2-M2. The difference between these two shaking events is shape of wave: one is positive tapered sinusoidal wave, and the other one is negative tapered sinusoidal wave. The vertical black solid line implied the 13th second of shaking duration of two shaking events in test NCU 2, which points out the trough and the crest of acceleration signal wave in NCU 2-M1 and NCU 2-M2, respectively. It shows that there is a phase difference of 180 degrees. In addition, the spikes are observed in the direction of positive acceleration both in these two events, which indicates that significant spikes occur when the direction of inertial force is in the direction of the dip of the slope.

### *18.5.2 Ground Surface Deformation*

The horizontal displacement of surface markers was recorded by cameras during centrifuge modeling tests, and the settlement of surface markers was measured after test. Figure 18.22a–c illustrates the movement of surface markers in tests NCU 1, NCU 2, and NCU 3, respectively. The *x*-axis and *y*-axis correspond to the boundary of the rigid container. The dimensions of container are 767 mm in the *x*-axis and 355 mm in the *y*-axis. The positive part of the *x*-axis indicates the uphill side of slope.

All the surface markers moved to downslope after Motion 1 and Motion 2 in each case. The No. A array and No. B array slightly moved to the positive *y*-axis due to the inertial force during centrifuge spinning; contrarily, the No. C array slightly moved to the negative *y*-axis due to the reflection of inertial force to the rigid boundary during centrifuge spinning.

To compare the *x*-component of displacement in cases NCU 1-M1, NCU 2-M1, and NCU 3-M1, the conditions of three motions have the same cycles and frequency, but the different peaks of base acceleration are 0.27 g, 0.22 g, and 0.18 g, respectively. The surface markers have the largest relative displacement at the downhill part in case NCU 3-M1 which is subjected to the lowest PBA. According to the average displacement in the *x*-axis direction of all surface markers in three tests, the displacements of surface markers of No. 3 array and No. 4 array are larger than those of others arrays.

Figure 18.22a shows the comparison of the displacements of surface markers induced by first main-shaking (M1) and second main-shaking (M2) in NCU 1. The difference between M1 and M2 is at the waveform: one is tapered sinusoidal wave (M1), and the other one is sinusoidal wave (M2). The *x*-axis displacement of surface markers induced by NCU 1-M2 is less than that induced by NCU 1-M1, perhaps because the relative density of strata increased after once destructive shaking. The largest displacement is recorded at No. 4 array. Figure 18.22b, c also illustrates displacement of each surface markers induced by first main-shaking (M1) and second main-shaking (M2) in both NCU 2 and NCU 3, correspondingly. The *x*-axis displacement of surface markers induced by NCU 2-M2 is less than that induced by NCU 2-M1, perhaps because the relative density of strata increased after once large shaking. The largest displacement happened at No. 3 array and No. 4 array. By different manner, the *x*-axis displacements of surface markers induced by NCU 3-M1 and NCU 3-M2 are almost the same.

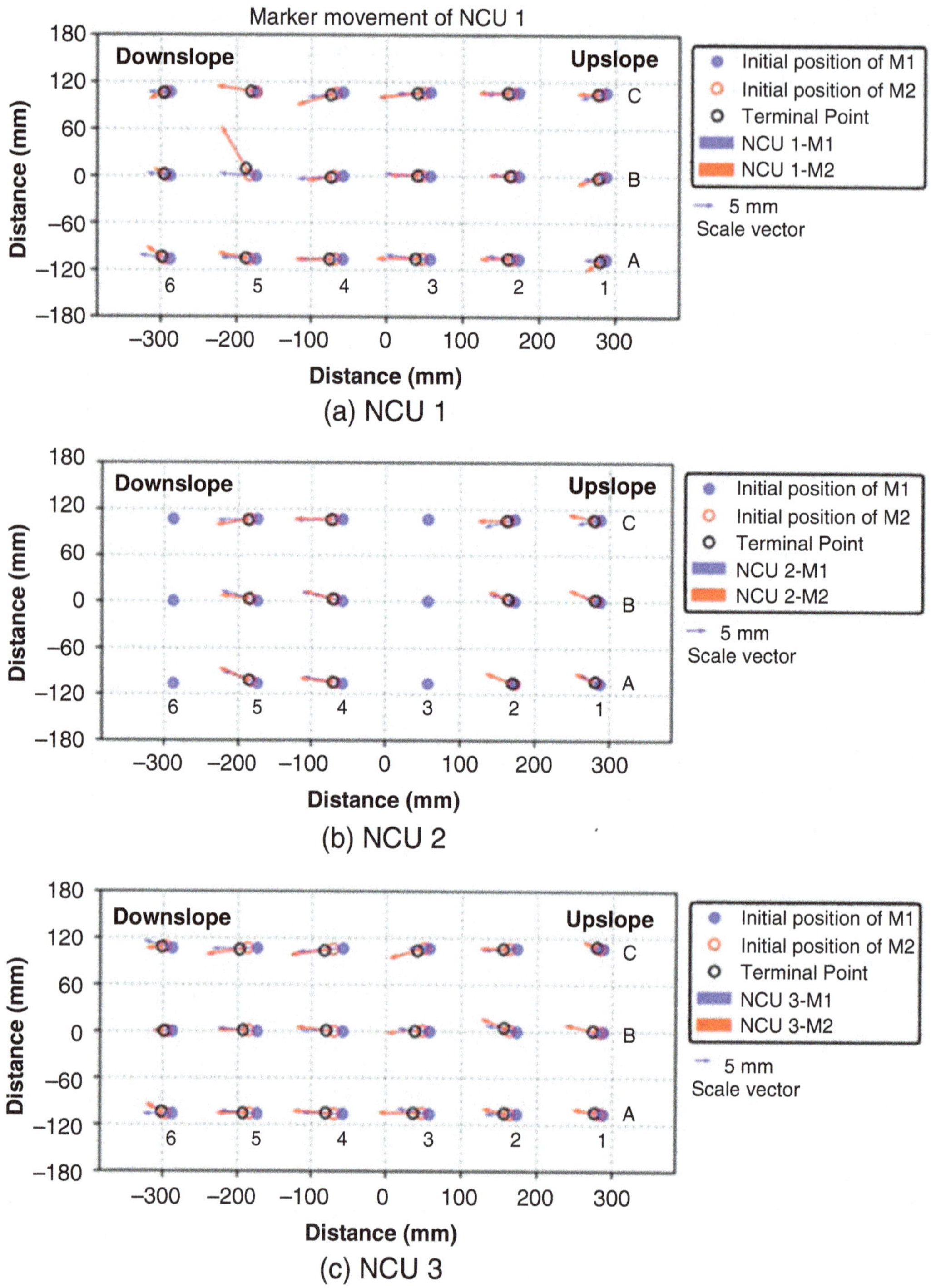


**Fig. 18.22** Movement of each surface markers after first main-shaking (M1) and the second main-shaking (M2) in tests of NCU 1 to NCU 3. (**a**) NCU 1, (**b**) NCU 2, and (**c**) NCU 3

### 18.5.3 Underground Deformation

The spaghetti noodles were inserted into the strata perpendicularly to the horizontal plane of rigid box. After test NCU 2, the profile of strata as in Figs. 18.23 and 18.24 shows the obvious deformation of noodles. The horizontal displacement of noodles in each test is listed in Tables 18.4a, 18.4b, and 18.4c. All of the spaghetti noodles are recorded to move in the downhill direction, and the horizontal displacement decreases along the depth.

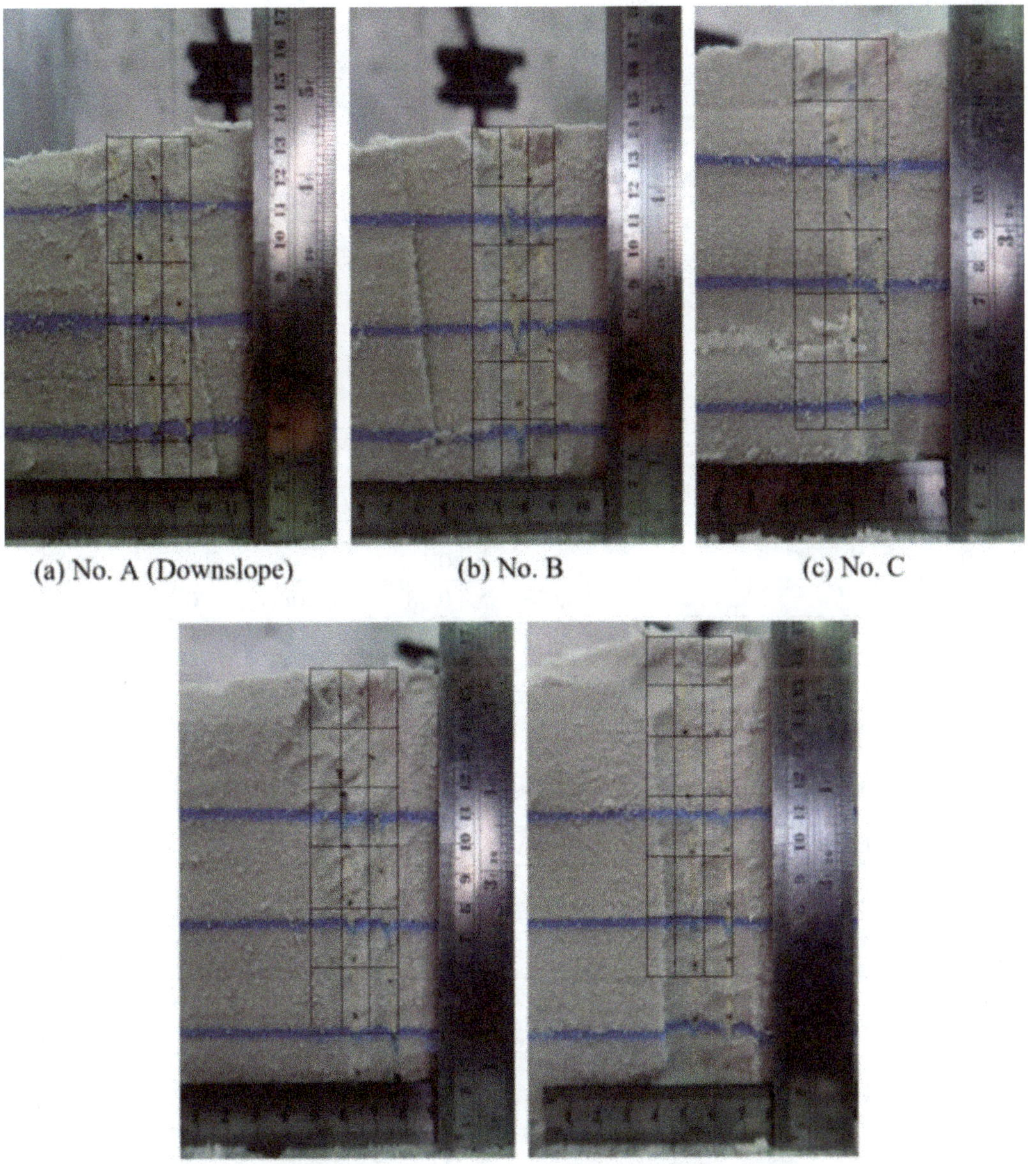

(a) No. A (Downslope) (b) No. B (c) No. C

(d) No. D (e) No. E (Upslope)

**Fig. 18.23** Deformation of spaghetti noodles at different position. (**a**) No. A (Downslope), (**b**) No. B, (**c**) No. C, (**d**) No. D, and (**e**) No. E (Upslope)

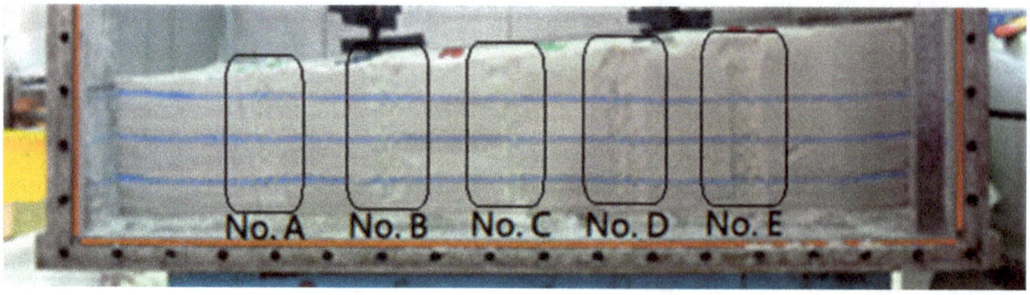


**Fig. 18.24** Results of spaghettis in NCU 2

**Table 18.4a** Horizontal displacement of NCU 1

| Depth under the ground surface (m) | No. A | No. B | No. C | No. D | No. E |
|---|---|---|---|---|---|
| 0 | 0.40 | 0.30 | 0.44 | 0.43 | 0.33 |
| −0.5 | 0.25 | 0.22 | 0.34 | 0.29 | 0.21 |
| −1.0 | 0.16 | 0.16 | 0.26 | 0.22 | 0.13 |
| −1.5 | 0.09 | 0.08 | 0.09 | 0.16 | 0.08 |
| −2.0 | 0.03 | 0.03 | 0.10 | 0.09 | 0.03 |
| −2.5 | 0.00 | 0.00 | 0.04 | 0.04 | 0 |
| −3.0 | – | – | 0 | 0 | 0 |

**Table 18.4b** Horizontal displacement of NCU 2

| Depth under the ground surface (m) | No. A | No. B | No. C | No. D | No. E |
|---|---|---|---|---|---|
| 0 | 0.51 | 0.35 | 0.30 | 0.36 | 0.21 |
| −0.5 | 0.26 | 0.21 | 0.18 | 0.18 | 0.10 |
| −1.0 | 0.14 | 0.10 | 0.12 | 0.14 | 0.08 |
| −1.5 | 0.07 | 0.08 | 0.08 | 0.12 | 0.04 |
| −2.0 | 0.04 | 0.04 | 0.05 | 0.07 | 0 |
| −2.5 | 0.00 | 0.03 | 0.03 | 0.04 | 0 |
| −3.0 | – | – | 0 | 0 | 0 |

**Table 18.4c** Horizontal displacement of NCU 3

| Depth under the ground surface (m) | No. A | No. B | No. C | No. D | No. E |
|---|---|---|---|---|---|
| 0 | 0.65 | 0.62 | 0.65 | – | 0.55 |
| −0.5 | 0.49 | 0.46 | 0.39 | – | 0.31 |
| −1.0 | 0.23 | 0.25 | 0.10 | – | 0.14 |
| −1.5 | 0.13 | 0.07 | 0.03 | – | 0.03 |
| −2.0 | 0.03 | 0.03 | 0.03 | – | 0.01 |
| −2.5 | 0.00 | 0.00 | 0.00 | – | 0.01 |
| −3.0 | – | – | – | – | 0.00 |

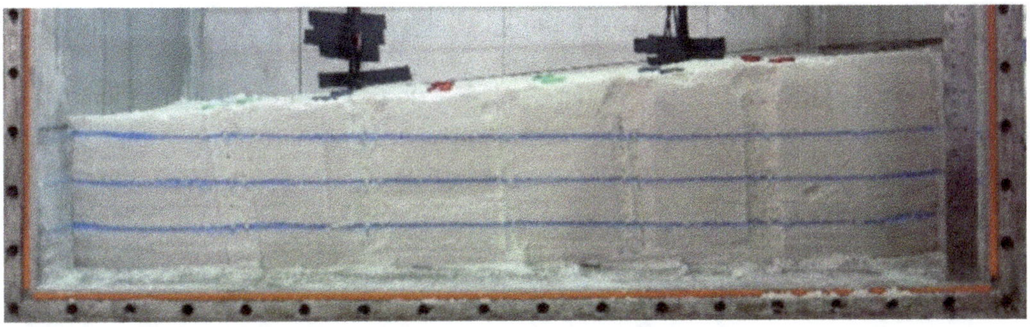

**Fig. 18.25** The profile of model after spinning

**Table 18.5** The settlement at each sand layer in prototype scale

| | NCU 1 | NCU 2 | NCU 3 |
|---|---|---|---|
| Top layer (m) | 0.0052 | 0.0078 | 0.0052 |
| Mid layer (m) | 0 | 0.0052 | 0.0052 |
| Bot layer (m) | 0 | 0 | 0 |

### *18.5.4 Blue Thin Sand Layers*

The three horizontal blue sand layers were pluviated; during the pluviation process, all of three sand layers are parallel to the plane of the bottom of rigid container. The depths under the ground surface of each color sand layer are 1 m, 2 m, and 3 m, respectively, which correspond to the depth of pore water pressure transducers P4, P3, and P2. The model would be cut after tests; as shown in Fig. 18.25, three color sand layers are still observed to be parallel to the plane of rigid container bottom, but the two upper layers have slight settlement after test, as shown in Table 18.5, and the pore water transducers were still attached within the color sand layers.

## 18.6 Conclusions

Three centrifuge modeling tests were conducted to simulate a 5-degree-inclined slope of 4-m-deep-saturated sandy ground subjected to different input motions. There are two kinds of main-shaking which includes Motion 1 and Motion 2 in each test. Motion 1 is following the exercise of LEAP-UC-2017, and Motion 2 is designed to observe the effect of different input motions for the responses of soil liquefaction. From the test results, the conclusions are fourfold: (1) the prominent amplitudes of spikes appear in the phase of the positive acceleration during both kinds of waveform which are tapered sinusoidal wave and uniform sinusoidal wave motions, which indicates that the pulses of acceleration are more prominent in the direction of the slope dip; (2) the surface markers move toward the dip direction along the *x*-axis, and additionally, a component of deformation in the transverse (*y*-direction) occurs in some events, which might be caused by a small bucket angle error or perhaps by Coriolis forces; and (3) the magnitude of lateral spreading decreases with depth in the strata, as there is negligible lateral spreading in the bottom meter of the 4 m thick soil deposit. (4) The pore water pressure transducers did not move relative to the colored sand layers after destructive shaking.

**Acknowledgments** The authors would like to express their gratitude for the financial support from the Ministry of Science and Technology, Taiwan (ROC) (MOST 106-2628-E-008 -004 -MY3) and the technical support from the Experimental Center of Civil Engineering, National Central University. Their support has made this study and future research possible and efficient.

## References

Carey, T. J., Stone, N., & Kutter, B. (2019). Grain size analysis and maximum and minimum dry density of Ottawa F-65 sand for LEAP-UCD-2017. In B. Kutter et al. (Eds.), *Model tests and numerical simulations of liquefaction and lateral spreading: LEAP-UCD-2017*. New York: Springer.

Kutter, B., Carey, T., Hashimoto, T., Zeghal, M., Abdoun, T., Kokalli, P., Madabhushi, G., Haigh, S., Hung, W.-Y., Lee, C.-J., Iai, S., Tobita, T., Zhou, Y. G., Chen, Y., & Manzari, M. T. (2017). LEAP-GWU-2015 experiment specifications, results, and comparisons. *International Journal of Soil Dynamics and Earthquake Engineering, 113*, 616–628. https://doi.org/10.1016/j.soildyn.2017.05.018.

Kutter, B. L., Carey, T. J., Stone, N., Bonab, M. H., Manzari, M., Zeghal, M., Escoffier, S., Haigh, S., Madabhushi, G., Hung, W.-Y., Kim, D.-S., Kim, N.-R., Okamura, M., Tobita, T., Ueda, K., & Zhou, Y.-G. (2019). In B. Kutter et al. (Eds.), *Model tests and numerical simulations of liquefaction and lateral spreading: LEAP-UCD-2017*. New York: Springer.

Manzari, M. T., Kutter, B. L., Zeghal, M., Iai, S., Tobita, T., Madabhushi, S. P. G., Haigh, S. K., Mejia, L., Gutierrez, D. A., & Armstrong, R. J. (2015). LEAP projects: Concept and challenges. In *Proceedings: Fourth International Conference on Geotechnical Engineering for Disaster Mitigation and Rehabilitation (4th GEDMAR), 2014 Sept 16–18*. Kyoto: Taylor & Francis.

Okamura, M., & Inoue, T. (2012). Preparation of fully saturated model for liquefaction study. *International Journal of Physical Modeling in Geotechnics, 12*(1), 39–46.

# Chapter 19
# Verification of the Repeatability of Soil Liquefaction Centrifuge Testing at Rensselaer

**Evangelia Korre, Tarek Abdoun, and Mourad Zeghal**

**Abstract** The Liquefaction Experiments and Analysis Projects (LEAP) is an international effort to use experimental data from physical modeling at different (international) centrifuge facilities to validate soil liquefaction numerical models and analysis tools. The goals of LEAP-2017 experimental efforts are to assess the repeatability potential at each facility, the reproducibility of centrifuge tests among different facilities, and the sensitivity of the experimental results to variation of testing parameters and conditions. A number of tests of the same (sloping deposit) centrifuge model were repeated at Rensselaer Polytechnic Institute in 2015 and 2017. This paper focuses on two specific tests to assess and demonstrate repeatability at this facility.

## 19.1 Introduction

The Liquefaction Experiments and Analysis Projects (LEAP) is an international effort to (1) investigate the dynamic response and liquefaction of a number of soil systems by means of centrifuge testing of small-scale physical models, and (2) use the experimental data to validate soil liquefaction numerical models and analysis tools.

In 2015, a LEAP was undertaken using a small-scale model of a sloping deposit. A number of tests were conducted at Cambridge University, Kyoto University, NCU of Taiwan, Rensselaer Polytechnic Institute (RPI), UC Davis, and Zhejiang University (Kutter et al. 2017, 2019). The data provided by these tests were used in a first validation exercise (Manzari et al. 2018). The LEAP 2017 expands and builds on the 2016 results and outcomes. Tests of the same sloping deposit were repeated in China (Zhejiang University), France (Institut français des sciences et technologies des transports, de l'aménagement et des réseaux), Japan (Ehime University and Kyoto University), Korea (KAIST University), Taiwan (National Taiwan University, UK (Cambridge University), and USA (UC Davis and Rensselaer). The tests are aimed

E. Korre · T. Abdoun · M. Zeghal (✉)
Department of Civil and Environmental Engineering, Rensselaer Polytechnic Institute, Troy, NY, USA
e-mail: zeghal@rpi.edu



B. Kutter et al. (eds.), *Model Tests and Numerical Simulations of Liquefaction and Lateral Spreading*, https://doi.org/10.1007/978-3-030-22818-7_19

at assessing the repeatability, reproducibility, and sensitivity of experimental results to variability in testing conditions among the different facilities.

A total of five tests of the sloping deposit were conducted at RPI. The first two tests (referred to as 2015_RPI01 and 2015_RPI02) were executed in 2015 (Kokkali et al. 2018; Abdoun et al. 2018). The LEAP 2017 three tests were aimed at repeating the LEAP 2015 tests and also assessing the soil system response to variations in input motion (i.e., presence of a 3 Hz component) and deposit relative density. Specifically, the first of this series (referred to as 2017_RPI01) was simply a repetition of the first test of 2015, and was conducted according to the same specification by two different researchers. This paper presents the 2017_RPI01 test and investigates the level of repeatability with the initial test performed in 2015 (2015_RPI01). The other LEAP 2017 tests will be published elsewhere.

## 19.2 Experimental Setup

### *19.2.1 Dry Model Preparation*

The dry 2017_RPI01 model was constructed according to the project specifications (Kutter et al. 2017) and following the methodology of 2015_RPI01, as described by Kokkali et al. (2018). A stratum of dry Ottawa F-65 sand (acquired from U. S. Silica) was deposited to form the specified 5-degree slope (shown in Fig. 19.1). The deposition was performed using dry pluviation in a rigid container made of aluminum and Plexiglas (Fig. 19.1). The series of experiments were all performed at a centrifugal level of 23 g. Hereafter, all experimental results are presented in prototype scale.

Pluviation was performed manually from a specific height and at constant (lateral) velocity. This height was determined after careful calibration of the utilized pluviator and was maintained throughout the entire process of model building. The slope surface was not curved as the model longitudinal direction and direction of shaking are parallel to the Rensselaer centrifuge axis of rotation (also, curving was deemed not necessary in the transverse direction in view of the relatively small model dimension in this direction).

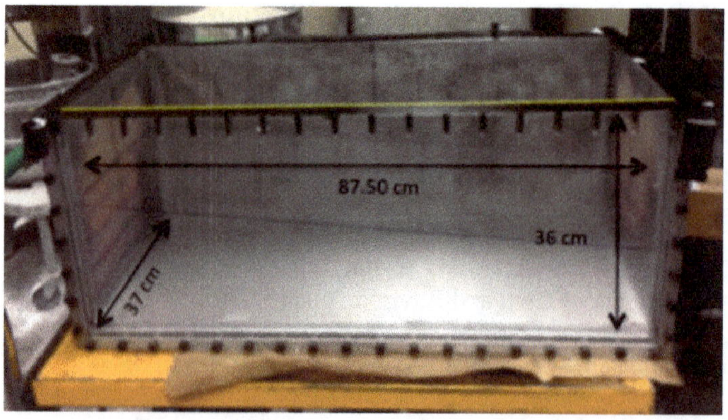


**Fig. 19.1** The rigid container utilized in the tests conducted at Rensselaer

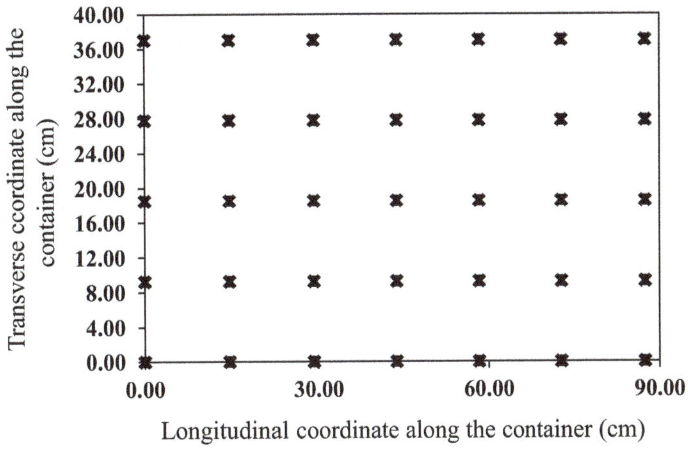


**Fig. 19.2** Plan view of the box and grid used to measure the achieved mass density (dimensions in model scale)

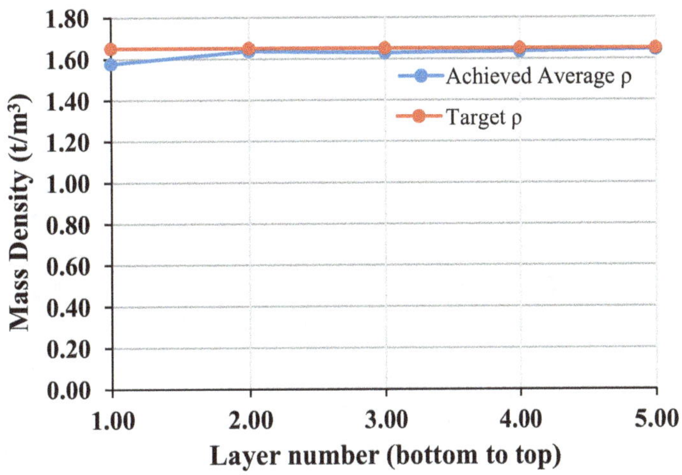


**Fig. 19.3** Mass density per horizontal layers

The specified target mass density of the LEAP 2015 and 2017 models was 1651 kg/m$^3$, corresponding to a relative density of 64–68% depending on which values of maximum and minimum density are assumed (Carey et al. 2019). To confirm the uniformity of the achieved deposit density, detailed mass and height measurements were taken on a grid (as shown in Fig. 19.2) at five levels along the deposit height. The measurements per layer (of about 1 m in thickness) are presented in Fig. 19.3 (showing the achieved and target mass densities). The first layer shows a slight discrepancy from the target. The conditions at the box boundary and the curvature around the edges (due to the applied silicone to seal the container) partly account for this discrepancy. While building the sloping surface of the deposit, it was not possible to take exact measurements until the final geometry was built. The obtained (slope) geometry is compared to the target geometry in Fig. 19.4. The achieved average density was 1.6508 g/cm$^3$, very close to the target value.

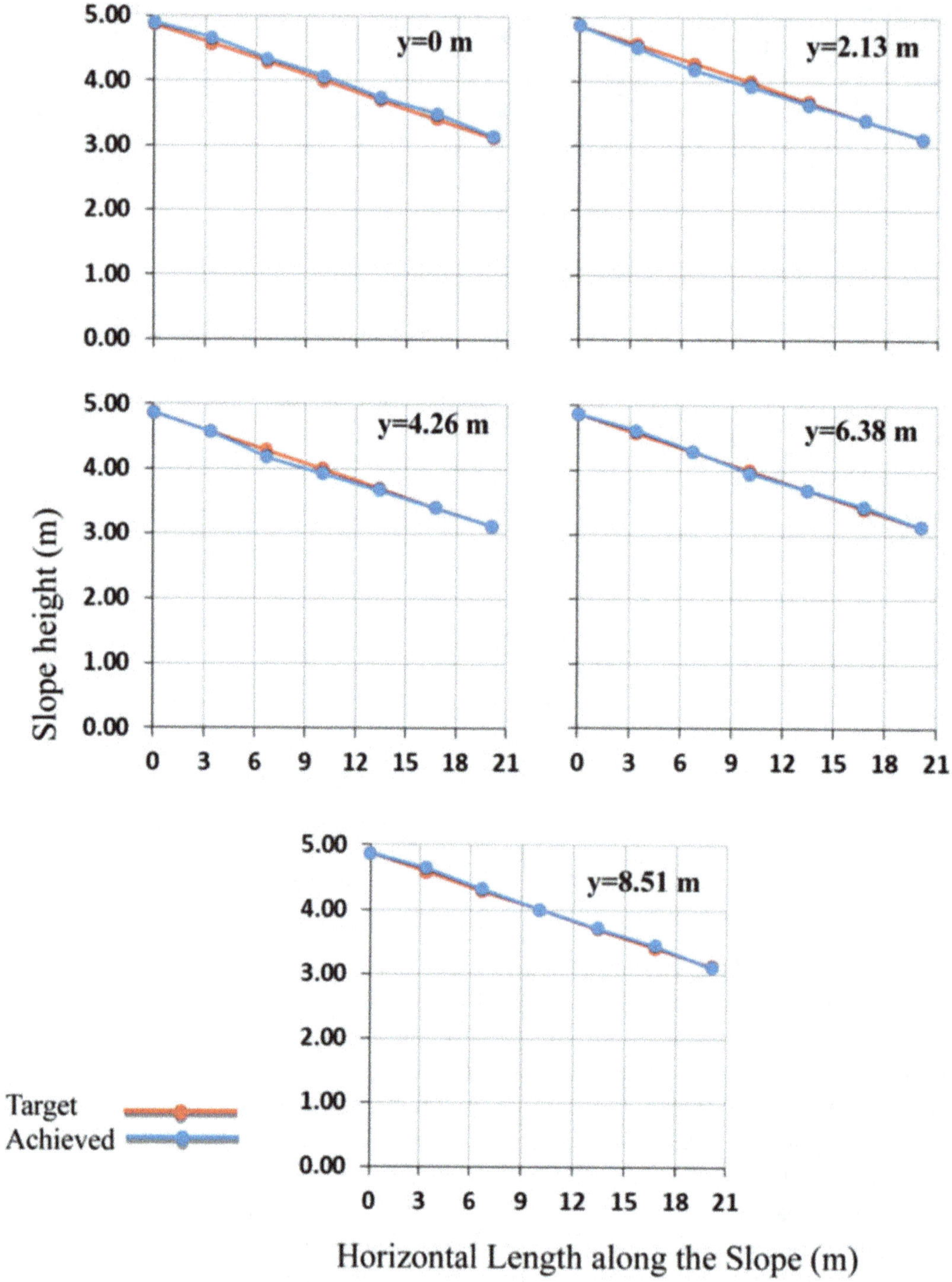


**Fig. 19.4** Comparison between the target and achieved slope geometry for selected locations along the transverse ($y$) direction

## 19.2.2 Instrumentation

The model was equipped with an extensive array of sensors that were installed according to the specified project locations (Kutter et al. 2017) while also matching the "as-built" coordinates of the sensors of 2015_RPI01. A sketch of the

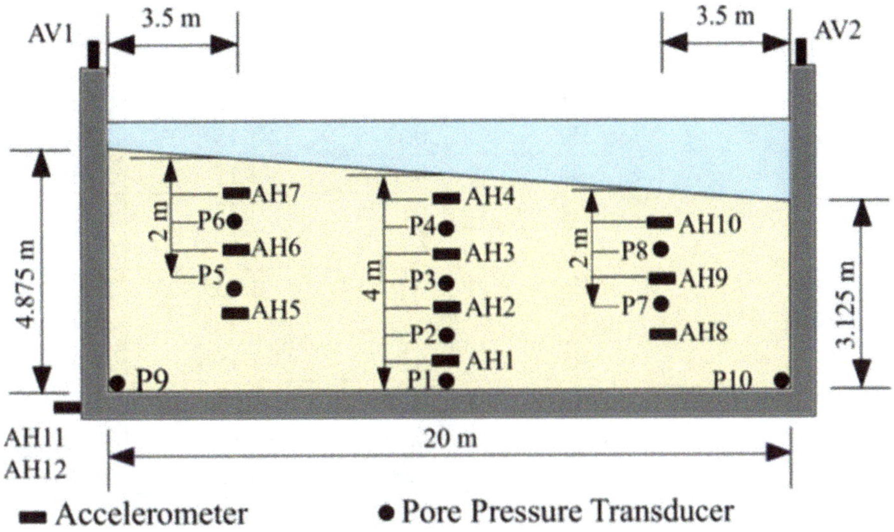


**Fig. 19.5** Experimental setup showing the installed model instrumentation

experimental setup is presented in Fig. 19.5. A total of 10 accelerometers (AH1–AH10) were buried within the sloping soil deposit (Fig. 19.5). Two accelerometers (AH11 and AH12) were installed at the bottom of the box to record the base longitudinal accelerations (at two opposite locations in the transverse direction), and two accelerometers (AV1 and AV2) were installed at the top of the box to record the vertical motion in the middle of the transverse side. The pore pressure was monitored by means of 10 buried pore water pressure transducers (P1–P10). In addition, zip-tie heads were inserted into the surface (of the slope) as targets which were tracked with a high-speed camera to measure the slope lateral spreading. The sensor locations inside the box (measured during construction of the model) are given in Table 19.1 in the form of *x-y-z* coordinates. The origin of the coordinate system is at the center of the container, as prescribed in Kutter et al. (2017). Table 19.2 provides the coordinates of the grid created with the (zip-tie head) targets.

### *19.2.3 Viscous Fluid Preparation and Saturation Process*

The dry model was first fixed on the centrifuge basket. Then, the container was sealed and put under vacuum. The dry model was thereafter saturated with $CO_2$ which pushes air out of the voids. The process of vacuum and $CO_2$ saturation was repeated twice, before starting saturation. The viscous fluid used for the saturation of the model was made with methylcellulose manufactured by Dow Chemical added at a specific ratio to the water to achieve a viscosity of 23.5 cp.

**Table 19.1** Sensor locations: (a) accelerometers and (b) pore water pressure transducers

| Sensor | *x* (m) | *y* (m) | *z* (m) |
|---|---|---|---|
| (a) | | | |
| AH1 | 0.2875 | −1.15 | 0.575 |
| AH2 | 0.1725 | −1.15 | 1.495 |
| AH3 | −0.0575 | −1.15 | 2.53 |
| AH4 | −0.6325 | 1.725 | 3.45 |
| AH5 | −6.6125 | −0.575 | 2.484 |
| AH6 | −6.1525 | 1.035 | 3.404 |
| AH7 | −5.4625 | 0.345 | 4.14 |
| AH8 | 6.7275 | −1.15 | 1.495 |
| AH9 | 6.7275 | −0.345 | 2.484 |
| AH10 | 6.7275 | −0.92 | 3.151 |
| AH11 | 10.0625 | −4.255 | 0 |
| AH12 | 10.0625 | 4.255 | 0 |
| AV1 | −10.0625 | 0 | Top of container |
| AV2 | 10.0625 | 0 | |
| (b) | | | |
| P1 | −0.5175 | −0.299 | 0.23 |
| P2 | −0.5175 | −0.345 | 1.012 |
| P3 | 0.0575 | −0.46 | 2.001 |
| P4 | 0.7475 | −0.345 | 2.99 |
| P5 | −7.8315 | 1.265 | 2.99 |
| P6 | −6.3825 | 0.115 | 4.002 |
| P7 | 6.4975 | 0.782 | 2.024 |
| P8 | 6.7275 | 0.851 | 2.99 |
| P9 | −8.0615 | −0.805 | 0.276 |
| P10 | 6.7965 | −0.046 | 0.23 |

### *19.2.4 Centrifuge Equipment*

The geotechnical centrifuge at RPI has a radius of 3 m and can withstand a total model weight of 150-g tons. The shaking table is installed within the centrifuge basket and can replicate dynamic (earthquake and artificial) motions in one or two lateral directions.

## 19.3 Recorded Response

### *19.3.1 Input Motion*

The sloping deposit model was subjected to two destructive excitations along the longitudinal direction. The excitations consisted of two sinusoidal ramped motions of 16 cycles, with a predominant frequency of 1 Hz. The sequence of shaking had an

**Table 19.2** Locations of tracking targets: (a) central rows along the slope and (b) central rows across the slope

| Target row | $x$ (m) | $y$ (m) | $z$ (m) |
|---|---|---|---|
| (a) | | | |
| | −66.125 | −1.15 | 48.07 |
| | −54.625 | −1.15 | 46.92 |
| A | −43.125 | −1.15 | 46.00 |
| B | −31.625 | −1.15 | 45.08 |
| C | −20.125 | −1.15 | 43.93 |
| D | −8.625 | −1.15 | 43.01 |
| E | 2.875 | −1.15 | 42.09 |
| F | 14.375 | −1.15 | 41.17 |
| G | 25.875 | −1.15 | 40.02 |
| H | 37.375 | −1.15 | 39.1 |
| I | 48.875 | −1.15 | 37.95 |
| | 60.375 | −1.15 | 36.57 |
| | 71.875 | −1.15 | 35.42 |
| (b) | | | |
| −3 | 2.875 | −35.65 | 41.17 |
| −2 | 2.875 | −24.15 | 39.79 |
| −1 | 2.875 | −12.65 | 40.25 |
| 0 | 2.875 | −1.15 | 40.02 |
| 1 | 2.875 | 10.35 | 39.10 |
| 2 | 2.875 | 21.85 | 39.10 |
| 3 | 2.875 | 33.35 | 39.10 |

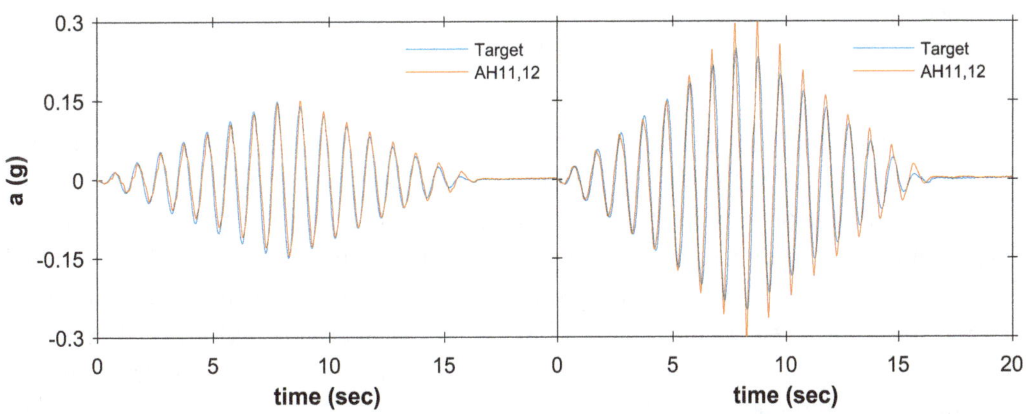


**Fig. 19.6** Comparison of the target and recorded input accelerations for the 0.15 and 0.25 g motions of 2017_RPI01. Left-side view refers to 0.15 g shake and right-side view to 0.25 g shake

increasing intensity, with the motions having a maximum acceleration of 0.15 g and then 0.25 g. Figure 19.6 shows a comparison between the target and achieved input motions and shows a remarkable level of fidelity in replicating the target in terms of both amplitude and frequency. Specifically, there is only a slight discrepancy in amplitude of the largest cycles of the 0.25 g motion; however, the frequency content was not affected. A non-destructive tremor was applied before and after each shake

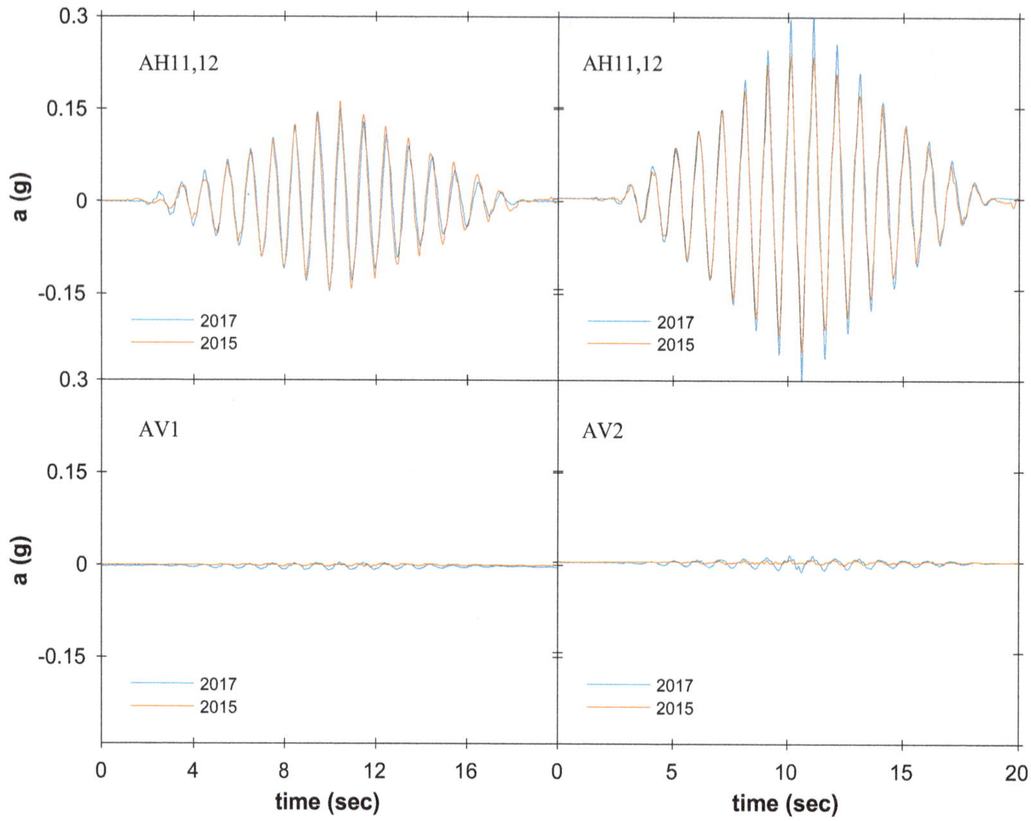


**Fig. 19.7** Comparison of the recorded horizontal and vertical input accelerations of the rigid container for the 2017_RPI01 and 2015_RPI01 tests. Left-side view refers to 0.15 g shake and right-side view to 0.25 g shake

for system identification purposes. The response associated with these tremors is not included in this article for reasons of brevity.

A comparison of the input motions that were recorded by the accelerometers AH11 and AH12 (mounted at the base of the rigid container) during the 2015_RPI01 and 2017_ RPI01 tests confirmed the successful repeatability of input motion for the 0.15 g and 0.25 g shakes (Fig. 19.7). Note also that the recorded vertical accelerations are negligible from a practical point of view (for both tests), assuring that no significant rocking had occurred.

### 19.3.2 Soil Acceleration

The accelerations of the soil deposit were recorded for the 0.15 and 0.25 g excitations of the 2015_RPI01 and 2017_RPI01 tests at the AH1 to AH4 locations (Fig. 19.8). The sensor AH2 malfunctioned during the second excitation of the 2017_ RPI01 test. There is a remarkable agreement between the accelerations of

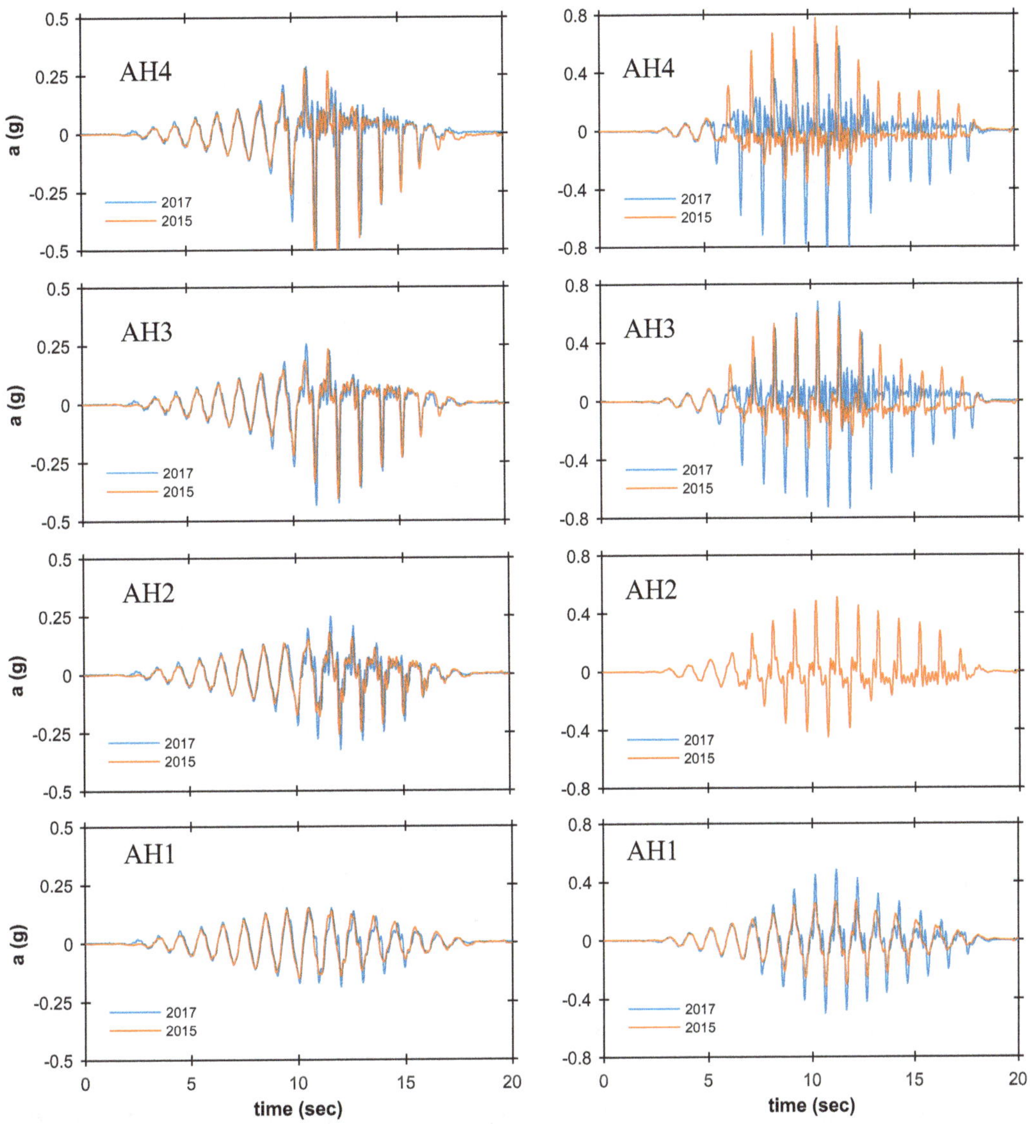


**Fig. 19.8** Acceleration time histories recorded along the sensors central array during the 2017_RPI01 and 2015_RPI01 tests. Left-side view refers to 0.15 g shake and right-side view to 0.25 g shake

the two tests in terms of both frequency content and intensity of the cyclic motion, including the spikes. This is especially the case for the 0.15 g excitation. However, the accelerations of 2017_ RPI01 during the 0.25 g excitation were higher than those of 2015_RPI01, especially within the shallow soil layers and for the prominent negative acceleration spikes. This is attributed partly to the larger amplitude of the 2017_ RPI01 input motion for the 7–13 s cycles. The larger negative acceleration spikes are indicative of a more intense dilation that occurred in the downslope direction at high strains of the 2017_ RPI01 test, as expected.

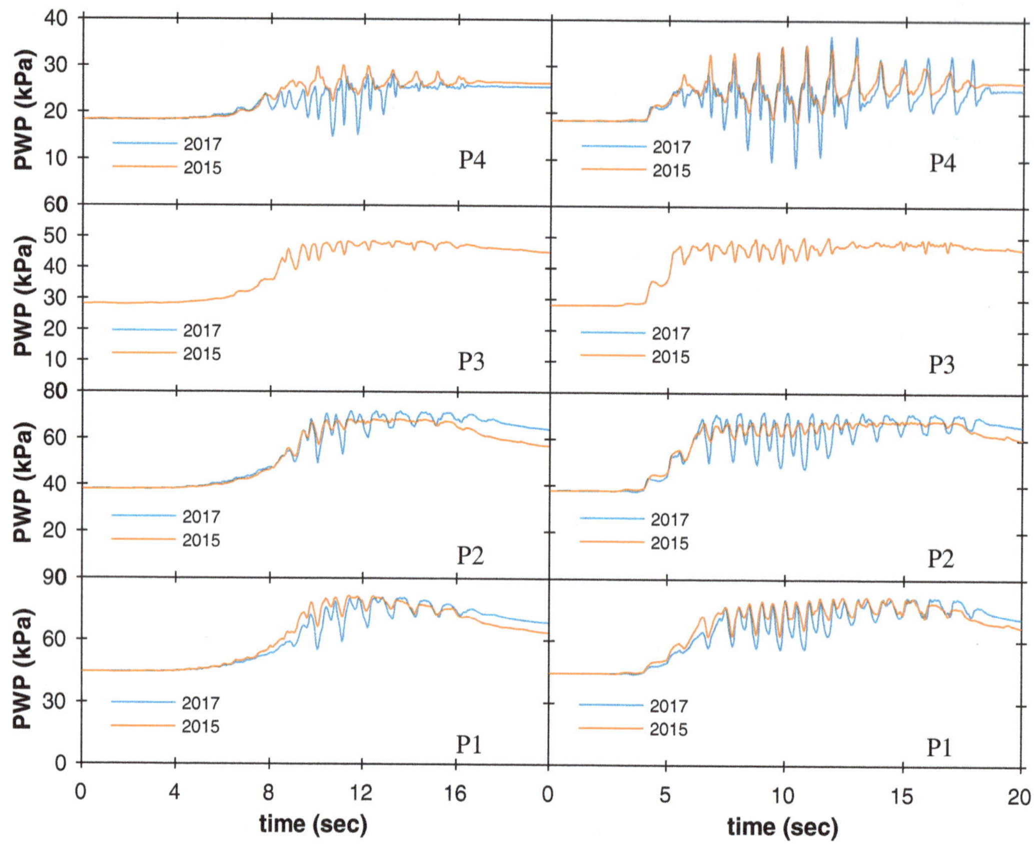


**Fig. 19.9** Pore water pressure (PWP) time histories recorded along the central array during the 2017_RPI01 and 2015_RPI01 tests. Left-side view refers to 0.15 g shake and right-side view to 0.25 g shake

## 19.3.3 Pore Water Pressure

The short-term pore water pressure time histories also showed a remarkable agreement between the 2015_RPI01 and 2017_RPI01 results, as depicted in Fig. 19.9 (note that the sensor P3 malfunctioned during the 2017 test). In general, the agreement is better for the lower layers of the deposit, while closer to the surface the agreement was affected by the dilative response (in consistency with the acceleration response discussed above). The dilative response was more prevalent for the 2017_RPI01 test, especially during the strongest phase of the 0.25 g shaking. Similar responses were obtained for the long-term pore water pressure time histories (Fig. 19.10). The pressure dissipation rates were noticeably close, and indicate that the achieved permeability and soil fabric were consistent for the two tests. The pore pressure results are also presented in terms of excess pore water pressure ratio in Fig. 19.11. The ratio was calculated as follows:

$$Ru = \frac{u - u_{\text{hydr}}}{\sigma'_v}$$

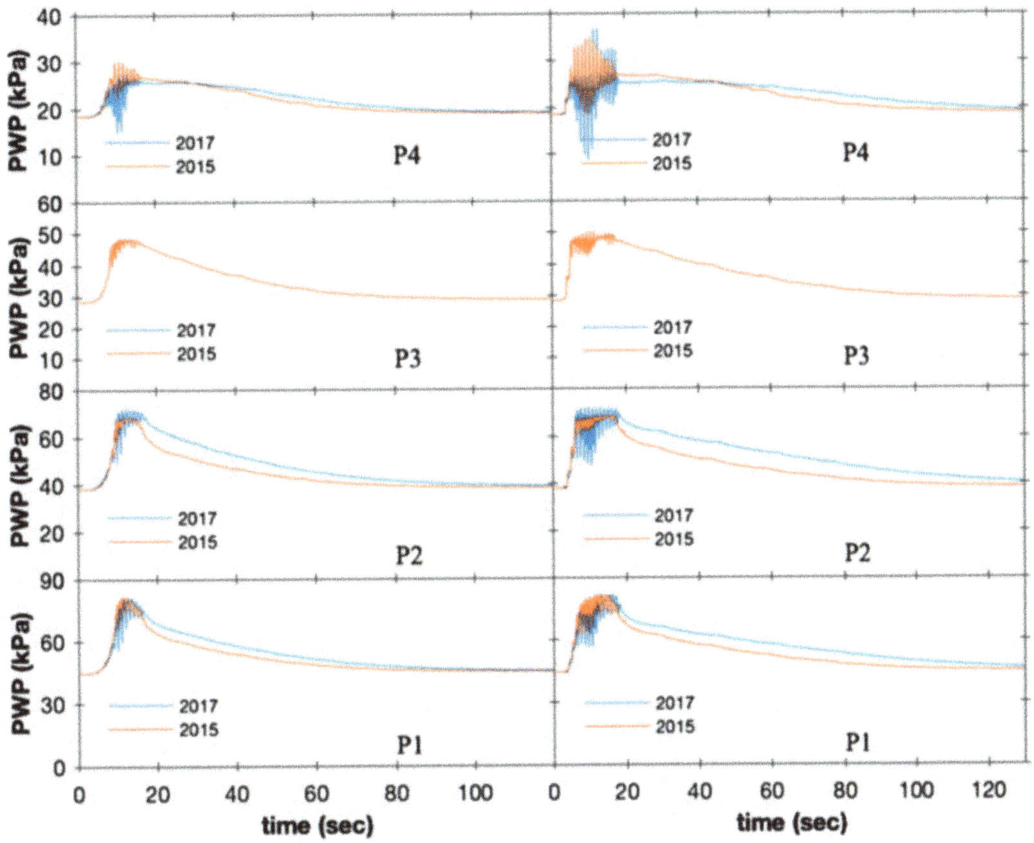


**Fig. 19.10** Long-term time histories of the pore pressures recorded along the central array during the 2017_RPI01 and 2015_RPI01 tests. Left-side view refers to 0.15 g shake and right-side view to 0.25 g shake

where $u$ is the excess pore water pressure, $u_{\text{hydr}}$ is the initial hydrostatic water pressure, and $\sigma'_v$ is the initial effective stress at the level of each pore water pressure transducer. The values of the $Ru$ ratio varied between 1 and 0, and that peak values of $Ru$ were close to 1.0 for the entire stratum in both tests.

## 19.4 Lateral Spreading

The permanent lateral (spreading) displacements along the slope were measured by means of a Phantom v5.1 HI-G high-speed camera, produced by Vision Research. The recorded videos had an actual sampling rate of 1000 frames/s, which corresponds to 43.5 frames/s in prototype scale (i.e., a Nyquest frequency of 21.7 Hz). The high-speed camera records at a resolution of 1024×1024 pixels, which corresponds to approximately 0.50 mm resolution in measured prototype-scale displacements at a 23 g centrifugal level (for the conducted tests). A more detailed description of the properties of the equipment is provided by Kokkali et al. (2018). The grid of the zip-tie heads, installed as targets on the surface of the sloping ground (Fig. 19.12), was created with a spacing of 1.15 m × 1.15 m, observing the

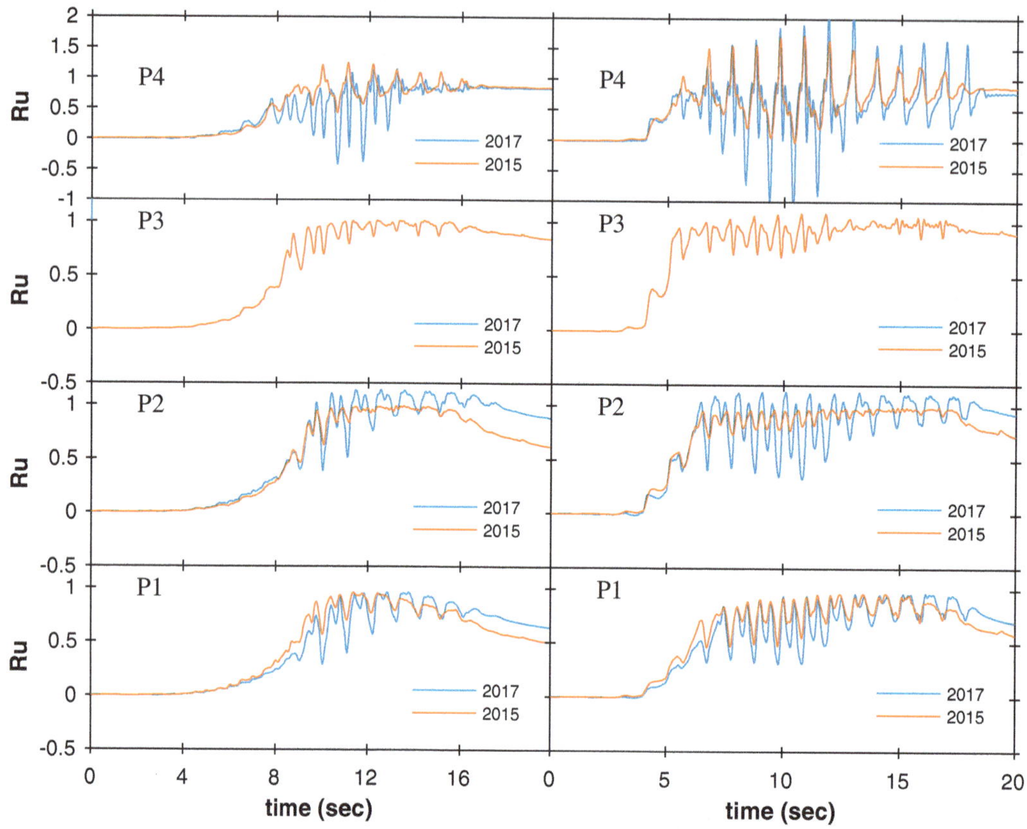


**Fig. 19.11** Time histories of the excess pore pressure ratios recorded along the central array during the 2017_RPI01 and 2015_RPI01 tests. Left-side view refers to 0.15 g shake and right-side view to 0.25 g shake

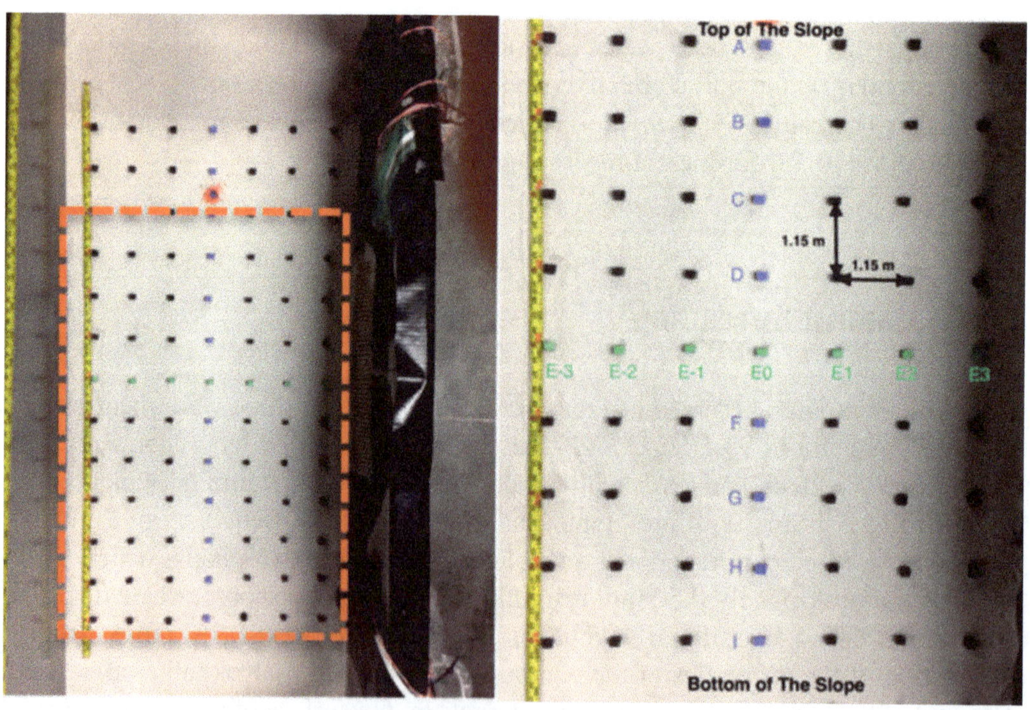


**Fig. 19.12** The grid of zip-tie heads used as targets for high-speed camera tracking

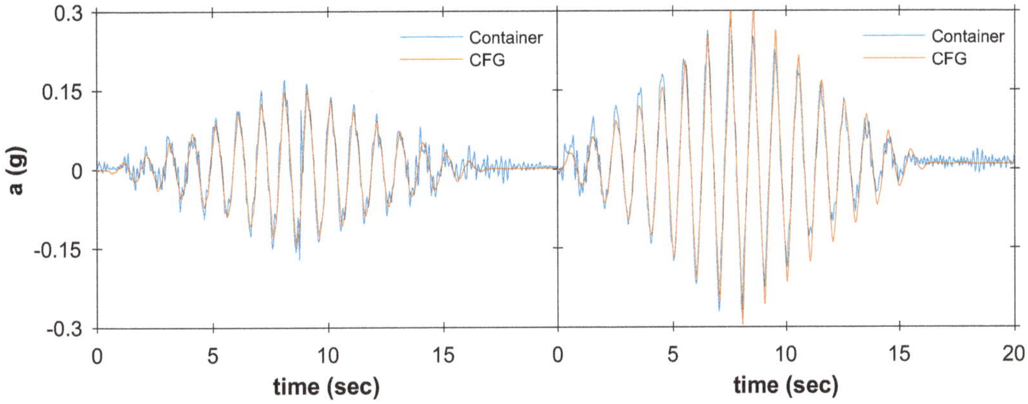


**Fig. 19.13** Acceleration time histories of the input motion obtained from image analysis of high-speed camera tracking and recorded by the accelerometers AH11 and AH12. Left-side view refers to 0.15 g shake and right-side view to 0.25 g shake

methodology adopted in 2015_RPI01. The red-dashed frame in Fig. 19.12 indicates the area that was actually tracked by the high-speed camera. Some target rows were not included because of light that is reflected on waves generated within the water above the slope making it difficult to track these targets. An image analysis was used to obtain the time histories of the soil total and relative displacements. The relative displacements are evaluated by subtracting the container displacement from the soil displacement provided by processing the zip-tie head images. A double differentiation was employed to obtain the corresponding acceleration time histories. Two (target) points were tracked on the container walls to verify the quality of obtained accelerations. The accelerations evaluated from these points were compared to the input recorded by the two accelerometers AH11 and AH12 at the base of the rigid container, as shown in Fig. 19.13. This figure shows that the magnitude and frequency content of the container target acceleration is consistent with the motions measured by the accelerometers. The agreement also confirms that the container behaves practically as rigid and apparently does not affect the input motion. Further verification of the tracking and image analysis was performed by comparing the acceleration AH4 (along the central array and the closest to the slope surface) with the target D0 (at the central target of the grid in Fig. 19.12), as shown in Fig. 19.14. The time histories for AH4 and D0 were very close and practically identical for both motions. The highly dilative spikes after full liquefaction (i.e., after about 8.0 s) are fully captured by target tracking. The displacement time histories for the central three target rows along the slope are exhibited in Fig. 19.15 for the 0.15 and 0.25 g excitation during the 2015_RPI01 and 2017_RPI01 experiments. The two tests were in remarkable agreement, showing a consistent value of permanent displacement approximately 12 cm in prototype scale at the end of shaking. Note that the displacements in Fig. 19.15 are shown in dotted lines after the 16 s instant to indicate a lower accuracy for these displacements (waves were generated within the water above the slope free surface and became more prominent after the end of shaking

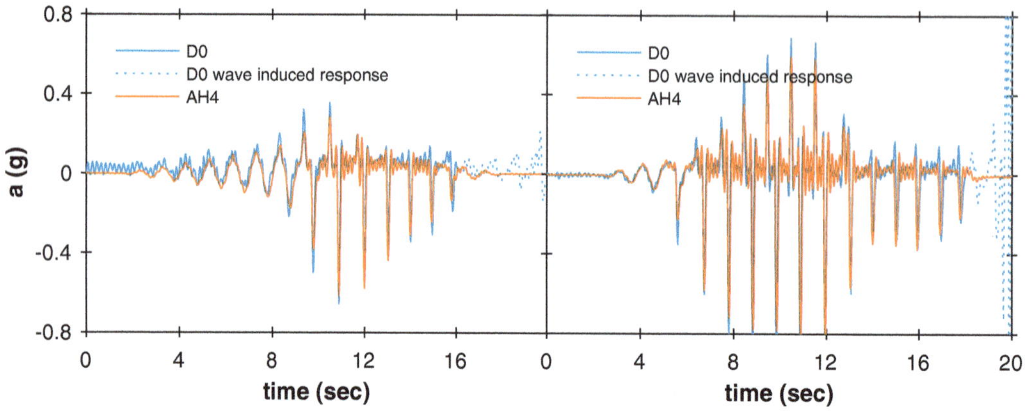


**Fig. 19.14** Acceleration time histories of the motion obtained from image analysis of the central target D0 and recorded by accelerometers AH4 (closest sensor to the surface along the central array). Left-side view refers to 0.15 g shake and right-side view to 0.25 g shake

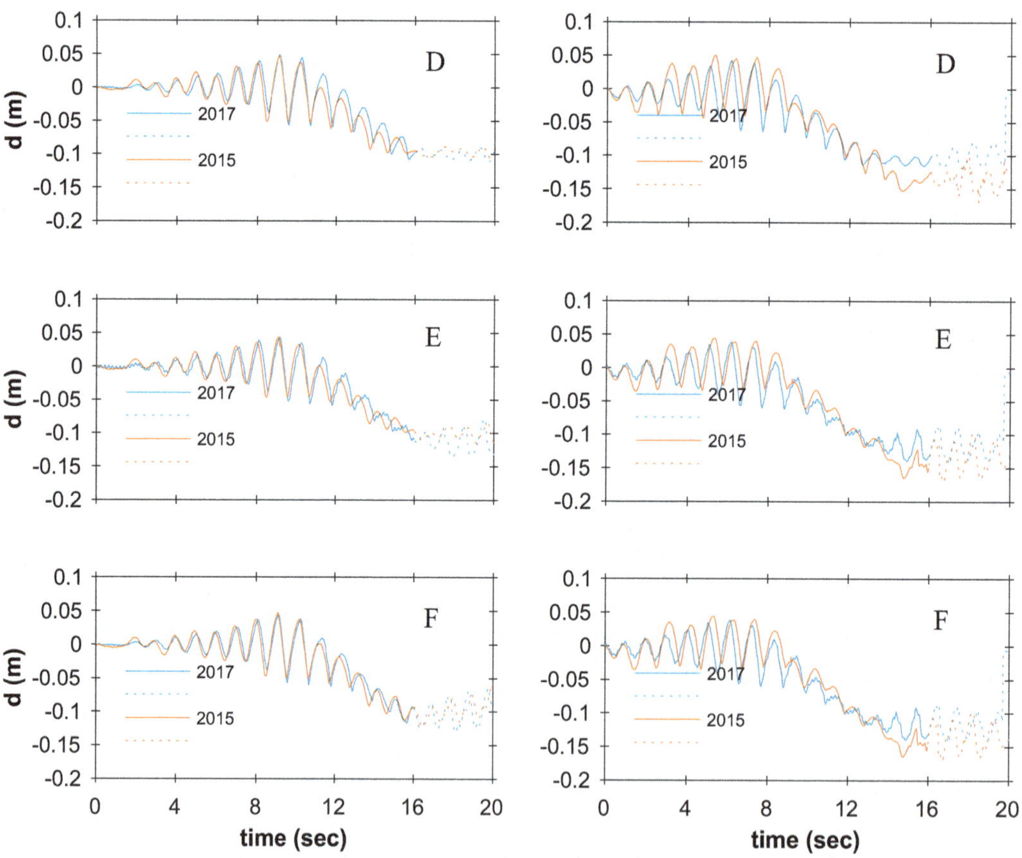


**Fig. 19.15** Time histories of lateral spreading displacements along the three central rows D, E, and F along the slope for the 0.15 g (left) and 0.25 g (right) shakes (the dotted portion of the graphs corresponds to lower accuracy due to water wave reflections)

which compromised the accuracy of the tracking process). Nevertheless, the overall trend of the permanent lateral displacement components remained consistent for the two tests.

## 19.5 Conclusions

Five LEAP centrifuge experiments were conducted at Rensselaer Polytechnic Institute in 2015 and 2017. This article presented, compared, and assessed the level of repeatability of two tests performed according to the same specification by two different researchers: the first test was conducted in 2015 (2015_RPI01) and the second was a repeat carried out in 2017 (2017_RPI01). Each test included a 0.15 and 0.25 g input excitations. The input motions were found to be highly repeatable and were consistent in frequency content and amplitude (especially for the 0.15 g motion) and showed that the two tests were subjected to practically the same loading conditions. The 2017 model building, instrumentation, preparation of the viscous pore fluid, and model saturation were performed following methodology of the 2015 test. The recorded accelerations and excess pore water pressures (of the two tests) were shown to be in very good agreement during the early phase of shaking and also during liquefaction and dilative response (including consistency in acceleration spikes and drops in pore pressure). The consistency in the pore pressure build-up between the two tests shows that the two models had comparable compressibility values. This finding along with the long-term time history of pore pressure dissipation confirmed that the two models have practically the same soil permeability. The lateral spreading displacements were also highly consistent not only in terms of the ultimate final values but also in time history of cyclic response. All these agreements are strong indicators that the two sloping soil deposits had consistent soil characteristics including stiffness and fabric.

**Acknowledgments** The authors would like to acknowledge the financial support of the National Science Foundation Geotechnical Engineering Program directed by Dr. Richard Fragaszy (Grant No. CMMI- 1635040). This support is gratefully acknowledged. The authors would also like to acknowledge the contribution of the staff of the Centrifuge Center at Rensselaer to the performance of the centrifuge tests.

## References

Abdoun, T., Kokkali, P., & Zeghal, M. (2018). Physical modeling of soil liquefaction: Repeatability of centrifuge experimentation at RPI. *Geotechnical Testing Journal, 41*(1), 141–163. https://doi.org/10.1520/GTJ20160192.

Carey, T. J., Stone, N., & Kutter, B. L. (2019). Grain size analysis and maximum and minimum dry density of Ottawa F-65 sand for LEAP-UCD-2017. In B. Kutter et al. (Eds.), *Model tests and numerical simulations of liquefaction and lateral spreading: LEAP-UCD-2017*. New York: Springer.

Kokkali, P., Abdoun, T., & Zeghal, M. (2018). Physical modeling of soil liquefaction: Overview of LEAP production test 1 at Rensselaer Polytechnic Institute. *Soil Dynamics and Earthquake Engineering, 113*, 629–649.

Kutter, B., Carey, T., Hashimoto, T., Zeghal, M., Abdoun, T., Kokalli, P., Madabhushi, G., Haigh, S., Hung, W.-Y., Lee, C.-J., Iai, S., Tobita, T., Zhou, Y. G., Chen, Y., & Manzari, M. T. (2017). LEAP-GWU-2015 experiment specifications, results, and comparisons. *International Journal of Soil Dynamics and Earthquake Engineering*. https://doi.org/10.1016/j.soildyn.2017.05.018.

Kutter, B. L., Carey, T. J., Stone, N., Bonab, M. H., Manzari, M., Zeghal, M., Escoffier, S., Haigh, S., Madabhushi, G., Hung, W., Kim, D., Kim, N., Okamura, M., Tobita, T., Ueda, K., & Zhou, Y. (2019). LEAP-UCD-2017 V. 1.01 model specifications. In B. Kutter et al. (Eds.), *Model tests and numerical simulations of liquefaction and lateral spreading: LEAP-UCD-2017*. New York: Springer.

Manzari, M., Ghoraiby, M. E., Kutter, B. L., Zeghal, M., Abdoun, T., Arduino, P., Armstrong, R. J., Beaty, M., Carey, T., Chen, Y.-M., Ghofrani, A., Gutierrez, D., Goswami, M., Haigh, S. K., Hung, W.-Y., Iai, S., Kokkali, P., Lee, C.-J., Madabhushi, S. P. G., Mejia, L., Sharp, M., Tobita, T., Ueda, K., Zhou, Y.-G., & Ziotopoulou, K. (2018). Liquefaction analysis and experiment projects (LEAP): Summary of observations from the planning phase. *International Journal of Soil Dynamics and Earthquake Engineering, 113*, 714–743. https://doi.org/10.1016/j.soildyn.2017.05.015.

# Chapter 20
# Specifications and Results of Centrifuge Model Test at Zhejiang University for LEAP-UCD-2017

**Kai Liu, Yan-Guo Zhou, Yu She, Di Meng, Peng Xia, Jin-Shu Huang, Gang Yao, and Yun-Min Chen**

**Abstract** Three dynamic centrifuge tests with different densities (corresponding to loose, medium dense, and dense deposits) were conducted at Zhejiang University for LEAP-UCD-2017 for an exercise of repeatability and reproducibility. The same model used in LEAP-GWU-2015 representing a 5-degree slope consisting of saturated Ottawa F-65 was repeated in 2017, but more rigorous protocols and new techniques (CPT and high-speed cameras) were included in Zhejiang University experiments. Test facilities and detailed modeling and testing procedures are presented; uncertainties in input parameters are also discussed. Preliminary results associated with selected experiment are presented.

## 20.1 Introduction

LEAP is an international collaboration aiming to establish rigorous protocols for validation of soil liquefaction numerical models for practical applications (Kutter et al. 2015; Manzari et al. 2015). LEAP has several exercises focusing on different aspects of liquefaction (Carey et al. 2017). In LEAP-Kyoto-2013 and LEAP-Kyoto-2014, some inconsistencies at different centrifuge facilities were noted due to laminar containers, which also caused challenges for numerical simulations (Tobita et al. 2015). Thus in LEAP-GWU-2015, a simple boundary condition (rigid container) and other specifications were proposed for more consistent results, and an identical 5-degree sloping deposit with medium dense Ottawa F-65 sand was repeated at six centrifuge facilities, and several predictors conducted numerical simulations for validation exercises. Kutter et al. (2017) compared the experimental results and found larger differences than ideal due to variability of input parameters

K. Liu · Y. She · D. Meng · P. Xia · J.-S. Huang · G. Yao
Institute of Geotechnical Engineering, Zhejiang University, Hangzhou, China

Y.-G. Zhou (✉) · Y.-M. Chen
Department of Civil Engineering, Zhejiang University, Hangzhou, China
e-mail: qzking@zju.edu.cn



B. Kutter et al. (eds.), *Model Tests and Numerical Simulations of Liquefaction and Lateral Spreading*, https://doi.org/10.1007/978-3-030-22818-7_20

(i.e., density, fabric, saturation, etc.) at different facilities. The Kyoto and GWU exercises promoted development of understanding in mechanism and protocols of specifications for later LEAP exercises. Several numerical simulations showed great consistencies with experimental tests, providing a good opportunity for uncertainty and sensitivities analysis. For the purpose of repeatability and reproducibility, better practical experimental technology and measuring techniques were adopted in LEAP-UCD-2017 for uncertainty quantification and quality control, including CPT tests for density estimation, high-speed camera for tracing surface lateral displacement, and shear-wave velocity for detecting initial state of the model (Zhou et al. 2017); other techniques like pressure sensors in RPI (Kokkali et al. 2017) were also encouraged to use for an abundant dataset.

Besides large geotechnical centrifuge ZJU-400, uniaxial hydraulic shaker, and advanced in-flight bender element (BE) system, other unique techniques, including a two-dimensional in-flight miniature CPT system and high-speed camera, were also used in Zhejiang University in LEAP-UCD-2017. Zhejiang University rigorously followed the specifications and procedures and conducted three important models with different densities, corresponding to loose, medium dense, and dense deposits. This paper describes the detailed facilities and test procedures; uncertainty analysis in input parameters and some preliminary experimental results are discussed as well, which is expected to help further researchers to understand the experimental benchmark data of Zhejiang University in LEAP-UCD-2017.

## 20.2 Test Facilities and Specifications

### *20.2.1 Test Facilities*

The LEAP-UCD-2017 tests of Zhejiang University were conducted using the ZJU-400 centrifuge with in-flight uniaxial shaker and BE testing system, which was described in detail in Zhou et al. (2017).

Significant uncertainty in soil density among six facilities in LEAP-GWU-2015 has been recognized, so standard cone penetration tests were specified in LEAP-UCD-2017, including identical procedure and cone with specified diameter, tip material, and penetration rate but different length with respect to container's size. A two-dimensional (i.e., moving horizontally and penetrating vertically) miniature CPT actuation system was developed for in-flight measurement of centrifuge model ground at Zhejiang University. The CPT system shown in Fig. 20.1 includes the cone, guiding frame, and driving apparatus. CPT tests have been conducted before each destructive motion to evaluate the uniformity and density of the soil model, and the velocity of penetration was 0.6 mm per second and sample rate was 1 Hz.

The inner dimensions of the rigid model container (Fig. 20.2) are 770 mm long, 400 mm wide, and 500 mm deep. The end walls are made of 2-cm-thick iron plates with pterygoid laminas weld around the outer wall at different levels to ensure high strength and stiffness in order to avoid little distortion during shaking.

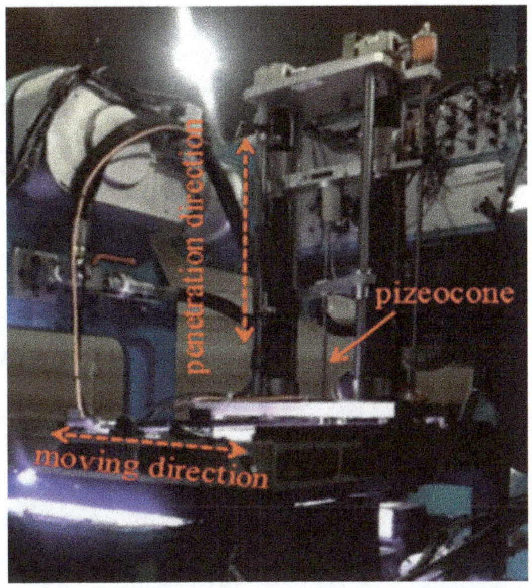


**Fig. 20.1** In-flight miniature CPTu system

**Fig. 20.2** Rigid container

### *20.2.2 Container Modifications and High-Speed Camera Installation*

The container was shortened to 666 mm in length to match the prototype specification of 20 m in length with g-level of 30 and to provide room for supporting block for supporting the thick rigid plastic cover. The supporting blocks are 52-mm-thick aluminum plate, which is braced at six locations and bolted to the end walls of the container. The block was sealed to prevent drainage along the aluminum-container interfaces.

Five high-speed cameras with good resolution were used to record the lateral displacement of surface markers on different regions of the model during spinning. A partially submersed thick rigid plastic bolted with the supporting blocks at opposite ends of the container was used to prevent surface wave of liquid.

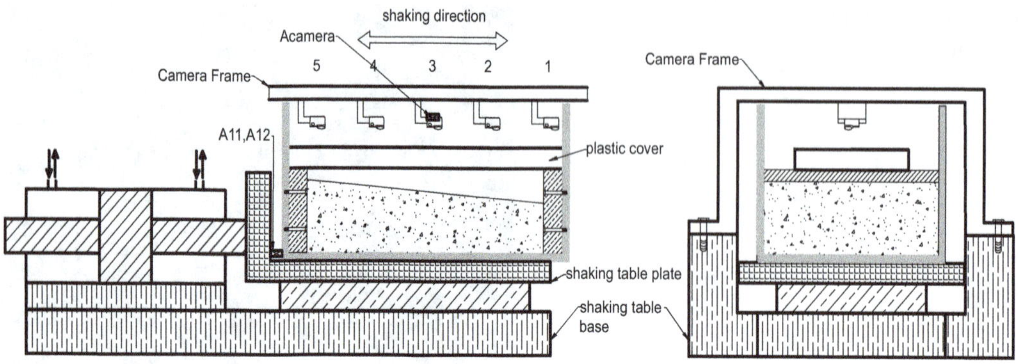


**Fig. 20.3** Structure with cameras and supporting block installation

The structure with cameras is as shown in Fig. 20.3. The cameras embraced with aluminum frame were bolted with camera frame, which is also bolted with shaking table base. The movement between the camera frame and the container bolted with shaking table plate is the same during shaking. The displacement obtained from video includes three components: movement of camera frame (d1), the container (d2), and the surface marks relative to the container (d3), above which the last component is of interest. Acceleration records of the accelerometer ($A_{camera}$) installed at the middle camera and A11 (A12) were used to obtain movement of camera frame and the container, respectively, by double integration.

## 20.2.3 Model Description

The same model used in LEAP-GWU-2015 was repeated in LEAP-UCD-2017. The model represents a 5-degree, 4-m-depth at midpoint, 20-m-length sand slope deposit of Ottawa F-65. It notes that the shaking direction is parallel to the axis of the centrifuge. The soil surface normal to slope direction was not curved according to the radius of the centrifuge. Considering the centrifuge radius to surface of the model and width of container are 4182 mm and 400 mm, respectively, the flat surface in the end view represented a gentle 0.14-m-high, 12-m-wide hill in prototype scale, with an average ground slope of 2.4% toward the side walls of the container.

The instruments are shown in Fig. 20.4. The required and highly recommended sensors were installed in the model. There were six accelerometers AH1 to AH9 inside the model recording horizontal acceleration and the other four (AH11, AH12, AV1, and AV2) on the container to monitor the input motion in both horizontal and vertical direction. Eight pore pressure transducers of P1 to P10 were also installed inside the model. Three pairs of bender elements, at the same depth with pore pressure transducers (elements S1-R1, S2-R2, and S3-R3 were located at the same

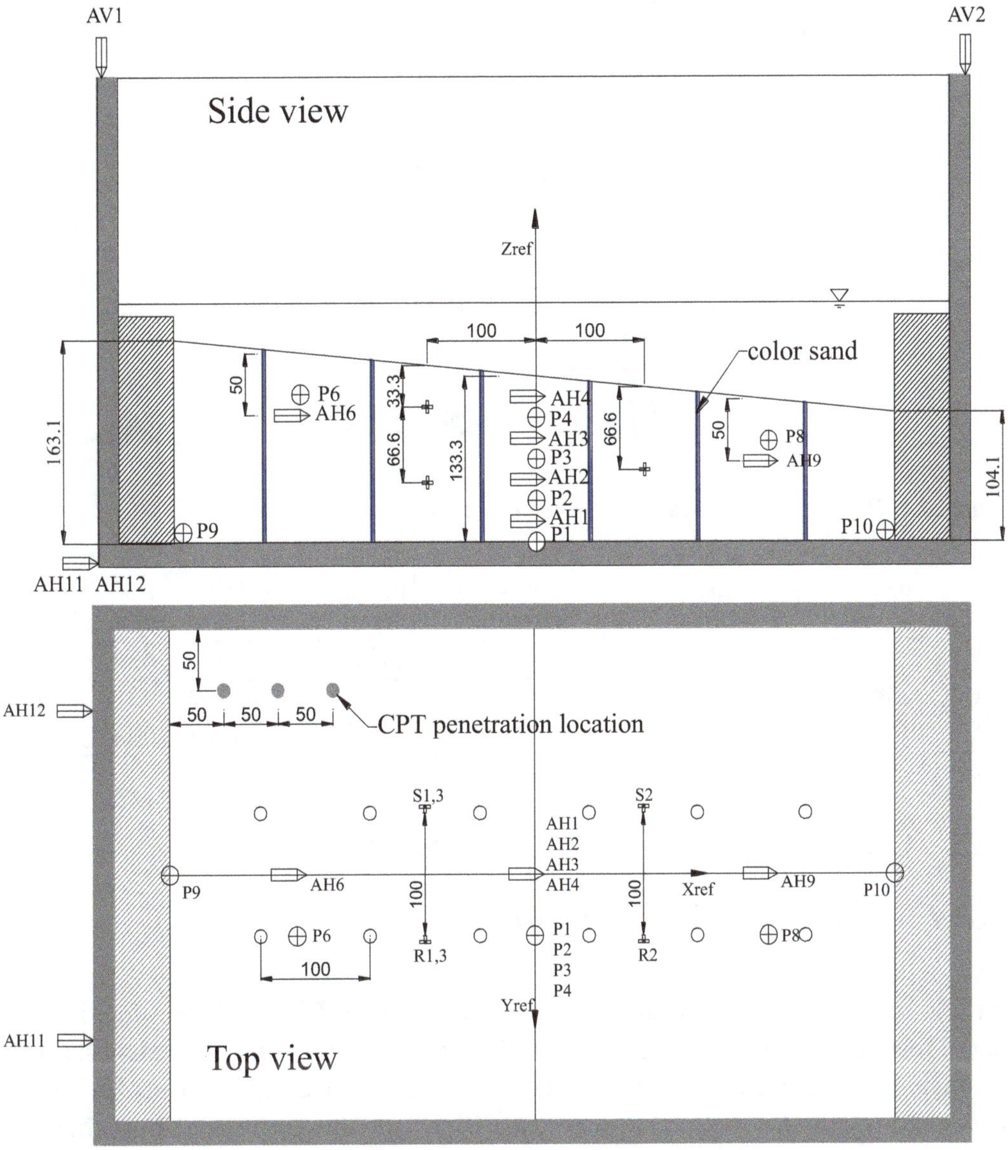


**Fig. 20.4** Centrifuge model setup and installation locations

depth as P4, P3, and P2, respectively), were placed to measure vertically polarized and horizontal travelling SV shear-wave velocity.

Surface markers were used to trace surface displacement. The specified surface markers are shown in Fig. 20.5 in red, which consists of a 10-mm-length, 25-mm-in-diameter PVC tube with an aluminum cross bar fixed in center. In addition, surface markers made of zip ties were also used, and all the surface markers were installed in a 0.05 m $\times$ 0.05 m grid (model scale). Colored sand columns following the same method as described in Zhou et al. (2017) were used to determine lateral spread profiles by excavation after finally spin-down.

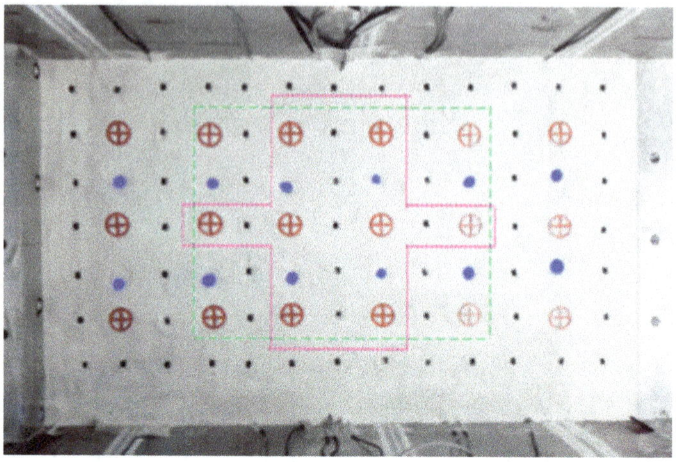

**Fig. 20.5** Surface markers and colored sand columns

**Table 20.1** Properties of Ottawa sand used in ZJU for each model

| Model | Target density (kg/m$^3$) | Achieved density (kg/m$^3$) | Min. density (kg/m$^3$) | Max. density (kg/m$^3$) | D10 (mm) | D30 (mm) | D60 (mm) |
|---|---|---|---|---|---|---|---|
| ZJU1 | 1651 | 1672 | 1496 | 1784 | 0.116 | 0.148 | 0.186 |
| ZJU2 | 1599 | 1606 | 1491 | 1786 | 0.102 | 0.136 | 0.186 |
| ZJU3 | 1703 | 1706 | 1490 | 1786 | 0.104 | 0.137 | 0.186 |

## 20.3 Centrifuge Experiment Preparation

### *20.3.1 Properties of Test Material*

For eliminating the effect of inconsistency of Ottawa F-65 at different facilities on model response, maximum and minimum dry density and grain analysis tests were conducted before each model according to "Chinese code for soil tests for hydropower and water conservancy engineering, 2006," and the results are summarized on Table 20.1. Figure 20.6 shows the grain-size distribution curve for three models, and there are little differences in grain-size distribution among three models. Additional material properties are available in Carey et al. (2016). In LEAP-UCD-2017 the density was specified instead of void ratio, and the target and achieved densities for ZJU1, ZJU2, and ZJU3 are summarized in Table 20.1 and correspond to loose, medium dense, and dense model.

### *20.3.2 Achieved Density*

Air pluviation method was adopted to ensure a high level of uniformity. The apparatus for air pluviation and construction procedures are available in Zhou et al. (2017). The calibration results using the specified spout were different from

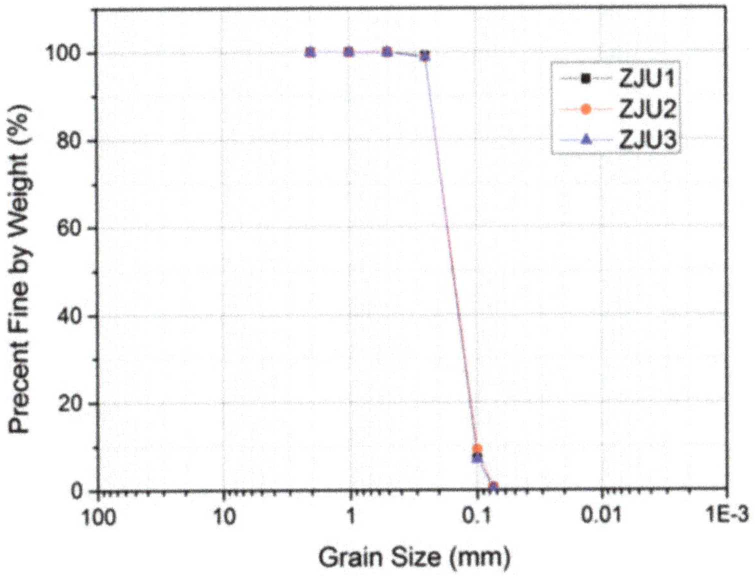


**Fig. 20.6** Grain-size distribution curve

**Fig. 20.7** Cone tip resistance profile results

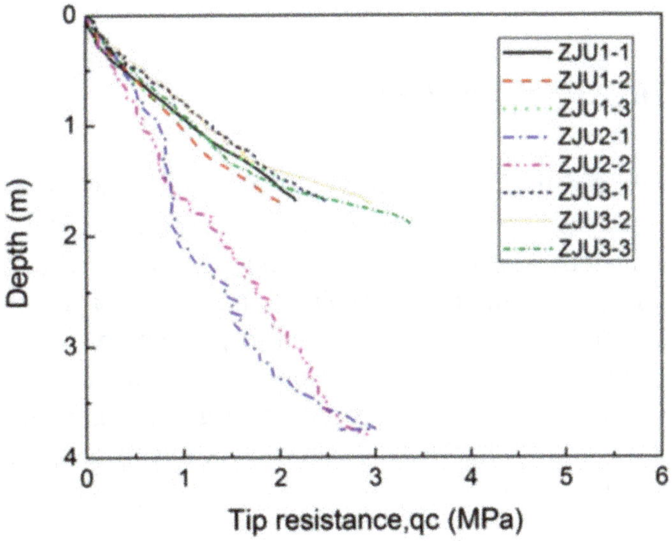


that specified. So a spout with one slot shape was adopted. ZJU1 and ZJU2 adopted the same calibration curve, which could not meet the specified density of ZJU3, so that a spout with a thinner slot was used. There was 3–4 kg/m$^3$ difference between two calibration results at the same drop height, besides the falling height changed less than 2 cm during pluviation, which causes a maximum deviation of 7 and 8 kg/cm$^3$ from the target density for ZJU1 (ZJU2) and ZJU3, respectively. Moreover the model settled after saturation. Overall, the achieved density is expected to deviate from the target within ±11 kg/m$^3$.

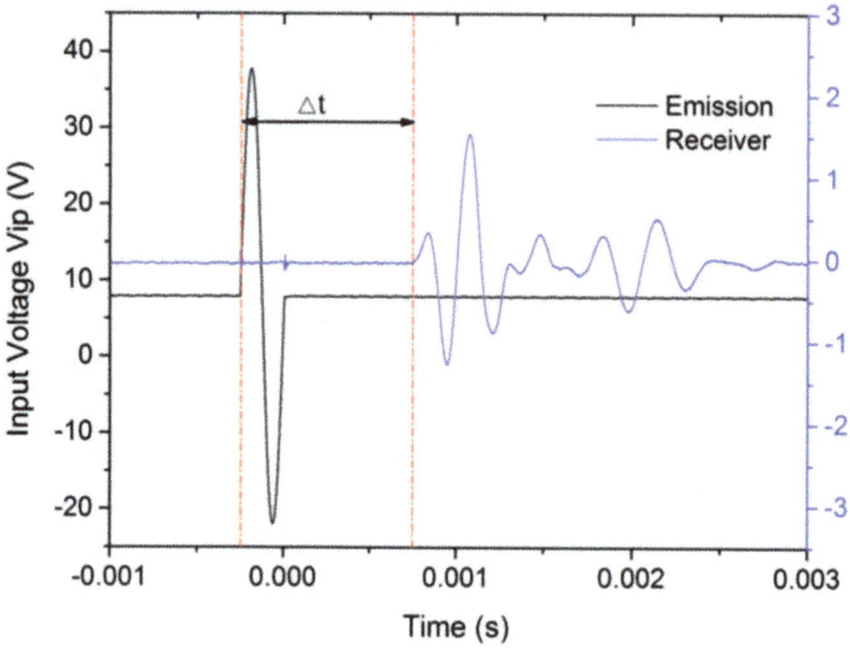


**Fig. 20.8** Typical signals of In-flight BE

### 20.3.3 *Saturation*

Silicone oil (RC0201-30cST, $\rho_d = 955$ kg/m$^3$ at 25 °C) with 30 times of viscosity of water was used as pore fluid to satisfy the scaling law. The oil tank and model container were kept under the same vacuum level and the oil was de-aired, and the transport from the reservoir to the container was driven by gravity feed. The saturation speed was controlled to prevent soil disturbance at the bottom of container. The degree of saturation for three models was not measured; however, the saturation devices and procedures were the same with these used in LEAP-GWU-2015, which provides a reference of degree of saturation for three models, indicating achieved $S_r > 99.5\%$.

## 20.4 Test Procedure and Achieved Motions

### 20.4.1 *Test Procedure*

The test procedure is shown in Table 20.2. A careful survey of the surface markers was made before spin-up. Then the centrifuge was spun up to 10 g, 20 g, and 30 g step by step. The $V_s$ was measured on each g-level when the pore pressure at this stage was stable. After consolidation in 30 g, the model was subjected to step-wave motion, which was too small to generate any excess pore pressure and was used to

Table 20.2 Test procedure and events of interest in the test

| No. | Name | N(g) | Description | No. | Name | N(g) | Description |
|---|---|---|---|---|---|---|---|
| 1 | S-1 | 1 | Surface marker survey | 18 | CPT-2 | 30 | CPT test |
| 2 | A | | Swing-up | 19 | E | 30 | Motion2 |
| 3 | BE-1 | 10 | $V_s$ measurement | 20 | BE-8 | 30 | $V_s$ measurement |
| 4 | BE-2 | 20 | $V_s$ measurement | 21 | SW-4 | 30 | Step wave |
| 5 | BE-3 | 30 | $V_s$ measurement | 22 | F | | Start of swing-down |
| 6 | SW-1 | 30 | Step wave | 23 | S-3 | 1 | Surface marker survey |
| 7 | CPT-1 | 30 | CPT test | 24 | G | | Start of swing-up |
| 8 | B | 30 | Motion1 | 25 | BE-9 | 10 | $V_s$ measurement |
| 9 | BE-4 | 30 | $V_s$ measurement | 26 | BE-10 | 20 | $V_s$ measurement |
| 10 | SW-2 | 30 | Step wave | 27 | BE-11 | 30 | $V_s$ measurement |
| 11 | C | | Start of swing-down | 28 | SW-5 | 30 | Step wave |
| 12 | S-2 | 1 | Surface marker survey | 29 | CPT-3 | 30 | CPT test |
| 13 | D | | Start of swing-up | 30 | H | 30 | Motion3 |
| 14 | BE-5 | 10 | $V_s$ measurement | 31 | BE-12 | 30 | $V_s$ measurement |
| 15 | BE-6 | 20 | $V_s$ measurement | 32 | SW-6 | 30 | Step wave |
| 16 | BE-7 | 30 | $V_s$ measurement | 33 | I | | Start of swing-down |
| 17 | SW-3 | 30 | Step wave | 34 | S-4 | 1 | Surface marker survey |

characterize the model. Then CPT test was conducted for determining the density of the model and then motion1 (destructive motion) followed. Enough time was needed for full dissipation of excess pore pressure in the soil, then step wave was used again to characterize the deposit, and the centrifuge was spun down and surface markers were surveyed. Thereafter the centrifuge spun up and later procedures were the same as before except for destructive motions. $V_s$ was measured by bender elements before and after each destructive motion when the pore pressure was stable at each stage. Three models followed the same procedures.

### 20.4.2 In-Flight Characterizations

Figure 20.7 shows the cone tip resistance results versus depth profile for three models. Because of design mistakes of CPT support device above the container, the penetration depths for ZJU1 and ZJU3 are less than 2 m in prototype scale. Unfortunately, due to time constraints associated with LEAP program, the deficient CPT support device was adopted in ZJU1 and ZJU3, the improved device was used in ZJU2, and the penetration depth was up to 4 m in prototype scale.

The bender elements (BE) used in tests are electrically shielded and grounded to avoid undesired electromagnetic interfere in centrifuge environment. Shown in Fig. 20.8 is a typical signal of BE during spinning, indicating the arrival of receiver is well distinguishable to ensure reliable results of BE. The $V_s$ data and associated analysis will be presented in a later publication.

Fig. 20.9 PIV analysis points of surface marker

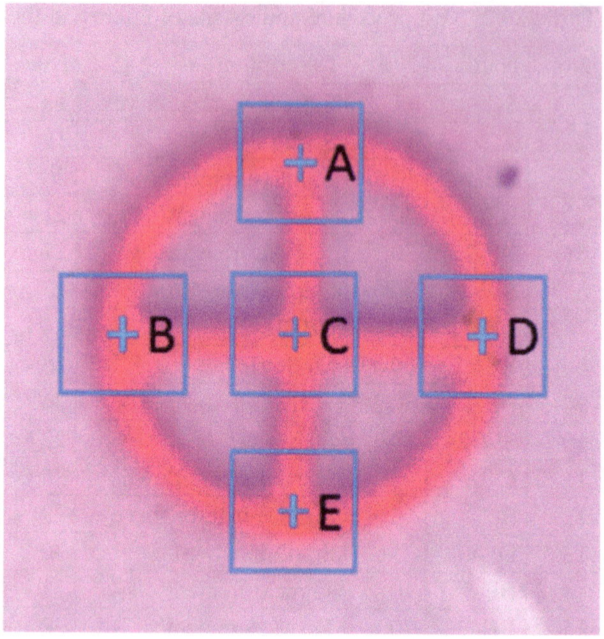


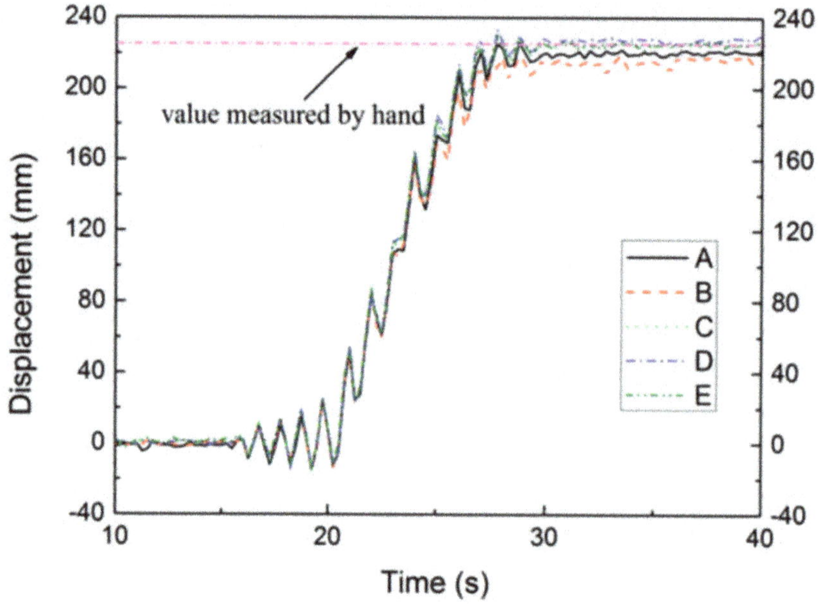


**Fig. 20.10** Typical results from PIV analysis

As mentioned above, five GoPro cameras were installed above the slope surface to record movement of surface markers. The videos were converted to displacement time history by GeoPIV analysis procedure. There are five points located at different regions of the surface marker used to ensure reliable results shown in Fig. 20.8. Figure 20.10 shows typical results of dynamic displacement of one surface marker from five points, showing high consistency within five points. And also, the displacement value obtained using videos is consistent with that measured by hand afterward.

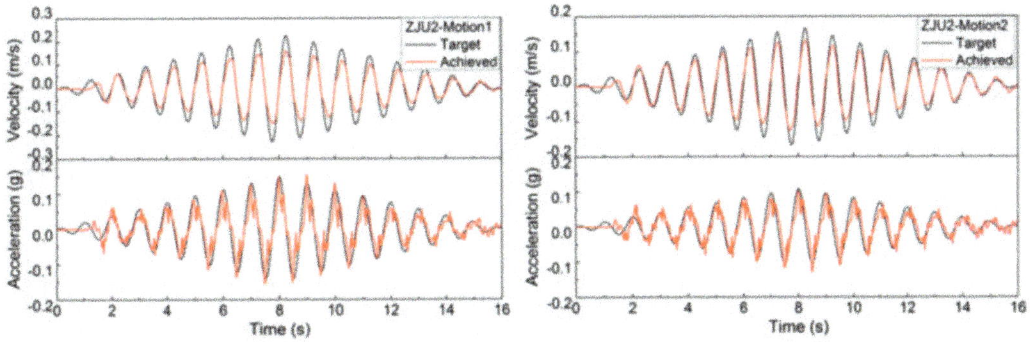


(a) (b)

**Fig. 20.11** Achieved and specified acceleration and velocity time history: (**a**) ZJU2-motion1; (**b**) ZJU2-motion2

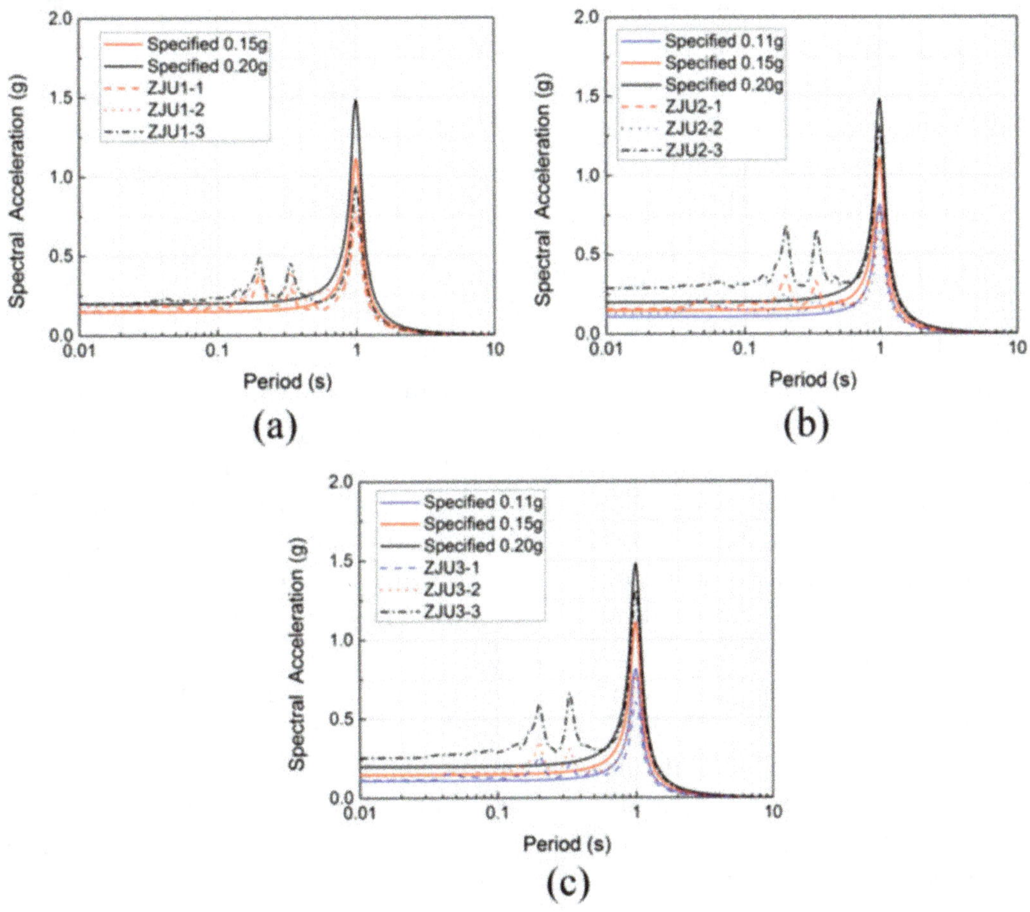


**Fig. 20.12** Achieved and specified 5% damped acceleration response spectra: (**a**) ZJU1; (**b**) ZJU2; (**c**) ZJU3

## 20.4.3 *Achieved Horizontal Components*

For length limitation, only selected achieved horizontal base accelerations are presented in Fig. 20.11 and compared with target accelerations. The shown

**Table 20.3** Ground motion sequence for LEAP-UCD-2017 experiments, unit:g

| Model | Motion | Target $PGA_{eff}$ | A11 | | | A12 | | |
|---|---|---|---|---|---|---|---|---|
| | | | $PGA_{1Hz}$ | $PGA_{hf}$ | $PGA_{eff}$ | $PGA_{1Hz}$ | $PGA_{hf}$ | $PGA_{eff}$ |
| ZJU1 | Motion111 | 0.15 | 0.093 | 0.070 | 0.128 | 0.093 | 0.071 | 0.128 |
| | Motion2 | 0.15 | 0.099 | 0.093 | 0.146 | 0.100 | 0.093 | 0.147 |
| | Motion3 | 0.20 | 0.118 | 0.101 | 0.169 | 0.118 | 0.095 | 0.166 |
| ZJU2 | Motion1 | 0.15 | 0.097 | 0.083 | 0.139 | 0.098 | 0.088 | 0.142 |
| | Motion2 | 0.11 | 0.079 | 0.061 | 0.110 | 0.08 | 0.066 | 0.113 |
| | Motion3 | 0.2 | 0.164 | 0.126 | 0.227 | 0.164 | 0.127 | 0.228 |
| ZJU3 | Motion1 | 0.11 | 0.077 | 0.051 | 0.103 | 0.077 | 0.045 | 0.100 |
| | Motion2 | 0.15 | 0.097 | 0.073 | 0.134 | 0.098 | 0.072 | 0.134 |
| | Motion3 | 0.2 | 0.164 | 0.101 | 0.215 | 0.165 | 0.097 | 0.214 |

acceleration time histories are filtered with a 0.2 Hz (prototype scale) high-pass filter, and the achieved acceleration matches the specified motion well. In LEAP-2017, a new concept of effective PGA was adopted to evaluate the accuracy and efficiency of the motions. The effective PGA is defined as below:

$$PGA_{effective} = PGA_{1Hz} + 0.5 \times PGA_{hf} \tag{20.1}$$

where $PGA_{hf}$ represents the peak acceleration of the high-frequency component of the motion and $PGA_{1Hz}$ denotes the peak acceleration with bandpass filtering of 0.9–1.1 Hz. The results of all the input motions for three models are summarized in Table 20.3. It is found that $PGA_{1Hz}$ values are almost the same between AH11 and AH12 for one input motion, indicating the inputs of shaking table plane are uniform.

The results showed that it is not easy to make the achieved motions match well with the target because the high-frequency components vary so abruptly and arbitrarily. However, the experiences learned from ZJU1 are important to accurately control the further base motions of ZJU2 and ZJU3. The achieved motions matched well with the target in amplitude and waveform, and the three models provide good benchmarks for numerical predictors.

Five percent damped acceleration response spectra (ARS) for all motions for three models are shown in Fig. 20.12. All achieved peak spectral accelerations were smaller than the specified at the main intended frequency of 1 Hz and showed high-frequency components. The achieved peak acceleration in 1 Hz (prototype scale) was almost identical for motions with the same target peak acceleration, which shows a good performance of the hydraulic shaker.

### *20.4.4 Achieved Vertical Components*

Figure 20.13 displays the selected vertical accelerations measured by AV1 and AV2 at opposite ends of container. Although zero vertical acceleration is desired during shaking, the hydraulic shaker produced unintended vertical component in

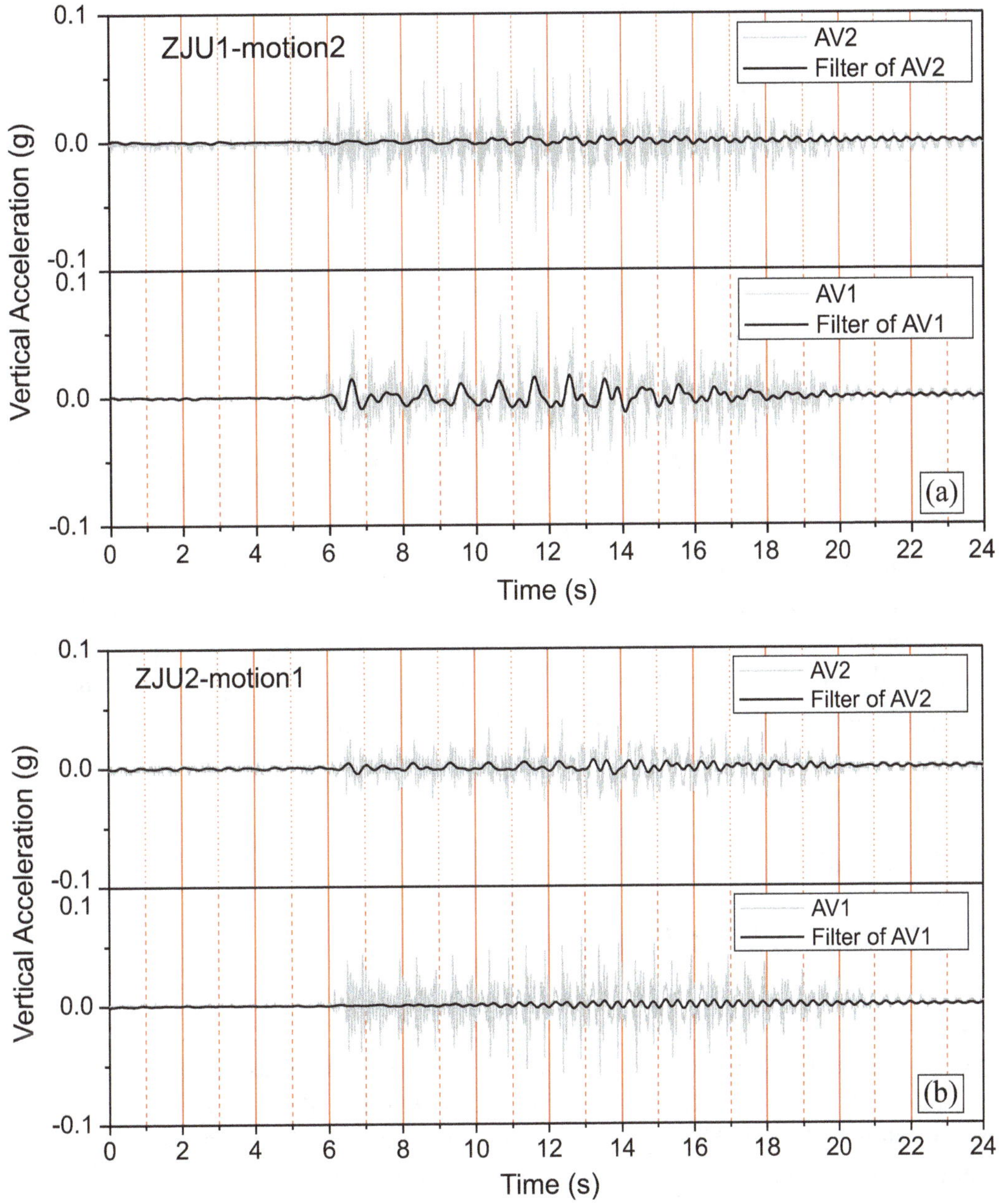


**Fig. 20.13** Vertical accelerations on container ends: (**a**) ZJU1-motion2; (**b**) ZJU2-motion1

addition to the horizontal accelerations. The gray lines indicate unfiltered vertical motions, and the black ones, bandpass filtering with 0.3–3 Hz in consistency with Kutter et al. (2017) to reduce high frequency noise, are superimposed as well. The amplitude of the filtered vertical component ranges from 4% to 17% of the achieved 1 Hz horizontal component. The rocking accelerations are not very large and the larger spikes are of relatively high frequency. It is worth noting that the angular displacement due to rocking is very small as shown in Fig. 20.14 for ZJU1-motion2

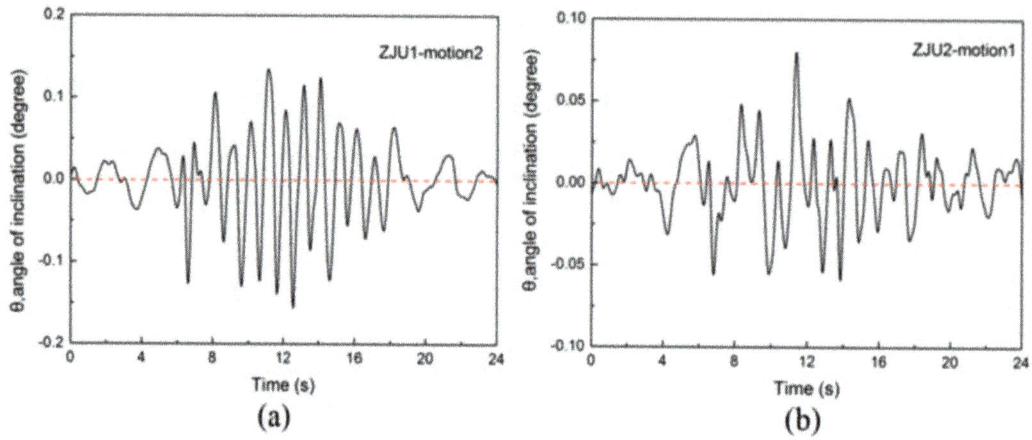


**Fig. 20.14** Angle of rocking displacement during shaking: (**a**) ZJU1-motion2; (**b**) ZJU2-motion1

and ZJU2-motion1, respectively. The angle of inclination was computed by integrating accelerations to obtain vertical displacements of AV1 and AV2. The difference in these displacements was divided by the length of the box to determine the rocking angle shown in Fig. 20.14. The maximum rocking angle during shaking was no more than 0.15 degrees along longitudinal direction of the container, indicating rocking displacements can be neglected during shaking.

## 20.5 Results

### *20.5.1 Acceleration Responses*

Figures 20.15 and 20.16 only show the acceleration time histories and Fourier spectrum of motion1 and motion3 in ZJU3 because of length restriction. For motion1 in ZJU3, the trace of accelerations recorded at base and central array along the depths is very similar during shaking, and obvious amplification effect was observed, indicating the soil moves almost as a rigid mass and the shear strain accumulated in the soil is small; however, for motion3 in ZJU3, the acceleration time history shows de-amplification in upslope direction and negative spikes in downslope direction, which was observed in LEAP-GWU-2015 and other previous studies (Zeghal and Elgamal 2015). Looking closely at the excess pore pressure time history during motion1 and motion3, there was little liquefaction occurring during motion1 and severe liquefaction happening during motion3, which accounts for different site responses of acceleration. It finds that the spikes mainly consist of high-frequency components when comparing Fourier spectrum of motion1 and motion3.

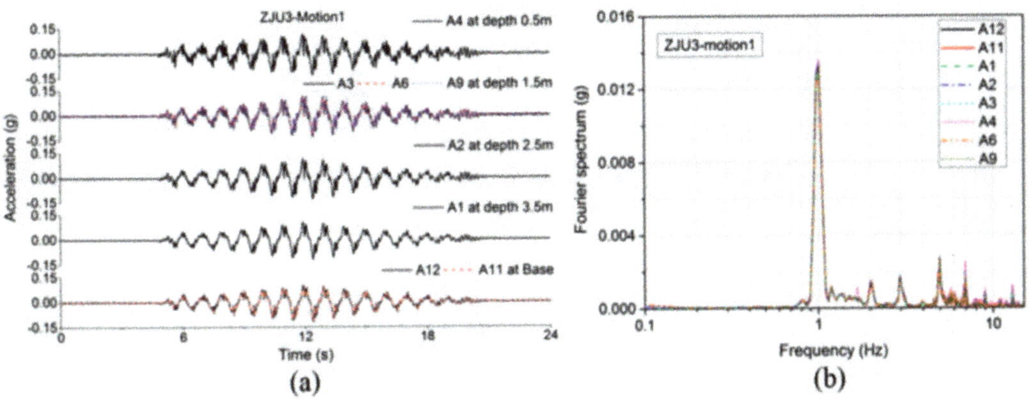


**Fig. 20.15** ZJU3-motion1: (**a**) acceleration time history; (**b**) Fourier spectrum

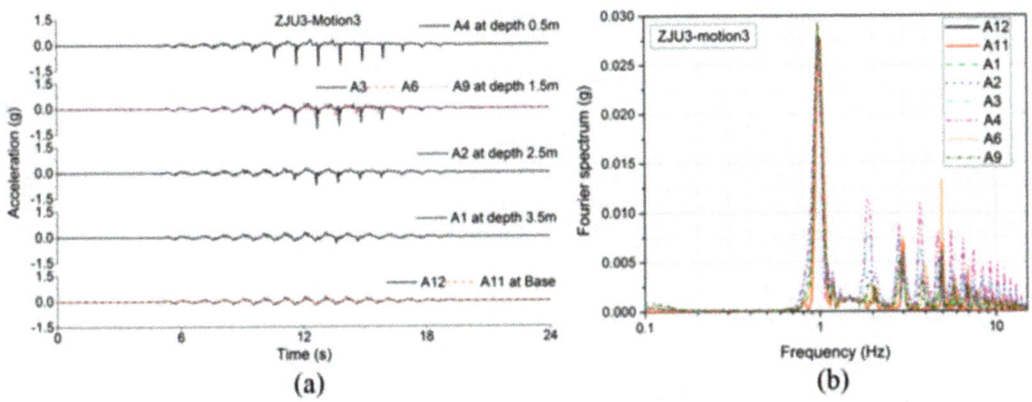


**Fig. 20.16** ZJU3-motion3: (**a**) acceleration time history; (**b**) Fourier spectrum

## *20.5.2 Pore Pressure Response*

Figure 20.17 shows the time history of excess pore pressure ratio for motion1 and motion3 in ZJU3. It notes that pore pressure sensors of P1 and P9 didn't work well due to without full saturation disposure before being placed in model, the same problem with P1 and P3 in ZJU1. For motion1 in ZJU3, excess pore pressure generated rapidly and started dissipating before the end of shaking; no liquefaction occurred even in shallow depth. However, severe liquefaction happened during motion3 in ZJU3; significant sudden drops in excess pore pressure due to dilatancy were observed, which corresponds to large spikes in acceleration. Figure 20.17 clearly depicts liquefaction state arose at shallow depth and then propagated downward. After maintaining for a while, then dissipation began after the end of shaking. The dissipation of excess pore pressure started early at deep depth, and then solidification front (red circle) moved upward.

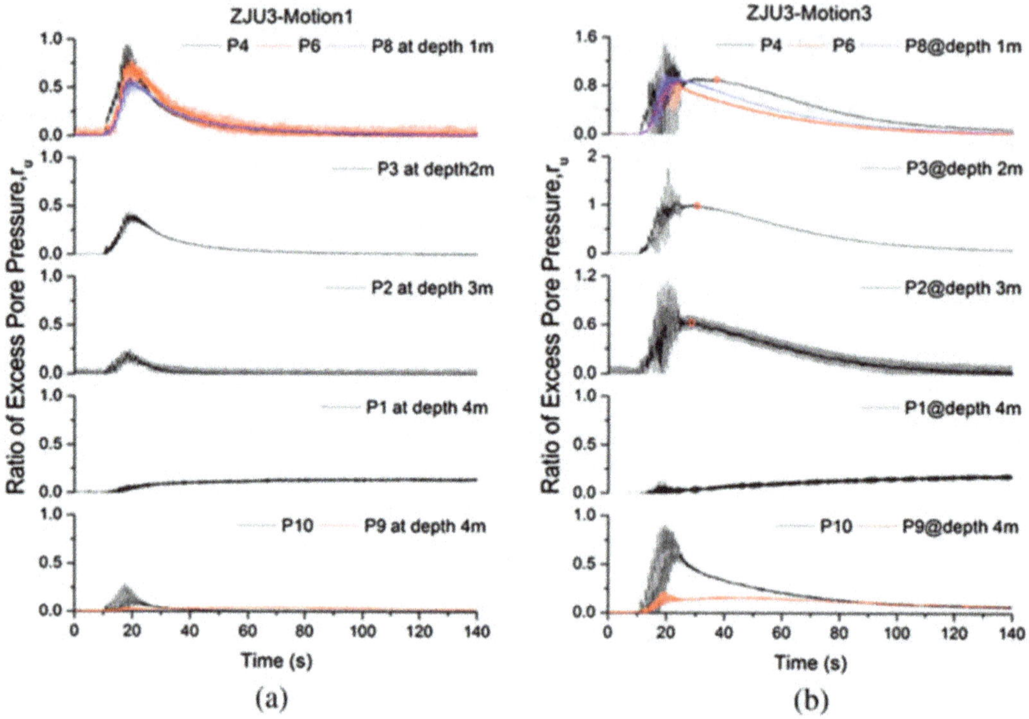


**Fig. 20.17** Pore pressure time history: (**a**) ZJU3-motion1; (**b**) ZJU3-motion3

## *20.5.3 Displacement Response*

### Vertical Displacement $D_v$ Response

The vertical displacements ($D_v$) were surveyed via the specified red surface markers after each motion, and the average values and standard deviation of vertical displacement for each motion calculated from only red surface in pink dashed frame shown in Fig. 20.5 are summarized in Table 20.4 in prototype scale. The standard deviations of $D_v$ of the surface markers are less than 20%, indicating the discreteness of the data is relatively small.

### Horizontal Displacement $D_v$ Response

Average values and standard deviation of lateral displacement ($D_L$) of surface markers in green dashed frame shown in Fig. 20.5 after each motion are summarized in Table 20.5 in prototype scale. The standard deviations of $D_v$ for the surface markers are within 63%, indicating it is acceptable that the average value represents the overall level of lateral displacement.

Figure 20.18 shows the lateral displacement profiles from excavation. The profiles showed that the displacement distributed over the whole depth and reached the maximum at the surface.

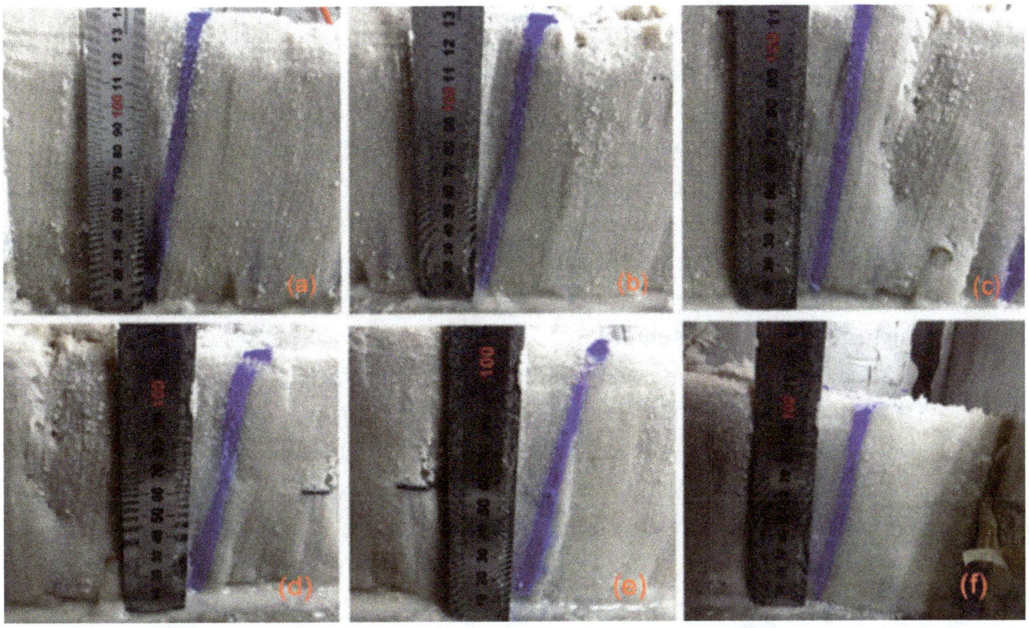


**Fig. 20.18** Typical lateral displacement profiles from excavation for ZJU2 model: pictures from (**a**) to (**e**) correspond to six colored sand columns from the top to the toe

**Table 20.4** Average vertical displacements after each motion

| | Motion1 | | Motion2 | | Motion3 | |
|---|---|---|---|---|---|---|
| Model | μ[a]/mm | $\sigma^b$ | μ/mm | $\sigma$ | μ/mm | $\sigma$ |
| ZJU1 | 42.4 | 13 | 30.0 | 16 | 24.0 | 5 |
| ZJU2 | 57.4 | 20 | 13.9 | 5 | 21.4 | 20 |
| ZJU3 | 22.9 | 10 | 7.9 | 7 | 39.0 | 14 |

[a]Average value
[b]Standard deviation of vertical displacement of the 12 red surface markers in green dashed frame shown in Fig. 20.5

**Table 20.5** Average values of lateral displacement after each motion

| | Motion1 | | Motion2 | | Motion3 | |
|---|---|---|---|---|---|---|
| Model | μ[a]/mm | $\sigma^b$ | μ/mm | $\sigma$ | μ/mm | $\sigma$ |
| ZJU1 | 151 | 51 | 55 | 63 | 36 | 28 |
| ZJU2 | 293 | 35 | 38 | 22 | 200 | 18 |
| ZJU3 | 22.9 | 10 | 7.9 | 7 | 39.0 | 14 |

[a]Average value and
[b]Standard deviation of horizontal displacement of surface markers in green dashed frame shown in Fig. 20.5

## 20.6 Summary and Conclusions

Three important centrifuge tests were conducted at Zhejiang University in LEAP-UCD-2017. Three models were designed in different densities corresponding to loose, medium dense, and dense Ottawa sand deposit for a comprehensive comparison of density effect on site response. The main purpose of this paper is to help further researcher to understand the data of tests uploaded in NEES website, so that information on test facilities, model setup and preparation, test procedures, and analysis of the results is presented.

Besides the centrifuge, shaking table, and in-flight BE system used in LEAP-GWU-2015, a two-dimensional miniature CPTu system was used to determine the density of the deposit following the specified and identical procedures and cone. High-speed cameras were installed above the model to trace the displacement of surface markers; the structures with cameras and the method to obtain displacement time history are introduced. The model preparation and tests procedures are also detailed. The pluviation devices and procedures lead to uncertainty in density within $\pm 11$ kg/m$^3$ with the achieved density. The degree of saturation for three models was not measured but based upon previous experience with this method (Zhou et al. 2017), the achieved $S_r > 99.5\%$.

The base input acceleration closely matched the target, but velocity was slightly smaller. The achieved effective PGA for each motion roughly match the targets. Five percent damped acceleration response spectra show the achieved motions were smaller than the target of 1 Hz components and some high-frequency components in input motion. The vertical accelerations at opposite ends of container were small, indicating a negligible rocking effect during shaking.

Results of motion1 and motion3 for ZJU3 are exampled to illustrate different responses to liquefaction and non-liquefaction. Motion1 didn't trigger liquefaction even at shallow depth but high excess pore pressure generating during shaking, the trace of accelerations for the base motion and central array is very similar, and an obvious amplification in amplitude was observed; there were no spikes for motion1. Motion3 led to severe liquefaction up to 2 m below the surface; spikes in acceleration time history due to dilatancy are consistent with drops in excess pore pressure. Lateral and vertical displacements for each motion were surveyed via some surface markers. Low standard deviation indicates high confidence with results.

**Acknowledgments** This study is partly supported by the National Natural Science Foundation of China (Nos. 51578501, 51778573), the National Program for Special Support of Top-Notch Young Professionals (2013), the Zhejiang Provincial Natural Science Foundation of China (No. LR15E080001), and the National Basic Research Program of China (973 Project) (No. 2014CB047005).

## References

Carey, T. J., Hashimoto, T., Cimini, D., & Kutter, B. L. (2017). LEAP-GWU-2015 centrifuge test at UC Davis. *International Journal of Soil Dynamics and Earthquake Engineering, 113*, 663–670. https://doi.org/10.1016/j.soildyn.2017.01.030.

Carey, T. J., Kutter, B. L., Manzari, M. T., Zeghal, M., & Vasko, A. (2016). *LEAP soil properties and element test data*. https://doi.org/10.17603/DS2WC7W.

Kokkali, P., Abdoun, T., & Zeghal, M. (2017). Physical modeling of soil liquefaction: Overview of LEAP production test 1 at Rensselaer Polytechnic Institute. *Soil Dynamics and Earthquake Engineering*. https://doi.org/10.1016/j.soildyn.2017.01.036.

Kutter, B. L., Manzari, M. T., Zeghal, M., Zhou, Y. G., & Armstrong, R. J. (2015). Proposed outline for LEAP verification and validation processes. In S. Iai (Ed.), *Geotechnics for catastrophic flooding events* (p. 99–108). London: Taylor & Francis.

Kutter B. L., Carey T. J., Hashimoto T., Zeghal M., Abdoun T., Kokkali P., Madabhushi G., Haigh S., d'Arezzo F. B., Madabhushi S., Hung W. Y., Lee C. J., Cheng H. C., Iai S., Tobita T., Ashino T., Ren J. F., Zhou Y. G., Chen Y. M., Sun Z. B., Manzari M. T. (2017). LEAP-GWU-2015 experiment specifications, results, and comparisons. *International Journal of Soil Dynamics and Earthquake Engineering, 113*, 618–28.

Manzari, M. T., Kutter, B. L., Zeghal, M., Iai, S., Tobita, T., Madabhushi, S. P. G., Haigh, S. K., Mejia, L., Gutierrez, D. A., & Armstrong, R. J. (2015). LEAP projects: Concept and challenges. In *Proceedings of the fourth international conference on geotechnical engineering for disaster mitigation and rehabilitation (4th GEDMAR): 2014 Sept 16–18*. Kyoto: Taylor & Francis.

Tobita, T., Manzari, M. T., Ozutsumi, O., Ueda, K., Uzuoka, R., & Iai, S. (2015). Benchmark centrifuge tests and analyses of liquefaction-induced lateral spreading during earthquake. In S. Iai (Ed.), *Geotechnics for catastrophic flooding events* (p. 127–82). London: Taylor & Francis.

Zeghal, M., & Elgamal, A. Site response and vertical seismic arrays (2015). *Progress in Structural Engineering and Materials*, 2(1):92–101.

Zhou, Y.-G., Sun, Z.-B., & Chen, Y.-M. (2017). Zhejiang University benchmark centrifuge test for LEAP-GWU-2015 and liquefaction responses of a sloping ground. *International Journal of Soil Dynamics and Earthquake Engineering, 113*, 698–713. https://doi.org/10.1016/j.soildyn.2017.03.010.

# Part III
# Numerical Simulation Papers

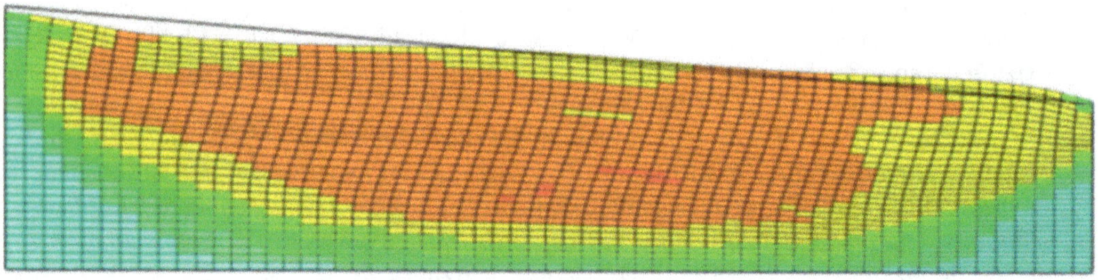

Image credit: Fukutake and Kiriyama

# Chapter 21
# Prediction of LEAP-UCD-2017 Centrifuge Test Results Using Two Advanced Plasticity Sand Models

**Long Chen, Alborz Ghofrani, and Pedro Arduino**

**Abstract** In accordance with the Liquefaction Experiments and Analysis Projects (LEAP)-UCD-2017 guidelines, two stress-dependent bounding surface constitutive models, Manzari-Dafalias and PM4Sand, were calibrated for Class-B prediction of centrifuge experiments of a sloped ground surface model of uniformly deposited Ottawa F-65 sand. Different calibration techniques and objectives were chosen for each material model. It was shown that both models were capable of simulating the behavior of cohesionless soils under liquefaction.

## 21.1 Introduction

Soil liquefaction induced by earthquakes can cause significant damage to adjacent structures and lead to considerable economic loss. The mechanism and effects of soil liquefaction have been studied extensively throughout the years. With the development of computational tools and advanced constitutive models which can capture complex soil behavior under various loading and drainage conditions, numerical modeling has become popular for predicting liquefaction-induced ground failure and deformations. For a numerical model to produce reasonable results, a comprehensive verification and validation study is necessary. Centrifuge tests, which can physically represent field conditions under earthquakes, have been used to provide benchmark studies for numerical analysis. VELACS (Verification of Liquefaction Analysis by Centrifuge Studies) (Arulanandan and Scott 1993) was a valuable initial attempt that identified the limitations of prevailing procedures. LEAP (Liquefaction Experiments and Analysis Projects) (Manzari et al. 2018) is a new effort to validate numerical models using centrifuge tests. In this work comparisons between experimental results and numerical

L. Chen · A. Ghofrani · P. Arduino (✉)
Department of Civil and Environmental Engineering, University of Washington, Seattle, WA, USA
e-mail: longchen@uw.edu; alborzgh@uw.edu; parduino@uw.edu



B. Kutter et al. (eds.), *Model Tests and Numerical Simulations of Liquefaction and Lateral Spreading*, https://doi.org/10.1007/978-3-030-22818-7_21

simulations obtained using two different constitutive models, i.e., Manzari-Dafalias (Dafalias and Manzari 2004) and PM4Sand (Boulanger and Ziotopolou 2015), implemented in the OpenSees finite element framework are presented and discussed.

## 21.2 Backgrounds

LEAP embodies a series of projects that attempt to extend understanding of the behavior of saturated granular soils subjected to seismic loading and at the same time provide high-quality experimental data sets for evaluation of constitutive models capable of reproducing liquefaction. Collectively these projects shed light on the performance of computational material models and numerical frameworks using benchmark centrifuge experiments representing different geotechnical settings. After an initial set of experiments in 2015 that resulted in the LEAP-GWU-2015 (Kutter et al. 2018) workshop, a second set of experiments was conducted at nine centrifuge facilities around the world, namely, Cambridge University (CU), Ehime University (Ehime), Korea Advanced Institute of Science and Technology (KAIST), Kyoto University (KyU), Taiwan National Central University (NCU), the University of California Davis (UCD), Rensselaer Polytechnic Institute (RPI), French Institute of Science & Technology for Transportation, development and networks (IFSTTAR), and Zhejiang University (ZJU). Blind simulations were completed and submitted to the LEAP organizers, and a second workshop (LEAP-UCD-2017) was organized at Davis at the end of 2017.

In all these studies, the well-characterized Ottawa F-65 sand (Bastidas 2016; Vasko 2015; El Ghoraiby et al. 2017, 2019) was used. During the second set of experiments, the sand was deposited uniformly in an instrumented 5-degree slope with a target density of 1652 kg/m$^3$. All centrifuge experiments used a rigid box container to eliminate complexities in boundary conditions. With small variations, depending on the configuration of each centrifuge facility, all models were prepared to represent a unique prototype size: 20 m in length, 4 m in height at the midpoint, and width greater than 9 m to reduce the effects of frictional boundary conditions on the container sidewalls. Four accelerometers (AH1–AH4) and four pore pressure transducers (P1–P4) were positioned along the centerline. A series of four 1 Hz ramped sine motions were imposed on each model. Motions 2 and 3 with a 0.15 g and 0.25 g PGA, respectively, were chosen strong enough to represent a destructive event. Motion 2 acceleration time history is shown in Fig. 21.1 and was used in all the Type B predictions presented in this paper.

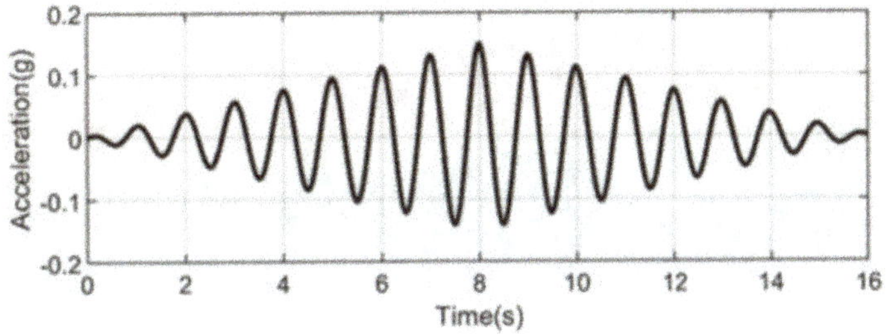


**Fig. 21.1** Target input motion 2 acceleration time history

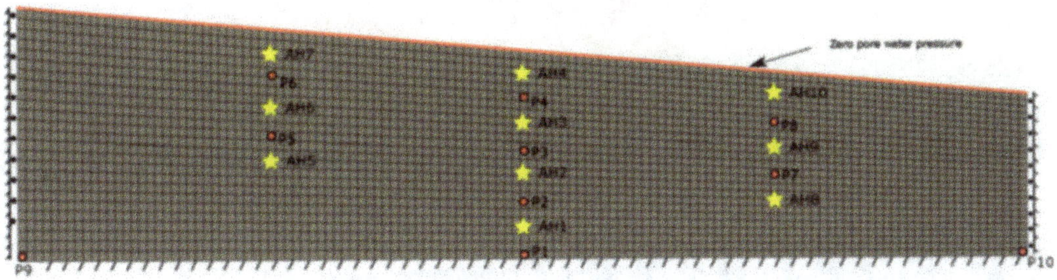


**Fig. 21.2** Finite element mesh and location of recorded pore water pressures (P) and accelerations (AH)

## 21.3 FE Model Development

In this work a 2D plane-strain representation of the problem was adopted, and numerical models were built in prototype scale using the OpenSees computational framework (McKenna 1997; OpenSees 2007). OpenSees (Open System for Earthquake Engineering Simulations) is an open-source, object-oriented finite element platform created and maintained by the Pacific Earthquake Engineering Research (PEER). The models consisted of 3125 four-node quadrilateral elements with an average size of 0.16 m × 0.16 m. The physically stabilized single-point integration and mixed displacement-pressure (u-p) element (SSPquadUP; see McGann et al. 2015) were used to capture the effective stress response of each simulated centrifuge test. A schematic of the mesh used in this work along with the location of recorded responses is depicted in Fig. 21.2. Appropriate mesh refinement was chosen to properly resolve the propagation of shear waves of up to 50 Hz in the soil domain.

### *21.3.1 Boundary and Loading Conditions*

The bottom boundary was fixed against vertical movement. The acceleration time history was applied to both vertical boundaries and the base of the model using the so-called *UniformExcitation* loading pattern in OpenSees. This was done to account for a rigid container. Pore pressures at the slope surface nodes were set to be zero during the analysis to ensure drainage and avoid generation of excess pore pressures

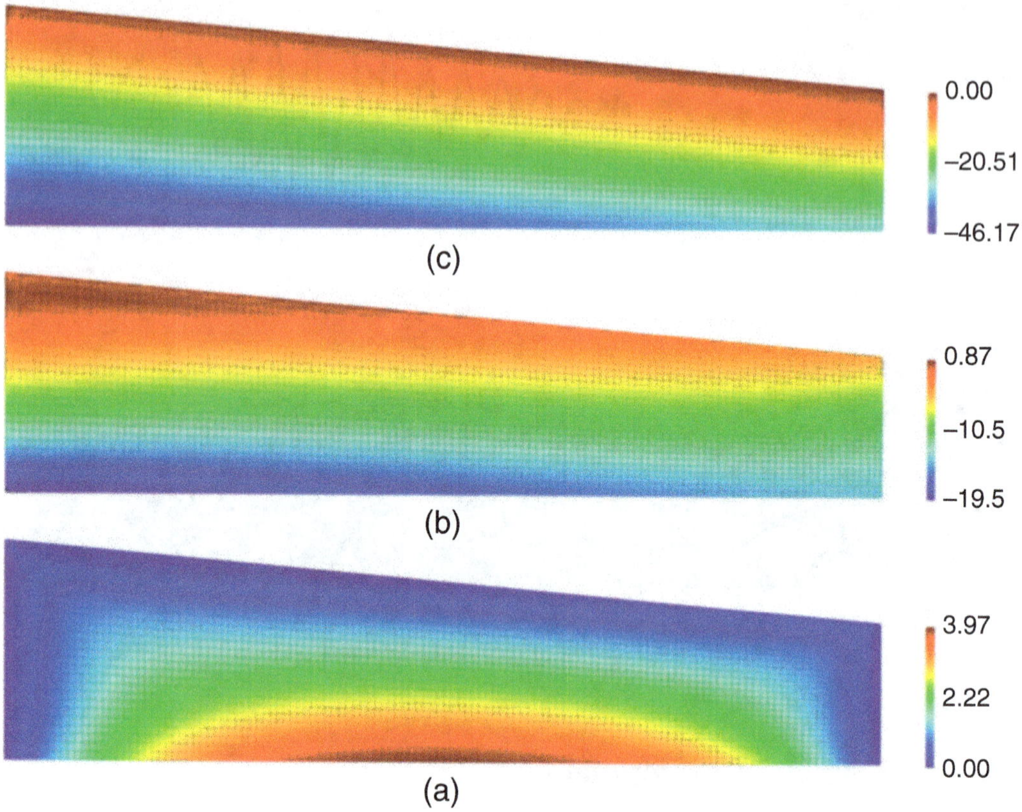


**Fig. 21.3** Initial stress field after applying gravity. (**a**) Vertical effective stress distribution. (**b**) Horizontal effective stress distribution. (**c**) Shear stress distribution. All units in kPa

at the surface. The free water on the slope was not modeled and effects of water sloshing (if any) were not considered. The soil was assumed to be always in contact with the container. A frequency-dependent Rayleigh damping was applied to compensate for small strain damping, which both MD and PM4Sand models lack in their formulation. The Rayleigh damping coefficients were chosen such that a 2% damping was obtained at 0.2 Hz and 20 Hz. The permeability was adopted from Ghofrani and Arduino (2018) and set to be $3.0 \times 10\ e^{-5}$m/s.

To apply gravity and centrifugal accelerations, the materials were set to be linear elastic, and elemental body force was increased gradually to reduce numerical instabilities in the model. Once gravity was in place, the materials were switched to have *elastoplastic* behavior, and enough extra steps were run to adjust their internal variables to any plastic behavior and maintain equilibrium. A Poisson's ratio of 0.3 was assumed to generate the initial stress state. This Poisson's ratio yielded a lateral earth pressure coefficient $K_0$ of 0.43 under level ground plane-strain conditions. Figure 21.3 shows the imposed initial state of stress in terms of vertical, horizontal, and shear stresses. After reaching the desired initial stress state, the acceleration time history was applied to the rigid boundaries. The constant average acceleration Newmark method ($\beta = 0.25$, $\gamma = 0.5$) was used in order to resolve the integration in time. To account for the material nonlinearity, both models used explicit modified Euler integration schemes with sub-stepping error control (Sloan et al. 2001). After the main shake portion of the motion was over, additional dynamic analysis steps were executed to dissipate any excess pore pressure generated during the shaking phase.

### *21.3.2 Material Constitutive Models*

The bounding surface constitutive models developed by Dafalias and Manzari (2004) (referred to as MD henceforth for brevity) and Boulanger and Ziotopolou (2015) (*PM4Sand*) were used for comparison purposes. The latter uses the basic framework of the model introduced by Dafalias and Manzari improved to better capture well-known trends in liquefiable soils. Both these models follow critical state soil mechanics concepts and are capable of capturing stress-strain relationships for denser-than-critical and looser-than-critical sands under different drainage and loading conditions. In the MD model, the relationship with critical state soil mechanics is through the so-called state parameter, defined as $\psi = e - e_c$ by Been and Jefferies (1985). Introducing a fabric tensor that evolves with accumulated plastic strain, MD can control contraction and dilation during cyclic loading.

Boulanger and Ziotopolou (2015) improved the capabilities of the MD model by developing a sand plasticity model for earthquake engineering applications that is simpler to calibrate and better represents some important aspects of liquefiable soils. Instead of the state parameter, a relative state parameter index, defined as $\xi_R = D_{R,\ \mathrm{cs}} - D_R$ by Boulanger (2003), is adopted and used together with the empirical relationship for critical state line proposed by Bolton (1986). Although the number of model parameters for PM4Sand is larger than MD, the number of parameters that need to be calibrated is reduced with considerations added internally for the model to follow general soil behavior. By changing three primary input parameters, namely, shear modulus coefficient, $G_0$; apparent relative density, $D_\mathrm{r}$; and contraction rate parameter, $h_\mathrm{po}$, the user can achieve reasonable approximations of desired behavior including pore pressure generation and dissipation, limiting strains, and cyclic mobility. Sending the "FirstCall" signal to the model allows users to initialize/reset secondary parameters (21 in total) and fabric history, while the "PostShake" signal allows the user to activate post-liquefaction functionality to improve simulation of reconsolidation strains. In the OpenSees analyses performed as part of this study, the "FirstCall" flag was called for material initialization just before switching to elastoplastic during gravity analysis. Obviously using the secondary parameters, the user can further fine-tune the response. Since its introduction, the PM4Sand model has drawn wide attention of geotechnical engineers and researchers due to its relatively easy calibration process and good agreement with field observations. More details on the model can be found at Boulanger and Ziotopolou (2015).

### 21.3.3 Calibration

In order to predict the response of centrifuge tests, material parameters were calibrated using available laboratory test data. Several drained and undrained cyclic triaxial as well as cyclic simple shear tests were used for this purpose (El Ghoraiby et al. 2017, 2019). To calibrate the model parameters, the constitutive evolution equations were integrated independently of the FE framework using a constitutive driver (referred to as MixedDriver) implemented in C++ based on the formulation proposed by Alawaji et al. (1992). The method uses an implicit backward Euler scheme to integrate the constitutive equations. Using this framework, it is possible to test any constitutive model under different stress paths and drainage conditions and independent of any finite element restriction. Also MixedDriver allows for simulating stress-controlled, strain-controlled, and mixed (stress-strain)-controlled loading conditions.

#### MD Parameters

Some of the MD constitutive model parameters can be estimated directly from drained and undrained monotonic triaxial tests. These parameters include $G_0$, $Mc$, $c$, $n^b$, $n^d$, $\lambda_c$, $e0$, and $\xi$. These parameters, in particular the last three, play an essential role in the response obtained from the MD model as the whole framework is built on the critical state surface concept. From the experimental standpoint, however, it is very difficult to obtain reliable data to represent the critical state of a soil. This has important implications in the MD model. Therefore, emphasis should be given to laboratory specimen sheared enough to reach critical state. The rest of the parameters can be calibrated using either trial and error or optimization techniques. Since the authors had experience working with Ottawa F65 sand from the previous LEAP 2015 exercise (Ghofrani and Arduino 2018), as well as few other projects, e.g., Ramirez et al. (2018), some of the same parameters used in previous calibration efforts were used for this study. Although it would have been ideal to calibrate the model for all of the characteristics revealed by the lab results at hand, due to intrinsic characteristics of the constitutive model, some specific objectives could not be achieved. In this context, the trend of number of cycles to reach liquefaction is one example. Based on experience, a number of cycles to liquefaction at smaller cyclic stress ratios do not play an important role in centrifuge experiments of this size and in particular are not observed with the loading conditions used in LEAP. Therefore, in this study the main objective when calibrating the MD model was to achieve a reasonable range of excess pore water pressure in cyclic tests as well as accumulation of shear strains, rather than trying to achieve perfect match between the number of cycles to reach liquefaction at different cyclic stress ratios. Table 21.1 summarizes the calibrated parameters for this phase of LEAP 2017.

**Table 21.1** Calibrated MD parameters and their values

| Parameter | | Value |
|---|---|---|
| Elastic | $G_0$ | 82.35 |
| | $\nu$ | 0.01 |
| Critical state | $M$ | 1.35 |
| | $c$ | 0.70 |
| | $\lambda c$ | 0.055 |
| | $e_0$ | 0.80 |
| | $\xi$ | 0.50 |
| Yield surface | $m$ | 0.02 |
| Plastic modulus | $h_0$ | 16.18 |
| | $c_h$ | 0.996 |
| | $n^b$ | 0.64 |
| Dilatancy | $A_0$ | 0.75 |
| | $n^d$ | 1.50 |
| Fabric tensor | $z_{max}$ | 12.50 |
| | $c_z$ | 500.0 |

## PM4Sand Parameters

The biggest advantage in using PM4Sand is that the calibration process is relatively straightforward. As mentioned before, only 3 primary parameters and 4 secondary parameters (out of 21) were used for calibration in this study. The remaining secondary parameters were kept at their default values, previously calibrated by Boulanger and Ziotopolou (2015) to represent typical soil behavior.

Since PM4Sand was developed for plane-strain conditions, undrained cyclic plane-strain compression (PSC) conditions were used for calibration purposes. Initially three sets of parameters were calibrated for three void ratios, namely, 0.515 ($D_r = 90.5\%$), 0.542 ($D_r = 79.6\%$), and 0.585 ($D_r = 62.2\%$), in accordance with LEAP phase 1 guidance. Maximum and minimum void ratios for the Ottawa F65 sand were adopted from previous test data and used to calculate relative densities. The critical state effective friction angle $\phi_{cv}$ was evaluated using shear and normal stresses at which the soil reached critical state during monotonic drained triaxial tests. The shear modulus coefficient $G_0$ was calibrated by matching the initial slope of the stress-strain curves in undrained cyclic tests. $H_{po}$, which controls the rate of pore pressure generation between contraction and dilation, and $c_z$, which controls the strain level at which fabric becomes relevant, were calibrated iteratively to match the experimental liquefaction strength curves. In contrast to the MD model, calibration of the PM4Sand model was done to capture the number of cycles to liquefaction observed in the triaxial tests. Table 21.2 presents the calibrated parameters for this phase of LEAP 2017.

Figure 21.4 depicts the case of a cyclic stress-controlled undrained triaxial test (El Ghoraiby et al. 2017, 2019) along with simulations obtained using MD and PM4Sand. The MD model was able to capture the asymmetry of the cyclic triaxial

**Table 21.2** Calibrated PM4Sand parameters and their values

| | | | $e_o = 0.585$ | |
|---|---|---|---|---|
| Primary parameters | $D_r$ | Relative density | 62.2% | Modified to 65% for Class-B prediction |
| | $G_0$ | Shear modulus coefficient | 350.0 | |
| | $h_{po}$ | Contraction rate parameter | 0.07 | |
| Secondary parameters | $e_{max}$ | Maximum void ratio | 0.7389 | Modified to match 65% relative density for Class-B prediction |
| | $e_{min}$ | Minimum void ratio | 0.4915 | |
| | $\varphi_{cv}$ | Critical state effective friction angle | 35.6 | |
| | $c_z$ | Controls the strain level at which fabric becomes important | 200.0 | |

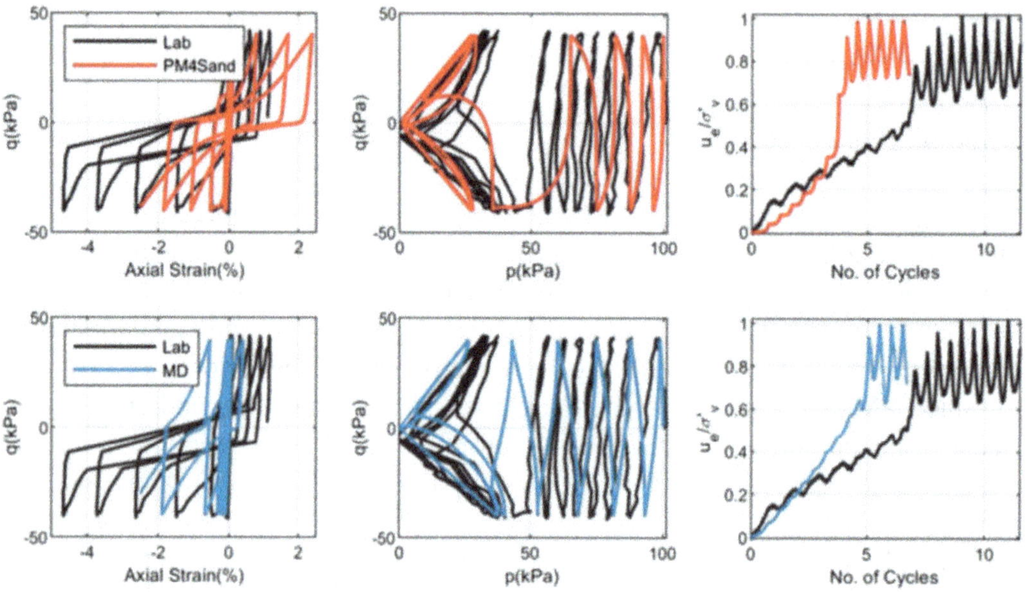


**Fig. 21.4** Elemental level calibration: comparison between simulations and experimental results. The results are presented in terms of deviatoric stress, $q$, vs. axial strain; deviatoric stress, $q$, vs. mean effective stress, $p$; and excess pore pressure ratio vs. number of cycles

**Fig. 21.5** Comparison of number of cycles required to reach 2.5% single-amplitude shear strain in simulations and laboratory tests

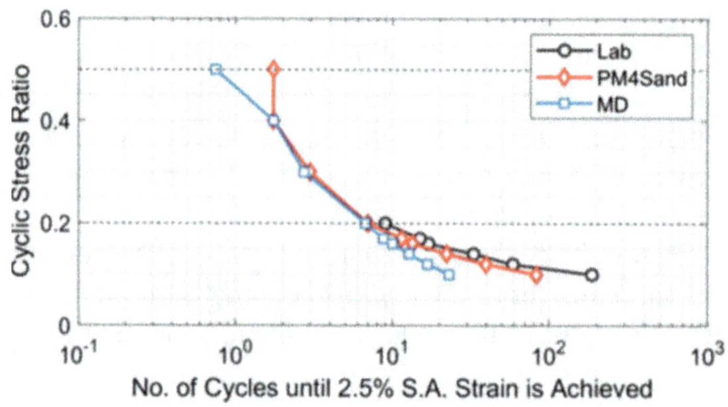

test and showed that axial strains accumulate in the extension direction, while the PM4Sand model results were symmetrical in compression and extension. Both the PM4Sand and MD models underpredicted excess pore pressure during the first cycle while overpredicted overall the rate of excess pore pressure generation. Figure 21.5 depicts comparisons between simulation results for both models and laboratory test data in terms of number of cycles to initial liquefaction for different cyclic stress ratios (CSR). The number of cycles to liquefaction was evaluated counting the number of cycles necessary to reach a 2.5% single-amplitude axial strain.

## 21.4 Type B Prediction Results

By definition Type B predictions are done with limited amount of data from completed experiments. In this study, the calibrated single set of parameters for MD was used without modification. For PM4Sand, the set of parameters calibrated for the void ratio of 0.585 was chosen, and the primary parameter $D_r$ was modified to 65%. The minimum void ratio achieved in the triaxial experiments was used in all cases, while the maximum void ratio was back calculated using the achieved void ratio in each centrifuge test to match the 65% relative density. In the following subsections, results and comparisons are presented in terms of acceleration response, pore pressure response, and surface displacement.

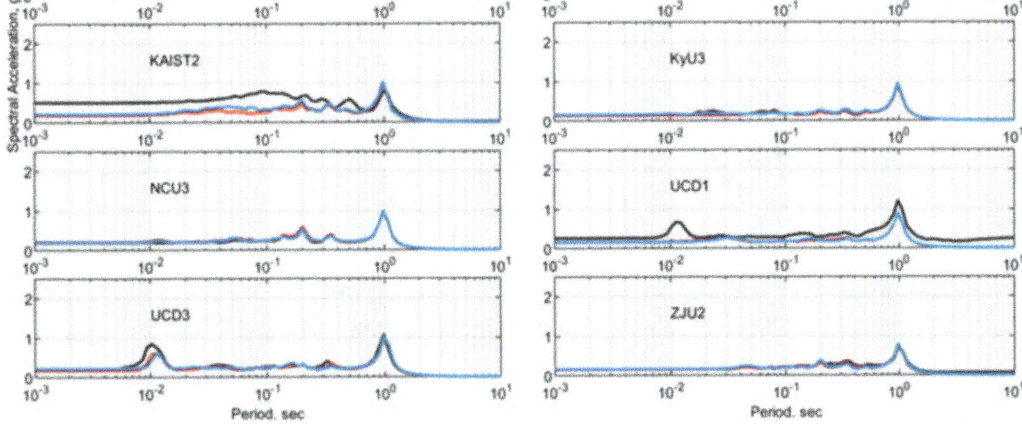


Fig. 21.6 Class-B prediction results—comparison of acceleration response spectra (5% damping) at AH1 in simulations and centrifuge tests

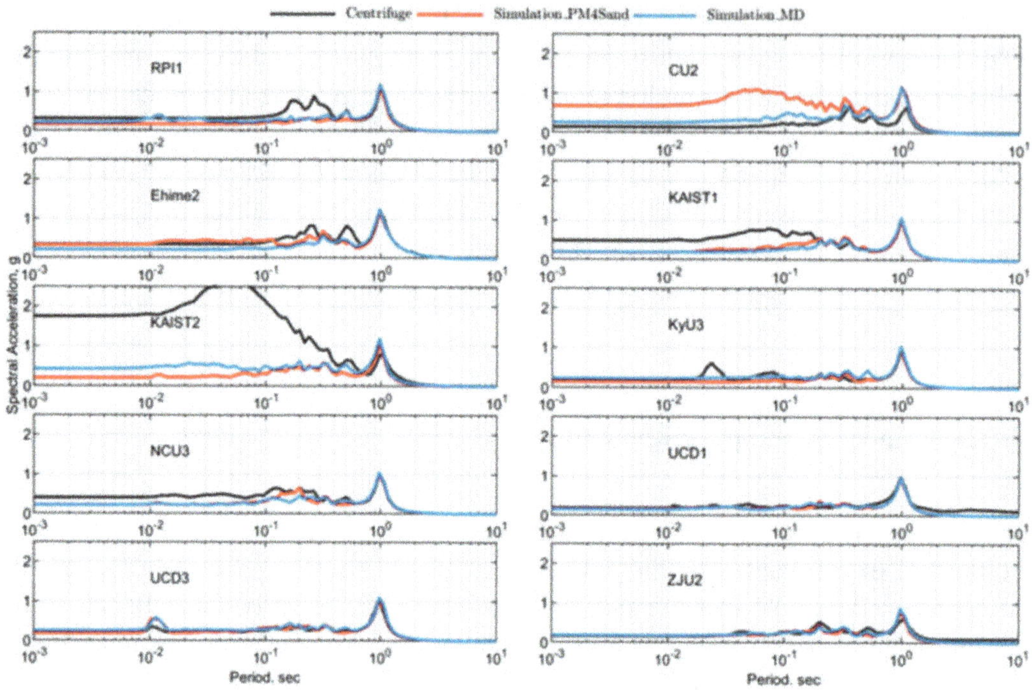


**Fig. 21.7** Class-B prediction results—comparison of acceleration response spectra (5% damping) at AH2 in simulations and centrifuge tests

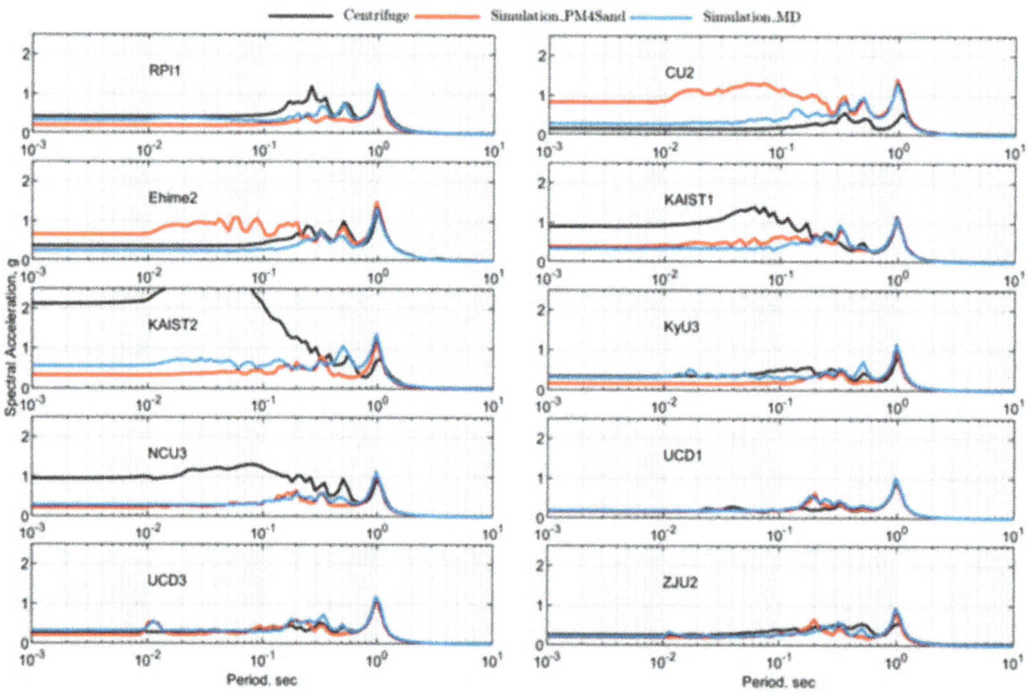


**Fig. 21.8** Class-B prediction results—comparison of acceleration response spectra (5% damping) at AH3 in simulations and centrifuge tests

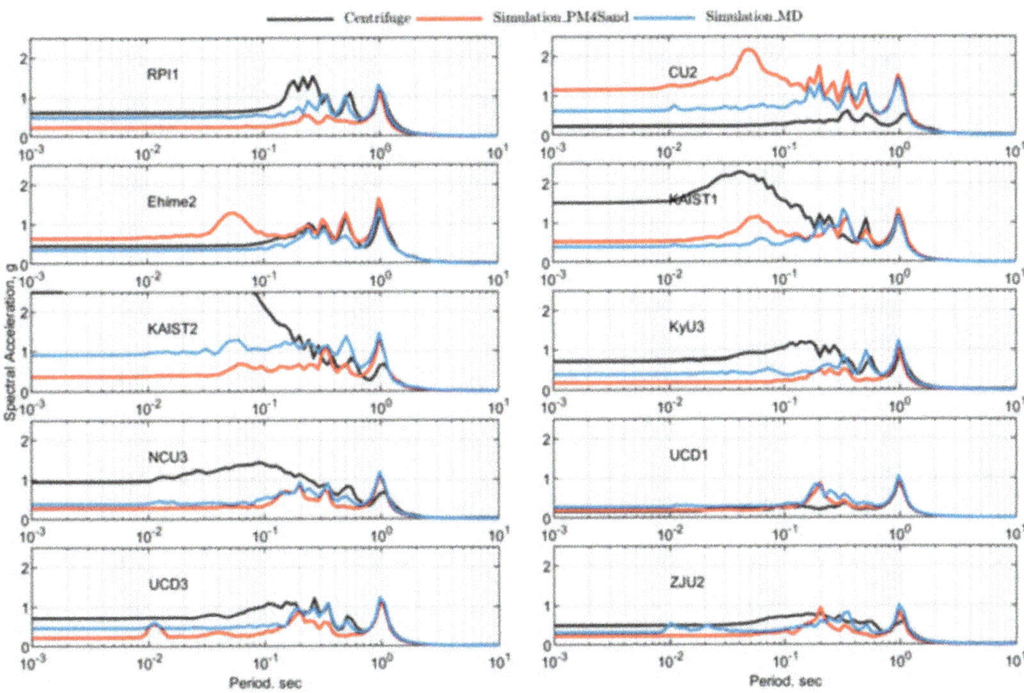


**Fig. 21.9** Class-B prediction results—comparison of acceleration response spectra (5% damping) at AH4 in simulations and centrifuge tests

### 21.4.1 Acceleration Response

Figures 21.6, 21.7, 21.8, and 21.9 show recorded and predicted acceleration response for all centrifuge experiments in terms of 5% damping response spectra at the AH1–AH4 sensors (see Fig. 21.2 for sensor location). The plots show the simulation results predicted very well the recorded experimental data at the intended 1 Hz input motion frequency. PM4Sand predicted higher PGAs for the CU2 and Ehime2 experiments and generally produced results with higher frequency content. This was most likely due to material overprediction of the soil stiffness under dilation which leads to stronger simulated dilation pulses. The acceleration time histories (not shown) show even a better match.

### 21.4.2 Pore Pressure Response

All centrifuge tests showed similar trends in the pore pressure response. Comparisons of predicted and recorded pore pressure response at pore pressure sensors along centerline are depicted in Figs. 21.10, 21.11, 21.12, and 21.13. The figures show both models could capture the trend of pore pressure generation observed in the experiments. As shown in Fig. 21.5, the MD model exhibited lower cyclic resistance under lower CSRs compared to the PM4Sand model. This manifested itself in different predicted responses of pore pressure for both models. In general, MD

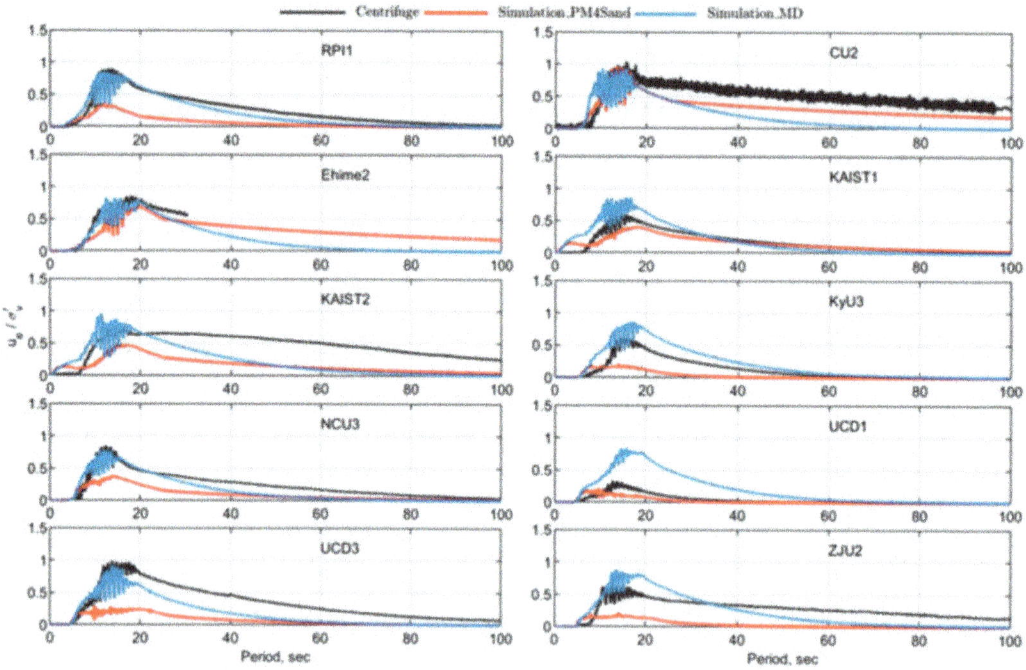


**Fig. 21.10** Class-B prediction results—comparison of excess pore water pressures at P1 in simulations and centrifuge tests

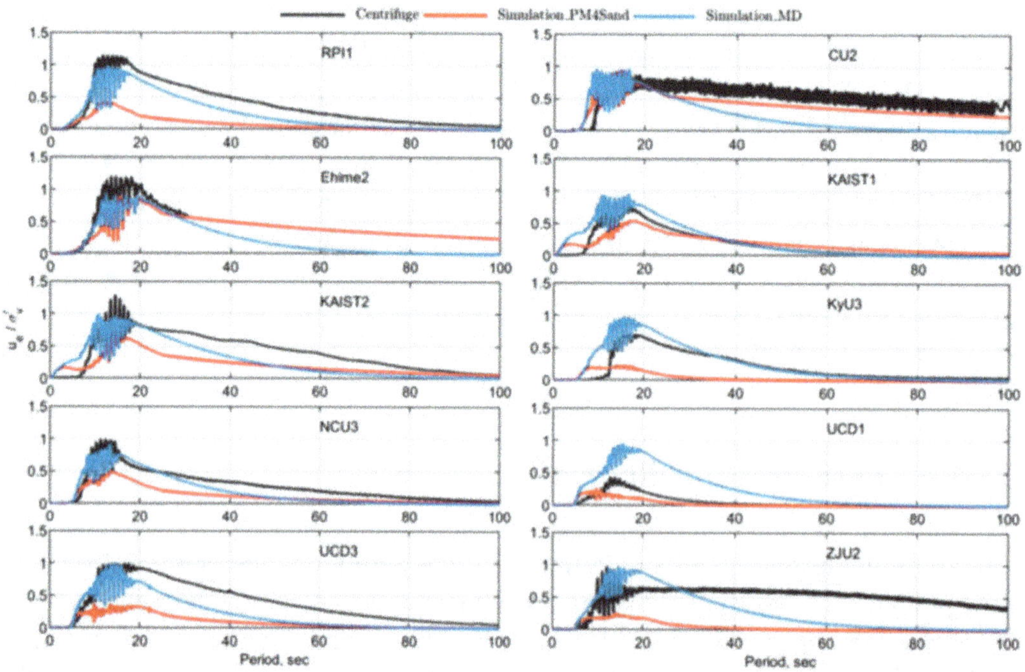


**Fig. 21.11** Class-B prediction results—comparison of excess pore water pressures at P2 in simulations and centrifuge tests

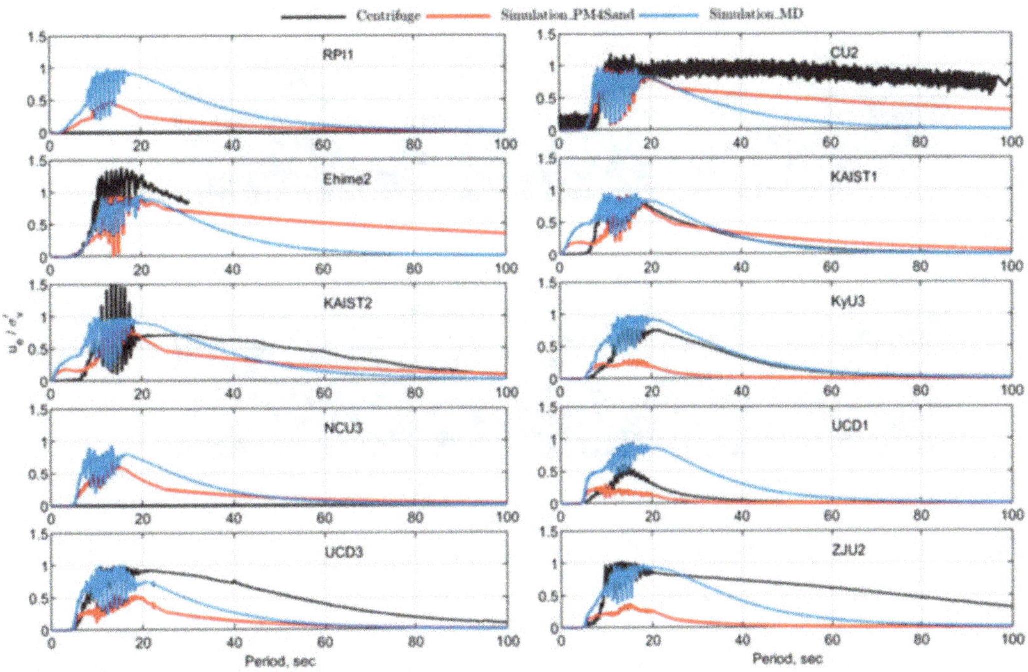


**Fig. 21.12** Class-B prediction results—comparison of excess pore water pressures at P3 in simulations and centrifuge tests

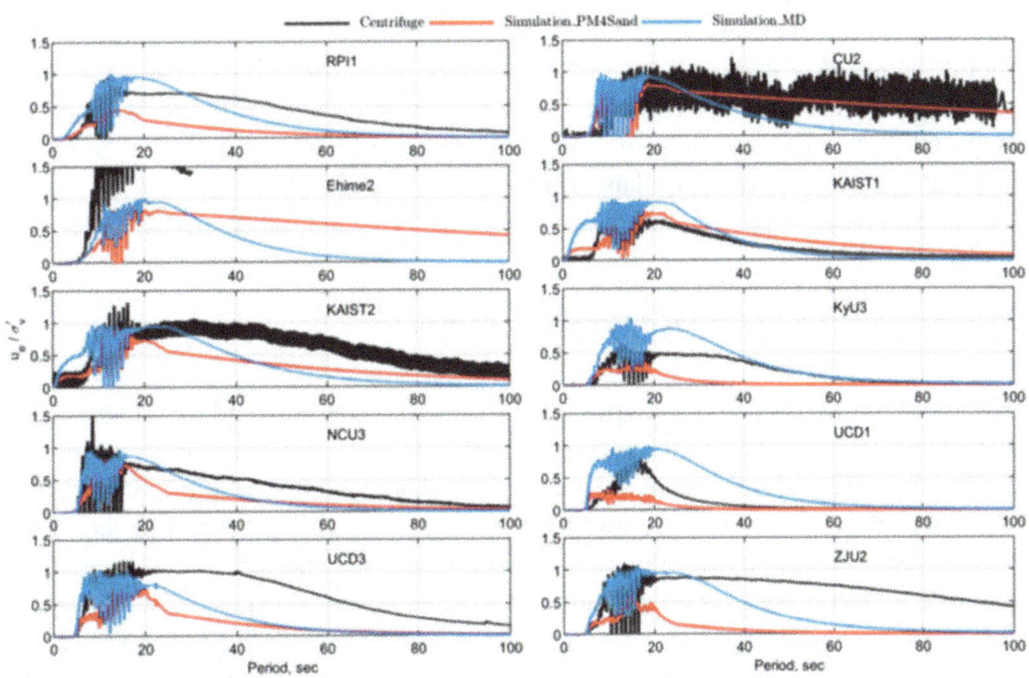


**Fig. 21.13** Class-B prediction results—comparison of excess pore water pressures at P4 in simulations and centrifuge tests

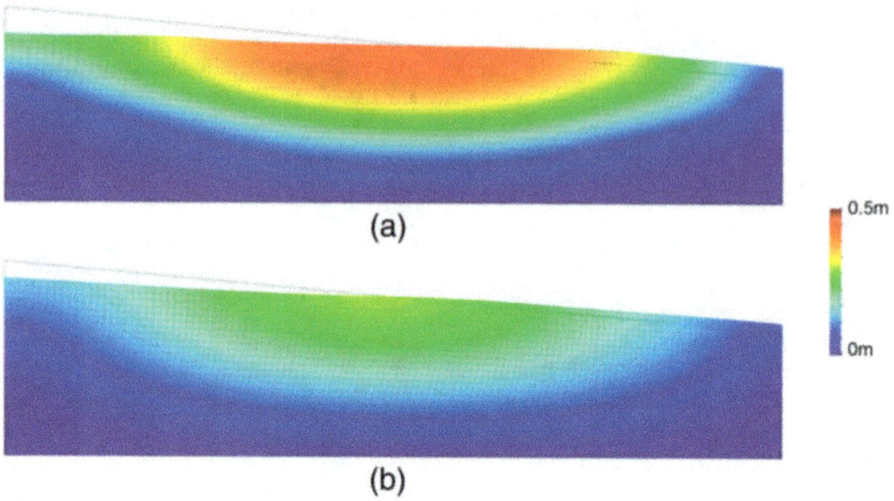


**Fig. 21.14** Class-B prediction results—contour plot of absolute displacement (Ehime2). (**a**) PM4Sand. (**b**) MD

model was able to predict excess pore pressure well, while PM4Sand model underpredicted excess pore pressure generation at deeper locations. Both models showed the same excess pore pressure dissipation pattern after the main shake and until 25 s. A slower dissipation rate of excess pore pressure shown in the simulation results after the main shake suggests that the permeability used in the numerical simulations was possibly larger than the actual value. It should be mentioned that after 25 s, the "PostShake" flag in PM4Sand model was turned "on" to achieve better post-shake settlements. Moreover, in PM4Sand the bulk modulus of the material was modified using the accumulated fabric history at each element. More details in PM4Sand can be found in Boulanger and Ziotopolou (2015).

### *21.4.3 Surface Displacement Response*

Figure 21.14 depicts nodal displacements contours at the end of shaking. As expected the upslope part of the soil settled, and the downslope part heaved resulting in the flattening of the slope. Figure 21.15 depicts the evolution of recorded horizontal displacements for a surface point at the center compared to simulated results. Large variability in both centrifuge experiments and numerical simulations are observed indicating that estimation of surface displacement continues to be a challenge for both physical and numerical models. Nevertheless, despite the observed variability, both models were able to capture the evolution of displacement reasonably well.

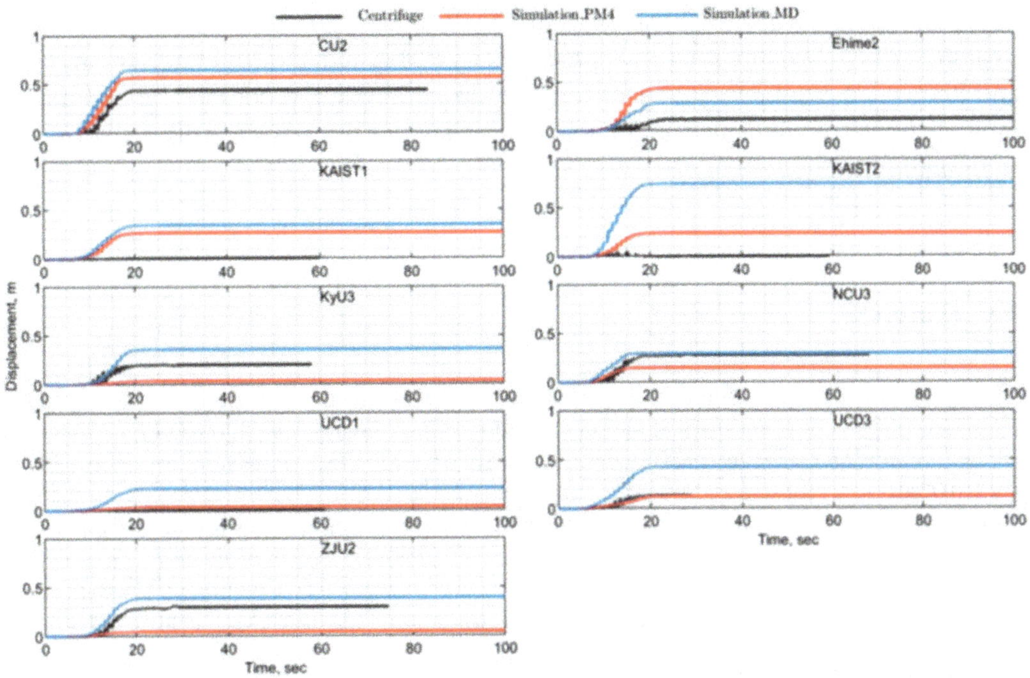


**Fig. 21.15** Class-B prediction results—comparison of horizontal displacements at centerline surface point in simulation and centrifuge tests

## 21.5 Conclusion

Two constitutive models based on critical state soil mechanics and bounding surface plasticity, MD and PM4Sand, were calibrated independently to simulate the behavior of Ottawa-F65 sand in accordance with the LEAP-UCD-2017 project guidelines. The calibrated models were then used to model the response of a boundary value problem using OpenSees. Besides using different material models, the same exact finite element model was used in all simulations. Type B predictions were obtained and simulations compared to recorded data from centrifuge experiments at several facilities around the world.

The MD model was calibrated for general soil behavior, and number of loading cycles to trigger liquefaction under small CSR was not prioritized. In the calibration of the PM4Sand model parameters, the number of cycles to liquefaction observed in the laboratory test results was emphasized and used as a calibration criterion.

Type B prediction results showed that although calibrated using different methods, both models were able to predict the results from centrifuge experiments, especially the acceleration response at the frequency of the input motion. The evolution of excess pore pressures predicted by PM4Sand and MD were comparable and showed the similarity and differences in calibration process. The amount of predicted horizontal displacement was in the range of the recorded displacements at each facility.

## References


Alawaji, H., Runesson, K., Sture, S., & Axelsson, K. (1992). Implicit integration in soil plasticity under mixed control for drained and undrained response. *International Journal for Numerical and Analytical Methods in Geomechanics, 16*(10), 737–756. https://doi.org/10.1002/nag.1610161004.

Arulanandan, K., & Scott, R. F. (1993). Verification of numerical procedures for the analysis of soil liquefaction problems. In *International Conference on the Verification of Numerical Procedures for the Analysis of Soil Liquefaction Problems (Davis, California)*. Rotterdam: AA Balkema.

Bastidas, A. M. P. (2016). *Ottawa F-65 sand characterization*. PhD thesis, University of California, Davis.

Been, K., & Jefferies, M. G. (1985). A state parameter for sands. *Géotechnique, 35*(2), 99–112.

Bolton, M. D. (1986). The strength and dilatancy of sands. *Géotechnique, 36*(1), 65578.

Boulanger, R. W. (2003). Relating $K_\alpha$ to relative state parameter index. *Journal of Geotechnical and Geoenvironmental Engineering, 129*(8), 770–773.

Boulanger, R. W., & Ziotopoulou, K. (2015). *PM4Sand (Version 3): A sand plasticity model for earthquake engineering applications. Technical Report No. UCD/CGM-15/xx, Center for Geotechnical Modeling*. Department of Civil and Environmental Engineering, University of California, Davis, CA.

Dafalias, Y. F., & Manzari, M. T. (2004). Simple plasticity sand model accounting for fabric change effects. *Journal of Engineering Mechanics, 130*(6), 622–634.

El Ghoraiby, M. A., Park, H., & Manzari, M. T. (2017). *LEAP 2017: Soil characterization and element tests for Ottawa F65 sand*. Washington, DC: The George Washington University.

El Ghoraiby, M. A., Park, H., & Manzari, M. T. (2019). Physical and mechanical properties of Ottawa F65 sand. In B. Kutter et al. (Eds.), *Model tests and numerical simulations of liquefaction and lateral spreading: LEAP-UCD-2017*. New York: Springer.

Ghofrani, A., & Arduino, P. (2018). Prediction of leap centrifuge test results using a pressure-dependent bounding surface constitutive model. *Soil Dynamics and Earthquake Engineering, 113*, 758–770. https://doi.org/10.1016/j.soildyn.2016.12.001.

Kutter, B. L., Carey, T. J., Hashimoto, T., Zeghal, M., Abdoun, T., Kokkali, P., Madabhushi, G., et al. (2018). LEAP-Gwu-2015 Experiment specifications, results, and comparisons. *Soil Dynamics and Earthquake Engineering, 113*, 616–628. https://doi.org/10.1016/j.soildyn.2017.05.018.

Manzari, M. T., El Ghoraiby, M., Kutter, B. L., Zeghal, M., Abdoun, T., Arduino, P., Armstrong, R. J., et al. (2018). Liquefaction experiment and analysis projects (Leap): Summary of observations from the planning phase. *Soil Dynamics and Earthquake Engineering, 113*, 714–743. https://doi.org/10.1016/j.soildyn.2017.05.015.

McGann, C. R., Arduino, P., & Mackenzie-Helnwein, P. (2015). A stabilized single-point finite element formulation for three-dimensional dynamic analysis of saturated soils. *Computers and Geotechnics, 66*(0), 126–141. https://doi.org/10.1016/j.compgeo.2015.01.002.

McKenna, F. T. (1997). *Object-Oriented Finite Element Programming: Frameworks for Analysis, Algorithms and Parallel Computing*. PhD thesis, University of California, Berkeley.

OpenSees. (2007). *Open System for Earthquake Engineering Simulation*. http://Opensees.berkeley.edu. University of California, Berkeley: Pacific Earthquake Engineering Research Center (PEER).

Ramirez, J., Barrero, A., Chen, L., Dashti, S., Ghofrani, A., Taiebat, M., & Arduino, P. (2018). Site response in a layered liquefiable deposit: Evaluation of different numerical tools and methodologies with centrifuge experimental results (accepted). *Journal of Geotechnical and Geoenvironmental, 144*, 10.

Sloan, S. W., Abbo, A. J., & Sheng, D. (2001). Refined explicit integration of elastoplastic models with automatic error control. *Engineering Computations, 18*(1/2). MCB UP Ltd), 121–194. https://doi.org/10.1108/02644400110365842.

Vasko, A. (2015). *An investigation into the behavior of Ottawa sand through monotonic and cyclic shear tests*. MS Thesis, The George Washington University.

# Chapter 22
# LEAP-UCD-2017 Centrifuge Test Simulation at UNINA

**Gianluca Fasano, Anna Chiaradonna, and Emilio Bilotta**

**Abstract** Within the framework of the LEAP-UCD-2017 exercise, Type B simulations of centrifuge tests were conducted assuming a hypoplastic constitutive model for sand. Differently from the most common elastoplastic approach, the hypoplasticity does not decompose the strain rate into elastic and plastic parts and does not use explicitly the notions of the yield surface and plastic potential surface. The process followed to calibrate the constitutive model is presented in detail. The initial state of stresses in the analyzed mesh, the key parameters used in the dynamic simulation phase, and a comparison of the simulation with some experimental results are reported. All the simulations were performed using the model parameters calibrated by using the laboratory test data. Finally, a sensitivity analysis of computed displacement to soil density and ground motion intensity show the influence of such factors on the seismic soil response of liquefiable soils.

## 22.1 Introduction

The aim of this paper is to describe the process followed to calibrate the constitutive model, to obtain the initial state of stresses in the analyzed soil-deposit mesh, to properly document the key parameters used in the seismic simulation phase, and to briefly present a comparison of the simulations with some experimental results of the LEAP-UCD-2017 exercise.

The content of the paper focuses on the following features of the performed numerical simulations: (1) the constitutive model used in the simulations, which is quite different from the most popular approaches that are currently adopted for studying liquefaction; (2) the calibration process, which is described in details; and (3) a sensitivity study.

The set of model parameters were calibrated on both triaxial and simple shear tests. The same parameters are used in the simulation of centrifuge tests, which are

G. Fasano · A. Chiaradonna (✉) · E. Bilotta
Department of Civil, Architectural and Environmental Engineering, University of Napoli Federico II, Naples, Italy
e-mail: anna.chiaradonna@unina.it



B. Kutter et al. (eds.), *Model Tests and Numerical Simulations of Liquefaction and Lateral Spreading*, https://doi.org/10.1007/978-3-030-22818-7_22

carried out at the prototype scale. Type B simulations section includes a description of the discretization, the numerical integration scheme, and numerical damping used in the analysis.

A sensitivity study (simulations NS1–NS7, exploring the sensitivity of displacement to soil density and ground motion intensity) has been carried out with the same parameters and using the same numerical model adopted in the Type B simulations.

The paper is divided into four sections. After this Introduction, Sect. 22.2 covers the essential features of the constitutive model, the final model parameters, the calibration philosophy, and the assumptions used in the calibration process. It also presents a comparison between the predicted and experimental cyclic strength curves. Section 22.3 discusses the main features of the numerical analysis platform used in the simulation, the model geometry and the discretization details, the boundary conditions of the numerical model, the solution algorithm employed, and some assumptions used in the reported analyses. Section 22.4 presents a comparison between the results of the Type B predictions and those of one of the centrifuge experiments. Finally, Sect. 22.5 shows the main results of the sensitivity study to explore the role of the soil density and ground motion intensity on slope behavior.

## 22.2 Calibration Phase

### *22.2.1 Hypoplastic Constitutive Model*

The constitutive model used in the simulation exercise is the hypoplastic model with intergranular strain concept (Von Wolffersdorff 1996; Niemunis and Herle 1997), implemented in the finite element code PLAXIS 2016 (Mašín 2010).

Hypoplasticity is a particular class of incrementally nonlinear constitutive models, developed specifically to predict the behavior of soils. The basic structure of the hypoplastic models has been developed during the 1990s at the University of Karlsruhe. In hypoplasticity, unlike in elastoplasticity, the strain rate is not decomposed into elastic and plastic parts, and the models do not use explicitly the notions of the yield surface and plastic potential surface. Still, the models are capable of predicting the important features of the soil behavior, such as the critical state, dependency of the peak strength on soil density, nonlinear behavior in the small and large strain range, and dependency of the soil stiffness on the loading direction.

Hypoplastic models can well represent deformations due to rearrangements of the grain skeleton. However, applications of the hypoplasticity to cyclic loading with small amplitudes revealed some shortcomings. In the basic hypoplastic model (Von Wolffersdorff 1996), the stiffness during loading and reloading is almost identical, which leads to an unrealistic large accumulation of deformation in cyclic loading. Therefore, for simulations of cyclic or seismic loading, it is necessary to enhance the model with an extension, considering an increased stiffness in a small strain range. Thus, the applied hypoplastic constitutive equation corresponds to the model by von

Wolffersdorff (1996) with the intergranular strain extension according to Niemunis and Herle (1997). In Plaxis, the constitutive model is integrated using explicit adaptive integration scheme with local sub-stepping (Mašín 2010).

The parameters of the hypoplastic model (Von Wolffersdorff 1996) are critical friction angle $\phi_c$, critical void ratio at zero pressure $e_{c0}$, void ratio at the maximum density at zero pressure $e_{d0}$, maximum void ratio at zero pressure $e_{i0}$, granular hardness $h_s$, stiffness exponent $n$, and the exponents $\alpha$ and $\beta$. The parameter $\alpha$ affects the peak friction angle in dependence of relative density, while the soil stiffness is affected by the parameter $\beta$. Moreover, the intergranular strain extension (Niemunis and Herle 1997) requires further five parameters: $R$, $m_R$, $m_T$, $\beta_r$, and $\chi$. The stiffness increase after a change in the loading direction of 180° and 90° is controlled by the parameters $m_R$ and $m_T$, respectively, with the loading direction referring to the direction of the strain rate. If $m_R = 0$, the intergranular strain concept is switched off, and the problem is simulated using the basic hypoplastic model (Von Wolffersdorff 1996).

The reduction in the shear modulus with increasing shear strain is controlled by the parameters $R$, $\beta_r$, and $\chi$.

Finally, in addition to the initial void ratio, $e_0$, further six state parameters may be specified, $\delta_{11}$, $\delta_{22}$, $\delta_{33}$, $\delta_{12}$, $\delta_{13}$, $\delta_{23}$, which are the initial values of the intergranular strain tensor $\delta = (\delta_{11}, \delta_{22}, \delta_{33}, 2\delta_{12}, 2\delta_{13}, 2\delta_{23})$. They describe the accumulation of permanent deformation. Their initial values have been set to zero. A more detailed description of the model can be found in the referenced publications.

Table 22.1 reports the input parameters used in the hypoplastic model, as listed in Plaxis 2016 (Brinkgreve et al. 2016). The model requires overall 21 input parameters.

**Table 22.1** Input parameters of the hypoplastic model

| | |
|---|---|
| $\phi_c$ [°] | Critical state friction angle |
| $p_t$ [kPa] | Shift of mean stress due to cohesion |
| $h_s$ [kPa] | Overall slope of compression curve |
| $n$ | Curvature of compression curve |
| $e_{d0}$ | Densest particle packing at zero mean stress |
| $e_{c0}$ | Critical particle packing at zero mean stress |
| $e_{i0}$ | Loosest particle packing at zero mean stress |
| $\alpha$ | Peak friction angle as function of relative density |
| $\beta$ | Shear stiffness as function of relative density |
| $m_R$ | Initial shear stiffness for initial and reverse condition |
| $m_T$ | Stiffness upon neutral condition |
| $R$ | Size elastic range |
| $\beta_r$ | Rate stiffness degradation |
| $\chi$ | Rate stiffness degradation |
| SV: $e_0$ | Initial void ratio |
| $\delta_{ij}$ $i, j = 1,2,3$ | Initial value of the intergranular strain tensor |

### 22.2.2 Model Calibration

The approach used in the calibration of the model parameters is hereafter explained. The calibration made use of the results of the provided cyclic triaxial stress-controlled tests and additional element test data (i.e., monotonic triaxial tests, simple shear tests).

Shear strength parameters are computed from the monotonic triaxial test data, available on the NEES Hub (https://nees.org/dataviewer/view/1064:ds/1189). Drained triaxial compression tests, carried out by Vasko (2015) on loose and dense specimens, were used to define the critical state line in the plane $q{:}p'$ and the constant volume friction angle, $\phi'_c$. As well known, the evaluation of critical state conditions in triaxial tests is a very complex issue, being such a test intrinsically affected by a number of experimental limitations (localization, bulging, shear stresses on the rough porous stones, difference between local and external displacements, etc.). One of the best ways to evaluate the final state is therefore by looking to dilatancy trend at the end of the tests. Based on all the elaborations of the available experimental data and considering that the hypoplastic model assumes the critical state as an asymptotic state at infinite strains, in this case the best fit of this parameter is the following:

$$\phi_c = 32^{\circ} \tag{22.1}$$

The critical void ratio at zero mean stress, $e_{c0}$, has been assumed, by extrapolation of experimental values as

$$e_{c0} = 0.746 \tag{22.2}$$

Hence, the other two parameters representing the initial values at $p' = 0$ of maximum and minimum void ratios have been set equal to

$$e_{d0} = e_{min} \tag{22.3}$$

$$e_{i0=1.2}e_{c0} \tag{22.4}$$

The parameter $\alpha$ was computed as a function of the void ratios, $e$, $e_c$, $e_d$, related to the peak mean stress, according to the equation proposed by Herle and Gudehus (1999):

$$\alpha = \frac{\ln\left[6\frac{(2+K_p)^2+a^2K_p(K_p-1-\tan\nu_p)}{a(2+K_p)(5K_p-2)\sqrt{4+2(1+\tan\nu_p)^2}}\right]}{\ln\left[\frac{e-e_d}{e_c-e_d}\right]} \tag{22.5}$$

where the peak ratios are function of the peak friction angle, $\phi_p$,

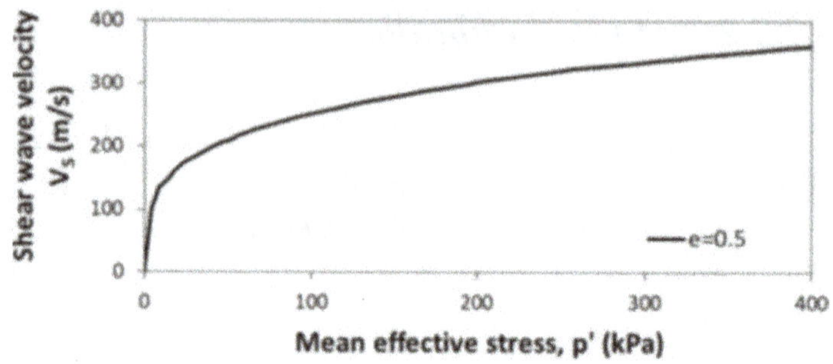


**Fig. 22.1** Shear wave velocity measurements, $V_S$, vs. mean effective stress, $p'$ during isotropic consolidation for Ottawa C-109 sand (modified after Robertson et al. 1995)

$$K_p = \frac{1 + \sin\phi_p}{1 - \sin\phi_p} \tag{22.6}$$

$$\tan\nu_p = 2\frac{K_p - 4 + 5AK_p{}^2 - 2AK_p}{(5K_p - 2)(1 + 2A)} - 1 \tag{22.7}$$

$$A = \frac{a^2}{(2 + K_p)^2}\left[1 - \frac{K_p(4 - K_p)}{4K_p - 2}\right] \tag{22.8}$$

while the parameter $a$ is function of the critical friction angle, $\phi_c$,

$$a = \frac{\sqrt{3}(3 - \sin\phi_c)}{2\sqrt{2}\sin\phi_c} \tag{22.9}$$

Finally, the parameter $\beta$ was calibrated on the results of the available drained triaxial tests on dense specimens (Vasko 2015).

The parameters $h_s$ and $n$ were calibrated by interpolating the $G_0$:$p'$ relationship proposed by Robertson et al. (1995) for dense Ottawa C-109 sand (Fig. 22.1) with the equation

$$G_0 = m_R \frac{h_s}{n}\left(\frac{3p}{h_s}\right)^{1-n}\left(\frac{e_{c0}}{e}\right)^{\beta} \times f_a \times f\langle K_0\rangle \tag{22.10}$$

where the constant of proportionality $m_R$ has been set to 5 and

$$f_a = \left(\frac{e_{i0}}{e_{c0}}\right)^{\beta}\frac{1 + e_{i0}}{e_{i0}}\left[3 + a^2 - a\sqrt{3}\frac{e_{i0} - e_{d0}{}^{\alpha}}{e_{c0} - e_{d0}}\right]^{-1} \tag{22.11}$$

$$f\langle K_0\rangle = \frac{1}{2}\frac{\langle 1 + 2K^2\rangle + a^2\langle 1 - K\rangle}{1 + 2K^2} \tag{22.12}$$

with $K = \sigma_2/\sigma_1$, hence for isotropic stress $f(K) = 1.5$.

In Eq. (22.10), a shear wave velocity of 250 m/s was assumed for a reference value of mean stress, $p' = 100$ kPa (Fig. 22.1). Then, the stiffness at small strains, $G_0$ was computed as

$$G_0 = \rho V_S^2 \tag{22.13}$$

After that, the parameters $h_s$ and $n$ were calibrated to fit the $G_0$:$p'$ relationship.

According to Mašín (2015), the parameter $m_T$ was considered equal to a fraction of $m_R$:

$$m_T = 0.7 m_R \tag{22.14}$$

Figures 22.2a, b and 22.3a, b show the simulation of two monotonic triaxial tests (at 100 and 200 kPa of confining effective stress, respectively) on dense samples with dry density, $\rho_d = 1673$ kg/m$^3$ in the $q$:$\varepsilon_a$ plane. The peak deviatoric values of both tests are well-predicted, while the simulated curves underestimate the stiffness and the dilatative behavior of the soil samples.

The implemented model requires also the definition of the parameter $p_t$, which is a shift of the mean stress due to cohesion. For the basic hypoplastic model, $p_t$ is set equal to zero, but a nonzero value of $p_t$ is needed to overcome problems with stress-free state (Mašín 2010).

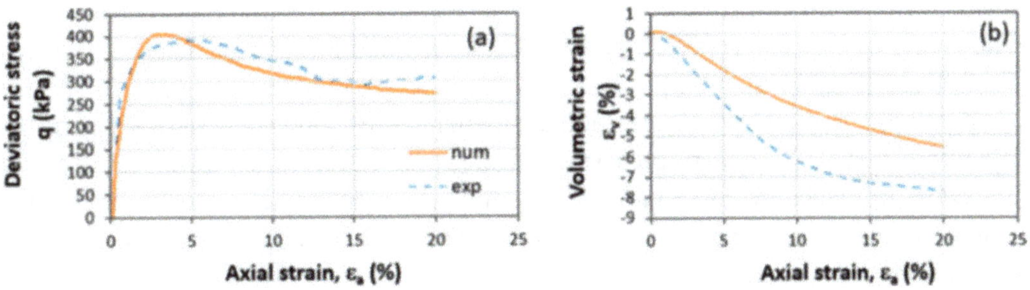


**Fig. 22.2** Simulation of monotonic triaxial tests on dense Ottawa sand ($\rho_d = 1673$ kg/m$^3$) with $p_0'$ equal to 100 kPa in the $\varepsilon_a$: $q$ (**a**) and $\varepsilon_a$: $\varepsilon_v$ (**b**) plane

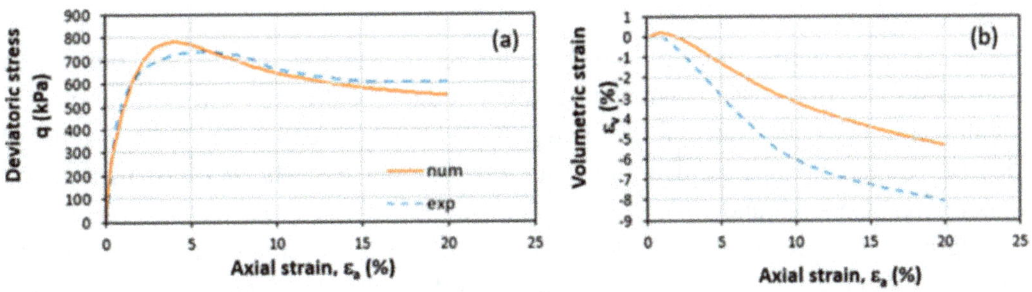


**Fig. 22.3** Simulation of monotonic triaxial tests on dense Ottawa sand ($\rho_d = 1673$ kg/m$^3$) with $p_0'$ equal to 200 kPa in the $\varepsilon_a$: $q$ (**a**) and $\varepsilon_a$: $\varepsilon_v$ (**b**) plane

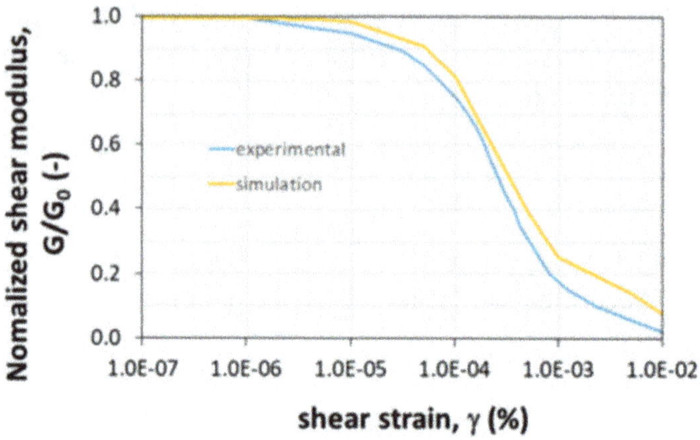


**Fig. 22.4** Simulated shear modulus degradation curve vs. experimental curve for Ottawa sand 20–30 (modified after Alarcon-Guzman et al. 1989)

**Table 22.2** Parameters of the hypoplastic model

| $\phi_c$ [°] | $p_t$ [kPa] | $h_s$ [kPa] | $n$ | $e_{d0}$ | $e_{c0}$ | $e_{i0}$ | $\alpha$ | $\beta$ | $m_R$ | $m_T$ | $R_{max}$ | $\beta_r$ | $\chi$ |
|---|---|---|---|---|---|---|---|---|---|---|---|---|---|
| 32 | 1 | 304,836 | 0.51 | 0.507 | 0.746 | 0.896 | 0.18 | 1.5 | 5 | 3.5 | 0.0001 | 0.4 | 1 |

In order to identify the parameters $R$, $\beta_r$, and $\chi$, simulations of strain-controlled cyclic shear tests were carried out, and the secant shear modulus $G$ was evaluated from them. The numerical results have been compared with the experimental data for ($G/G_0$, $\gamma$) reported in literature for Ottawa sand 20–30 (Alarcon-Guzman et al. 1989). The values of $R$ and $\chi$ were assumed as $R = 0.0001$ and $\chi = 1.0$. Hence $\beta_r = 0.4$ was determined (Fig. 22.4).

The cyclic triaxial tests were simulated by using the "soil test" tool of Plaxis 2016 (Brinkgreve et al. 2016). The prescribed cyclic stress ratio, $CSR_{CTX}$, was imposed for a number of cycles, $N$, large enough to induce liquefaction. The model parameters that were finally obtained are summarized in Table 22.2. They were adopted for the simulation of the centrifuge tests.

### 22.2.3 *Liquefaction Strength Curves*

For each group of tests with the same void ratio, the relative density was defined as the mean value of the relative densities calculated by the different experimenters (see Table 22.3). For the sake of simplicity, the obtained mean values were rounded down. Table 22.3 shows that the available laboratory tests were performed on specimens of dense sand.

Figure 22.5 reports an example of the stress-strain and stress path curves simulated by the model for a cyclic stress-controlled test ($e_0 = 0.515$, $p'_0 = 100$ kPa, and

**Table 22.3** Relative density according to the reported $e_{min}$ and $e_{max}$

| Tested by | Institute | $e_{max}$ | $e_{min}$ | $e_o = 0.515$ | $e_o = 0.542$ | $e_o = 0.585$ |
|---|---|---|---|---|---|---|
| Eduardo Cerna | UC Davis | 0.8375 | 0.5116 | 98.96% | 90.67% | 77.48% |
| Cooper Labs | UC Davis | 0.7162 | 0.4977 | 92.08% | 79.73% | 60.05% |
| Wen-Yi Hung | NCU | 0.7544 | 0.4599 | 81.29% | 72.12% | 57.52% |
| GeoComp (for RPI) | RPI | 0.7403 | 0.479 | 86.22% | 75.89% | 59.43% |
| Yan-Guo (Eagle) ZHOU | Zhejiang University | 0.7857 | 0.5003 | 94.85% | 85.39% | 70.32% |
| Ana Maria Parra Bastidas | UC Davis | 0.791 | 0.4897 | 91.60% | 82.64% | 68.37% |
| Andrew Vasko | GWU | 0.7389 | 0.4915 | 90.50% | 79.59% | 62.21% |
| Mean value | | | | 90% | 80% | 65% |

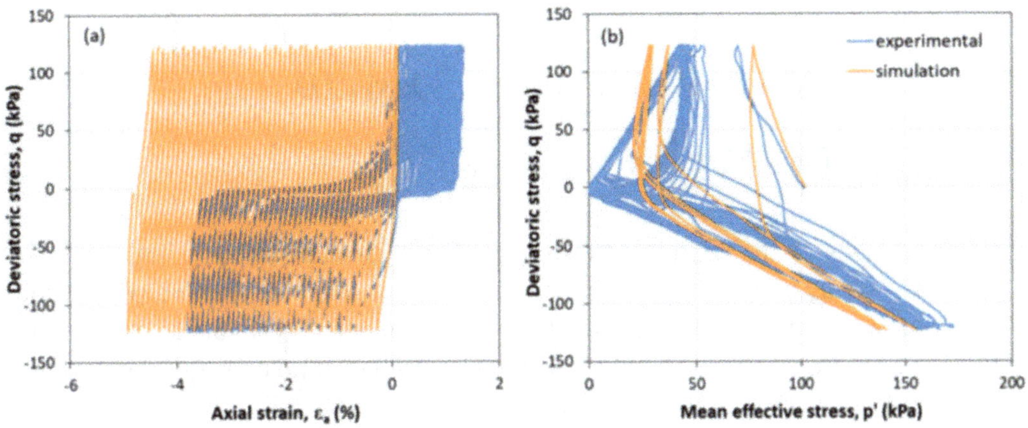


**Fig. 22.5** Simulated stress-strain (**a**) and stress path (**b**) vs. experimental curve for a cyclic triaxial test with $e_0 = 0.515$, $p'_0 = 100$ kPa, and CSR = 0.6

CSR = 0.6) provided in El Ghoraiby et al. (2017) and compares them with the experimental data. The simulated stress-strain cycles are quite different from the experimental results, while the prediction of the stress path is satisfied especially during the extension phase of the test. Liquefaction has been assumed when a computed axial strain of 2.5% was achieved. It is worth noting that such a value is affected by ratcheting that is a common issue for many constitutive models among those implemented in commercial codes.

The liquefaction strength curves, obtained from the simulated cyclic stress-controlled triaxial tests, are plotted and compared with the experimental results (Fig. 22.6).

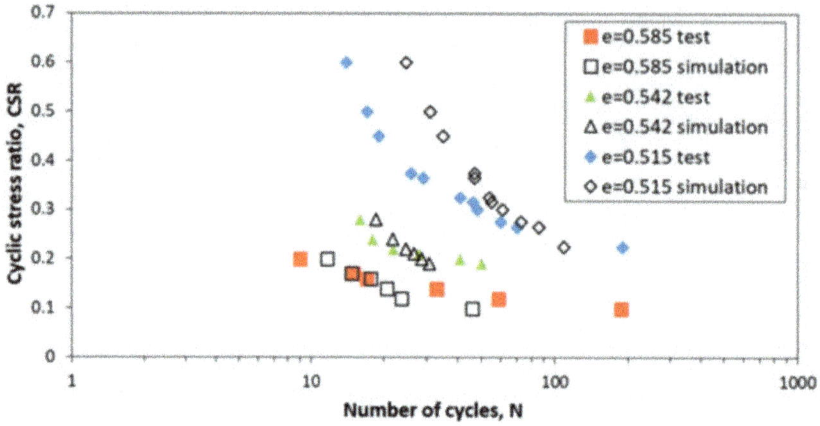


**Fig. 22.6** Liquefaction strength curves obtained from experimental (El Ghoraiby et al. 2017, 2019) and simulated cyclic stress-controlled triaxial tests on Ottawa F65 Sand

## 22.3 Numerical Analysis Platform: Plaxis 2D

### *22.3.1 Analysis Platform*

The simulations of the centrifuge tests have been carried out by using Plaxis 2D (Brinkgreve et al. 2016). This is a 2D commercial finite element method (FEM) code that includes several constitutive models. Many of them are available as user-defined model, such as the hypoplastic model with intergranular strain concept, implemented and made available in the platform by Mašín (2010).

The main reason to use Plaxis rather than other platforms is that, although not specifically oriented to solve boundary value problems in earthquake geotechnical engineering, this numerical code is quite well widespread in the community of geotechnical practitioners. Hence, it was interesting to check the possible benefit of a rigorous validation of numerical simulation procedures implemented in Plaxis through experimental data, in order to apply those procedures to a boundary value problem involving soil liquefaction.

Unfortunately, modelling the pore pressure dissipation during a dynamic analysis was not possible in Plaxis at the time when simulations were carried out. Therefore, pore pressure buildup needed to be calculated by imposing undrained conditions during shaking. This is obviously a limitation of the selected calculation method, since it affects the prediction of pore pressure buildup.

### *22.3.2 Model Geometry/Mesh and Boundary Conditions*

Figure 22.7 shows the mesh density and the boundary conditions. The mesh consists of 428 15-noded triangular elements with an average size of 0.6 m. The nodes located

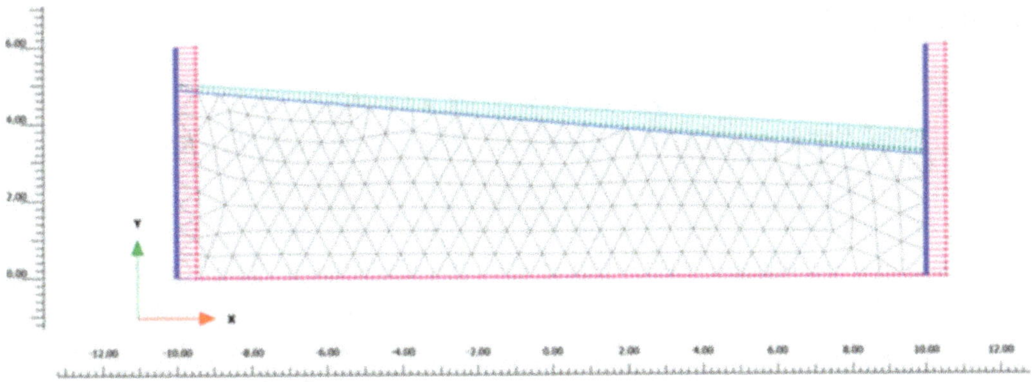

Fig. 22.7 Finite element model

at the base are fully constrained in $x$- and $y$-direction, while the nodes on the side walls are constrained laterally. The nodes on the ground surface allow full drainage.

### 22.3.3 Solution Algorithm and Assumptions

The Newmark time integration scheme is used in the simulations where the time step is constant and equal to the critical time step during the whole analysis. The proper critical time step for dynamic analyses is estimated in order to model accurately wave propagation and reduce error due to integration of time history functions. First, the material properties and the element size are taken into account to estimate the time step, and then the time step is adjusted based on the time history functions used in the calculation. During each calculation step, the Plaxis calculation kernel performs a series of iterations to reduce the out-of-balance errors in the solution. To terminate this iterative procedure when the errors are acceptable, it is necessary to establish the out-of-equilibrium errors at any stage during the iterative process automatically. Two separate error indicators are used for this purpose, based on the measure of either the global equilibrium error or the local error. The "global error" is related to the sum of the magnitudes of the out-of-balance nodal forces. The term "out-of-balance nodal forces" refers to the difference between the external loads and the forces that are in equilibrium with the current stresses. Such a difference is made nondimensional dividing it by the sum of the magnitudes of loads over all nodes of all elements. The "local error" is related to a norm of the difference between the equilibrium stress tensor and the constitutive stress tensor. It is made nondimensional dividing by the maximum value of the shear stress as defined by the failure criterion. The values of both indicators must be below a tolerated error set to 0.01 for the iterative procedure to terminate. In general, the solution procedure restricts the number of iterations that take place to 60, in order to ensure that computer time does not become too high.

A full Rayleigh damping formulation has been considered in the simulation, and the coefficients $\alpha_{RAY}$ and $\beta_{RAY}$ are equal to 0.025 and $0.63 \times 10^{-3}$, respectively.

## 22.4 Type B Simulations

### 22.4.1 *State of Stresses and Internal Variables of the Constitutive Model in the Pre-shaking Stage*

All the centrifuge models are subjected to centrifugal accelerations that are increased from 1 g to a designated value during the spin-up process. The soil specimen is subjected to an increasing centrifugal acceleration that affects the state of stresses and the subsequent seismic response. The initialization of stresses and internal variables of the constitutive model after the centrifuge spin-up and just before the start of shaking was addressed in the simulation through a static analysis aimed at computing the initial effective stress state in the soil profile.

### 22.4.2 *Results of the Dynamic Analysis*

Relevant results of the simulations are hereafter discussed, while all the data of the numerical analysis can be found in the platform DesignSafe (www.designsafe-ci.org). As described in Kutter et al. (2019), the centrifuge model is monitored with accelerometers and pore pressure transducers as reported in Fig. 22.8.

Figure 22.9 shows the comparisons among simulated and predicted time histories of acceleration, pore pressure, and horizontal displacement at the surface for the

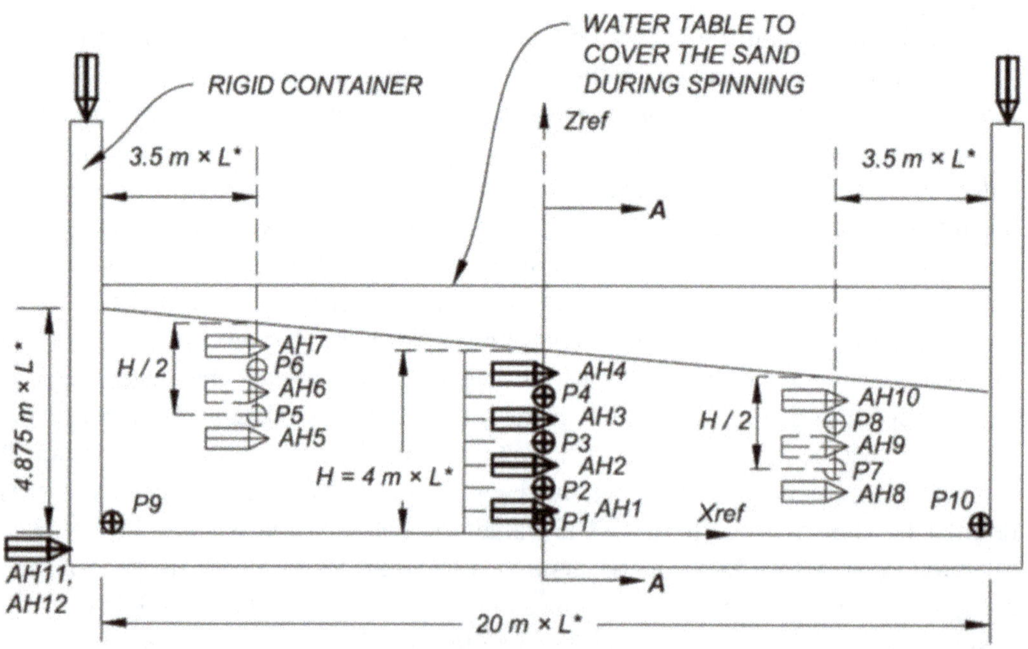


**Fig. 22.8** Baseline schematic for centrifuge experiment for shaking parallel to the axis of the centrifuge (Kutter et al. 2019)

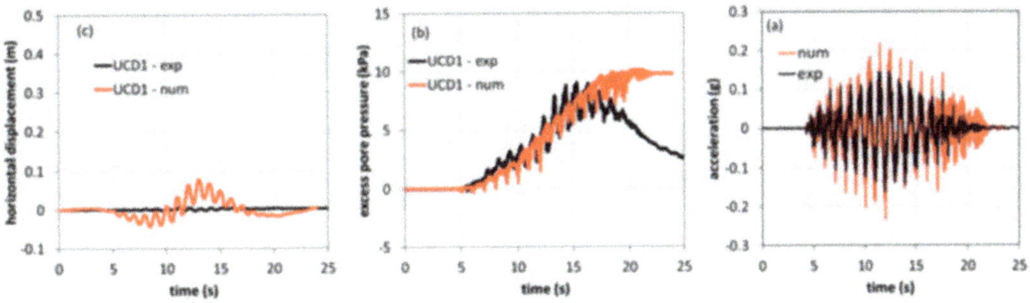


**Fig. 22.9** Computed vs. measured acceleration time history of AH4 (**a**), pore pressure time history of P4 (**b**), and horizontal displacement at surface (**c**) in UCD 1 test

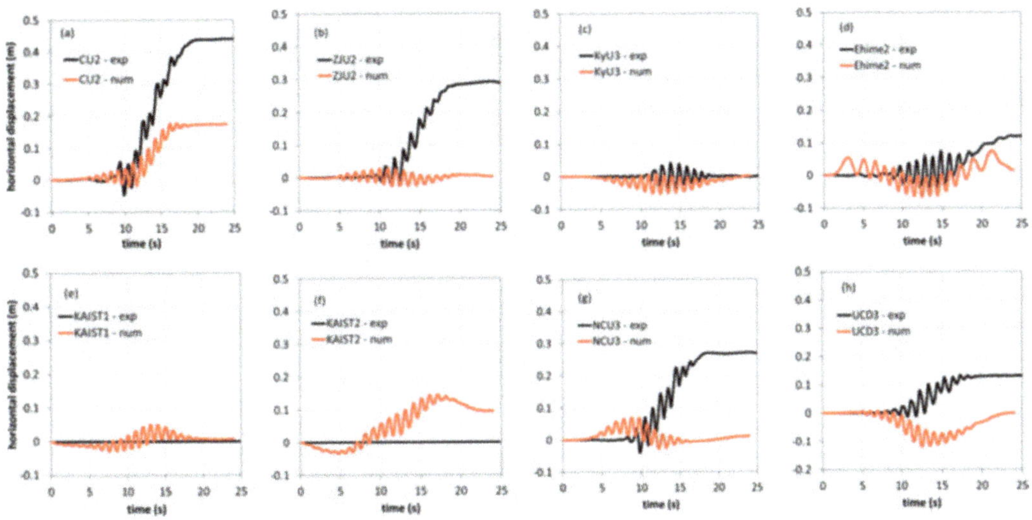

**Fig. 22.10** Computed vs. measured horizontal displacement at surface in CU2 (**a**), ZJ2 (**b**), KyU3 (**c**), Ehime2 (**d**), KAIST1 (**e**), KAIST2 (**f**), NCU3 (**g**), and UCD3 (**h**) centrifuge test

UCD1 centrifuge test. A reasonable prediction of acceleration and pore pressure is provided by the numerical simulation, while simulated displacements strongly underestimate the experimental time history.

Since the horizontal displacements result to be the most critical aspect to be reproduced, the simulated displacements at the surface are compared to the experimental ones in all the other centrifuge tests (Fig. 22.10). The comparison shows that the amplitude of the simulated time histories is quite similar to the experimental one in centrifuge test KyU3 and Ehime 2 (Fig. 22.10c, d).

## 22.5 Sensitivity Analysis

The considered key input parameters of the sensitivity study are relative density and ground motion intensity, as described below. Table 22.4 shows the characteristics of each analysis.

**Table 22.4** Characteristics of the analyses carried out for the sensitivity study

| Simulation # | NS-1 | NS-2 | NS-3 | NS-4 | NS-5 | NS-6 | NS-7 |
|---|---|---|---|---|---|---|---|
| Dry density (kg/m$^3$) | 1651 | 1608 | 1683 | 1651 | 1651 | 1651 | 1651 |
| Soil | Ottawa F65 | Ottawa F65 | Ottawa F65 | Ottawa F65 | Ottawa F65 | Ottawa F65 | Ottawa F65 |
| $D_r$ | 65% | 50% | 75% | 65% | 65% | 65% | 65% |
| Input ground motion | Achieved RPI-1, Motion 1 | Anticipated RPI-3 Motion 1 | Achieved RPI-1, Motion 1 | Achieved RPI-1, Motion 1 scaled up | Achieved RPI-1, Motion 1 scaled down | Achieved RPI-2 Motion 1 | Achieved RPI-2 Motion 1 scaled up |
| PGA (g) | 0.150 | 0.150 | 0.150 | 0.250 | 0.110 | 0.14 | 0.20 |
| PGA—1 Hz comp. (g) | 0.135 | 0.135 | 0.135 | 0.270 | 0.099 | 0.11 | 0.16 |
| PGA—high freq (g) | 0.021 | 0.021 | 0.021 | 0.035 | 0.015 | 0.08 | 0.11 |

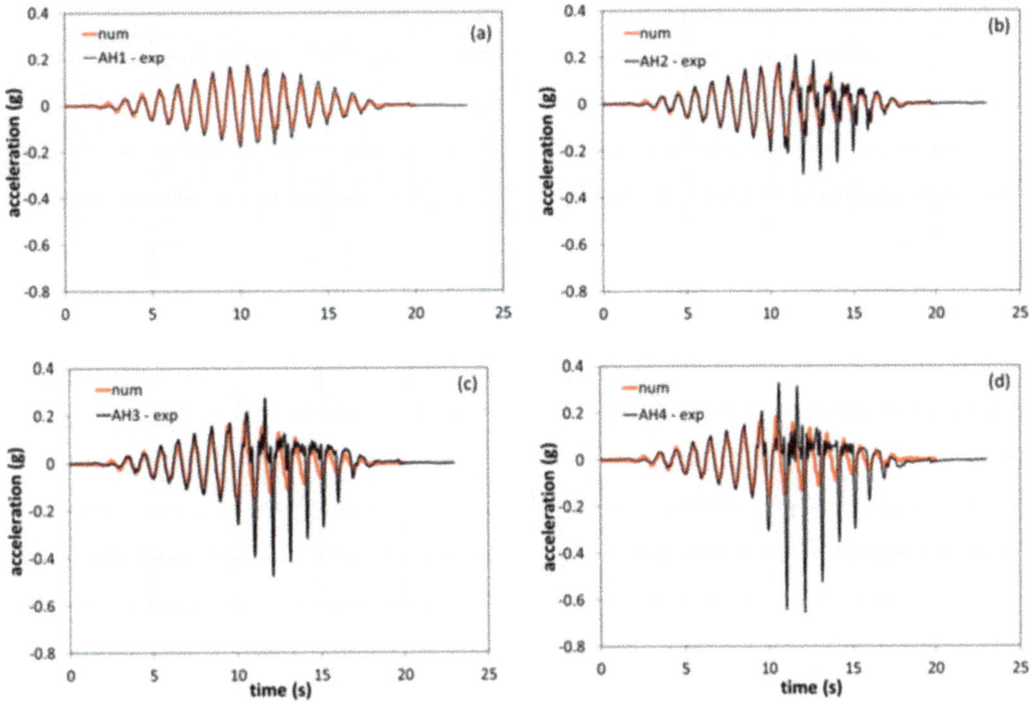


**Fig. 22.11** Computed vs. measured acceleration time histories in AH1 (**a**), AH2 (**b**), AH3 (**c**), and AH4 (**d**)—RPI01 test

The sensitivity study incorporates Type C simulation of the RPI-1 test (simulation NS-1), Type B simulation of RPI-2 (NS-6) test, and an almost Type A simulation of RPI-3 (NS-2). NS-1 to NS-3 has been used to deduce the sensitivity of simulations to relative density. NS-1, NS-4, and NS-5 allowed deduction of the sensitivity of simulations to motion intensity. NS-6 and NS-7 allowed assessment of the influence of superimposed high frequencies on simulation results.

Figures 22.11, 22.12, and 22.13 report the comparison between computed and measured acceleration and pore pressure time histories related to the RPI01 centrifuge test. Experimental acceleration time histories are well-reproduced until liquefaction condition is attained; after that, the simulations lead to an underestimation of the observed trend. Underestimation is also related to the prediction of the pore pressure in the center of the model, while pore pressure closer to the boundaries of the laminar box are better simulated.

Table 22.5 summarizes the results of the analysis in terms of computed horizontal displacement at the middle point of the model surface. It must be noted that shaking was done imposing undrained conditions and a consolidation phase was run only after the end of the shaking; hence the analyses did not provide information about the duration of liquefaction after the end of shaking. In the following sections, the effects of relative density and motion intensity are shown.

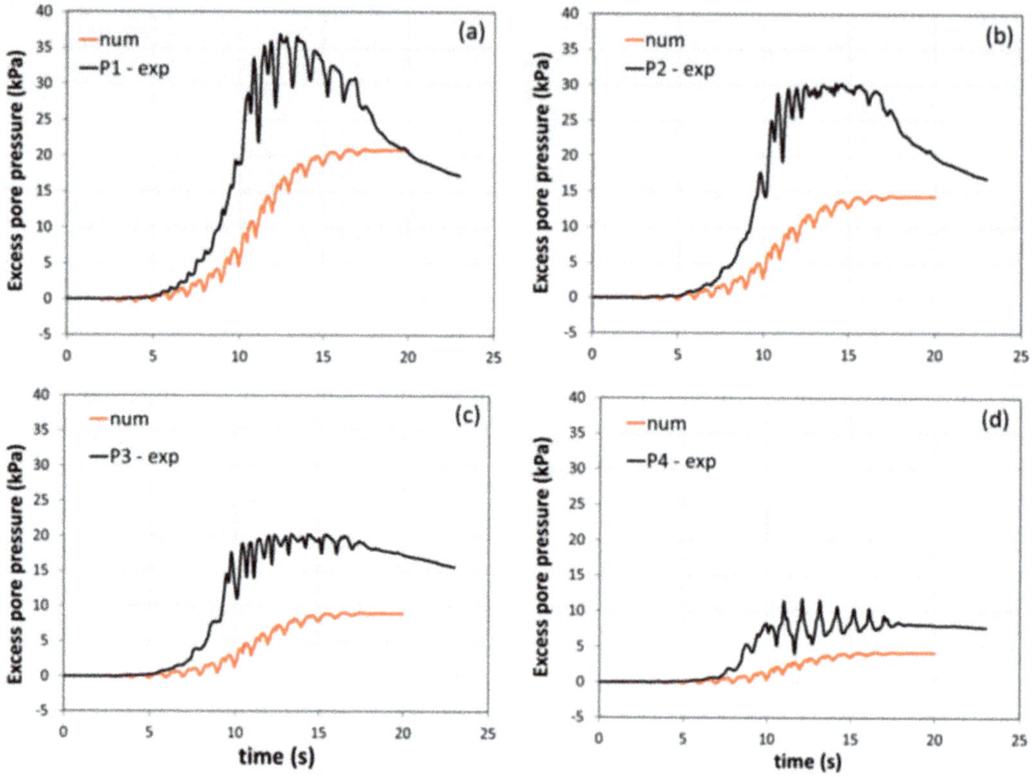


**Fig. 22.12** Computed vs. measured EPP time histories in P1 (**a**), P2 (**b**), P3 (**c**), and P4 (**d**)—RPI01 test

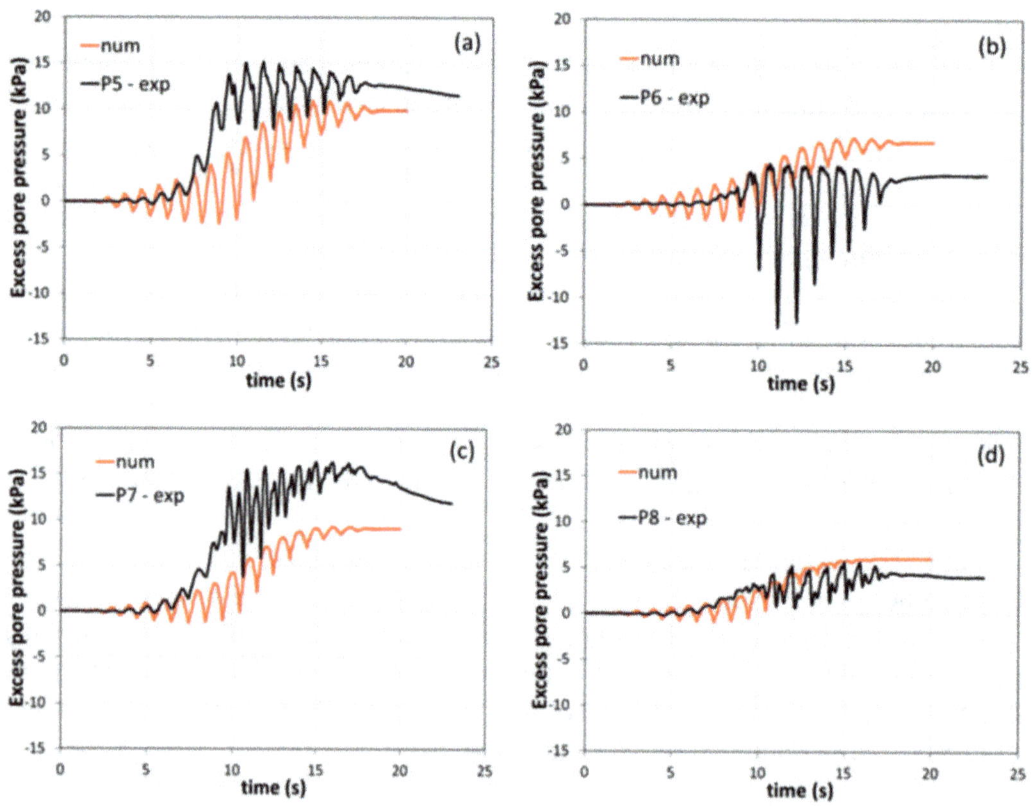


**Fig. 22.13** Computed vs. measured EPP time histories in P5 (**a**), P6 (**b**), P7 (**c**), and P8 (**d**)—RPI01 test

**Table 22.5** Main results of the sensitivity study

| Simulation # | NS-1 | NS-2 | NS-3 | NS-4 | NS-5 | NS-6 | NS-7 |
|---|---|---|---|---|---|---|---|
| X-displacement (m) at middle point on the model surface | $7.1 \times 10^{-4}$ | $1.7 \times 10^{-3}$ | $5.8 \times 10^{-4}$ | $1.4 \times 10^{-1}$ | $4.4 \times 10^{-4}$ | $2.2 \times 10^{-3}$ | $2.5 \times 10^{-1}$ |

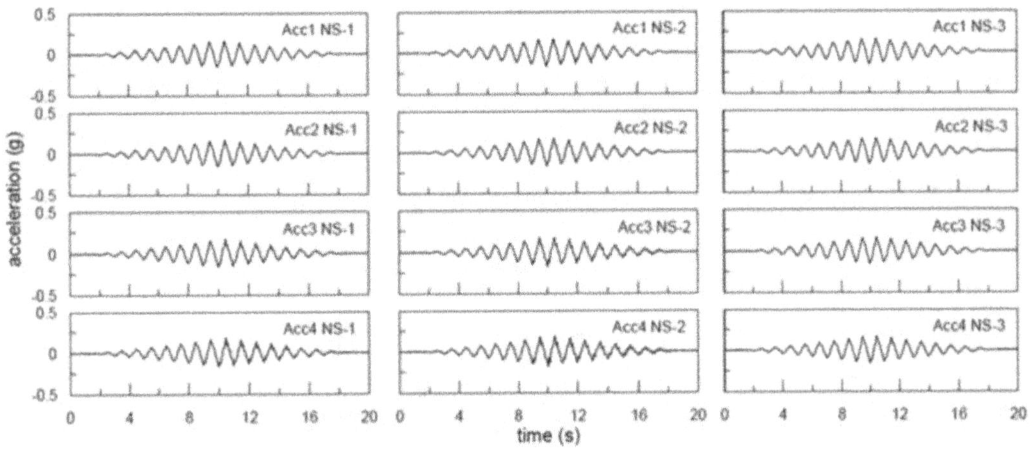


**Fig. 22.14** Influence of relative density: simulated acceleration time histories

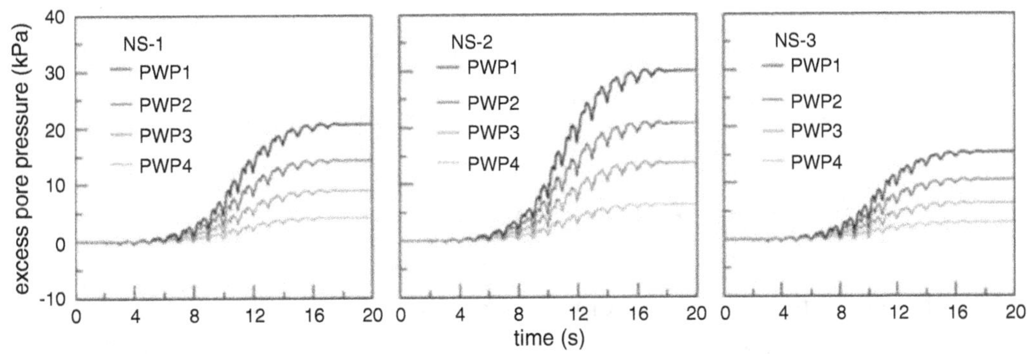


**Fig. 22.15** Influence of relative density: simulated pore pressure time histories

## 22.5.1 *Influence of Relative Density*

With reference to the influence of relative density, three different analyses were carried out with a relative density of 65% (NS-1), 50% (NS-2), and 75% (NS-3), as reported in Fig. 22.14.

Small differences can be observed in the simulated acceleration time histories (Fig. 22.14), while higher pore pressure are generated for decreasing values of relative density (Fig. 22.15).

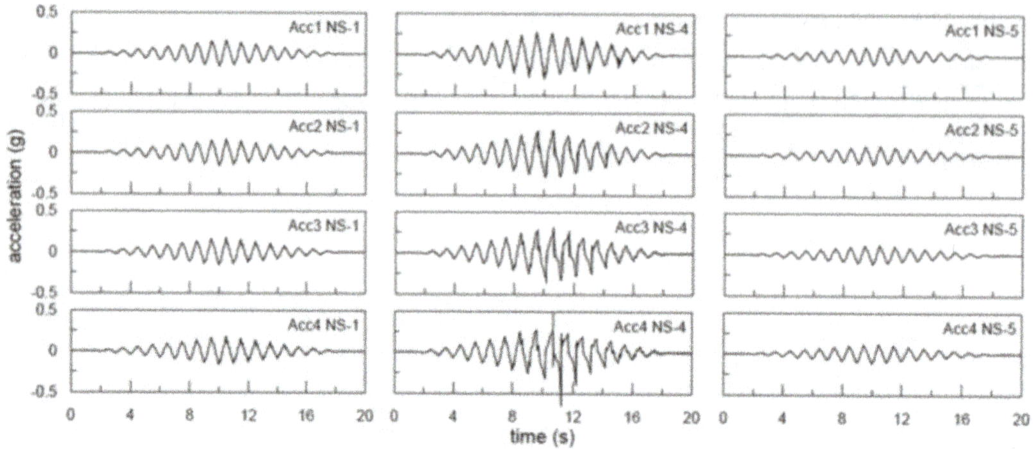


**Fig. 22.16** Influence of motion intensity: simulated acceleration time histories

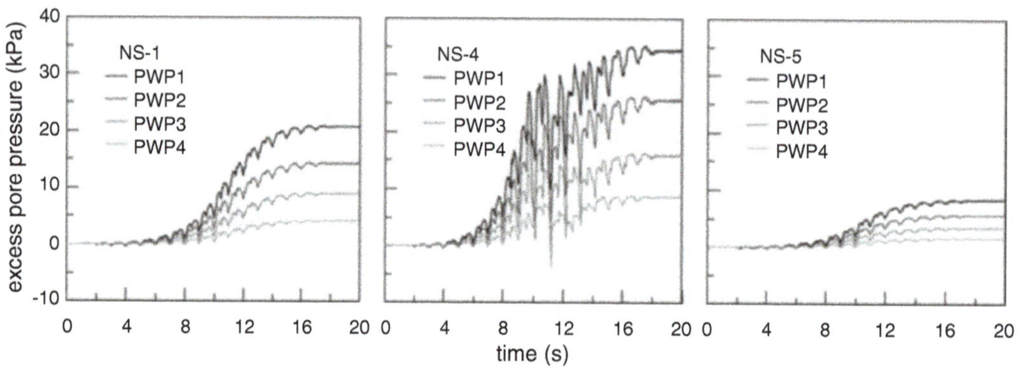


**Fig. 22.17** Influence of motion intensity: simulated pore pressure time histories

### 22.5.2 *Influence of Motion Intensity*

With reference to the influence of input motion, three different analyses were carried out with a maximum acceleration of 0.15 g (NS-1), 0.25 g (NS-4), and 0.11 g (NS-5), as reported in Fig. 22.16. As expected, significant differences can be observed in the simulated acceleration time histories along the vertical profile (Fig. 22.16), while higher pore pressure are generated for increasing values of the maximum acceleration (Fig. 22.17).

## 22.6 Conclusion

This paper described the model calibration and numerical simulations of centrifuge test for LEAP-UCD-2017, carried out at the University of Napoli Federico II (Italy).

The adopted hypoplastic model allowed a relatively easy calibration of the model parameters compared to other advanced models. Cyclic resistance curves are well-predicted although with some drawbacks on the shear strain histories, where "ratcheting" effects can be identified.

Modelling pore pressure dissipation during dynamic analysis was not possible in the version of Plaxis available at the time of simulations. Therefore, pore pressure buildup needed to be calculated by imposing undrained conditions during shaking. Nevertheless, in most cases, the buildup of excess pore pressure was reasonably captured. However, the magnitude of calculated horizontal displacements was in most cases unsatisfactory.

The sensitivity study highlighted the crucial role of relative density and the intensity of the applied input motion on the liquefaction behavior of the slope.

**Acknowledgments** The authors would like to thank the graduate students Giuseppe Colamarino, Giovanna Landolfo, and Gerardo Mascolo that carried out part of the simulations. Further appreciation goes to Prof Alessandro Flora and Prof Francesco Silvestri for their constructive comments and fruitful discussion. The Authors also wish to thank Prof David Masin for making available the Plaxis implementation of the sand hypoplasticity model used in the paper through the website www.soilmodels.com.

## References

Alarcon-Guzman, A., Chameau, J. L., Leonards, G. A., & Frost, J. D. (1989). Shear modulus and cyclic undrained behavior of sands. *Soils and Foundations, 29*(4), 105–119.

Brinkgreve, R. B. J., Kumaeswamy, S., & Swolfs, W. M. (2016). *PLAXIS 2016 User's manual*. Retrieved from PLAXIS Website, https://www.plaxis.com/kb-tag/manuals/

El Ghoraiby, M. A., Park, H., & Manzari, M. T. (2017). *LEAP 2017: Soil characterization and element tests for Ottawa F65 sand*. Washington, DC: The George Washington University.

El Ghoraiby, M. A., Park, H., & Manzari, M. T. (2019). Physical and mechanical properties of Ottawa F65 sand. In B. Kutter et al. (Eds.), *Model tests and numerical simulations of liquefaction and lateral spreading: LEAP-UCD-2017*. New York: Springer.

Herle, I., & Gudehus, G. (1999). Determination of parameters of a hypoplastic constitutive model from properties of grain assemblies. *Mechanics of Cohesive-Frictional Materials, 4*, 461–486.

Kutter, B. L., Carey, T., Stone, N., Zheng, B.L., Gavras, A., Manzari, M., et al. (2019). LEAP-UCD-2017 comparison of centrifuge test results. In B. Kutter et al. (Eds.), *Model tests and numerical simulations of liquefaction and lateral spreading: LEAP-UCD-2017*. New York: Springer.

Mašín, D. (2010). *PLAXIS implementation of hypoplasticity*. Retrieved March 2010 from PLAXIS Website.

Mašín, D. (2015). *Hypoplasticity for Practical Applications*. PhD course on hypoplasticity at Zhejiang University, June 2015. Retrieved from http://web.natur.cuni.cz/uhigug/masin/hypocourse/download/Hypoplasticity-course-handouts-public.pdf

Niemunis, A., & Herle, I. (1997). Hypoplastic model for cohesionless soils with elastic strain range. *Mechanics of Cohesive-Frictional Materials, 2*, 279–299.

Robertson, P. K., Sasitharan, S., Cunning, J. C., & Sego, D. C. (1995). Shear-wave velocity to evaluate in-situ state of Ottawa sand. *Journal of Geotechnical Engineering, 121*, 262–273.

Vasko, A. (2015). *An investigation into the behavior of Ottawa sand through monotonic and cyclic shear tests*. Master Thesis, The George Washington University.

Von Wolffersdorff, P. A. (1996). A hypoplastic relation for granular materials with a predefined limit state surface. *Mechanics of Cohesive-Frictional Materials, 1*, 251–271.

# Chapter 23
# LEAP-2017 Centrifuge Test Simulation Using HiPER

**Kiyoshi Fukutake and Takatoshi Kiriyama**

**Abstract** Simulations by effective stress analysis were carried out for nine centrifuge experiments of a gentle sloped ground. The constitutive equation is a hyperbolic model as a stress–strain relationship and a bowl model as a strain–dilatancy relationship. The experimental results largely vary depending on density and input acceleration. The analyses results can explain such a tendency of variability.

## 23.1 Outline of the Analysis Method

Analysis is done by FEM and the software used is "HiPER." The integration scheme is implicit and the integration time interval is 0.002 s. The Newmark $\beta$ a numerical integration method is adopted ($\beta = 1/4$). The method used for convergence is a modified Newton Raphson method. Convergence is determined as a relative force imbalance of 1.0E-3 or less. The maximum number of iterations is four; if convergence is not reached, the unbalanced force is carried over to the next step.

To ensure analytical stability, a low stiffness proportional damping ($C = \beta K$, where $K$ is stiffness matrix and $\beta = 1.0\text{E}{-3}$) is used.

The u-p formulation is used and excess pore water pressure is evaluated at nodes (the Sandhu method (Sandhu and Wilson 1969)).

## 23.2 Outline of the Constitutive Equations

In this section, the three-dimensional stress–strain-dilatancy relationship is explained. A hyperbolic model extending in three dimensions is used for the stress–strain relationship, while the strain–dilatancy relationship is modeled with a bowl

K. Fukutake (✉) · T. Kiriyama
Institute of Technology, Shimizu Corporation, Tokyo, Japan
e-mail: kiyoshi.fukutake@shimz.co.jp



B. Kutter et al. (eds.), *Model Tests and Numerical Simulations of Liquefaction and Lateral Spreading*, https://doi.org/10.1007/978-3-030-22818-7_23

function. The hyperbolic stress–strain model parameters are determined from dynamic deformation tests ($G/G_{max}$–$\gamma$, $h$–$\gamma$ relationships). The bowl model parameters are determined from liquefaction resistance tests (from the relationship between stress ratio and number of cycles).

### *23.2.1 Hyperbolic Model and Its Parameters*

In multidimensionalizing the hyperbolic model, the shear stress versus shear strain relationships for the shear component and the axial difference component, respectively, are defined by the following equations.

$$\tau_{xy} = \frac{G_{max} \cdot \gamma_{xy}}{1 + \frac{\gamma_{xy}}{\gamma_r}}, \quad \tau_{yz} = \frac{G_{max} \cdot \gamma_{yz}}{1 + \frac{\gamma_{yz}}{\gamma_r}}, \quad \tau_{zx} = \frac{G_{max} \cdot \gamma_{zx}}{1 + \frac{\gamma_{zx}}{\gamma_r}} \tag{23.1a}$$

$$\frac{\sigma_x - \sigma_y}{2} = \frac{G_{max} \cdot (\varepsilon_x - \varepsilon_y)}{1 + \frac{(\varepsilon_x - \varepsilon_y)}{\gamma_r}}, \quad \frac{\sigma_y - \sigma_z}{2} = \frac{G_{max} \cdot (\varepsilon_y - \varepsilon_z)}{1 + \frac{(\varepsilon_y - \varepsilon_z)}{\gamma_r}}, \quad \frac{\sigma_z - \sigma_x}{2} = \frac{G_{max} \cdot (\varepsilon_z - \varepsilon_x)}{1 + \frac{(\varepsilon_x - \varepsilon_x)}{\gamma_r}} \tag{23.1b}$$

When solving two-dimensional problems, the hyperbolic model is used for the shear components $\tau_{xy} - \gamma_{xy}$ and the axial difference components $(\sigma_x - \sigma_y)/2 - \varepsilon_x - \varepsilon_y$.

Here, $G_{max}$ is the initial shear modulus, and $\gamma_r$ is the reference strain. $\gamma_r$ is obtained from the shear strength $\tau_f$ by the following equation.

$$\gamma_r = \frac{\tau_f}{G_{max}} \tag{23.2}$$

The $h$–$\gamma$ relationship is given by the following equation.

$$h = h_{max} \cdot \left(1 - \frac{G}{G_{max}}\right) \tag{23.3}$$

where, $h$ is hysteretic damping parameter, $h_{max}$ is the maximum damping ratio and $G$ is shear modulus.

Three parameters are required to construct the hyperbolic model; $G_{max}$, $h_{max}$, and $\gamma_r$. Of these, $G_{max}$ and $\gamma_r$ are functions of effective stress. If $G_{max}$ and $\gamma_r$ at a certain reference effective stress $\sigma'_{mi}$ are $G_{maxi}$ and $\gamma_{ri}$, then $G_{max}$ and $\gamma_r$ satisfy the following equations.

**Table 23.1** Parameters of the hyperbolic model

| Parameters | Physical meaning of parameters |
|---|---|
| $G_{max}$ | Initial shear modulus. $G_{max} = \rho V_s^2$ |
| $h_{max}$ | Maximum damping ratio. As $h_{max}$ increases the nonlinearity becomes stronger |
| $\gamma_r$ | Reference strain. $\gamma_r = \tau_f/G_0$<br>The shear strain when $G/G_{max} = 0.5$ |

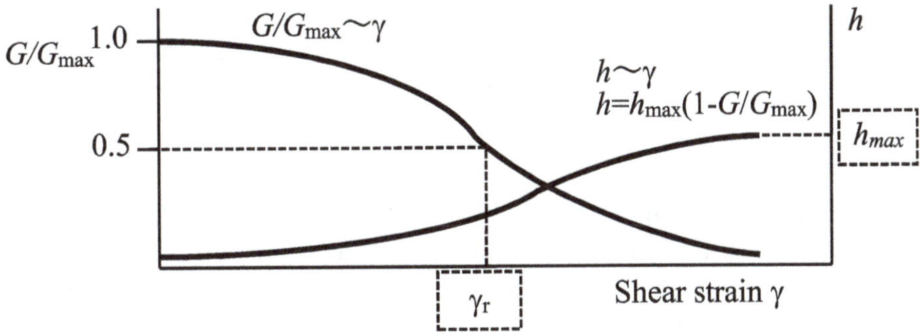


**Fig. 23.1** Parameters of hyperbolic model: $h_{max}$, $\gamma_r$

$$G_{max} = G_{maxi}\left(\frac{\sigma'_m}{\sigma'_{mi}}\right)^{0.5}, \quad \gamma_r = \gamma_{ri}\left(\frac{\sigma'_m}{\sigma'_{mi}}\right)^{0.5} \tag{23.4}$$

As the effective stress varies, at each incremental calculation step, these parameters are calculated as they vary with time based on the above equations. At the same time, the shear stress versus shear strain relationship in Eq. 23.1 varies with time in accordance with the effective stress.

Applying the Masing rule to the hyperbolic model results in excessive hysteretic damping. Therefore, hysteretic damping $h$ is adjusted using the method described by Ishihara et al. (1985).

The three parameters of the hyperbolic model ($G_{maxi}$, $h_{max}$, $\gamma_{ri}$) are defined in Table 23.1 and Fig. 23.1. As shown in Eq. 23.4 above, $G_{max}$ and $\gamma_r$ depend on the effective stress. The formulation $G_{maxi}$, $\gamma_{ri}$ is used when the mean effective stress $\sigma'_m$ is 1.0 kN/m$^2$. These values can be simply determined from the stiffness reduction curve ($G/G_{max}$ versus $\gamma$ relationship) or the damping increase curve ($h$–$\gamma$ relationship) obtained from the dynamic deformation tests.

### 23.2.2 *Bowl Model (Dilatancy Model) and Its Parameters*

To model deformation in three dimensions, in addition to the shear strains ($\gamma_{zx}$, $\gamma_{yx}$, $\gamma_{xy}$) representing simple shearing deformation and the axial shear strain differentials ($\varepsilon_x - \varepsilon_y$, $\varepsilon_y - \varepsilon_z$, $\varepsilon_z - \varepsilon_x$) representing axial deformation differential, definitions of resultant shear strain $\varGamma$ and cumulative shear strain $G^*$ are needed. They are defined by the following equations.

$$\Gamma = \sqrt{\gamma_{zx}^2 + \gamma_{zy}^2 + \gamma_{xy}^2 + (\varepsilon_x - \varepsilon_y)^2 + (\varepsilon_y - \varepsilon_z)^2 + (\varepsilon_z - \varepsilon_x)^2} \tag{23.5}$$

$$\begin{aligned} G^* &= \sum \Delta G^* \\ &= \sum \sqrt{\Delta\gamma_{zx}^2 + \Delta\gamma_{zy}^2 + \Delta\gamma_{xy}^2 + \Delta(\varepsilon_x - \varepsilon_y)^2 + \Delta(\varepsilon_y - \varepsilon_z)^2 + \Delta(\varepsilon_z - \varepsilon_x)^2} \end{aligned} \tag{23.6}$$

The bowl model (as proposed by Fukutake and Matsuoka 1989, 1993) focuses on the resultant shear strain $\varGamma$ and the cumulative shear strain $G^*$. Dilatancy $\varepsilon^s{}_v$ results from a certain soil particle repeatedly dropping into the valley between other soil particles (negative dilatancy), and rising up on other soil particles (positive dilatancy). This mechanism is represented by the superposition of positive and negative dilatancy using the following equation.

$$\varepsilon_v^s = \varepsilon_\Gamma + \varepsilon_G = \mathrm{A} \cdot \Gamma^{1.4} + \frac{\mathrm{G}^*}{\mathrm{C} + \mathrm{D} \cdot \mathrm{G}^*} \tag{23.7}$$

Here $A$, $C$, and $D$ are parameters. $\varepsilon_G$ is monotonic negative dilatancy (compressive strain). It is irreversible and represented as a hyperbolic function with respect to $G^*$. $\varepsilon_\varGamma$ is cyclic positive dilatancy (swelling strain) and is reversible and represented as an exponential function with respect to $\varGamma$. The $\varepsilon_G$ component is the master curve and is the component that determines the basic dilatancy during cyclic shearing, while the $\varepsilon_\varGamma$ component is a oscillating component associated with it. $1/D$ is the asymptotic line to the hyperbolic curve, corresponding to a relative density of 100%.

The mechanism of such a bowl model is the movement of soil particles in seven-dimensional strain space with $\gamma_{xy}$, $\gamma_{yz}$, $\gamma_{zx}$, $(\varepsilon_x - \varepsilon_y)$, $(\varepsilon_y - \varepsilon_z)$, $(\varepsilon_z - \varepsilon_x)$, and $\varepsilon^s{}_v$ as axes.

Figure 23.2 illustrates the case of unidirectional cyclic shearing.

Next, the consolidation term is taken into consideration in the stress–strain relationship and effective stress is modeled under undrained (constant volume) conditions.

The volumetric strain $\varepsilon_v{}^s$ represented by Eq. 23.7 is the dilatancy component due to shearing, but an additional volumetric strain due to change in the effective stress $\sigma'_m$ has to be considered. The total volumetric strain increment of the soil $d\varepsilon_v$ is then given by the following equation, where the shear component is $d\varepsilon_v{}^s$ and the consolidation component is $d\varepsilon_v{}^c$.

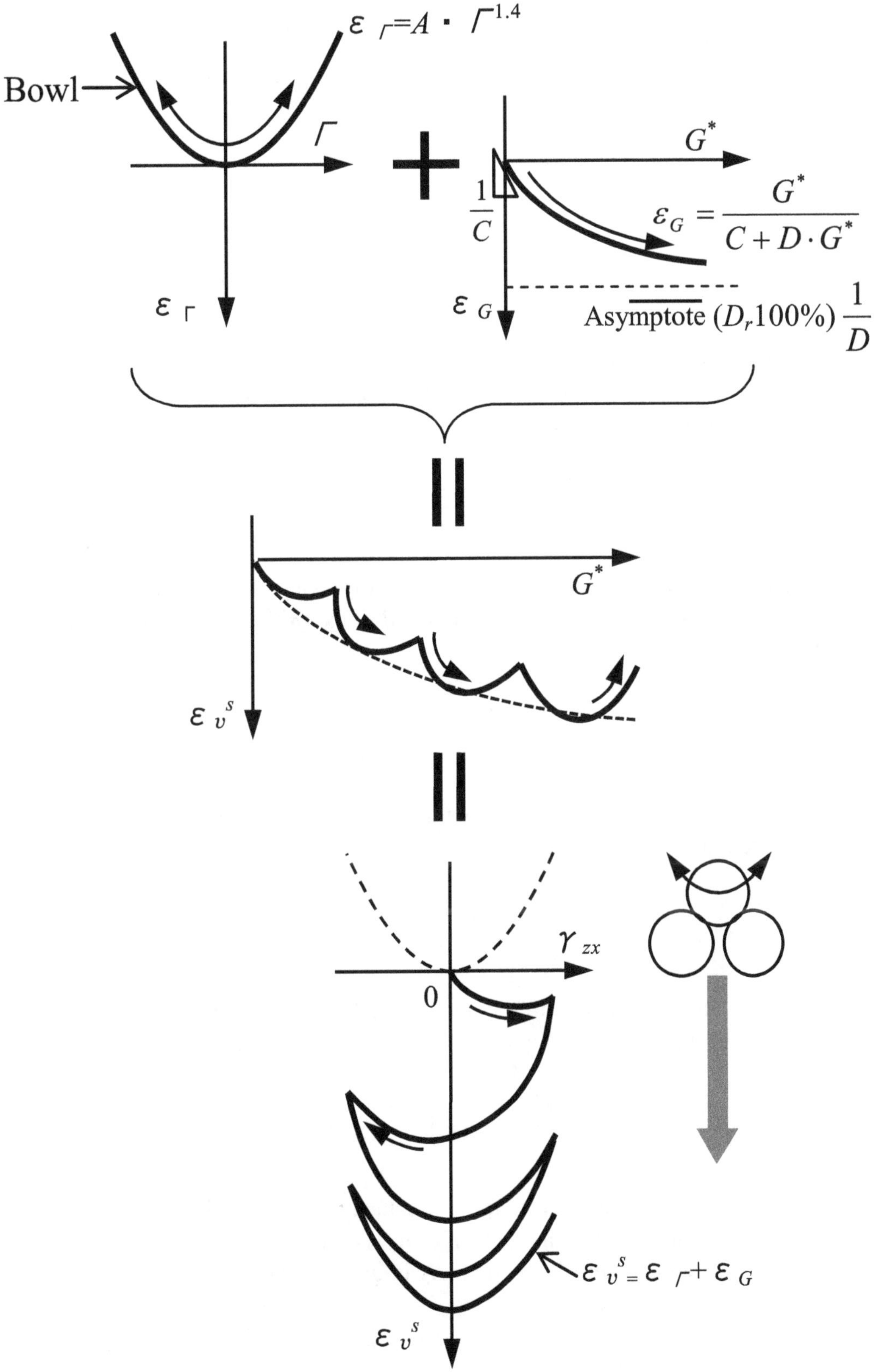


**Fig. 23.2** Dilatancy in unidirectional cyclic shearing

$$d\varepsilon_v = d\varepsilon_v{}^s + d\varepsilon_v{}^c \tag{23.8}$$

The consolidation term $d\varepsilon_v{}^c$ is given by the following equations, assuming one-dimensional consolidation conditions.

$$d\varepsilon_v{}^c = \frac{0.434 \cdot C_s}{1 \cdot e_0} \cdot \frac{d\sigma'_m}{\sigma'_m} \quad (\text{for } d\sigma'_m < 0) \tag{23.9}$$

$$d\varepsilon_v{}^c = \frac{0.434 \cdot C_c}{1 \cdot e_0} \cdot \frac{d\sigma'_m}{\sigma'_m} \quad (\text{for } d\sigma'_m > 0) \tag{23.10}$$

Here, $C_s$ is the swelling index, $C_c$ is the compression index, and $e_0$ is the initial void ratio. If undrained conditions ($d\varepsilon_v = 0$) are assumed in Eqs. 23.9 and 23.10, the following equation is obtained.

$$d\varepsilon_v{}^s + \frac{0.434 \cdot C_s}{1 + e_0} \cdot \frac{d\sigma'_m}{\sigma'_m} = 0 \tag{23.11}$$

If the mean effective stress in the initial shearing stage is $\sigma'_{m0}$, and if the above equation is integrated under the initial conditions ($\sigma'_m = \sigma'_{m0}$), the following equation is obtained.

$$\varepsilon_v{}^s + \frac{C_s}{1 + e_0} \cdot \log \frac{\sigma'_m}{\sigma'_{m0}} = 0 \tag{23.12}$$

From Eq. 23.12, the effective stress under undrained shear $\sigma'_m$ is given by the following equations.

$$\sigma'_m = \sigma'_{m0} \cdot 10^{\alpha}, \quad \alpha \equiv \frac{-\varepsilon_v{}^s}{C_s/(1 + e_0)} \tag{23.13}$$

Therefore, substituting α from the above equation, the effective stress reduction ratio is given by the following equation:

$$\left( \frac{\sigma'_{m0} - \sigma'_m}{\sigma'_{m0}} \right) = 1 - 10^{\alpha} \tag{23.14}$$

In order to suppress the occurrence of dilatancy under small shear amplitude, a spherical region with shear strain radius $\Gamma = R_e$ is considered within the strain space, and within this region there is no $d\varepsilon_G$. $R_e$ is given as follows, with reference to Fig. 23.3.

Figure 23.3a shows the liquefaction resistance curve and the relationship between liquefaction resistance and the lower limit value $X_l$. $X_l$ is the liquefaction resistance

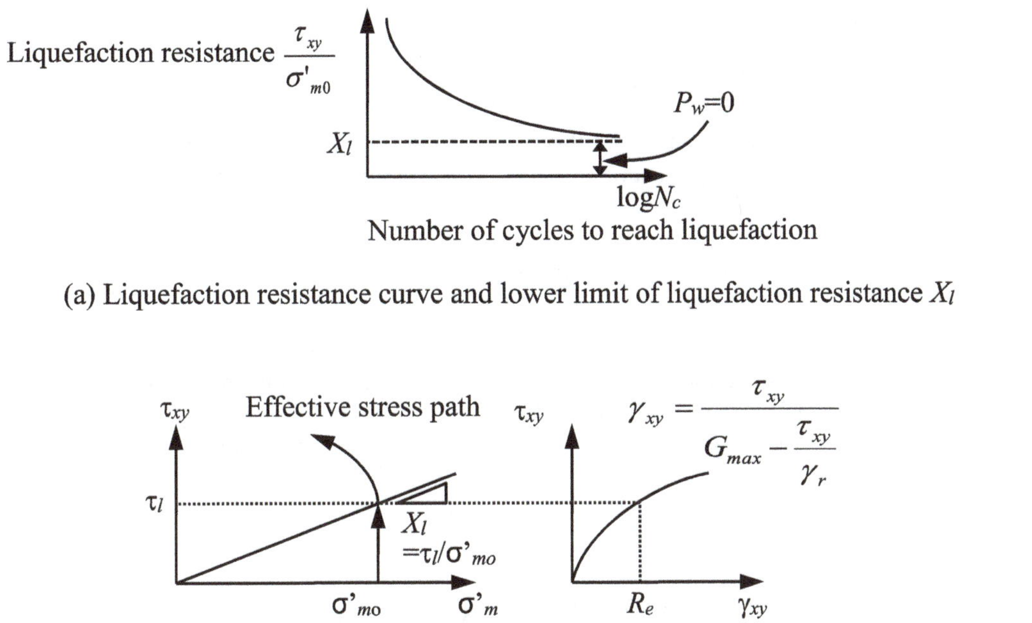


(a) Liquefaction resistance curve and lower limit of liquefaction resistance $X_l$

(b) Effective stress path and $X_l$ (c) Skeleton curve of hyperbolic model and $R_e$

**Fig. 23.3** Lower limit of liquefaction resistance $X_l$ and shear strain $R_e$. (**a**) Liquefaction resistance curve and lower limit of liquefaction resistance $X_l$. (**b**) Effective stress path and $X_l$. (**c**) Skeleton curve of hyperbolic model and $R_e$

after a very large number of cycles, in other words, it represents the stress ratio at which liquefaction does not occur even however many cycles occur. Also, for simplicity, it is assumed that the excess pore water pressure does not arise ($P_w = 0$) for repeated stress ratios less than $X_l$. From Fig. 23.3b, c, setting the shear strain when the stress ratio is $X_l$ equal to $R_e$, then $R_e$ is given by the following equation.

$$R_e = \frac{X_l \sigma'_{m0}}{G_{\max} - \frac{X_l \sigma'_{m0}}{\gamma_r}} \tag{23.15}$$

Here $\sigma'_{m0}$ is the mean effective stress in initial shear. When the amplitude of the stress ratio is $X_l$ or less, positive excess pore water pressure does not arise.

There are six parameters in the bowl model ($A$, $C$, $D$, $C_s/(1 + e_0)$, $C_c/(1 + e_0)$ and $X_l$) as shown in Table 23.2. The values of these parameters are determined from a liquefaction resistance curve.

Table 23.2 and Fig. 23.4 give the meanings of the bowl model parameters. These parameters are determined by fitting to the liquefaction resistance curve.

**Table 23.2** Parameters of bowl model

| Parameter | Physical meaning of parameters |
|---|---|
| $A$ | Parameter representing the swelling component $\varepsilon_r$ of the dilatancy components. The larger the absolute value of $A$, the greater the cyclic mobility |
| $C$, $D$ | Parameters representing the compression component $\varepsilon_G$ of the dilatancy components. $1/C$ is the slope of the dilatancy in the initial stage of shear. $1/D$ is calculated from the minimum void ratio $e_{\min}$ on the hyperbolic asymptotic line (maximum amount of compression) |
| $C_s/(1 + e_0)$ | $C_s$ is the swelling index; $e_0$ is the initial void ratio |
| $C_c/(1 + e_0)$ | $C_s$ is the compression index; $e_0$ is the initial void ratio |
| $X_l$ | The lower limit value of the liquefaction resistance. In the relationship between stress ratio $\tau/\sigma'_{m0}$ and the number of cycles $N_c$, it is represented by $\tau/\sigma'_{m0}$ when $N_c$ is sufficiently large. When $\tau/\sigma'_{m0} > X_l$ excess pore water pressure arises |

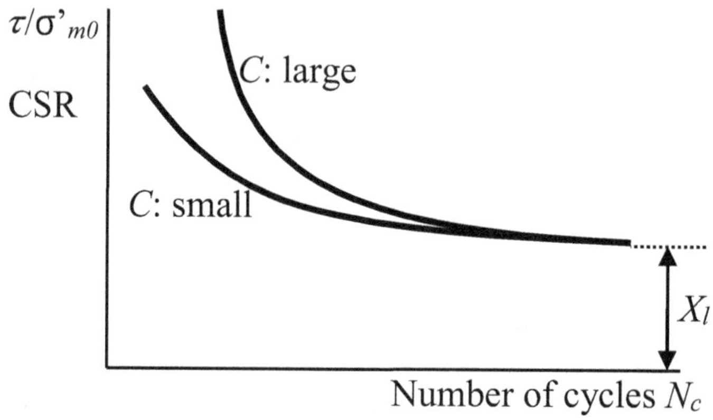


**Fig. 23.4** Liquefaction resistance curves and bowl model parameters $C$ and $X_l$

## 23.3 Element Test Simulation

### 23.3.1 Determining Parameters by Test Simulation

#### Hyperbolic Model

The initial shear stiffness $G_{\max}$ is calculated from the following equation based on research by Liu et al. (2017).

$$G_{\max} = 6.35 \frac{1}{0.3 + 0.7e^2} \sigma_m^{0.5} \tag{23.16}$$

Here $\gamma_r$ and $h_{\max}$ are standard values for sand based on the shear strain dependence of stiffness ratio and damping for Toyoura sand ($G/G_{\max}$–$\gamma$, $h$–$\gamma$ relationship). Since the experiment of Ottawa sand was not carried out, it replaced it with Toyoura sand experiment. Figure 23.5 shows the $G/G_{\max}$–$\gamma$, $h$–$\gamma$ relationship used in calculations.

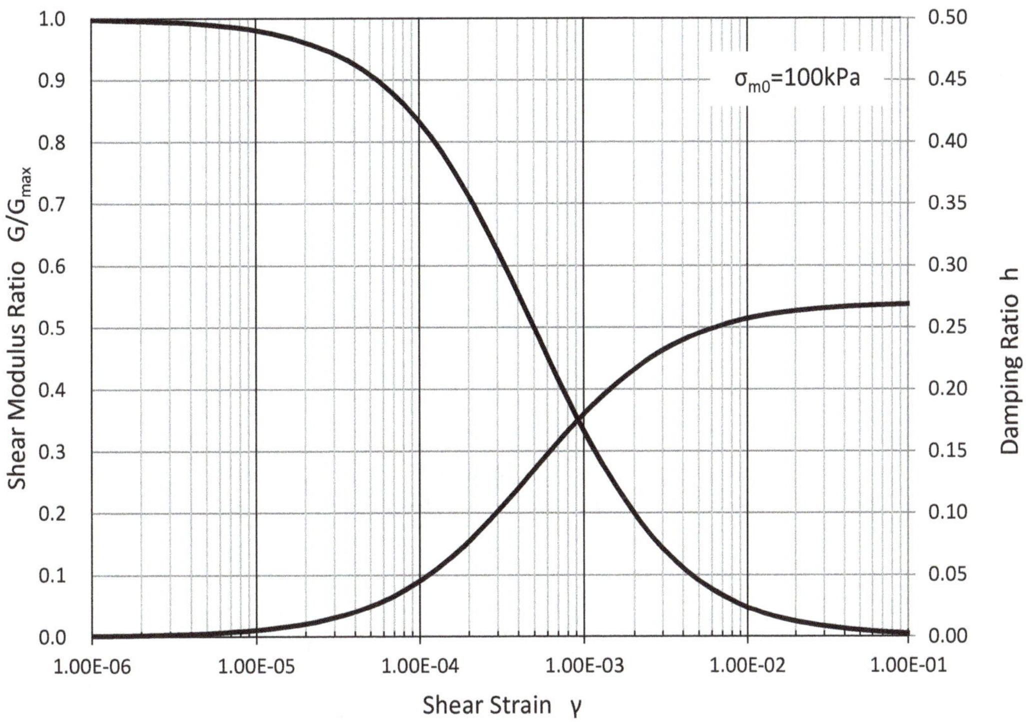


**Fig. 23.5** $G/G_{max}$–$\gamma$, $h$–$\gamma$ relationship used in calculations (hyperbolic model)

### Bowl Model

The parameter $D$ is determined from the following equation since $1/D$ is the asymptote of dilatancy. The average of $e_{min}$ is 0.49.

$$\frac{1}{D} = \frac{e_0 - e_{min}}{1 + e_0} \tag{23.17}$$

The lower limit value of liquefaction resistance $X_l$ is determined based on the liquefaction resistance curve. The liquefaction resistance curve is calculated using values for standard sand for the other parameters. Parameter $C$ is adjusted while comparing the test results and the calculation results so as to match the whole liquefaction resistance curve with the measured values.

The parameters for Ottawa sand set by this procedure are given in Table 23.3.

## *23.3.2 Results of Element Test Simulations*

Figure 23.6 shows the simulations of liquefaction resistance curves. They explain the experiment results for the three selected densities.

Figure 23.7 shows the effective stress path and stress–strain relationship at these three different densities. In the dense case, strong cyclic mobility is present.

**Table 23.3** Parameters for Ottawa sand

| | | | | | | | Parameter of constitutive equation | | | | | | | | |
|---|---|---|---|---|---|---|---|---|---|---|---|---|---|---|---|
| | | | | | | | Hypabolic model | | | Bowl model | | | | | |
| No | $\sigma_{m0}$(kN/m$^2$) | $e$ | $\rho_d$ (t/m$^3$) | $V_s$ (m/s) | $G_0$ (kN/m$^2$) | Reference strain ($\gamma_r$) | $G_{0i}$ (kN/m$^2$) | $\gamma_{ri}$ | $h_{max}$ (%) | $A$ | $C$ | $D$ | $C_s$ (1 + $e_0$) | $C_c$ (1 + $e_0$) | $X_l$ |
| 1 | 100 | 0.515 | 1.7442 | 274 | 130751 | 0.0005 | 13075 | 5.0E−5 | 27 | −1.5 | 35.0 | 61 | 0.0060 | 0.0061 | 0.20 |
| 2 | 100 | 0.542 | 1.7126 | 271 | 125585 | 0.0005 | 12558 | 5.0E−5 | 27 | −0.8 | 9.0 | 30 | 0.0060 | 0.0061 | 0.16 |
| 3 | 100 | 0.585 | 1.6656 | 266 | 117689 | 0.0005 | 11769 | 5.0E−5 | 27 | −0.6 | 6.0 | 17 | 0.0060 | 0.0061 | 0.09 |

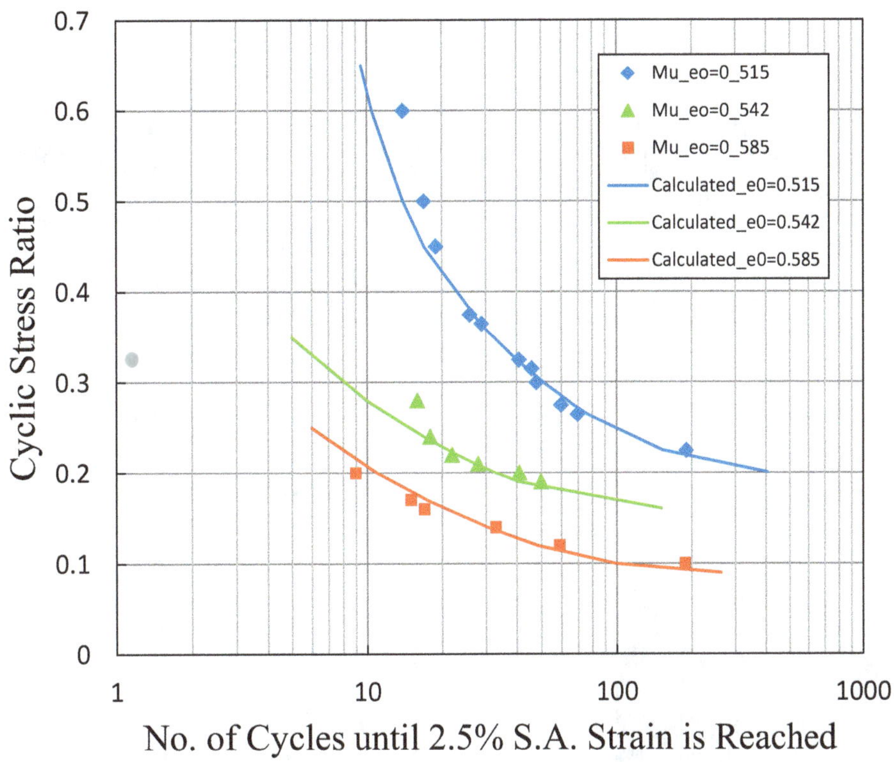


**Fig. 23.6** Liquefaction resistant curves

## 23.4 Centrifuge Simulation

### 23.4.1 Determination of Parameters

The model parameters (that is to say the $G/G_0$–$\gamma$, $h$–$\gamma$ relationship and the liquefaction resistance curve) are determined by element test simulations of Phase I except for $G_0$ and $D$. $G_0$ and $D$ are reset by the following procedure for each centrifuge test. From the density of sand (Table 23.4), the relative density $D_r$ is obtained, and the void ratio $e$ is calculated from $D_r$. The initial shear stiffness $G_0$ is calculated from Eq. 23.16 based on the research of Liu et al. 2017. $G_{0i}$ is the value at unit mean stress ($\sigma_m$ = 1.0 kN/m$^2$). Parameter $D$ is determined from Eq. 23.17 since 1/$D$ is the asymptote of dilatancy. The average of $e_{min}$ is 0.49.

Table 23.5 shows the parameters set by this method. The Poisson ratio of the ground is set to 0.33. The bulk modulus of water is $K_w$ = 2.2E + 6 kN/m$^2$. For soil permeability, $k$ = 0.015 cm/s is used from experimental results.

### 23.4.2 Summary of the Numerical Simulations

The FEM mesh is shown in Fig. 23.8. The FEM model has 2013 nodes and 1920 elements. A plane strain four-noded quadrilateral element with displacements and pore pressure degrees of freedom at each node is employed. Four integration points

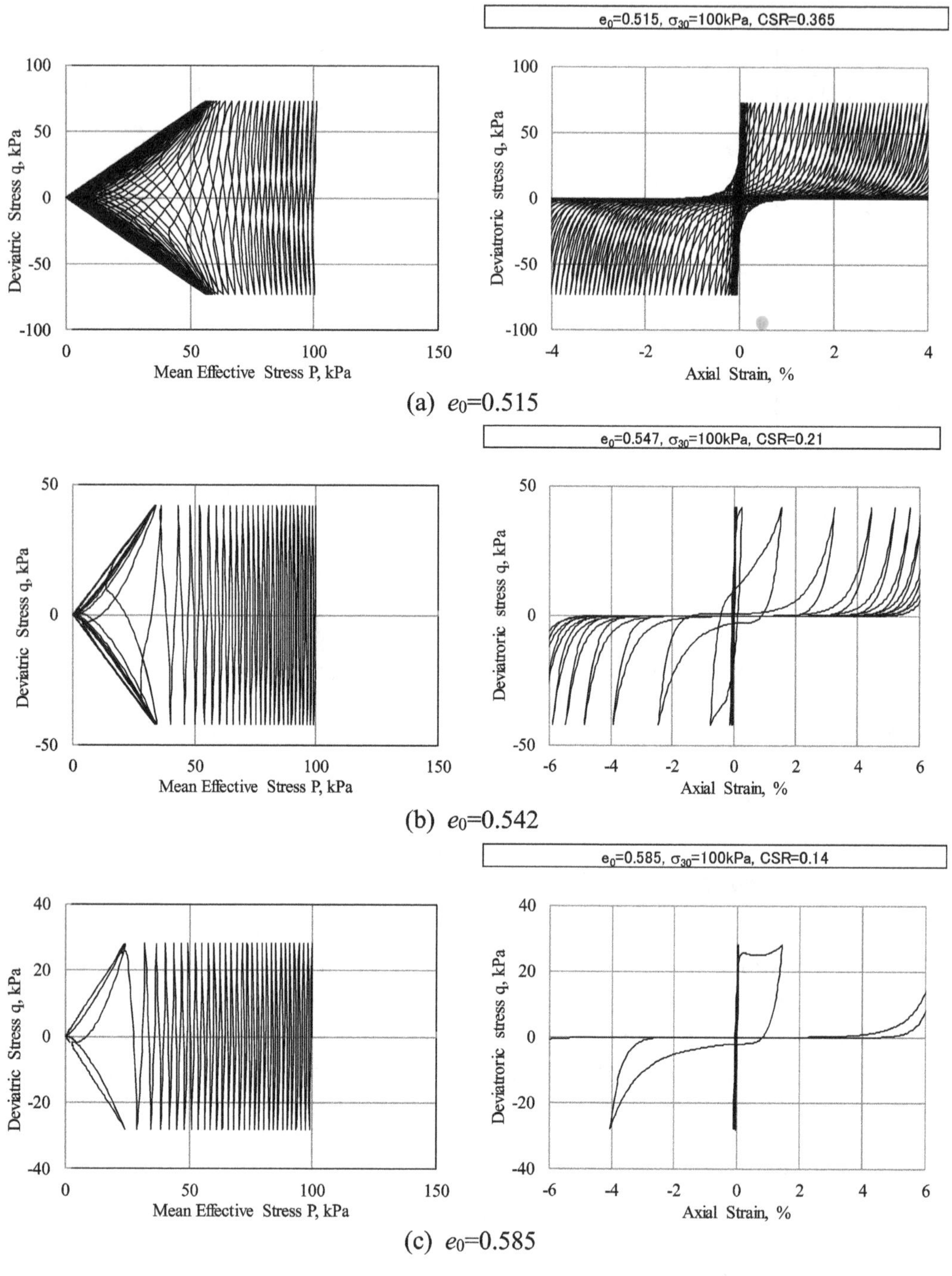


**Fig. 23.7** Effective stress path and stress strain relationship. (**a**) $e_0 = 0.515$. (**b**) $e_0 = 0.542$. (**c**) $e_0 = 0.585$

**Table 23.4** Soil density and void ratio of Ottawa sand

| $\rho$dmax (kg/m$^3$) | 1765 | $\rho$dmin (kg/m$^3$) | 1480 |
|---|---|---|---|
| *e*max | 0.766 | *e*min | 0.49 |

**Table 23.5** Parameters of constitutive equation

| | | ρd | Dr | e | Parameter of constitutive equation | | | | | | | | |
|---|---|---|---|---|---|---|---|---|---|---|---|---|---|
| | | | | | Hyperbolic model | | | Bowl model | | | | | |
| | | ρd (kg/m3) | Dr | e | $G_{0i}$ kN/m$^2$ | $\gamma_{ri}$ | $h_{max}$ % | $A$ | $C$ | $D$ | $\frac{Cs}{(1+e_0)}$ | $\frac{Cc}{(1+e_0)}$ | $X_l$ |
| 1 | CU-2 | 1605.8 | 0.485 | 0.632 | 10954 | 5.0E-5 | 27 | -0.6 | 6.0 | 11 | 0.0060 | 0.0061 | 0.09 |
| 2 | Ehime-2 | 1656.55 | 0.660 | 0.584 | 11790 | 5.0E-5 | 27 | -0.6 | 6.0 | 17 | 0.0060 | 0.0061 | 0.09 |
| 3 | KAIST-1* | 1701.2 | 0.805 | 0.544 | 12525 | 5.0E-5 | 27 | -0.8 | 9.0 | 29 | 0.0060 | 0.0061 | 0.16 |
| 4 | KAIST-2 | 1592.5 | 0.437 | 0.645 | 10736 | 5.0E-5 | 27 | -0.6 | 6.0 | 11 | 0.0060 | 0.0061 | 0.09 |
| 5 | KyU-3 | 1637 | 0.594 | 0.602 | 11467 | 5.0E-5 | 27 | -0.6 | 6.0 | 14 | 0.0060 | 0.0061 | 0.09 |
| 6 | NCU-3 | 1652 | 0.645 | 0.588 | 11715 | 5.0E-5 | 27 | -0.6 | 6.0 | 16 | 0.0060 | 0.0061 | 0.09 |
| 7 | UCD-1 | 1665 | 0.688 | 0.576 | 11929 | 5.0E-5 | 27 | -0.6 | 6.0 | 18 | 0.0060 | 0.0061 | 0.09 |
| 8 | UCD-2 | 1658 | 0.665 | 0.582 | 11814 | 5.0E-5 | 27 | -0.6 | 6.0 | 17 | 0.0060 | 0.0061 | 0.09 |
| 9 | ZJU-2 | 1606 | 0.486 | 0.632 | 10958 | 5.0E-5 | 27 | -0.6 | 6.0 | 12 | 0.0060 | 0.0061 | 0.09 |

∗ dense

$$G_{0i} = 6.35 \frac{1}{0.3 + 0.7e^2} 1.0^{0.5}$$

$$\frac{1}{D} = \frac{e - e_{min}}{1 + e}$$

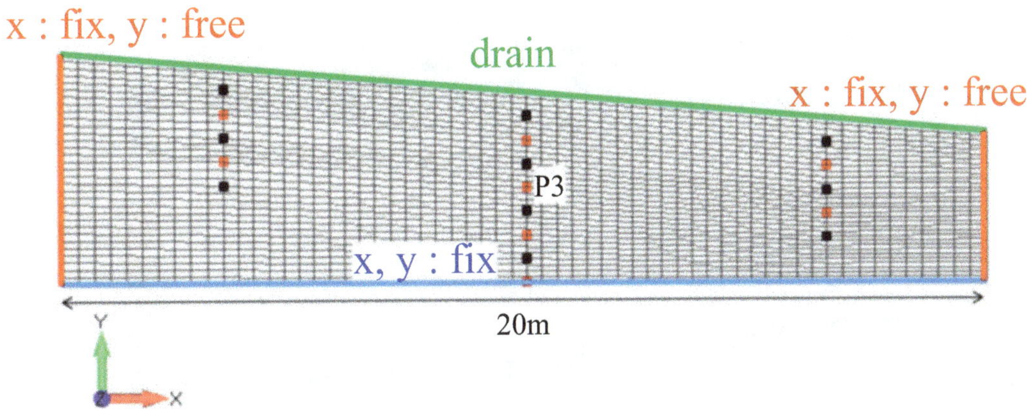


**Fig. 23.8** Boundary conditions and time history output points (black square box acceleration, red square box pore water pressure)

are used in each element. The *u-p* formulation is used. The excess pore water pressure is evaluated at the nodes (the Sandhu method).

The nodes located at the base of the model are fully constrained in the $x$ and $y$ directions while the nodes on the side walls are constrained laterally. The nodes on the free surface allow full drainage and have a fixed pore water pressure.

Figure 23.9 shows the initial stress calculated by linear analysis. The initial stress is determined by linear self-weight analysis. For stiffness, the value at 4 m of depth is used.

In the dynamic analysis, horizontal acceleration and vertical acceleration were input simultaneously.

### 23.4.3 *Results of Centrifuge Simulations*

Horizontal displacement time histories of the ground surface at the center are shown in Fig. 23.10. The amount of lateral spreading of the ground surface was of the same

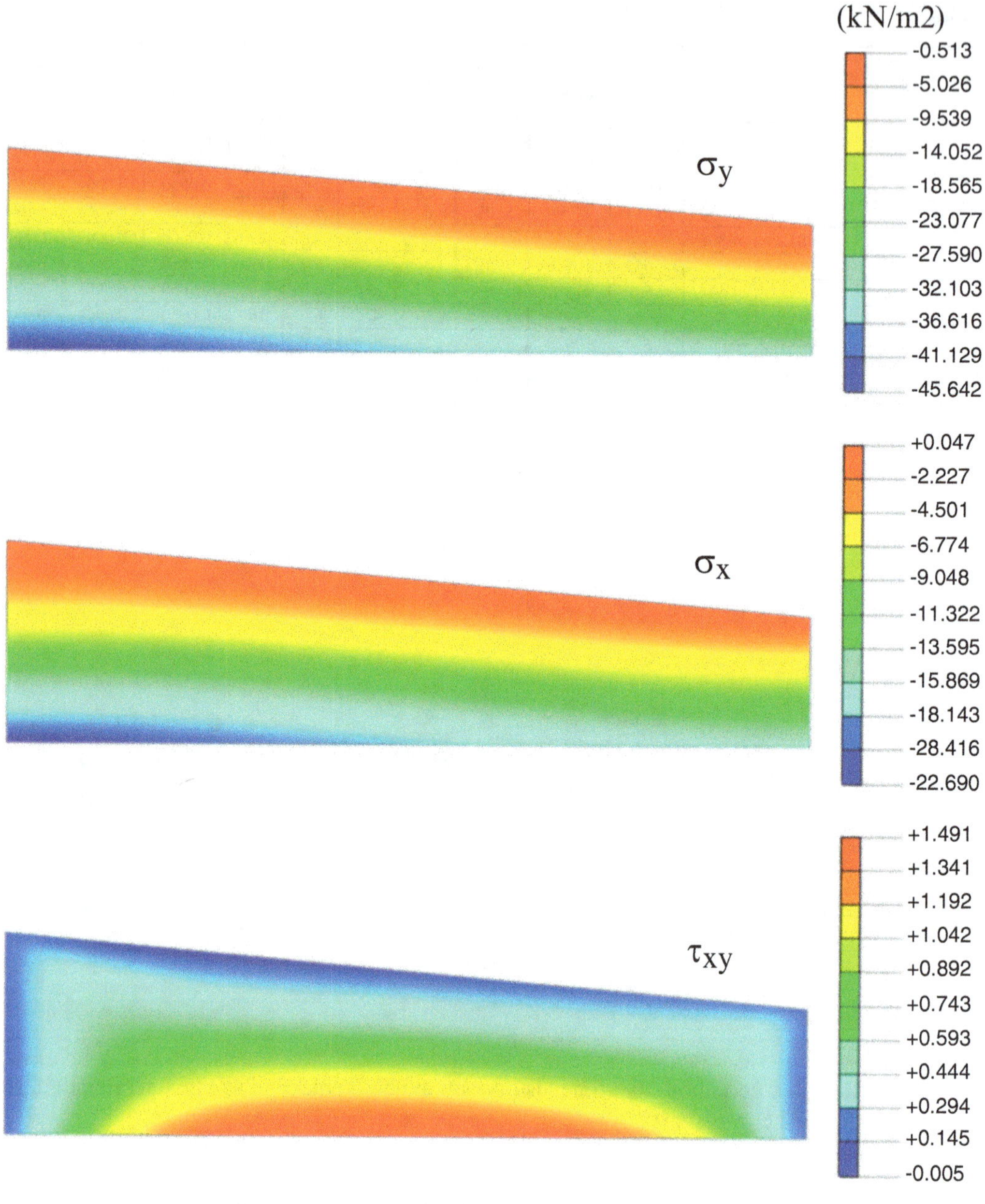


**Fig. 23.9** Initial stress calculated by linear analysis

order in experiment and analysis. For UCD3, Ehime2, and CUD1, the simulated values and experimental values show good correspondence. The analysis values are rather smaller for CU 2, ZJU 2, and NCU 3. The experimental show large variation depending on density and input acceleration. The simulations successfully explain this tendency.

Figure 23.11 shows fivefold enlarged deformation plots and contours of excess pore water pressure ratio. For KAIST1, KyU3, and UCD1, liquefaction does not occur and deformations are small. In the other six cases, the ground liquefies and deforms in the downstream direction.

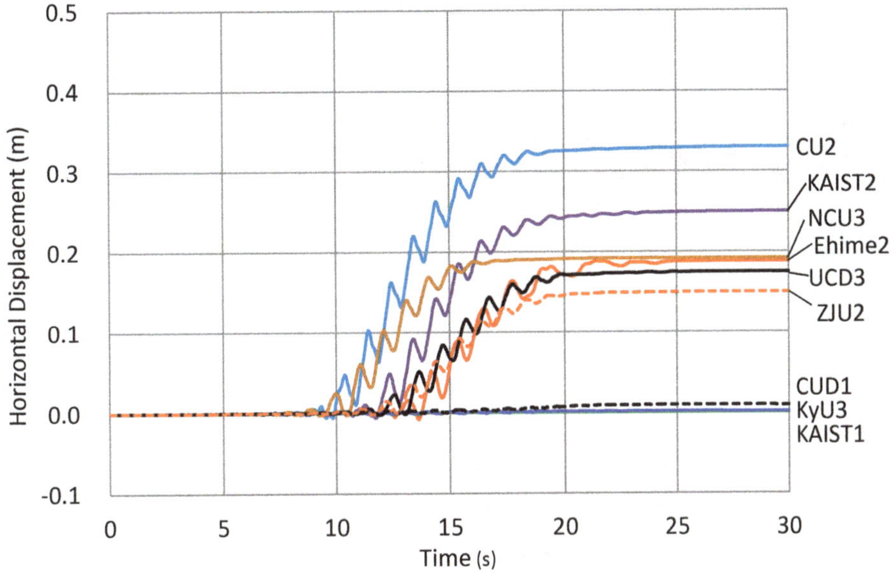


(a) Simulations

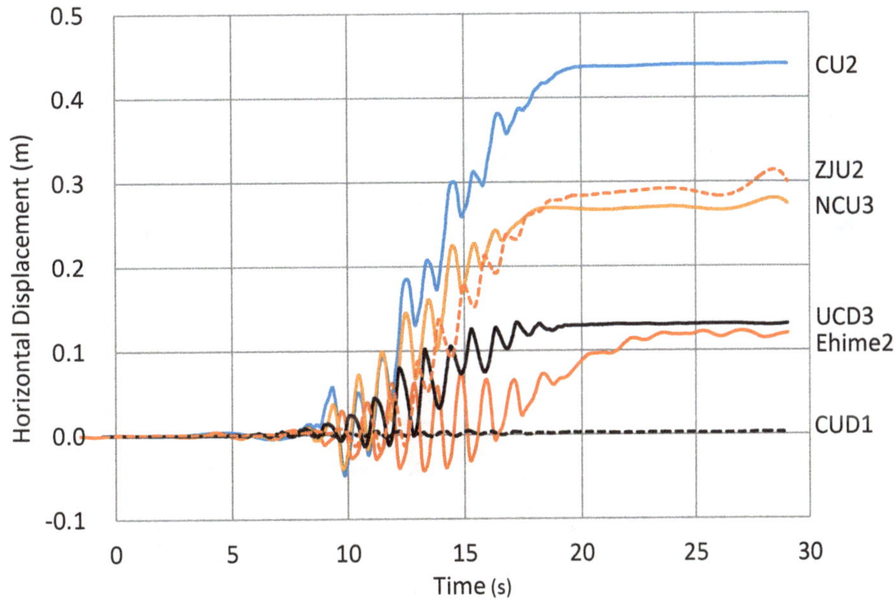


**Fig. 23.10** Horizontal displacement time histories of ground surface at the center. (**a**) Simulations, (**b**) experiments (displacements from ACC and surface marker)

According to Fig. 23.12, shear strain accumulates in the direction in which the initial shear stress acts. The reverse warping tendency of the stress–strain relationship due to cyclic mobility is not noticeable.

Generally, horizontal deformation is large in the upper layer at the center of the slope. Vertical deformation is dominated by sinking on the upper side of the slope. Deformation peaks at the end of the excitation, and then recovers slightly during the process of dissipating excess pore water pressure. Excess pore water pressure is fully dissipated after 600 s.

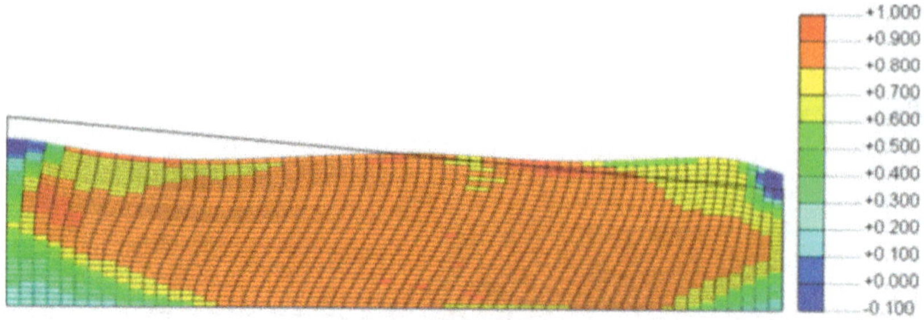


(a) CU2

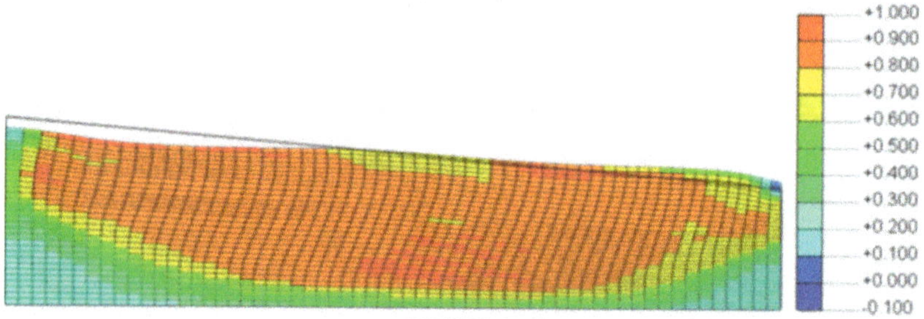


(b) Ehime2

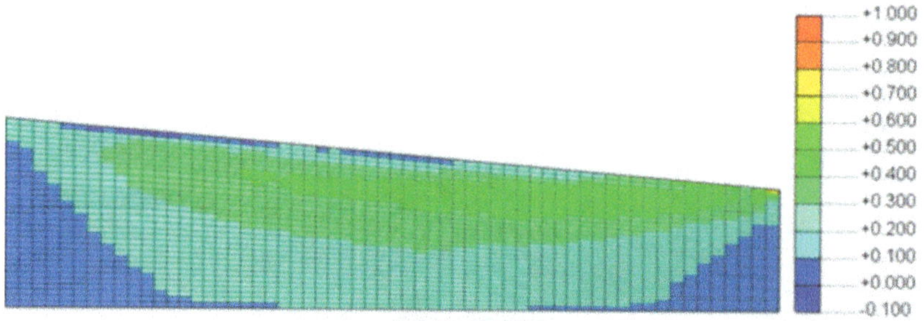


(c) KAIST1

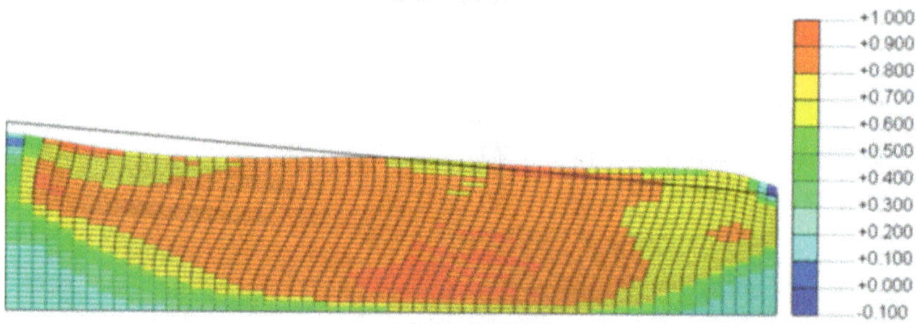


(d) KAIST2

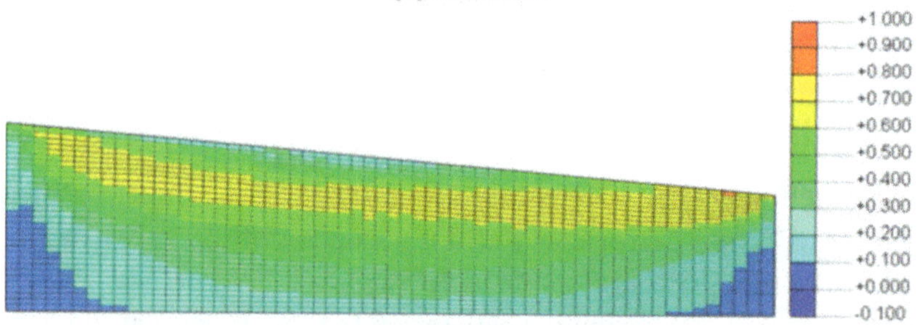


(e) KyU3

**Fig. 23.11** Fivefold enlarged deformation and excess pore pressure ratio distribution (contour) at 25 s. (**a**) CU2, (**b**) Ehime2, (**c**) KAIST1, (**d**) KAIST2, (**e**) KyU3, (**f**) NCU3, (**g**) UCD1, (**h**) UCD3, and (**i**) ZJU2

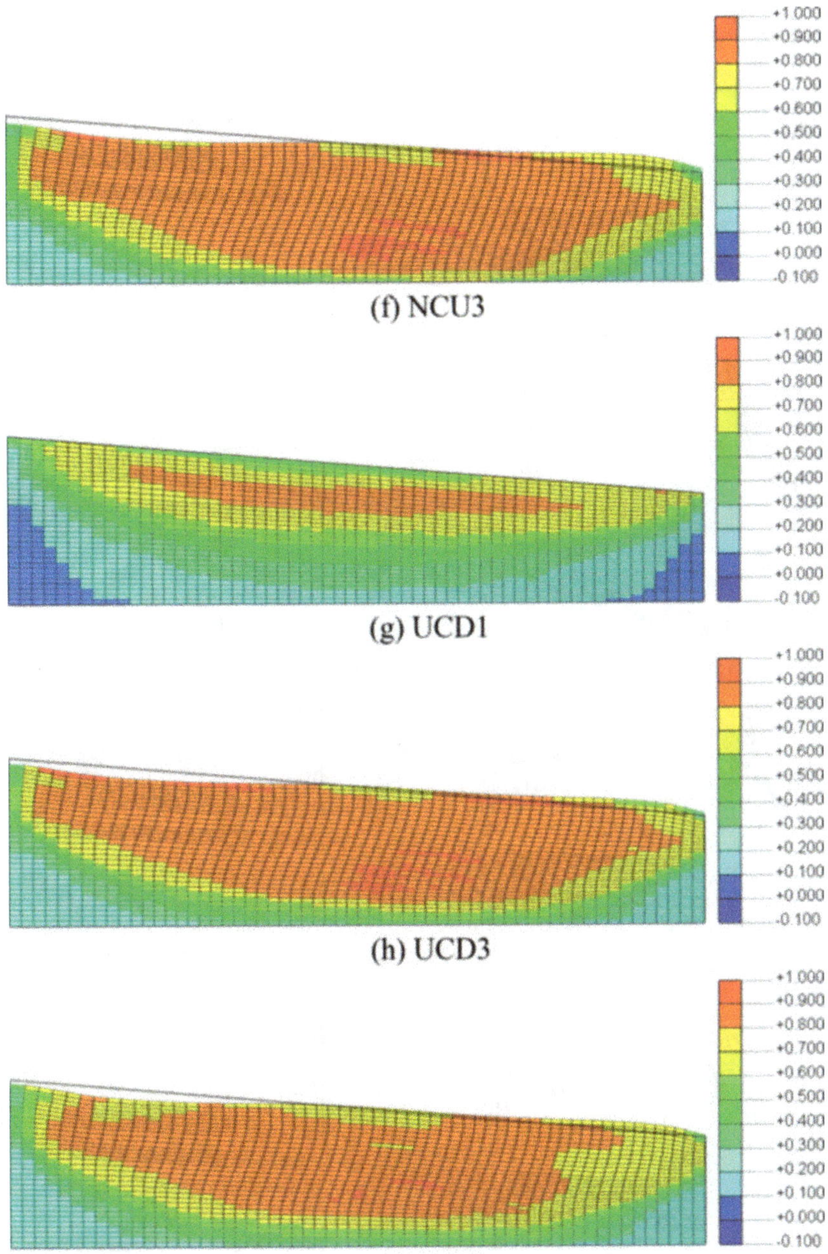


(f) NCU3

(g) UCD1

(h) UCD3

(i) ZJU2

**Fig. 23.11** (continued)

## 23.5 Conclusion

We simulated the lateral spreading of sloping ground using a hyperbolic model and a bowl model for comparison against experimental results. The parameters of the constitutive equation were determined based on simulation of element tests. Experimental results show large variation with density and input acceleration. The simulations successfully explain this tendency to variability. The lateral spreading of the ground surface was of the same order in the experiments and simulations.

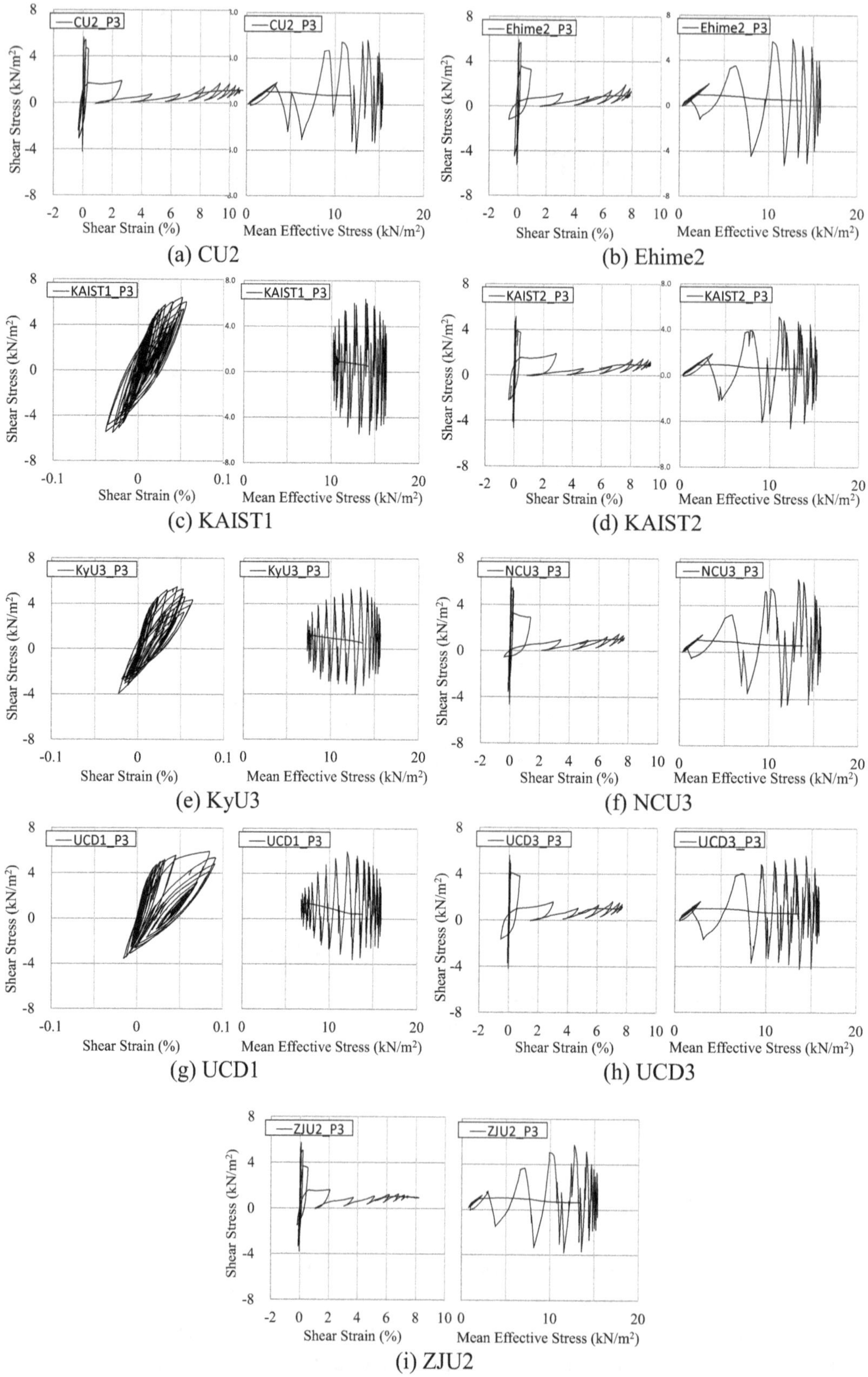


**Fig. 23.12** Stress–strain and effective stress path at P3. (**a**) CU2, (**b**) Ehime2, (**c**) KAIST1, (**d**) KAIST2, (**e**) KyU3, (**f**) NCU3, (**g**) UCD1, (**h**) UCD3, and (**i**) ZJU2

## References

Fukutake, K., & Matsuoka, H. (1989). A unified law for dilatancy under multi-directional simple shearing. In *Proceedings of Japan Society of Civil Engineers, No. 412/III-12* (pp. 143–151).

Fukutake, K., & Matsuoka, H. (1993). Stress-strain relationship under multi-directional cyclic simple shearing. In *Proceedings of Japan Society of Civil Engineers, No. 463/III-22* (pp. 75–84).

Ishihara, K., Yoshida, N., & Tsujino, S. (1985). Modelling of stress-strain relations of soils in cyclic loading. In *Proceedings of 5th International Conference for Numerical Method in Geomechanics, Nagoya* (Vol. 1, pp. 373–380).

Liu, K., Zhou, Y., & Chen, Y. (2017). *Hardin equation of Ottawa sand-F65 for LEAP*.

Sandhu, R. S., & Wilson, E. L. (1969). Finite-element analysis of seepage in elastic media. *Proceedings of ASCE, 95*(EM3), 641–652.

# Chapter 24
# Numerical Simulations of Selected LEAP Centrifuge Experiments with PM4Sand in FLAC

**Jack Montgomery and Katerina Ziotopoulou**

**Abstract** Increasing the confidence in estimates of liquefaction-induced deformations from numerical models requires careful validation of both constitutive models and numerical simulation approaches. This validation should be performed with proper consideration of the intended use of the simulation results and the uncertainty inherent in comparing numerical simulations to laboratory or physical models. This paper describes the comparison of a set of calibrated numerical simulations to the results of centrifuge experiments with a submerged, liquefiable slope. These simulations were performed as part of the 2017 LEAP simulation exercise using the constitutive model PM4Sand and the numerical platform FLAC. The simulation results were compared to those from the centrifuge to identify which aspects of the experimental response the simulations were adequately able to capture. This paper provides a description of the simulation approach including the constitutive model and numerical platform. The process of calibrating PM4Sand to the results of laboratory experiments is described. This is followed by a description of the system-level simulations and a comparison of the simulation results with the experimental results. A sensitivity study was performed to examine trends in the simulation response and a discussion of the findings from this study is presented.

## 24.1 Introduction

Liquefaction-induced ground failure continues to be a major source of damage to civil infrastructure during earthquakes. The ability of engineers to predict the response of infrastructure to liquefaction requires analysis techniques that can provide reliable estimates of the response of soils to seismic loading. For important structures, engineers are increasingly using nonlinear effective stress analyses that combine a constitutive model capable of reproducing important aspects of soil

J. Montgomery (✉)
Department of Civil Engineering, Auburn University, Auburn, AL, USA
e-mail: jmontgomery@auburn.edu

K. Ziotopoulou
Department of Civil and Environmental Engineering, University of California, Davis, CA, USA



B. Kutter et al. (eds.), *Model Tests and Numerical Simulations of Liquefaction and Lateral Spreading*, https://doi.org/10.1007/978-3-030-22818-7_24

behavior with a finite element or finite difference platform that can capture the system response subjected to the applicable boundary and loading conditions. The quality of these analyses will at a minimum depend on the selection and calibration of the constitutive model, the ability of the numerical simulation approach to capture the important system-level responses, the quality of the soil characterization, and the quality of the documentation and review processes. Validation of the constitutive model and numerical platform is often accomplished by comparing results from numerical simulations to carefully performed and documented centrifuge tests. These comparisons provide a means to estimate the current capabilities and limitations of the numerical tools commonly used in liquefaction analyses, to identify major sources of uncertainty in both the numerical and experimental techniques, and to provide insights for future pathways.

This paper documents a set of numerical simulations performed as part of the 2017 LEAP simulation exercise. The simulations described in this paper used the constitutive model PM4Sand (Version 3, Boulanger and Ziotopoulou 2015) and the numerical platform FLAC (Version 8.0, Itasca 2016). The simulations are compared to selected experimental results with the goal of identifying aspects of the experimental response the numerical simulations were able to reasonably capture and which may require further investigation. A brief description of the simulation approach is presented, including the constitutive model and numerical platform used for both the calibration and system-level simulations. The calibration of the model to the results of laboratory experiments conducted on the designated soil for LEAP (Ottawa F-65 sand) is described. This is followed by a description of the simulations performed for the Type-B (Kutter et al. 2015) predictions of the centrifuge model tests, and a critical comparison between simulation and experimental results. A sensitivity study was performed to identify which of the input parameters had the largest effect on the response. A discussion of some of the findings from this study is presented.

## 24.2 Modeling Approach

The numerical platform used for these simulations was FLAC 8.0 (Itasca 2016). FLAC is a two-dimensional, explicit, finite difference program for engineering mechanics computation. The program simulates the behavior of structures built of soil, rock, or other materials that may undergo plastic flow when their yield limits are reached. Materials are represented by elements, or zones, which form a grid that is adjusted by the user to fit the shape of the object to be modeled. Each element behaves according to a prescribed linear or nonlinear constitutive stress/strain law in response to the applied forces and boundary conditions. FLAC uses an explicit, Lagrangian calculation scheme and a mixed-discretization zoning technique that allows for accurate modeling of plastic collapse and flow. Details of the formulation and validation of FLAC can be found in the software manual (Itasca 2016).

The nonlinear constitutive model used for this study is PM4Sand Version 3 (formulation and implementation described in Boulanger and Ziotopoulou 2015, compiled for FLAC 8.0 in 2017) which was specifically developed for earthquake engineering applications. PM4Sand is a stress-ratio controlled, critical state compatible, bounding-surface plasticity model, based on the plasticity model initially developed by Dafalias and Manzari (2004) and described in detail by Boulanger and Ziotopoulou (2015). It is cast in terms of the relative state parameter ($\xi_R$), which measures the difference between the relative density ($D_R$) and the relative density at critical state ($D_{Rcs}$) for the current confining pressure (Boulanger 2003). This formulation allows soil properties to change during the simulation as a function of the change in state (i.e., changes in mean effective stress and/or in void ratio). PM4Sand Version 3 has 22 input parameters, from which only three are considered primary and are required as model inputs. The other 19 (2 flags and 17 secondary parameters) can be left with their preset default values if no other information is available or calibrated to the desired response based on the available lab data. Pertinent information on the parameters of PM4Sand and their calibrated values are given in the following section, while extensive information on the model is provided by Boulanger and Ziotopoulou (2015) and Ziotopoulou and Boulanger (2016).

## 24.3 Calibration Approach

The goal of the constitutive model calibration performed for this study was to select parameters that can reasonably capture the experimental results, while considering the uncertainty inherent in using laboratory tests to predict centrifuge model test results. The calibration of any constitutive model must focus on the aspects of the model behavior which are important to the problem being analyzed (i.e., the constitutive behaviors that are reasonably expected to be activated by the loading paths of the problem at hand). The most critical aspects of the model response for the current study are expected to be the generation of pore pressure and triggering of liquefaction, the accumulation of strains during shaking, and the subsequent reconsolidation of the soil after shaking has stopped. The current calibration intentionally focused on capturing liquefaction triggering curves provided by the LEAP organizers (Fig. 24.1) as well as some general aspects of the individual stress–strain and stress path responses for the cyclic triaxial test results (El Ghoraiby et al. 2019). For all other aspects of model behavior, PM4Sand parameters retained their default values that are generally functions of an index property ($D_R$) chosen to reasonably approximate design correlations.

Calibration of PM4Sand was performed by using single-element simulation of undrained cyclic stress-controlled plane strain compression (PSC) tests and comparing the results to the provided results from undrained cyclic stress-controlled triaxial tests on Ottawa F-65 sand (El Ghoraiby et al. 2017, 2019). The response of cyclic triaxial tests is, as expected, asymmetric with respect to deviatoric stresses. This is

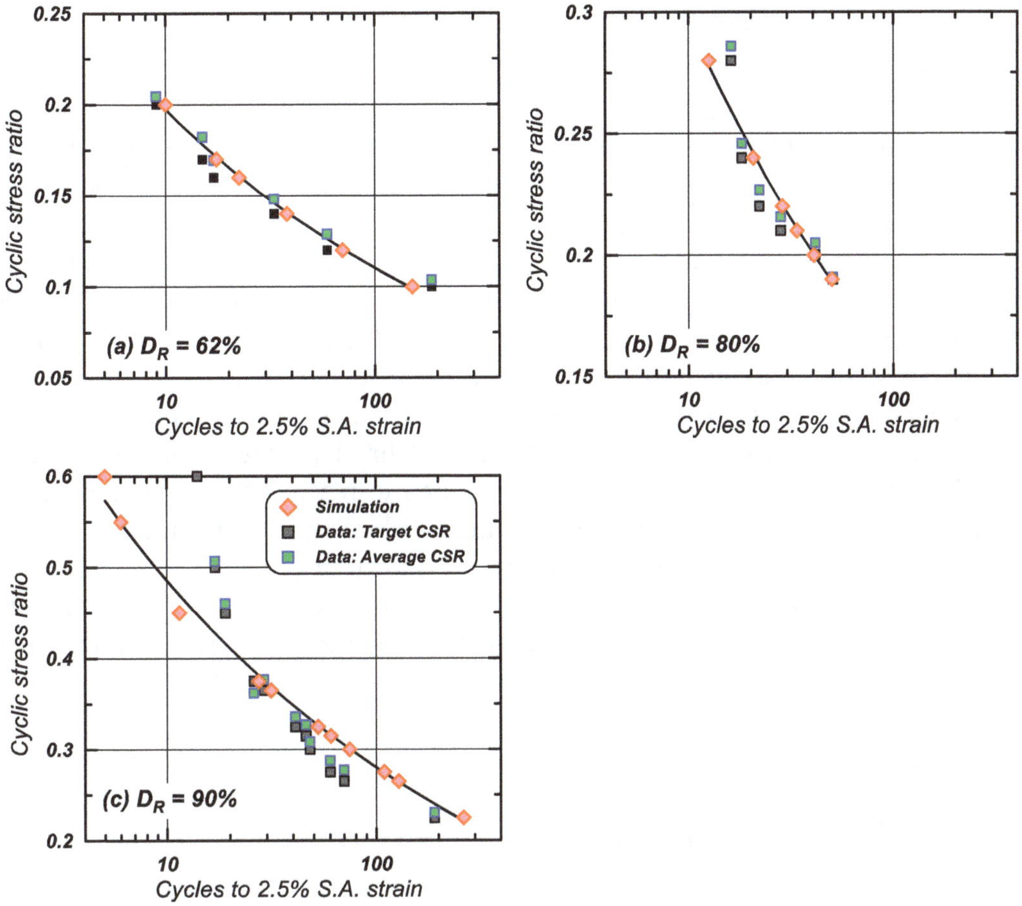


**Fig. 24.1** Cyclic stress ratio versus number of cycles to reach single amplitude strain of 2.5% in single-element cyclic plane strain simulations, for the calibration developed for (**a**) $D_R$ of 62%, (**b**) $D_R$ of 80%, and (**c**) $D_R$ of 92%. Confinement is set to 100 kPa. *[Note different y-axis scales]*

not the case with undrained cyclic PSC, so the goal of the calibration was to capture a reasonable overall match within each group of cyclic triaxial tests (each density) without looking to precisely match details from the individual tests.

Triaxial test results were provided for three void ratios ($e_o = 0.585$, 0.542, and 0.515). PM4Sand is cast in terms of $D_R$, so it was necessary to estimate corresponding $D_R$ values for each of the tests. The organizers provided a database of index test results for Ottawa F-65 sand, which included minimum and maximum void ratios ($e_{min}$ and $e_{max}$) from various authors. There was considerable scatter in some of these properties and so some judgment was necessary in selecting single values. The authors chose to use an $e_{max}$ of 0.739 and an $e_{min}$ of 0.492 based on the results from George Washington University, which also performed the triaxial tests. This selection had a major impact on the comparison to the centrifuge results as will be discussed later. Using these values, the corresponding $D_R$ values for the experimental results were 62%, 80%, and 90%, respectively.

### 24.3.1 Single-Element Simulations

The undrained cyclic PSC simulations were performed using a single element (or zone) in FLAC 8.0. PM4Sand was assigned to the zone using the desired properties at the beginning of the simulation. Stresses were then initialized in the zone to replicate isotropic consolidation at an effective stress of 100 kPa. Cyclic loading was applied by applying a constant vertical velocity to the top boundary of the zone while fixing the bottom boundary against vertical movement. The left boundary of the zone was fixed against horizontal displacement while the right boundary had a constant horizontal force applied to represent the cell pressure. The direction of the velocity of the top boundary was reversed when the internal deviatoric stress reached the desired level as defined by the cyclic stress ratio (CSR $= q/2\sigma'_{vo}$). The number of cycles required to reach 2.5% single amplitude axial strain and an excess pore pressure ratio ($r_u$) of 98% were tracked. Cyclic strength curves for each of the experiments are shown in Fig. 24.1. Two values of CSR are shown for the laboratory data in this figure. The first is the reported CSR provided by the organizers and the second is the average of the peak CSR achieved during each cycle of loading.

### 24.3.2 Calibrated PM4Sand Parameters

Calibration of PM4Sand requires, at a minimum, the three primary input parameters: (1) the apparent relative density ($D_R$) which controls the dilatancy and stress–strain response characteristics of the soil; (2) the contraction rate parameter ($h_{po}$) which controls the cyclic strength of the soil; and (3) the shear modulus coefficient ($G_o$) which controls the small strain stiffness of the model ($G_{max}$). For the current study, the parameter $h_{po}$ was iteratively adjusted for each $D_R$ until the liquefaction strength curves produced from the simulation reasonably matched the experimental data to the extent possible (Fig. 24.1). The calibration reasonably tracked the pattern of responses across all the reported tests. The simulation results from the individual CSR levels were fit with a power function (CSR $= aN^{-b}$) which is illustrated as a solid line in the figures. The CSR to reach 2.5% single amplitude axial strain in 15 cycles was approximated for all three $D_R$ values (Fig. 24.2a) using the results in Fig. 24.1. This curve is often called a liquefaction triggering curve and can also be used to calibrate $h_{po}$ for untested density levels. This relationship is later used to account for the variations with $D_R$ between the different centrifuge model tests.

The most challenging parameter to calibrate was the shear modulus coefficient ($G_o$), because no information on small strain stiffness or shear wave velocity was provided and $G_o$ was not well constrained by calibrating to the results from cyclic triaxial tests. Values for this parameter were obtained by comparing the simulation results to the stress–strain response from the cyclic triaxial tests during the first few cycles of loading. $G_o$ was iteratively changed until a reasonable approximation was

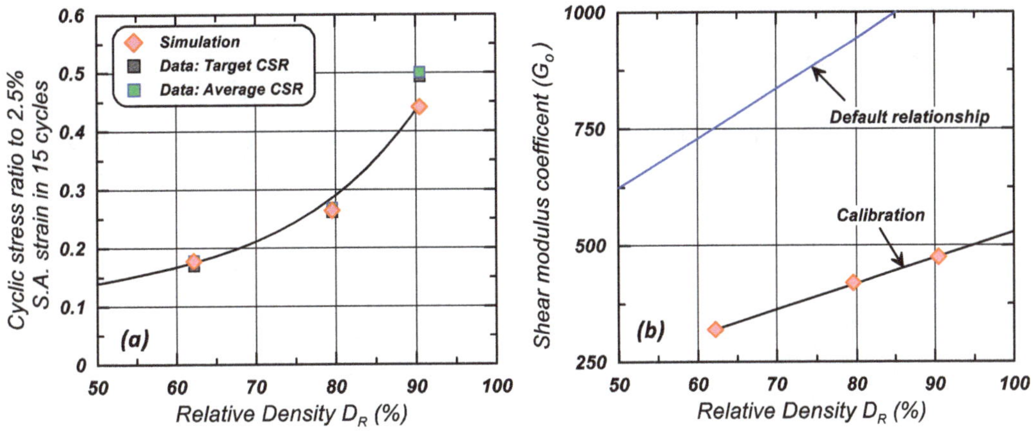


**Fig. 24.2** (**a**) Relationship between cyclic stress ratio required to reach a single amplitude strain of 2.5% and $D_R$; (**b**) relationship between calibrated shear modulus coefficient ($G_o$) and $D_R$ (assigned in single-element drivers) compared to the default relationship for PM4Sand

achieved. Ziotopoulou (2018), in the predictions for LEAP 2015, had selected a $G_o$ of 240 for a $D_R$ of 65%, which was used as a starting point for this calibration. The selected $G_o$ values using this approach are shown in Fig. 24.2b. These values are lower than the default relationship used by PM4Sand would suggest (Fig. 24.2b), but follow the same linear trend with $D_R$ as expected. The effect of this difference will be examined in subsequent simulations.

Two of the secondary model parameters were adjusted to improve the comparison of the simulations results to the stress–strain and stress path results from the cyclic triaxial tests. The critical state friction angle $\varphi_{cv}$ was reduced to 30° from the default value of 33°, to better match the slope of the frictional envelopes (bounding line) in the stress path plots. This value is also more consistent with the values reported by Parra Bastidas (2016) based on a review of available literature. The second default parameter that was modified was $n^b$ which controls the bounding ratio and therefore dilatancy and peak effective friction angles. The default value of 0.5 was increased to 0.7 to reduce the rate of strain accumulation in the cyclic mobility regime following the triggering of liquefaction.

Table 24.1 lists all model parameters for PM4Sand that were given values other than their default during the calibration and simulations. Default values for the remaining parameters are provided by Boulanger and Ziotopoulou (2015). The index properties that were utilized are:

- The as-placed dry density or void ratio of as indicated by the reported results.
- The maximum and minimum void ratios of 0.739 and 0.492.
- The average specific gravity $G_s = 2.66$ as determined through tests (Vasko et al. 2014).

**Table 24.1** Calibrated parameters

| Parameter | Description | $e_o = 0.585$ | $e_o = 0.542$ | $e_o = 0.515$ |
|---|---|---|---|---|
| $D_R$ | Apparent relative density | 62% | 80% | 90% |
| $G_o$ | Shear modulus coefficient | 320.0 | 420.0 | 475.0 |
| $h_{po}$ | Contraction rate parameter | 0.13 | 0.05 | 0.11 |
| $e_{max}$ | Maximum void ratio | 0.739 | 0.739 | 0.739 |
| $e_{min}$ | Minimum void ratio | 0.492 | 0.492 | 0.492 |
| $n_b$ | Bounding ratio constant<br>Default value is 0.50 | 0.7 | 0.7 | 0.7 |
| $\varphi_{cv}$ | Critical state friction angle default value is 33° | 30° | 30° | 30° |

## 24.4 System-Level Numerical Simulations

Simulations were performed for each of the nine centrifuge tests completed as part of the LEAP 2017 exercise. All simulations used the same geometry which was based on the prototype dimensions (Fig. 24.3). The numerical mesh consists of one uniform layer of sand, 80 zones wide and 16 zones high, yielding 81 gridpoints in the *x*- and 17 gridpoints in the *y*-direction (Fig. 24.4). Zones in the simulations presented herein had an average size of 0.25 × 0.25 m, and a maximum aspect ratio of 2:1 at the bottom right corner of the grid. At the time of the Type-B predictions, no information was provided by the centrifuge facilities on the actual location of the instruments, so the goal of the discretization was to have grid points at the prescribed instrument locations.

### *24.4.1 Boundary Conditions*

The mechanical boundary conditions in the simulations replicated the boundary conditions imposed by the rigid container used in the centrifuge model tests, without explicitly simulating the rigid box that surrounded the soil. All bottom nodes were fixed in the *x*- and *y*-directions and all side nodes were fixed in the *x*-direction. The side walls of the rigid container were assumed to have no friction and thus the soil was able to freely slide vertically during all stages of the simulation.

Flow of water was allowed across the top surface of the model and restricted across the container boundaries. The experiments were submerged, so pore pressures and saturation were fixed at the top nodes and a pressure was applied to simulate the weight of the fluid. The values of the surface pressure from the water (both the external pressure and boundary pore pressure) were updated every 0.2 s during the simulation to account for any settlement of the soil surface. The elevation of the free water surface was assumed to be at 5 m. This value was determined using the initial hydrostatic values of the pore pressure histories reported by the centrifuge facilities.

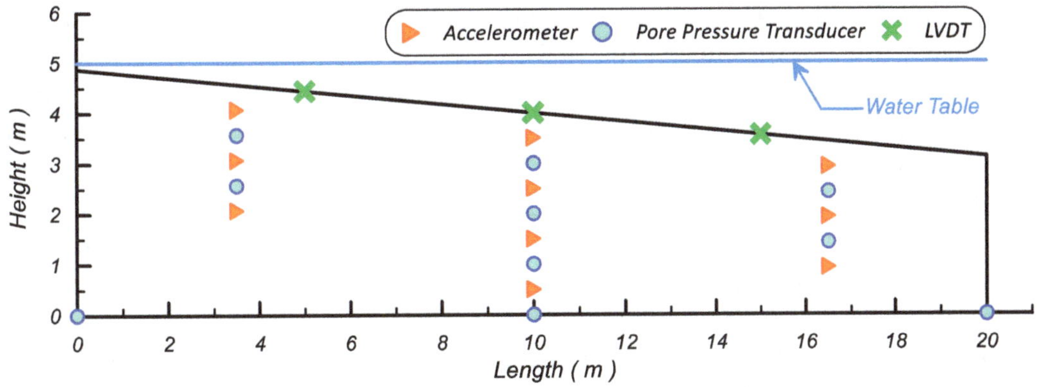


**Fig. 24.3** Interpreted geometry of LEAP centrifuge tests (prototype units). The boundary of the soil model as well as the locations of the 23 instruments (10 accelerometers, 10 pore pressure transducers and 3 LVDTs) are shown

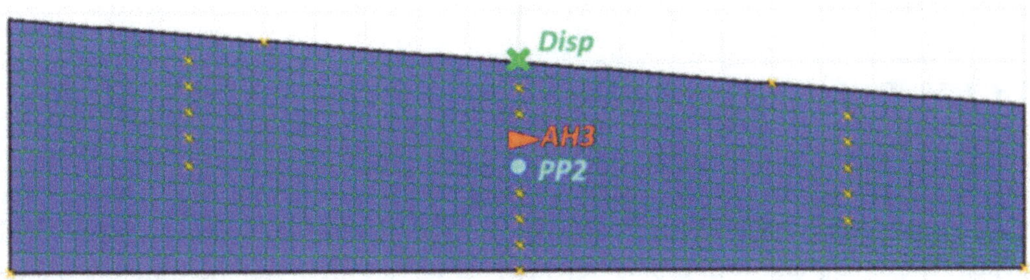


**Fig. 24.4** FLAC numerical grid used in the simulations. The markings showing the locations where time histories are recorded to match the instrument locations shown in Fig. 24.3

The input motion was applied as a horizontal acceleration time history to the base and sides of the model, and as a vertical acceleration time history to the bottom of the model. Time histories were provided for each experiment at two locations and so an average time history was computed. This did not allow for consideration of container rocking which appeared to be significant for some of the experiments. The average time histories were baseline corrected to remove any residual displacement drifts. A quadratic baseline correction was applied to all the motions of all facilities by fitting a quadratic curve with the least squares method and subtracting from the recordings.

### *24.4.2 Solution Scheme*

The numerical platform utilized (FLAC) solves the full dynamic equations of motion for each zone and follows an explicit integration scheme where the equations of motion are used to derive new velocities and displacements from stresses and forces. Strain rates are then derived from velocities, and new stresses from strain rates. This calculation loop is performed at each time increment or time step. The time step is calculated to be small enough that information cannot physically pass from one element to another in that interval, and this way, the computational information is always ahead from the physical information. No iteration process is necessary when computing stresses from strains in a zone, even if the constitutive law is highly nonlinear (as is the case with PM4Sand). The disadvantage of the explicit method is that the required dynamic time step is very small resulting in significant computational time for a simulation. Time steps for the simulations in the current study fluctuated around 7.3e−6 s. All simulations were conducted with "large" deformations enabled, which allowed the mesh nodes to update their coordinates during dynamic shaking, and the geometry progressively to change. Rayleigh damping was set to 0.5% at a center frequency of 1 Hz (frequency of input motion) in order to reduce noise in the simulations and to account for small strain damping which is not captured by the constitutive model.

### *24.4.3 Soil Input Parameters*

The achieved densities in the centrifuge tests varied between the different tests and none exactly matched the target value (62%). The model parameters $h_{po}$ and $G_o$ are dependent on $D_R$ and would thus be expected to vary between the different experiments. To address this variation, four separate sets of simulations were performed with each one taking a slightly different approach in the selection of $h_{po}$ and $G_o$. All other model parameters retained the values from Table 24.1. The first two sets of simulations were considered as the "best estimate" approaches and are discussed in detail below. The third set of simulations examined the effect of using the properties in Table 24.1 without adjusting for any relative density variation and essentially assuming the target density was achieved, while the fourth set of simulations examined the effect of using the CPT data provided by the organizers to estimate the $D_R$. Only the results from the first two sets of simulations (referred to as Calibration 1 and 2) are presented here, as these represented the authors' best estimate of the soil behavior.

*Calibration 1*: For this calibration, the cyclic resistance ratio (CRR) of the soil was estimated using the relationship shown in Fig. 24.2a. $G_o$ was estimated using the calibrated relationship shown in Fig. 24.2b. $h_{po}$ was then estimated using single-element simulations following the procedure described above.

*Calibration 2*: This calibration was performed in an identical manner to Calibration 1, except for $G_o$ which was calculated using the default PM4Sand relationship shown in Fig. 24.2b. This calibration was performed to examine the sensitivity of the simulation results to $G_o$ which, as noted, was poorly constrained by the experimental data.

### 24.4.4 Simulation Procedure

The simulations were performed using the reported centrifugal acceleration from each facility to establish the pre-shaking stress conditions. The sequence of centrifuge model construction was simulated to establish a reasonable initial stress state for the soil. The soil was sequentially placed (or constructed) in layers under a gravity of 1/*N* *g*, which simulates the stresses in the model before centrifugal spinning, where *N* is the centrifugal acceleration from the experiment. During construction, the sand was assigned a Mohr-Coulomb material model with appropriate stress-dependent stiffness. A $K_o$ value close to 0.45 was expected for normally consolidated sand, so the Poisson's ratio for the Mohr-Coulomb model was chosen to produce this value under one-dimensional conditions. The sand was placed layer-by-layer to avoid any stress arching at the container walls. Saturation was performed by setting the saturation value to 1 everywhere and allowing pore pressures to reach equilibrium. Next, gravity was increased from 1/*N* *g* to 1 *g* in 10 increments to simulate the spin-up of the centrifuge. Following each increase in gravity, the stress-dependent stiffness moduli were updated, and the coefficient of earth pressure at rest ($K_o$) was evaluated. The slight slope of the model produced $K_o$ values that differed only slightly from 0.45. Since the modeling took place for the prototype scale, the shape of the ground surface (which during shaking depends on the direction of shaking relative to the curved g-field) was not curved.

After establishing the final static stresses, the water table was applied at an elevation of 5 m. The fluid boundary conditions were set as described in the *Boundary Conditions* section of this paper. The hydraulic conductivity of the sand was set at 1.18e–4 m/s based on the results presented by Vasko et al. (2014). The centrifuge experiments all used an appropriate viscous fluid, so the laboratory permeability was used without scaling. Example stress profiles at this stage are shown in Figs. 24.5, 24.6, 24.7, and 24.8. After initial conditions had been established, the PM4Sand was assigned to all zones using the selected properties. The entire mesh was simulated with the same material model since it was all made of uniform sand with a uniform permeability.

The dynamic shaking was applied to the model using the procedures described in the *Boundary Conditions* section of this paper. The shaking was continued for the

**Fig. 24.5** Initial (pre-shaking) pore pressures for experiment ZJU-2 (units in kPa)

**Fig. 24.6** Initial vertical effective stresses ($\sigma'_{yy}$) for experiment ZJU-2 (units in kPa)

**Fig. 24.7** Initial horizontal effective stress ($\sigma'_{xx}$) for experiment ZJU-2 (units in kPa)

**Fig. 24.8** Initial shear stress ($\tau_{xy}$) for experiment ZJU-2 (units in kPa)

duration of the recorded event for each facility (approximately 26 s). After the end of shaking, the motion of the model base was stopped by applying a vertical and horizontal acceleration opposite in direction to the average velocities along this boundary and with a magnitude to reduce the average velocity to zero in 0.5 s. This is equivalent to the centrifuge container being stopped, but the ramping procedure described above was used to avoid sudden changes in velocity. After the velocity of the model base was brought to zero, the model was allowed to reconsolidate. During reconsolidation, the permeability of the sand was increased by 10 times to account for an increase in permeability of the sand layer which has been observed by others during liquefaction (e.g., Shahir et al. 2012, 2014). During reconsolidation, the Post_Shake flag of PM4Sand was activated (set equal to 1) with $G_{sed} = 0.03$ and $p_{sedo} = -17{,}000$ (Ziotopoulou and Boulanger 2013).

## 24.5 Comparison to Experimental Results

Simulations were performed for each experiment using the two calibrations described above (Table 24.2). Results were compared in terms of horizontal accelerations, excess pore pressures, and vertical and horizontal displacements at the locations shown in Fig. 24.3. The results generally showed that the two calibrations enveloped the observed responses for many of the tests in terms of excess pore pressures and accelerations. The results also generally overpredicted displacements at the center of the model.

Excess pore pressures are compared between the numerical and experimental results for eight of the experiments in Figs. 24.9 and 24.10. Pore pressures are shown for piezometer PP2 which is located near the center of the model (Fig. 24.4). Figure 24.9 shows the time histories for the duration of shaking and shows that both calibrations were able to match the rate and magnitude of pore pressure generation for most tests. Some discrepancies are noted between the maximum excess pore pressure between the simulation and experiments which is attributed to differences between the actual location of instruments and the assumed location in the simulations. Figure 24.10 illustrates an extended version of these time histories to show the dissipation process. Calibration 2 does a better job of matching the dissipation pattern for the majority of the simulations.

**Table 24.2** Model parameters for Calibrations 1 and 2

| Experimental values | | | | | Calibration 1 | | Calibration 2 | |
|---|---|---|---|---|---|---|---|---|
| Centrifuge test | Dry density (kg/m$^3$) | Void ratio | $D_R$ | CRR[a] | $G_o$[b] | $h_{po}$[c] | $G_o$[d] | $h_{po}$[c] |
| CU-2 | 1605.8 | 0.656 | 33% | 0.099 | 161 | 0.330 | 460 | 0.305 |
| Ehime-2 | 1656.55 | 0.606 | 54% | 0.149 | 274 | 0.187 | 664 | 0.143 |
| KAIST-1 | 1701.2 | 0.564 | 71% | 0.216 | 368 | 0.090 | 845 | 0.050 |
| KAIST-2 | 1592.5 | 0.670 | 28% | 0.089 | 130 | 0.325 | 410 | 0.311 |
| KyU-3 | 1637 | 0.625 | 46% | 0.129 | 231 | 0.268 | 585 | 0.204 |
| NCU-3 | 1652 | 0.610 | 52% | 0.145 | 264 | 0.210 | 646 | 0.163 |
| UCD-1 | 1665 | 0.598 | 57% | 0.159 | 292 | 0.165 | 699 | 0.120 |
| UCD-3 | 1658 | 0.604 | 54% | 0.151 | 277 | 0.187 | 670 | 0.140 |
| ZJU-2 | 1606 | 0.656 | 33% | 0.099 | 162 | 0.330 | 461 | 0.300 |

[a]CRR is computed using the relationship in Fig. 24.2a and the corresponding $D_R$ value
[b]$G_o$ is computed using the "Calibration" relationship from Fig. 24.2b and the corresponding $D_R$ value
[c]$h_{po}$ is found using single-element simulations
[d]$G_o$ is computed using the "Default" relationship from Fig. 24.2b and the corresponding $D_R$ value

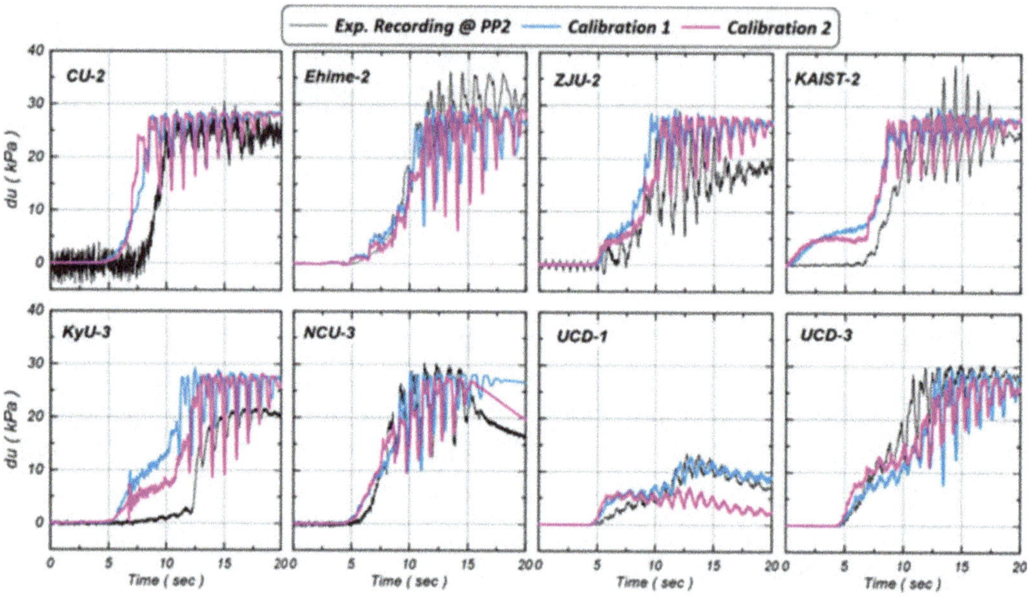


**Fig. 24.9** Comparison between simulation and experimental results for excess pore pressure generation at piezometer PP2 using Calibrations 1 and 2

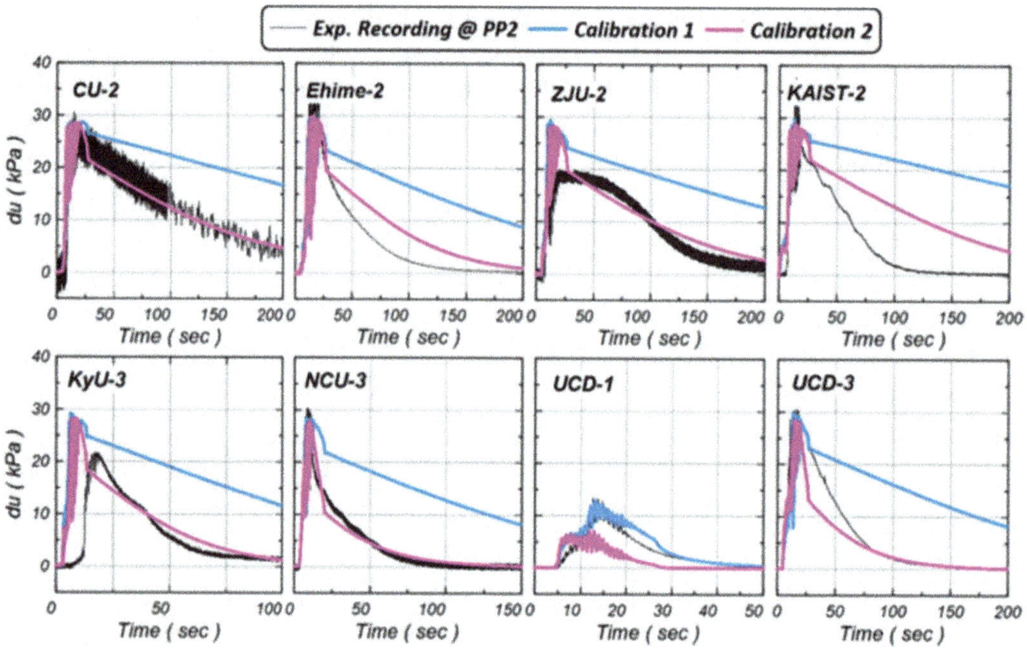


**Fig. 24.10** Comparison between simulation and experimental results for excess pore pressure dissipation at piezometer PP2 using Calibrations 1 and 2

Horizontal acceleration response spectra are compared for the numerical and experimental results for eight of the experiments in Fig. 24.11. The response spectra are calculated based on the horizontal accelerations at accelerometer AH3 which is located near the center of the model (Fig. 24.4). The two calibrations produced

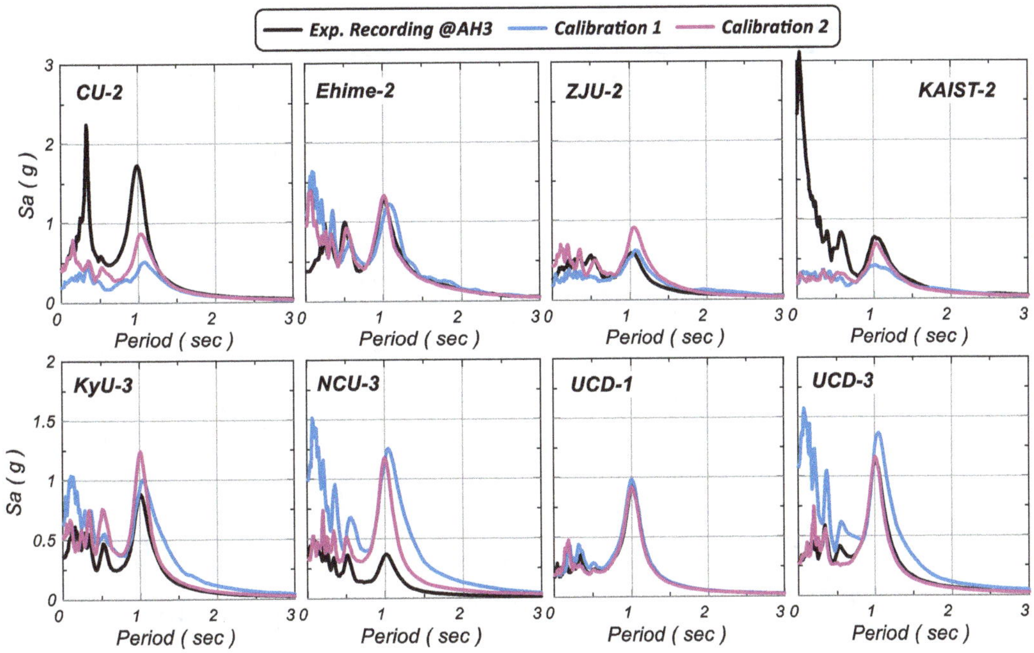


**Fig. 24.11** Comparison between simulation and experimental results for acceleration response spectra at accelerometer AH3 using Calibrations 1 and 2

similar results in terms of acceleration spectra. Overall, simulations were able to reproduce the frequency content, but the magnitudes were over predicted in some simulations and under predicted in others, making the extraction of a discernable trend challenging. These discrepancies would be most likely attributed to experimental variation and uncertainty, since the same numerical modeling protocol was used in all simulations and any achieved and reported properties from the centrifuge facilities were honored. It is expected however, that a case-by-case calibration and simulation would resolve these issues.

A markedly lesser degree of agreement was observed in the comparison between simulations and experimental results in terms of deformations (i.e., lateral displacements). The simulations tended to overpredict displacements for almost all of the experiments. Calibration 2 tends to show lower displacements, but still tends to overpredict the observations. An example of the horizontal displacement time histories for the center of the profile (location shown in Fig. 24.3) is shown in Fig. 24.12. Reasons for the discrepancies in the displacements became apparent after the release of the experimental data which included estimates of $D_R$ values for each centrifuge test that had previously not been provided. These estimates were based on $e_{max}$ and $e_{min}$ values that were selected by the LEAP organizers and differed from the values selected by the authors during the calibration phase. This difference led to a systematic underprediction of the $D_R$ for most of the experiments and an associated overprediction of the displacements, which are very sensitive to small changes in $D_R$ (Fig. 24.13).

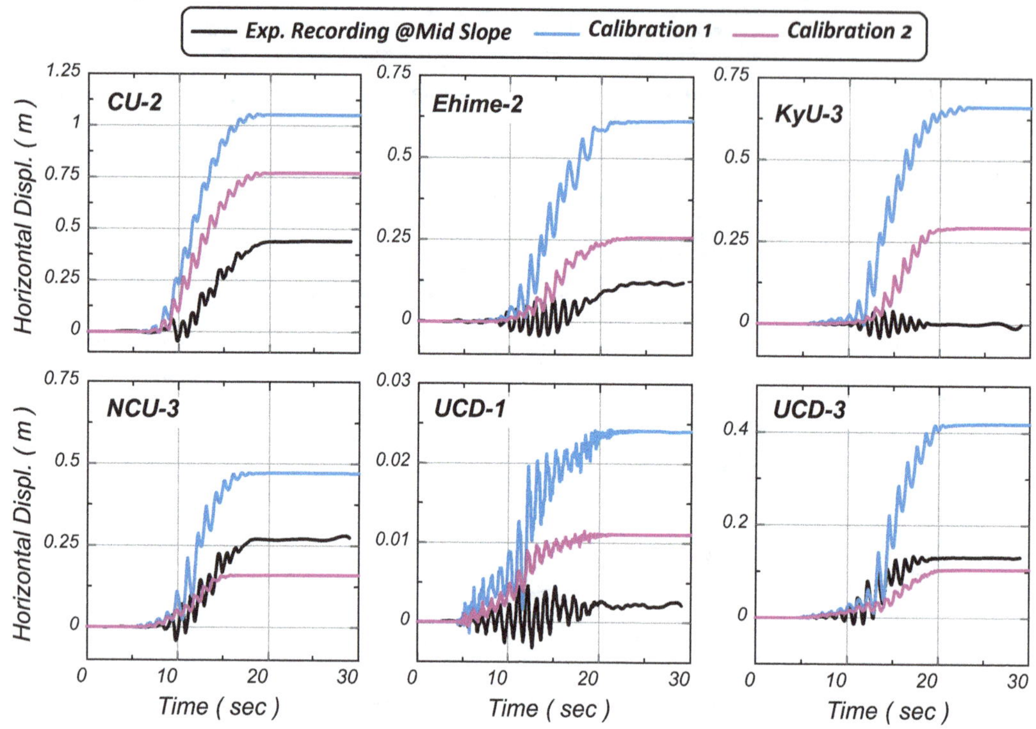


**Fig. 24.12** Comparison between simulation and experimental results for lateral displacements at the middle of the model surface using Calibrations 1 and 2 (note different y-axis scales)

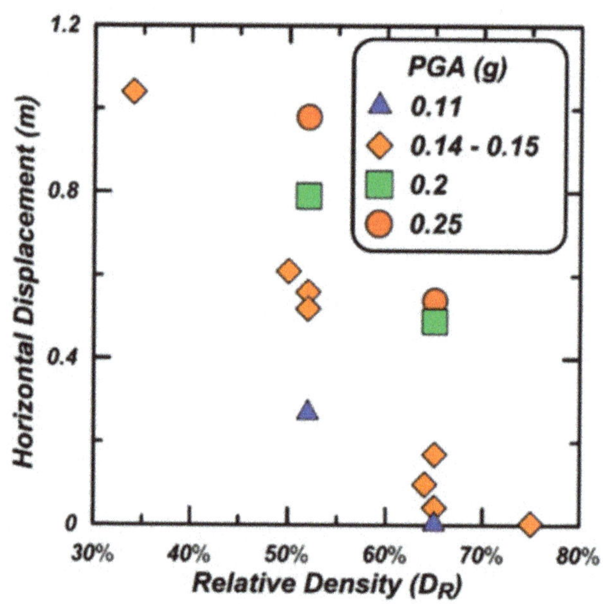


**Fig. 24.13** Results of the sensitivity study showing horizontal surface displacements at the center of the model as a function of $D_R$ and peak ground acceleration (PGA)

## 24.6 Sensitivity Study

A sensitivity study was performed at the request of the LEAP organizers to identify how the simulation results varied with changes in density and input motion. This sensitivity study included five different input motions, which had different magnitudes (as represented by peak ground acceleration) and different frequency contents, and three dry densities. PM4Sand requires $D_R$ values, so these were calculated using the two sets of $e_{max}$ and $e_{min}$ values to account for the uncertainty in these parameters discussed previously. This led to simulations being performed at six different $D_R$ values (Fig. 24.13). The sensitivity study was only performed using Calibration 1 (Table 24.2). Lateral displacements for each of the simulations versus $D_R$ are shown in Fig. 24.13. The displacements in this figure represent final horizontal surface displacements at the center of the model (Fig. 24.4). The results show that the amount of displacement is very sensitive to changes in $D_R$ and peak ground acceleration (PGA). Displacements are especially sensitive to changes in $D_R$ when values are between 40 and 60% which is where most of the experiments for this exercise fall. These sensitivity results demonstrate that a systematic underprediction in $D_R$ could lead to an overprediction in displacement as was observed in the previous comparisons. It also emphasizes the need for reliable estimates of $e_{max}$ and $e_{min}$ when interpreting void ratios from experimental data.

## 24.7 Summary

Results have been presented from a set of numerical simulations using the constitutive model PM4Sand and the numerical platform FLAC. PM4Sand was calibrated to results from cyclic triaxial tests on Ottawa F-65 sand. The calibrations were performed for three $D_R$ values and relationships were developed between the input parameters and $D_R$ to allow for simulations to be performed at any $D_R$ value. The calibrated model was used to simulate nine centrifuge experiments that examined the response of a gently sloping saturated sand deposit. The simulations were able to envelope most of the recorded responses, but some discrepancies were observed. Recordings of excess pore pressure and accelerations generally matched well with results falling in the range of experimental uncertainty. The simulations tended to overpredict lateral displacements which is likely due to a systematic underestimation of $D_R$ for the centrifuge experiments. This underestimation was caused by the choice of $e_{max}$ and $e_{min}$ during the model calibration process.

A sensitivity study was performed to investigate the effects of $D_R$ and input motion on the displacement results. The sensitivity study showed that displacements are very sensitive to small changes in $D_R$. This result suggests that a more accurate assessment of the index parameters for the soil could have led to better agreement between the simulations and experiments. For future simulations using PM4Sand and FLAC, careful measurement of $D_R$ should be performed for the experiments to avoid the need for the numerical modeler to make assumptions regarding the experimental results. Overall, the simulation results suggest that the combination of FLAC with PM4Sand can give reliable predictions of the different response metrics with proper calibration.

## References

Boulanger, R. W. (2003). High overburden stress effects in liquefaction analyses. *J. Geotechnical and Geoenvironmental Eng.*, ASCE *129*(12), 1071–082.

Boulanger, R. W., & Ziotopoulou, K. (2015). *PM4Sand (Version 3): A sand plasticity model for earthquake engineering applications*. Technical Report No. UCD/CGM-15/xx, Center for Geotechnical Modeling. Department of Civil and Environmental Engineering, University of California, Davis, CA.

Dafalias, Y. F. & Manzari, M. T. (2004). Simple plasticity sand model accounting for fabric change effects. *Journal of Engineering Mechanics*, ASCE, *130*(6), 622-634.

El Ghoraiby, M. A., Park, H., & Manzari, M. T. (2017). *LEAP 2017: Soil characterization and element tests for Ottawa F65 sand*. Washington, DC: The George Washington University.

El Ghoraiby, M. A., Park, H., & Manzari, M. T. (2019). Physical and mechanical properties of Ottawa F65 sand. In B. Kutter et al. (Eds.), *Model tests and numerical simulations of liquefaction and lateral spreading: LEAP-UCD-2017*. New York: Springer.

Itasca. (2016). *FLAC – Fast Lagrangian Analysis of Continua, Version 8.0*. Minneapolis, MN: Itasca Consulting Group, Inc.

Kutter, B. L., Manzari, M. T., Zeghal, M., Zhou, Y. G., & Armstrong, R. J. (2015). Proposed outline for LEAP verification and validation processes. In S. Iai (Ed.), *Geotechnics for catastrophic flooding events*. London: Taylor & Francis.

Parra Bastidas, A. M. (2016). *Ottawa F-65 Sand Characterization*. PhD Dissertation, University of California, Davis.

Shahir, H., Pak, A., Taiebat, M., & Jeremić, B. (2012). Evaluation of variation of permeability in liquefiable soil under earthquake loading. *Computers and Geotechnics, 40*, 74–88.

Shahir, H., Mohammadi-Haji, B., & Ghassemi, A. (2014). Employing a variable permeability model in numerical simulation of saturated sand behavior under earthquake loading. *Computers and Geotechnics, 55*, 211–223.

Vasko, A., El Ghoraiby, M. A., & Manzari, M. (2014). *An Investigation Into the Behavior of Ottawa Sand Under Monotonic and Cyclic Shear Tests*. Washington, DC: Department of Civil and Environmental Engineering, George Washington University.

Ziotopoulou, K. (2018). Seismic response of liquefiable sloping ground: Class A and C numerical predictions of centrifuge model responses. *Soil Dynamics and Earthquake Engineering, 113*, 744–757. https://doi.org/10.1016/j.soildyn.2017.01.038.

Ziotopoulou, K., & Boulanger, R. W. (2013). Numerical modeling issues in predicting post-liquefaction reconsolidation strains and settlements. In *10th International Conference on Urban Geotechnical Engineering (CUEE), March 1–2, Tokyo, Japan*.

Ziotopoulou, K., & Boulanger, R. W. (2016). Plasticity modeling of liquefaction effects under sloping ground conditions and irregular cyclic loading. *Soil Dynamics and Earthquake Engineering, 84*, 269–283. https://doi.org/10.1016/j.soildyn.2016.02.013.

# Chapter 25
# LEAP-UCD-2017 Numerical Simulation at Meisosha Corp

**Osamu Ozutsumi**

**Abstract** As a part of the LEAP-UCD-2017 project, type B simulations were performed to predict the results of nine selected centrifuge model tests using a two-dimensional effective stress analysis program incorporating the cocktail glass model. In Phase I of the numerical simulation, three parameter sets of the cocktail glass model were determined for three different dry densities. In Phase II, numerical models of the finite element method were made based on the data of the centrifuge model. At that time, for each centrifuge model test, we decided which parameter set of the cocktail glass model to apply by referring to the dry density of the analyzed ground. There was a tendency that the response values of the numerical simulation were slightly larger than the response values expected from the difference between the achieved dry density of the centrifuge model and the dry density on which the applied parameter set depends.

## 25.1 Introduction

It is considered advantageous that the constitutive model of sand for numerical simulation is based on the interaction between the sand particles. The strain space multiple mechanism model developed by Iai (Iai et al. 2011), which originated from the multi-inelastic spring model proposed by Towhata and Ishihara (1985), is such a model based on the soil particle interaction (Iai and Ueda 2016). The strain space multiple mechanism model has excellent extensibility. The model can be expanded to a three-dimensional constitutive model (Iai and Ozutsumi 2005). The model also can be expanded to a constitutive model for clay (Iai et al. 2015).

We participated as one of the numerical simulation teams of the LEAP-UCD-2017 project and got the opportunity to perform type B simulation targeted at the centrifuge model tests of liquefiable ground with a gently inclined surface. In the numerical simulations, we applied "the constitutive cocktail glass model" (Iai et al.

O. Ozutsumi (✉)
Meisosha Corporation, Tokyo, Japan
e-mail: ozutsumi@meisosha.co.jp



B. Kutter et al. (eds.), *Model Tests and Numerical Simulations of Liquefaction and Lateral Spreading*, https://doi.org/10.1007/978-3-030-22818-7_25

2011) to reproduce the behavior of sand. The cocktail glass model is one of the achievements of the strain space multiple mechanism model.

The cocktail glass model was incorporated in a two-dimensional dynamic effective stress analysis program, "FLIP ROSE," based on the finite element method. The program is able to analyze earthquake responses accompanied by liquefaction by considering the permeability of the ground. We used this program to perform type B LEAP simulation to confirm the performance of this constitutive model and program.

## 25.2 Constitutive Model

### 25.2.1 Model Description

The "cocktail glass model" was proposed for sandy soil based on the framework of a strain space multiple mechanism model for granular materials (Iai et al. 2011; Iai et al. 2013).

The macroscopic effective stress in granular materials is given by the average over contact forces **P** within a representative volume element with volume $V$ as follows (Iai et al. 2013).

$$\boldsymbol{\sigma}' = \frac{1}{V}\sum l\mathbf{P}\otimes\mathbf{n} \tag{25.1}$$

See Fig. 25.1 for a definition of the variables.

The framework of the strain space multiple mechanism model is used to upscale the micromechanical structure into the macroscopic structure (Iai et al. 2013).

$$\boldsymbol{\sigma}' = -p\mathbf{I} + \int_0^{\pi} q\langle\mathbf{t}\otimes\mathbf{n}\rangle d\omega(\omega = 2\theta) \tag{25.2}$$

$$\langle\mathbf{t}\otimes\mathbf{n}\rangle = \begin{bmatrix} \cos\omega & \sin\omega \\ \sin\omega & -\cos\omega \end{bmatrix} \tag{25.3}$$

Here, $p$ represents isotropic stress and $q$ virtual simple shear stress (Iai et al. 2013). In order to define the relationship between macroscopic strain tensor $\boldsymbol{\varepsilon}$ and macroscopic effective stress tensor $\boldsymbol{\sigma}'$ through the structure defined by Eq. (25.2), let volumetric strain $\varepsilon$ and virtual simple shear strain $\gamma(\omega)$ be defined as follows (Iai et al. 2013).

$$\varepsilon = \mathbf{I} : \boldsymbol{\varepsilon} \tag{25.4}$$

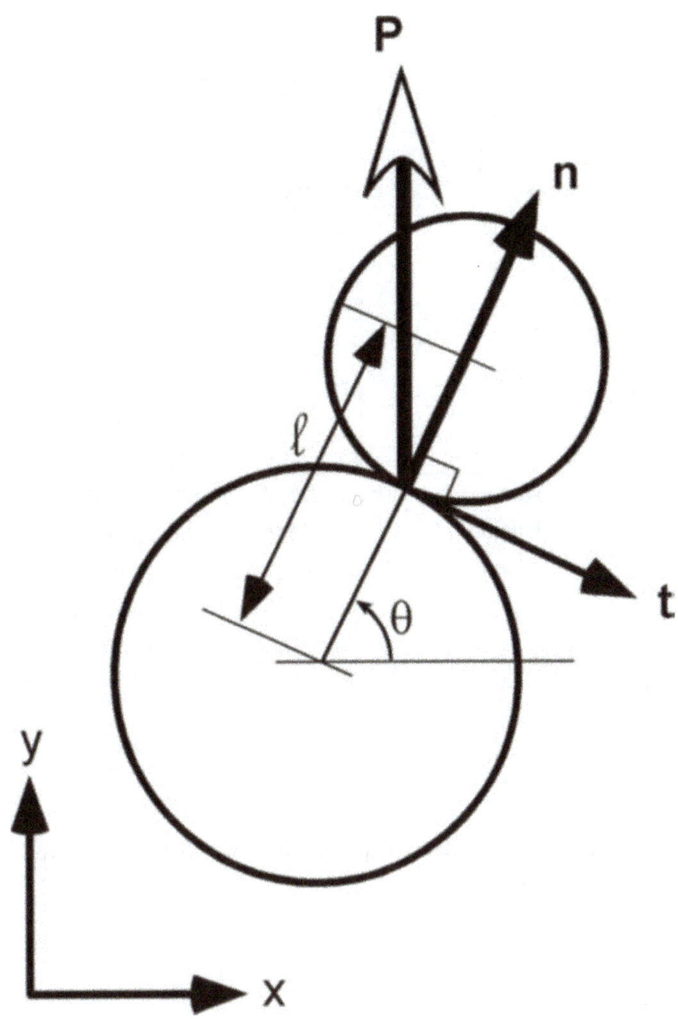


**Fig. 25.1** Contact between two particles: normal **n**, tangential direction **t**, and contact force **P** (Iai et al. 2013)

$$\gamma(\omega) = \langle \mathbf{t} \otimes \mathbf{n} \rangle : \boldsymbol{\varepsilon} \tag{25.5}$$

where the double dot symbol denotes a double contraction. To account for the effect of the volumetric strain due to dilatancy ($\varepsilon_{\mathrm{d}}$), effective volumetric strain $\varepsilon'$ is introduced as follows (Iai 1993).

$$\varepsilon' = \varepsilon - \varepsilon_{\mathrm{d}} \tag{25.6}$$

Effective volumetric strain $\varepsilon'$ and virtual simple shear strain $\gamma$ are used respectively to define isotropic stress $p$ and virtual simple shear stress $q$ as follows (Iai et al. 2013).

$$p = p(\varepsilon') \tag{25.7}$$

**Table 25.1** Main model parameters (based on Iai et al. 2011)

| Symbol | Mechanism | Parameter designation |
|---|---|---|
| $K_{L/Ua}$ | Volumetric | Bulk modulus corresponding to mean effective confining pressure $p_a$ |
| $r_K$ | Volumetric | Reduction factor of bulk modulus for liquefaction analysis |
| $l_K$ | Volumetric | Power index of bulk modulus for liquefaction analysis |
| $G_{ma}$ | Shear | Shear modulus corresponding to mean effective confining pressure $p_a$ |
| $\phi_f$ | Shear | Internal friction angle |
| $h_{max}$ | Shear | Upper bound for hysteretic damping factor |
| $\phi_p$ | Dilatancy | Phase transformation angle |
| $r_{\varepsilon_d}$ | Dilatancy | Parameter controlling dilative and contractive components |
| $r_{\varepsilon_d^c}$ | Dilatancy | Parameter controlling contractive component |
| $q_1$ | Dilatancy | Parameter controlling initial phase of contractive component |
| $q_2$ | Dilatancy | Parameter controlling final phase of contractive component |
| $q_4$ | Dilatancy | Parameter controlling shear stiffness under liquefaction consideration |
| $\varepsilon_d^{cm}$ | Dilatancy | Limit of contractive component |
| $S_1$ | Dilatancy | Small positive number to avoid zero confining pressure |
| $c_1$ | Dilatancy | Parameter controlling elastic range for contractive component |
| $q_{us}$ | Dilatancy | Undrained shear strength (for steady-state analysis) |

$$q = q(\gamma(\omega)) \tag{25.8}$$

Dilatancy $\varepsilon_d$ is decomposed into contractive dilatancy $\varepsilon_d^c$ and dilative dilatancy $\varepsilon_d^d$ as follows (Iai et al. 2011).

$$\varepsilon_d = \varepsilon_d^c + \varepsilon_d^d \tag{25.9}$$

The rate of $\varepsilon_d^c$ is controlled by the irreversible (plastic) portion of virtual simple shear strain rate ($\dot{\gamma}_p$) and the rate of $\varepsilon_d^d$ is controlled by virtual simple shear strain rate ($\dot{\gamma}$).

### *25.2.2 Model Parameters*

Table 25.1 shows main model parameters for cocktail glass model.

## 25.3 Analysis Platform

### *25.3.1 Overview*

We used a two-dimensional dynamic effective stress analysis program, "FLIP ROSE," in the type B predictions. "FLIP ROSE" (Finite element analysis program of Liquefaction Process/Response of Soil structure systems during Earthquakes) is based on the finite element method and infinitesimal strain formulation.

The program provides a multistage analysis function that enables the analysis to reproduce the initial stress state accurately by considering the construction process.

### 25.3.2 Elements

The program has 20 kinds of element. We used the pore water elements (partially drained) for modeling the behavior of the pore water and the cocktail glass model elements for modeling the behavior of the sand soil skeleton. The pore water element is based on u-p formulation by Zienkiewicz and Bettss (1982). The cocktail glass model element is based on the model proposed by Iai et al. (2011). For simulation of soil behavior below the ground water table, the pore water element (partially drained) is superposed with the cocktail glass model element by sharing the same nodes.

### 25.3.3 Governing Equations

Following Zienkiewicz and Bettss (1982), the governing equations for porous media saturated with pore water are given as a combination of equilibrium equation and mass balance equation of pore water as follows (u-p formulation).

$$\sigma_{ij,j} + \rho g_i = \rho \ddot{u}_i \ \text{ on } V_{\mathrm{a}}, \tag{25.10}$$

$$\left(k_{ij} p_{,j}\right)_{,i} - \dot{\varepsilon}_{ii} - \left(k_{ij} \rho_f g_j\right)_{,i} = -\left(k_{ij} \rho_f \ddot{u}_j\right)_{,i} + n\dot{p}/K_f \ \text{ on } V_{\mathrm{b}}, \tag{25.11}$$

where

$\sigma_{ij}$: stress tensor, $k_{ij}$: coefficient of permeability, $\varepsilon_{ij}$: strain tensor,
$\rho$: mass density (saturated density), $\rho_f$: mass density of pore water,
$g_i$: gravity acceleration vector, $u_i$: displacement vector, $n$: porosity,
$K_f$: bulk modulus of pore water.

### 25.3.4 Boundary Conditions

The following boundary conditions are assigned.

$$u_i = \bar{u}_i \ \text{ on } \Gamma_1, \tag{25.12}$$

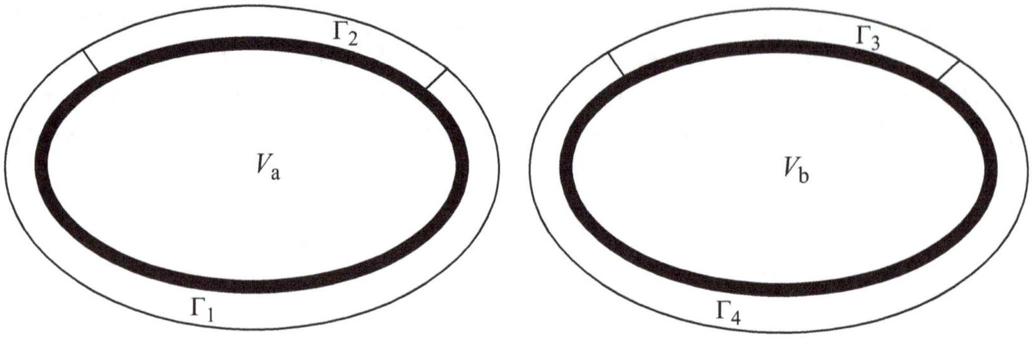


**Fig. 25.2** Boundaries for equilibrium equation and mass balance equation of pore water

$$\sigma_{ij}\nu_j = \bar{T}_i \ \text{on}\,\Gamma_2, \tag{25.13}$$

$$p = \bar{p} \ \text{on}\,\Gamma_3, \tag{25.14}$$

$$-\dot{w}_j\nu_j = k_{ij}\left(p_{,i} - \rho_f g_i + \rho_f \ddot{u}_i\right)\nu_j = \bar{q} \ \text{on}\,\Gamma_4, \tag{25.15}$$

where

$\bar{u}_i$—Displacement specified on boundary $\Gamma_1$.

$\bar{T}_i$—Traction specified on boundary $\Gamma_2$.

$\bar{p}$—Pore water pressure specified on $\Gamma_3$.

$\bar{q}$—Pore water inflow specified on $\Gamma_4$.

$\dot{w}_j$ —Velocity of pore water relative to soil skeleton (average over total sectional area).

The union of the boundaries $\Gamma_1$ and $\Gamma_2$ is equal to the total area $\Gamma$ of the analysis domain $V_a$. as shown in Fig. 25.2. The union of the boundaries $\Gamma_3$ and $\Gamma_4$ is equal to the total area $\Gamma$ of the analysis domain $V_b$.

## 25.4 Model Calibration (Phase I)

### *25.4.1 Overview*

The characterization tests, including cyclic undrained triaxial tests, of Ottawa F65 sand were conducted (El Ghoraiby et al. 2017, 2019). We performed the calibration of the model parameters for the constitutive model, that is, the cocktail glass model, as a part of the LEAP 2017 simulation exercise (Manzari et al. 2019). We are concerned here with the calibration procedures and the model parameters obtained. Also, we will show a comparison between the results of the numerical element simulations and the results of the corresponding cyclic undrained triaxial tests.

**Table 25.2** Model parameter values determined through first step

| $e_0$ | $\rho_d$ (kg/m$^3$) | $\rho_{sat}$ (t/m$^3$) | $n$ | $G_{ma}$ (kPa) | $p_a$ (kPa) | $\nu$ | $K_{L/Ua}$ (kPa) | $\phi_f$ (deg.) | $\phi_p$ (deg.) | $h_{max}$ |
|---|---|---|---|---|---|---|---|---|---|---|
| 0.515 | 1744.2 | 2.089 | 0.340 | 125,326 | 100 | 0.33 | 326,831 | 36.0 | 28.0 | 0.24 |
| 0.542 | 1712.6 | 2.070 | 0.351 | 119,147 | 100 | 0.33 | 310,716 | 36.0 | 28.0 | 0.24 |
| 0.585 | 1665.6 | 2.041 | 0.369 | 109,872 | 100 | 0.33 | 286,529 | 36.0 | 28.0 | 0.24 |

$p_a = 100$ kPa

### 25.4.2 *Calibration Process (First Step)*

We decided on the values of the parameters that can be determined without trial and error. Table 25.2 shows model parameter values determined through the first step. These parameter values were obtained through the following process:

1. Saturated mass density $\rho_{sat}$ was obtained from void ratio $e_0$, specific gravity $\rho_s$, and mass density of pore water $\rho_w$.
2. Porosity $n$ was calculated from void ratio $e_0$.
3. Mean effective confining pressure $p_a$ corresponding to $G_{ma}$, $K_{L/Ua}$ was set to 100 kPa. This is an arbitrary value.
4. Initial shear stiffness $G_{ma}$ for mean effective confining pressure $p_a$ was calculated using the following equation (Hardin and Richart Jr 1963).

$$G_{ma} = 700\frac{(2.17 - e_0)^2}{1 + e_0}(p_a)^{1/2} \quad (\text{kgf/cm}^2)$$

5. Poisson's ratio $\nu$ was assumed to be 0.33.
6. Bulk modulus $K_{L/Ua}$ was calculated by the relationship between bulk modulus, initial shear modulus, and Poisson's ratio for isotropic elastic medium.
7. Using the following equation, internal friction angle $\phi_f$ was calculated from the gradient $M$ measured on the compression side of the effective stress path diagram.

$$M = 6\sin\phi_f/(3 - \sin\phi_f)$$

8. Phase transformation angle $\phi_p$ was assumed to be 28 degrees.
9. Upper bound for hysteretic damping factor $h_{max}$ was assumed to be 0.24.

### 25.4.3 *Calibration Process (Second Step)*

We decided on the values of the parameters that should be determined by trial and error. During the trial and error procedure, we kept the values determined through the first step. Table 25.3 shows the model parameter values determined through the

**Table 25.3** Model parameter values determined through second step

| $e_0$ | $\rho_d$ (kg/m$^3$) | $r_K$ | $l_K$ | $r_{\varepsilon_d}$ | $r_{\varepsilon_d^c}$ | $q_1$ | $q_2$ | $q_4$ | $\varepsilon_d^{cm}$ | $S_1$ | $c_1$ | $q_{us}$ (kPa) |
|---|---|---|---|---|---|---|---|---|---|---|---|---|
| 0.515 | 1744.2 | 0.225 | 1.4 | 0.444 | 0.20 | 20 | 0.9 | 1.0 | 0.0366 | 0.005 | 2.7 | 0 |
| 0.542 | 1712.6 | 0.3 | 1.4 | 0.333 | 0.95 | 20 | 1.3 | 1.0 | 0.0561 | 0.005 | 2.1 | 0 |
| 0.585 | 1665.6 | 0.3 | 1.4 | 0.333 | 1.70 | 20 | 1.3 | 1.0 | 0.0599 | 0.005 | 1.5 | 0 |

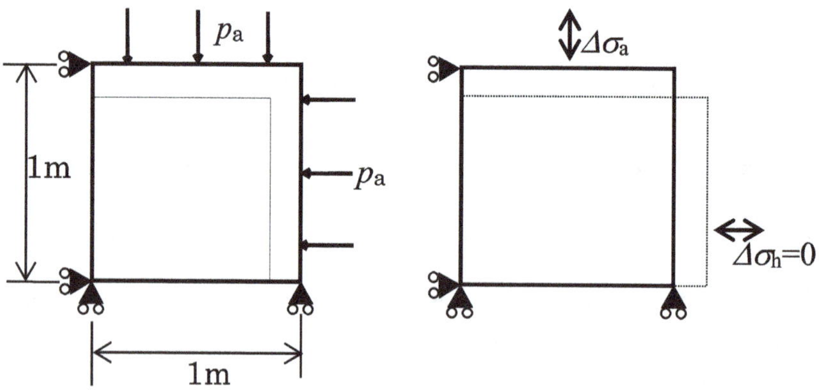


(1) Isotropic consolidation under complete drainage condition with confining pressure $p_a$ (100 kPa)

(2) Cyclic loading under undrained triaxial condition

**Fig. 25.3** Procedure of element simulation for performing calibration

second step. In order to obtain these parameters, element simulations under plane strain condition were carried out by the procedure shown in Fig. 25.3.

According to the characterization test results, the axial strain time histories were unidirectional, but this tendency was not seen in the element simulations. Therefore, we did not select the liquefaction resistance curves determined with 2.5% single amplitude axial strain, shown in the report (El Ghoraiby et al. 2017, 2019), as targets of the calibration. Instead, we adjusted the parameters by fitting mainly axial strain double amplitude shown in the analytical time histories to those obtained from the triaxial tests.

### *25.4.4 Results of the Simulation of the Cyclic Triaxial Tests*

Figure 25.4 shows comparisons of the liquefaction resistance curves between the test results and the corresponding element simulation results. Element simulations were carried out using the parameters shown in Tables 25.2 and 25.3. The liquefaction resistances plotted in Fig. 25.4 were determined with 2.5% single amplitude axial strain. Figure 25.5 shows sample comparisons of the effective stress path, the deviatoric stress–axial strain relation, the mean effective stress–axial strain relation and the time history of the excess pore water pressure ratio. Figure 25.6 shows comparisons of the time histories of mean effective stress and axial strain. In

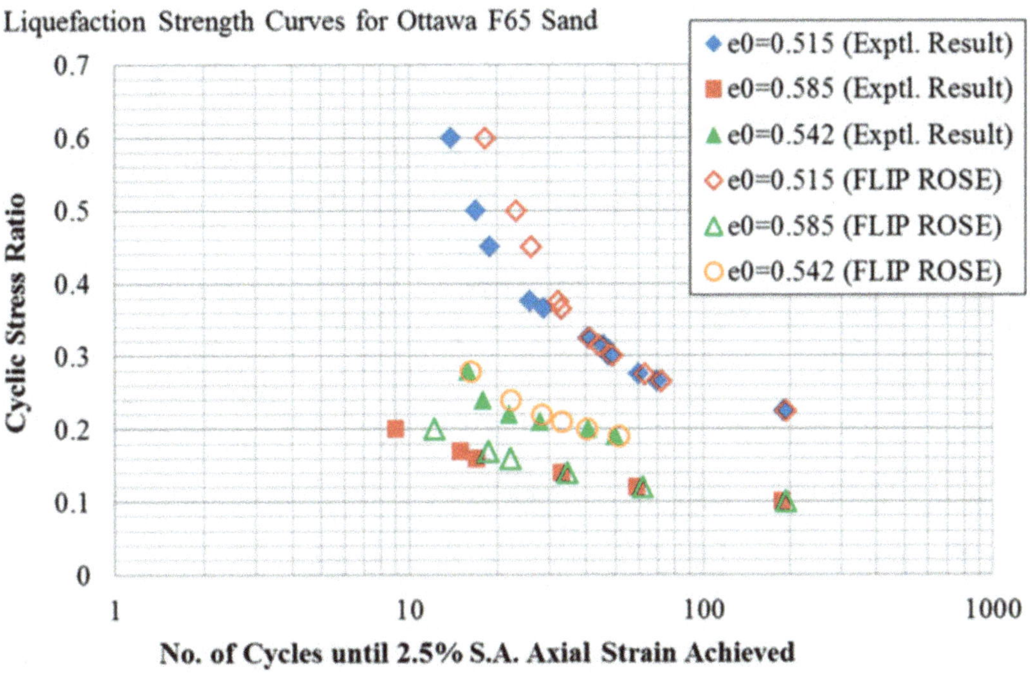


**Fig. 25.4** Liquefaction strength curves obtained from cyclic stress-controlled triaxial tests (El Ghoraiby et al. 2017, 2019) and FLIP ROSE element simulations (Filled markers correspond to test results. Open markers correspond to simulation results.)

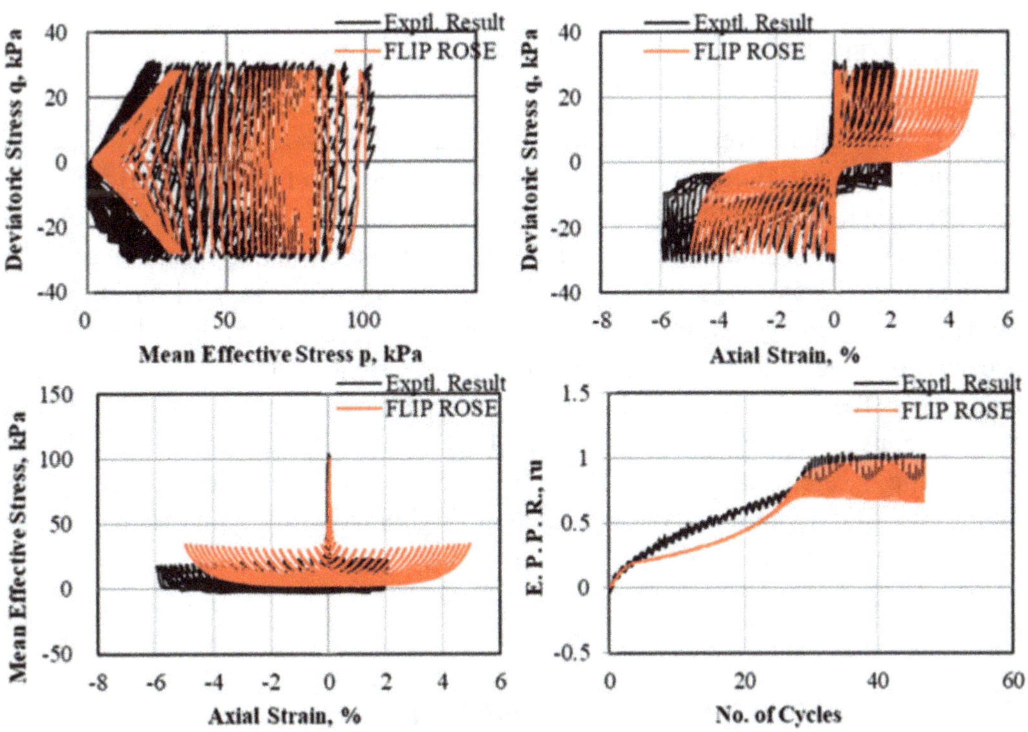


**Fig. 25.5** A sample comparison of the simulated versus measured stress–strain responses and effective stress path ($e_0 = 0.585$, $\rho_d = 1665.6$ kg/m$^3$; CSR = 0.14)

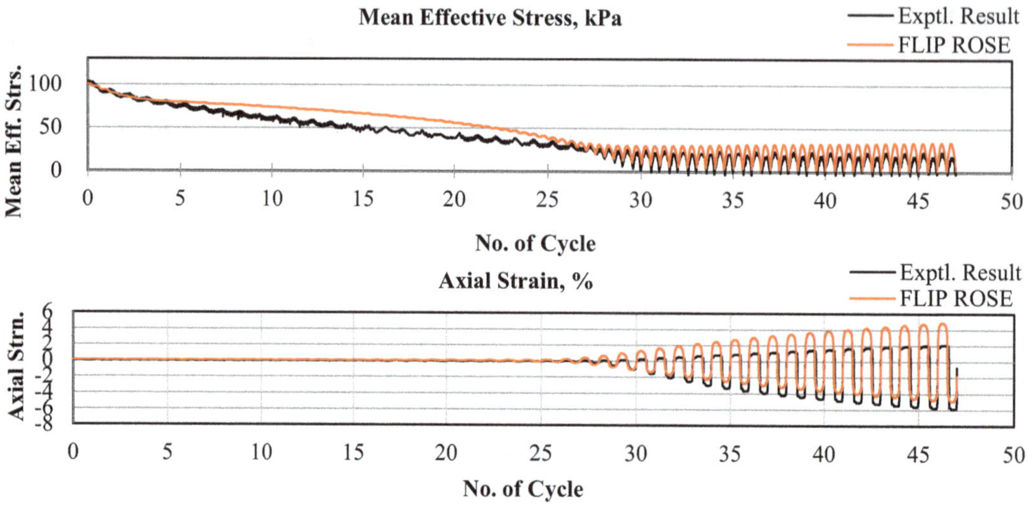


**Fig. 25.6** A sample comparison of the simulated versus measured mean effective stress and axial strain histories ($e_0 = 0.585$, $\rho_d = 1665.6$ kg/m$^3$; CSR = 0.14)

Figs. 25.5 and 25.6, the responses in the case of $e_0 = 0.585$, $\rho_d = 1666$ kg/m$^3$ and CSR = 0.14 are shown as an example.

## 25.5 Type B Simulation (Phase II)

### *25.5.1 Overview*

Type B simulations were carried out to predict the response of nine cases of centrifuge model tests conducted in seven organizations in the LEAP 2017 project. The centrifuge model was a ground made of Ottawa F65 sand whose surface was gently inclined. The ground dimensions were 20 m in the shaking direction and 4 m average depth (prototype scale). Numerical simulations were carried out using 1 g field equivalent analysis model. The input ground motion used in the analysis was an acceleration time history with maximum amplitude of about 0.15 g and duration of about 20 s. The code names of the nine centrifuge model tests were CU2, Ehime2, Kaist1, Kaist2, KyU3, NCU3, UCD1, UCD3, and ZJU2. See "LEAP 2017 Simulation Exercise Phase II: Type B Prediction of Centrifuge Experiments Guidelines for submission of the numerical simulations." for details (George Washington University et al. 2017).

### *25.5.2 Model Geometry/Mesh and Boundary Conditions*

Figure 25.7 shows details of the finite element model. We divided the whole area into four sub-areas so as to be able to assign the different material parameters to each

**Fig. 25.7** Finite element model: soil structure

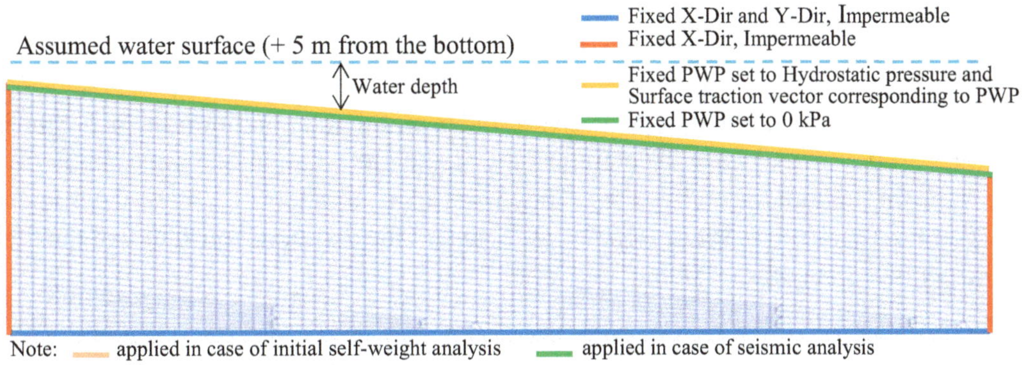


**Fig. 25.8** Finite element model: mesh and boundary conditions

area based on the soil density. However, we set the same material parameters to each area as discussed later. Figure 25.8 shows the finite element mesh and the boundary conditions.

Each element shown in Fig. 25.8 is a cocktail glass model element. Every cocktail glass model element is accompanied by a pore water element (partially drained) by sharing the same nodal points. Total number of the cocktail glass model elements was 2686; hence, total number of the pore water elements was also 2686. Total number of the nodal points was 2800.

The boundary conditions we applied are shown in Fig. 25.8.

### 25.5.3 *Analysis Conditions*

1. Time integration method.
   The coupled set of equations consisting of equilibrium equation and mass balance equation of pore water was integrated in time by using generalized time integration method (SSpj method, Zienkiewicz et al. (2005)). We set the time integration interval $\Delta t$ to 0.001 sec so as to avoid the numerical instability in all cases. We also performed numerical analyses for 50 s, therefore, the number of time steps was 50,000 in each case.

2. Rayleigh damping.
   Rayleigh damping was used instead of the intrinsic viscous damping of the soil (Tatsuoka et al. 2002). We applied Rayleigh damping shown in Eq. (25.16) to the equilibrium equation.

$$\mathbf{C} = \alpha\mathbf{M} + \beta\mathbf{K} \tag{25.16}$$

where **C** is a Rayleigh damping matrix, **M** is a mass matrix. **K** is an initial tangential stiffness matrix at the beginning of dynamic analysis and is constant throughout the entire analysis time. We set parameters $\alpha, \beta$ as follows in all cases.

$$\alpha = 0.0, \;\; \beta = 1.555 \times 10^{-4} \tag{25.17}$$

We performed eigenmode analysis under undrained condition in each centrifuge experimental case. The frequency of the first eigenmode was almost the same throughout the entire cases. They were approximately 10.24 Hz. Coefficient $\beta$ in Eq. (25.16) was set to $1.555 \times 10^{-4}$ so that the damping constant $h$ of the first eigenmode becomes 0.005.

### *25.5.4 Material Properties and Constitutive Model Parameters*

Model parameters determined in Phase I were used as they were, except for mass density, porosity and permeability coefficient.

According to the following procedure, we calculated the dry density ($\rho_d$) of the ground for each case of the centrifuge model test.

1. We read the cone penetration resistance ($q_c$) at three or four depths (1.5 m, 2.0 m, 2.5 m, 3.0 m) from the graph representing the depth dependence of $q_c$ submitted by each institution conducting centrifuge model test.
2. The dry density ($\rho_d$) at each of these depths was read from the correlation diagram (Manzari 2017) of the cone penetration resistance ($q_c$) to the dry density ($\rho_d$).
3. The dry densities obtained at these depths were averaged.

Table 25.4 shows mean dry density and coefficient of variation of dry densities for each centrifuge model experiment. As shown in Table 25.4, the coefficients of variation are mainly smaller than or equal to 1.0%; therefore, we decided to apply these mean values to the entire soil layer in each centrifuge model test.

In the Phase I, the parameter sets of the cocktail glass model were determined for three cases of dry densities; $\rho_d = 1665.6, 1712.6, 1744.2$ kg/m$^3$. Each centrifuge test case had to be assigned with one of these parameter sets having the nearest dry density, because the liquefaction characteristics of soil with any other dry density except those of the three cases in Phase I was unknown.

**Table 25.4** Mass density, porosity, and permeability coefficient

| Centrifuge test | Dry density (Mean) $\rho_d$ (kg/m$^3$) | Dry density (CV) | Void ratio $e_0$ | Saturated mass density $\rho_{sat}$ (t/m$^3$) | Porosity $N$ | Permeability, $T = 20$ °C $k$ (m/s) | $\mu{*}/g{*}$ (cSt) | Permeability, $T = 20$ °C adjusted $k{*}$ (m/s) |
|---|---|---|---|---|---|---|---|---|
| CU2 | 1581 | 0.010 | 0.677 | **1.984** | **0.404** | 1.108E−4 | 1.175 | **9.434E−5** |
| Ehime2 | 1666 | 0.007 | 0.590 | **2.037** | **0.371** | 1.020E−4 | 1.0 | **1.020E−4** |
| KAIST1 | 1681 | 0.006 | 0.577 | **2.047** | **0.366** | 1.005E−4 | 0.897 | **1.121E−4** |
| KAIST2 | 1591 | 0.007 | 0.666 | **1.990** | **0.400** | 1.098E−4 | 0.936 | **1.173E−4** |
| KyU3 | 1652 | 0.012 | 0.604 | **2.028** | **0.377** | 1.034E−4 | 0.991 | **1.043E−4** |
| NCU3 | 1651 | 0.010 | 0.606 | **2.028** | **0.377** | 1.035E−4 | 1.0 | **1.035E−4** |
| UCD1 | 1659 | 0.005 | 0.597 | **2.033** | **0.374** | 1.027E−4 | 1.0 | **1.027E−4** |
| UCD3 | 1626 | 0.007 | 0.630 | **2.012** | **0.386** | 1.060E−4 | 1.0 | **1.060E−4** |
| ZJU2 | 1585 | 0.009 | 0.671 | **1.987** | **0.402** | 1.103E−4 | 1.0 | **1.103E−4** |

Note 1: Dry density (Mean) is the average value of $\rho_d$ based on the penetration resistance $q_c$ at 4 or 3 depths (1.5 m, 2.0 m, 2.5 m, 3.0 m) obtained by the cone penetration test (CPT) during the centrifugal experiment
Note 2: Dry density (CV) represents coefficient of variation of dry densities calculated at three or four depths
Note 3: Void ratio ($e_0$) was calculated using the equation of $e_0 = \rho_s/\rho_d - 1$
Note 4: Saturated density ($\rho_{sat}$) was calculated using the equation of $\rho_{sat} = (\rho_s + e_0)/(1 + e_0\rho_w)$
Note 5: Porosity ($n$) was calculated using the equation of $n = e_0/(1 + e_0)$
Note 6: Coefficient of permeability at $T = 20$ °C ($k$) was set as a value depending on the void ratio $e_0$ with reference to El Ghoraiby et al. (2017)
Note 7: Adjusted coefficient of permeability at $T = 20$ °C ($k{*}$) was calculated using the equation of $k{*} = k/(\mu{*}/g{*})$
Note 8: $\rho_s$(=2.65 t/m$^3$) is mass density of soil particle, $\rho_w$(=1.0 t/m$^3$) is mass density of pore water

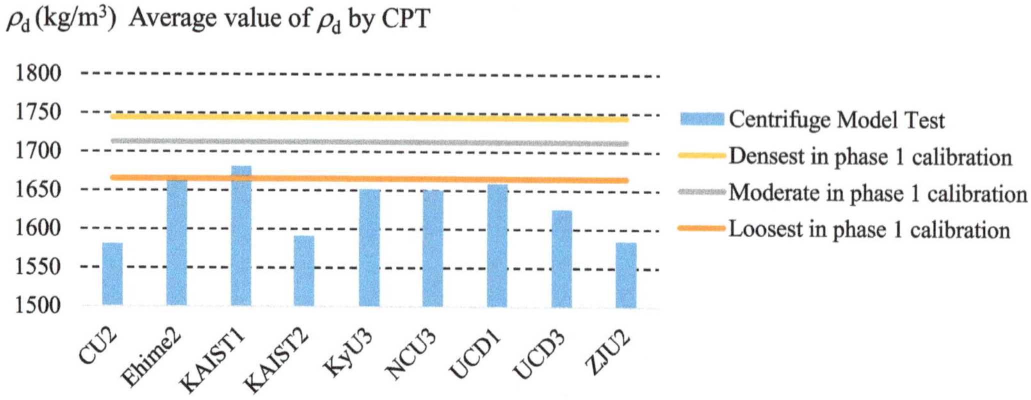


**Fig. 25.9** Dry densities in the centrifuge model tests and in Phase I

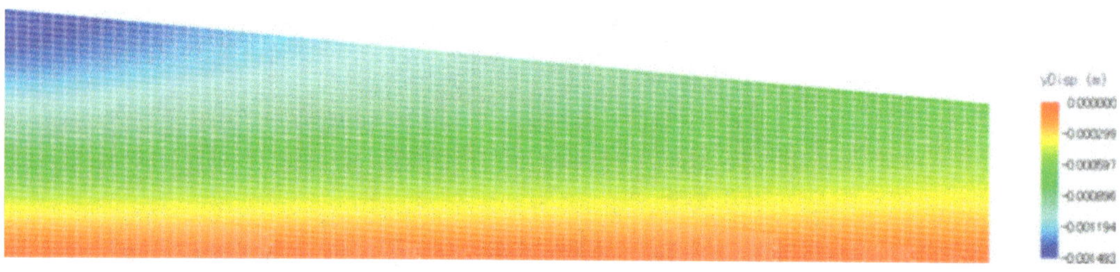

**Fig. 25.10** Initial state of the model for UCD1—vertical displacement (m) due to gravity

As shown in Fig. 25.9, the dry density of each centrifuge test was closest to the dry density of the loosest case in Phase I. Therefore, in the type B simulation, we applied a parameter set of the loosest specimen, whose $\rho_d$ was 1665.6 kg/m$^3$, to any case in the centrifuge model test. This implies that the liquefaction strength in all centrifuge model test cases becomes the same with each other, and then the results of the seismic response analyses in the type B simulations may become similar.

As shown in Table 25.4, saturated mass density ($\rho_{sat}$), porosity ($n$) and coefficient of permeability at $T = 20$ °C (k$*$) were determined for each centrifugal test case based on their mean value of mass density $\rho_d$.

### 25.5.5 *State of Stresses and Internal Variables of the Constitutive Model in the Pre-shaking Stage*

Figures 25.10, 25.11, 25.12, 25.13, 25.14, and 25.15 show the contour plots obtained by initial self-weight analysis for the model of UCD1. Since the plot diagrams of the other cases are almost the same as the diagrams for UCD1, these drawings of the other cases are omitted. In these figures, compression of stress is represented by negative sign, and tension is represented by positive sign.

Figures 25.10 and 25.11 show the distributions of vertical displacement and horizontal displacement, respectively. Figure 25.12 shows the distribution of pore

**Fig. 25.11** Initial state of the model for UCD1—horizontal displacement (m) due to gravity

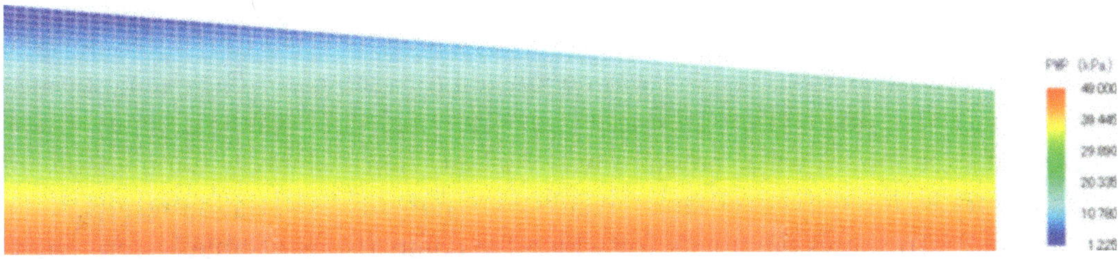


**Fig. 25.12** Initial state of the model for UCD1—pore water pressure due to gravity

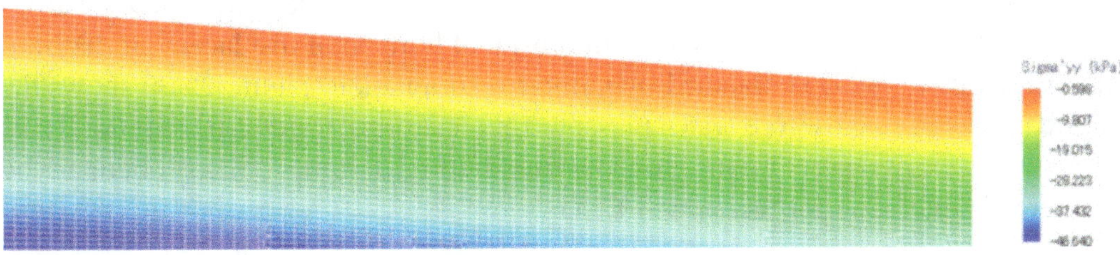


**Fig. 25.13** Initial state of the model for UCD1—vertical effective stress $\sigma_{yy}'$ (kPa) due to gravity

**Fig. 25.14** Initial state of the model for UCD1—horizontal effective stress $\sigma_{xx}'$ (kPa) due to gravity

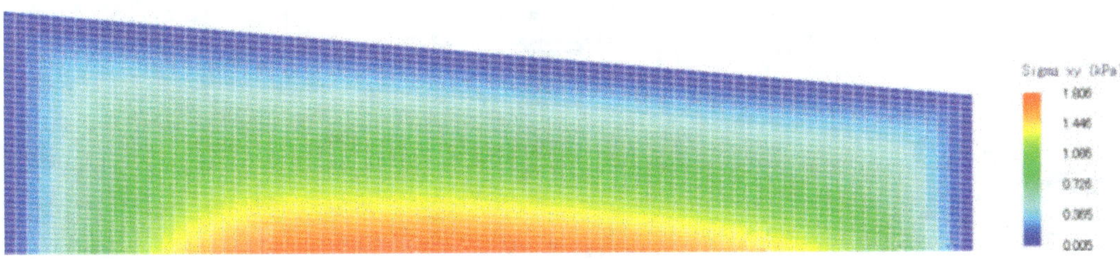


**Fig. 25.15** Initial state of the model for UCD1—shear stress $\tau_{xy}$ (kPa) due to gravity

water pressure. According to this figure, if the depth from the assumed water surface (see Fig. 25.8) is the same, the pore water pressure will be the same. The effective vertical stress distribution shown in Fig. 25.13 is a bedded distribution according to the depth. Figures 25.14 and 25.15 show the distributions of effective horizontal stress and shear stress, respectively.

### 25.5.6 Results of the Dynamic Analysis

Seismic response analysis of 50 s was conducted for each case. The excitation was carried out in the first half for 20 s, and in the latter half for 30 s the drainage of the increased excess pore water pressure was tracked. Here, we show comparisons of time history between the experiment results and numerical simulation results.

Figure 25.16 shows the points for which the comparisons of time history charts between experiment results and numerical simulation results were made. Figure 25.17 shows the comparisons of horizontal acceleration time histories at point AH4. Figure 25.18 shows the comparisons of horizontal displacement time histories at point Disp_0. Figure 25.19 shows the residual horizontal displacement distribution of the case "UCD3" obtained by numerical simulation. Figure 25.20 shows the comparisons of excess pore water pressure time histories at point P1. Figure 25.21 shows the excess pore water pressure distribution at time 15.1 s obtained by numerical simulation for the case "UCD3."

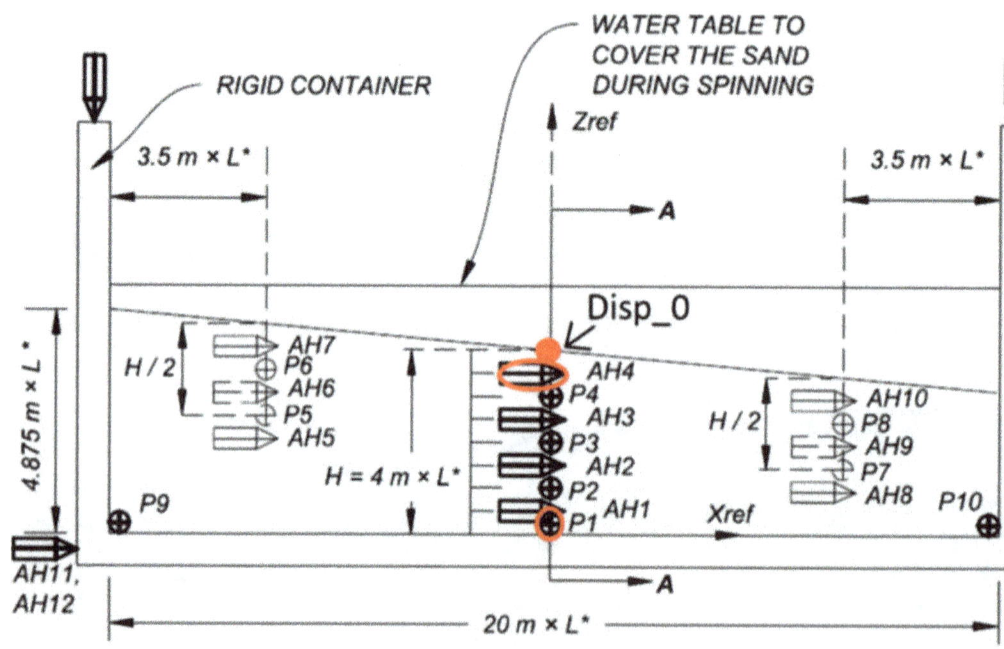


**Fig. 25.16** Output points, added to "Baseline schematic for LEAP 2017 centrifuge experiment.", for time histories: AH4 (Horizontal Acc.), Disp_0 (Horizontal Disp.), P1 (E. P. W. P.)

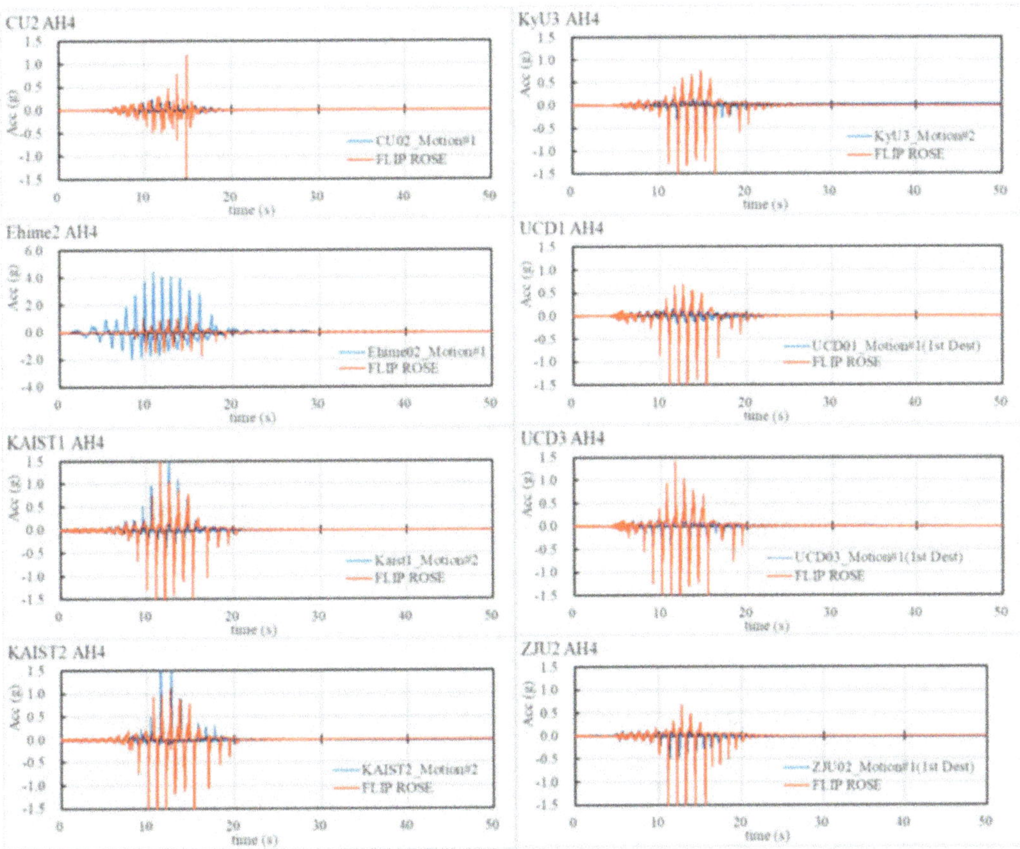


**Fig. 25.17** Acceleration time histories at AH4; center of the model, 0.5 m below the ground surface

Figure 25.17 shows that the acceleration time histories by numerical simulations trace the results of experiments well, except spikes, which can be seen in the results of numerical simulations. These spikes seem to indicate that there is attenuation that is not taken into consideration in the analysis. For example, attenuation due to sliding friction between sand and rigid wall can be mentioned. In any case, if there is attenuation that is not considered in the analysis, the analysis result will have a larger displacement.

Figure 25.18 shows that the horizontal displacement time histories by numerical simulations trace well the results of experiments in case of Ehime2, KAIST2, KyU3, and UCD3 except oscillations, which can be seen in the simulation results. The cause of these oscillations seems to be the same as the cause of the spikes in the acceleration time histories. If we take into account the attenuation, then the amplitude of the oscillations will decrease.

On the other hand, according to Fig. 25.9, the results of the numerical simulation may be underestimated for cases of "CU2", "KAIST2," "UCD3" and "ZJU2," and expected to be correct in other cases. Because, the dry densities of these four cases are below the dry density of the loosest sand in Phase I and the dry densities of the other cases are roughly equal to the dry density of the loosest sand in Phase I.

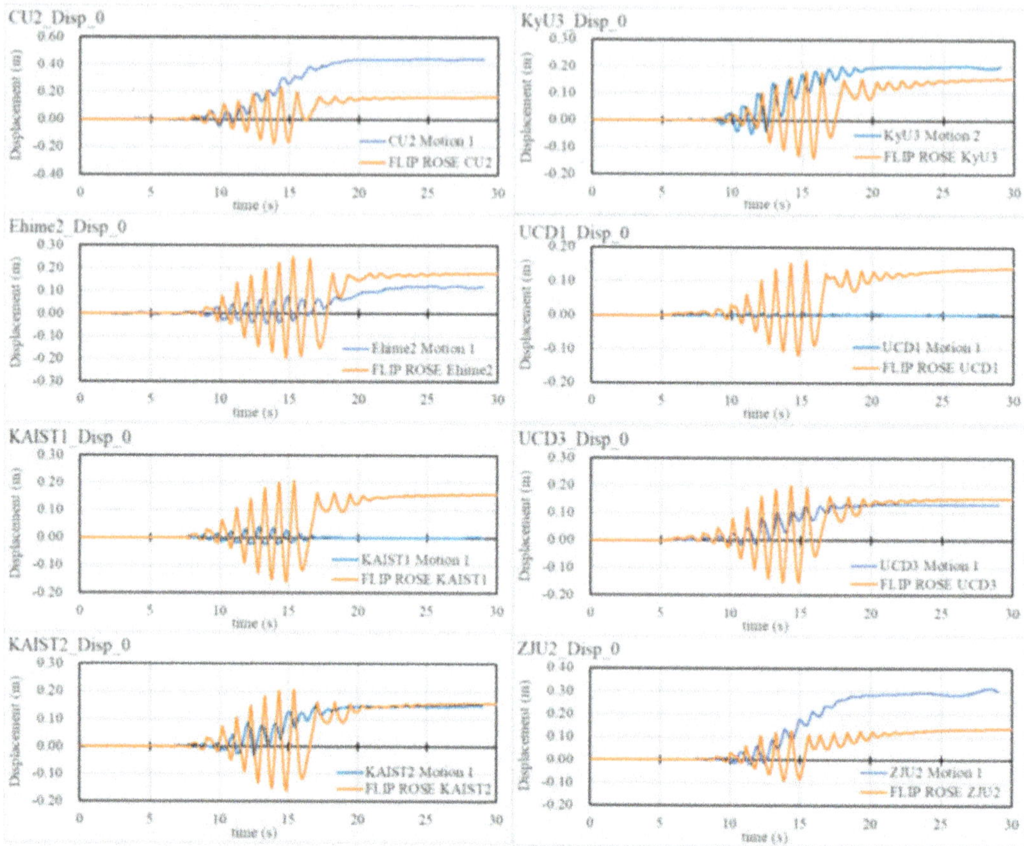


**Fig. 25.18** Horizontal displacement time histories at Disp_0; center of the model, on the surface

**Fig. 25.19** Final state of the model for UCD3 at 50 s—horizontal displacement

If we take into account the unconsidered damping in the numerical analyses, then the residual displacements in Fig. 25.18 will decrease and the tendency of these residual displacements appears to become close to the above expectation based on Fig. 25.9.

According to the comparison of the time history of the excess pore water pressure shown in Fig. 25.20, the results of numerical simulation generally correlate with the test results except for the water pressure dissipation phase. However, in terms of maximum value of excess pore water pressure, the numerical simulation results in many cases exceed the test results.

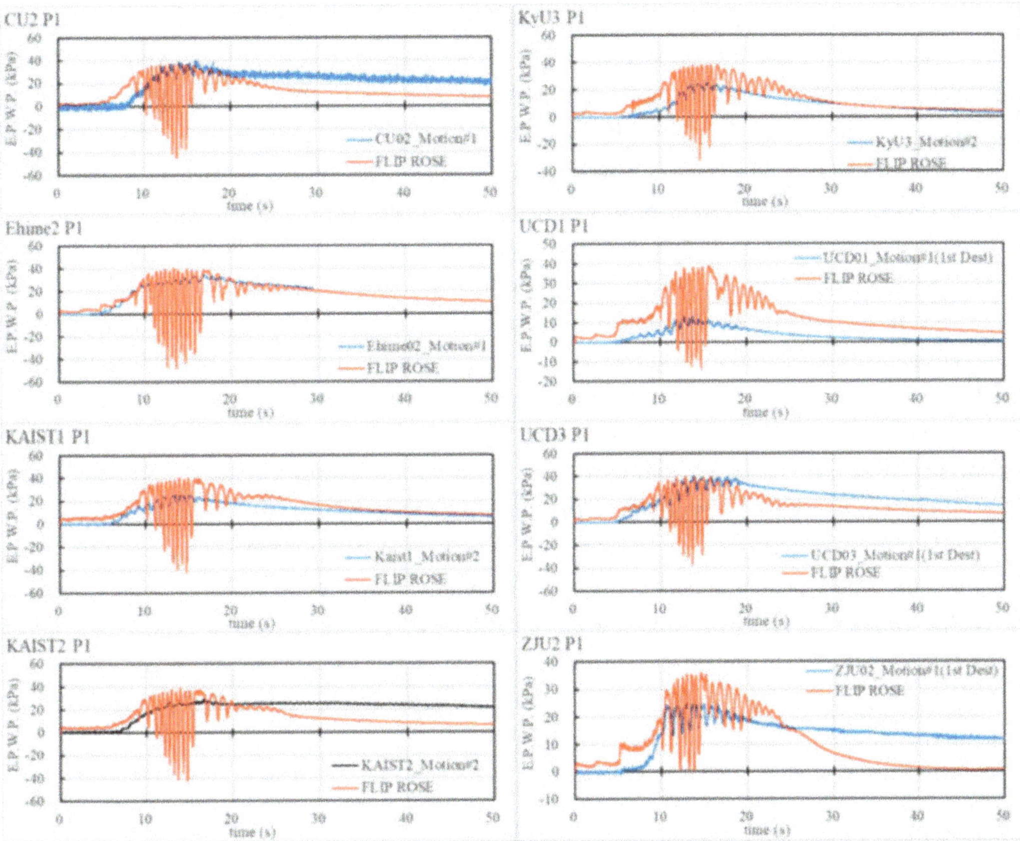


**Fig. 25.20** Excess pore water pressure time histories at P1; center of the model, depth 4 m

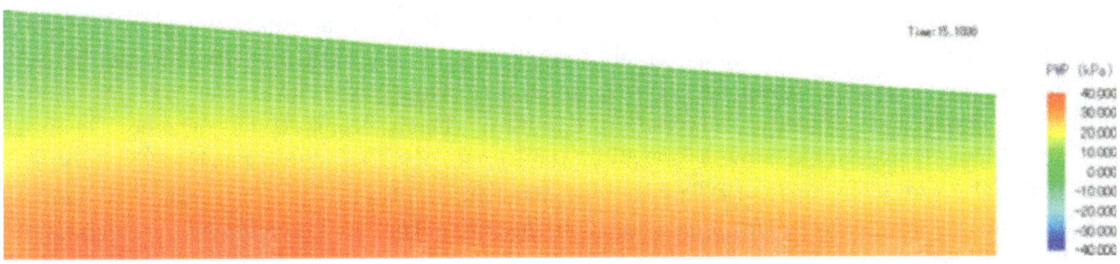


**Fig. 25.21** State of the model for UCD3 at 15.1 s—Excess pore water pressure distribution

## 25.6 Conclusion

As a part of the LEAP-UCD-2017 project, type B simulations were performed for centrifuge model tests using a two-dimensional effective stress analysis program "FLIP ROSE" incorporating a cocktail glass model (Iai et al. 2011). The centrifuge model tests were conducted as a part of the LEAP-UCD-2017 project to investigate the behavior of the liquefiable ground with gentle slope.

In Phase I of the numerical simulation, three parameter sets of the cocktail glass model were determined by trial and error so as to reproduce the results of the undrained cyclic triaxial tests for the specimens of three different dry densities. In Phase II, models were made based on the data of the centrifuge model tests. For each centrifuge model test, we decided which parameter set of the cocktail glass model to apply by referring to the dry density of the ground. As a result, for all of the centrifuge model tests, the parameter set obtained based on the triaxial test result of the specimen with the loosest dry density was applied. It seems that it is difficult to adjust the dry density of the centrifuge model and in many of the models the density was lower than that of the loosest specimen.

Since it was a type B simulation, numerical simulations were performed without knowing the experiment results. Acceleration time histories, displacement time histories and excess pore water pressure time histories simulation results were generally consistent with the results of the centrifugal model tests. There was a tendency that the response values of the numerical simulation were slightly larger than the response values expected from the difference between the dry density of the ground of the centrifuge model and the dry density of the specimen on which the applied parameter set depends. These differences seem to be related to the attenuation due to sliding friction between the ground and the rigid wall, which is not taken into account in the numerical simulation.

## References

El Ghoraiby, M. A., Park, H., & Manzari, M. T. (2017). *leap 2017: soil characterization and element tests for Ottawa F65 sand*. Washington, DC: The George Washington University.

El Ghoraiby, M. A., Park, H., & Manzari, M. T. (2019). Physical and mechanical properties of Ottawa F65 sand. In B. Kutter et al. (Eds.), *Model tests and numerical simulations of liquefaction and lateral spreading: LEAP-UCD-2017*. New York: Springer.

George Washington University, University of California, Davis, Rensselaer Polytechnic Institute, Cambridge University, Ehime University, IFSTTAR, Kyoto University, KAIST, K-Water, National Central University of Taiwan, Zhejiang University. (2017). *LEAP 2017 Simulation Exercise Phase II: Type B Prediction of Centrifuge Experiments Guidelines for submission of the numerical simulations*.

Hardin, B. O., & Richart, F. E., Jr. (1963). Elastic wave velocities in granular soils. *Journal of Soil Mechanics and Foundations Division, ASCE, 89*(SM1), 33–66.

Iai, S. (1993). Concept of effective strain in constitutive modeling of granular materials. *Soils and Foundations, 33*(2), 171–180.

Iai, S., & Ozutsumi, O. (2005). Yield and cyclic behaviour of a strain space multiple mechanism model for granular materials. *International Journal for Numerical and Analytical Methods in Geomechanics, 29*(4), 417–442.

Iai, S., & Ueda, K. (2016). Energy-less strain in granular materials – Micromechanical background and modeling. *Soils and Foundations, 56*(3), 391–398.

Iai, S., Tobita, T., Ozutsumi, O., & Ueda, K. (2011). Dilatancy of granular materials in a strain space multiple mechanism model. *International Journal for Numerical and Analytical Methods in Geomechanics, 35*(3), 360–392.

Iai, S., Tobita, T., & Ozutsumi, O. (2013). Evolution of induced fabric in a strain space multiple mechanism model for granular materials. *International Journal for Numerical and Analytical Methods in Geomechanics, 37*(10), 1326–1336.

Iai, S., Ueda, K., & Ozutsumi, O. (2015). Strain space multiple mechanism model for clay under monotonic and cyclic loads. In *6th International Conference on Earthquake Geotechnical Engineering, Christchurch, New Zealand, 264, CD-ROM.*

Manzari, M. T. (2017). *Supplemental information for numerical simulation of LEAP-2017 centrifuge tests*. Attached to an email from Majid Manzari on 2017.08.10.

Manzari, M. T., El Ghoraiby, M. A., Kutter, B. L., Zeghal, M., Wang, R., Chen, R., Zhang, J.-M., Ozutsumi, O., Fukutake, K., Kiriyama, T., Fasano, G., Chiaradonna, A., Bilotta, E., Montgomery, J., Ziotopoulou, K., Chen, L., Ghofrani, A., Arduino, P., Wada, T., Ueda, K., Mercado, V., Fuentes, W., Lascarro, C., Yang, M., Barrero, A. R., Taiebat, M., Tsiaousi, D., Ugalde, J., Thaleia Travasarou, T., Ichii, K., Uemura, K., Orai, N., Hyodo, M., Abdoun, T., Haigh, S., Madabhushi, M. S. P. G., Tobia, T., Hung, W.-Y., Kim, D. S., Okamura, M., Zhou, Y.-G., & Escoffier, S. (2019). LEAP-2017 simulation exercise: Calibration of constitutive models and simulation of the element test. In B. Kutter et al. (Eds.), *Model tests and numerical simulations of liquefaction and lateral spreading: LEAP-UCD-2017*. New York: Springer.

Tatsuoka, F., Ishihara, M., Di Benedetto, H., & Kuwano, R. (2002). Time dependent shear deformation characteristics of geomaterials and their simulation. *Soils and Foundations, 42* (2), 103–129.

Towhata, I., & Ishihara, K. (1985). Modelling soil behavior under principal stress axes rotation. In *Proceedings of 5th international conference on numerical method in geomechanics, Nagoya* (pp. 523–530).

Zienkiewicz, O. C., & Bettss, P. (1982). Soil and other saturated media under transient, dynamic conditions, general formulation and the validity various simplifying assumptions. In *Soil Mechanics – Transient and Cyclic Loads*. Hoboken, NJ: Wiley.

Zienkiewicz, O. C., Taylor, R. L., & Zhu, J. H. (2005). *Finite element method its basis and fundamentals* (6th ed.). Amsterdam: Elsevier.

# Chapter 26
# Numerical Simulations of LEAP Dynamic Centrifuge Model Tests for Response of Liquefiable Sloping Ground

**Zhijian Qiu and Ahmed Elgamal**

**Abstract** This paper presents numerical simulations (modified Type-B) related to LEAP-UCD-2017 (Liquefaction Experiments and Analysis Projects) dynamic centrifuge model tests for a liquefiable sloping ground conducted by various institutions. The numerical simulations are performed using a pressure-dependent constitutive model implemented with the characteristics of dilatancy and cyclic mobility. The soil parameters are determined based on a series of available stress-controlled cyclic triaxial tests during the Type-A simulation phase for matching the liquefaction strength curves of Ottawa F-65 sand. The computational framework for the dynamic response analysis is discussed. Computed results are presented for the selected centrifuge experiments (modified Type-B simulations). Measured time histories (e.g., displacement, acceleration and excess pore pressure) of these experiments are reasonably captured. Comparisons between the numerical simulations and measured results showed that the pressure-dependent constitutive model as well as the overall employed computational framework have the potential to predict the response of the liquefiable sloping ground, and subsequently realistically evaluate the performance of an equivalent soil system subjected to seismically induced liquefaction.

## 26.1 Introduction

LEAP (Liquefaction Experiments and Analysis Projects) is an effort to facilitate validation and verification of numerical procedures for liquefaction-induced lateral spreading analysis of a liquefiable sloping ground (Kutter et al. 2014, 2015; Manzari et al. 2014). In order to obtain a set of reliable centrifuge test data with high quality among different centrifuge facilities, a centrifuge experiment was repeated at six centrifuge facilities in LEAP-GWU-2015 (Kutter et al. 2017; Manzari et al. 2018), as one project within LEAP. In addition, the associated numerical simulations conducted by several predictors (Ueda and Iai 2018; Ziotopoulou 2018; Zeghal

Z. Qiu · A. Elgamal (✉)
Department of Structural Engineering, University of California, La Jolla, CA, USA
e-mail: zhq009@eng.ucsd.edu; aelgamal@ucsd.edu



B. Kutter et al. (eds.), *Model Tests and Numerical Simulations of Liquefaction and Lateral Spreading*, https://doi.org/10.1007/978-3-030-22818-7_26

et al. 2018; Ghofrani and Arduino 2018; Armstrong 2018) were compared with the measured response from the conducted experiments.

As part of the ongoing LEAP, i.e., LEAP-UCD-2017 (Kutter et al. 2018), a new set of dynamic centrifuge tests have been performed to simulate the liquefaction-induced lateral spreading phenomenon in a fully saturated sloping ground. A major goal of LEAP-UCD-2017 is to quantify repeatability of the centrifuge test as well as sensitivity of the response with respect to variations of the soil density and base input excitation.

On this basis, this paper presents the results of numerical simulations for the dynamic centrifuge tests and the emphasis is on calibration based on the Type-B simulation phase, and subsequent minor modification based on knowledge of experiment results during the Type-C phase. All the numerical simulations are performed using a pressure-dependent constitutive model (Yang 2000; Elgamal et al. 2003; Parra 1996; Yang and Elgamal 2002) implemented with the characteristics of dilatancy and cyclic mobility. The soil parameters are determined based on a series of stress-controlled cyclic triaxial tests provided during the Type-A simulation phase for matching the liquefaction strength curves of Ottawa F-65 sand. To better capture the overall dynamic response of each selected centrifuge test in the modified Type-B simulations, only one parameter was adjusted by observations from all selected centrifuge test results provided in the Type-C simulation phase (to increase the rate of pore pressure build-up).

The following sections of this paper outline: (1) computational framework, (2) specifics and calibration processes, (3) details of the employed FE modeling techniques, and (4) computed results of the selected centrifuge tests. Finally, a number of conclusions are presented and discussed.

## 26.2 Brief Summary of the Centrifuge Tests

A schematic representation of the centrifuge tests is shown in Fig. 26.1. The soil specimen is a sloping layer of Ottawa F-65 sand with 5° slope (target dry density is 1652 kg/m$^3$, $D_r = 65\%$). The soil layer has a length of 20 meters (in prototype scale) and a height of 4 m (in prototype scale) at the center. The specimen is built in a container with rigid walls. Three arrays of accelerometers and pore pressure transducers are placed in the central section and at 3.5 m away from the side walls on the right and left of the centrifuge model. In the vertical direction, the sensors are 0.5 m apart. The sensors in the central section are required sensors and those of side arrays (away from central section) are recommended for all centrifuge facilities. Table 26.1 shows a summary of the experiments selected for LEAP-UCD-2017 Type-B simulations including centrifugal acceleration and the achieved density of the soil. All centrifuge models were subjected to a target motion of ramped, 1 Hz sine wave base motion with amplitude 0.15 g. Figure 26.2 shows the target motion and achieved motions for the selected centrifuge experiments.

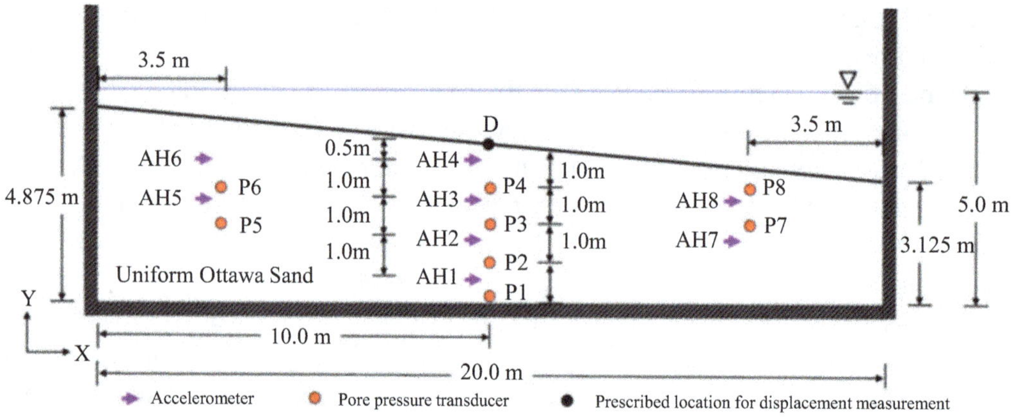


**Fig. 26.1** Schematic representation of the centrifuge test layout (after Kutter et al. 2018)

## 26.3 Finite Element Model

A two-dimensional FE mesh with element size 0.5 m (Fig. 26.3) is created to represent the centrifuge model with rigid walls. All numerical simulations for the selected nine centrifuge experiments (Table 26.1) are conducted using the computational platform OpenSees. The Open System for Earthquake Engineering Simulation (OpenSees, McKenna et al. 2010, http://opensees.berkeley.edu) developed by the Pacific Earthquake Engineering Research (PEER) Center, is an open source, object-oriented finite element platform. Currently, OpenSees is widely used for the simulation of structural and geotechnical systems (Yang 2000; Yang and Elgamal 2002) under static and seismic loading.

Four-node plane-strain elements with two-phase material following the *u-p* (Chan 1988) formulation were employed for simulating saturated soil response, where *u* is the displacement of the soil skeleton and *p* is the pore water pressure. Implementation of the *u-p* element is based on the following assumptions: (1) small deformation and rotation; (2) solid and fluid density remain constant in time and space; (3) porosity is locally homogeneous and constant with time; (4) soil grains are incompressible; (5) solid and fluid phases are accelerated equally. Hence, the soil layers represented by effective stress fully coupled *u-p* elements are capable of accounting for soil deformations and the associated changes in pore water pressure.

### 26.3.1 Soil Constitutive Model

The employed soil constitutive model (Yang 2000; Elgamal et al. 2003; Parra 1996; Yang and Elgamal 2002) were developed based on the multi-surface-plasticity theory (Mroz 1967; Iwan 1967; Prevost 1978, 1985). In this employed soil constitutive model, the shear–strain backbone curve was represented by the hyperbolic relationship with the shear strength based on simple shear (reached at an octahedral

**Table 26.1** Summary of centrifuge experiments selected for LEAP-UCD-2017 Type-B simulations (Kutter et al. 2018)

| Centrifuge test | CU-2 | Ehime-2 | KAIST-1 | KAIST-2 | KyU-3 | NCU-3 | UCD-1 | UCD-3 | ZJU-2 |
|---|---|---|---|---|---|---|---|---|---|
| $g^*$ | 40 | 40 | 40 | 40 | 44.4 | 26.0 | 43.0 | 43.0 | 30.0 |
| Density (kg/m$^3$) | 1605.8 | 1656.55 | 1701.2 | 1592.5 | 1637 | 1652 | 1665 | 1658 | 1606 |

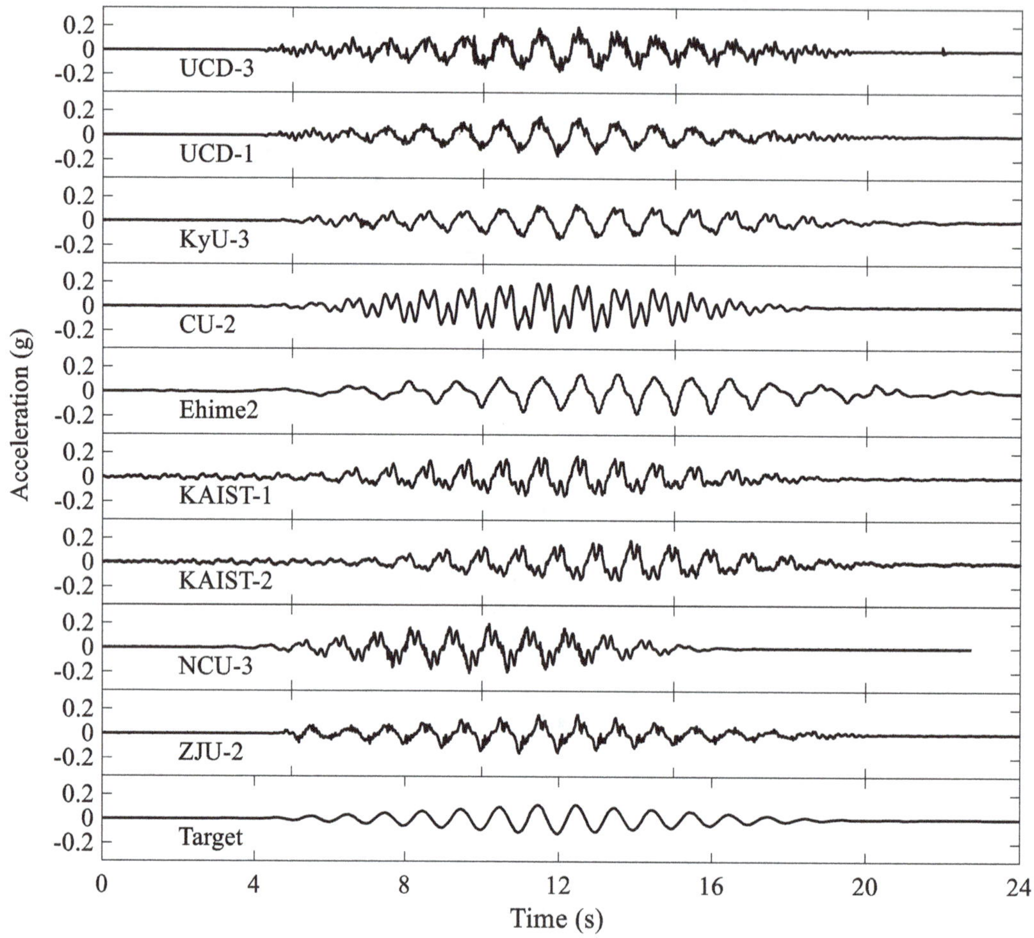


**Fig. 26.2** Selected input motions for LEAP-UCD-2017 Type-B simulations from different facilities (Kutter et al. 2018)

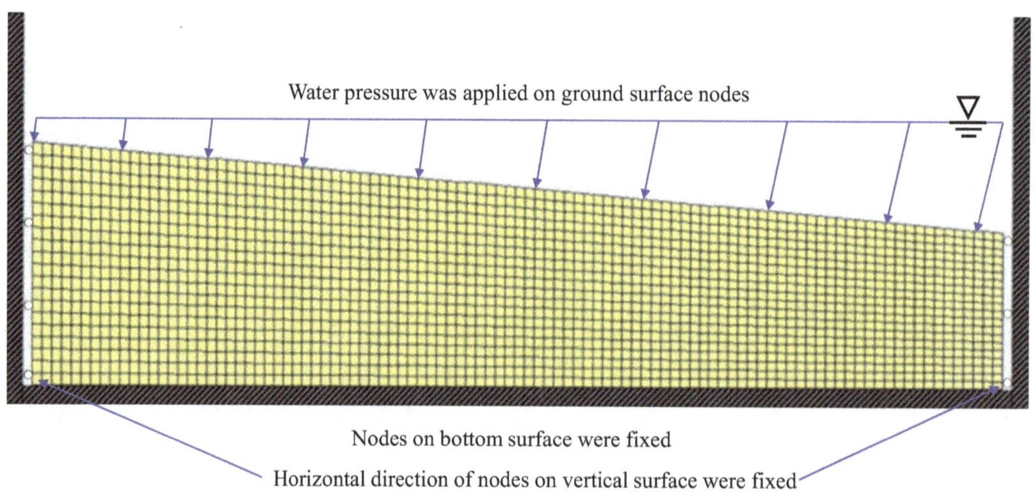


**Fig. 26.3** Finite element mesh (element size = 0.5 m)

shear strain of 10%). The small strain shear modulus under a reference effective confining pressure $p'_r$ is computed using the equation$G = G_0(p'/p'_r)^n$, where $p'$ is effective confining pressure. The dependency of shear modulus on confining pressure is taken as ($n = 0.5$). The critical state frictional constant $M_f$ (failure surface) is related to the friction angle $\phi$ (Chen and Mizuno 1990) and defined as $M_f = 6\sin\phi/(3\text{-}\sin\phi)$. As such, the soil is simulated by the implemented OpenSees material PressureDependMultiYield02. Brief descriptions of this soil constitutive model are included below.

### Yield Function

The yield function is defined as a conical surface in principal stress space (Prevost 1985; Lacy 1986; Yang and Elgamal 2002):

$$f = \frac{3}{2}\left(\boldsymbol{s} - \left(p' + p_0'\right)\boldsymbol{a}\right) : \left(\boldsymbol{s} - \left(p' + p_0'\right)\boldsymbol{a}\right) - M^2\left(p' + p_0'\right)^2 = 0 \tag{26.1}$$

where, $\boldsymbol{s} = \boldsymbol{\sigma}' - p'\boldsymbol{\delta}$ is the deviatoric stress tensor, $\boldsymbol{\sigma}'$ is the effective Cauchy stress tensor, $\boldsymbol{\delta}$ is the second-order identity tensor, $p'$ is mean effective stress, $p_0'$ is a small positive constant (0.3 kPa in this paper) such that the yield surface size remains finite at $p' = 0$ for numerical convenience and to avoid ambiguity in defining the yield surface normal to the yield surface apex. $\boldsymbol{a}$ is a second-order deviatoric tensor defining the yield surface center in deviatoric stress subspace, $M$ defines the yield surface size, and "**:**" denotes doubly contracted tensor product.

### Contractive Phase

Shear-induced contraction occurs inside the phase transformation (PT) surface ($\eta < \eta_{PT}$), as well as outside ($\eta > \eta_{PT}$) when $\dot{\eta} < 0$, where, $\eta$ is the stress ratio and $\eta_{PT}$ is the stress ratio at phase transformation surface. The contraction flow rule is defined as (Yang et al. 2008):

$$P'' = \left(1 - \frac{\dot{\boldsymbol{n}} : \dot{\boldsymbol{s}}}{\|\dot{\boldsymbol{s}}\|}\frac{\eta}{\eta_{PT}}\right)^2 \left(c_1 + c_2\gamma_d\right)\left(\frac{p'}{p_a}\right)^{c_3}, \tag{26.2}$$

where $c_1$, $c_2$ and $c_3$ are non-negative calibration constants, $\gamma_d$ is octahedral shear strain accumulated during previous dilation phases, $p_a$ is atmospheric pressure for normalization purpose, and $\dot{\boldsymbol{s}}$ is the deviatoric stress rate. The $\dot{\boldsymbol{n}}$ and $\dot{\boldsymbol{s}}$ tensors are used to account for general 3D loading scenarios, where, $\dot{\boldsymbol{n}}$ is the outer normal to a surface. The parameter $c_3$ is used to represent the dependence of pore pressure buildup on initial confinement (i.e., $K_\sigma$ effect).

#### Dilative Phase

Dilation appears only due to shear loading outside the PT surface ($\eta > \eta_{PT}$ with $\dot{\eta} > 0$), and is defined as (Yang et al. 2008):

$$P'' = \left(1 - \frac{\dot{n} : \dot{s}}{\|\dot{s}\|} \frac{\eta}{\eta_{PT}}\right)^2 d_1 (\gamma_d)^{d_2} \left(\frac{p'}{p_a}\right)^{-d_3} \tag{26.3}$$

where $d_1$, $d_2$ and $d_3$ are non-negative calibration constants, and $\gamma_d$ is the octahedral shear strain accumulated during all dilation phases in the same direction as long as there is no significant load reversal. It should be mentioned that $\gamma_d$ accumulates even if there are small unload–reload phases, resulting in increasingly stronger dilation tendency and reduced rate of shear strain accumulation.

#### Neutral Phase

When the stress state approaches the PT surface ($\eta = \eta_{PT}$) from below, a significant amount of permanent shear strain may accumulate prior to dilation, with minimal changes in shear stress and $p'$ (implying $p'' = 0$). For simplicity, $P'' = 0$ is maintained during this highly yielded phase until a boundary defined in deviatoric strain space is reached, and then dilation begins. This yield domain will enlarge or translate depending on load history (Yang et al. 2008).

It should be noted that PressureDependMultiYield02 has been improved with new flow rules in order to better capture contraction and dilation in sands and the model parameters were calibrated with established guidelines on the liquefaction triggering logic, i.e., cyclic stress ratio versus number of equivalent uniform loading cycles in undrained (direct simple shear) DSS loading to cause single-amplitude shear strain of 3% (Khosravifar et al. 2018).

### 26.3.2 Boundary and Loading Conditions

The boundary and loading conditions for the dynamic analysis of the sloping ground under an input motion are implemented in a staged fashion as follows:

1. Gravity was applied to activate the initial static state (Fig. 26.4) for the sloping ground with: (a) linear elastic properties (Poisson's ratio of 0.47), (b) nodes on both side boundaries (vertical faces) of the FE model were fixed against longitudinal translation, (c) nodes were fixed along the base against vertical translation, (d) water table was specified (Fig. 26.2) with related water pressure and nodal forces specified along ground surface nodes. At the end of this step, the static soil state was imposed and displacements under own-weight application were re-set to zero using the OpenSees command InitialStateAnalysis.

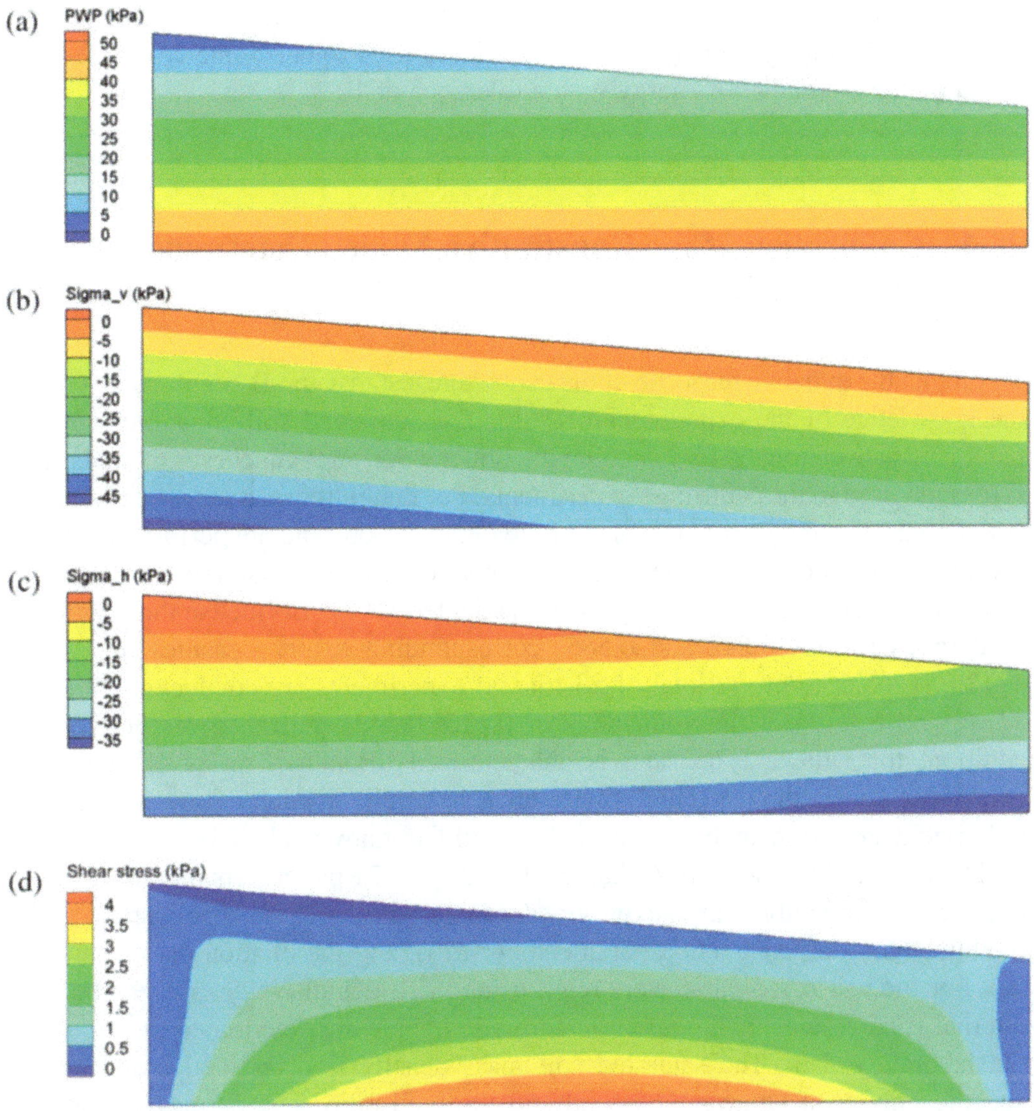


**Fig. 26.4** Initial state of soil due to gravity: (**a**) Pore water pressure; (**b**) vertical effective stress; (**c**) horizontal effective stress; (**d**) shear stress

2. Soil properties were switched from elastic to plastic.
3. Nodes were fixed along the base against longitudinal translation.
4. Dynamic analysis is conducted by applying an acceleration time history to the base of the FE model.

The FE matrix equation is integrated in time using a single-step predictor multi-corrector scheme of the Newmark type (Chan 1988; Parra 1996) with integration parameters $\gamma = 0.6$ and $\beta = 0.3025$. The equation is solved using the modified Newton-Raphson method, i.e., Krylov subspace acceleration (Carlson and Miller 1998) for each time step. A relatively low level of stiffness proportional damping (coefficient = 0.003) with the main damping emanating from the soil nonlinear shear stress–strain hysteresis response was used to enhance the numerical stability of the liquefiable sloping system. The tolerance criteria used to check the convergence is based on the increment of energy with a tolerance of $10^{-6}$.

## 26.4 Calibration of the Constitutive Model: Determination of Soil Model Parameters

To predict the dynamic response of the centrifuge test in Type-B simulations without the knowledge of the experimental results, the employed soil constitutive model parameters are calibrated for matching the liquefaction strength curves of the Ottawa F-65 sand used in the centrifuge experiments. For that purpose, a series of undrained stress-controlled cyclic triaxial tests for the Ottawa F-65 sand are performed and the test results are provided in the Type-A simulation phase (El Ghoraiby et al. 2017, 2019; Vasko 2015; Bastidas 2016). The triaxial tests are performed at three different void ratios (0.515, 0.542 and 0.585) corresponding to dry densities of 1744.2, 1712.6, 1665.6 kg/m$^3$. A total of 23 triaxial experiments are conducted, in which 11, 6, and 6 tests for the void ratios of 0.515, 0.542, and 0.585 respectively. In addition, the Ottawa F-65 sand is characterized through a series of laboratory experiments, including specific gravity tests, hydraulic conductivity tests, and particle size distribution analysis (Vasko 2015; El Ghoraiby et al. 2019).

The PressureDependMultiYield02 soil model has 21 parameters in total. The key parameters for calibrating the provided triaxial data are the small strain shear modulus ($G_o$), the contraction parameters ($c_1$, $c_2$), and the dilation parameters ($d_1$, $d_2$). For the rest of 16 parameters (e.g., $c_3$ and $d_3$), default values were used. The internal friction angle and phase transformation angle are determined directly from the laboratory data. The saturated soil mass density is detemined based on the specific gravity of Ottawa F-65 sand and the void ratio. The parmeability of the the Ottawa F-65 sand is $1.1 \times 10^{-4}$ m/s determined from El Ghoraiby et al. (2019). On this basis, the calibrated soil model parameters for the Ottawa F-65 sand at three different void ratios are presented in Table 26.2.

Figure 26.5 shows the plot of the liquefaction strength curves for Ottawa F-65 sand with initial void ratios of 0.515, 0.542 and 0.585 ($\rho_d$ = 1744.2, 1712.6, 1665.6 kg/m$^3$). The data plotted is composed of the number of cycles until a 2.5% single amplitude of strain is achieved versus the applied cyclic stress ratio. It can be seen that a comparatively good match is reached between simulation results and laboratory test data for the Ottawa sand F-65 at three different void ratios. Figure 26.6 shows an example of comparison results between simulation and experiment for Ottawa F-65 sand with void ratio 0.585, which will be further used in the centrifuge tests of Type-B simulations. It is noted that the yield surface of PressureDependMultiYield02 is conical in principal stress space, resulting in symmetric response for deviatoric stress versus axial strain (Fig. 26.6b). However, it still can reasonably capture the stress path, maximum axial strain, and pore pressure generation (Fig. 26.6a, c, and d).

**Table 26.2** Sand model parameters employed in Type-A simulations

| Model parameters for different void ratios $e_o$ | 0.585 | 0.542 | 0.515 |
|---|---|---|---|
| Reference mean effective pressure, $p'_r$ (kPa) | 101 | | |
| Mass density $\rho$ (t/m$^3$) | 2.04 | 2.07 | 2.09 |
| Maximum shear strain at reference pressure, $\gamma_{\max,r}$ | 0.1 | | |
| Shear modulus at reference pressure, $G_r$ (atm) | 250 | 300 | 350 |
| Stiffness dependence coefficient $d$, $G = G_r\left(\frac{p'}{p'_r}\right)^d$ | 0.5 | | |
| Poisson's ratio $v$ (for dynamics) | 0.4 | | |
| Shear strength at zero confinement, $c$ (kPa) | 0.3 | | |
| Friction angle $\phi$ | 36° | 36° | 36° |
| Phase transformation angle | 22° | 22° | 22° |
| Contraction coefficient, $c_1$ | 0.015 | 0.012 | 0.005 |
| Contraction coefficient, $c_2$ | 3.0 | | |
| Contraction coefficient, $c_3$ | 0.15 | 0.15 | 0.15 |
| Dilation coefficient, $d_1$ | 0.08 | 0.1 | 0.4 |
| Dilation coefficient, $d_2$ | 3.0 | | |
| Dilation coefficient, $d_3$ | 0.2 | 0.1 | 0.00 |
| Damage parameter, $Liq1$ | 3.0 | 2.0 | 0.5 |
| Damage parameter, $Liq2$ | 3.0 | 0.0 | 0.0 |
| Permeability (m/s) | $1.1 \times 10^{-4}$ | | |
| Number of yield surfaces | 20 | | |

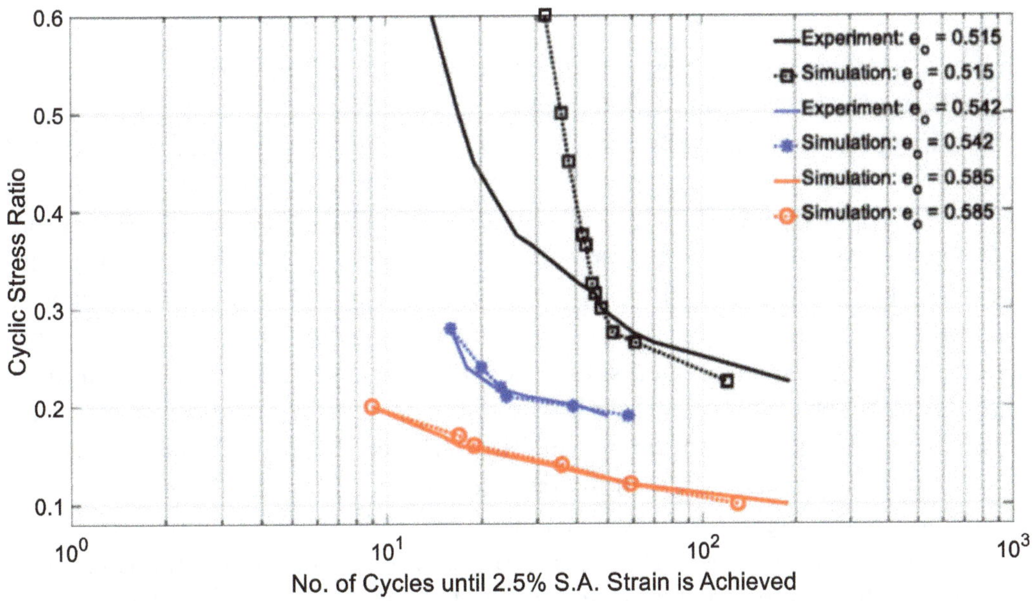


**Fig. 26.5** Simulated liquefaction strength curves with measured data obtained from the cyclic stress-controlled triaxial tests on Ottawa F-65 Sand (Kutter et al. 2018)

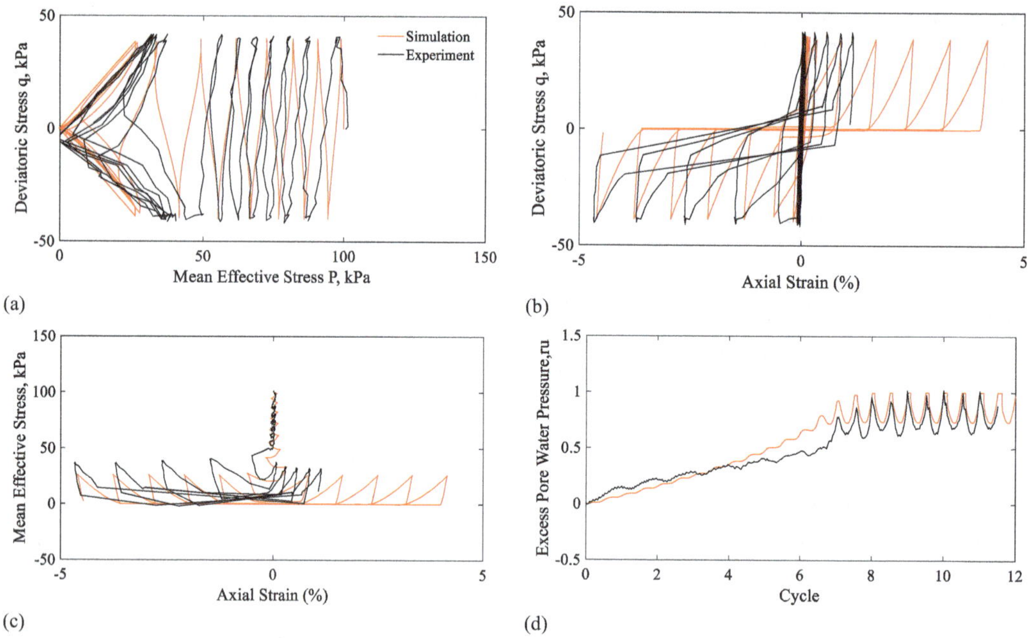


**Fig. 26.6** Simulation of cyclic undrained stress-controlled triaxial test for Type-A simulations (confining stress = 100 kPa, cyclic stress ratio = 0.2): (**a**) Mean effective stress-deviatoric stress; (**b**) axial strain- deviatoric stress; (**c**) axial strain-mean effective stress; (**d**) load cycles-excess pore water pressure

## 26.5 Computed Results of Type-B Simulations

In this section, the numerical results of selected nine centrifuge tests in the modified Type-B simulations for Ottawa F-65 sand with void ratio $e_o = 0.585$ are presented. The achieved motion for each centrifuge test is shown in Fig. 26.2. As discussed above, the model parameters are calibrated for matching the liquefaction strength curves for the Ottawa F-65 sand with different void ratios from the Type-A simulation phase. Although the measured and simulated results are in good agreement for the case of void ratio $e_o = 0.585$, it was found that the simulation was unable to capture the overall response for the selected nine centrifuge tests in Type-B simulations (once the centrifuge test results were made available for Type-C simulations). In order to better capture the overall dynamic response for all selected centrifuge tests, the contraction parameter $c_1$ equal to 0.015 in Table. 26.2 was adjusted to $c_1 = 0.08$ based on the observations from the provided test results in the Type-C simulation phase. Figure 26.7 shows comparison results between numerical simulation and triaxial experiment for Ottawa F-65 sand with void ratio 0.585 (Fig. 26.6) after the contraction parameter $c_1$ was adjusted. As such, some discrepancies of the soil response can be found, e.g., the pore pressure in Fig. 26.7d generates faster than that in Fig. 26.6d, and the axial strain in Fig. 26.7b is overestimated compared to that in Fig. 26.6b.

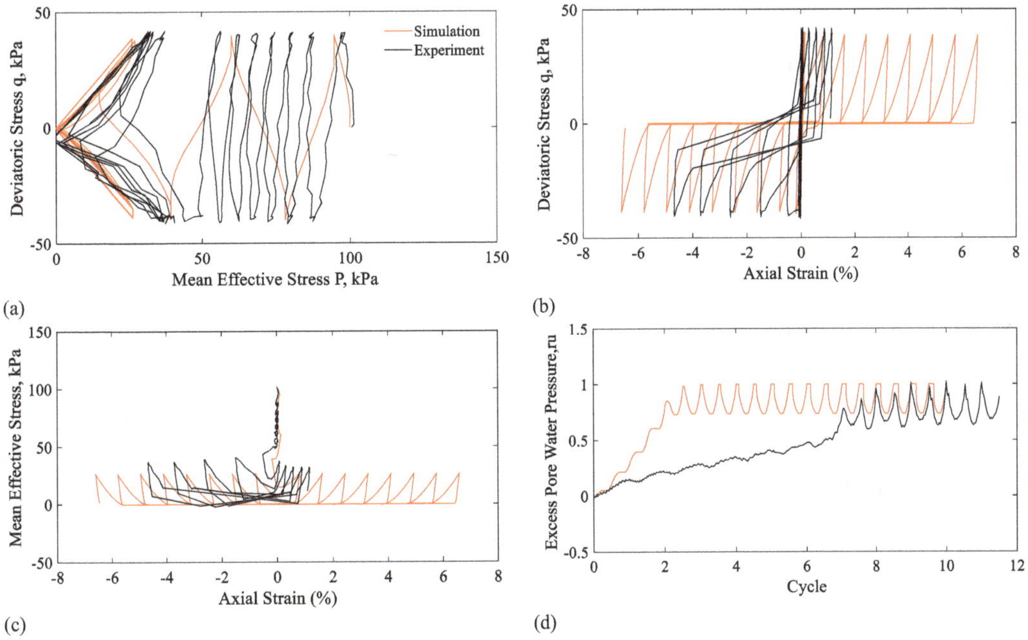


**Fig. 26.7** Simulation of cyclic undrained stress-controlled triaxial test for Type-A simulations (confining stress = 100 kPa, cyclic stress ratio = 0.2): (**a**) Mean effective stress-deviatoric stress; (**b**) axial strain- deviatoric stress; (**c**) axial strain-mean effective stress; (**d**) load cycles-excess pore water pressure

As discussed above, the contraction parameter $c_1$ was adjusted to better capture the overall response of the centrifuge tests. Therefore, the following numerical results for each centrifuge test are computed based on the parameters in Table 26.2 with the adjusted $c_1$ parameter as described above.

The numerical results for UCD-3 are presented in Fig. 26.8. The computed time histories of acceleration (AH1-AH4) and excess pore pressure (P1-P4) reasonably match with those from measurements (Fig. 26.8a, b). Both the computed results and the measurements showed a consistent trend for negative spikes of pore pressure generation and acceleration due to dilation. The computed lateral displacement in Fig. 26.8c at the midpoint of the ground surface shows a trend similar to the measurement. Figure 26.9 shows the computed horizontal and vertical displacement contours at the end of shaking. It is noted that the arrows in Fig. 26.9a shows the soil movement and the length represents the total displacement. Figure 26.10 shows the computed soil response including the shear stress versus mean effective stress (confinement) and shear stress versus shear strain for the integration points near the locations of pore pressure transducer (P1–P4). It is observed that the pore pressure dips are consistent with the spikes in shear stress due to dilation.

The numerical results for UCD-1 are shown in Fig. 26.11. Comparison of the results in Fig. 26.11a shows that the computed acceleration is in good agreement with that of measurement. In Fig. 26.11b, the time histories of pore pressure are overestimated by numerical simulations. Figure 26.11c shows the comparison of displacement between numerical and measured computed results. It can be seen that

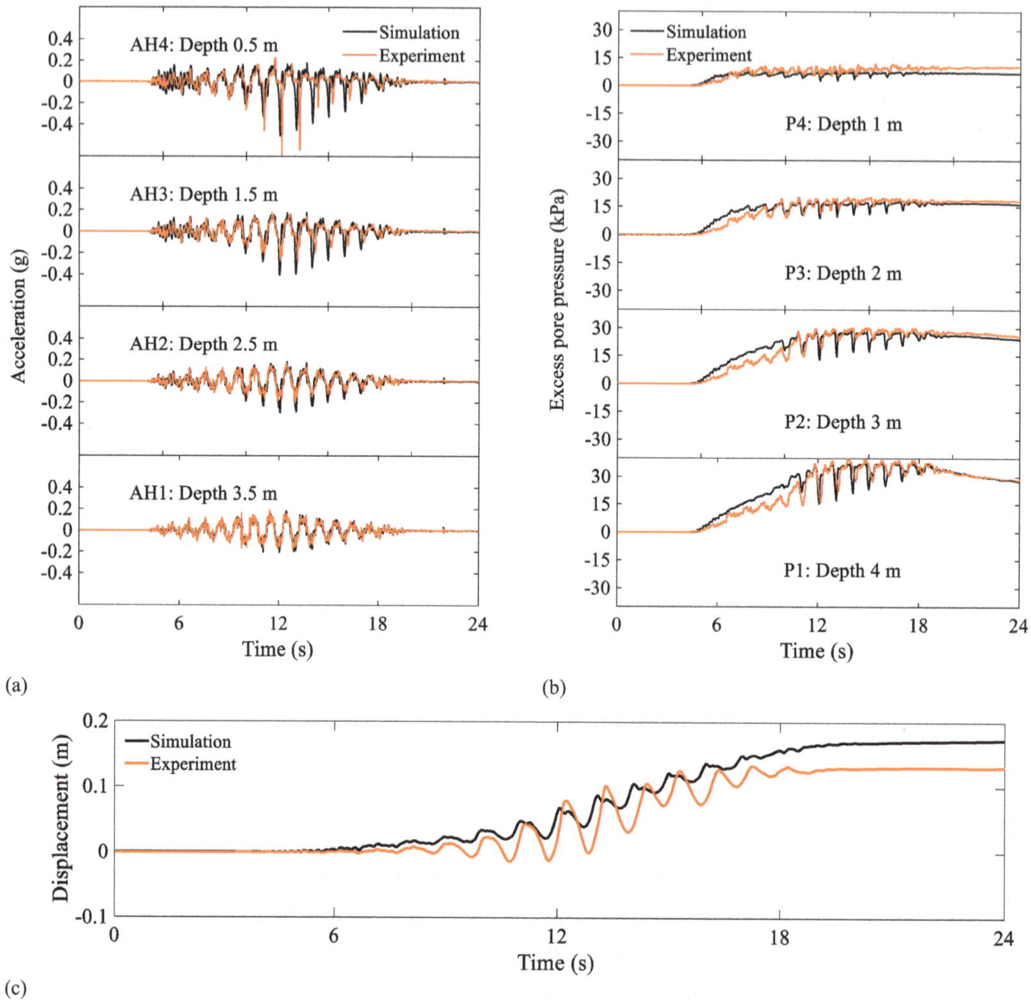


**Fig. 26.8** Measured and computed time histories of UCD-3: (**a**) Acceleration; (**b**) excess pore water pressure; (**c**) displacement at the middle of ground surface

the measured displacement is significantly smaller than that of simulation. Comparing the achieved response of UCD-1 with those of other centrifuge tests, the reason for discrepancies between the simulations might stem from the fact that the soil near the ground surface was built denser than what was expected.

Figure 26.12 shows the results of NCU-3. The computed time histories of acceleration (AH3-AH4) at deeper depth are in good agreement with those from measurements (Fig. 26.12a). The acceleration at AH1 and AH2 show very high negative spikes due to dilation, that the computed results did not reproduce. By comparing time histories of pore pressure, a relatively good agreement is shown in Fig. 26.12b in which, time histories of P3 at depth 2 m was not available. Figure 26.12c indicates that the computed lateral displacement at midpoint of the ground surface is accumulating cycle-by-cycle with residual lateral displacement around 0.055 m. Since the measured time history of lateral displacement not being available, the dynamic component computed from the acceleration history at A4 is shown (no residual lateral displacement induced by the inclination of the ground).

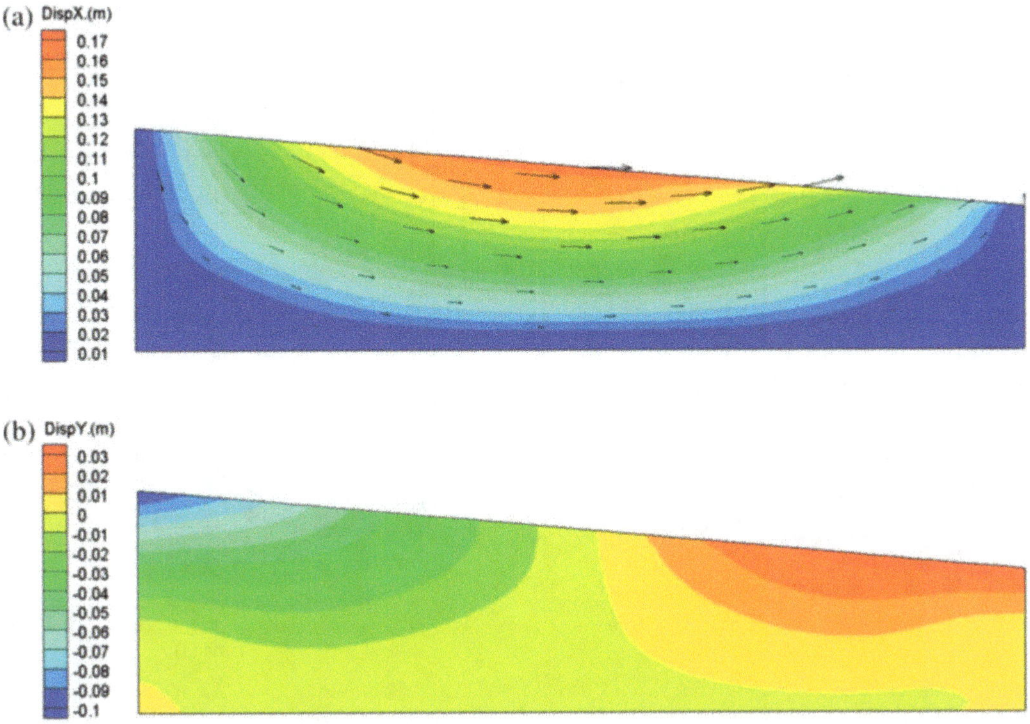


**Fig. 26.9** Computed displacement contours of UCD-3: (**a**) Horizontal (arrow shows the total displacement); (**b**) vertical

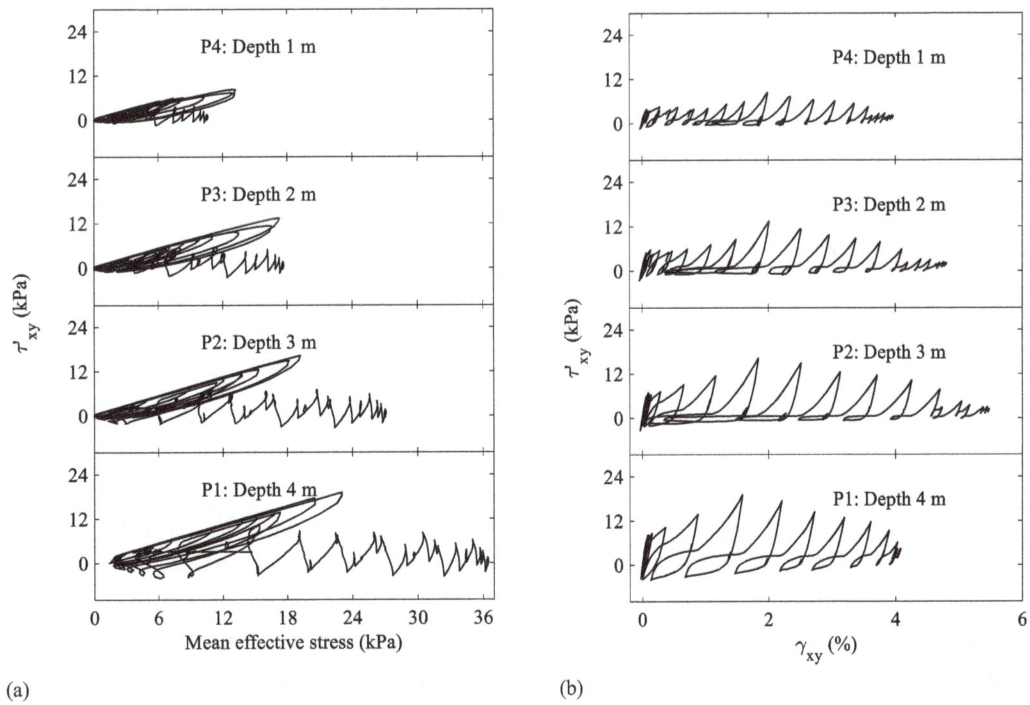


**Fig. 26.10** Computed soil responses of UCD-3: (**a**) Mean effective stress-shear stress; (**b**) shear strain–stress

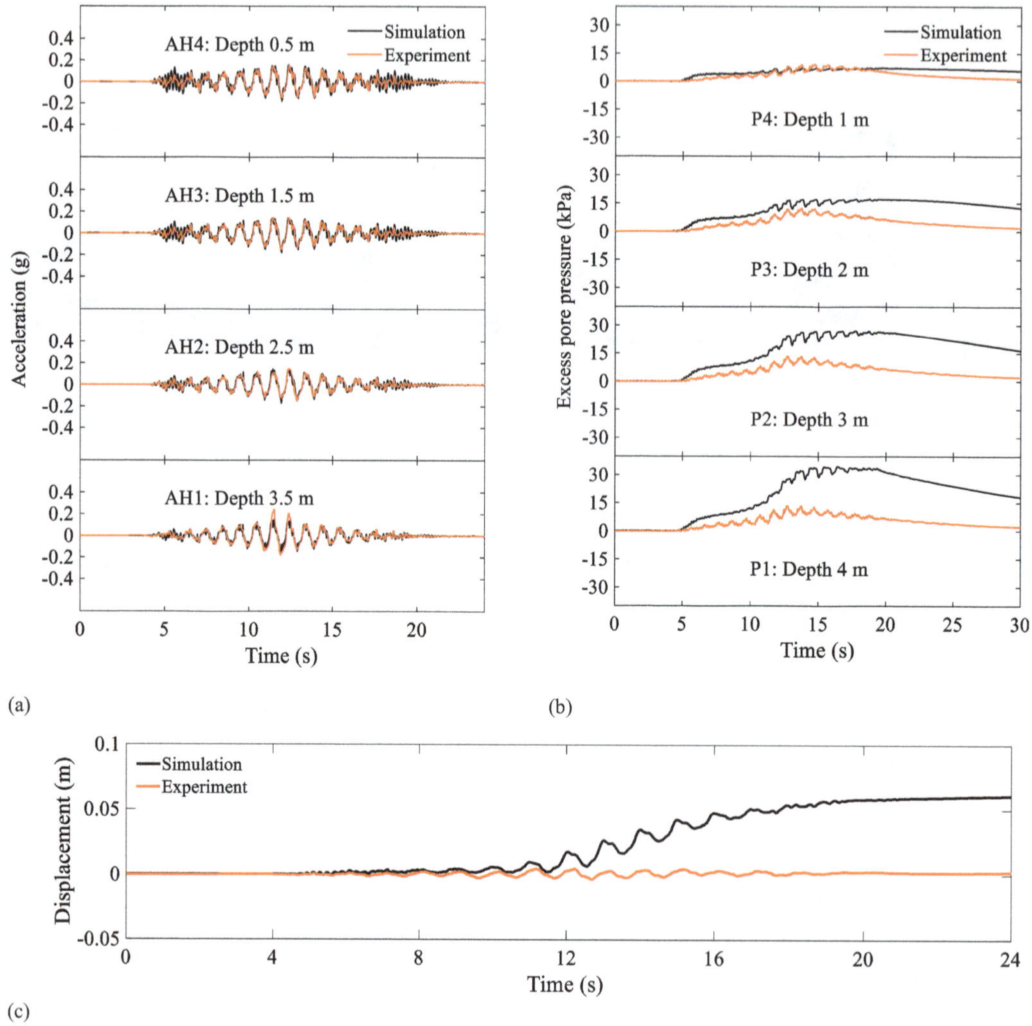


**Fig. 26.11** Measured and computed time histories of UCD-1: (**a**) Acceleration; (**b**) excess pore water pressure; (**c**) displacement at the middle of the ground surface

The results for CU-2 are shown in Fig. 26.13. The simulation results provide a similar response for AH1 and AH2 (Fig. 26.13a). Compared to the time histories of the numerical simulations, the experiment results show shifted negative spikes in acceleration due to dilation. The overall trend of the time histories of the excess pore pressure is generally consistent with the measurements (Fig. 26.13b). In Fig. 26.13c, the computed lateral displacement is around 0.18 m, however, the lateral displacement at the midpoint of the ground in the experiment shows around 0.42 m at the end of shaking. By comparing the displacement with those of the other testing facilities, the induced lateral displacement of CU-2 is relative higher. This might be due to that the soil near the ground surface was built looser than what was expected.

Figure 26.14 presents the results of Ehime-2. The computed time histories of acceleration and excess pore pressure are in good agreement with those from measurements (Fig. 26.14a, b). Although, both the computed results and the measurements showed a consistent trend for the negative spikes of pore pressure

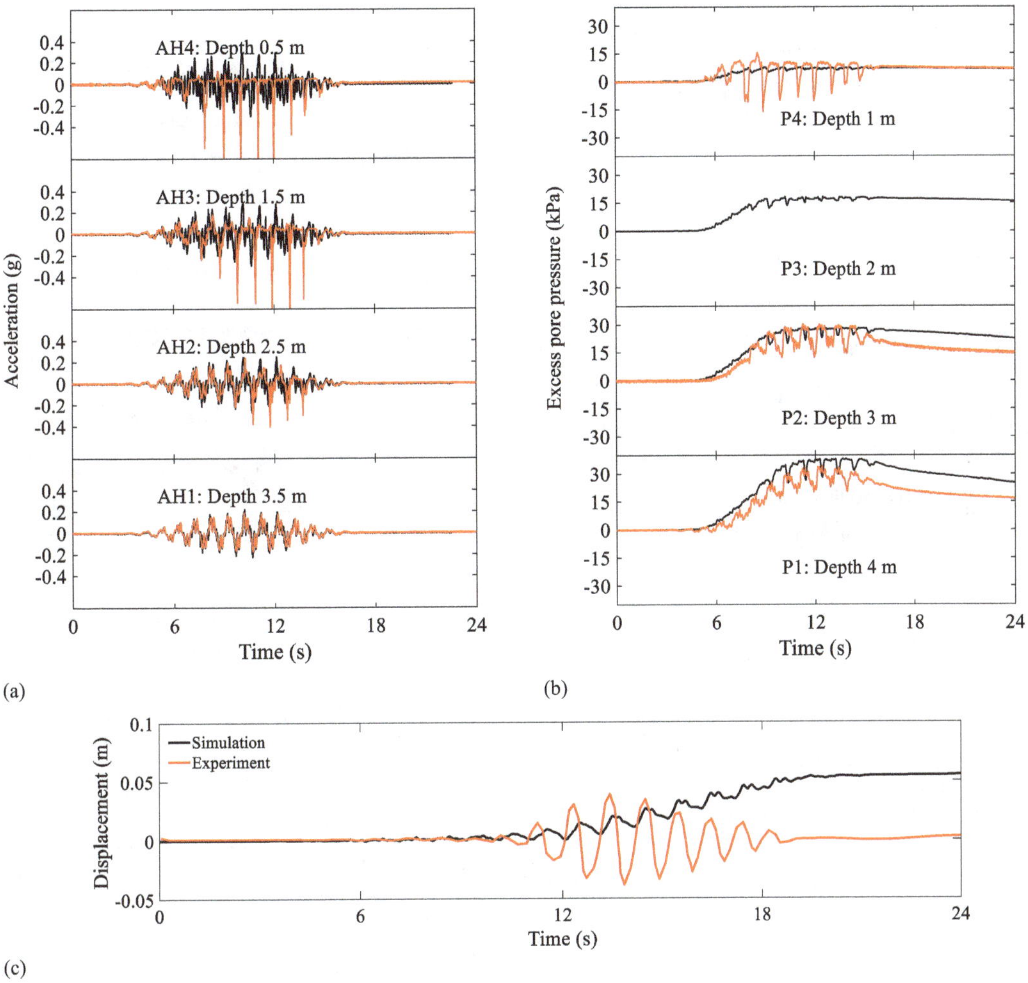


**Fig. 26.12** Measured and computed time histories of NCU-3: (**a**) Acceleration; (**b**) excess pore water pressure; (**c**) displacement at the middle of ground surface

generation and acceleration due to dilation, some discrepancies are found in the values of excess pore pressure between the measured and computed results. At the locations of P2–P4, the values of excess pore pressure for the numerical results are slightly lower than those of the measurement, but P1 is slightly higher. The computed lateral displacement in Fig. 26.14c at the midpoint of the ground surface shows a good trend similar to the measurement and with similar residual displacement at the end of shaking.

The results for ZJU-2 are shown in Fig. 26.15. The simulation results provide a similar response for AH1 and AH2 (Fig. 26.15a). The comparison results for ZJU-2 are similar to CU-2 (i.e., the experiment results show shifted negative spikes in acceleration compared to the numerical results). The overall trend of the time histories of excess pore pressure is generally consistent with the measurements (Fig. 26.15b). Discrepancies are also found in the values of excess pore pressure between the measured and computed results. At the locations of P1–P2, the values of

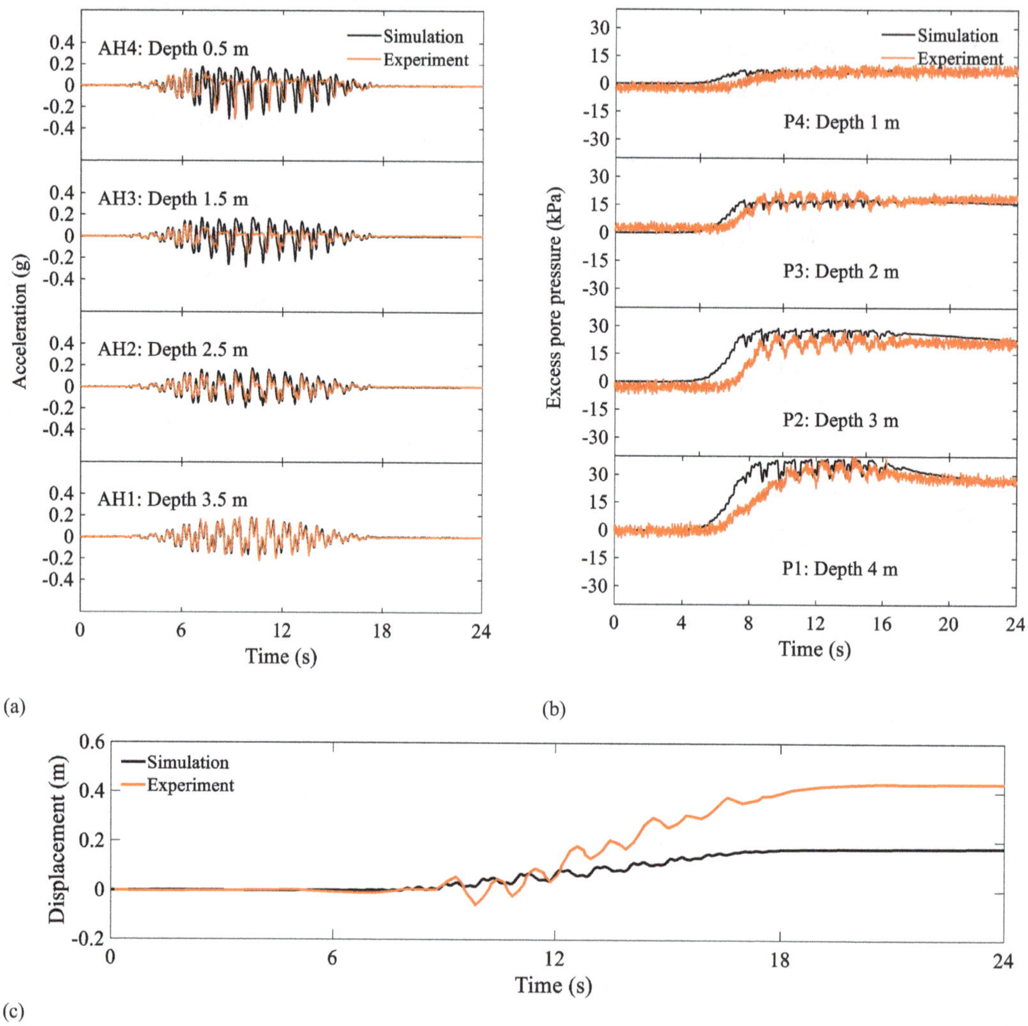


**Fig. 26.13** Measured and computed time histories of CU-2: (**a**) Acceleration; (**b**) Excess pore water pressure; (**c**) Displacement at the middle of ground surface

excess pore pressure for the numerical results are slightly higher than those of the measurement. In Fig. 26.15c, the computed lateral displacement is around 0.12 m and the lateral displacement at the midpoint of the ground in the experiment shows around 0.3 m at the end of shaking.

The numerical results for KAIST-1 are presented in Fig. 26.16. The computed time histories of acceleration and excess pore pressure are in good agreement with those from measurements (Fig. 26.16a, b). The time histories of AH3 and AH4 in measurement show more negative spikes due to dilation. Both the computed results and the measurements showed a consistent trend for the negative spikes of pore pressure generation due to dilation. In addition, the computed time histories of excess pore pressure at P1 and P2 reasonably captured the pore pressure dispassion at the end of shaking from 18 to 24 s. Figure 26.16c indicates that the computed lateral displacement at the midpoint of the ground surface is accumulating cycle-by-

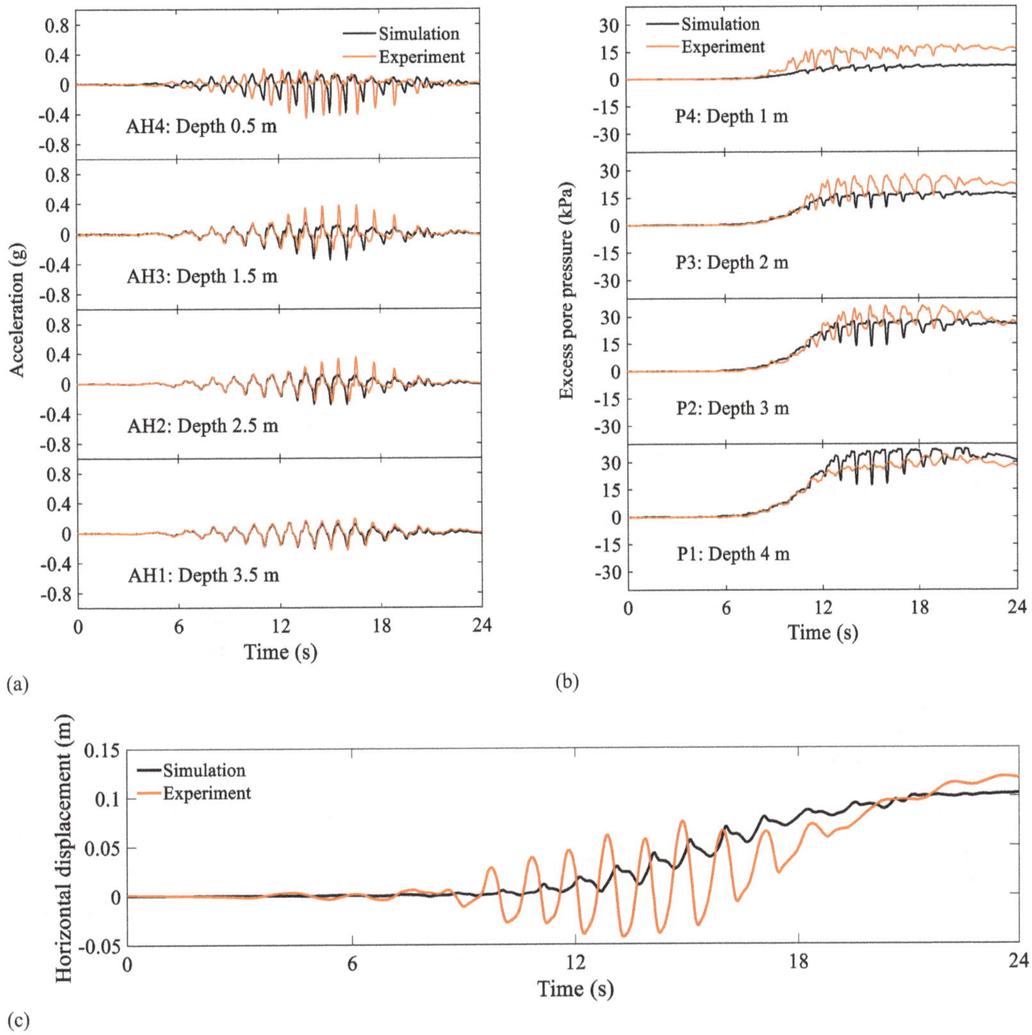


**Fig. 26.14** Measured and computed time histories of Ehime-2: (**a**) Acceleration; (**b**) excess pore water pressure; (**c**) displacement at the middle of ground surface

cycle and with residual lateral displacement around 0.05 m. Similar to NCU-3, the dynamic component of displacement computed from the acceleration history at AH4, is shown in Fig. 26.16c.

Figure 26.17 presents the results of KAIST-2. The computed time histories of acceleration and excess pore pressure are in good agreement with those from measurements (Fig. 26.17a, b). Compared to the time histories of the numerical simulations, the experiment results show shifted negative spikes in acceleration due to dilation. The overall trend of time histories of excess pore pressure is generally consistent with the measurements (Fig. 26.17b). The computed lateral displacement in Fig. 26.17c at the midpoint of the ground surface shows a good trend similar to the measurement and with similar residual displacement at the end of shaking.

Figure 26.18 shows the results of KyU-3. The computed time histories of acceleration (AH1-AH3) are in good agreement with those from measurements (Fig. 26.18a). The acceleration at AH4 near the ground surface showed very high

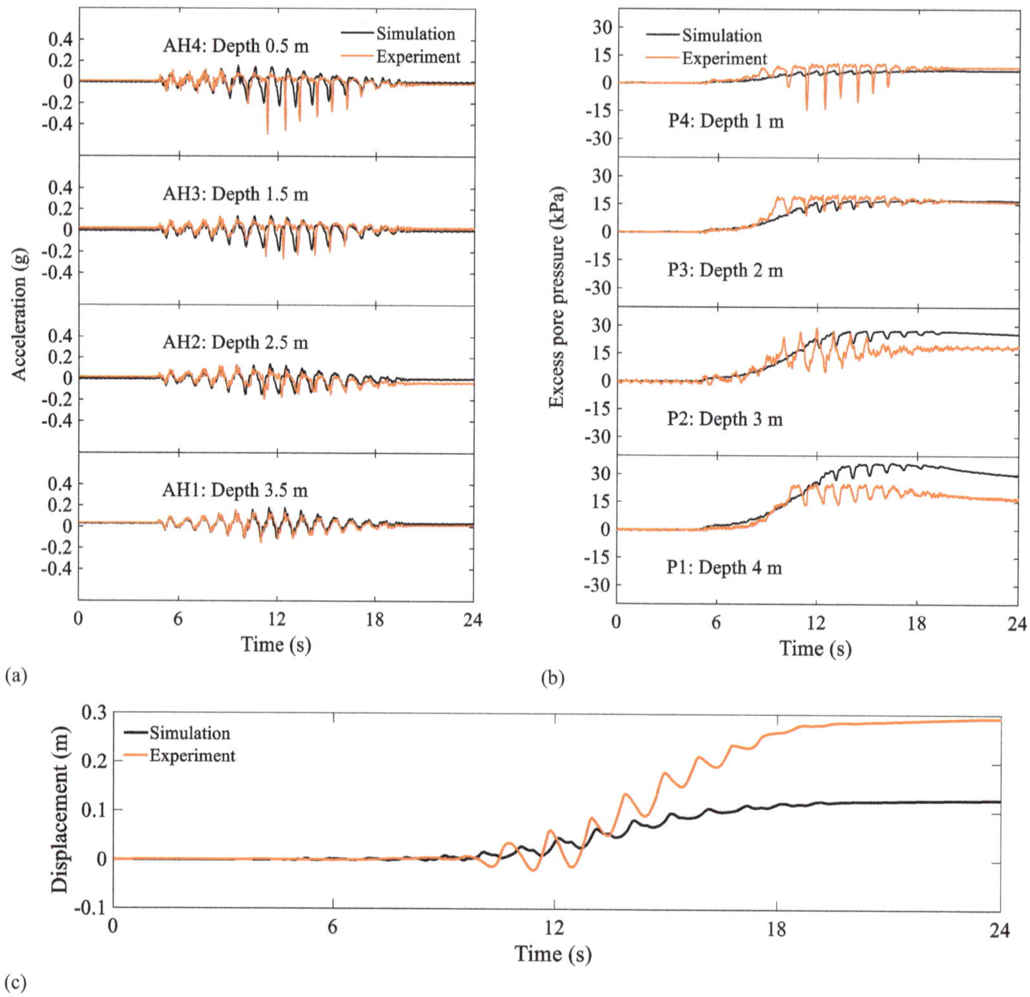


**Fig. 26.15** Measured and computed time histories of ZJU-2: (**a**) Acceleration; (**b**) excess pore water pressure; (**c**) displacement at middle of ground surface

negative spikes due to dilation and was not captured by the computed result. By comparing the time histories of pore pressure, a relatively good agreement is shown. Figure 26.18c indicates that the computed lateral displacement at the midpoint of the ground surface is accumulating cycle-by-cycle and with residual lateral displacement around 0.055 m. Similar to NCU-3 and KAIST-1, the dynamic component of displacement computed from the acceleration history at AH4 is shown in Fig. 26.18 (no residual lateral displacement).

## 26.6 Conclusions

The numerical simulations (modified Type-B) in LEAP-UCD-2017 centrifuge model tests for a liquefiable sloping ground conducted by various institutions are presented in this paper. The computations are performed using a pressure-dependent

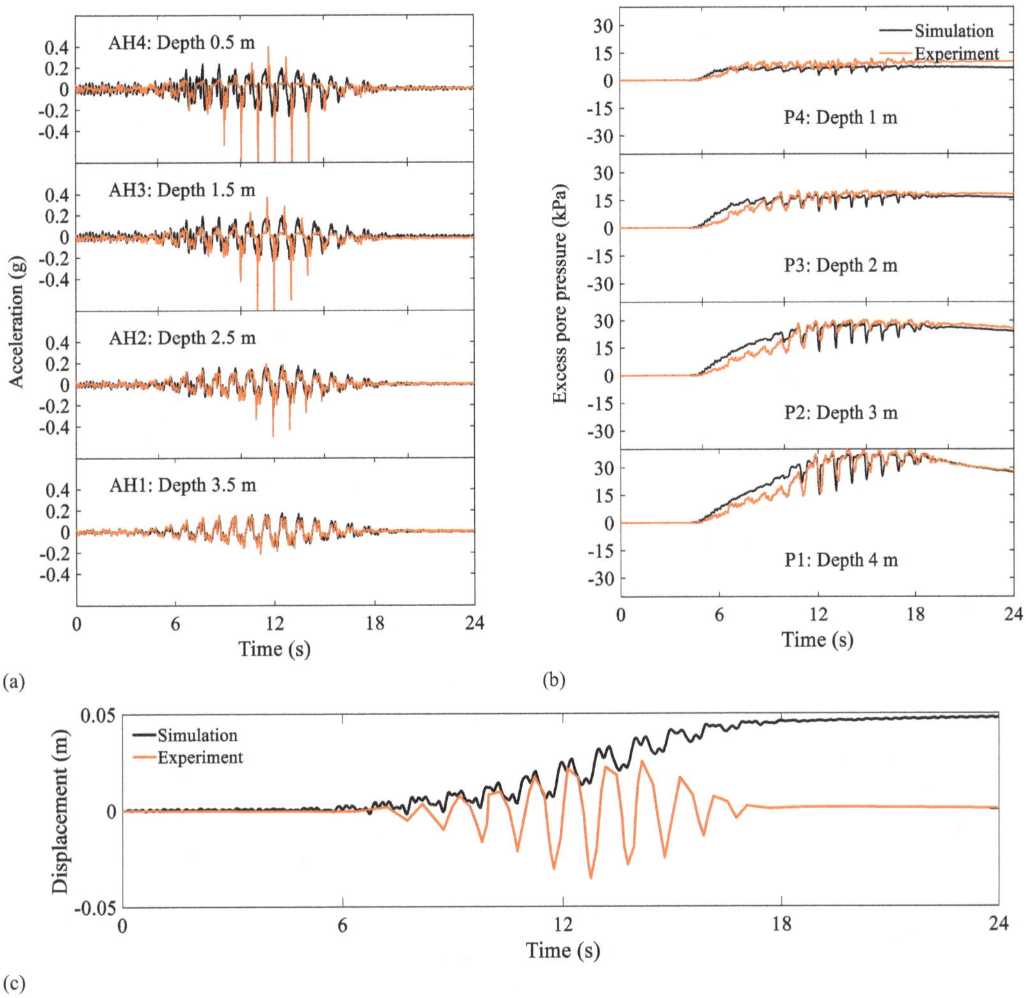


**Fig. 26.16** Measured and computed time histories of KAIST-1: (**a**) Acceleration; (**b**) excess pore water pressure; (**c**) displacement at the middle of ground surface

constitutive model (PressureDependMultiYield02) implemented with the characteristics of dilatancy and cyclic mobility. The soil parameters are determined based on a series of available stress-controlled cyclic triaxial tests (provided during the Type-A simulation phase) for matching the liquefaction strength curves. The computational framework and staged analysis procedure are presented as well. The primary conclusions can be drawn as follows:

1. Determination of soil model parameters in the Type-A simulation phase showed that the pressure-dependent model, i.e., PressureDependMultiYield02 material, generally resonably captured the liquefaction strength curves for the cyclic stress-controlled triaxial tests of the Ottawa F-65 sand at three different void ratios. Although the curve of deviatoric stress versus axial strain is symmetric due to the fact that the yield surface defined in principal stress space is conical, the soil model still can resonably capture the stress path, maximum axial strain, and the pore presssure generation for each triaxial cyclic stress-controlled test.

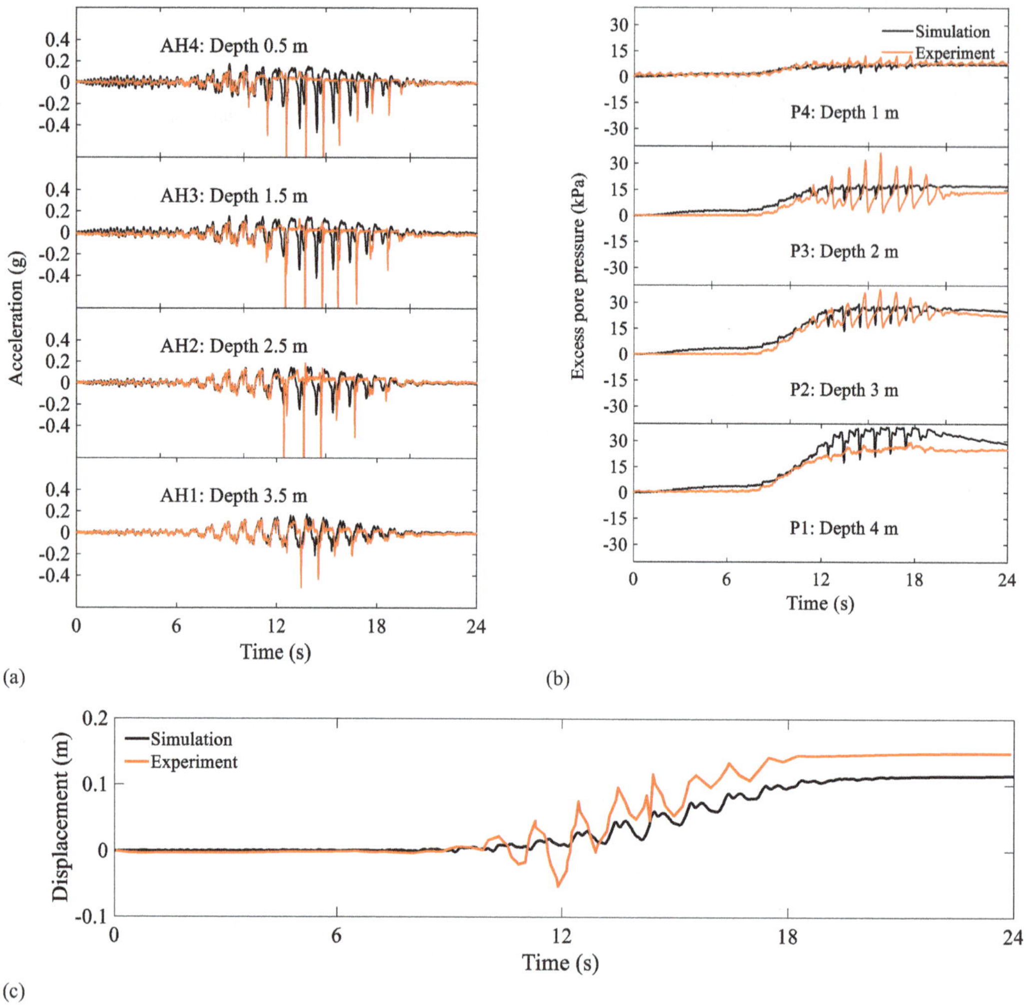


**Fig. 26.17** Measured and computed time histories of KAIST-2: (**a**) Acceleration; (**b**) excess pore water pressure; (**c**) displacement at the middle of ground surface

2. With the knowledge of centrifuge test results provided in the Type-C simulation phase, only the contraction parameter $c_1$ (controlling the pore pressure generation) was adjusted to better capture the dynamic response of the selected centrifuge tests, leading to what is presented herein as the modified Type-B simulations. The adjustment of this parameter is based on observations for all selected centrifuge test results provided in the Type-C simulation phase. This indicates that, to achieve better fits to the experimental data of the liquefiable sloping ground, the soil constitutive model parameters should be calibrated with some additional knowledge from the response of the centrifuge tests, as well as the triaxial tests on the soils.
3. Although the centrifuge tests in various institutions are not providing completely consistent results, measured time histories of acceleration and pore pressure for these experiments are reasonably captured by the numerical simulations, using the same soil constitutive model parameters. In addition, time histories of

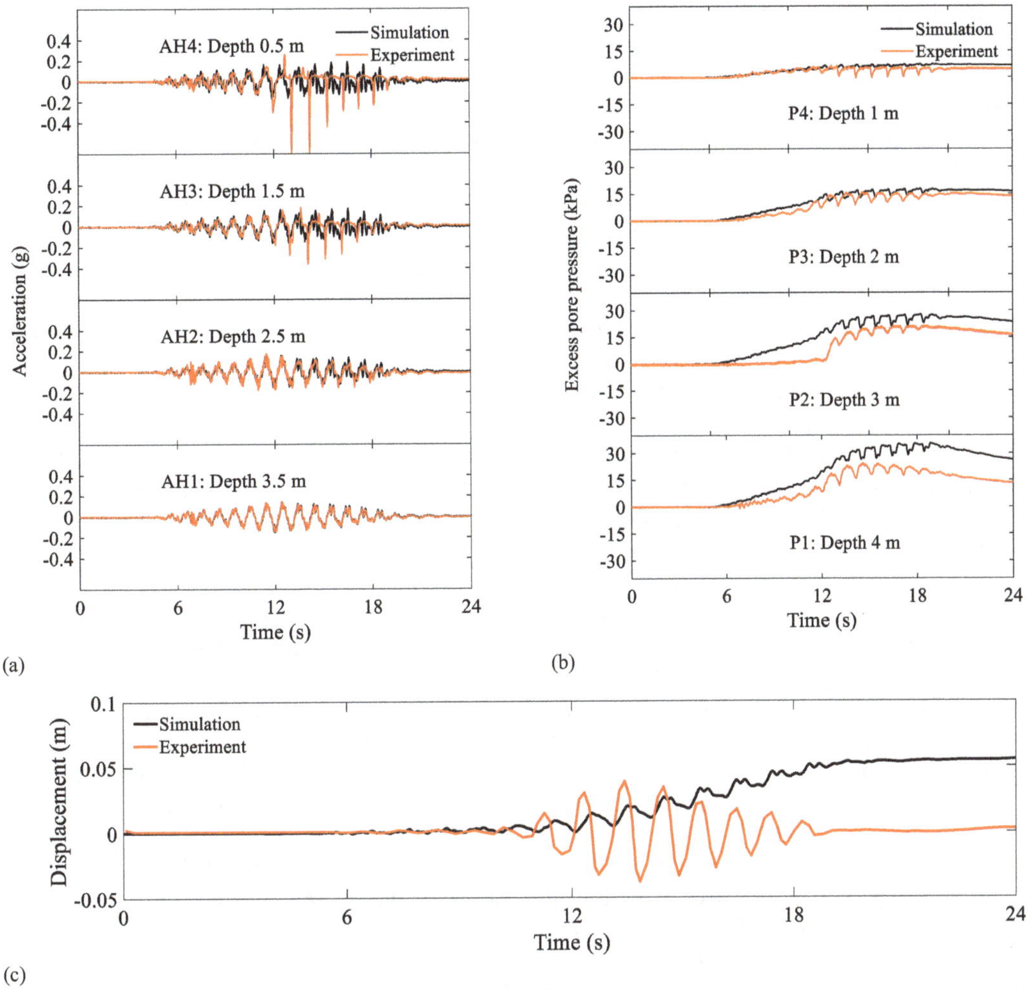


**Fig. 26.18** Measured and computed time histories of KyU-3: (**a**) Acceleration; (**b**) excess pore water pressure; (**c**) displacement at middle of ground surface

accumulated displacement are also captured to a reasonable level. Comparisons between these numerical simulations and measured results for each centrifuge test demonstrated that the PressureDependMultiYield02 soil model as well as the overall employed computational methodology have the potential to predict the response of liquefiable sloping ground, and subsequently realistically evaluate the seismic performance of analogous soil systems subjected to seismically induced liquefaction.

**Acknowledgments** The authors are grateful for the kind invitation by Professors Majid T. Manzari, Mourad Zeghal, and Bruce L. Kutter to participate in LEAP-UCD-2017. Partial funding of this effort was provided by Caltrans through a project in which Ottawa sand is being used for experimentation.

## References

Armstrong, R. J. (2018). Numerical analysis of LEAP centrifuge tests using a practice-based approach. *Soil Dynamics and Earthquake Engineering, 113*, 793–803.

Bastidas, A. M. P. (2016). *Ottawa F-65 Sand Characterization*. University of California, Davis.

Carlson, N. N., & Miller, K. (1998). Design and application of a gradient-weighted moving finite element code II: In two dimensions. *SIAM Journal on Scientific Computing, 19*(3), 766–798.

Chan, A. H. C. (1988). *A unified finite element solution to static and dynamic problems in geomechanics*. PhD Thesis, University College of Swansea.

Chen, W. F., & Mizuno, E. (1990). *Nonlinear analysis in soil mechanics, theory and implementation*. New York, NY: Elsevier.

El Ghoraiby, M. A., Park, H., & Manzari, M. T. (2017). *LEAP 2017: Soil Characterization and Element Tests for Ottawa F65 Sand*. Washington, DC: The George Washington University.

El Ghoraiby, M. A., Park, H., & Manzari, M. T. (2019). Physical and mechanical properties of Ottawa F65 sand. In B. Kutter et al. (Eds.), *Model tests and numerical simulations of liquefaction and lateral spreading: LEAP-UCD-2017*. New York: Springer.

Elgamal, A., Yang, Z., & Parra, E. (2003). Modeling of cyclic mobility in saturated cohesionless soils. *International Journal of Plasticity, 19*(6), 883–905.

Ghofrani, A., & Arduino, P. (2018). Prediction of LEAP centrifuge test results using a pressure-dependent bounding surface constitutive model. *Soil Dynamics and Earthquake Engineering, 113*, 758–770.

Iwan, W. D. (1967). On a class of models for the yielding behavior of continuous and composite systems. *Journal of Applied Mechanics, ASME, 34*, 612–617.

Khosravifar, A., Elgamal, A., Lu, J., & Li, J. (2018). A 3D model for earthquake-induced liquefaction triggering and post-liquefaction response. *Soil Dynamics and Earthquake Engineering, 110*, 43–52.

Kutter, B. L., Carey, T. J., Hashimoto, T., Manzari, M. T., Vasko, A., Zeghal, M., & Armstrong, R. J. (2015). LEAP databases for verification, validation, and calibration of codes for simulation of liquefaction. In *Sixth International Conference on Earthquake Geotechnical Engineering, Christchurch, New Zealand*.

Kutter, B. L., Carey, T. J., Hashimoto, T., Zeghal, M., Abdoun, T., Kokkali, P., Madabhushi, G., Haigh, S. K., d'Arezzo, F. B., Madabhushi, S., & Hung, W. Y. (2017). LEAP-GWU-2015 experiment specifications, results, and comparisons. *Soil Dynamics and Earthquake Engineering, 113*, 616–628.

Kutter, B. L., Carey, T. J., Zheng, B. L., Gavras, A., Stone, N., Zeghal, M., Abdoun, T., Korre, E., Manzari, M., Madabhushi, G. S., & Haigh, S. (2018). Twenty-four centrifuge tests to quantify sensitivity of lateral spreading to $D_r$ and PGA. *Geotechnical Earthquake Engineering and Soil Dynamics V GSP, 293*, 383–393.

Kutter, B. L., Manzari, M. T., Zeghal, M., Zhou, Y. G., & Armstrong, R. J. (2014). Proposed outline for LEAP verification and validation processes. In *Safety and Reliability: Methodology and Applications* (p. 99).

Lacy, S. (1986). *Numerical procedures for nonlinear transient analysis of two-phase soil system*. Ph.D. dissertation, Princeton University, Princeton, NJ.

Manzari, M. T., El Ghoraiby, M., Kutter, B. L., Zeghal, M., Abdoun, T., Arduino, P., Armstrong, R. J., Beaty, M., Carey, T., Chen, Y., & Ghofrani, A. (2018). Liquefaction experiment and analysis projects (LEAP): Summary of observations from the planning phase. *Soil Dynamics and Earthquake Engineering, 113*, 714–743.

Manzari, M. T., Kutter, B. L., Zeghal, M., Iai, S., Tobita, T., Madabhushi, S. P. G., Haigh, S. K., Mejia, L., Gutierrez, D. A., Armstrong, R. J., & Sharp, M. K. (2014). LEAP projects: Concept and challenges. In *Proceedings of 4th International Conference on Geotechnical Engineering for Disaster Mitigation and Rehabilitation* (p. 109–116).

McKenna, F., Scott, M. H., & Fenves, G. L. (2010). Nonlinear finite-element analysis software architecture using object composition. *Journal of Computing in Civil Engineering, 24*(1), 95–107.

Mroz, Z. (1967). On the description of anisotropic work hardening. *Journal of the Mechanics and Physics of Solids, 15*(3), 163–175.

Parra, E. (1996). *Numerical Modeling of Liquefaction and Lateral Ground Deformation Including Cyclic Mobility and Dilation Response in Soil Systems*. PhD Thesis. Rensselaer Polytechnic Institute.

Prevost, J. H. (1978). Plasticity theory for soil stress-strain behavior. *Journal of the Engineering Mechanics Division, 104*(5), 1177–1194.

Prevost, J. H. (1985). A simple plasticity theory for frictional cohesionless soils. *Soil Dynamics and Earthquake Engineering, 4*(1), 9–17.

Ueda, K., & Iai, S. (2018). Numerical predictions for centrifuge model tests of a liquefiable sloping ground using a strain space multiple mechanism model based on the finite strain theory. *Soil Dynamics and Earthquake Engineering, 113*, 771–792.

Vasko, A. (2015). *An Investigation into the Behavior of Ottawa Sand through Monotonic and Cyclic Shear Tests*. Washington, DC: The George Washington University.

Wotherspoon, L., Bradshaw, A., Green, R., Wood, C., Palermo, A., Cubrinovski, M., & Bradley, B. (2011). Performance of bridges during the 2010 Darfield and 2011 Christchurch earthquakes. *Seismological Research Letters, 82*(6), 950–964.

Yang, Z. (2000). *Numerical modeling of earthquake site response including dilation and liquefaction*. PhD Thesis, Columbia University.

Yang, Z., & Elgamal, A. (2002). Influence of permeability on liquefaction-induced shear deformation. *Journal of Engineering Mechanics, 128*(7), 720–729.

Yang, Z., Lu, J., & Elgamal, A. (2008). *OpenSees Soil Models and Solid-Fluid Fully Coupled Elements User Manual*. University of California, San Diego, La Jolla, CA.

Zeghal, M., Goswami, N., Kutter, B. L., Manzari, M. T., Abdoun, T., Arduino, P., Armstrong, R., Beaty, M., Chen, Y. M., Ghofrani, A., & Haigh, S. (2018). Stress-strain response of the LEAP-2015 centrifuge tests and numerical predictions. *Soil Dynamics and Earthquake Engineering, 113*, 804–818.

Ziotopoulou, K. (2018). Seismic response of liquefiable sloping ground: Class A and C numerical predictions of centrifuge model responses. *Soil Dynamics and Earthquake Engineering, 113*, 744–757.

# Chapter 27
# LEAP-UCD-2017 Simulation Team Fugro

**Dimitra Tsiaousi, Jose Ugalde, and Thaleia Travasarou**

**Abstract** Fugro participated in the Liquefaction Experiment and Analysis Projects (LEAP) by performing numerical simulations using two different constitutive models implemented in the software FLAC. Fugro developed a calibration framework based on soil-specific laboratory test data considering multiple elements of dynamic response such as liquefaction triggering criteria (i.e., number of cycles to a specified strain and pore pressure ratio threshold) as well as post-triggering strain accumulation rate. The calibrated model parameters were subsequently used to obtain "Type B (blind)" predictions of the centrifuge experiments with the opportunity to refine after the centrifuge results were provided (Type C simulations). Overall, Fugro's blind predictions captured the centrifuge test responses well with small refinements needed during Type C simulations. Estimated deformations were within a factor of about two compared to the observed. The overall good comparison provides confidence in the proposed calibration framework, which can be implemented and used for different sand types and project conditions.

## 27.1 Introduction

The LEAP 2017 simulation exercise consisted of the following four stages: (1) constitutive model calibration, (2) Type-B (blind) predictions, (3) Type-C simulations, and (4) sensitivity analyses. A series of stress-controlled cyclic triaxial laboratory tests were available (El Ghoraiby et al. 2017, 2019) and used in the calibration process. Next, the model parameters obtained during model calibration were used to analyze the centrifuge experiments, providing the base excitation and the soil density without any knowledge of the actual results (Type-B predictions). The simulation teams were subsequently provided with the centrifuge experiment results and simulations were refined, if necessary, to obtain better agreement (Type-C simulations).

D. Tsiaousi (✉) · J. Ugalde · T. Travasarou
Fugro, Walnut Creek, CA, USA
e-mail: dtsiaousi@fugro.com





B. Kutter et al. (eds.), *Model Tests and Numerical Simulations of Liquefaction and Lateral Spreading*, https://doi.org/10.1007/978-3-030-22818-7_27

Last, a series of sensitivity analyses were performed to assess the influence of soil relative density ($D_R$), motion intensity, and motion high frequency content on the simulation results.

## 27.2 Constitutive Model Calibration

### *27.2.1 Introduction*

Fugro calibrated two constitutive models, PM4SAND and UBCSAND, to the available cyclic stress-controlled triaxial laboratory tests on Ottawa F65 sand (El Ghoraiby et al. 2017, 2019). The cyclic triaxial tests were performed for three different soil densities. The three groups of specimens tested had void ratios of approximately 0.585, 0.542, and 0.515 corresponding to relative densities of about 65%, 80%, and 90%, respectively, based on a maximum void ratio ($e_{max}$) of 0.74 and a minimum void ratio ($e_{min}$) of 0.49 per Vasko as tabulated in Table 6 of the El Ghoraiby et al. (2017) report. At the time of this calibration process, no other test data was available.

Constitutive model calibration was performed considering multiple elements of response such as number of cycles to specified strain and pore pressure ratio thresholds as well as post-triggering strain accumulation rate.

### *27.2.2 Model Description, Parameters, and Implementation*

#### PM4SAND Constitutive Model

PM4SAND Version 3 constitutive model developed by Boulanger and Ziotopoulou (2015) was initially calibrated. Stress-controlled plane strain compression (PSC) test simulations were performed using PM4SAND and calibrated against the available stress-controlled triaxial test data. The soil elements were initially confined with vertical and horizontal stresses of one atmosphere, considering coefficient of earth pressures at rest of one ($K_o = 1$). Plane strain compression loading conditions were stress controlled by imposing a velocity until a specific stress ratio was reached and then the sign of velocity was changed.

The model parameters are grouped into two categories: a primary and a secondary set of parameters that may be modified from their default values in special circumstances. The three primary input parameters are the sand's apparent relative density $D_R$, the shear modulus coefficient $G_o$, and the contraction rate parameter $h_{po}$. The contraction rate parameter $h_{po}$ was used to calibrate the liquefaction triggering resistance, and the shear modulus parameter $G_o$ was mainly used to calibrated the strain rate accumulation. PM4SAND calibrated model parameters are tabulated in Table 27.1. A detailed description of the primary and secondary input parameters

**Table 27.1** PM4SAND calibrated model parameters

| Parameter | Function | Values |
|---|---|---|
| $D_R$ | Relative density | 65%, 80%, and 90% for $e_o$ of 0.585, 0.542, and 0.515, respectively based on $e_{max}$ ~ 0.74 and $e_{min}$ ~ 0.49 |
| $G_o$ | Shear modulus coefficient | 625, 1321, and 2001 for void ratio $e_o$ of 0.585, 0.542, and 0.515, respectively based on the equation: $G_o = 167 \times (N_{1,60} + 2.5)^{0.5}$ and using multiplier of 0.8, 1.4, and 1.9 for $e_o$ of 0.585, 0.542, and 0.515, respectively |
| $h_{po}$ | Contraction rate parameter | 0.07, 0.038, and 0.03 for void ratio $e_o$ of 0.585, 0.542, and 0.515, respectively |
| $e_{max}$ and $e_{min}$ | Maximum and minimum void ratios | $e_{max}$ ~ 0.74 and $e_{min}$ ~ 0.49 based on lab test data per Table 6 of the El Ghoraiby et al. (2017) report |

and default values is provided in Tables 4.1 and 4.2 in Boulanger and Ziotopoulou (2015).

Plane strain compression tests were simulated for the relative densities of 65%, 80%, and 90% for various cyclic stress ratios (CSR) in order to compare the plane strain simulation results with the triaxial test results. Each plane strain compression simulation was compared to a triaxial test of a soil sample with similar relative density and stress ratio (CSR) in terms of liquefaction triggering and strain accumulation rate. Liquefaction strength curves were plotted for both the actual triaxial tests and the simulated plane strain compression tests in terms of numbers of cycles to 98% excess pore pressure ratio, and 2 and 5% double amplitude axial strain.

Figure 27.1 presents the observed (from the tests—red) and simulated (with calibrated soil properties—blue) liquefaction strength curve for the case of sand with $D_R$ of 65%. Figure 27.2 presents the observed and simulated response in terms of (a) deviatoric stress versus mean effective stress, (b) deviatoric stress versus axial strain, (c) mean effective stress versus axial strain, and (d) pore pressure ratio versus number of cycles for all the plane strain compression simulations and superimposed with the respective triaxial tests for comparison purposes for the case of $D_R$ of 65% and CSR of 0.2.

The constitutive model was also calibrated based on the average axial strain accumulation rate (SAR). SAR is defined in units of percent of strain per cycle and indicates the half double-amplitude strain accumulation per cycle (average of triaxial extension and compression) after liquefaction triggering. Figure 27.3 presents a comparison of SAR observed during the triaxial tests and that from the simulations using the calibrated parameters for PM4SAND for the case of $D_R$ of 65%. For the PSC simulations, the strain accumulation rate (SAR) increases consistently with larger CSR. However, this trend is not consistent in the actual triaxial tests. Given this discrepancy, our goal was to approximate the average rate of SAR for CSR values between 0.14 and 0.2, which would be more relevant for the subsequent centrifuge tests. The observed versus simulated average rate of strain accumulation over these CSR values are 0.39% versus 0.42% strain per cycle, which are relatively in good agreement.

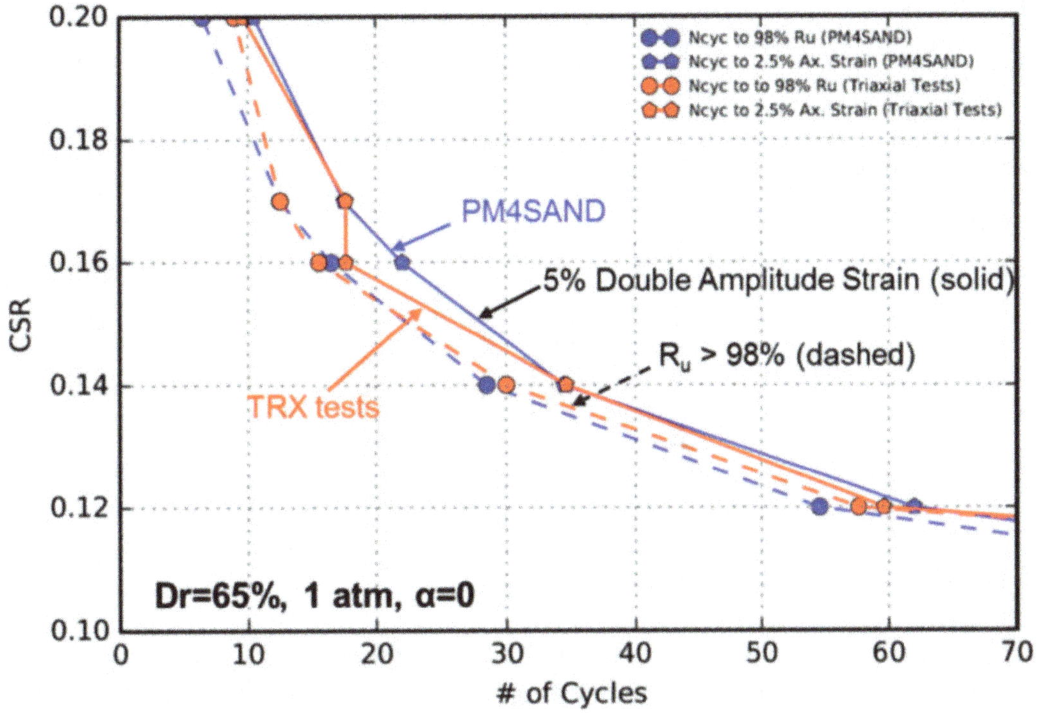


**Fig. 27.1** Liquefaction strength curve for $D_R \sim 65\%$: triaxial test data versus plane strain compression simulation results using PM4SAND

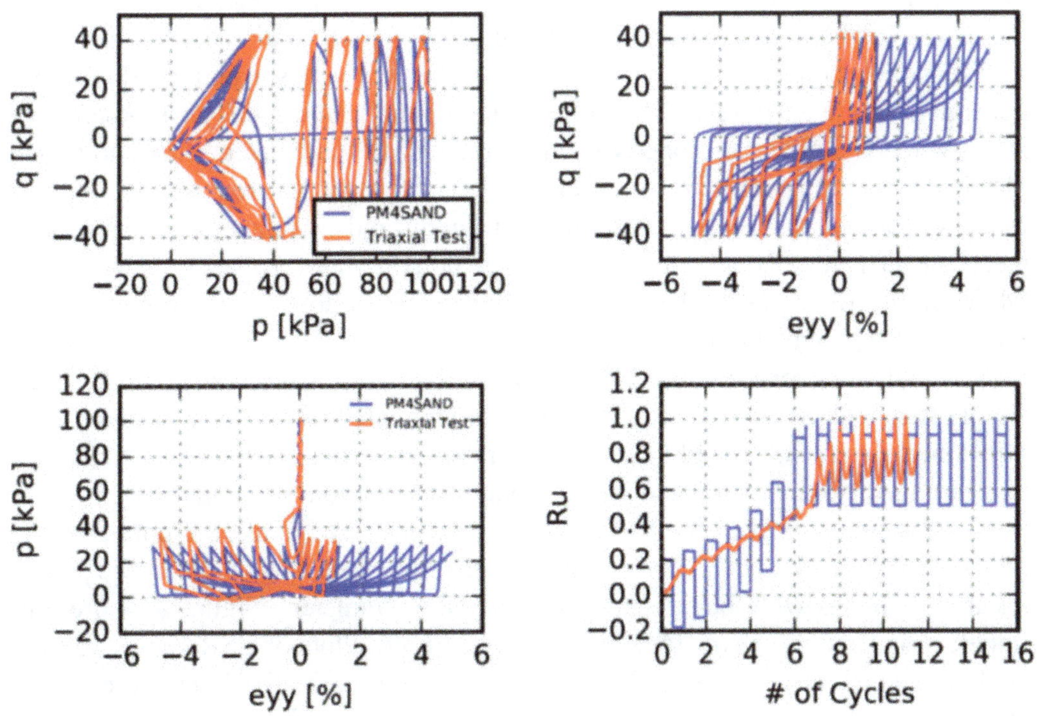


**Fig. 27.2** Deviatoric stress versus mean effective stress, deviatoric stress versus axial strains, mean effective stress versus axial strains and pore pressure ratio versus number of cycles for $D_R \sim 65\%$ and CSR = 0.2, triaxial test data versus plane strain compression simulation using PM4SAND

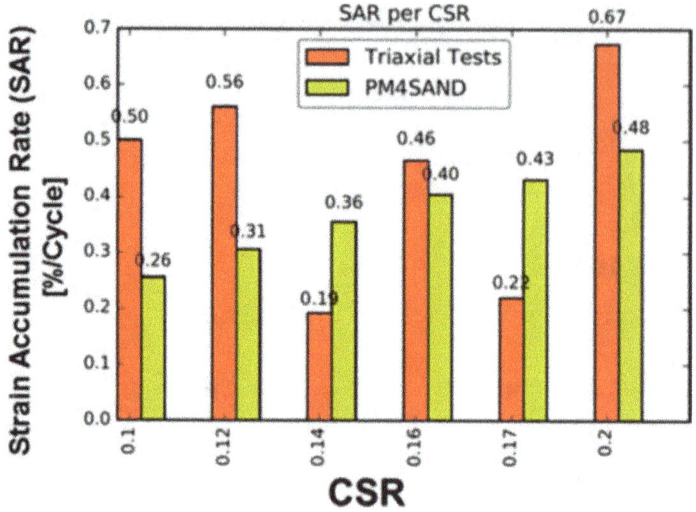


**Fig. 27.3** Strain accumulation rate [%Strain/Ncyc] for $D_R \sim 65\%$ and different CSR values, triaxial test data versus plane strain compression simulation using PM4SAND—calibration process example

## UBCSAND Constitutive Model

The UBCSAND constitutive model, which has been modified from the commonly available 904ar version by Fugro and Peter Burn, was also used in this study. This model has been modified, subsequently calibrated and validated by Fugro (Giannakou et al. 2011), and used in major projects such as the BART Offshore Transbay Tube Seismic Retrofit (Travasarou et al. 2011). The primary modification consists of introducing one additional model parameter (hfac4), which controls the rate of shear strain accumulation after liquefaction triggering. Detailed description of the model can be found in Beaty and Byrne (1998) and Byrne et al. (2004).

The model is fully defined by means of nine elastic and plastic parameters that can generally capture the observed response from monotonic and cyclic tests. Apart from $N_{1,60}$ which is a physical property of sand, four model parameters (hfac1 to hfac4; plastic hardeners) which are not related to physical properties are available to allow model calibration to observed behavior of different sands. Among these four parameters, the first three control the triggering of liquefaction (primarily the first two) while the fourth controls the post-trigger response. Table 27.2 summarizes the UBCSAND calibrated model parameters.

For calibration of the UBCSAND constitutive model, undrained cyclic direct simple shear (DSS) single element tests were simulated and compared against the DSS simulations using the PM4SAND calibrated properties. The UBCSAND model was not directly calibrated against the triaxial tests due to time limitations.

Liquefaction strength curves in terms of number of cycles to 98% pore pressure ratio, 1% shear strain, and 3% shear strain were compared against the cyclic simple shear simulations with the calibrated PM4SAND parameters. Figure 27.4 presents liquefaction strength curves for sand with $D_R$ of 65%. Plots of CSR versus normalized vertical effective stress, CSR versus shear strain, normalized vertical effective stress versus shear strain, and pore pressure ratio versus number of cycles are also provided in Fig. 27.5. For comparison purposes, the results from the cyclic DSS simulations using PM4SAND with the calibrated parameters are superimposed with

**Table 27.2** UBCSAND calibrated model parameters

| Parameter | Function | Values |
|---|---|---|
| $N_{1,60}$ | Normalized and corrected SPT Blowcount | 19, 29, and 37 for $D_R$ of 65%, 80%, and 90%, respectively |
| KGE | Elastic shear modulus multiplier | 1166, 1339, and 1448 for $D_R$ of 65%, 80%, and 90%, respectively based on the equation: $KGE = 21.7 \times 20 \times (N_{1,60})^{0.333}$ |
| me | Elastic shear exponent | 0.5 |
| KB | Elastic bulk modulus multiplier | 816, 937, and 1013 for $D_R$ of 65%, 80%, and 90%, respectively based on the equation: $KB = KGE \times 0.7$ |
| ne | Elastic bulk exponent | 0.5 |
| KGP | Plastic bulk modulus multiplier | 1421, 3581, and 6130 for $D_R$ of 65%, 80%, and 90%, respectively based on the equation: $KGP = KGE \times (N_{1,60})^2 \times 0.003 + 100.0$ |
| np | Plastic bulk exponent | 0.4 |
| $\phi_{cs}$ (degrees) | Critical state friction angle | 33 |
| $\phi_{peak}$ (degrees) | Peak friction angle | 36.9, 38.9, and 40.5 for $D_R$ of 65%, 80%, and 90%, respectively |
| $R_f$ | Failure ratio | 0.81, 0.71, and 0.63 for $D_R$ of 65%, 80%, and 90%, respectively |
| hfac1 | Controls liquefaction triggering | 0.45 |
| hfac2 | Controls liquefaction triggering | 0.5 |
| hfac3 | Controls liquefaction triggering | 1.0 |
| hfac4 | Controls post-trigger response | 2.2, 2.0, and 1.8 for $D_R$ of 65%, 80%, and 90%, respectively |

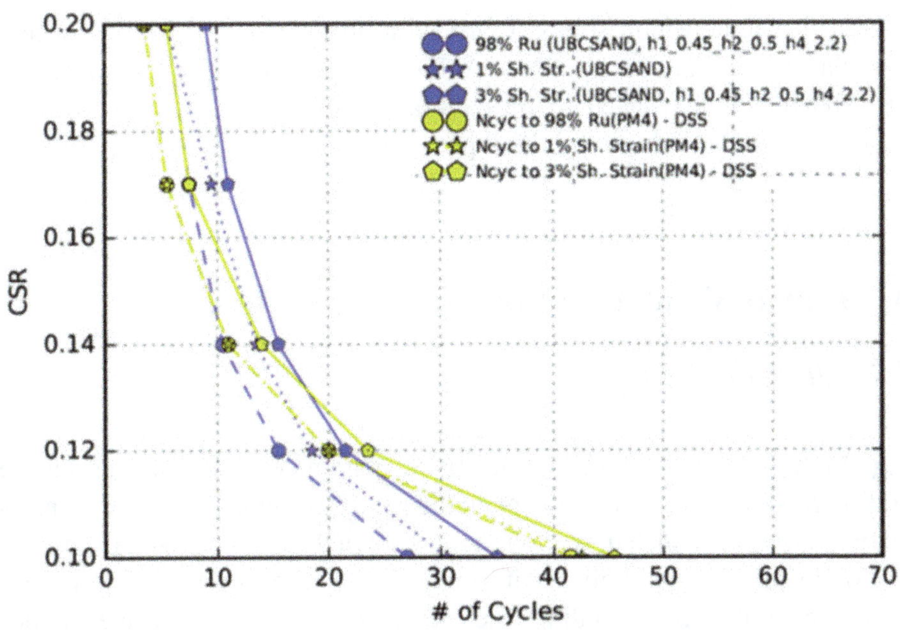


**Fig. 27.4** Liquefaction strength curve in DSS conditions for $D_R \sim 65\%$, and cyclic direct simple shear simulations using PM4SAND and UBCSAND versus triaxial test data and plane strain compression simulations using PM4SAND

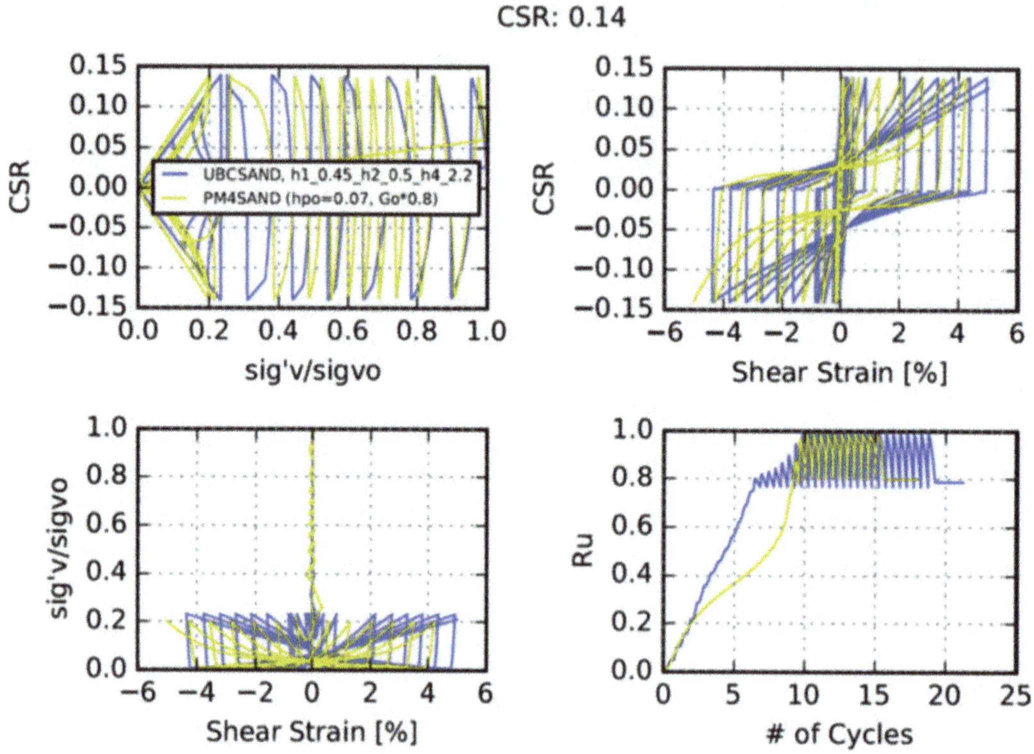


**Fig. 27.5** CSR versus normalized vertical effective stress, CSR versus shear strain, normalized vertical effective stress versus shear strain and excess pore pressure ratio versus number of cycles for $D_R \sim 65\%$ and CSR = 0.14, DSS simulations using PM4SAND and UBCSAND

the results from the DSS simulations using UBCSAND with the calibrated parameters.

Overall, the liquefaction strength curves implied by UBCSAND are generally steeper than those implied by PM4SAND and the laboratory tests. Hence, the UBCSAND model may need to be calibrated with a tighter target CSR range, corresponding to that induced by the input motions.

## 27.3 Type B Simulations (Blind Predictions)

### *27.3.1 Introduction*

During this phase of the project, the model parameters obtained as part of the calibration process were used to simulate the centrifuge experiments. The numerical simulations were conducted without any knowledge of the actual results. The simulation teams were only given information about the main characteristics of the centrifuge experiments, including the geometry of the centrifuge model, the

**Table 27.3** Summary of centrifuge experiments selected for LEAP-2017 Type-B simulations

| Experiment | Cent. Acc. ($g$) | $\mu/g$ | $\rho_d$ (kg/m$^3$) | $D_R$ (%) |
|---|---|---|---|---|
| CU-2 | 40 | 1.175 | 1605.8 | 46 |
| Ehime-2[a] | 40 | 1 | 1656.55 | 64 |
| KAIST-1[a] | 40 | 0.897 | 1701.2 | 78 |
| KAIST-2 | 40 | 0.936 | 1592.5 | 42 |
| KyU-3 | 44.4 | 0.991 | 1637 | 57 |
| NCU-3[a] | 26 | 1 | 1652 | 62 |
| UCD-1[a] | 43 | 1 | 1665 | 67 |
| UCD-3[a] | 43 | 1 | 1658 | 64 |
| ZJU-2[a] | 30 | 1 | 1606 | 46 |

[a]Selected subset of centrifuge tests considered for Type-B simulations

centrifugal acceleration, the viscosity of the pore fluid ($\mu$), the base excitation (horizontal and vertical components) and the achieved soil density.

Nine centrifuge experiments were selected by the organizing committee for LEAP-2017 Type-B simulations. The main characteristics of these experiments such as the centrifugal acceleration in $g$, the ratio of the pore fluid viscosity over the gravity ($\mu/g$), and the achieved dry soil density ($\rho_d$) are summarized in Table 27.3 This table also contains the relative density ($D_R$) which is calculated for each test according to Eq. 27.1:

$$D_R = \left( \frac{1/\rho_{d,\min} - 1/\rho_d}{1/\rho_{d,\min} - 1/\rho_{d,\max}} \right) \tag{27.1}$$

where $\rho_{d,\min}$ and $\rho_{d,\min}$ are considered equal to 1485 and 1773 kg/m$^3$, respectively based on the average values of LEAP-2017 laboratory data on Ottawa F65 sand (El Ghoraiby et al. 2017). Six centrifuge experiments were selected by Fugro to be simulated at this phase due to time limitations: Tests Ehime-2, NCU-3, UCD-1, and UCD-3 with a relative density ($D_R$) of approximately 65%, test KAIST-1 with a $D_R$ of approximately 80%, and test ZJU-2 with a $D_R$ of approximately 50%.

### 27.3.2 Analysis Platform

The analysis platform used for the numerical simulation of the centrifuge experiments is FLAC2D, Version 7.0 for PM4Sand and Version 8.0 for UBCSAND, respectively (Itasca 2011, 2017). FLAC is a two-dimensional explicit finite difference program for engineering mechanics computation, following a “mixed discretization” scheme (Marti and Cundall 1982).

### 27.3.3 Numerical Modeling

Numerical modeling in FLAC was performed at the prototype scale. The model is 20 m long and 4 m high at the centerline. The soil surface forms a 5-degree slope. A horizontal ground water table with coordinates (−10, 0.875) and (10, 0.875) was assumed so that the slope is submerged. A baseline schematic of the centrifuge in prototype scale, illustrating the locations of pore pressure transducers, accelerometers and prescribed locations of displacement measurements, is presented in Fig. 27.6a. The grid and its dimensions of the numerical model are shown in Fig. 27.6b. The grid was built so that the grid points correspond to the locations of the accelerometers and the pore water pressure sensors illustrated in Fig. 27.6a.

A triangular pressure was applied at the soil surface simulating the load due to the water mass. Initially, during the gravity loading, Mohr-Coulomb properties were assigned to soil. A gravitational field with a vertical acceleration equal to 9.81 m/s$^2$ was applied and once equilibrium was reached, PM4SAND or UBCSAND constitutive models were assigned to soil elements and equilibrium was reached again. In the next step, the groundwater flow mode was set on and equilibrium was reached once more. During this process of gravity loading, the error tolerance was on the order of $10^{-4}$. The pore pressure was fixed at the soil surface. The horizontal displacement was also fixed at the vertical boundaries of the model. Last, the base of the model was constrained to move neither horizontally nor vertically.

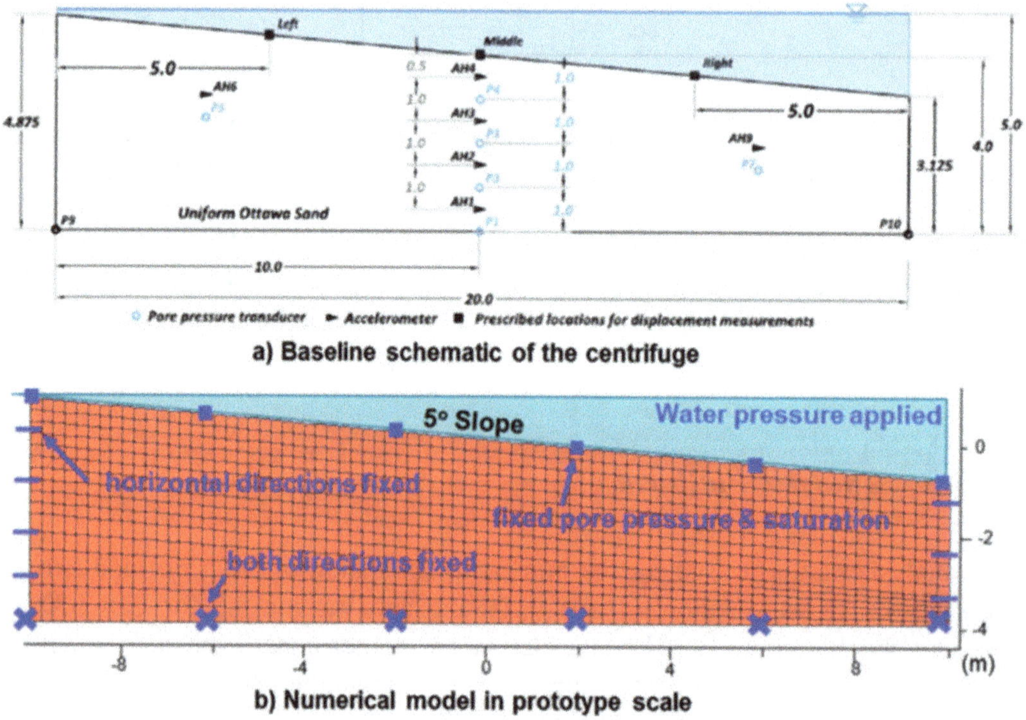


**Fig. 27.6** (**a**) Baseline schematic of the centrifuge experiments, (**b**) Grid, model geometry, and boundary conditions used for numerical simulations of centrifuge experiments

After the end of the gravity loading phase, horizontal and vertical mean input motions, obtained from recorders AH11–12 and AV1–2, respectively, were applied at the base and the vertical boundaries of the model. Rayleigh damping (stiffness- and mass-proportional) with a minimum value of 0.5% at 3.3 Hz was assigned to the soil elements. After the end of shaking, the analyses continued until the excess pore water pressures dissipated fully. This procedure was followed for each of the simulated centrifuge tests. For each simulation, the corresponding input motions were baseline corrected and then applied to the model.

### 27.3.4 Material Properties and Constitutive Model Parameters

The material properties assumed in the numerical simulations, such as permeability, the water bulk modulus, the soil relative density and the dry soil density, are given in Table 27.4. These model properties were adopted following sensitivity analyses evaluating the effect of changes in water bulk modulus, permeability anisotropy, and Raleigh damping.

After shaking, the model was allowed to reconsolidate until excess pore pressures dissipated and initial effective stresses were re-established. After end of shaking, permeability was scaled by ten to reduce calculation time, and the time was also scaled in the post-processing phase to ensure that the results are not affected compared to the solutions that would have been obtained by not scaling the permeability.

As aforementioned, the gravity loading was applied in two stages. In the first stage, Mohr-Coulomb properties were assigned to the soil elements. Basically, the gravity loading was initially applied elastically since no failure occurred.

In the second stage of gravity loading, as well as the dynamic part of the numerical analyses, advanced constitutive models were assigned to the soil elements, such as PM4SAND and UBCSAND models. Constitutive models PM4SAND and UBCSAND were calibrated for soil relative densities of 65%, 80%, and 90%. For the Type-B simulations of centrifuge tests with $D_R$ of about 65% (Ehime-2, NCU-3, UCD-1, and UCD-3) and 80% (KAIST-1), the initial calibrated parameters for a $D_R$ of 65% and 80% were used, respectively. There

**Table 27.4** Material properties used for numerical simulations

| Centrifuge test | Permeability (cm/s) | Water bulk modulus (kPa) | $D_R$ (%) | $\rho_d$ (kg/m$^3$) |
|---|---|---|---|---|
| Ehime-2 | 0.0118 | 480,000 | 65 | 1657 |
| NCU-3 | | | | 1652 |
| UCD-1 | | | | 1665 |
| UCD-3 | | | | 1658 |
| KAIST-1 | | | 80 | 1701 |
| ZJU-2 | | | 50 | 1606 |

was no calibration performed for a $D_R$ of 50%; hence, for Type-B simulation of ZJU-2 with a $D_R$ of about 50%, the calibrated parameters corresponding to a $D_R$ of 65% were used while a $D_R$ of 50% was used as input to the simulation. For PM4SAND and UBCSAND constitutive models, the calibrated model parameters presented in Tables 27.1 and 27.2, respectively, were used based on the idealized relative density of each centrifuge test per Table 27.4.

### 27.3.5 Simulation Results

Figure 27.7 illustrates a comparison of the horizontal displacement and excess pore pressure ratio time histories at the middle surface of the model for Type-B simulation results using constitutive models UBCSAND, and PM4SAND versus Ehime 2 centrifuge experiment results. Both constitutive models and the centrifuge test results developed excess pore pressure ratio in excess of 80% while the rate of excess pore pressure dissipation recorded in the centrifuge is bounded by the two constitutive model predictions. Figure 27.8 presents a comparison of permanent horizontal displacement at the middle surface of the numerical models for Type-B simulation results using PM4SAND and UBCSAND versus the centrifuge experiment results for all cases simulated. Simulations with PM4SAND result in displacements closer to those measured and within a factor of 2 from the actual measurements with the exception of NCU3. Simulations with UBCSAND resulted in generally larger displacements than the actual measurements by a factor of about 2. This is likely attributed to the model calibration which was initially targeted at larger CSR ranges compared to those induced on average by the input ground motions. Overall, the blind predictions captured the centrifuge tests' behavior well and bounded the centrifuge response in most cases, indicating the benefits of using more than one constitutive model when performing numerical modeling.

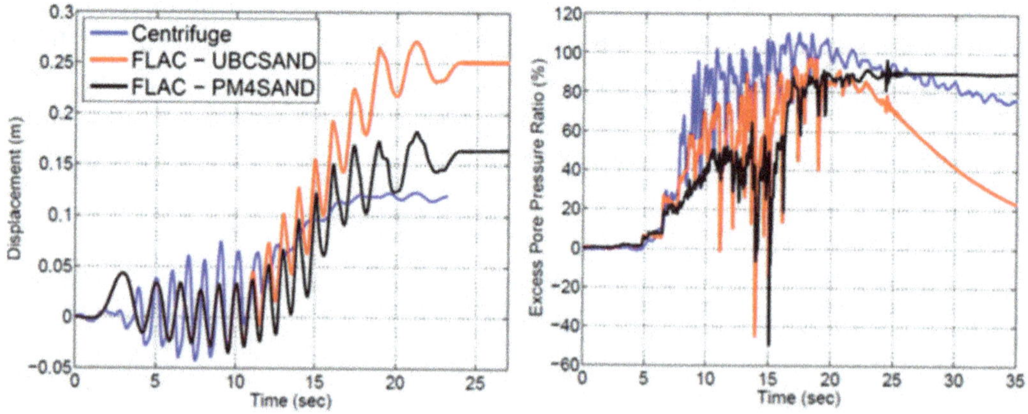


**Fig. 27.7** Comparison of horizontal displacement and excess pore pressure ratio time histories at the middle surface of the model for Type-B simulation results using PM4SAND and UBCSAND versus Ehime 2 centrifuge experiment results

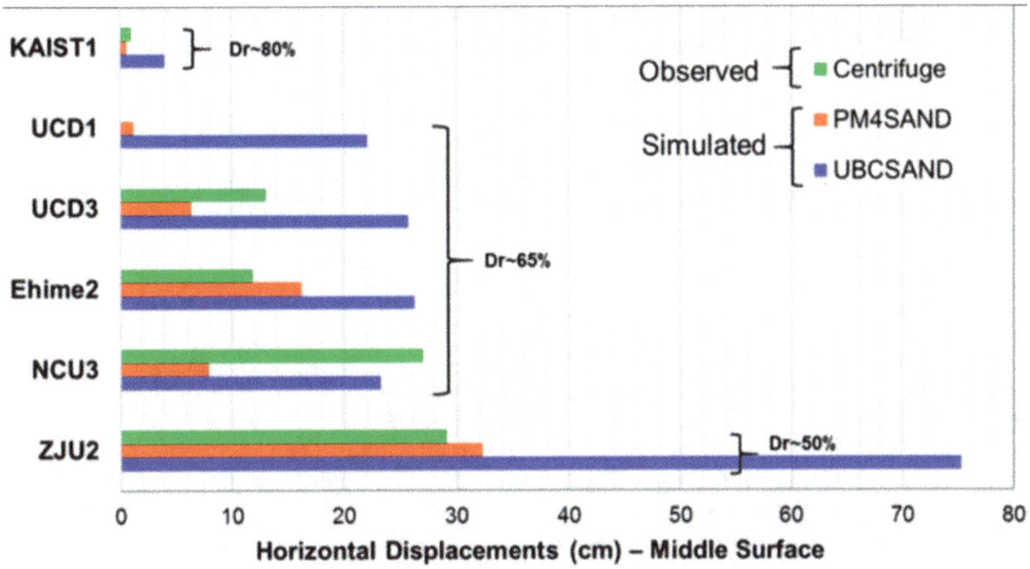


**Fig. 27.8** Comparison of permanent horizontal displacement at the middle surface of the model for Type-B simulation results using PM4SAND and UBCSAND versus centrifuge experiment results for all cases simulated

## 27.4 Type-C Simulations

### *27.4.1 Introduction and Constitutive Model Parameters*

At this stage, the centrifuge experiment results were released to the teams who were provided with the opportunity to refine the simulations, if necessary. In the Type-C simulations, the UBCSAND constitutive model parameters hfac1 and hfac2 were increased by 40% (hfac1 $= 0.45 \times 1.4 = 0.63$ & hfac2 $= 0.5 \times 1.4 = 0.7$) and the rest of the parameters are the same with the ones presented in Table 27.2. The PM4SAND constitutive model parameters were not modified and are the same with the ones presented in Table 27.1. As discussed before, the liquefaction strength curves implied by the UBCSAND model are steeper than the ones implied by PM4SAND and the laboratory tests. Hence, the UBCSAND model had to be recalibrated for a specific CSR range. Figure 27.9 illustrates the liquefaction strength curves from DSS simulations at a vertical effective stress of 0.2 atm for constitutive models PM4SAND (red) and UBCSAND using Type-B (blue) and Type-C (cyan) calibration parameters.

### *27.4.2 Simulation Results*

Recalibrating UBCSAND constitutive model parameters resulted in permanent displacements closer to the actual measurements. Figure 27.10 illustrates a

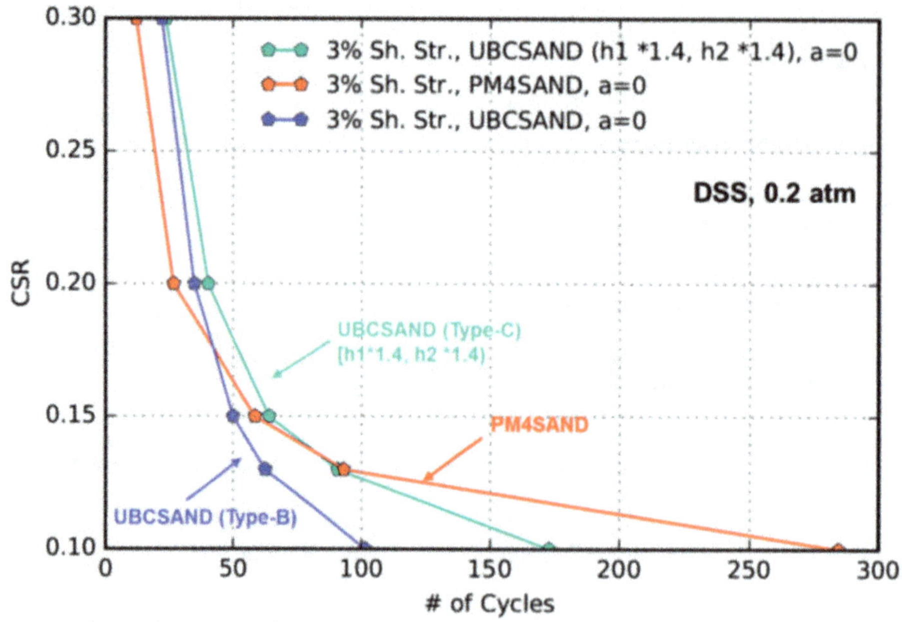


**Fig. 27.9** Liquefaction strength curve for $D_R \sim 65\%$ and normal effective stress of 0.2 atm, direct simple shear simulations for PM4SAND and UBCSAND using calibration parameters from Type-B and Type-C calibrations

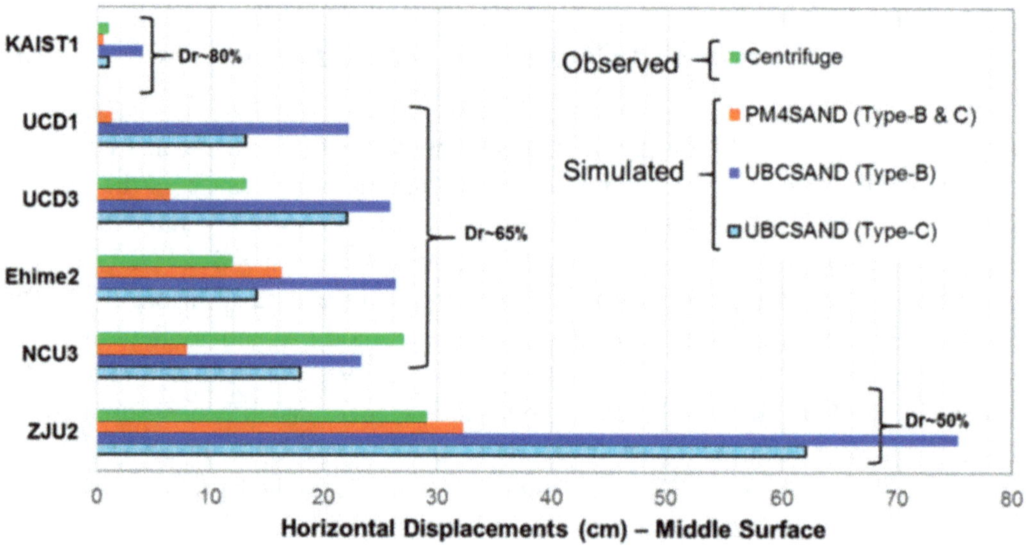


**Fig. 27.10** Comparison of permanent horizontal displacement at the middle surface of the model for Type-B and Type-C simulation results using PM4SAND and UBCSAND versus centrifuge experiment results for all cases simulated

comparison of permanent horizontal displacement at the middle surface of the model for Type-B and Type-C simulation results for all cases simulated. The permanent displacements predicted from both constitutive models are within a factor of 2 from the actual measurements.

## 27.5 Sensitivity Study

### 27.5.1 Introduction

A sensitivity analysis study was performed to illustrate the sensitivity of the simulations to key input parameters such as relative density, ground motion intensity and the influence of superimposed high frequencies in the ground motions. The properties of the seven sensitivity analyses (NS-1 to NS-7) are tabulated in Table 27.5. All sensitivity analyses were performed using the constitutive model parameters that were used in Type-C simulations discussed in the previous section. Simulations NS-1, NS-4, NS-5, NS-6, and NS-7 with soil relative density of 65% were performed using the Type-C calibrated parameters for the case of $D_R$ of 65%. Simulation NS-2 was performed using the Type-C calibrated parameters for the case of $D_R$ of 65% but using as input a $D_R$ of 50%. Simulation NS-3 was performed using the Type-C calibrated parameters for the case of $D_R$ of 80% but using as input a $D_R$ of 75%.

### 27.5.2 Sensitivity Analyses Results

Sensitivity analyses NS-1 to NS-3 were used to explore the sensitivity of simulations to relative density. Sensitivity analyses NS-1, 4, and 5 were used to explore the sensitivity of simulations to motion intensity. Sensitivity analyses NS-6 and NS-7 were used to assess the influence of superimposed ground motion high frequencies on simulation results. Table 27.5 presents the resulting horizontal displacement at the middle of the model surface and the duration of liquefaction at point P4 (P4 location: middle at 1-m depth) after the end of shaking.

Figure 27.11 presents a comparison of sensitivity analyses using PM4SAND in terms of excess pore pressure ratio, horizontal displacement time history and amplification ratio at the middle surface of the model for NS1, NS6, and NS7. Input ground motions for analyses NS6 and NS7 have superimposed high frequencies, and input motion NS7 is equal to input motion NS6 multiplied by 1.43. This figure indicates that superimposed high frequencies in the input ground motion as well as motion intensity can have a significant effect on the model response.

The predicted displacements were evaluated against various intensity parameters, and meaningful trends were discovered between permanent displacements and input ground motion Arias intensity. Figure 27.12 presents the predicted horizontal displacements at the middle surface of the models versus Arias intensity for sensitivity analyses with $D_R$ of 65%. As expected, for higher Arias intensity motions, the predicted permanent displacements are higher. It appears that motions with high frequency content (NS-6 and NS-7) can result in greater displacements than motions with lower high frequency content and similar Arias intensity.

The sensitivity analyses also show that soil relative density has a significant impact on the model behavior.

**Table 27.5** Sensitivity analyses: inputs and results

| Simulation # | | NS-1 | NS-2 | NS-3 | NS-4 | NS-5 | NS-6 | NS-7 |
|---|---|---|---|---|---|---|---|---|
| Dry density (kg/m$^3$) | | 1651 | 1608 | 1683 | 1651 | 1651 | 1651 | 1651 |
| Soil | | Ottawa F65 | Ottawa F65 | Ottawa F65 | Ottawa F65 | Ottawa F65 | Ottawa F65 | Ottawa F65 |
| $D_R$ (assuming $p_{max} = 1765$ & $p_{min} = 1476$ kg/m$^3$) | | 65% | 50% | 75% | 65% | 65% | 65% | 65% |
| Motion to be used for simulation provided in Excel sheet | | Achieved RPI-1, Motion 1 | Anticipated RPI-3 Motion 1 | Achieved RPI-1, Motion 1 | Achieved RPI-1, Motion 1 scaled up | Achieved RPI-1, Motion 1 scaled down | Achieved RPI-2 Motion 1 | Achieved RPI-2 Motion 1 scaled up |
| PGA (g) | | 0.15 | 0.15 | 0.15 | 0.25 | 0.11 | 0.14 | 0.2 |
| PGA of 1 Hz component (g) | | 0.135 | 0.135 | 0.135 | 0.27 | 0.099 | 0.11 | 0.16 |
| PGA of the high frequency component (g) | | 0.021 | 0.021 | 0.021 | 0.035 | 0.015 | 0.08 | 0.11 |
| Simulation result: X-displacement at middle point on the specimen surface (cm) | PM4SAND | 0.7 | 48 | 0 | 28.9 | 0.2 | 10.6 | 32.6 |
| | UBCSAND | 4.9 | 63.6 | 0.2 | 13.9 | 0.2 | 12.4 | 25.1 |
| Simulation result: Duration of liquefaction at P4 after end of shaking (cm) | PM4SAND | 0 | 130 | 0 | 80 | 0 | 20 | 60 |
| | UBCSAND | 0 | 0 | 0 | 0 | 0 | 0 | 0 |

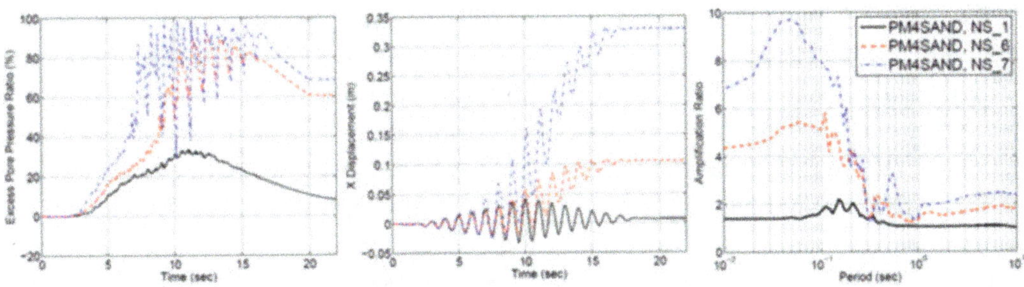


**Fig. 27.11** Comparison of sensitivity analyses using PM4SAND in terms of excess pore pressure ratio, horizontal displacement time history and amplification ratio at the middle surface of the model for NS1, NS6, and NS7. (Input ground motions for analyses NS6 and NS7 have superimposed high frequencies. Input motion NS7 is input motion NS6 multiplied by 1.43)

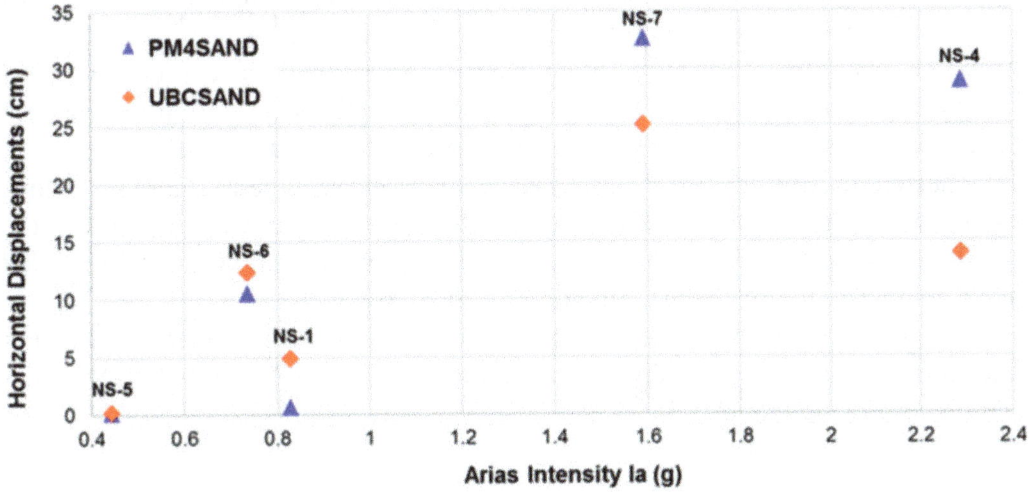


**Fig. 27.12** Horizontal displacements at middle surface of the models versus Arias intensity for sensitivity analyses with $D_R$ of 65%

## 27.6 Conclusions

We used two different constitutive models (PM4SAND and UBCSAND), implemented in the software FLAC, to simulate liquefaction-induced slope instability from six centrifuge tests.

The methodology adopted for the constitutive model calibration process considered both liquefaction triggering and post-liquefaction accumulation of shear strains. Firstly, constitutive models, PM4SAND and UBCSAND, were calibrated against laboratory tests to comply with liquefaction triggering criteria such as the number of cycles required to reach specific strain and pore pressure ratio thresholds. Secondly, the strain accumulation rate (SAR) following liquefaction triggering was used as a target for model parameter calibration.

Overall, the blind predictions captured the centrifuge tests behavior well and bounded the centrifuge response in most cases, indicating the benefits of using more than one constitutive model when performing numerical modeling. The predicted permanent displacements based on both constitutive models were generally within a factor of 2 from the actual measurements. The liquefaction strength curves implied by the UBCSAND model are steeper than the ones implied by PM4SAND and the laboratory test data. Hence, during the Type-C simulations, the UBCSAND model calibration was adjusted considering a CSR range corresponding to the CSR induced by the input motions. This should be considered when model calibrations are performed in practice, especially when multiple hazard levels are involved.

Differences between the predicted and measured responses are partially associated with limitations in the laboratory test data available, which affect the results of the numerical model calibration. For example, direct simple shear tests rather than triaxial tests, tests with static bias in addition to the no-bias conditions as well as tests at different CSR levels, representative of the CSR induced by the input motions, would be more relevant and are recommended. Differences between observed and computed responses are also attributed to simplifications in the prototype-scale numerical model (e.g., simulating possibly heterogeneous actual conditions with an idealized uniform relative density), accuracy of measurements and conversions from model to prototype scale, and differences between the input motions in the simulations ("average" of recorded at the centrifuge base) and the actual recorded input motions in the centrifuge which may vary across its base.

The sensitivity analyses suggest that horizontal displacements increase when high frequencies are introduced in the ground motion and for smaller soil relative density and higher ground motion intensity.

## References

Beaty, M., & Byrne, P. M. (1998). An effective stress model for predicting liquefaction behavior of sand. In *Proceedings of a specialty conference, geotechnical earthquake engineering and soil dynamics* (pp. 766–777). Seattle: ASCE.

Boulanger, R. W., & Ziotopoulou, K. (2015). *PM4Sand (Version 3): A sand plasticity model for earthquake engineering applications. Technical Report No. UCD/CGM-15/xx, Center for Geotechnical Modeling*. Department of Civil and Environmental Engineering, University of California, Davis, CA

Byrne, P. M., Park, S. S., Beaty, M., Sharp, M., Gonzalez, L., & Abdoun, T. (2004). Numerical modeling of liquefaction and comparison with centrifuge tests. *Canadian Geotechnical Journal, 41*(2), 193–211.

El Ghoraiby, M. A., Park, H., & Manzari, M. T. (2017). *LEAP 2017: Soil Characterization and Element Tests for Ottawa F65 Sand*. Washington, DC: The George Washington University.

El Ghoraiby, M. A., Park, H., & Manzari, M. T. (2019). Physical and mechanical properties of Ottawa F65 sand. In B. Kutter et al. (Eds.), *Model tests and numerical simulations of liquefaction and lateral spreading: LEAP-UCD-2017*. New York: Springer.

Giannakou, A., Travasarou, T., Ugalde, J., Chacko, J., & Byrne, P. (2011). Calibration methodology for liquefaction problems considering level and sloping ground conditions. In *5th International Conference on Earthquake Geotechnical Engineering, Paper No. CMFGI, January*.

Itasca Consulting Group Inc. (2011). *Fast Lagrangian Analysis of Continua (FLAC2D)*. V. 7.0.
Itasca Consulting Group Inc. (2017). *Fast Lagrangian Analysis of Continua (FLAC2D)*. V. 8.0.
Marti, J., & Cundall, P. A. (1982). Mixed discretisation procedure for accurate solution of plasticity problems. *International Journal Numerical Methods and Analytical Methods in Geomechanics, 6*, 129–139.
Travasarou, T., Chen W. Y., & Chacko, J. (2011). Liquefaction-induced uplift of buried structures insights from the study of an immersed railway tunnel. In *5th International Conference on Earthquake Geotechnical Engineering, Paper No. LITUR, January*.

# Chapter 28
# LEAP-UCD-2017 Type-B Predictions Through FLIP at Kyoto University

**Kyohei Ueda and Toma Wada**

**Abstract** This study reports the results of type-B predictions for dynamic centrifuge model tests of a liquefiable sloping ground conducted at various centrifuge facilities within a framework of the LEAP-UCD-2017. The simulations are carried out with a finite strain analysis program, called "FLIP TULIP," which incorporates a strain space multiple mechanism model based on the finite strain theory (including both total and updated Lagrangian formulations). The program can take into account the effect of geometrical nonlinearity as well as material nonlinearity's effect. Soil parameters for the constitutive model are determined referring to the results of laboratory experiments (e.g., cyclic triaxial tests) and some empirical formulae. This chapter describes the parameter identification process in details as well as the computational conditions (e.g., geometric modeling, initial and boundary conditions, numerical schemes such as time integration technique). Type-B prediction results are compared with the centrifuge test results to examine the applicability of the program and constitutive model.

## 28.1 Introduction

To estimate liquefaction-induced damage to soil-structure systems during huge earthquakes, analytical techniques (e.g., effective stress analysis) have been actively studied as well as experimental ones (e.g., laboratory soil test, centrifuge model test) since 1970s. In particular, research on constitutive laws of sandy (or granular) materials has been dramatically developed by academic researchers since 1990s for its application in practice; an effective stress analysis technique incorporating the constitutive laws is being used more and more in seismic design to estimate the

K. Ueda (✉)
Disaster Prevention Research Institute, Kyoto University, Uji, Kyoto, Japan
e-mail: ueda.kyohei.2v@kyoto-u.ac.jp

T. Wada
Department of Civil and Earth Resources Engineering, Kyoto University, Kyoto, Japan
e-mail: wada.touma.67a@st.kyoto-u.ac.jp


B. Kutter et al. (eds.), *Model Tests and Numerical Simulations of Liquefaction and Lateral Spreading*, https://doi.org/10.1007/978-3-030-22818-7_28

damage level of soil-structure systems due to liquefaction. Laboratory and model experimental results are often used for validation of the analysis technique, and such constitutive models have been modified and updated if need arises. However, a practical way for the validation process (e.g., how to assess the validity of constitutive models) has not yet been established, in particular for dynamic liquefaction-induced behavior of soil-structure systems, as well known in geotechnical engineer/researcher community.

The importance of validation was already pointed out in the VELACS project more than 20 years ago (Arulanandan and Scott 1993/1994). The project was an opportunity for geotechnical researchers to realize the need for improvement of numerical modeling on liquefiable ground. In addition, the project has revealed that reliable data for the validation process were difficult to obtain from laboratory and/or centrifuge experimental results because such experimental results generally have some variation among different facilities, particularly in the case of experiments on complicated phenomena such as liquefaction. To answer the problems that were pointed out but could not be solved in the VELACS project, some researchers have proposed a new international collaborative effort called "LEAP" (Liquefaction Experiment and Analysis Projects) (Iai 2015; Kutter et al. 2015; Manzari et al. 2015; Zeghal et al. 2015). One of the main goals is to establish a practical way for validating the capabilities of existing analytical techniques for liquefaction-induced behavior, including constitutive laws of granular materials, through comparison with laboratory and centrifuge experiments.

As part of LEAP exercises, LEAP-GWU-2015 aimed to obtain a set of high-quality experimental data among different centrifuge facilities; a sloping liquefiable ground was built up in a rigid model container, having a simpler lateral boundary condition rather than a laminar container, in order to avoid difficulties in numerical modeling associated with complex boundary conditions in the latter case. Model specifications in LEAP-GWU-2015 are presented by Kutter et al. (2018), in which the results of centrifuge experiments performed at Cambridge University (CU) in the UK, Kyoto University (KU) in Japan, National Central University (NCU) in Taiwan, Rensselaer Polytechnic Institute (RPI) and University of California Davis (UCD) in the USA, and Zhejiang University (ZU) in China are compared. More recently, LEAP-UCD-2017 has started to obtain a sufficient number of centrifuge experimental results for a sloping liquefiable ground with higher accuracy and to quantify the magnitude of variability among different centrifuge facilities.

This chapter presents results of numerical simulations for the dynamic centrifuge model tests, within a framework of type-B prediction (e.g., Lambe (1973)) phases of the LEAP-UCD-2017. The simulations are carried out with a finite strain, effective stress finite element (FE) program, called "FLIP TULIP," which incorporates a strain space multiple mechanism model based on the finite strain theory (including both total and updated Lagrangian formulations) (Ueda 2009; Iai et al. 2013). The program can take into account the effect of geometrical nonlinearity as well as that of material nonlinearity. This chapter describes the identification procedure of input model parameters in details as well as the computational conditions such as geometrical modeling, initial, and boundary conditions.

## 28.2 Brief Overview of Centrifuge Experiments

Vargas et al. (2019) describe the centrifuge modeling equipment used at Disaster Prevention Research Institute, Kyoto University, and Kutter et al. (2019a, b) present the detailed specifications for the LEAP-UCD-2017 experiments conducted at all of the centrifuge facilities. This section briefly describes a summary of the centrifuge model tests. A sketch of the model is shown in Fig. 28.1. The experiments modeled submerged Ottawa F65 sand (a target relative density of about 65%) with a 5-degree sloping ground surface in a rigid box subjected to ramped sinusoidal wave motions (1 Hz, 16 cycles). At some facilities (e.g., Kyoto) in which the direction of horizontal shaking is in the plane of spinning of the centrifuge, the slope in the shaking direction was built up as a curved surface considering the effective radius from the axis of rotation of the centrifuge (Fig. 28.1b). The sloping ground had the width of

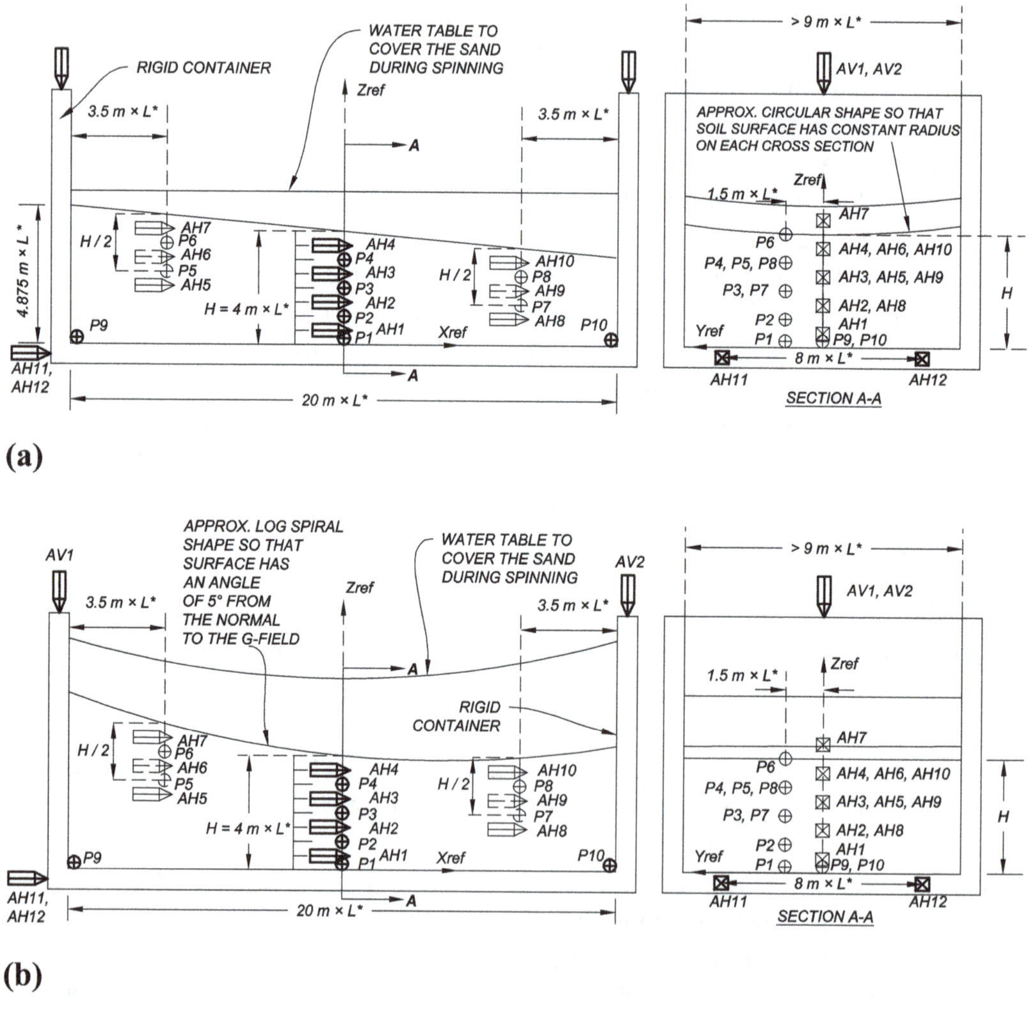


**Fig. 28.1** Schematic for LEAP-UCD-2017 centrifuge model tests (Kutter et al. 2019a): (**a**) Sectional drawing for shaking parallel to the axis of the centrifuge (ZU, RPI, NCU). (**b**) Sectional drawing for shaking in the plane of spinning of the centrifuge (UCD, Kyoto, Cambridge)

20 m and the height of 4 m at midpoint in prototype scale. The input motions used for LEAP-UCD-2017 validation experiments consisted of three nondestructive motions (i.e., motions 1, 3, and 5) and two destructive motions (i.e., motions 2 and 4): the first destructive motion, which corresponded to the second motion of the sequence (motion 2), had an acceleration amplitude of 0.15 g. The nondestructive motions were applied to evaluate the ground properties (e.g., shear wave velocity) after each destructive motion. In this chapter, we focus on the dynamic responses under the first destructive motion (motion 2) for validation of the effective stress FE program (i.e., FLIP TULIP) for liquefaction problems incorporating the strain space multiple mechanism model.

## 28.3 Constitutive Model of Soils

As is the case with the type-B and -C simulations for LEAP-GWU-2015 (Ueda and Iai 2018), a strain space multiple mechanism model based on the finite strain theory (Ueda 2009; Iai et al. 2013) is used for LEAP-UCD-2017 type-B predictions. The original version of the model was proposed within the context of the infinitesimal strain theory (Iai et al. 1992), and implemented in a FE program called "FLIP ROSE" (Finite Element Analysis Program of LIquefaction Process/Response Of Soil-structure Systems during Earthquakes). The program has been frequently used in design practice for evaluating the seismic performance of soil-structure systems, particularly port structures, in Japan (Iai et al. 1992; Iai et al. 1995; Ozutsumi et al. 2002). On the basis of a multitude of virtual simple shear mechanisms oriented in an arbitrary direction, the model has an ability to simulate the evolution of induced fabric under various complicated loading conditions (e.g., the rotation of principal stress axis direction). About 10 years ago, a new stress-dilatancy relationship was introduced in the constitutive model for controlling dilative (or positive) and contractive (or negative) components of dilatancy in a more sophisticated manner (Iai et al. 2011).

With an aim to take into account the effect of geometrical nonlinearity in addition to material nonlinearity's effect, the model was extended within the context of the finite strain (or large deformation) theory (Ueda 2009; Iai et al. 2013). Incorporating the extended model, a finite strain FE program called "FLIP TULIP" (Finite Element Analysis Program of LIquefaction Process/Total and Updated Lagrangian Program of LIquefaction Process) has been developed based on the original infinitesimal strain program (i.e., FLIP ROSE). As might be surmised from the abbreviation, two different methods—the total Lagrangian (TL) and updated Lagrangian (UL) approaches—are available in the program. The TL approach is based on the reference (or undeformed) configuration corresponding to a fixed reference time, in which the Lagrangian (or material) description is used. On the other hand, the Eulerian (or spatial) description based on the current (or deformed) configuration is used in the UL approach. However, both approaches are theoretically equivalent to each other: the only difference is a standing position (i.e., the reference or current configuration). One of the major advantages in conducting both the TL and UL analyses is to be able to directly compare the numerical results obtained from different numerical schemes for validating the reliability of each numerical approach. The finite strain program begins to be used in research and design practice for evaluating the seismic performance of a soil-structure system considering its large deformation behavior (Ueda et al. 2011; Ueda et al. 2015; Ueda and Iai 2018). In this chapter, only the TL approach is applied because the UL analysis has been found to give an almost identical simulation result with the TL analyses.

## 28.4 Numerical Simulation (FE Analysis)

### *28.4.1 Definition of Type-A, -B, and -C Predictions*

In this chapter, type-B predictions are carried out using the strain space multiple mechanism model extended within the context of the finite strain theory for validation of the model capability. Before proceeding to explain the analytical condition in detail, the definition of type-A, -B, and -C predictions is briefly given following Lambe (1973) as follows:

*Type-A prediction*: Type-A prediction is done toward a planned experiment, not targeted at an actual experiment. This means type-A is a true prediction of an event made prior to the event.

*Type-B prediction*: Type-B prediction is targeted at an actual experiment after the experiment is conducted. The predictors can get information on as-built properties and measured input data, but no knowledge of the results is available for them.

*Type-C prediction*: Type-C prediction is performed after the experiment is completed, with results known to the predictors. They are able to iteratively adjust the model parameters, if necessary, to enhance the quality of their simulation results compared with observations.

### *28.4.2 Model Parameters*

Most of the input parameters of the strain space multiple mechanism model were determined in the same manner as in the type-B simulations for LEAP-GWU-2015 (Ueda and Iai 2018). This section briefly describes how to set the required model parameters for liquefaction analyses.

The parameters for defining the characteristics of volumetric and shear deformation are shown in Table 28.1. In the type-B prediction, the measured density at each facility should be used. However, a significant difference in the density was not

**Table 28.1** Model parameters for deformation characteristics

| Symbol | Mechanism | Parameter designation | |
|---|---|---|---|
| $\rho_t$ | – | Mass density | 2.041 t/m$^3$ |
| $p_a$ | – | Reference confining pressure | 100.0 kPa |
| $K_{L/Ua}$ | Volumetric | Bulk modulus | $2.86 \times 10^5$ kPa |
| $r_K$ | Volumetric | Reduction factor of bulk modulus for liquefaction analysis | 0.5 |
| $l_K$ | Volumetric | Power index of bulk modulus for liquefaction analysis | 2.0 |
| $G_{ma}$ | Shear | Shear modulus | $1.10 \times 10^5$ kPa |
| $\phi_f^{PS}$ | Shear | Internal friction angle for plane strain | 40.0° |
| $h_{max}$ | Shear | Upper bound for hysteretic damping factor | 0.24 |

found among different facilities, and thus the value shown in Table 28.1 was used for all facilities.

The initial (small-strain) shear modulus $G_m$ under an arbitrary confining pressure $p$ is automatically calculated in the program following the equation below:

$$G_m = G_{ma}(p/p_a)^{m_G} \tag{28.1}$$

by specifying the reference effective confining pressure $p_a$ and the index $m_G$, which was set to be 0.5 in this study. The index controls the dependency of shear modulus on confining pressure. The initial shear modulus $G_{ma}$ under the confining pressure $p_a$ was determined based upon the void ratio by applying an empirical relation (Ishihara 1996) as follows:

$$G_{ma} = 7000\frac{(2.17 - e)^2}{1 + e}p_a^{0.5}. \tag{28.2}$$

As is the case with the density, the shear modulus in Table 28.1 was applied to all facilities because the difference in the shear modulus estimated from the void ratio was very small among the different facilities.

The internal friction angle $\phi_f^{PS}$ for plane strain shown in Table 28.1 was determined as follows:

$$sin\phi_f^{PS} = \frac{1}{2}\frac{1}{cos(\pi/6)}\mathrm{M} \tag{28.3}$$

in which the critical state frictional constant M was estimated from a monotonic triaxial compression test under undrained condition of Ottawa F-65 sand. As shown in Table 28.1, the maximum damping constant $h_{max}$ was set to be the standard value for sands (= 0.24).

In analogy with Eq. (28.1) for the shear modulus, the initial bulk modulus $K_{L/U}$ under an arbitrary confining pressure $p$ is automatically evaluated in the program as follows:

$$K_{\mathrm{L/U}} = K_{\mathrm{L/Ua}}(p/p_{\mathrm{a}})^{n_{\mathrm{K}}} \tag{28.4}$$

where $K_{\mathrm{L/Ua}}$ was estimated from the shear modulus $G_{\mathrm{ma}}$ in Eq. (28.1) with the Poisson ratio of 0.33, and the index $n_{\mathrm{K}}$, which controls the dependency of bulk modulus on confining pressure, was set to be 0.5 in this study. In the case of liquefaction analyses, the above equation is extended as

$$K_{\mathrm{L/U}} = r_{\mathrm{K}} K_{\mathrm{U0}}(p/p_0)^{l_{\mathrm{K}}} \tag{28.5}$$

where $p_0$ denotes the initial confining pressure, $r_{\mathrm{K}}$ is a reduction factor of bulk modulus, and the power index $l_{\mathrm{K}}$ represents the confining pressure dependency of bulk modulus (Iai et al. 2011). By changing the parameters $r_{\mathrm{K}}$ and $r_{\varepsilon_{\mathrm{d}}}$, the latter of which is a parameter for controlling both dilative and contractive components, with keeping the product (i.e., $r_{\varepsilon_{\mathrm{d}}} \times r_{\mathrm{K}}$) constant, volumetric characteristics due to the dissipation of excess pore water pressure (EPWP) following liquefaction can be independently controlled without altering liquefaction resistance. In this study, the standard value (i.e., 0.5) shown in Table 28.1 was used for $r_{\mathrm{K}}$ due to the lack of knowledge about the volumetric characteristics. As suggested by Ishihara and Yoshimine (1992), the characteristics can be given as a relationship between the volumetric strain due to consolidation following liquefaction and the maximum amplitude of shear strain during shaking by conducting laboratory experiments such as triaxial or torsional tests.

In addition to the centrifuge experiments, a series of laboratory tests (e.g., stress-controlled cyclic undrained triaxial tests) were carried out by a research group from the George Washington University (El Ghoraiby et al. 2017, 2019). Figure 28.2

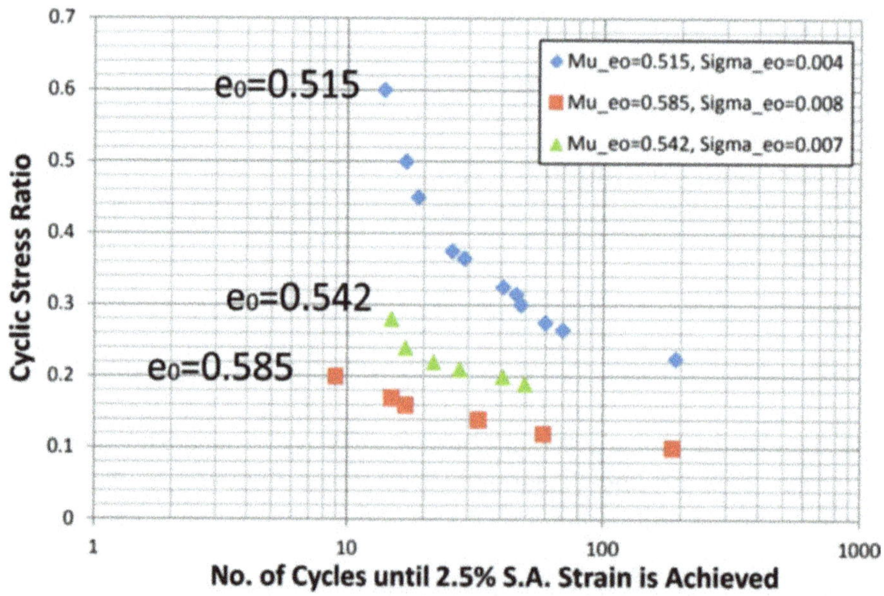


**Fig. 28.2** Liquefaction strength curves from stress-controlled cyclic undrained triaxial tests (El Ghoraiby et al. 2017, 2019)

**Table 28.2** Model parameters for dilatancy

| Symbol | Parameter designation | |
|---|---|---|
| $\phi_p$ | Phase transformation angle | 28.0° |
| $\varepsilon_d^{cm}$ | Limit of contractive component | 0.15 |
| $r_{\varepsilon_d^c}$ | Parameter controlling contractive component | 0.72 |
| $r_{\varepsilon_d}$ | Parameter controlling dilative and contractive components | 0.85 |
| $q_1$ | Parameter controlling initial phase of contractive component | 1.0 |
| $q_2$ | Parameter controlling final phase of contractive component | 0.1 |
| $S_1$ | Small positive number to avoid zero confining pressure | 0.005 |
| $c_1$ | Parameter controlling elastic range for contractive component | 1.39 |
| $q_{us}$ | Undrained shear strength (for steady-state analysis) | – |

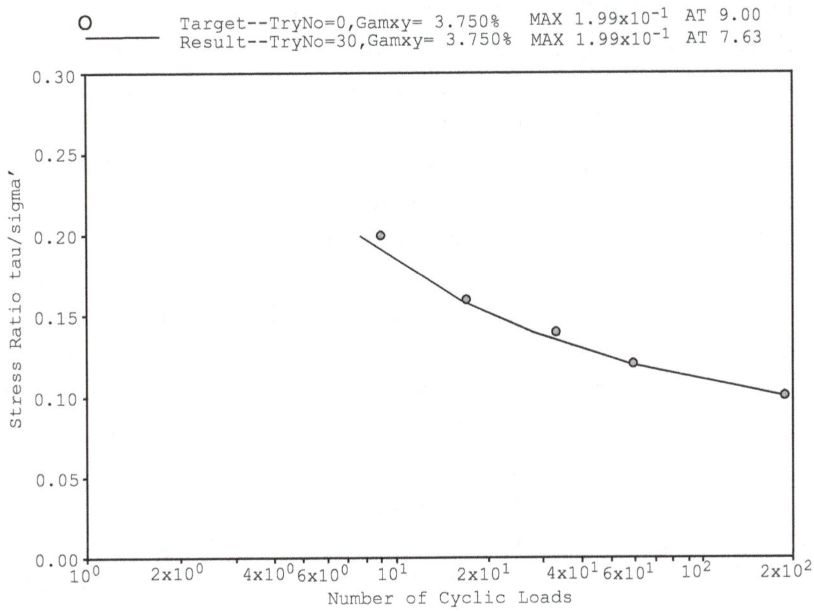


**Fig. 28.3** Comparison of simulated liquefaction strength with experimental results

shows liquefaction strength curves for different three void ratios obtained from the cyclic triaxial tests. Model parameters controlling the characteristics of liquefaction and dilatancy in Table 28.2 were determined by referring to the liquefaction resistance curve for the loosest sample (i.e., $e_0 = 0.585$) as shown in Fig. 28.3. The simulated liquefaction strength curve was able to well capture the measured relationship between the shear stress ratio and the number of cyclic loads, including at lower stress levels. The undrained shear strength $q_{us}$ for steady-state analysis was not specified, which means $q_{us}$ was set to be an infinite value, because the strength is normally very large for clean sands such as Ottawa F-65 sand. The permeability of the ground was set to a constant value of $1.26 \times 10^{-4}$ m/s to take into account the effect of pore water flow and migration; the value was determined based on permeability tests conducted at the George Washington University.

### 28.4.3 Initial/Boundary Conditions and Input Motions

The type-B FE simulations were performed using a 2-dimensional FE mesh shown in Fig. 28.4, which has the same prototype dimension as the centrifuge experiments. The number of nodes and elements (including pore water elements) is 861 and 1600, respectively. In order to avoid shear locking and hourglass modes, the selective reduced integration (SRI) techniques (Hughes 1980) were applied to four-node quadrilateral soil elements. The FE mesh size was determined by considering the wavelength corresponding to the highest frequency of interest as suggested by Alford et al. (1974); in this study, 15 Hz was used for the highest frequency because the value is large enough compared to the natural frequency (i.e., 1 Hz) of the input motion. For simulating the boundary conditions of the rigid box in the centrifuge experiments, displacement boundary conditions for the type-B predictions were determined as follows: the degree of freedom at the side boundaries was fixed only horizontally (i.e., vertical roller), and both horizontal and vertical displacements were fixed at the base. When it comes to the degree of freedom of pore water pressure, a hydrostatic condition was specified at the ground surface whereas the side and bottom boundaries were set to be impermeable.

Before proceeding to a seismic response (or dynamic) analysis, a self-weight analysis was carried out by applying the gravity acceleration to the model for obtaining the initial stress and strain distributions before shaking. In the seismic response analysis, the first destructive motions (motion 2) recorded at the bottom of the container during centrifuge experiments were used as an input motion for the type-B predictions; the motions were intended to be identical among all facilities, but somewhat different from the target motion depending on facilities. The numerical time integration was carried out by the SSpj method (Zienkiewicz et al. 2000) with the standard parameters $\theta_1 = 0.6$ and $\theta_2 = 0.605$ for the equation of motion and $\theta_1 = 0.6$ for the mass balance equation of pore water flow, using a time step of 0.005 s; both equations are written in the context of the finite strain formulation

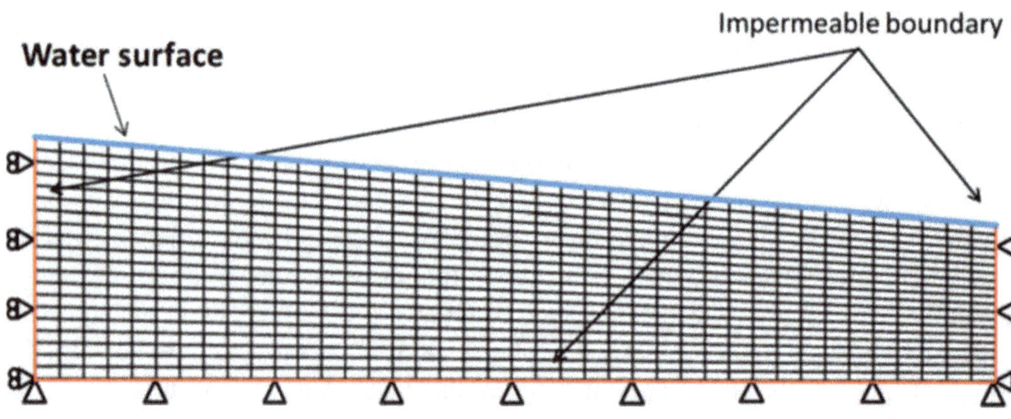


**Fig. 28.4** Finite element mesh for numerical analysis

(i.e., the Lagrangian or Eulerian description) (Ueda 2009). In the dynamic simulation, Rayleigh damping ($\alpha = 0.0$, $\beta = 0.0002$) was applied for stabilizing the numerical solution process.

## 28.5 Type-B Predictions (for Motion 2)

The results of the type-B predictions are compared to centrifuge experiments in Figs. 28.5, 28.6, 28.7, 28.8, 28.9, 28.10, 28.11, 28.12, and 28.13. In each figure, simulation results, in which model parameters for dilatancy were determined referring to the liquefaction strength for $e_0 = 0.542$, are also shown. The measured responses for KyU3 in Fig. 28.5 are generally well simulated using appropriate model parameters (i.e., parameters for the closest void ratio, $e_0 = 0.585$) rather than the case with the parameters for $e_0 = 0.542$: the simulated lateral displacement becomes closer to the measurement with the increase in the void ratio. The negative spikes in the measured acceleration are reasonably simulated in the type-B prediction ($e_0 = 0.585$). However, the dynamic amplitude of simulated EPWP is overestimated to some extent. Thus, type-C simulations should be required to capture the measured behavior with higher precision.

Following is a summary of the comparison between the type-B predictions and centrifuge model tests. By setting the model parameters referring to liquefaction

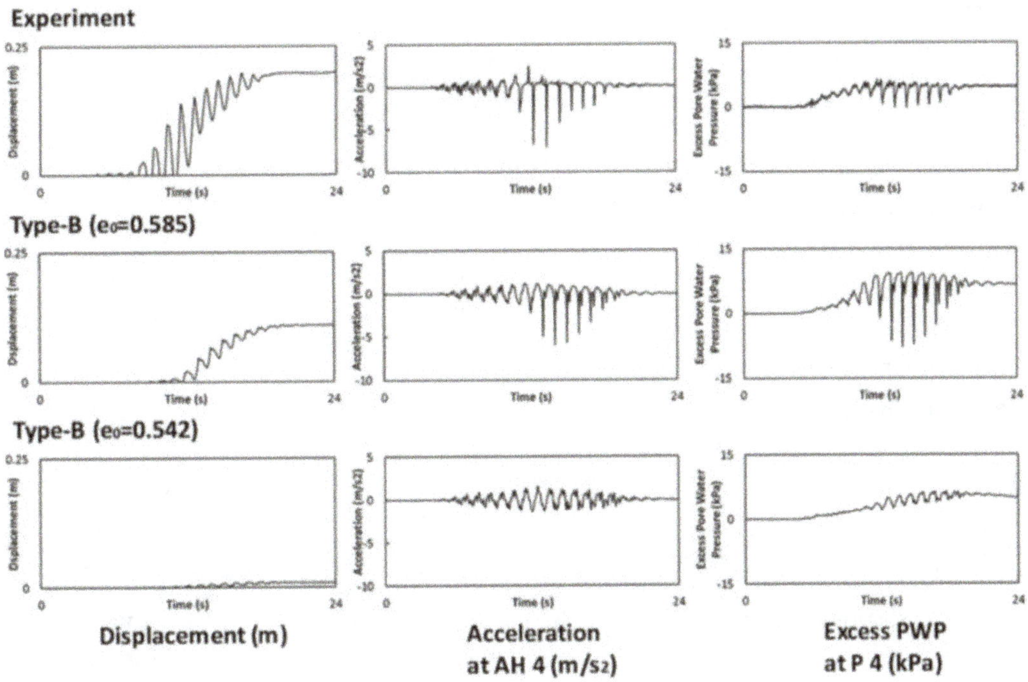


**Fig. 28.5** Computed time histories (type-B prediction for KyU3, $e_0 = 0.619$) with experimental results

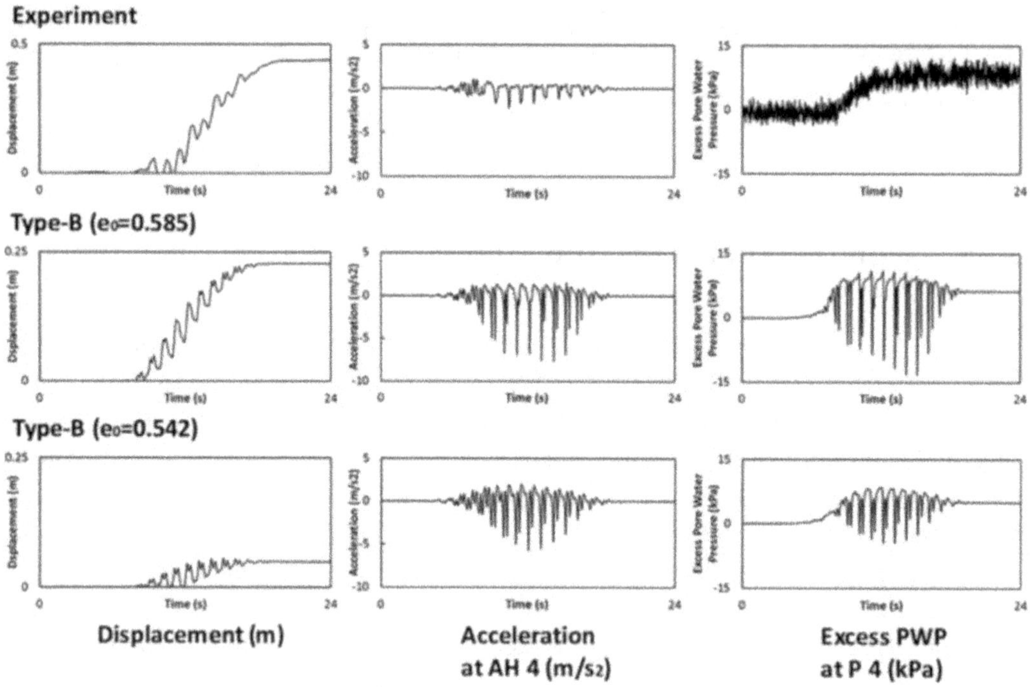


**Fig. 28.6** Computed time histories (type-B prediction for CU2, $e_0 = 0.650$) with experimental results

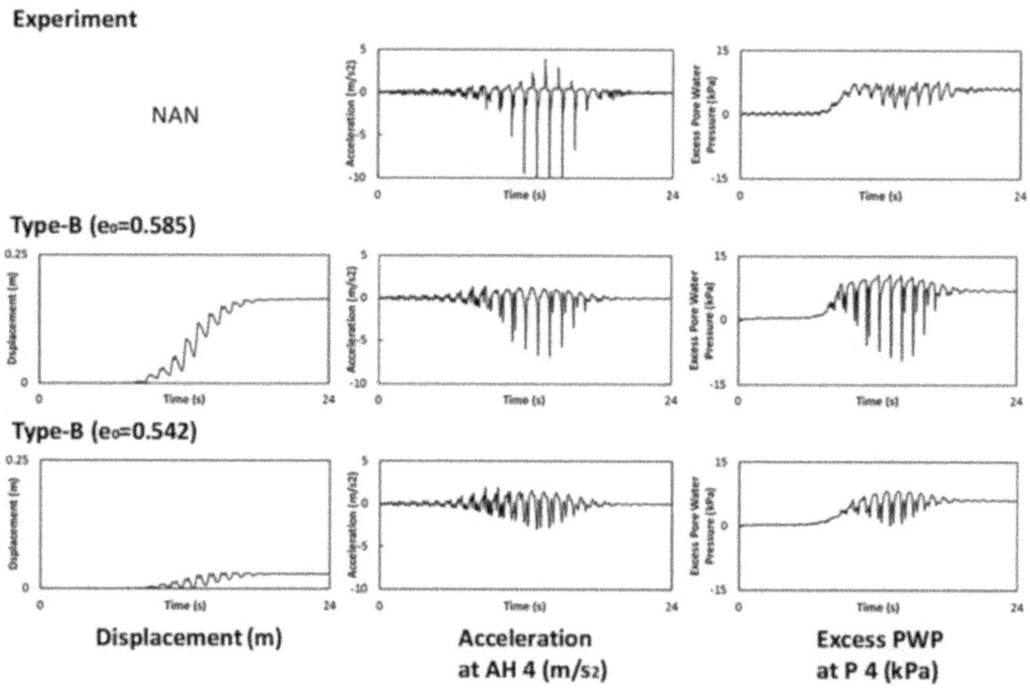


**Fig. 28.7** Computed time histories (type-B prediction for KAIST1, $e_0 = 0.558$) with experimental results

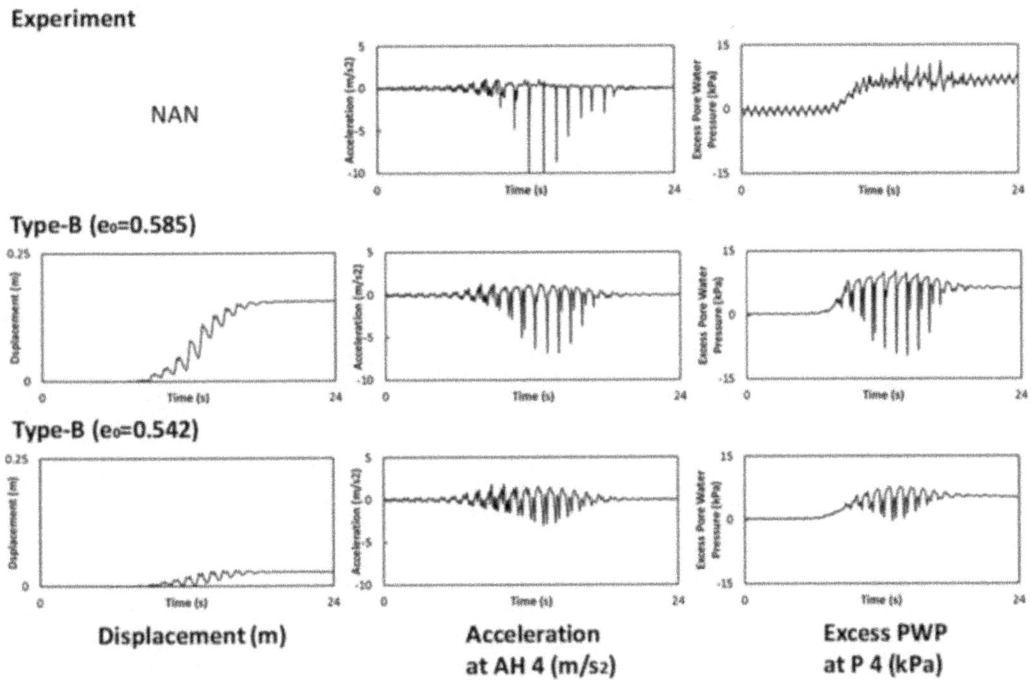


**Fig. 28.8** Computed time histories (type-B prediction for KAIST2, $e_0 = 0.664$) with experimental results

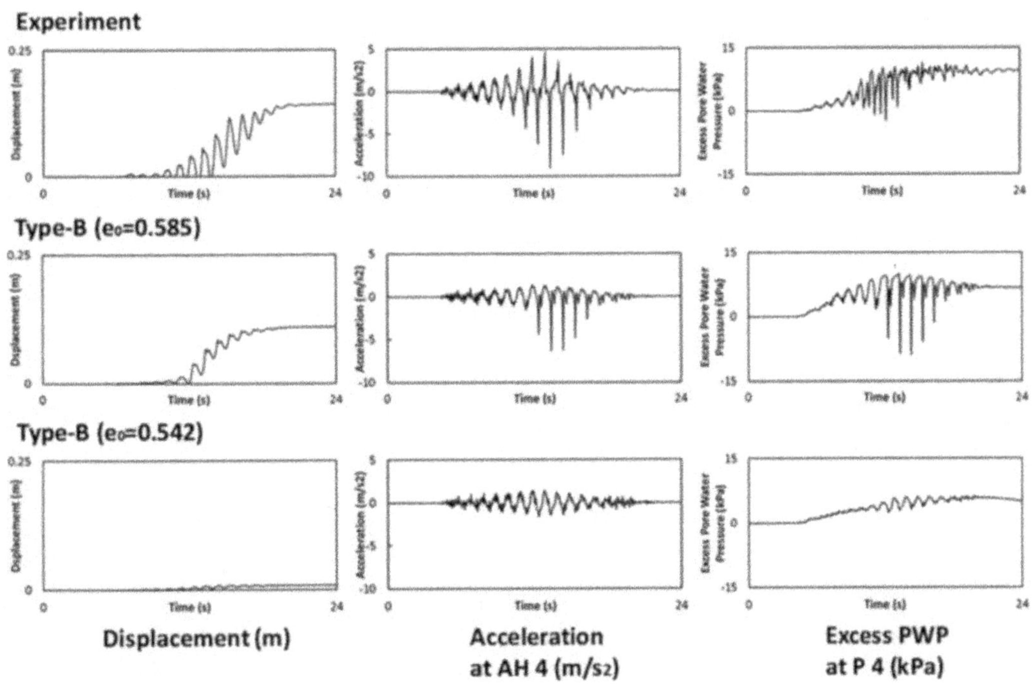


**Fig. 28.9** Computed time histories (type-B prediction for UCD1, $e_0 = 0.592$) with experimental results

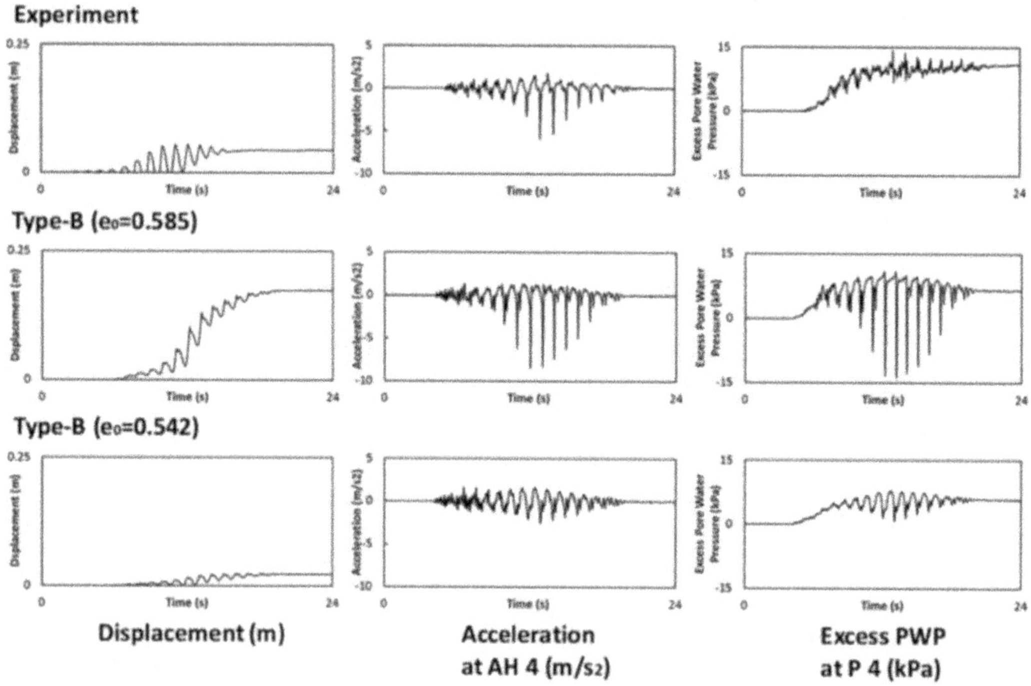


**Fig. 28.10** Computed time histories (type-B prediction for UCD3, $e_0 = 0.598$) with experimental results

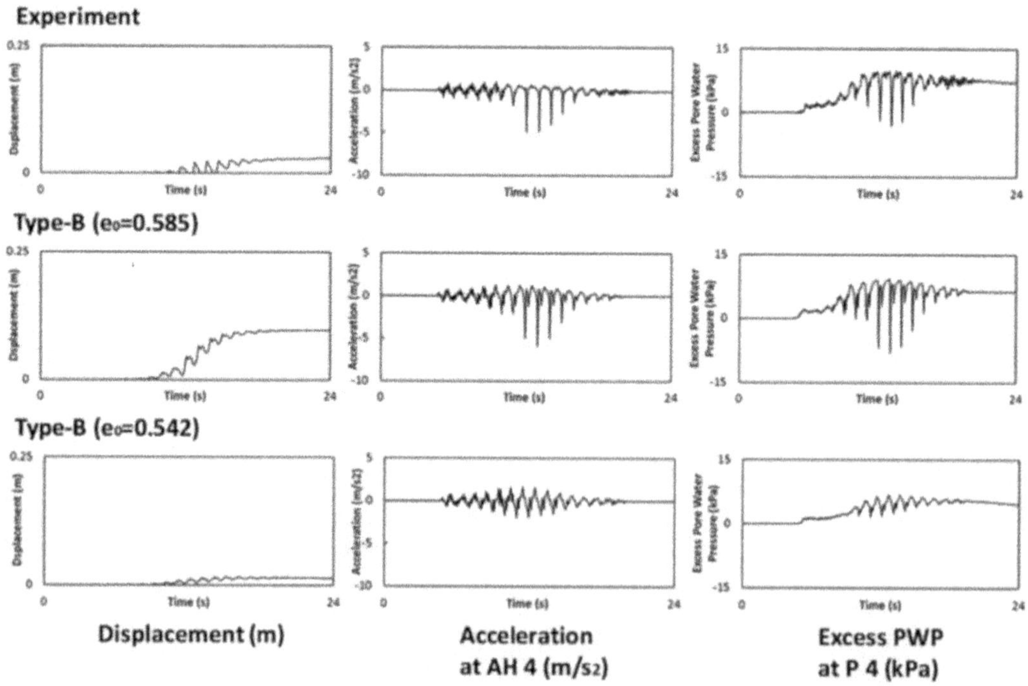


**Fig. 28.11** Computed time histories (type-B prediction for ZJU2, $e_0 = 0.650$) with experimental results

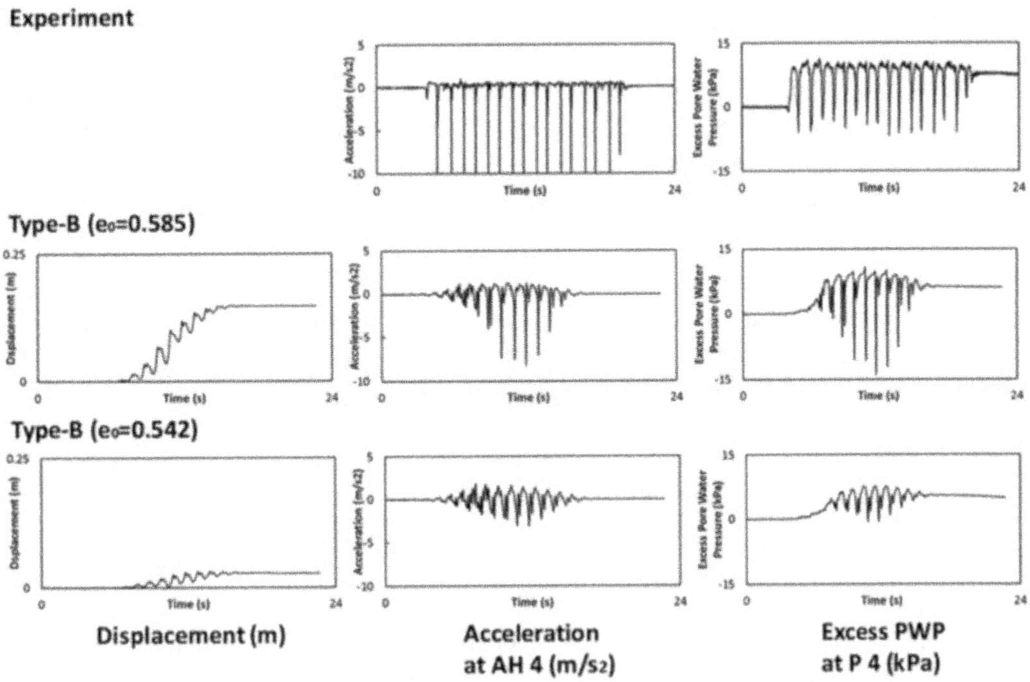


**Fig. 28.12** Computed time histories (type-B prediction for NCU3, $e_0 = 0.604$) with experimental results

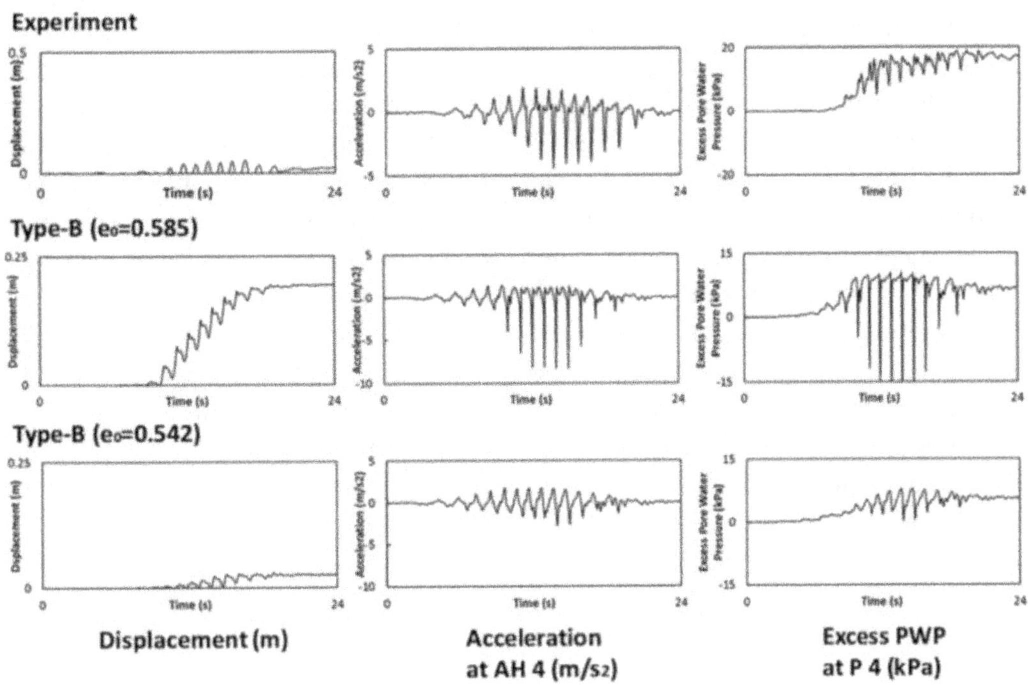


**Fig. 28.13** Computed time histories (type-B prediction for Ehime2, $e_0 = 0.560$) with experimental results

strength curves (Figs. 28.2 and 28.3), the prediction is able to reasonably simulate the measured lateral displacement, horizontal acceleration, and EPWP. However, it seems difficult to achieve a perfect agreement between predictions and experiments without knowledge of experimental results, even though input motions recorded during experiments can be used for the prediction. The series of centrifuge model tests conducted for LEAP-UCD-2017 show that the degree of variation in measured dynamic responses, particularly lateral displacements, among the facilities was larger than that in the input motions. This suggests that non-negligible differences in the test procedure among the facilities (e.g., how to build up the sloping ground) should be taken into account in addition to the input motion difference, if the predictor wants to simulate the experimental results more realistically.

## 28.6 Conclusions

This chapter reported the results of Type-B predictions for dynamic centrifuge model tests of a liquefiable sloping ground conducted at various centrifuge facilities within a framework of LEAP-UCD-2017. The simulations were performed with a finite strain analysis program, called "FLIP TULIP," which incorporates a strain space multiple mechanism model based on the finite strain theory (including both total and updated Lagrangian formulations). The program can consider the effect of geometrical nonlinearity as well as material nonlinearity's effect. By setting soil parameters for the constitutive model referring to the results of laboratory experiments (e.g., cyclic triaxial tests), the prediction was able to reasonably simulate the measured lateral displacement, horizontal acceleration, and excess pore water pressure. However, it was found that a perfect agreement between predictions and experiments seems difficult to achieve without knowledge of experimental results, even though input motions recorded during experiments are available for the prediction. This may be because the variation in measured dynamic responses, particularly lateral displacements, among the facilities was more or less influenced by a series of test procedures (e.g., how to build up the sloping ground) as well as the differences of input motions. However, the variation is hard to quantify in advance of predictions at the current moment, although the influence is non-negligible for the prediction accuracy. Thus, how to estimate the degree of variation in advance is left for future work as well as how to reduce experimental errors. To simulate the experimental results more realistically, type-C simulations may be required by identifying model parameters with trial and error for each facility's results.

## References

Alford, R. M., Kelly, K. R., & Boore, D. M. (1974). Accuracy of finite difference modeling of the acoustic wave equation. *Geophysics, 39*(6), 834–842.

Arulanandan, K., & Scott, R. F. (1993/1994). Verification of numerical procedures for the analysis of soil liquefaction problems. In *Proceedings of the International Conference on the Verification of Numerical Procedures for the Analysis of Soil Liquefaction Problems* (Vols. 1 and 2). Rotterdam: A. A. Balkema.

El Ghoraiby, M. A., Park, H., & Manzari, M. T. (2017). LEAP 2017: Soil Characterization and 339 Element Tests for Ottawa F65 Sand. Washington, DC: The George Washington University.

El Ghoraiby, M. A., Park, H., & Manzari, M. T. (2019). Physical and mechanical properties of Ottawa F65 Sand. In B. Kutter et al. (Eds.), *Model tests and numerical simulations of liquefaction and lateral spreading: LEAP-UCD-2017*. New York: Springer.

Hughes, T. J. R. (1980). Generalization of selective integration procedures to anisotropic and nonlinear media. *International Journal for Numerical Methods in Engineering, 15*, 1413–1418.

Iai, S. (2015). Liquefaction experiment and analysis projects (LEAP) through a generalized scaling relationship. In S. Iai (Ed). *Geotechnics for catastrophic flooding events* (pp. 95–97). London: CRC Press.

Iai, S., Matsunaga, Y., & Kameoka, T. (1992). Strain space plasticity model for cyclic mobility. *Soils and Foundations, 32*(2), 1–15.

Iai, S., Morita, T., Kameoka, T., Matsunaga, Y., & Abiko, K. (1995). Response of a dense sand deposit during 1993 Kushiro-Oki earthquake. *Soils and Foundations, 35*(1), 115–131.

Iai, S., Tobita, T., Ozutsumi, O., & Ueda, K. (2011). Dilatancy of granular materials in a strain space multiple mechanism model. *International Journal for Numerical and Analytical Methods in Geomechanics, 35*(3), 360–392.

Iai, S., Ueda, K., Tobita, T., & Ozutsumi, O. (2013). Finite strain formulation of a strain space multiple mechanism model for granular materials. *International Journal for Numerical and Analytical Methods in Geomechanics, 37*(9), 1189–1212.

Ishihara, K. (1996). *Soil behaviour in earthquake geotechnics*. New York, NY: Oxford University Press.

Ishihara, K., & Yoshimine, M. (1992). Evaluation of settlements in sand deposits following liquefaction during earthquakes. *Soils and Foundations, 32*(1), 173–188.

Kutter, B. L., Carey, T. J., Hashimoto, T., Zeghal, M., Abdoun, T., Kokkali, P., Madabhushi, G., Haigh, S., d'Arezzo, F. B., Madabhushi, S., Hung, W. Y., Lee, C. J., Cheng, H. C., Iai, S., Tobita, T., Ashino, T., Ren, J., Zhou, Y. G., Chen, Y., Sun, Z. B., & Manzari, M. T. (2018). LEAP-GWU-2015 experiment specifications, results, and comparisons. *Soil Dynamics and Earthquake Engineering, 113*, 616–628.

Kutter, B. L., Carey, T. J., Stone, N., Bonab, M. H., Manzari, M., Zeghal, M., Escoffier, S., Haigh, S., Madabhushi, G., Hung, W. Y., Kim, D.S., Kim N. R., Okamura, M., Tobita, T., Ueda, K., & Zhou, Y. G. (2019a). LEAP-UCD-2017 V. 1.01 model specifications. In B. Kutter et al. (Eds.), *Model tests and numerical simulations of liquefaction and lateral spreading: LEAP-UCD-2017*. New York: Springer.

Kutter, B. L., Carey, T. J., Stone, N., Zheng, B. L., Gavras, A., Manzari, M., Zeghal, M., Abdoun, T., Korre, E., Escoffier, S., Haigh, S., Madabhushi, G., Madabhushi, S. S. C., Hung, W. Y., Liao, T. W., Kim, D. S., Kim, S. N., Ha, J. G., Kim, N.R., Okamura, M., Sjafuddin, A. N., Tobita, T., Ueda, K., Vargas, R., Zhou, Y. G., & Liu, K. (2019b). LEAP-UCD-2017 comparison of centrifuge test results. In B. Kutter et al. (Eds.), *Model tests and numerical simulations of liquefaction and lateral spreading: LEAP-UCD-2017*. New York: Springer.

Kutter, B. L., Manzari, M. T., Zeghal, M., Zhou, Y. G., & Armstrong, R. J. (2015). Proposed outline for LEAP verification and validation processes. In S. Iai (Ed). *Geotechnics for catastrophic flooding events* (pp. 99–108). London: CRC Press.

Lambe, T. W. (1973). Predictions in soil engineering. *Geotechnique, 23*(2), 151–201.

Manzari, M. T., Kutter, B. L., Zeghal, M., Iai, S., Tobita, T., Madabhushi, S. P. G., Haigh, S. K., Mejia, L., Gutierrez, D. A., Armstrong, R. J., Sharp, M. K., Chen, Y. M., & Zhou, Y. G. (2015). LEAP projects: Concept and challenges. In S. Iai (Ed). *Geotechnics for catastrophic flooding events* (pp. 109–116). London: CRC Press.

Ozutsumi, O., Sawada, S., Iai, S., Takeshima, Y., Sugiyama, W., & Shimazu, T. (2002). Effective stress analyses of liquefaction-induced deformation in river dikes. *Soil Dynamics and Earthquake Engineering, 22*(9–12), 1075–1082.

Ueda, K. (2009). Finite Strain Formulation of a Strain Space Multiple Mechanism Model for Granular Materials and Its Application. Doctoral Thesis, Kyoto University, (in Japanese).

Ueda, K., & Iai, S. (2018). Numerical predictions for centrifuge model tests of a liquefiable sloping ground using a strain space multiple mechanism model based on the finite strain theory. *Soil Dynamics and Earthquake Engineering, 113*, 771–792.

Ueda, K., Iai, S., & Ozutsumi, O. (2015). Finite deformation analysis of dynamic behavior of embankment on liquefiable sand deposit considering pore water flow and migration. In *Proceedings of the 6th International Conference on Earthquake Geotechnical Engineering, Christchurch, New Zealand*. Paper No. 215.

Ueda, K., Iai, S., & Tobita, T. (2011). Finite deformation analysis of earthquake induced damage to breakwater during 1995 Hyogoken-Nanbu Earthquake. In *Proceedings of the 8th International Conference on Urban Earthquake Engineering, Tokyo, Japan* (pp. 261–270).

Vargas, R. R., Tobita, T., Ueda, K., & Yatsugi, H. (2019). LEAP-UCD-2017 Centrifuge Test at Kyoto University. In B. Kutter et al. (Eds.), *Model tests and numerical simulations of liquefaction and lateral spreading: LEAP-UCD-2017*. New York: Springer.

Zeghal, M., Manzari, M. T., Kutter, B. L., & Abdoun, T. (2015). LEAP: Data, calibration and validation of soil liquefaction models. In *Proceedings of the 6th International Conference on Earthquake Geotechnical Engineering*, *Christchurch, New Zealand.* Paper No. 432.

Zienkiewicz, O. C., Taylor, R. L., & Zhu, J. Z. (2000). *The finite element method: Its basis and fundamentals* (6th ed.). Amsterdam: Elsevier.

# Chapter 29
# LEAP-UCD-2017 Simulations at Tsinghua University

**Rui Wang, Renren Chen, and Jian-Min Zhang**

**Abstract** Under the LEAP-UCD-2017 project framework, simulations of centrifuge shaking table tests on gently sloping ground are conducted in this study. A unified plasticity model for large post-liquefaction shear deformation of sand is used in the simulations. The model is able to provide a unified description of sand behaviour under different states from the pre- to post-liquefaction regimes. The model parameters are calibrated against undrained cyclic triaxial test results. Using the calibrated parameters, a series of Type-B simulations are performed to predict the results of nine centrifuge tests conducted at various facilities with different settings without previously knowing the test results. Using the same simulation setup, a sensitivity study is conducted to investigate the influence of soil density and input motion on the response of the slope.

## 29.1 Introduction

The LEAP (Liquefaction Experiments and Analysis Projects, Manzari et al. 2015) project is a collaborative effort to verify and validate numerical liquefaction models, following the influential work of the VELACS project (Arulanandan and Scott 1993–1994), the results of which are still often being used in validating numerical simulation methods for soil liquefaction.

Over the past two decades, significant developments in the constitutive modelling of soil liquefaction behaviour have taken place. The phenomenon and physics of soil liquefaction is better understood, and the state dependency of sand is better described. Hence, it is important and timely that such developments be validated against actual experimental data, aided with new developments in experimental technology.

This paper presents the simulations of geotechnical centrifuge tests on gently sloping ground under the LEAP-UCD-2017 project framework. A unified plasticity model for large post-liquefaction shear deformation of sand proposed by Wang et al.

R. Wang (✉) · R. Chen · J.-M. Zhang
Department of Hydraulic Engineering, Tsinghua University, Beijing, China
e-mail: wangrui_05@mail.tsinghua.edu.cn; zhangjm@mail.tsinghua.edu.cn



B. Kutter et al. (eds.), *Model Tests and Numerical Simulations of Liquefaction and Lateral Spreading*, https://doi.org/10.1007/978-3-030-22818-7_29

(2014) is used in the simulations of this study. The model is calibrated and used in Type-B simulations of centrifuge tests and sensitivity studies.

## 29.2 Constitutive Model

### 29.2.1 Model Formulation

The constitutive model used for the simulations of the centrifuge tests in LEAP-UCD-2017 is based on the unified plasticity model for large post-liquefaction shear deformation of sand developed by Wang et al. (2014) of Tsinghua University with modifications made to take better consideration of the influence of material state. This constitutive model has been validated against many laboratory element tests and has previously been applied to simulations of centrifuge tests (Chen et al. 2018; Wang et al. 2016; Wang 2016). This paper only presents a brief description of the specific modifications of the original model for the calibration and simulation of the cyclic triaxial data obtained for LEAP-UCD-2017 (El Ghoraiby et al. 2017, 2019). Readers should refer to Wang et al. (2014) for the full formulation of the original model.

The function $g(\theta)$ used for the description of the shape of the critical, maximum stress ratio and reversible dilatancy surfaces in the octahedral stress plane is modified based on Zhang's (1997) original proposition to be:

$$g(\theta)=\left(\frac{1}{1+M_p(1+\sin 3\theta-\cos^2 3\theta)/6+\left(M_p-M_{p,o}\right)\cos^2 3\theta/M_{p,o}}\right)^{0.7} \tag{29.1}$$

In Eq. 29.1, $M_p$ is the peak mobilized stress ratio at triaxial compression, $\phi_f$ is the corresponding friction angle, and $M_{p,\ o}$ is the peak mobilized stress ratio under torsional shear after isotropic consolidation. The original model does not have the exponent 0.7 in Eq. (29.1), which is introduced here specifically to better represent the difference in the $p$–$q$ curves during the "butterfly orbit" between triaxial compression and extension observed in the reported triaxial tests. Compared with the triaxial test data, the original formulation overestimates the difference of peak mobilized stress ratio between triaxial compression and extension for Ottawa F65, and hence the exponent was introduced to reduce such a difference.

According to the propositions made by Shamoto and Zhang (1997) and Zhang (1997), the dilatancy of sand is decomposed into a reversible and an irreversible component through which the dilatancy during load reversal and cyclic loading can be properly reflected. The generation rate, $D_{\mathrm{re,gen}}$, of reversible dilatancy is:

$$D_{\text{re,gen}} = \sqrt{\frac{2}{3}} d_{\text{re},1} (\exp(1 + n_{\text{d}}\psi))^{-d_{\text{re},3}} (\mathbf{r_d} - \mathbf{r}) : \mathbf{n} \tag{29.2}$$

The state parameter $\psi$, proposed by Been and Jefferies (1985), is introduced to consider the dependency of sand behaviour on the current state. $d_{\text{re},1}$ and $d_{\text{re},3}$ are model parameters. $\mathbf{r}$ is the deviatoric stress ratio tensor, $\mathbf{r_d}$ is the projection of the current stress ratio on the reversible dilatancy surface, and n is a unit deviatoric tensor serving as the loading direction in deviatoric stress space in the model. The term $(\exp(1 + n_{\text{d}}\psi))^{-d_{\text{re},3}}$ in Eq. (29.2) is newly introduced compared with the original formulation here based on recent observations from undrained cyclic torsional tests on the increase of post-liquefaction shear strain as sand becomes looser.

The modified constitutive model has a total of 15 parameters, which can be calibrated using drained and undrained triaxial and torsional shear tests (El Ghoraiby et al. 2017, 2019).

### *29.2.2 Model Calibration*

Undrained cyclic triaxial tests (El Ghoraiby et al. 2017, 2019) were presented to us for the calibration of the constitutive model prior to the simulations of the centrifuge tests. The four critical state parameters are hence directly obtained from Vasko (2015) without any alteration. The elastic modulus, plastic modulus, and dilatancy parameters are fitted based on three triaxial tests with various void ratios (0.585, 0.542, and 0.515) and CSRs (0.2, 0.28, and 0.6). Figure 29.1 shows the calibration

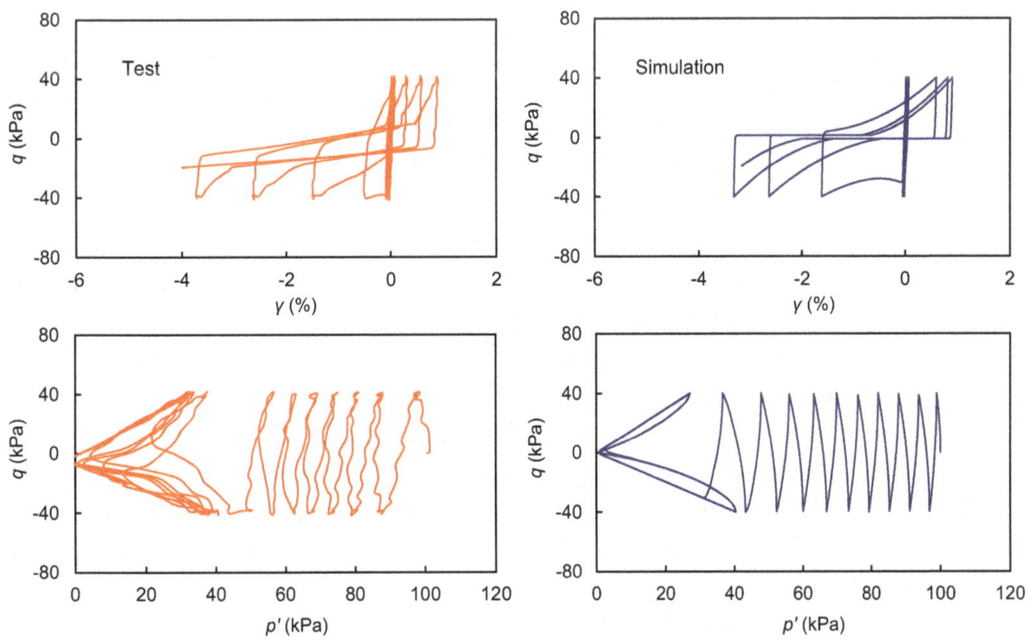


**Fig. 29.1** Calibration of an undrained cyclic triaxial test on Ottawa F65 sand with void ratio of 0.585 and CSR of 0.2

**Table 29.1** Model parameters obtained from calibration of the cyclic triaxial tests

| Sand | $G_o$ | $\kappa$ | $h$ | $M$ | $d_{\mathrm{re},1}$ | $d_{\mathrm{re},2}$ | $d_{\mathrm{re},3}$ | $d_{ir}$ |
|---|---|---|---|---|---|---|---|---|
| Ottawa F65 | 300 | 0.008 | 3.0 | 1.45 | 0.3 | 60 | 3 | 3 |
| Sand | $\alpha$ | $\gamma_{d,r}$ | $n^p$ | $n^d$ | $\lambda_c$ | $e_0$ | $\xi$ | |
| Ottawa F65 | 180 | 0.05 | 0.1 | 5.0 | 0.0112 | 0.78 | 0.715 | |

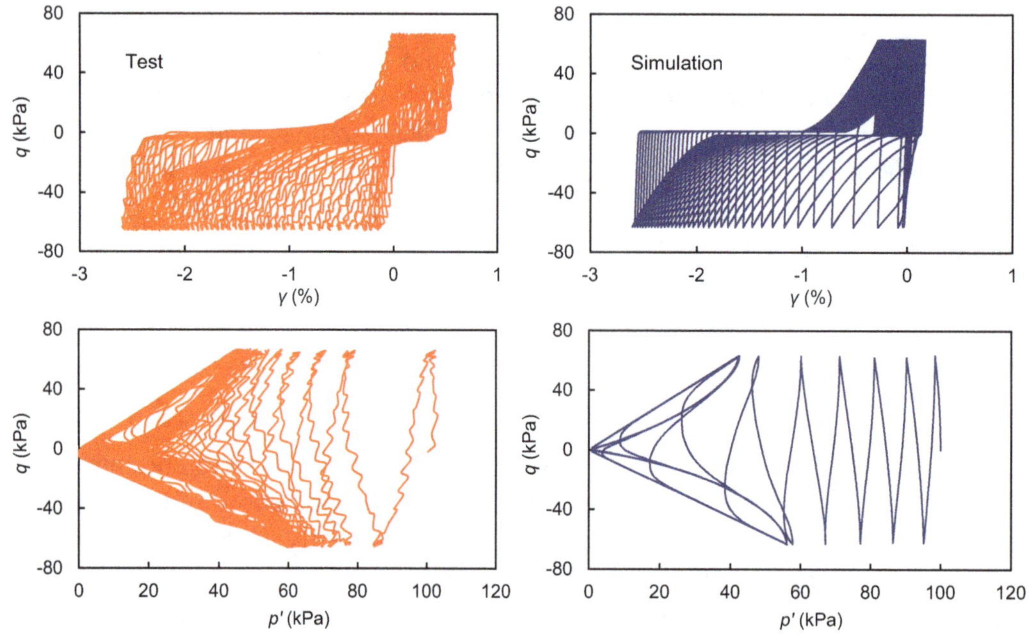


**Fig. 29.2** Simulation of an undrained cyclic triaxial test on Ottawa F65 sand with void ratio of 0.515 and CSR of 0.315

results of the undrained cyclic triaxial test on Ottawa F65 sand with void ratio of 0.585 and CSR of 0.2 (El Ghoraiby et al. 2017, 2019). The model parameters obtained from this calibration process are listed in Table 29.1. Simulations of the rest of the triaxial tests are then carried out without any further changes to the model parameters.

The simulation of a typical undrained cyclic triaxial test on Ottawa F65 sand with void ratio of 0.515 and CSR of 0.315 is plotted in Fig. 29.2. The calibration and simulation results of typical undrained cyclic triaxial tests highlights the model's ability to capture the dilatancy of sand under cyclic loading and especially the accumulation of shear strain after the sample reaches initial liquefaction. With the introduction of state dependency, the model is able to provide a unified description of sand with different densities using the same set of parameters.

The results for liquefaction strength curves obtained from the simulations are plotted in Fig. 29.3 in comparison with the test results. The number of cycles until 2.5% single amplitude axial strain is achieved for each test is used to determine the liquefaction strength curves. In general, the strength curves are well represented by the numerical simulations though the discrepancy at low cyclic stress ratio is somewhat significant.

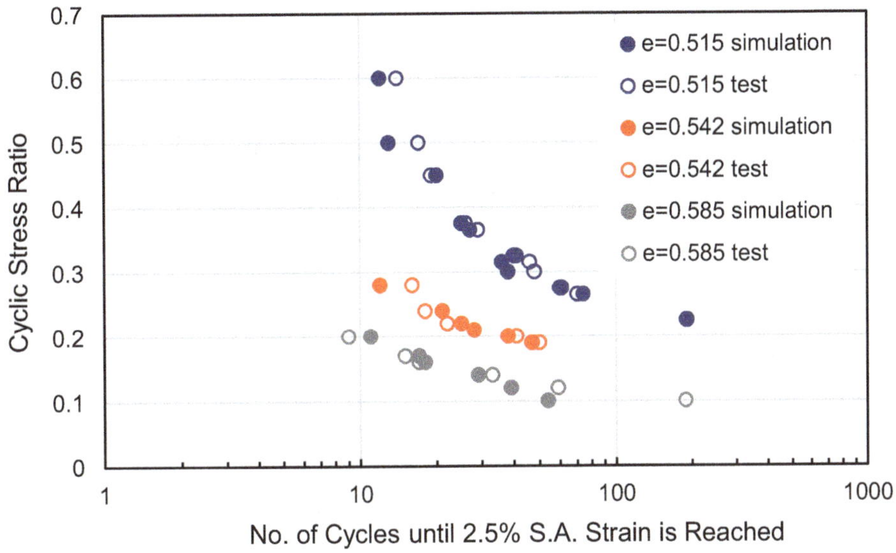


**Fig. 29.3** Liquefaction strength curves for Ottawa F64 sand from the tests and simulations

## 29.3 Setup for Numerical Simulations of Centrifuge Tests

The simulations in this study are conducted using the open source finite element framework, OpenSees (McKenna and Fenves 2001). Solid-fluid coupled elements needed for the undrained and partially drained analysis of sand, which is essential for liquefaction analysis, are already incorporated into OpenSees (e.g. u-p elements by Yang et al. (2008)).

quadUP elements in the OpenSees framework is used for the dynamic analysis conducted. The element is a four-noded quadrilateral plane strain element that has two displacement degrees of freedom and one pore pressure degree of freedom at each node and follows the u-p formulation proposed by Zienkiewicz and Shiomi (1984).

The configuration of the mesh is illustrated in Fig. 29.4, which consists of 1280 elements and 1377 nodes. The simulations are conducted in the prototype scale. The bottom boundary of the model is constrained in both $x$ and $y$ directions and is undrained. The lateral boundaries are undrained and constrained in the $x$ direction. The model container itself is not explicitly simulated. Constant pore pressure is applied on the surface of the model. Since the water table is not reported for the centrifuge tests, we assume that the water table is level with the top of the slope to guarantee that the model is fully submersed. The curvature of the centrifuge model ground surface to compensate for the non-parallel nature of the centrifugal force is not considered in the simulations.

The Krylov Newton solution algorithm along with a ProfileSPD approach is used to solve the system of equations (OpenSees manual). Hilber-Hughes-Taylor (HHT) time integration scheme is used. The boundary conditions are enforced using the

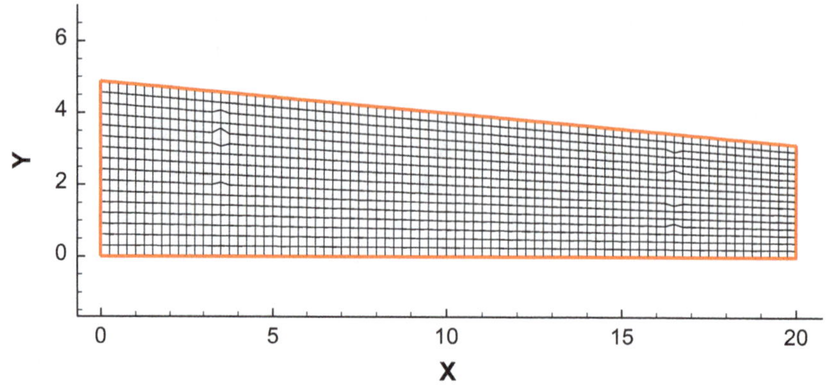


**Fig. 29.4** Finite element mesh for LEAP-UCD-2017 simulations

**Table 29.2** Initial void ratio and density of each centrifuge test provided for the LEAP-UCD-2017 simulations

| Test | $e_{in}$ | $\rho$ |
|---|---|---|
| CU-2 | 0.650268 | 1.999838 |
| Ehime-2 | 0.59971 | 2.031437 |
| KAIST-1 | 0.557724 | 2.059238 |
| KAIST-2 | 0.66405 | 1.991557 |
| KyU-3 | 0.618815 | 2.019264 |
| NCU-3 | 0.604116 | 2.028604 |
| UCD-1 | 0.591592 | 2.036698 |
| UCD-3 | 0.598311 | 2.03234 |
| ZJU-2 | 0.650062 | 1.999962 |

penalty method with the penalty number of $10^{12}$. For the test of convergence, a norm of the displacement increment test is used with a $10^{-3}$ tolerance and maximum number of iterations of 50.

The constitutive model described in the previous section is implemented in OpenSees as described by Wang et al. (2014). The model parameters used are as listed in Table 29.1, which are kept constant during the entire simulation process. The same constant permeability of $1.4 \times 10^{-4}$ m/s is used for all the analyses in this study. The initial void ratio and density used for each test are listed in Table 29.2.

The input motions for the Type-B simulations follow the exact motions provided to us from the centrifuge tests, including both horizontal and vertical motions. The input motions used for the NCU-3 and UCD-3 centrifuge test simulations are plotted in Fig. 29.5 as typical examples. For the sensitivity analysis, the motions designated for LEAP-UCD-2017 are used. Prior to the seismic event simulations, each simulation involves a gravity step to obtain the initial state of stresses in the model under centrifugal acceleration. The initial pore pressure, and the vertical effective, horizontal, and shear stress after spin-up and before shaking in a typical simulation are illustrated in Figs. 29.6, 29.7, 29.8, and 29.9. Note that due to the constant pore pressure boundary condition soil surface, the pore pressure contour in Fig. 29.6 is horizontally layered.

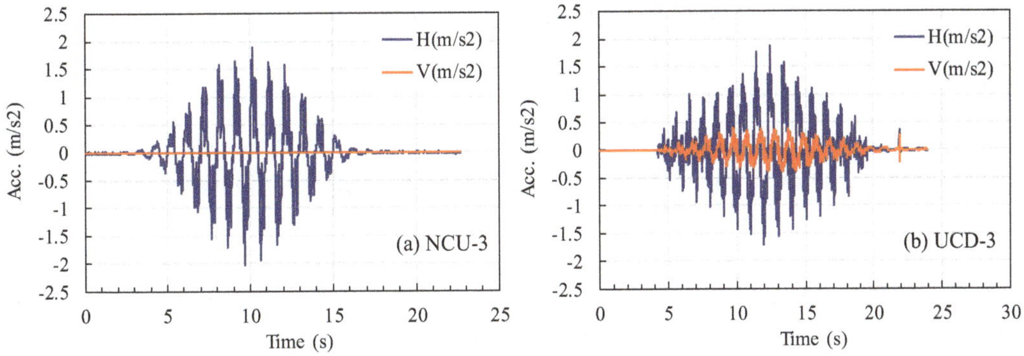


**Fig. 29.5** Examples of input motions used in the Type-B simulations (**a**) NCU-3; (**b**) UCD-3

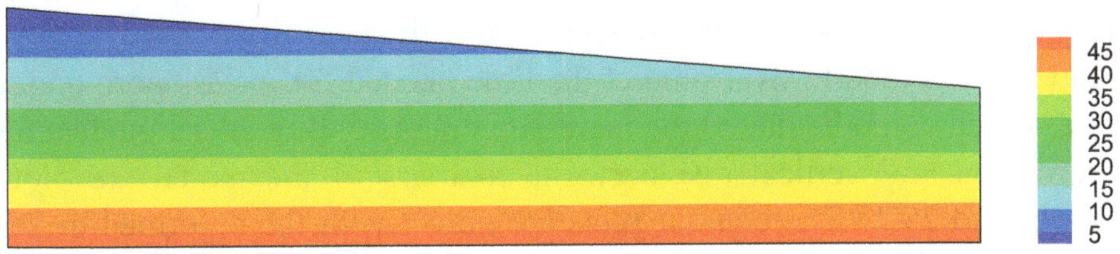


**Fig. 29.6** Initial pore water pressure (kPa) in a typical simulation

**Fig. 29.7** Initial vertical effective stress (kPa) in a typical simulation

**Fig. 29.8** Initial horizontal effective stress (kPa) in a typical simulation

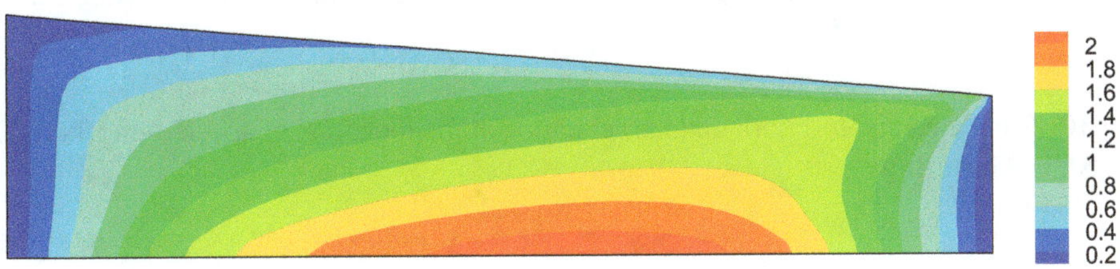


**Fig. 29.9** Initial shear stress $\sigma_{xy}$ (kPa) in a typical simulation

## 29.4 Type-B Simulation Results

### *29.4.1 Typical Results*

This section exhibits the results of a typical dynamic analysis on the NCU-3 centrifuge test. Figure 29.10 shows the simulated and experimentally measured horizontal accelerations at the locations of accelerometers AH1-AH4. At greater depths (AH1), the numerical and experimentally obtained accelerations match well. However, the Type-B simulations do not capture the spikes in the recorded acceleration time histories. The spikes are most likely caused by the sudden increase in soil stiffness as the soil leaves liquefaction during shear due to dilatancy. This suggests that the parameters used in these simulations likely underestimates the reversible dilatancy component. As for the pore pressure time histories in Fig. 29.11, the simulation results tend to over-predict the time needed for excess pore pressure to dissipate, which could be due to the increase in permeability after liquefaction. The final horizontal and vertical displacement contours of the model are plotted in Figs. 29.12 and 29.13.The contour plots are drawn with the deformed mesh, with deformation amplified by 4 times. The maximum horizontal displacement in the simulation appears slightly downslope of the center and reaches 0.398 m, which is greater than the maximum horizontal displacement measured in test NCU-3 of 0.349, and is also greater than the average horizontal displacement of the two centermost markers at 0.279 m.

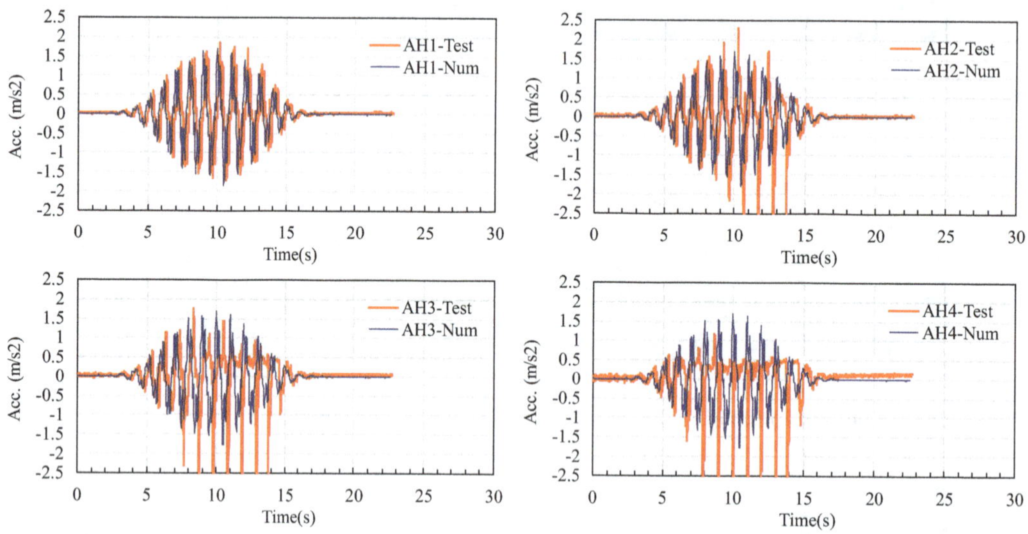


**Fig. 29.10** Simulated and test horizontal acceleration at the locations of accelerometers AH1-AH4 for the NCU-3 centrifuge test

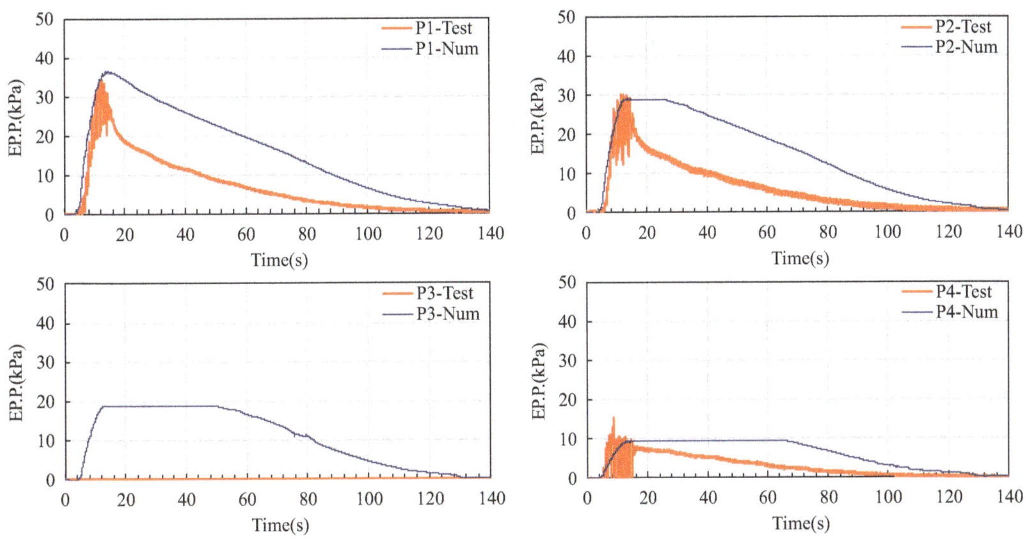


**Fig. 29.11** Simulated and test excess pore pressure at the locations of pore pressure transducers P1-P4 for the NCU-3 centrifuge test

**Fig. 29.12** Final horizontal displacement contour (deformation of the model is amplified by 4 times, unit: m)

**Fig. 29.13** Final vertical displacement contour (deformation of the model is amplified by 4 times, unit: m)

### *29.4.2 Overall Comparison Between Simulation and Test Results*

The simulated maximum horizontal displacement in all nine centrifuge tests used for the Type-B predictions are presented and compared with the measured maximum horizontal displacement and average horizontal displacement of the two centermost markers in Table 29.3. If the test results are used in a non-biased comparison with the

**Table 29.3** Simulated maximum horizontal displacement compared with the measured horizontal displacement in all nine centrifuge tests used for the Type-B predictions

| Test No | Test ID | $D_r$ | Motion | $PGA_{eff}$ (g) | Average horizontal displacement of center 2 markers in test (m) | Maximum horizontal displacement in test (m) | Maximum horizontal displacement in simulation (m) |
|---|---|---|---|---|---|---|---|
| 1 | CU-2 | 0.463 | 1 | 0.195 | 0.490 | 0.520 | 0.462 |
| 2 | Ehime-2 | 0.638 | 1 | 0.158 | 0.100 | 0.160 | 0.418 |
| 3 | KAIST-1 | 0.782 | 2 | 0.168 | 0.004 | 0.009 | 0.312 |
| 4 | KAIST-2 | 0.416 | 2 | 0.166 | 0.007 | 0.205 | 0.456 |
| 5 | KyU-3 | 0.572 | 2 | 0.133 | 0.200 | 0.266 | 0.406 |
| 6 | NCU-3 | 0.622 | 1 | 0.176 | 0.279 | 0.349 | 0.398 |
| 7 | UCD-1 | 0.666 | 1 | 0.149 | −0.002 | 0.079 | 0.343 |
| 8 | UCD-3 | 0.642 | 1 | 0.183 | 0.160 | 0.306 | 0.372 |
| 9 | ZJU-2 | 0.464 | 1 | 0.148 | 0.263 | 0.315 | 0.415 |

simulation results, the displacements predicted in the Type-B simulations in this study generally overestimate lateral deformation of the soil slope.

However, it should be pointed out that the inconsistencies in the centrifuge test results may in fact hinder such comparisons. Take tests NCU-3 and UCD-3 as two examples. The sand in the UCD-3 test is slightly denser than that in the NCU-3 test, while the peak input acceleration of UCD-3 is slightly greater than that in the NCU-3 test. Intuitively, the horizontal displacement in these two tests should be similar in terms of both value and distribution. The measured results in Fig. 29.14 contradicts such understandings. The measured horizontal displacement in test UCD-3 is significantly smaller than that in the NCU-3 test. The measured maximum horizontal displacement in the UCD-3 test of 0.306 m is only an anomaly compared with the results of the other markers in the same test; the displacement of the markers are generally less than half of those in the NCU-3 test. The average horizontal displacement of the two centermost markers is 0.349 m in the NCU-3 test, while being only 0.160 m in the UCD-3 test. Such drastic discrepancies in the deformation of the slopes of two experiments with very similar settings is very hard to explain and is certainly not possible to be simulated by the same procedure using the same parameters. The simulation results show that the Type-B prediction of maximum horizontal displacement for the two tests are 0.398 m and 0.372 m, respectively. In a non-biased comparison, one can say that the displacement in the UCD-3 test is significantly "overestimated" though such an overestimation is mostly due to the variation in test results rather than being due to deficiencies in the numerical simulations.

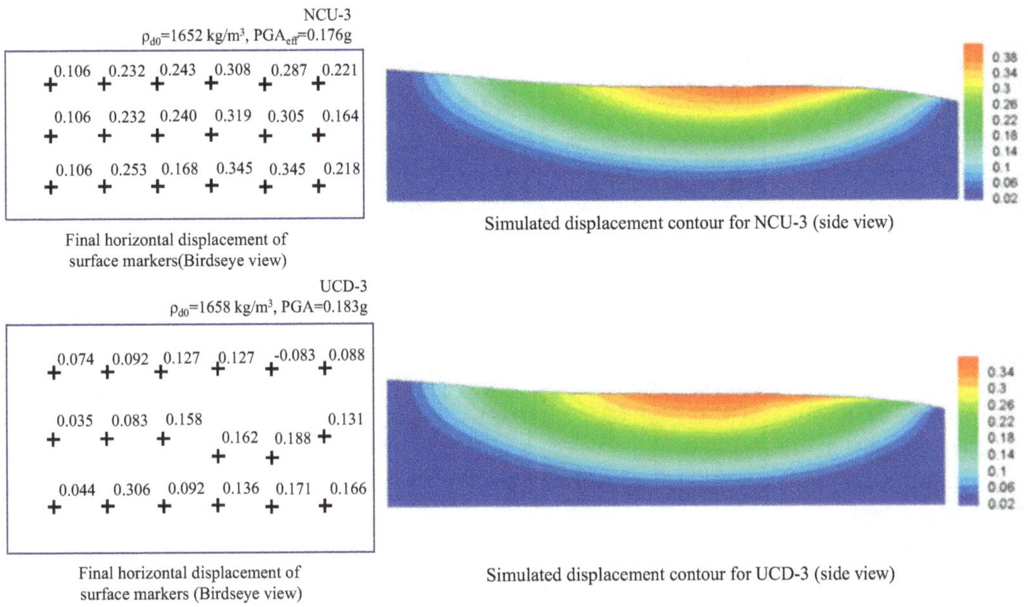


**Fig. 29.14** Comparison of simulated and measured horizontal displacement for NCU-3 and UCD-3 tests. The measured test results are presented in Birdseye views of the surface markers. The simulation results are presented in side view contours (m)

## 29.5 Sensitivity Study

Using the same material parameters and model geometry setup, a sensitivity analysis is carried out in this study to analyze the influence of soil density, input motion intensity, high frequency component of the input motion on the displacement, and liquefaction of the sloping ground. The soil density and input motion information are listed in Table 29.4. Three different relative densities, including 50%, 65%, and 75%, are adopted in this study. The same input motion with the PGA scaled up or down is used for simulations NS-1 to NS-5; another motion with greater high frequency component is used for simulations NS-6 and NS-7.

The simulation results show that at high relative density, i.e. 65–75%, the relative density significantly influences the soil displacement; this is due to the fact that as the sand becomes denser, the soil becomes more difficult to reach liquefaction, which is evident from the duration of liquefaction after the end of shaking. For the same reason, the influence of input motion PGA is significant at low PGA levels. For example, the soil displacement in NS-5 is only around half of that in NS-1.

The high frequency component of the input motion is shown to have little influence on the liquefaction and displacement of the soil. The high frequency component of the input motion in NS-6 is significantly greater than that in NS-1, and the 1 Hz component is slightly smaller. The resulting displacement of NS-6 is slightly smaller than that of NS-1, suggesting that the 1 Hz component plays a much more significant role on the soil response than the high frequency component.

Table 29.4 Information and results of the sensitivity study

| Simulation # | NS-1 | NS-2 | NS-3 | NS-4 | NS-5 | NS-6 | NS-7 |
|---|---|---|---|---|---|---|---|
| Dry density (kg/m$^3$) | 1651 | 1608 | 1683 | 1651 | 1651 | 1651 | 1651 |
| Soil | Ottawa F65 | Ottawa F65 | Ottawa F65 | Ottawa F65 | Ottawa F65 | Ottawa F65 | Ottawa F65 |
| $D_r$ (assuming $\rho_{max} = 1765$ & $\rho_{min} = 1476$ kg/m$^3$) | 65% | 50% | 75% | 65% | 65% | 65% | 65% |
| PGA (g) | 0.150 | 0.150 | 0.150 | 0.250 | 0.110 | 0.14 | 0.20 |
| PGA of 1 Hz component (g) | 0.135 | 0.135 | 0.135 | 0.270 | 0.099 | 0.11 | 0.16 |
| PGA of the high frequency component (g) | 0.021 | 0.021 | 0.021 | 0.035 | 0.015 | 0.08 | 0.11 |
| Simulation result: X-displacement at middle point on the specimen surface (m) | 0.365 | 0.384 | 0.283 | 0.425 | 0.184 | 0.328 | 0.398 |
| Simulation result: Duration of liquefaction at P4 after end of shaking (s) | 72 | 68 | 55 | 74 | 36 | 65 | 74 |

## 29.6 Conclusion

This study describes the simulations conducted at Tsinghua University for LEAP-UCD-2017. The calibration of the constitutive model, Type-B simulation of nine centrifuge tests, and sensitivity studies are presented in this study.

A unified plasticity model for large post-liquefaction shear deformation of sand is used in the simulations of this study; the model is able to provide a unified description of sand behaviour under different states from the pre- to post-liquefaction regimes. The constitutive model is calibrated against undrained cyclic triaxial tests with various void ratios and CSRs with the results highlighting the models' advantage in capturing the accumulation of shear strain during cyclic loading after initial liquefaction. In general, the strength curves of the tests are well represented by the numerical simulation results though the discrepancy at low cyclic stress ratio is somewhat significant.

Applying the model parameters obtained from the calibration process, the open source finite element framework, OpenSees, is used to conduct Type-B simulations of nine centrifuge tests with different soil densities and input motions. The simulation results show that the deformation of the model slopes of some tests are well predicted while there are significant differences between simulation and test slope displacement for a few other tests. This is to some extent due to the discrepancies between the tests conducted at various facilities. The dilatancy spikes observed in the tests are not captured in the simulations, possibly due to a relatively small dilatancy parameter used. Another notable difference between simulation and test is that the post-shaking dissipation of excess pore pressure in the simulations takes much longer, possibly due to the increase in permeability after liquefaction in actual tests.

Sensitivity studies on the influence of soil density, input motion intensity, high frequency component of the input motion on the displacement, and liquefaction of the sloping ground were carried out. As expected, the relative density and input motion intensity are found to be influential factors. The high frequency component of the input motion is shown to have little influence on the liquefaction and displacement of the soil.

**Acknowledgements** The authors would like to thank the National Natural Science Foundation of China (No. 51678346 and No. 51708332) and the State Key Laboratory of Hydroscience and Engineering Project (2018-KY-04) for funding the work presented in this chapter.

## References

Arulanandan, K., & Scott, R. F. (Eds.). (1993). *Verification of Numerical Procedures for the Analysis of Soil Liquefaction Problems*. Rotterdam: A.A. Balkema.

Been, K., & Jefferies, M. G. (1985). A state parameter for sands. *Geotechnique, 35*(2), 99–112.

Chen, R., Taiebat, M., Wang, R., Zhang J. M. (2018). Effects of layered liquefiable deposits on the seismic response of an underground structure. *Soil Dynamics and Earthquake Engineering*, 113, 124–135.

El Ghoraiby, M. A., Park, H., & Manzari, M. T. (2017). *LEAP 2017: Soil Characterization and Element Tests for Ottawa F65 Sand*. Washington, DC: The George Washington University.

El Ghoraiby, M. A., Park, H., & Manzari, M. T. (2019). Physical and mechanical properties of Ottawa F65 sand. In B. Kutter et al. (Eds.), *Model tests and numerical simulations of liquefaction and lateral spreading: LEAP-UCD-2017*. New York: Springer.

Manzari, M. T., Kutter, B. L., Zeghal, M., Iai, S., Tobita, T., Madabhushi, S. P. G., Haigh, S. K., Mejia, L., Gutierrez, D. A., & Armstrong, R. J. (2015). LEAP projects: Concept and challenges. In *Proceedings: Fourth International Conference on Geotechnical Engineering for Disaster Mitigation and Rehabilitation (4th GEDMAR), 2014 Sept 16–18*. Kyoto, Japan: Taylor & Francis.

McKenna, F., & Fenves, G. L. (2001). *OpenSees Manual*. PEER Center. http://OpenSees.berkeley.edu.

Shamoto, Y., & Zhang, J. M. (1997). Mechanism of large post-liquefaction deformation in saturated sands. *Soils and Foundations, 2*(37), 71–80.

Vasko, A. (2015). *An Investigation into the Behavior of Ottawa Sand Through Monotonic and Cyclic Shear Tests*. Masters Thesis, The George Washington University.

Wang, R. (2016). *Single Piles in Liquefiable Ground: Seismic Response and Numerical Analysis Methods*. Springer.

Wang, R., Fu, P., & Zhang, J. M. (2016). Finite element model for piles in liquefiable ground. *Computers and Geotechnics, 72*, 1–14.

Wang, R., Zhang, J. M., & Wang, G. (2014). A unified plasticity model for large post-liquefaction shear deformation of sand. *Computers and Geotechnics, 59*, 54–66.

Yang, Z., Lu, J., & Elgamal, A. (2008). *OpenSees Soil Models and Solid-Fluid Fully Coupled Elements User Manual*. UCSD. http://cyclic.ucsd.edu/opensees/OSManual_UCSD_soil_models_2008.pdf.

Zhang, J. M. (1997). *Cyclic Critical Stress State Theory of Sand With Its Application to Geotechnical Problems*. PhD thesis, Tokyo Institute of Technology, Tokyo.

Zienkiewicz, O. C., & Shiomi, T. (1984). Dynamic behaviour of saturated porous media; the generalized biot formulation and its numerical solution. *International Journal for Numerical and Analytical Methods in Geomechanics, 8*(1), 71–96.

# Chapter 30
# Application of a SANISAND Model for Numerical Simulations of the LEAP 2017 Experiments

**Ming Yang, Andres R. Barrero, and Mahdi Taiebat**

**Abstract** Numerical simulations of LEAP-UCD-2017 were performed to validate the numerical modeling approach and provide insight to capabilities and limitations of the adopted constitutive model. This chapter focuses on using an extended version of the SANISAND constitutive model implemented in $FLAC^{3D}$ program at UBC. The constitutive model was calibrated based on the available laboratory element tests on Ottawa F65 sand. It was then used for simulation of the centrifuge tests on a mildly sloping liquefiable ground of the same soil subjected to dynamic loading. The study covered the Types B and C simulations and the sensitivity analyses. Type B simulations were successful in capturing some aspects of measurements from the experiments. A simplified approach for changing the soil permeability was adopted in Type C simulations, and the improved simulation results were again compared with those measured in the experiments. In the numerical sensitivity analyses, the model appeared to provide reasonable trends for simulation of different sample densities, and ground motion intensities and frequency contents.

## 30.1 Introduction

In geotechnical engineering, numerical modeling has become significantly more popular with the fast development of modeling tools and techniques. The key component of various continuum mechanics based numerical platforms for solving boundary value problems (BVP) is proper capturing of the material stress-strain response, or the material constitutive model. In recent decades, many advanced and sophisticated constitutive models have been developed, evaluated against laboratory element tests, and used in the analysis of dynamic problems related to soil liquefaction. Soil constitutive models are in a wide spectrum of sophistication. The developers often seek a balance between including sufficient features for capturing the key

M. Yang · A. R. Barrero · M. Taiebat (✉)
Department of Civil Engineering, University of British Columbia, Vancouver, BC, Canada
e-mail: mtaiebat@civil.ubc.ca



B. Kutter et al. (eds.), *Model Tests and Numerical Simulations of Liquefaction and Lateral Spreading*, https://doi.org/10.1007/978-3-030-22818-7_30

aspects of the mechanical response of soil and also keeping the models and, more importantly, their calibration sufficiently simple and straightforward, not an easy task particularly for a complex natural material like soil. The laboratory element tests are essential in the development of the constitutive models as they provide insight into the mechanical response of soil and can be used in evaluation of the range of applicability of the constitutive models. In the big picture, the goal of modeling in this field is the successful simulation of BVPs to provide an analysis-based insight for practical cases of interest in geotechnical engineering. As such, the experimental test on representative BVPs is also very essential in evaluation of the numerical modeling methods including (but not limited to) the role of the material constitutive models and their range of applicability and limitations. For geotechnical problems, centrifuge tests are particularly of interest in providing insights into the response of soil system BVPs. They have evolved and improved over the last three decades and can be sources of useful data for specific problems related to soil response subjected to dynamic loading. The Liquefaction Experiments and Analysis Project (LEAP) aims specifically at the combined use of advanced centrifuge testing and numerical modeling as described by Manzari et al. (2014).

The LEAP-UCD-2017 focused on evaluating the seismic response of a submerged, medium dense, clean sand with a mildly sloping surface subjected to base excitation. The exercise consisted of three phases. Phase 1 is on the constitutive model calibration based on the provided laboratory element tests on Ottawa F65 sand. Phase 2 is on the simulation of nine centrifuge experiments without knowing the experiment results; this is also referred to as Type B simulation. Phase 3 is re-evaluating the model simulation of the centrifuge experiments but now after knowing the centrifuge tests results; this is referred to as Type C simulation. In Type C simulations, the numerical modeling teams had the option to tune the constitutive model or the other simulation parameters or settings so as to achieve a closer match between the experiments and simulations. Along with Type C simulation settings, the simulation teams were requested to model another seven centrifuge tests without the corresponding experimental results provided, just to illustrate the simulation sensitivity to the key input parameters including the soil relative density, and the intensity and frequency content of the ground motion; this is referred to as sensitivity analysis.

This study focuses on explaining the numerical simulations conducted by the University of British Columbia (UBC) team for the LEAP-UCD-2017. It includes using an extended version of the SANISAND constitutive model following the simulation guidelines of the project. In Sect. 30.2, the constitutive model is briefly introduced and selected results of its calibration and performance are presented based on the available laboratory element tests on Ottawa F65 sand. Section 30.3 explains the key aspects of the finite difference model used for simulation of the LEAP centrifuge experiments. The results of Type B and Type C simulations and the sensitivity analyses on the selected input parameters are presented in Sect. 30.4. Brief conclusions and outlooks from the study are presented in Sect. 30.5.

## 30.2 Constitutive Model

This section includes a brief description of the constitutive model along with the calibration process and selected element test simulation results based on the available experimental data of cyclic triaxial and direct simple shear tests under undrained conditions on Ottawa F65 sand.

### *30.2.1 Extended SANISAND Model*

The material constitutive model used by the UBC team for simulating both the element tests and the centrifuge tests in the LEAP-UCD-2017 is an extended version of the SANISAND model. The term SANISAND stands for Simple ANIsotropic SAND constitutive model. This class of sand constitutive models follows the elegant formulation of the original two-surface plasticity model developed by Manzari and Dafalias (1997). The model formulation is based on the bounding surface plasticity within the framework of critical state soil mechanics for unifying the description for different pressures and densities. The SANISAND class of models includes various extensions developed by Dafalias and Manzari (2004), Dafalias et al. (2004), Taiebat and Dafalias (2008), where the name SANISAND was first adopted, Li and Dafalias (2012) and Dafalias and Taiebat (2016). The work by Manzari and Dafalias (1997) represents the core of the constitutive model and the above-referenced subsequent works include various additional constitutive ingredients.

The version of SANISAND model with the fabric dilatancy effects by Dafalias and Manzari (2004), an overshooting correction scheme described in Dafalias and Taiebat (2016), and a work in progress extending the model formulation to capture the post-liquefaction response were considered as the constitutive model for this exercise. Inheriting 15 input parameters from the version of Dafalias and Manzari (2004), two overshooting correction parameters, and three post-liquefaction modeling parameters, this extended version includes 20 parameters as listed in Table 30.1 in detail. This model is focused on capturing stress-strain characteristics in the semi-fluidized regime during the liquefaction process. Its details are the subject of a forthcoming publication. This extended version is simply referred to as SANISAND hereafter. The model has been numerically implemented as a user-defined material model for application in the form of a dynamic link library (DLL) in the Finite Difference (FD) program *FLAC*$^{3D}$ (Barrero 2019), which was used in the present study.

### *30.2.2 Calibration Process*

In preparation for the Type B simulations, the SANISAND model was calibrated for simulation of a series of laboratory element tests on the designated Ottawa F65 sand

**Table 30.1** Calibrated model parameters for SANISAND constitutive model

| Parameter | Symbol | Value | Parameter | Symbol | Value |
|---|---|---|---|---|---|
| Elasticity | $G_0$ | 125 | Kinematic | $n^b$ | 1.15 |
| | $\nu$ | 0.05 | Hardening | $h_0$ | 5.0 |
| Critical state line | $M$ | 1.26 | | $c_h$ | 0.968 |
| | $c$ | 0.99 | Fabric dilatancy | $z_{max}$ | 18 |
| | $e_0$ | 0.78 | | $c_z$ | 500 |
| | $\lambda_c$ | 0.029 | Overshooting correction | $e_{eq}^{-p}$ | 0.01% |
| | $\xi$ | 0.7 | | $n$ | 1.0 |
| Yield surface | $m$ | 0.02 | Post-liquefaction | $x_1$ | 0.05 |
| Dilatancy | $n^d$ | 1.25 | | $x_2$ | 50 |
| | $A_0$ | 0.35 | | $x_3$ | 0.05 |

for the LEAP-UCD-2017. The experimental database includes a number of stress-controlled cyclic triaxial tests under undrained conditions, conducted and provided as part of the project to be used in the calibration process (El Ghoraiby et al. 2017, 2019). Besides, two additional sets of element test data on this type of sand were provided as supplementary data for the constitutive model calibration: monotonic and cyclic triaxial tests by Vasko (2015), and monotonic and cyclic direct simple shear tests by Bastidas (2016).

Details of the calibration process for the base model have been elaborated in the related reference as mentioned in Sect. 30.2.1, and also in Taiebat et al. (2010). The specific calibration of the model for the Ottawa F65 sand is presented in Ramirez et al. (2019) for all model parameters except for the last three that are related to the modeling of post-liquefaction. These latter parameters that are used for modeling the accumulation of post-liquefaction shear strains were approximated from the cyclic direct simple shear test data. Then the model parameters including $c$, $n^b$, $h_0$, $n^d$, $A_0$, $c_z$, $z_{max}$ were slightly adjusted to better capture the stress-strain response of both the cyclic triaxial tests of the LEAP-UCD-2017 and the cyclic direct simple shear tests of Bastidas (2016). The final values of calibrated model parameters are listed in Table 30.1. The model performance for selected undrained cyclic triaxial tests at void ratios of 0.585 and 0.515 and also for selected undrained cyclic direct simple shear tests at void ratios of 0.802 and 0.539 are presented in Figs. 30.1 and 30.2, respectively.

## 30.3 Numerical Model Development

This section includes a brief introduction to the numerical modeling platform used in the present study. It follows by description of the model set up including the spatial discretization and mechanical and fluid boundary conditions, and simulation procedures related to initial condition set up, dynamic analysis parameters such as temporal discretization and numerical damping, and other provisions followed in the Type B and Type C simulations and the sensitivity analyses.

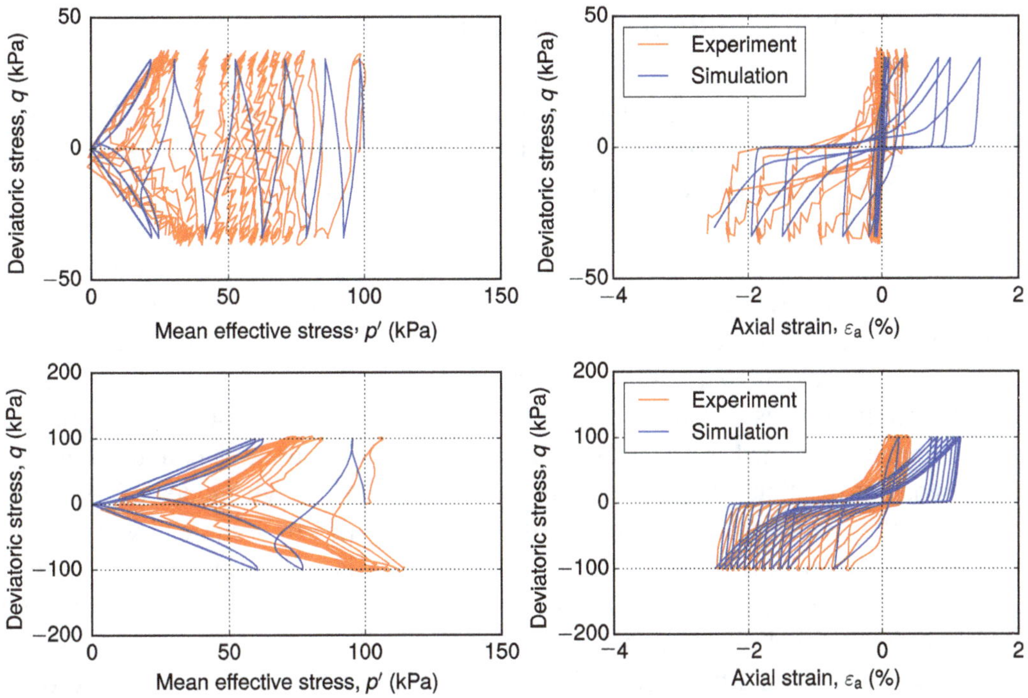


**Fig. 30.1** Simulations versus experiments in stress-controlled cyclic triaxial tests on Ottawa F65 sand (El Ghoraiby et al. 2017, 2019) with initial void ratios 0.585 (top) and 0.515 (bottom)

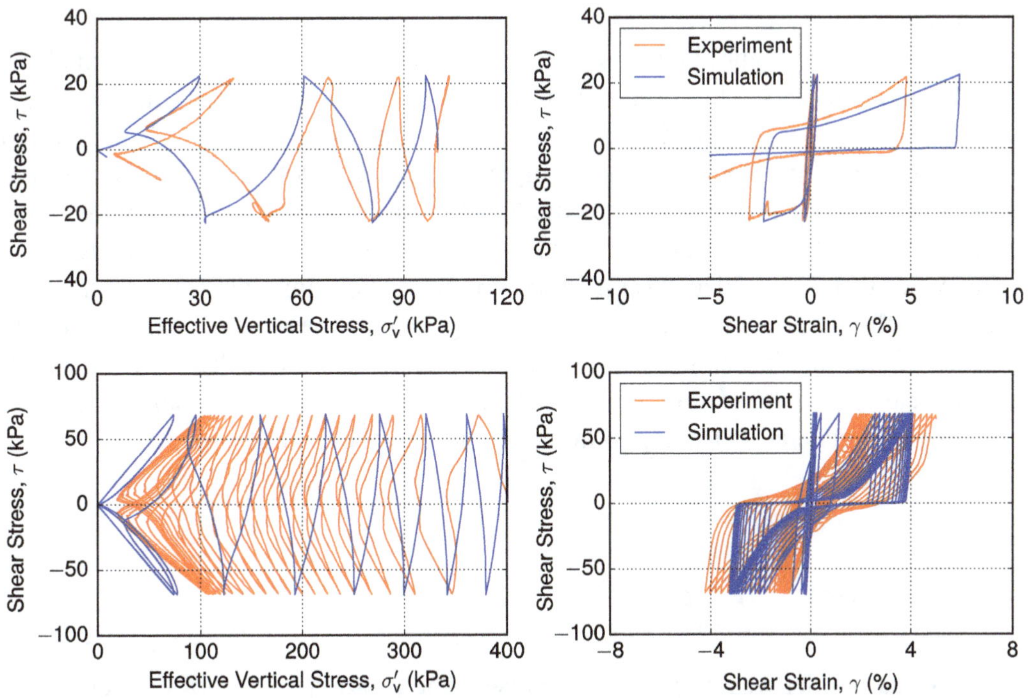


**Fig. 30.2** Simulations versus experiments in cyclic direct simple shear tests on Ottawa F65 sand with initial void ratios 0.802 (top) and 0.539 (bottom); experimental data from Bastidas (2016)

### 30.3.1 General Numerical Platform

The analysis platform used in simulating centrifuge experiments is $FLAC^{3D}$ (Itasca Consulting Group 2013). $FLAC^{3D}$ is a three-dimensional explicit finite-difference program for engineering mechanics computation, extended from *FLAC* to simulate the behavior of three-dimensional structures built of soil, rock, and other materials that undergo plastic flow when their yield limits are reached. Materials are represented by polyhedral elements within a three-dimensional grid that is adjusted by the user to fit the shape of the object to be modeled. Each element behaves according to a prescribed linear or nonlinear stress/strain law in response to the applied boundary restraints.

This platform uses constant strain rate tetrahedral elements (sub-zones) that do not generate unwanted hour-glassing modes of deformation. When used in the framework of plasticity, these elements do not provide enough modes of deformation; e.g., they cannot deform individually without change of volume. To overcome this problem, a so-called mixed-discretization process is applied in $FLAC^{3D}$. In this procedure, the discretization for the isotropic part of stress and strain tensors are different from the discretization for the deviatoric part. The explicit Lagrangian calculation scheme and the mixed-discretization zoning technique used in $FLAC^{3D}$ are to ensure proper modeling of the plastic collapse and flow of the material. Each zone (hexahedron) in this program consists of two overlays of five sub-zones (tetrahedron), and in each sub-zone a constant strain rate is assumed. Thus, quantities such as displacements and pore pressures are accessible at the grid points while stresses and strains are only accessible at the sub-zone.

Two analysis configurations exist in $FLAC^{3D}$: fluid-mechanical interaction analysis and dynamic analysis. The formulation of coupled deformation–fluid diffusion processes in $FLAC^{3D}$ is done within the framework of the quasi-static Biot's theory and can be applied to problems involving single-phase Darcy flow in a porous medium. The dynamic analysis option extends the $FLAC^{3D}$ analysis capability to a wide range of dynamic problems in disciplines such as earthquake engineering, seismology, and mine rock bursts. The calculation is based on the explicit finite difference scheme to solve the full equations of motion using lumped grid point masses derived from the real density of surrounding zones. The dynamic feature can be coupled with the fluid-mechanical interaction feature, which makes $FLAC^{3D}$ capable of modeling dynamic pore pressure generation leading to liquefaction in cyclic loading process.

### 30.3.2 Numerical Model Description

A numerical model was prepared to represent the prototype scale of the centrifuge tests conducted in LEAP-UCD-2017. This included a three-dimensional finite difference mesh with only one zone along the slope strike direction to model all the centrifuge experiments. The mesh was not allowed to deform in the slope strike direction, and, therefore, worked in the same way as a plane strain condition. The mesh density, boundary conditions, and locations of the recording sensors/control

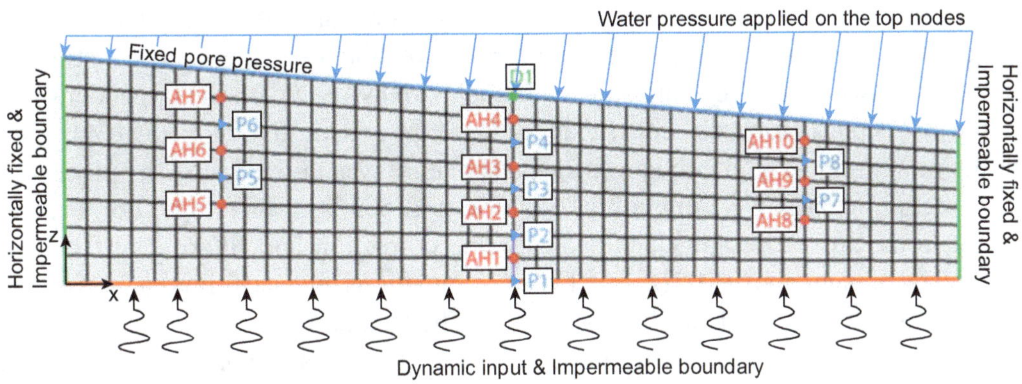


**Fig. 30.3** Two-dimensional view of the $FLAC^{3D}$ mesh showing the boundary conditions, and the recordings locations for the pore pressures (P), accelerations (AH), and displacements (D)

points are all presented in the two-dimensional view of the model geometry as shown in Fig. 30.3. In this figure AH1-AH10, P1-P8, and D1 refer to the acceleration, pore pressure, and displacement control points, respectively.

The mesh consists of 40 zones in the slope dip direction, 8 zones in the height direction, and 1 zone in the slope strike direction. The sizes of the zones are 0.5 m in the slope dip direction and 0.39–0.61 m in the height direction. In total, there are 320 zones and 738 grid points in this model. The grid points on the model base were fully constrained along all the three directions, while the grid points on the side walls were constrained laterally and the grid points on the top surface allowed full drainage with fixed values of pore pressure to model the submerged surface of the slope and replicated the situation in the centrifuge tests. After establishing the self-weight and the related initial conditions, the input motions were applied at the base grid points of the model for simulating the dynamic response of the slope, with the details presented in the following section.

### 30.3.3 Simulation Procedure

In order to establish a reasonable initial stress state of the model in the prototype scale, the numerical simulation started with stage construction of the slope where the dry soil deposit was constructed in layers and was allowed to establish the stress state under a gravity of 1*g*. In this process the mechanical boundary conditions are as shown in Fig. 30.3 except for the stress boundary condition on the top surface of the slope. A simple Mohr-Coulomb material model with bulk modulus 0.622 GPa, shear modulus 0.2385 GPa, and friction angle 33° was assigned to the sand layers to represent their stress-strain response. After the mechanical equilibrium was achieved for all layers of the slope deposit, the fluid-mechanical interaction was switched on, and a normal stress gradient representing the target submerged pressures of water was applied on the top surface of the slope, as shown in Fig. 30.3. The pore pressures at the top surface were fixed to the submerged pressures of water during the analysis. When both mechanical and fluid equilibria were obtained, the sand constitutive model was switched to the elastoplastic model SANISAND with the target densities of the soil deposits for each centrifuge test. Once the constitutive model was adjusted to the new stress state, the dynamic analysis feature was utilized, and the processed acceleration records were applied at the base grid points of the slope. For Type B simulations, this processing included baseline correction and filtering of the high frequency content above 10 Hz. For Type C simulations and sensitivity analyses, the processing only included baseline correction. The simulation continued well beyond the end of the base excitation to dissipate excess pore pressure and allowed for the resulting settlement of the soil deposit.

$FLAC^{3D}$ adopts the central finite difference approximation method for time integration, and for the simulations of the present study a mechanical ratio $10^{-6}$ was chosen as the convergence criterion in this process. In solid-fluid interaction, the water bulk modulus was chosen as $2.2 \times 10^9$ Pa, and the soil permeability was fixed to $k = 1.53 \times 10^{-4}$ m/s during the static stage of the analysis for establishing self-weight. In dynamic loading, a Rayleigh damping of 0.5% with the central frequency of 1 Hz was adopted. The soil permeability $k$ was kept unchanged during the dynamic loading in Type B simulations. Following the suggestion of Shahir et al. (2012), the permeability can be argued to increase during the liquefaction state of material. To approximately account for that suggestion, the Type C simulations and sensitivity analyses were conducted using a time-dependent permeability that was increased to 10 $k$ between $t = 6$ s and $t = 16$ s (an approximate time period representing the nearly liquefied state of the slope). It should be noted that the constitutive model parameters were kept the same for the Type B and Type C simulations and the sensitivity analyses.

## 30.4 Simulation Results

Results of the Type B and Type C numerical simulations are presented and compared with the experimental data for the nine centrifuge tests in terms of acceleration response spectra, excess pore pressures, and horizontal displacements at selected control points. Similar results are also reported from the numerical sensitivity analyses conducted on seven additional simulations that aim for variations of sand density and ground motion intensity and frequency content. Because of the limited space allowed for this chapter, a detailed description and analysis of the results will be presented in a more expanded forthcoming manuscript.

### *30.4.1 Type B and Type C Simulations*

For the Type B simulations, the numerical modeling teams were provided with the information of laboratory element tests to be used for calibration of the constitutive models, the configuration for the centrifuge tests, and the input motions. Based on this information, Type B simulations of the nine centrifuge tests (Madabhushi et al. 2019; Okamura and Nurani Sjafruddin 2019; Kim et al. 2019; Vargas Tapia et al. 2019; Hung and Liao 2019; Carey et al. 2019; Liu et al. 2019) were completed and the results for the accelerations pore pressures and horizontal displacements at the control points shown in Fig. 30.3 were reported. The corresponding experimental data were released after this stage. In Type C simulations, the simulation teams were allowed to adjust their parameters to get a better match to the centrifuge test results. The changes considered by this team in Type C simulations were limited to the use of variable permeability and not applying the high-frequency filter to the input motions, as described in the previous section.

The comparisons, at different control points, between the experimental recordings from the nine centrifuge tests and the corresponding Type B and Type C simulations are presented in this section. In the comparisons between the experiments and simulations, among the nine centrifuge tests, NCU3 is chosen to illustrate details of one centrifuge test at AH1-AH10 in the form of acceleration response spectra (Fig. 30.4) and at P1-P8 in the form of excess pore pressure time histories (Fig. 30.5). The simulation results match the experiments very well around the predominant

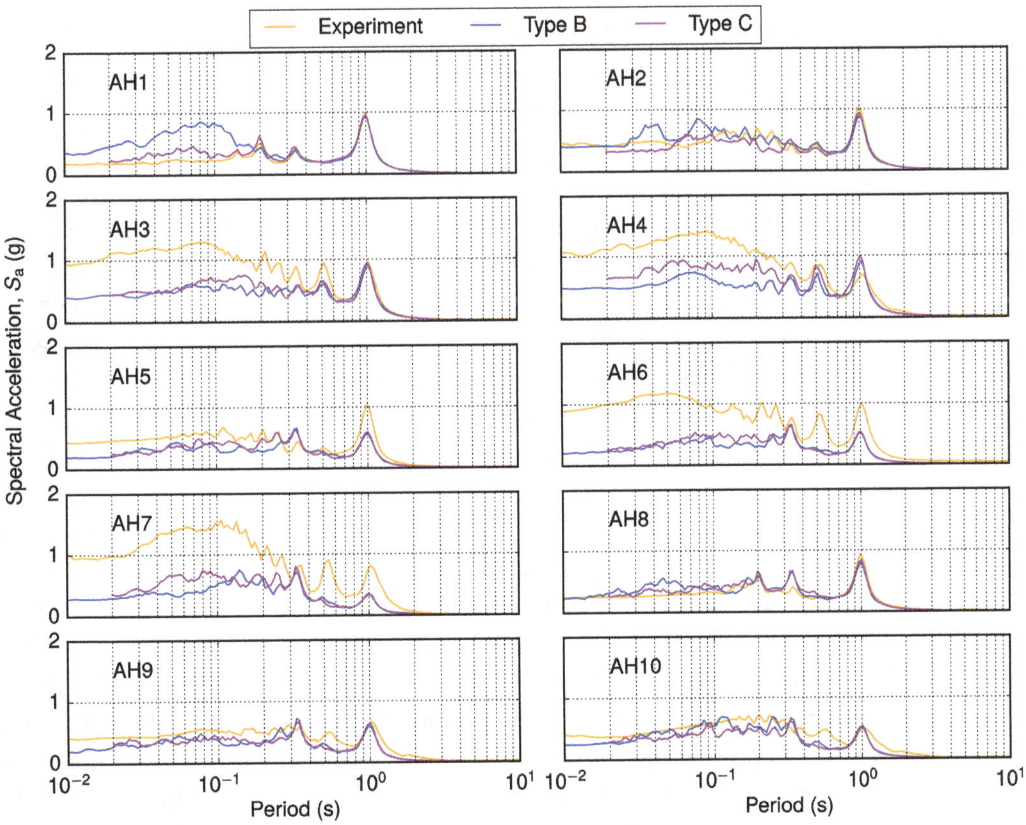


**Fig. 30.4** Simulation results for the NCU3 centrifuge experiment: acceleration response spectra

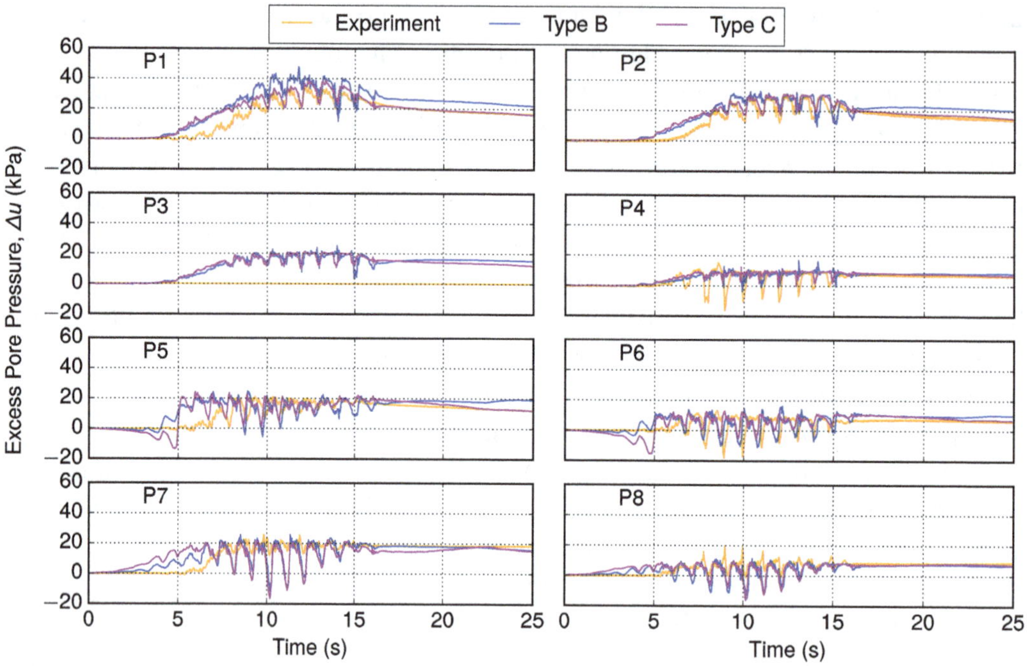


**Fig. 30.5** Simulation results for the NCU3 centrifuge experiment: excess pore pressures

frequency of the input motion (1 Hz) except the accelerometers AH5, AH6, and AH7 located on the upslope side where the spectral accelerations are underestimated. Both experiments and simulations show presence of dilation pulses in the plots of excess pore pressure. The simulated dilation pulses in P7 appear to be stronger than those reported from the centrifuge test while in P4 the simulated ones appear to be weaker. Following this detailed comparisons of spectral acceleration and excess pore pressure in all control points of NCU3, results of all the other eight tests, but only at selected points, are presented to reveal the comparative performance of numerical models in simulating the centrifuge experiments. These results are presented only at AH1 (Fig. 30.6) and AH4 (Fig. 30.7) for accelerations, and P1 (Fig. 30.8) and P4 (Fig. 30.9) for excess pore pressures. Finally results of all nine tests for the horizontal displacement time histories at D1 are compared with the corresponding ones from the simulations (Fig. 30.10).

The results suggest that in most cases the comparisons between the experiments and simulations are reasonable for the acceleration response spectra and the excess pore pressure time histories. Simulation of the acceleration response spectra at higher frequencies and also the peak values of the excess pore pressure time histories appear to have improved in Type C simulations for a number of cases. While the simulations properly capture the trend of evolution of horizontal displacement at D1, in all cases the magnitude of simulated displacement appears to be larger or considerably larger than those derived from the experiments. Again the Type C simulations have improved the simulation of the displacements at D1.

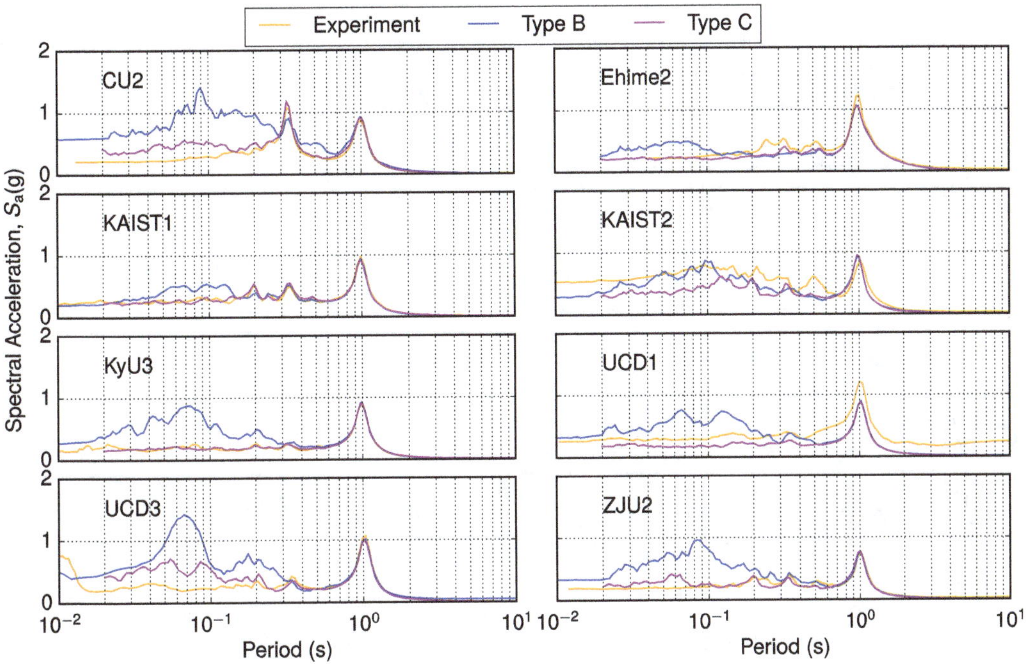


**Fig. 30.6** Simulation results for the remaining eight centrifuge experiments: acceleration response spectra at AH1

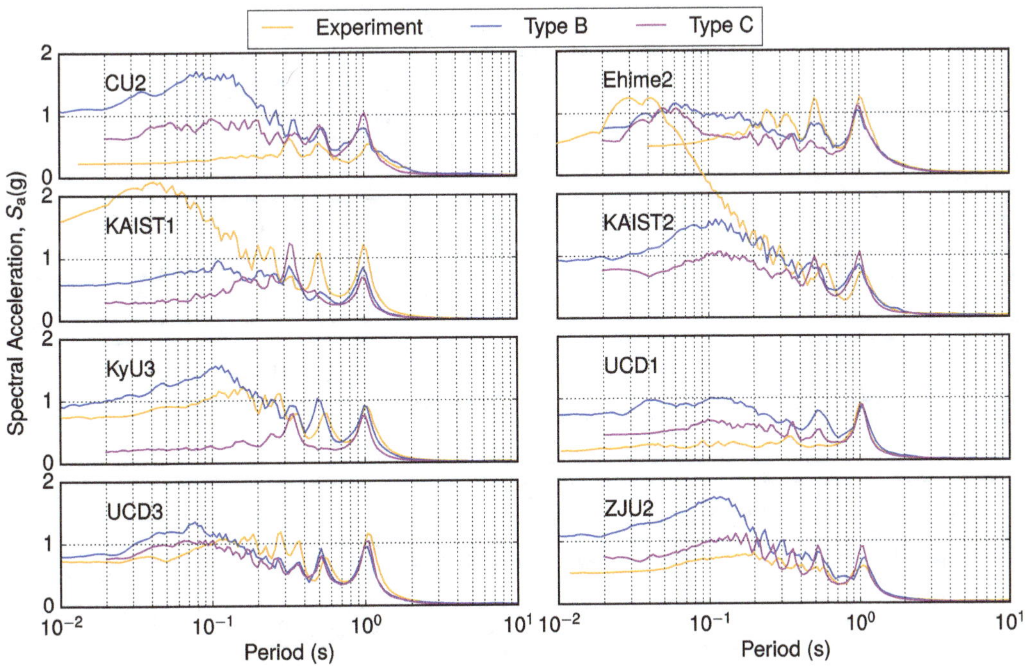


**Fig. 30.7** Simulation results for the remaining eight centrifuge experiments: acceleration response spectra at AH4

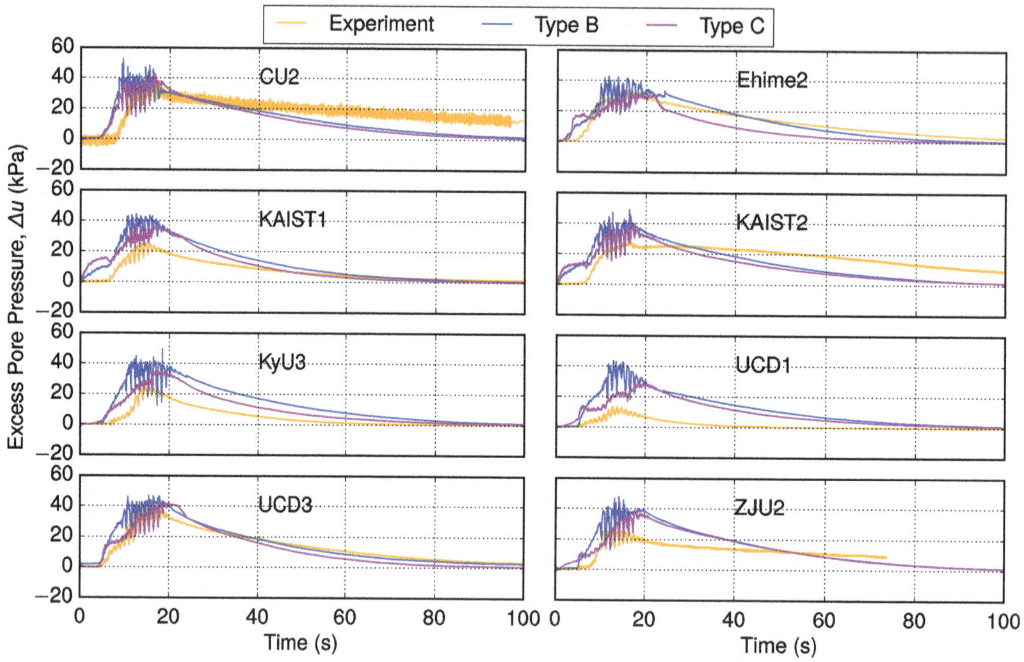


**Fig. 30.8** Simulation results for the remaining eight centrifuge experiments: excess pore pressure at P1

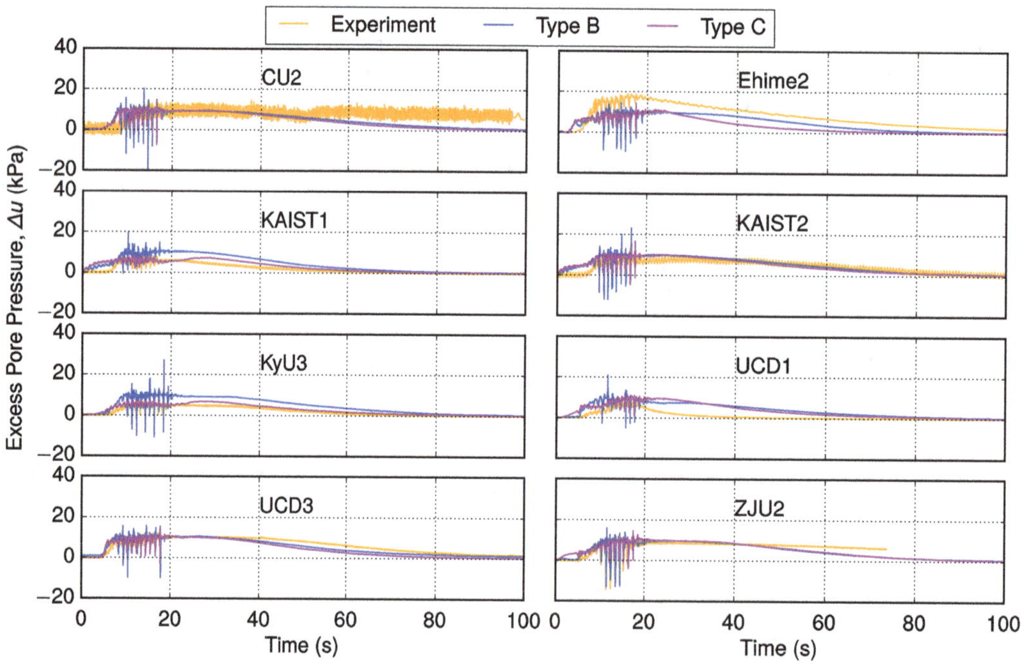


**Fig. 30.9** Simulation results for the remaining eight centrifuge experiments: excess pore pressure at P4

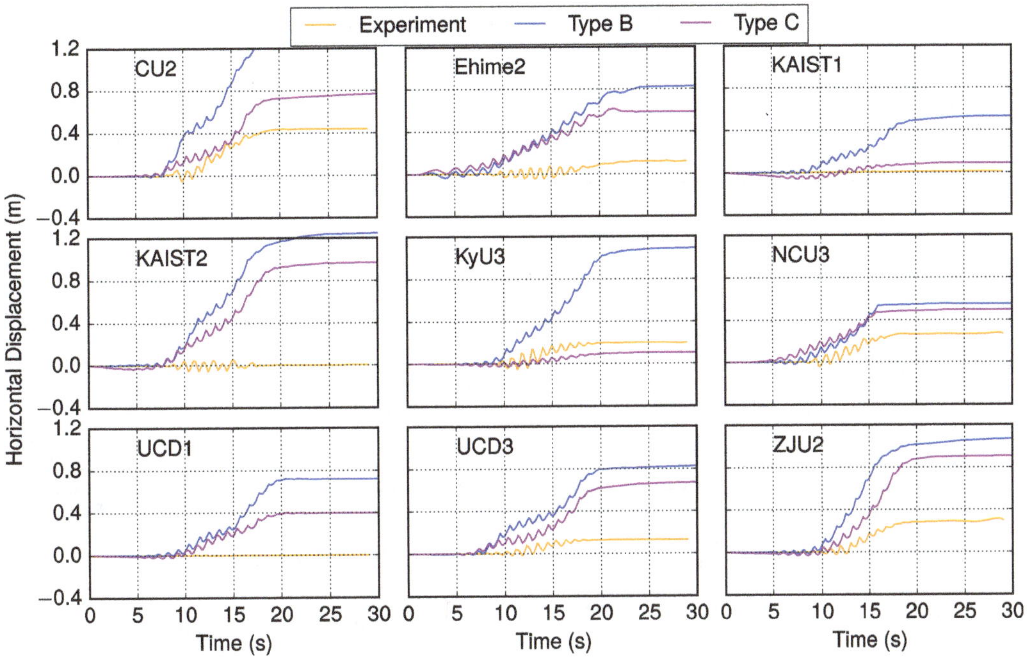


Fig. 30.10 Simulation results for all nine centrifuge experiments: horizontal displacements at D1

### *30.4.2 Sensitivity Analyses*

With the same configurations as Type C simulations, numerical sensitivity analyses were also conducted on seven planned centrifuge tests, NS1-NS7, based on the provided densities of the sand layer and characteristics of the input motion, but the corresponding experimental results were not provided. In particular, these sensitivity analyses aim at variations of sand density (NS1, NS2, and NS3), ground motion intensity (NS1, NS4, and NS5), and frequency content (NS1, NS6, and NS7). Comparative simulation results at different control points are summarized in Figs. 30.11, 30.12, 30.13, and 30.14. The trends observed in the results appear to be reasonable and are illustrative of the features embedded in the constitutive model such as density dependency and small yield locus; the effects will be discussed and elaborated in details in a more extended manuscript.

## 30.5 Summary and Outlook

The extended version of SANISAND for the semi-fluidized regime of response is expected to have improved the predictive capabilities in capturing stress-strain loops beyond the onset of liquefaction. Several aspects of response in the centrifuge testing appear to have been captured by the numerical model. The lateral displacements, however, have been overpredicted in a number of cases. This is believed to be due to the performance of the calibrated model as can be observed also in the element level.

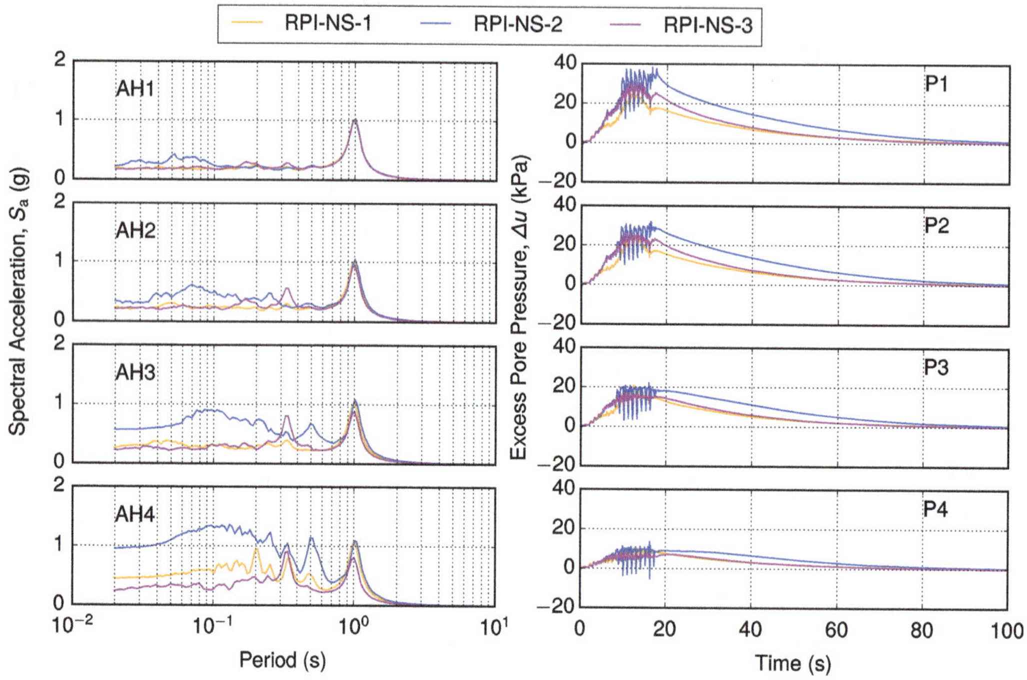


**Fig. 30.11** Simulation results for sensitivity analysis on initial density: acceleration response spectra and excess pore pressures

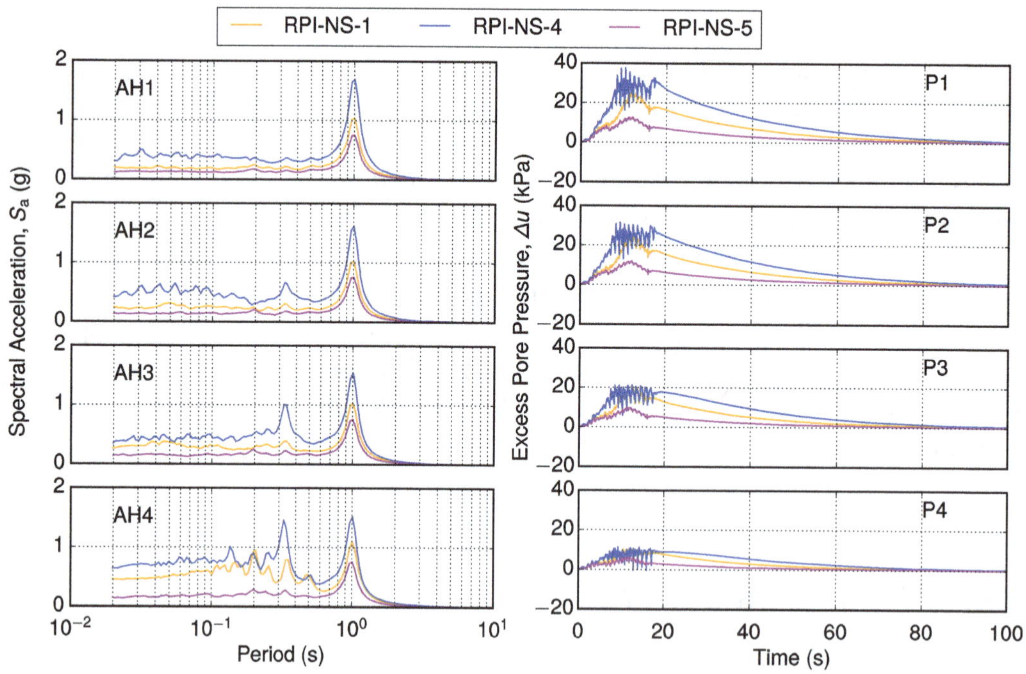


**Fig. 30.12** Simulation results for sensitivity analysis on ground motion intensity: acceleration response spectra and excess pore pressures

A lower shear-induced volumetric stiffness in the pre-liquefaction stage, and a higher shear stiffness in the post-liquefaction stage are expected to improve the predictive capability of the model. The simulation results can be revisited by using such revised group of model parameters.

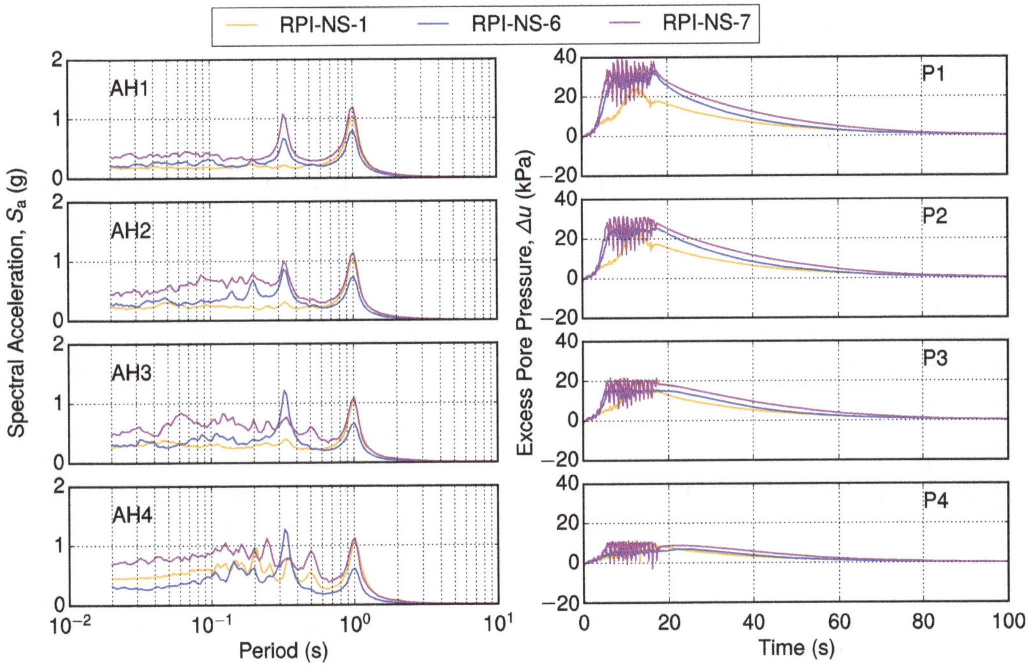


**Fig. 30.13** Simulation results for sensitivity analysis on ground motion frequency content: acceleration response spectra and excess pore pressures

**Fig. 30.14** Simulation results for sensitivity analysis: horizontal displacement at D1

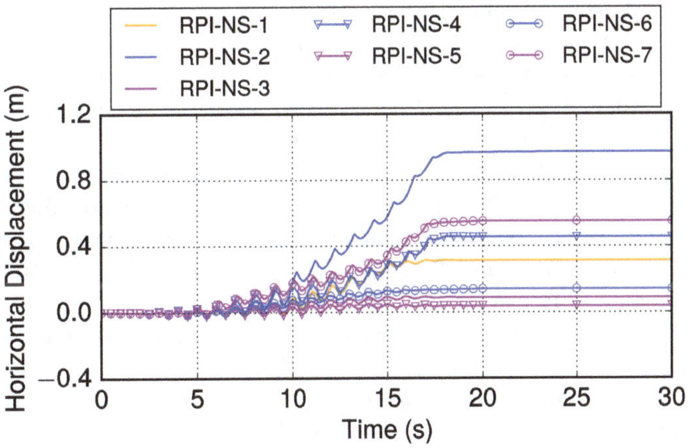


## References


Barrero, A. R. (2019). *Multi-Scale Modeling of the Response of Granular Soils Under Cyclic Shearing*. Ph.D. thesis, University of British Columbia (in preparation).

Bastidas, A. M. P. (2016). *Ottawa F-65 Sand Characterization*. Ph.D. thesis, University of California, Davis.

Carey, T. J., Stone, N., Hajialilue Bonab, M., & Kutter, B. L. (2019). LEAP-UCD-2017 centrifuge test at University of California, Davis. In B. Kutter et al. (Eds.), *Model tests and numerical simulations of liquefaction and lateral spreading: LEAP-UCD-2017*. New York: Springer.

Dafalias, Y. F., & Manzari, M. T. (2004). Simple plasticity sand model accounting for fabric change effects. *Journal of Engineering Mechanics, 130*(6), 622–634.

Dafalias, Y. F., Papadimitriou, A. G., & Li, X. S. (2004). Sand plasticity model accounting for inherent fabric anisotropy. *Journal of Engineering Mechanics, 130*(11), 1319–1333.

Dafalias, Y. F., & Taiebat, M. (2016). SANISAND-Z: Zero elastic range sand plasticity model. *Géotechnique, 66*(12), 999–1013.

El Ghoraiby, M. A., Park, H., & Manzari, M. T. (2017). *LEAP 2017: Soil Characterization and Element Tests for Ottawa F65 Sand*. Washington, DC: The George Washington University.

El Ghoraiby, M. A., Park, H., & Manzari, M. T. (2019). Physical and mechanical properties of Ottawa F65 sand. In B. Kutter et al. (Eds.), *Model tests and numerical simulations of liquefaction and lateral spreading: LEAP-UCD-2017*. New York: Springer.

Hung, W.-Y., & Liao, T.-W. (2019). LEAP-UCD-2017 centrifuge tests at NCU. In B. Kutter et al. (Eds.), *Model tests and numerical simulations of liquefaction and lateral spreading: LEAP-UCD-2017*. New York: Springer.

Itasca Consulting Group, Inc. (2013). *FLAC3D: Fast Lagrangian Analysis of Continua in 3 Dimensions, Ver. 5.01*. Minneapolis: Itasca.

Kim, S.-N., Ha, J.-G., Lee, M.-G., & Kim, D.-S. (2019). LEAP-UCD-2017 centrifuge test at KAIST. In B. Kutter et al. (Eds.), *Model tests and numerical simulations of liquefaction and lateral spreading: LEAP-UCD-2017*. New York: Springer.

Li, X. S., & Dafalias, Y. F. (2012). Anisotropic critical state theory: Role of fabric. *Journal of Engineering Mechanics, 138*(3), 263–275.

Liu, K., Zhou, Y.-G., She, Y., Peng Xia, D.-M., Huang, J.-S., Yao, G., & Chen, Y.-M. (2019). Specifications and results of centrifuge model test at Zhejiang University for LEAP-UCD-2017. In B. Kutter et al. (Eds.), *Model tests and numerical simulations of liquefaction and lateral spreading: LEAP-UCD-2017*. New York: Springer.

Madabhushi, S. S. C., Dobrisan, A., Beber, R., Haigh, S. K., & Madabhushi, S. P. G. (2019). LEAP-UCD-2017 centrifuge tests at Cambridge. In B. Kutter et al. (Eds.), *Model tests and numerical simulations of liquefaction and lateral spreading: LEAP-UCD-2017*. New York: Springer.

Manzari, M. T., & Dafalias, Y. F. (1997). A critical state two-surface plasticity model for sands. *Géotechnique, 47*(2), 255–272.

Manzari, M. T., Kutter, B. L., Zeghal, M., Iai, S., Tobita, T., Madabhushi, S. P. G., Haigh, S. K., Mejia, L., Gutierrez, D. A., Armstrong, R. J., Sharp, M. K., Chen, Y. M., & Zhou, Y. G. (2014). *LEAP projects: Concept and Challenges* (p. 109–116).

Okamura, M. & Nurani Sjafruddin, A. (2019). LEAP-2017 centrifuge test at Ehime University. In B. Kutter et al. (Eds.), *Model tests and numerical simulations of liquefaction and lateral spreading: LEAP-UCD-2017*. New York: Springer.

Ramirez, J., Barrero, A. R., Chen, L., Dashti, S., Ghofrani, A., Dafalias, Y. F., Taiebat, M., & Arduino, P. (2019). Site response in a layered liquefiable deposit: Evaluation of different numerical tools and methodologies with centrifuge experimental results. *ASCE Journal of Geotechnical and GeoEnvironmental Engineering, 144*, 10.

Shahir, H., Pak, A., Taiebat, M., & Jeremić, B. (2012). Evaluation of variation of permeability in liquefiable soil under earthquake loading. *Computers and Geotechnics, 40*, 74–88.

Taiebat, M., & Dafalias, Y. F. (2008). SANISAND: Simple anisotropic sand plasticity model. *International Journal for Numerical and Analytical Methods in Geomechanics, 32*, 915–948.

Taiebat, M., Jeremić, B., Dafalias, Y. F., Kaynia, A. M., & Cheng, Z. (2010). Propagation of seismic waves through liquefied soils. *Soil Dynamics and Earthquake Engineering, 30*(4), 236–257.

Vargas Tapia, R. R., Tobita, T., Ueda, K., & Yatsugi, H. (2019). LEAP-UCD-2017 centrifuge test at Kyoto University. In B. Kutter et al. (Eds.), *Model tests and numerical simulations of liquefaction and lateral spreading: LEAP-UCD-2017*. New York: Springer.

Vasko, A. (2015). *An investigation into the behavior of Ottawa sand through monotonic and cyclic shear tests*. Master thesis, The George Washington University.

# Chapter 31
# Numerical Simulation Trial by Cocktail Glass Model in FLIP ROSE for LEAP-UCD-2017

**Koji Ichii, Kazuaki Uemura, Naoki Orai, and Junichi Hyodo**

**Abstract** LEAP-UCD-2017 is a blind prediction exercise for numerical modelers with relatively simple centrifuge tests. However, the application of numerical simulation should be discussed from both modeler's and practitioner's viewpoints. The authors applied one of the most common dynamic analysis programs in Japan (FLIP ROSE) for this exercise. The program, FLIP ROSE, has been continuously updated for 20 years with voluntary supports from practitioners. Using this scheme, the authors discuss several issues including the possible variation in parameter determination in practice and the effect of artificial Rayleigh damping on predictions. Although only a limited number of cases of centrifuge tests were analyzed, the application of a common analysis program by practitioners pointed out some issues to be addressed by numerical modelers.

## 31.1 Introduction

LEAP (Liquefaction Experiments and Analysis Projects) is an international joint research project to discuss centrifuge modeling and numerical modeling of liquefaction. The name LEAP was proposed by one of the authors at an early phase of the project. However, the authors did not participate in the experimental program since Hiroshima University does not have a centrifuge. In other word, the authors present a numerical simulation perspective. Furthermore, the authors' main concern is not

K. Ichii (✉)
Faculty of Societal Safety Science, Kansai University (Formerly at Hiroshima University), Osaka, Japan
e-mail: ichiik@kansai-u.ac.jp

K. Uemura
Core Laboratory, Oyo Corporation, Saitama, Japan

N. Orai
Chuden Engineering Consultants, Hiroshima, Japan

J. Hyodo
Tokyo Electric Power Service Co, Tokyo, Japan



B. Kutter et al. (eds.), *Model Tests and Numerical Simulations of Liquefaction and Lateral Spreading*, https://doi.org/10.1007/978-3-030-22818-7_31

developing numerical constitutive models, but how to use numerical simulation adequately in practice. In short, the authors belong to the practitioner's side.

LEAP will contribute to verification and validation of the dynamic analysis program of soils. This is quite important for practitioners. Currently, we often used dynamic analysis programs in practice. The reliability and possible degree of variation in the computed results are quite important issues for practitioners, and the participation of practitioners in simulation in LEAP will be beneficial to the project. Thus, the authors join the simulation exercise with one of the most common dynamic analysis program in Japan: FLIP ROSE (*F*inite element analysis program for *LI*quefaction *P*rocess, *R*esponse *O*f *S*oil-structure systems during *E*arthquakes).

FLIP ROSE is a FEM program with an advanced constitutive model of soil that considers liquefaction. Both the program and the constitutive model have been continuously updated for these 20 years. These updates have been supported by enthusiastic voluntary support by practitioners. This experience indicates the importance of the participation by practitioners. The trial of LEAP-UCD-2017 exercise explained here is within the same context. With the voluntary support of the practitioners related to FLIP ROSE (FLIP Consortium WG members), the authors discussed several issues including the possible variation in parameter determination in practice and the effect of artificial Rayleigh damping. These discussions may contribute to the future direction on exercise design.

## 31.2 Numerical Analysis Code and Model Features

FLIP ROSE (ver.7.3.0) with the cocktail glass model (Iai et al. 1992, 2011) is used for element test simulation. The main features of the constitutive model are as follows:

1. The model in FLIP ROSE is based on multiple shear mechanisms. The contact forces between particles in an arbitrary direction are considered as the stress states of the soil element.
2. There are two types of constitutive models for liquefaction in FLIP ROSE. The cocktail glass model is the recent advanced model. In the cocktail glass model, dilatancy is given as the sum of contractive part and dilative part. The increment of contractive part of dilatancy $\varepsilon_d^c$ is given by the strain increment in a multiple shear mechanism $\gamma_{xy}$, and dilative part of dilatancy $\varepsilon_d^d$ is given by the current shear strain $\gamma_{xy}$ in an arbitrary direction. The value of dilative part of dilatancy was determined so as to make no work under shear. As a result, the void ratio decreases from the initial value $e_0$ to a possible limiting value $e_{\mathrm{min}}$, but the current void ratio of the soil element depends on the current strain level. Thus the shape of possible points of void ratio is similar to the shape of a cocktail glass as shown in Fig. 31.1.

One of the most unique features of the code is that more than 100 groups are using the program in practice in Japan. Before the implementation of the cocktail glass model, which is used here, another model (only applicable for fully drained or fully

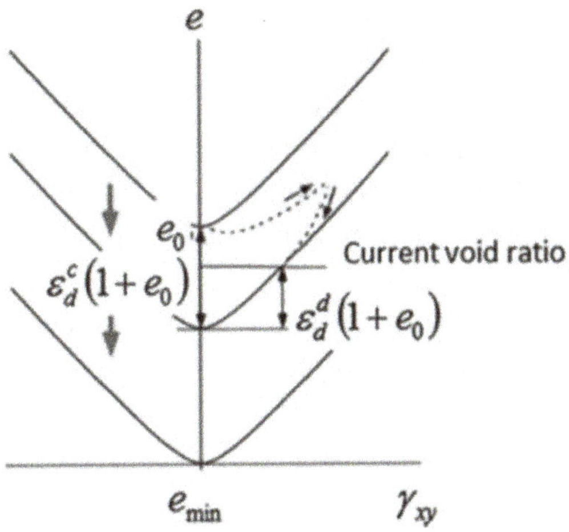


**Fig. 31.1** Schematic illustration of void ratio variation in cocktail glass model (Iai et al. 2011)

undrained conditions) was implemented in the FLIP ROSE program. This model was quite stable and showed good performance in the analysis of the structures during the 1995 Kobe earthquake. Therefore, many practitioners use it in seismic design of structures. These practitioners gradually started using the cocktail glass model in practice.

## 31.3 Parameter Calibration

### *31.3.1 Difference Between the Method Used in Practice and the Method of This Exercise*

In practice, the in-situ information should be referenced as much as possible. It is very important to have parameters with physical meaning. For example, the internal friction angle shall be determined by CD or CUB tests of in-situ samples. Shear modulus shall be determined by PS logging results. However, there are only re-constituted samples in this project. Thus our standard procedure in parameter setting cannot be used. The authors determined the parameters from the reported laboratory test results and/or standard values often used in practice in case of insufficient information.

On the other hand, there are different types of parameters, which have no apparent physical meaning. For example, the parameters controlling the dilatancy characteristics are quite model-dependent. Although these parameters may have a conceptual background, there is no explicit procedure to determine the value of the parameters. In practice, practitioners often conduct the trial-and-error fitting process to obtain the appropriate value for these parameters. Liquefaction resistance curves are often used as the target of trial-and-error to obtain the parameters for dilatancy. In this case, cyclic tests were simulated in the program to obtain the computed liquefaction resistance curves. (Iai et al. 1990)

In this exercise, the authors conducted the analysis in a two-dimensional plane strain condition. This is because 2D analysis is preferred in most of the practical cases for its simplicity. Therefore, parameter setting is also conducted in the two-dimensional plane strain condition in the simulation of the cyclic test of the soil element. In other words, a simple shear test or torsional shear test is more appropriate as the element test in this project. However, there were results of cyclic tri-axial tests, and the authors simulated these tests in the 2D condition. This constituted a difficulty in the application of our standard parameter setting procedures in this project.

### *31.3.2 Concept of Parameters and the Detailed Parameter Setting Procedure*

The followings are steps used in the project:

1. Reference confining pressure, $P_a$
   Parameter values are given at a certain value of confining stress called reference confining pressure. The reference confining pressure, $P_a$, was determined from experimental test conditions (=100 kPa). In practice, basically, it should be the in situ confining pressure.
2. Initial shear modulus, $G_{ma}$
   In practice, it should be given by the PS logging result. In the test, bender elements or the resonance column test may be good to obtain the exact value. Without these test results, initial shear modulus, $G_{ma}$, was calculated using the following equation (Kokusyo 1980):

$$G_{ma} = 840\frac{(2.17 - e_0)^2}{1 + e_0}\left(\sigma'_c\right)^{0.5} \tag{31.1}$$

   Here, $e_0$: initial void ratio, $\sigma'_c$: effective confining pressure.
   The equation is suitable for round-shaped particles. If the particle is not round shaped, a different equation shall be used. Considering the small value of friction angle in this case, the above equation was used.
3. Initial bulk modulus, $K_{La}$ and $K_{Ua}$
   Initial bulk modulus, $K_{La}$, for the loading process, and $K_{Ua}$ for the unloading process were calculated using the following equation:

$$K_{La} = K_{Ua} = \frac{2(1+\nu)}{3(1-2\nu)}G_{ma} \tag{31.2}$$

   Here, $\nu$: Poisson's ratio of soil skeleton and is assumed as 0.33.

4. Parameters mG, mK
   mG is the parameter to control the confining pressure dependency of the shear modulus, and mK is the parameter to control the confining pressure dependency of the bulk modulus. mG and mK were set 0.5 as the general values.
5. Bulk density, $\rho$
   Bulk density, $\rho$, was determined from the experimental test condition.
6. Porosity, $n$
   Porosity, $n$, was calculated using the following equation:

$$n = \frac{e_0}{1 + e_0} \tag{31.3}$$

7. Maximum damping coefficient, $h_{max}$
   Maximum damping coefficient, $h_{max}$, was set 0.24 as the general value of sand. In practice, cyclic test results should be used to obtain these dynamic characteristics of the soil. It relates to the hysteresis damping of soil. When the effect of dilatancy is neglected, i.e. fully drained condition, a hyperbolic relationship is assumed for the shear stress-shear strain relationship. The extended Masing rule is applied to obtain the stress-strain in unloading and reloading processes. In the extension of the Masing rule in the model, the limit value of hysteresis damping in quite large strain is adjusted to agree with the input value of the maximum damping coefficient.
8. Soil strength parameters: cohesion, $c$, and internal friction angle, $\phi_f$
   Cohesion, $c$, and internal friction angle, $\phi_f$, are determined from laboratory test results (stress path shown in Fig. 31.2 as an example). The average value of the test results was used.
9. Permeability, $k$
   Permeability is determined from experimental results (=0.01 cm/s). However, the process of pore water pressure dissipation and consolidation of liquefied soil were skipped in the computation exercises. This is because the deformation of

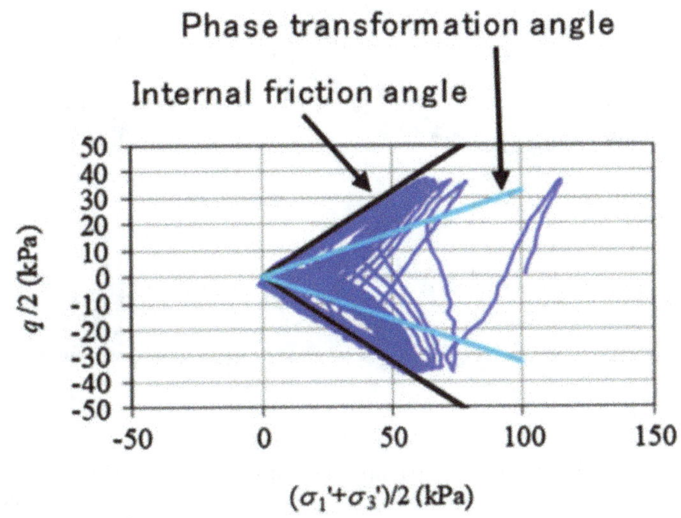


**Fig. 31.2** Example of internal friction angle determination from stress paths

simple soil layers' model without any clayey layers often stop immediately after the cessation of shaking as shown in the LEAP2014 exercise (Tobita et al. 2015), and the settlement due to consolidation is not significant (no more than 3–5%) as indicated in the experimental chart (Ishihara and Yoshimine 1992) in many cases. Thus, from a practical point of view, the authors are afraid that the computation process after shaking stopped is time consuming and less important. The authors assume that the effect of the permeability in this computation exercise is not significant and skip the detailed consideration in this exercise.

10. Phase transformation angle, $\phi_p$
    Phase transformation angle, $\phi_p$, is determined from laboratory test results (stress paths) as shown in example in Fig. 31.2.
11. Liquefaction parameters: other parameters for dilatancy
    Liquefaction parameters are necessary for successful numerical simulations of cyclic shear tests. The simulations were done for direct shear tests in a 2D plane strain condition, and the liquefaction strength curves were the main targets to be simulated. Furthermore, not only the liquefaction strength curves, but also the characteristics of strain accumulation rates with the increase in number of loading cycles were carefully considered.

Since the target of the centrifuge tests is an inclined slope, soil behaviors under initial shear stress may have important effects on the result. In other words, $K\alpha$ effect may be important. As shown in Fig. 31.3, the cocktail glass model can consider $K\alpha$ effect with appropriate choice in the parameters, which depend on the relative densities. However, due to insufficient test results of the soil behaviors under initial shear, the detailed consideration of $K\alpha$ effect was skipped in the parameter determination in this exercise.

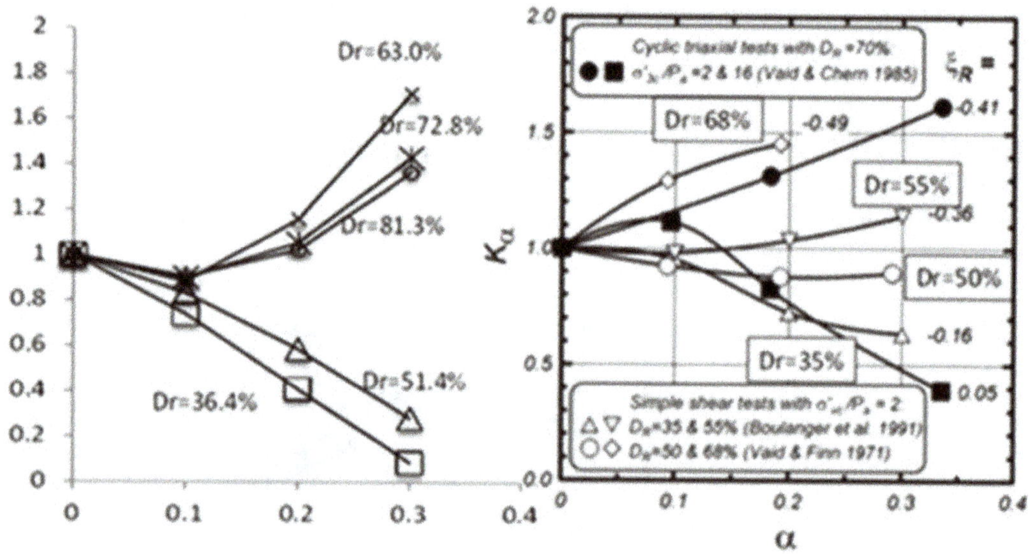


**Fig. 31.3** Example of parameter determination considering $K\alpha$ effect (left figure: results with appropriate parameter in the cocktail glass model, right figure: the target after Boulanger 2003)

### 31.3.3 Parameter Setting Results Considering the Variation in Densities

The parameters for Ottawa F65 in loose ($e_0 = 0.585$), medium ($e_0 = 0.542$), and dense conditions ($e_0 = 0.515$) were determined. The values of parameters are shown in Table 31.1 and in Table 31.2. The comparison of target liquefaction resistance

**Table 31.1** List of parameters for Ottawa F65 sand (physical parameters)

| Name of the parameters | Loose condition | Medium condition | Dense condition | Meaning of the parameters |
|---|---|---|---|---|
| Initial void ratio $e_0$ | 0.585 | 0.542 | 0.515 | Just a reference |
| Reference confining pressure $P_a$ (kPa) | 100 | 100 | 100 | Relates to shear modulus input |
| Initial shear modulus $G_{ma}$ (kPa) | 1.33E+04 | 1.44E+04 | 1.52E+04 | |
| Initial bulk modulus $K_{La}$, $K_{ua}$ (kPa) | 3.47E+04 | 3.76E+04 | 3.96E+04 | Assuming Poisson's ratio $\nu = 0.33$ |
| Poisson's ratio $\nu$ | 0.33 | 0.33 | 0.33 | For soil skeleton |
| Parameters for confining stress dependency, mG, mK | 0.5 | 0.5 | 0.5 | |
| Bulk density $\rho$ (t/m$^3$) | 1.666 | 1.713 | 1.744 | |
| Porosity | 0.369 | 0.351 | 0.340 | |
| Maximum damping coefficient $h_{max}$ | 0.24 | 0.24 | 0.24 | |
| Cohesions $c$ (kPa) | 0 | 0 | 0 | |
| Internal friction angle $\phi_f$ (deg.) | 29.5 | 30.2 | 33.0 | |
| Permeability $k$ (cm/s) | 1.1E-02 | 1.0E-02 | 9.8E-03 | |

**Table 31.2** List of parameters for Ottawa F65 sand (liquefaction parameters)

| Name of the parameters | Loose condition | Medium condition | Dense condition | Meaning of the parameters |
|---|---|---|---|---|
| Initial void ratio $e_0$ | 0.585 | 0.542 | 0.515 | Just to give relative density |
| Phase transformation angle $\phi_p$ (deg.) | 14.0 | 14.0 | 17.8 | NOT control the stress path directly |
| $r_{\varepsilon dc}$ | 0.73 | 0.40 | 0.25 | Control negative dilatancy |
| $r_{\varepsilon d}$ | 0.50 | 0.50 | 0.50 | Control overall dilatancy |
| $q_1$ | 1.00 | 1.00 | 1.00 | Control negative dilatancy in initial phase |
| $q_2$ | 0.50 | 1.00 | 2.00 | Control negative dilatancy in final phase |
| $r_k$ | 1.00 | 1.00 | 1.00 | Control negative dilatancy in final phase and $K_\alpha$ effect |
| $c_1$ | 1.53 | 2.00 | 2.50 | Threshold for dilatancy |

The values for the following parameters are generally unique: $-\varepsilon_{dcm} = 0.10$, STOL = 1.0E-06, $r_K'' = 1.0$, $S_1 = 0.005$, $q_{us} = 0$, $q_4 = 1.0$, rmtmp =0.5, I865SW = 0

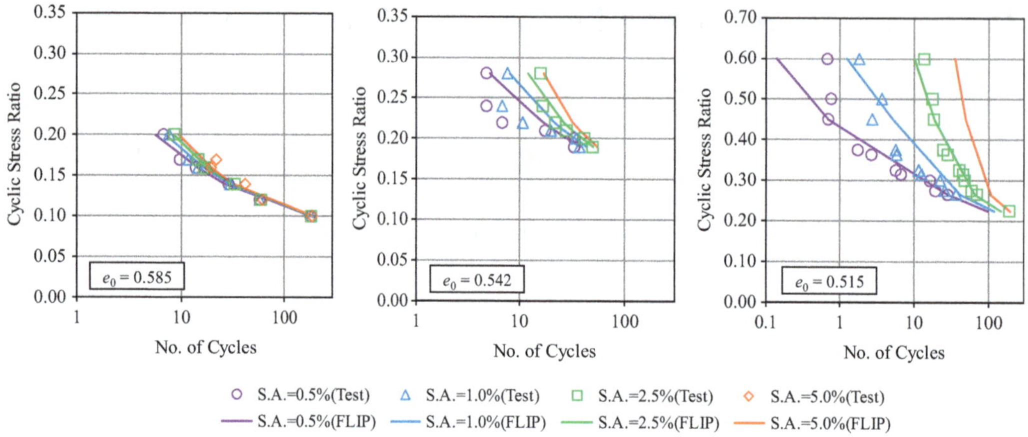


**Fig. 31.4** Comparison of target liquefaction resistance curves and computed curves with the parameters for Ottawa F95 sand in loose (left), medium (center), and dense conditions (right)

curves (El Ghoraiby et al. 2017, 2019) and computed curves are shown in Fig. 31.4. Multiple definitions of liquefaction, 0.5%, 1.0%, 2.5%, and 5.0% strain in single amplitude, were considered as the liquefaction resistance curves.

### *31.3.4 Parameter Setting Results by Multiple Practitioners*

The program FLIP ROSE has been continuously updated for the last 20 years with voluntary supports from practitioners. Using this scheme, the authors tried to discuss the possible variation in parameter determination in practice.

Currently, the FLIP Consortium organizes a working group (WG) to discuss the appropriate procedure of parameter determination in practice. The WG members are practitioners with the experience of using FLIP ROSE in practice. The authors asked them to determine the possible parameters independently. The results from nine practitioners from four companies were obtained.

The values of determined parameters for Ottawa F95 sand under medium dense conditions are shown in Table 31.3. The differences in the determined values were caused by the following reasons:

1. The methods to determine the initial shear stiffness were different. In reality, there are more variations in the procedure to determine the shear stiffness, i.e., from the result of small shaking prior to the main shaking, or from the result of inflight cone penetration in the centrifuge tests.
2. Regarding the method to determine the internal friction angle $\phi_f$, some practitioners determine the friction angle from the observed stress path, and other practitioners apply an empirical formula.
3. In the model, the behavior of soil is controlled by several parameters. There are multiple combinations of parameter set to successfully simulate the target

**Table 31.3** Parameters determined by multiple practitioners A to D-4

| Parameters | A | B-1 | B-2 | C-1 | C-2 | D-1 | D-2 | D-3 | D-4 |
|---|---|---|---|---|---|---|---|---|---|
| $P_a$ (kPa) | 100 | | | | | | | | |
| $G_{ma}$ (kPa) | 1.44E+05 | 1.57E+04 | | 1.25E+05 | | 1.19E+05 | | | |
| $K_{La}$, $K_{ua}$ (kPa) | 3.76E+05 | 4.10E+5 | | 3.33E+05 | | 3.10E+05 | | | |
| $\nu$ | 0.33 | | | | | | | | |
| Porosity | 0.351 | | | | | | | | |
| $h_{max}$ | 0.24 | | | | | | | | |
| $c$ (kPa) | 0.0 | | | | | | | | |
| $\phi_f$ (deg.) | 34.0 | 42.0 | | | | 40.2 | | | |
| $k$ (cm/s) | 1.0E-04 | N/A | | | | 1.21E-04 | | | |
| $\phi_p$ (deg.) | 17.0 | 28.0 | | | | | | | |
| $r_{\varepsilon dc}$ | 0.90 | 1.00 | 0.90 | 1.50 | 1.50 | 1.00 | 0.60 | 1.50 | 2.50 |
| $r_{\varepsilon d}$ | 0.50 | | | 1.00 | 0.50 | 0.70 | 1.00 | 0.20 | 0.50 |
| $q_1$ | 1.00 | 8.00 | | 6.00 | | 10.00 | 8.00 | 10.00 | 1.00 |
| $q_2$ | 0.85 | 1.00 | 1.10 | 1.25 | | 1.00 | 0.80 | 1.00 | 1.25 |
| $r_k$ | 1.0 | 0.5 | | | 1.0 | | 0.5 | 6.0 | 2.2 |
| $c_1$ | 1.95 | 2.00 | | 2.05 | | 2.10 | 1.90 | 2.20 | 2.10 |
| $-\varepsilon_{dcm}$ [a] | 0.10 | 0.20 | | | 0.10 | 0.20 | | 0.10 | 0.20 |

Ottawa F65 sand, medium dense condition: $e_0 = 0.542$

[a] $-\varepsilon_{dcm}$ controls the ultimate state of liquefaction

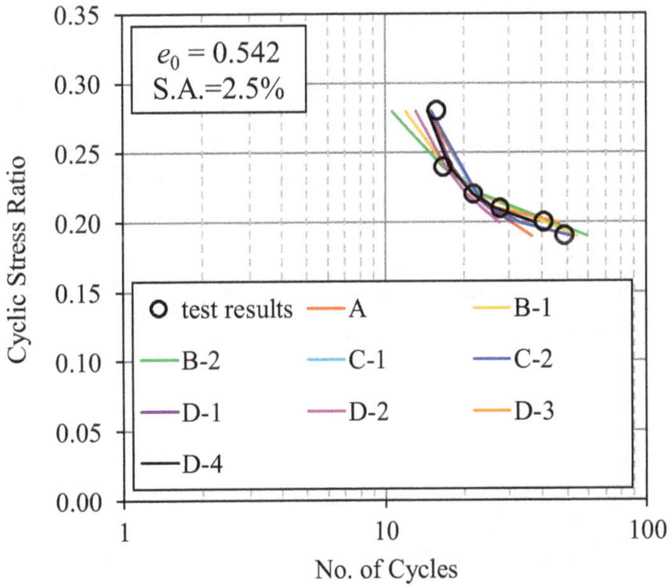


**Fig. 31.5** Variation in the computed liquefaction resistance curves and the target. (A to D-4 are the nine practitioner groups)

behavior of soil. In other words, there is no theoretical background for a unique combination of parameter set to be obtained.

4. The degree of agreement to the target is different. Not only for the overall agreement, but the stress level and/or strain level, which were the focus of the practitioner, may be different. For example, Fig. 31.5 shows the variation of computed liquefaction resistance curves defined by single amplitude of strain 2.5%. Generally, the computed results are similar but there still remains slight differences.

## 31.4 Type B Simulation of the Centrifuge Tests

### *31.4.1 Target Test Case and Parameter Determination*

A total of nine centrifuge tests were reported as the target of Type B simulation. There are some differences in test conditions, but the most important difference is the soil density.

As mentioned above, the parameters for loose ($e_0 = 0.585$), medium dense ($e_0 = 0.542$), and dense ($e_0 = 0.515$) conditions were determined. However, the actual void ratio is far larger than the loose condition in the parameter determination except for one case (KAIST-1). It means the parameter determination for eight cases shall be extrapolations from three reported cases, and the reliability of the parameter is quite low. Note the situation above is related to the significant variations of the reported values of the maximum and minimum void ratios for Ottawa F-65 sand. The reason for these variations is not clear, but the authors guess the difference in the sand lot may result in the difference of soil behavior. Anyway, this variation resulted in a significant uncertainty on the achieved relative density of the soil. Considering the target tests were done at George Washington University (GWU), the relative densities for loose ($e_0 = 0.585$), medium dense ($e_0 = 0.542$), and dense ($e_0 = 0.515$) conditions using the reported values of the maximum and minimum void ratios from GWU are 62%, 80%, and 91%, respectively.

Figure 31.6 show the schematics of parameter determination with interpolation and extrapolation. The left figure is for the parameter $r_{\varepsilon dc}$. The variation is quite linear as a function of void ratio, and both interpolation and extrapolation work well. However, the right figure is for the parameter $q_2$. The variation of this parameter is quite non-linear, and the extrapolated value for large void ratio is negative; nevertheless the value of the parameter is theoretically positive.

Therefore, the authors picked up only the KAIST-1 case for the Type B simulation, and the exact value of the parameters are determined by linear interpolation. Table 31.4 shows the summary of the values of the parameters for KAIST-1 case. Other parameters are same as the values shown in Tables 31.1 and 31.2.

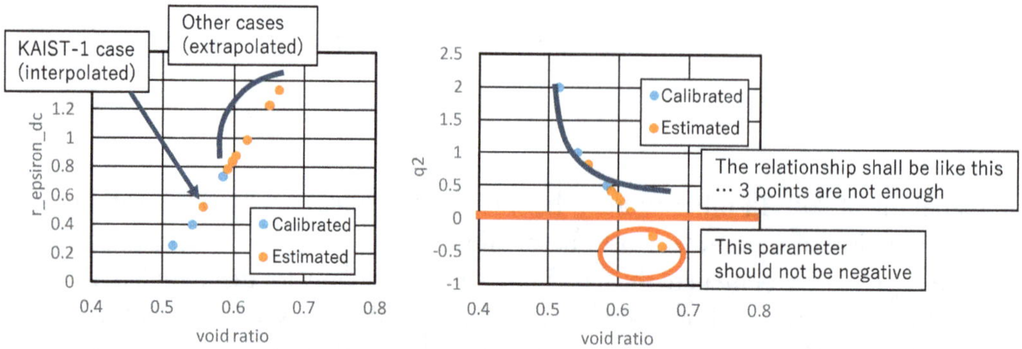


**Fig. 31.6** Schematics of interpolation and extrapolation in parameter determination

**Table 31.4** Summary of the determined parameters for KAIST-1 case

| **Physical parameters** | | | | | | |
|---|---|---|---|---|---|---|
| **Void ratio** | $G_{ma}$ **(kPa)** | $K_{La}$, $K_{ua}$ **(kPa)** | **Bulk density** $\rho$ **(t/m$^3$)** | **Porosity** | **Permeability** $k$ **(cm/s)** | **Internal friction angle** $\phi_f$ **(deg.)** |
| 0.577 | 1.4E+04 | 3.654E+04 | 1.70 | 0.36 | 1.0E-02 | 29.94 |
| **Liquefaction parameters** | | | | | | |
| **Phase transformation angle** $\phi_p$ **(deg.)** | | $r_{\varepsilon dc}$ | $r_{\varepsilon d}$ | $q_1$ | $q_2$ | $c_1$ |
| 14.0 | | 0.52 | 0.50 | 1.00 | 0.817 | 1.828 |

### 31.4.2 Analysis Conditions

An average of horizontal accelerations is used as the input motion. The original data showed some fluctuations on the time stamp; the authors estimate the fluctuation is just because of a small error in digitalization (cancellation of significant unit). Time step for all data is assumed as 0.005 s. Although there are small spikes on the recorded motion, no artificial manipulation was added to the input motion. Since the vertical motion is smaller than the horizontal motion, the authors ignored it due to time limitation to submit the predictions. The computation stopped when the input motion ends, and no computation on the consolidation process was conducted. This is because of time limitation and no interest of the authors in this issue. The pore water pressure dissipation process can be computed by FLIP ROSE/cocktail glass model (please refer the results of other teams in the LEAP exercises).

As for the mesh geometry, the same mesh indicated as an example in the guideline is used (a total of 2470 nodes and 4480 elements including pore water elements and water elements above the ground surface). The bottom and side boundaries are fixed. The undrained boundary was given for pore water.

### 31.4.3 Artificial Rayleigh Damping

For numerical stability, the authors applied artificial Rayleigh damping of very small value. Rayleigh damping is given as $\alpha[M] + \beta[K]$, and we usually assume $\alpha = 0$ for FLIP application in practice.

The remaining Rayleigh damping parameter $\beta$ for the FLIP program is carefully determined by parametric studies in 1D analysis. The idea of determination of Rayleigh damping is to avoid large values, which artificially reduce the deformation (Fig. 31.7).

Following the practical scheme of FLIP application, the parametric analysis for the left side column as shown in Fig. 31.8 were conducted. Dynamic analysis with a non-liquefaction condition (not considering dilatancy effect) with various values of Rayleigh damping parameter $\beta$ was done.

With a larger value in damping parameter, a smaller value of the computed displacement in response of soil layers will be given. These kinds of reductions in response can be regarded as the result of unnecessary artificial damping. Therefore, the threshold value not to minimize the computed displacement is chosen as the Rayleigh damping parameter in practice. Figure 31.10 is the results of the case with KAIST-1 input motion. From this result, the authors chose $\beta = 0.0002$ ($\alpha = 0$) as the Rayleigh damping parameter.

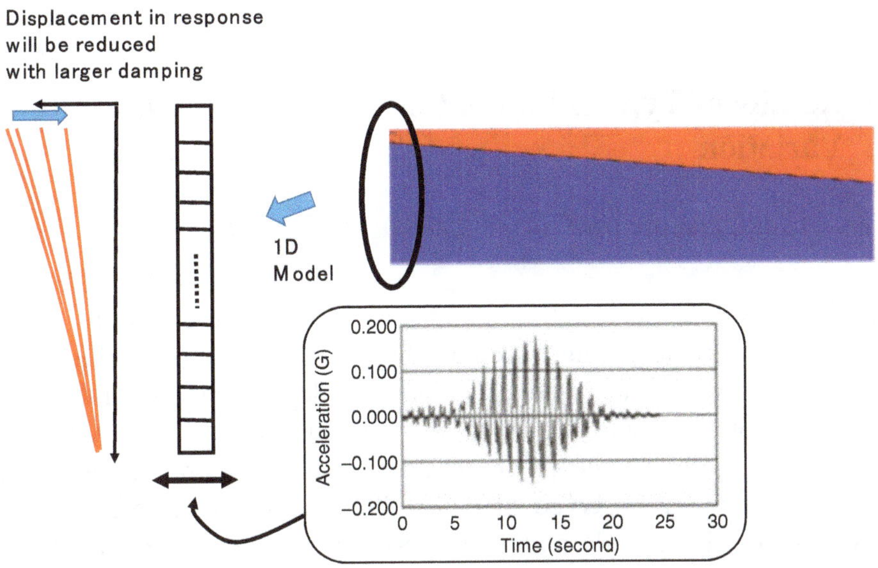


**Fig. 31.7** Schematics of Rayleigh damping parameter determination in FLIP application

**Fig. 31.8** Results of 1D analysis for Rayleigh damping parameter determination

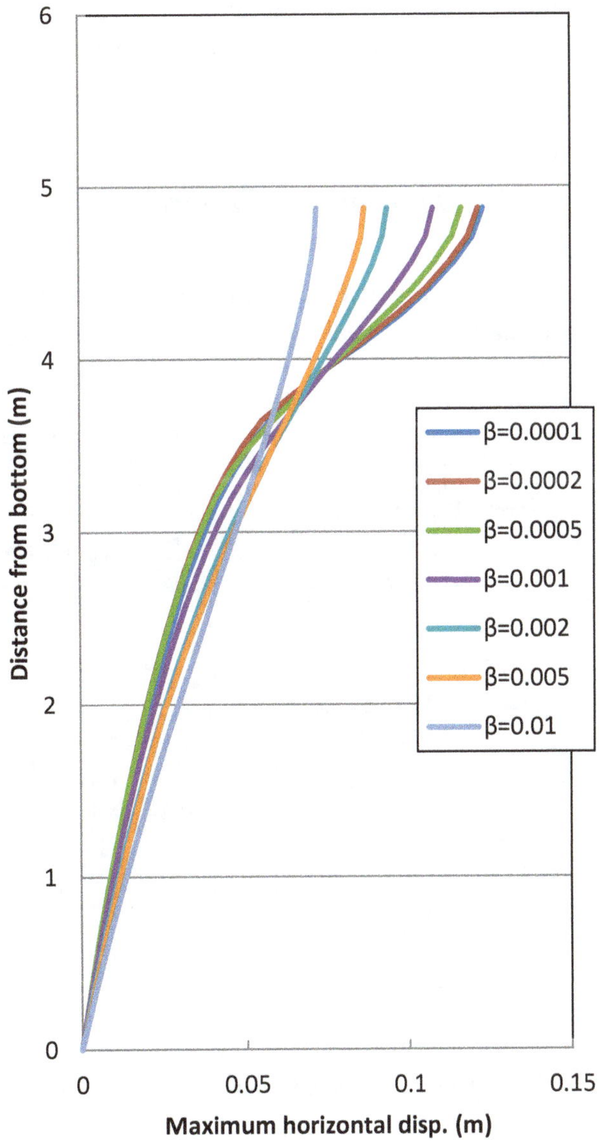

## 31.5 Results of Type B Simulation and Its Possible Variation

### 31.5.1 Acceleration and Pore Pressure Response

Figure 31.9 shows the acceleration and excess pore pressure response of the left array. There was no significant difference between the results in left, center, and right-side arrays. Spiky peaks were observed at all depth locations, and more spikes were observed at the points closer to the ground surface. Pore water pressure increased in the beginning; however, strong reduction of pore water pressure due to positive dilatancy was observed after 10 s. Slight reduction of pore water pressure was observed after 18 seconds at all points, and this may be related with dissipation.

Many spiky peaks were observed in acceleration time histories. These spiky peaks may be related to instability of the analysis mainly induced by significant change of the soil status due to dilatancy. In many cases, these spikes can be reduced with artificial Rayleigh damping. However, using quite a small value of Rayleigh damping parameter ($\beta = 0.0002$, $\alpha = 0$) in this case, the spiky peaks remained.

### 31.5.2 Deformation Pattern

A slight level of lateral spread of the ground is observed. Figure 31.10 shows the deformation after shaking ($t = 24.55$ s). The direction of the deformation agree well

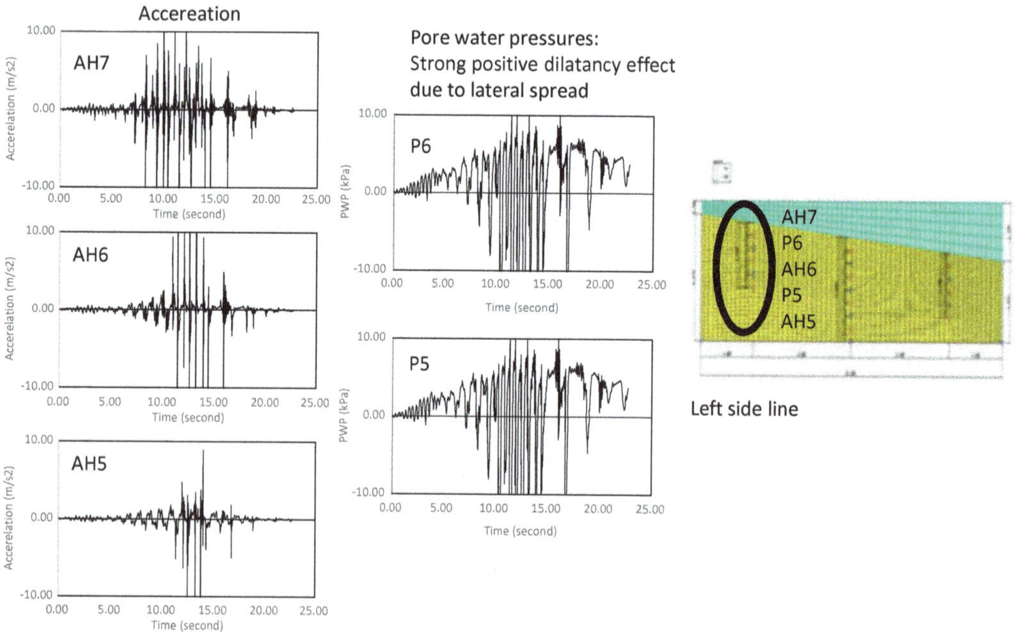


**Fig. 31.9** Time histories of the acceleration/pore water pressure at the left side line

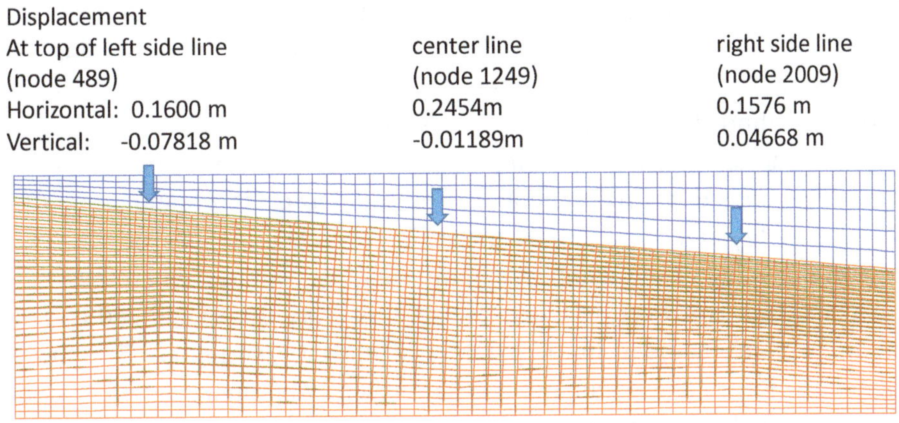


**Fig. 31.10** Computed deformation after shaking ($t = 24.55$ s)

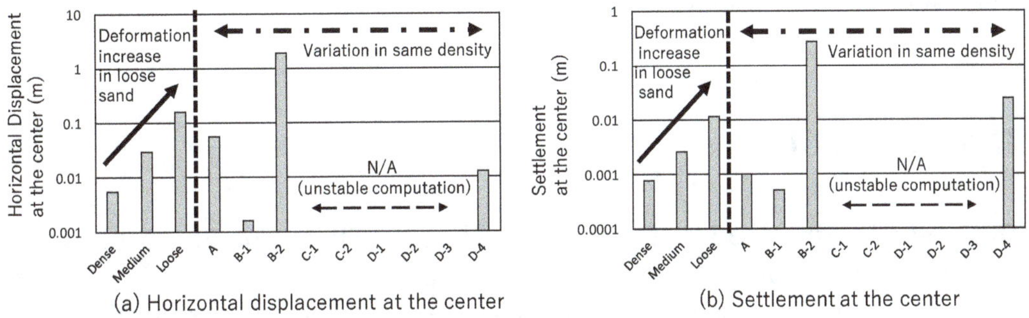


**Fig. 31.11** Deformation at the center with various parameter settings

with the direction to make the ground flat: left side goes down, and right side goes up. However, as shown in Fig. 31.10, the ground was not flat at the end of shaking. This may be due to the positive dilatancy to restrain the deformation. Note the deformation in pore water pressure dissipation process is not included in this result.

### *31.5.3 Possible Variation on Estimated Deformation*

Figures 31.11 shows a summary of the deformation at the center in all numerical cases. With increase in relative density, both horizontal displacement and settlement decrease. This trend is quite reasonable.

With the cases of various parameters determined by many practitioners, the results are quite scattered. Although the parameters are determined for the target density ($e_0 = 0.558$: between medium: $e_0 = 0.542$ to loose: $e_0 = 0.585$), the results are less than 1/10 of the medium sand result to 10 times of the loose sand results. It indicates that the magnitude of the deformation can be controlled by parameter setting, and comparison of the magnitude of deformation between the test results and computed results is not meaningful. In other words, the magnitude of deformation is not only the issue of constitutive modeling but also of parameter setting. A more comprehensive discussion on parameter determination procedure including laboratory testing method, $K\alpha$ effect, etc., is necessary.

Another problem is unstable computation with the parameter sets C-1 to D-3. This may be induced by insufficient Rayleigh damping in computation. As mentioned above, Rayleigh damping is determined by 1D analysis of the target soil layers, and the value of the damping should be re-determined when the soil profile has been changed. However, in this exercise, this process was skipped and only the soil parameters are modified in the parametric study for simplicity. In other words, with the variation of the parameter for soil layers, computations were done with inappropriate Rayleigh damping, and it induced variations in the computed results. Careful discussion on the damping effect in the computation should be done. Note, not only the artificial damping but also the numerical damping effect in the time integration scheme should be examined.

## 31.6 Conclusion

This study describes the LEAP-UCD-2017 exercise with one of the most common dynamic analysis programs in Japan (FLIP ROSE). Although only a limited number of cases of centrifuge tests were analyzed, the demonstration of a practical application of a common analysis program to this exercise clarified some issues that should be addressed by numerical modelers. First, there is a possible variation in parameter determination in practice. Second, the possible variation of parameters can cause a large variation in computed deformations. Third, the effect of artificial Rayleigh damping should be carefully examined.

More detailed discussions on the variation of the results and comparison with the centrifuge test results remain for future study.

**Acknowledgments** The analyses of the LEAP project reported here were supported by JSPS Kakenhi. This report was prepared with volunteer works of the WG members of the FLIP consortium. The parameter calibration part was done by Mr.Kazuaki Uemura and WG members (Mr. Sato, Mr. Hyodo, Mr. Watabe, Mr. Nakagama, Mr. Niina, Mr. Nakajima, Mr. Ise, Mr. Matsumoto, Mr. Igawa, and Mr. Masuda), mesh generation and data preparation were done by Mr. Naoki Orai, and drawing up of computed results were contributed by Mr. Jyunichi Hyodo.

## References

Boulanger, R. W. (2003). Relating K$\alpha$ to relative state parameter index. *Journal of Geotechnical and Geoenvironmental Engineering, ASCE, 129*(8), 770–773.

El Ghoraiby, M. A., Park, H., & Manzari, M. T. (2017). *LEAP 2017: Soil Characterization and Element Tests for Ottawa F65 Sand.* Washington, DC: The George Washington University.

El Ghoraiby, M. A., Park, H., & Manzari, M. T. (2019). Physical and mechanical properties of Ottawa F65 sand. In B. Kutter et al. (Eds.), *Model tests and numerical simulations of liquefaction and lateral spreading: LEAP-UCD-2017*. New York: Springer.

Iai, S., Matsunaga, Y., & Kameoka, T. (1990). Parameter identification for a cyclic mobility model. *Report of the Port and Harbour Research Institute, 29*(4), 3.

Iai, S., Matsunaga, Y., & Kameoka, T. (1992). Strain space plasticity model for cyclic mobility. *Soils and Foundations, 32*(2), 1–15.

Iai, S., Tobita, T., Ozutsumi, O., & Ueda, K. (2011). Dilatancy of granular materials in a strain space multiple mechanism model. *International Journal for Numerical and Analytical Methods in Geomechanics, 35*(3), 360–392.

Ishihara, K., & Yoshimine, M. (1992). Evaluation of settlements in sand deposits following liquefaction during earthquakes. *Soils and Foundations, 32*(1), 173–188.

Kokusyo, T. (1980). Cyclic triaxial test of dynamic soil properties for wide strain range. *Soils and Foundations, 20*(2), 45–60.

Tobita, T., Manzari, M. T., Ozutsumi, O., Ueda, K., Uzuoka, R., & Iai, S. (2015). Benchmark centrifuge tests and analyses of liquefaction-induced lateral spreading during earthquake. In S. Iai (Ed.), *Geotechnics for catastrophic flooding events* (pp. 127–182). Leiden: CRC Press., ISBN:978-1-138-02709-1.

# Part IV
# Workshop Essays

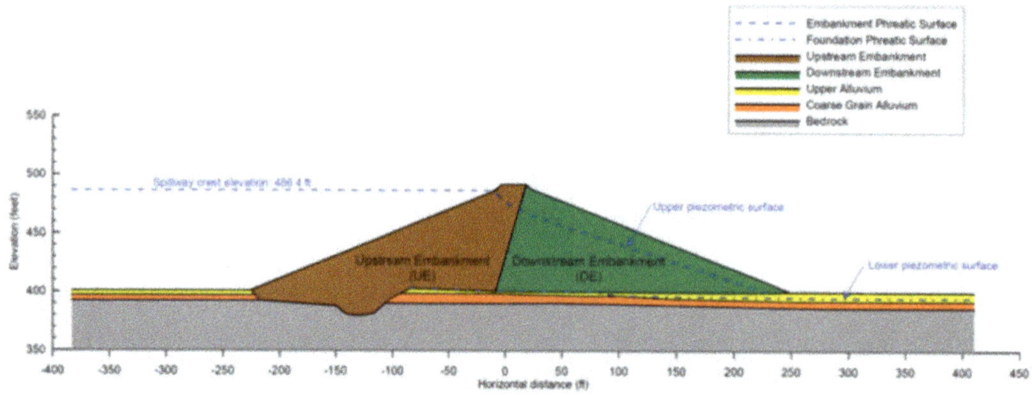


Image credit: Mejia

# Chapter 32
# Preliminary Seismic Deformation and Soil-Structure Interaction Evaluations of a Caisson-Supported Marine Terminal Wharf Retaining and Founded on Liquefiable Soils

**Arul K. Arulmoli**

**Abstract** The seismic deformation and soil-structure interaction analyses completed to date (i.e., preliminary design) for the project would benefit from more rigorous validation and calibration of the constitutive models and numerical procedures used in future analyses. The results from the Liquefaction Experiments and Analysis Project (LEAP) are expected to benefit future phases of the project (i.e., detailed design).

## 32.1 Thesis Statement

The seismic deformation and soil-structure interaction analyses completed to date (i.e., preliminary design) for the project would benefit from more rigorous validation and calibration of the constitutive models and numerical procedures used in future analyses. The results from the Liquefaction Experiments and Analysis Project (LEAP) are expected to benefit future phases of the project (i.e., detailed design).

## 32.2 Essay

The Port of Vancouver (Port) is proposing to build a new marine terminal in Delta, British Columbia, Canada. The project calls for a caisson-supported wharf founded on potentially liquefiable foundation soils and retaining soils that are also potentially liquefiable. Soils immediately below and behind the caisson will be replaced with materials that will be densified after placement. The total retained height of the fill exceeds 25 m.

A. K. Arulmoli (✉)
Earth Mechanics, Inc., Fountain Valley, CA, USA
e-mail: arulmoli@earthmech.com



B. Kutter et al. (eds.), *Model Tests and Numerical Simulations of Liquefaction and Lateral Spreading*, https://doi.org/10.1007/978-3-030-22818-7_32

The wharf will be designed to the most current seismic design requirements: i.e., three levels of earthquake ground motions (ASCE 2014). The wharf performance requirements for the three levels of earthquake limit lateral displacements at the base and top of the caisson (i.e., horizontal displacements and rotations are limited).

As the Port's Owner's Engineer (OE), the OE team performed preliminary seismic evaluations to develop initial optimum wharf configuration meeting the lateral displacement requirements. Initial evaluations included uncoupled geotechnical and structural analyses. Simplified lateral seismic deformation evaluations using available procedures (Bray and Travasarou 2007; Transportation Research Board 2008) and pseudo-static structural analysis using the computer program SAP2000 (Computers, and Structures, Inc. 2013) were completed to develop the initial wharf configuration. Seismic soil-structure interaction evaluations were performed for this configuration using the finite difference computer program FLAC (Itasca Consulting Group 2011) and using two constitutive models for sandy soils, UBCSAND (Beaty and Byrne 2011) and PM4SAND (Boulanger and Ziotopoulou 2015).

Significant differences were noted between the results from the UBCSAND and PM4SAND constitutive models. Since there was no available data on past performance of caissons in liquefiable soils from well-documented case histories or experiments, it was a challenge to reconcile these differences. The OE team members used their collective experience and judgment to develop the preliminary wharf configuration that could be refined during the subsequent detailed design phase with additional analyses.

In conclusion, the seismic deformation and soil-structure interaction analyses completed by the OE team to date would benefit from more rigorous validation and calibration of the analytical models, especially for adequately predicting liquefaction triggering, post-liquefaction soil strengths, and capturing post-liquefaction behavior of retaining structures in liquefiable soils. It is the author's hope that the Liquefaction Experiments and Analysis Project (LEAP) would include experiments using waterfront earth-retaining structures as well as evaluation of soil-structure interaction in liquefied soils (Kutter et al. 2014), which is expected to benefit the project during its future design phases.

**Acknowledgement** This abstract is based on the project that is funded by the Port of Vancouver. The Owner's Engineer team is led by Moffat & Nichol and supported by Stantec and Earth Mechanics, Inc.

## References

ASCE 61-14. (2014). *Seismic design of piers and wharves*. American Society of Civil Engineering (ASCE) Standard.

Beaty, M. H., & Byrne, P. M. (2011). *UBCSAND Constitutive Model, Version 904aR*. Documentation Report on Itasca UDM Website, February.

Boulanger, R. W., & Ziotopoulou, K. (2015). *PM4Sand (Version 3): A sand plasticity model for earthquake engineering applications. Technical Report No. UCD/CGM-15/xx, Center for*

*Geotechnical Modeling*. Department of Civil and Environmental Engineering, University of California, Davis, CA.

Bray, J. D., & Travasarou, T. (2007). Simplified procedure for estimating earthquake-induced deviatoric slope displacements. *Journal of Geotechnical and Geoenvironmental Engineering, ASCE, 133*(4), 381–392.

Computers & Structures, Inc. (2013). *SAP2000 v15.2.1.*

Itasca Consulting Group. (2011). *Fast Lagrangian analysis of continua*. FLAC Version 7.0.

Kutter, B. L., Manzari, M. T., Zeghal, M., Zhou, Y. G., & Armstrong, R. J. (2014). Proposed outline for LEAP verification and validation processes. In *Proceedings of the Fourth International Conference on Geotechnical Engineering for Disaster Mitigation and Rehabilitation*. Kyoto: Kyoto University.

Transportation Research Board. (2008). Seismic analysis and design of retaining walls, buried structures, slopes, and embankments. In *National Cooperative Highway Research Program Report 611, Project 12–70, Washington, DC.*

# Chapter 33
# Significance of Calibration Procedure Consistency

**Zhao Cheng**

**Abstract** Significance of calibration procedure consistency for geotechnical earthquake engineering is emphasized. The material parameters should be calibrated consistently for specimen preparation during the tests. Particularly, the laboratory-based parameters should not be mixed up with those for in-situ conditions.

Reliable prediction for nonlinear dynamic problems using an advanced constitutive model remains challenging either for real-site response analyses or centrifuge tests. Uncertainties come from many factors, such as calibration procedure consistency, which is stressed here for the numerical prediction. The material parameters should be calibrated consistently for specimen preparation during the tests. Particularly, the laboratory-based parameters should not be mixed up with those for in situ conditions.

There are many different behaviors for sands in the field and in the laboratory, essentially because the sand samples prepared in the laboratory usually have been disturbed or the fabric has been damaged or reconstituted, even though some material parameters, e.g., critical state parameters, are believed to be independent of sample preparation approaches. Yoshmi et al. (1989) showed that the cyclic resistance ratio (CRR) of frozen (assumed undisturbed) samples from the in situ deposit is about 200% of that of the samples from the new deposit sand fill, despite very small differences of sample densities. The material parameters calibrated from the disturbed laboratory samples should be adjusted for in situ deposit simulations.

The slope of cyclic stress ratio (CSR) versus number of cycles ($N$) is also observed steeper for undisturbed samples compared to that for disturbed samples (Yoshmi et al. 1989). The results of CSR versus $N$ can be approximately fitted with a power law, CSR $= aN^{-b}$. The $b$-value for the naturally deposited sands in the field could be much higher than those obtained from the laboratory tests, depending on the sample preparation methods. Applying an underestimated b-value based on the laboratory tests for field case studies may lead to an increased calculated liquefaction

Z. Cheng (✉)
Itasca Consulting Group, Inc., Minneapolis, MN, USA
e-mail: zcheng@itascacg.com



B. Kutter et al. (eds.), *Model Tests and Numerical Simulations of Liquefaction and Lateral Spreading*, https://doi.org/10.1007/978-3-030-22818-7_33

hazard for magnitudes less than 7.5, but reduce calculated hazards for magnitudes greater than 7.5, if the CSR has been adjusted to the correct value at magnitude 7.5.

Even for the material parameters that define the model elasticity in constitutive models, calibrated parameters are quite different for in situ and laboratory samples. Cheng (2018) illustrated that the slope of elastic shear modulus ($G_e$) versus sand relative density ($D_r$) calibrated from field data is much steeper than that calibrated from empirical relations based on laboratory data, while the intercept of $G_e$ versus $D_r$ is much smaller for in situ sands compared to that for laboratory-based sands. This probably partially explains the higher *b*-values for in situ sands.

One key calibration procedure for some constitutive models is to match an accepted empirical curve of CRR versus a field measure, e.g., normalized blow-counts, or $(N_1)_{60}$, which is one of the most welcomed features by practice engineers, assuming that CRR for a magnitude 7.5 earthquake and effective overburden stress of 1 atm is approximately equal to the CRR during a DSS (direct simple shear) test with 15 uniform loading. However, this match implies that the calibrated material parameters should be used for field sands and not for laboratory sands because the empirical CRR versus $(N_1)_{60}$ curve is based on the field observations. These parameters are probably going to predict a steeper curve of CSR versus *N*, or other discrepancies, if used for laboratory-based sands. Note that the liquefaction triggering in the empirical curve of CRR versus $(N_1)_{60}$ is based on case histories with any of the ground observations, e.g., boils, failures, large settlement or lateral deformations, so the liquefaction triggering criteria based on this curve for a constitutive model should be consistently based on a certain excess pore pressure ratio or shear strain value, whichever is earlier satisfied, but not just one of them. Some constitutive models are using a single liquefaction criterion, e.g., 3% shear strain, for simplicity, which is more or less acceptable, but the users should be aware of the possible intrinsic discrepancy.

Another note is that because the target empirical relation was based on in situ data covering a wide range of sands, it represents only general sand behavior in a statistic sense, but may have considerable bias if evaluating some specific sands. For example, for a relative density of 80% Fraser River sand, to match the target CRR determined by the NCEER empirical curve of CRR versus $(N_1)_{60}$, the laboratory test showed liquefaction in approximately 11.5 cycles rather than 15 cycles (Beaty and Byrne 2011).

The CSR versus *N* curve is also quite sensitive to the test types, e.g., triaxial compression tests, DSS tests, or torsion shear tests, and sample preparation methods. The calibrated parameters should be used only for predictions with similar conditions, or careful adjustments should be incorporated.

In conclusion, to obtain satisfactory prediction between the numerical modeling and the centrifuge tests, the material parameter calibration procedure should be consistent such that it is based on laboratory samples considering disturbances, not directly or indirectly based on the in situ materials. In the other direction, some parameters that are calibrated from the laboratory tests should be used cautiously for case studies if not correctly adjusted.

## References

Beaty, M. H., & Byrne, P. M. (2011). *UBCSAND constitutive model version 904aR*. Document report on Itasca UDM website. Retrieved March 8, 2016 from https://www.itascacg.com/software/udm-library.

Cheng, Z. (2018). A practical 3D bounding surface plastic sand model for geotechnical earthquake engineering application. In *Geotechnical Earthquake Engineering and Soil Dynamics V: Numerical Modeling and Soil Structure Interaction* (pp. 37–47). Reston, VA: American Society of Civil Engineers.

Yoshmi, I., Tokimatsu, K., & Hosaka, Y. (1989). Evaluation of liquefaction resistance of clean sands based on high-quality undisturbed samples. *Soils and Foundations, 29*(1), 93–104.

# Chapter 34
# Paths Forward for Evaluating Seismic Performance of Geotechnical Structures

**Susumu Iai**

**Abstract** Two suggestions are provided for expanding the scope of LEAP project for the next stage. The first suggestion is to perform undrained cyclic shear tests with addditional static deviator stress to represent more realistic stress conditions in soil-structure systems during earthquakes. The other suggestion is to perform centrifuge model tests with non-homogeneous soil layer such as mildly sloping ground of liquefiable sand overlain by a less permeable capping crust layer, leading to a unlimited flow failure of surface crust. Effects of void ratio redistribution and seepage flow will be studied in detail in this type of cetrifuge tests.

## 34.1 Thesis Statement

The paths forward for evaluating seismic performance of geotechnical structures along the framework of LEAP project are to expand the scope of soil element tests from the conventional cyclic undrained tests toward the undrained shear with additional static deviator stress and to expand the scope of centrifuge tests from the behavior of homogenous sand deposit toward the non-homogenous sand deposit involving the effects of void redistribution due to pore water flow.

## 34.2 Essay

The laboratory tests of soil specimens have been typically performed through drained/undrained monotonic and cyclic shear tests. However, soil element behavior in the typical soil-structure system in two or three dimension includes the effect of initial static deviator stress in addition to the cyclic shear during shaking. The paths forward to improving the understanding of the source of uncertainty in evaluating

S. Iai (✉)
FLIP Consortium, Kyoto, Japan
e-mail: iai.susumu@flip.or.jp



B. Kutter et al. (eds.), *Model Tests and Numerical Simulations of Liquefaction and Lateral Spreading*, https://doi.org/10.1007/978-3-030-22818-7_34

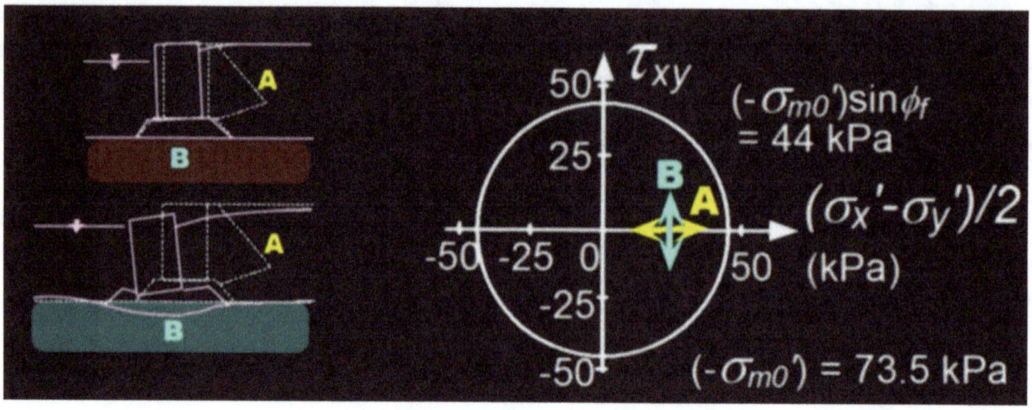


**Fig. 34.1** Laboratory soil element tests with combination of static and cyclic undrained shear

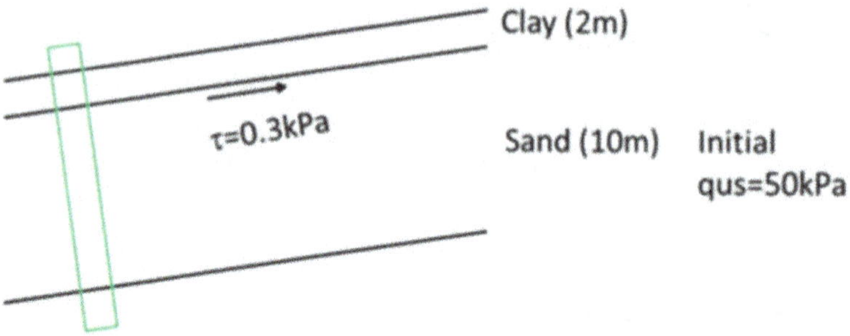


**Fig. 34.2** Centrifuge tests of mildly sloping ground of liquefiable sand deposit overlain by surface crust layer of less permeable nature

the seismic behavior of soil-structure systems are to include this type of soil element tests. Figure 34.1 indicates two of the possible combinations of the static and cyclic shear.

The stress path A is easier to perform using the triaxial testing device. The stress path B is more difficult to perform and needs a testing device such as a hollow cylinder testing device. However, the stress path B has wider applications such as for evaluating behavior of shallow foundations.

The centrifuge tests of soil-structure systems have been often performed using homogenous sand deposit. In these tests, the excess pore water pressure in the sand deposit typically shows initial increase during shaking and then dissipation afterwards associated with settlement. The paths forward along the line of the LEAP project are to expand the scope toward the non-homogenous sand deposit involving the effects of void redistribution due to pore water flow.

The simplest example of the non-homogenous sand deposit can be a combination of liquefiable sand layer overlain by a surface crust layer of less permeable nature as shown in Fig. 34.2 (Iai 2017). With the slight inclination, there may be a potential for unlimited sliding along the boundary of the two layers depending on the void ratio and thus the phenomenon involved in this type of tests can be the basis for evaluating

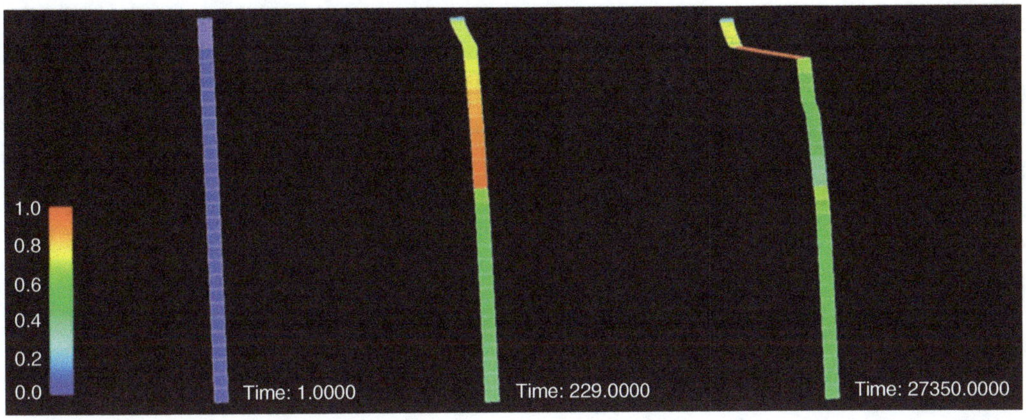


**Fig. 34.3** Expected examples of numerical analysis

the combined effects of steady state, void ratio dependency, and rate of void redistribution associated with the seepage flow of pore water toward the boundary of two layers. The application of this degree of complexity in practice can be proven to be useful (Fig. 34.3).

## Reference

Iai, S. (2017). Performance of port structures during earthquakes. In *Proceedings of the Third International Conference on Performance-Based Design of Geotechnical Structures, Vancouver*.

# Chapter 35
# Selected Issues in the Seismic Evaluation of Embankment Dams for Possible Investigation by LEAP

**Lelio H. Mejia**

**Abstract** The post-liquefaction residual strength of soils often plays a key role in the seismic stability analysis of embankment dams and other earth structures. A commonly held view is that back analyses of field case histories of liquefaction flow failures offer at present the most practical approach for estimating the residual strength of liquefied soils. Thus, it is common in engineering practice to estimate this strength parameter based on empirical correlations. However, the available data for such correlations are very limited and represent only a small range of conditions encountered in practice. Thus, use of the correlations in practice generally involves considerable uncertainty. A case history is presented that illustrates questions that are often encountered in practice and that would benefit from additional research, perhaps by means of centrifuge testing and numerical modeling.

## 35.1 Background and Thesis Statement

The post-liquefaction residual strength of soils often plays a key role in the seismic stability analysis of embankment dams and other earth structures. A commonly held view is that back analyses of field case histories of liquefaction flow failures offer at present the most practical approach for estimating the residual strength of liquefied soils. Thus, it is common in engineering practice to estimate this strength parameter based on empirical correlations.

Published relationships have been developed by relating the residual strength of soils back-calculated from case histories of liquefaction flow failure to soil indices such as the standard penetration test (SPT) blow count or the cone penetration test (CPT) tip resistance (and the vertical effective consolidation stress). However, the available data are very limited and exhibit large scatter, and the case histories represent only a small fraction of the range of soil types, site conditions, consolidation stresses, and geologic settings encountered in practice. Thus, application of the

L. H. Mejia (✉)
Geosyntec Consultants, Oakland, CA, USA
e-mail: LMejia@Geosyntec.com



B. Kutter et al. (eds.), *Model Tests and Numerical Simulations of Liquefaction and Lateral Spreading*, https://doi.org/10.1007/978-3-030-22818-7_35

correlations to any one practical problem generally involves a great deal of uncertainty and requires the exercise of judgment.

It is now recognized by many geotechnical engineers that the residual strength of liquefied soil is a characteristic of a specific soil-structure system (i.e., a system parameter) as opposed to a property of the soil, and that it depends on many factors. One key factor is the potential for void redistribution in the soil after liquefaction triggering. Void redistribution results from the collapse of the soil structure and the tendency for densification of the soil upon triggering and has been shown to be a function of the thickness of a liquefied soil layer and the density (initial void ratio) of the liquefied soil (e.g. Malvick et al. 2006).

Many cases arise in practice for which additional information on the effects of soil layer thickness and of density on void redistribution would be helpful in evaluating the range of values of residual strength to be used in a stability analysis. Furthermore, the analysis of seismic deformations and of potential gross instability in such cases would benefit considerably from insight into the ability of available numerical analysis procedures and constitutive models to simulate void redistribution of soils after liquefaction triggering. One example case is the seismic evaluation of Calero Dam, which is briefly described below. A more complete description of the case history is presented by Mejia et al. (2014).

## 35.2 Calero Dam Seismic Evaluation

Built in 1935 and located in Santa Clara County, California, Calero Dam is a 100-foot-high compacted earthfill embankment that straddles a wide valley blanketed by 15–20 ft of alluvium overlying bedrock. As shown in Fig. 35.1, the embankment consists of upstream and downstream zones and a seepage cutoff trench excavated through the alluvium. The alluvium was stripped from beneath the embankment upstream of the cutoff trench but was left in place downstream of the trench.

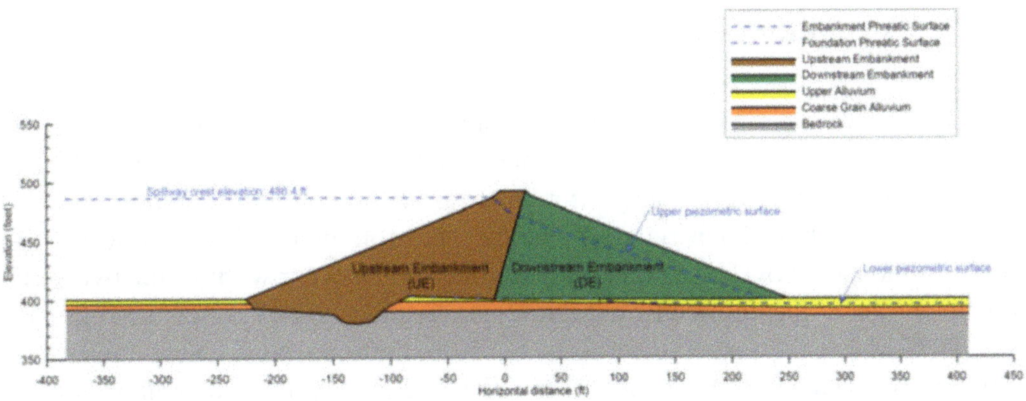


Fig. 35.1 Transverse cross-section of Calero Dam

The lower 5–10 ft of the alluvium consists of medium dense gravelly sands with 10–20% low-plasticity fines, whereas the upper 5–10 ft consist of clayey sands and sandy clays of medium plasticity. The upstream and downstream embankment materials are generally similar and consist of dense clayey sands with gravel.

Two piezometric surfaces are discerned from piezometer data at the site. The lower piezometric surface, shown in Fig. 35.1, applies to the foundation alluvium downstream of the cutoff trench, whereas the upper surface applies to the embankment.

Although the dam withstood the 1989 Loma Prieta Earthquake with only minor effects, the dam could be vulnerable to large earthquakes on the San Andreas and Calaveras faults, located within 20 km to the west and east, and earthquakes on other active faults within a few kilometers of the site. Large earthquakes on the nearby faults could generate ground motions characterized by peak horizontal rock accelerations exceeding 1 g, whereas the Loma Prieta Earthquake generated peak accelerations of about 0.3 g.

Extensive field investigations with the SPT and the Becker Penetration Test (BPT) showed that the corrected, normalized SPT blow count, $(N_1)_{60}$, in the lower alluvium ranges between about 6 and 30 and averages about 17, suggesting an overall medium dense condition for that soil. Thus, while the embankment and upper alluvium materials are not susceptible to liquefaction, the potential for liquefaction triggering in the lower alluvium during intense earthquake shaking was a key consideration in the seismic evaluation of the dam.

Likewise, assessment of the post-liquefaction residual strength of the lower alluvium, should liquefaction trigger in that layer, was a key issue in the dam stability analysis and required careful consideration of uncertainty and exercise of judgment. The question of uncertainty was further underscored by the fact that key characteristics of the dam and its foundation are outside the range represented by case histories in the database of liquefaction flow failures. A key question was the potential for liquefaction triggering and for void redistribution in a 10-foot-thick layer of gravelly sand with an average blow count of about 17.

## 35.3 Concluding Remarks

The Calero Dam case history illustrates questions that are often encountered in practice in the seismic stability analysis of embankment dams and that would benefit from additional research, possibly through physical modeling by centrifuge tests and numerical modeling. Such questions include the following: Can significant void redistribution and flow failure (i.e., large deformations) occur within a relatively thin confined layer of medium-dense cohesionless soil under significant vertical effective stress? How are the potential for void redistribution and flow failure under such conditions affected by the thickness of the liquefied soil layer? How are they affected by soil relative density? Are numerical analyses with available constitutive models able to simulate void redistribution and large deformations in those types of situations?

The above questions might be answered, at least in part, through a program of centrifuge testing and numerical modeling to simulate development of flow failure of earth embankments underlain by varying layer thicknesses of liquefaction-susceptible soil of varying relative densities. Perhaps such a program could be undertaken within the LEAP framework.

## References

Malvick, E. J., Kutter, B. L., Boulanger, R. W., & Kulasingam, R. (2006). Shear localization due to liquefaction-induced void redistribution in a layered infinite slope. *Journal of Geotechnical and Geoenvironmental Engineering, ASCE, 132*(10), 1293–1303.

Mejia, L., Wu, J., Newman, E., & Mooers, M. (2014). Seismic stability evaluation of dam underlain by coarse-grained alluvium. In *Proceedings of U.S. Society on Dams Annual Meeting and Conference, San Francisco, CA*.

# Chapter 36
# Soil Permeability in Centrifuge Modeling

**Inthuorn Sasanakul**

If the same prototype soil and water are used for a centrifuge model, as it is generally accepted, the seepage velocity in the centrifuge is increased by *N* times earth gravity produced by the centrifuge environment. The increased seepage velocity produces a time scaling conflict between diffusion time and dynamic time. Many centrifuge researchers prefer not to scale soil particle size; thus, the viscous fluid is used to resolve the time conflict. The purpose of viscous fluid is to maintain the same seepage velocity as in the prototype condition. In other words, the hydraulic conductivity of model soil in centrifuge environment is same as in the prototype condition. For a parametric study, especially, a comparison of test results produced by multiple centrifuge facilities such as the LEAP project, the soil permeability should be consistent. Viscous fluid should be prepared carefully and accurately measured fluid viscosity used for each centrifuge test should be reported because it could affect the soil permeability significantly. It is ideal that the permeability of soil with viscous fluid is measured in the laboratory following ASTM standards before each test. The permeability test should be conducted in such a way that it validates the Darcy's flow behavior for the hydraulic gradients (and seepage velocity) anticipated in the centrifuge test. Permeability of the sample using the viscous fluid should be reported for each test conducted by each facility. It is unlikely that the permeability will match perfectly between each test but numerical validation can account for the variation of the soil permeability.

I. Sasanakul (✉)
Department of Civil and Environmental Engineering, University of South Carolina, Columbia, SC, USA
e-mail: sasanaku@cec.sc.edu



B. Kutter et al. (eds.), *Model Tests and Numerical Simulations of Liquefaction and Lateral Spreading*, https://doi.org/10.1007/978-3-030-22818-7_36

# Chapter 37
# Variation of Permeability of Viscous Fluid During Liquefaction Model Testing

**Tetsuo Tobita**

## 37.1 Summary

In dynamic centrifuge modelling, to resolve conflicts in scaling of time between dynamic and diffusion events, viscous fluids are used to reduce a model's permeability. Recently, with the ease of its handling, methylcellulose (MC) solution is commonly used (Stewart et al. 1998). The author conducted constant head permeability tests of Toyoura sand by varying the fluid viscosity, paying attention to transient changes in its permeability and found that the permeability of sand with MC solution continuously decreased with the fluid passing, whereas it remained constant for purified water. Then, by modelling of models of the liquefaction model in centrifuge testing, the effect of permeability reduction is monitored and was confirmed that the effect is relatively minor. However, this is thought to be one of the sources of epistemic errors in centrifuge modelling.

## 37.2 Essay

Constant head permeability tests were conducted with a particular focus on the transient changes of permeability with viscous fluids made of methylcellulose (MC) solution. Tested fluid viscosities were 20, 44 and 60 times that of purified water (Fig. 37.1). Results of the constant head tests showed that the permeability tends to degrade with the increasing volume of passing fluid. The effect might seriously affect the physical modelling in practice. Results of the constant head

T. Tobita (✉)
Department of Civil Engineering, Kansai University, Osaka, Japan
e-mail: tobita@kansai-u.ac.jp



B. Kutter et al. (eds.), *Model Tests and Numerical Simulations of Liquefaction and Lateral Spreading*, https://doi.org/10.1007/978-3-030-22818-7_37

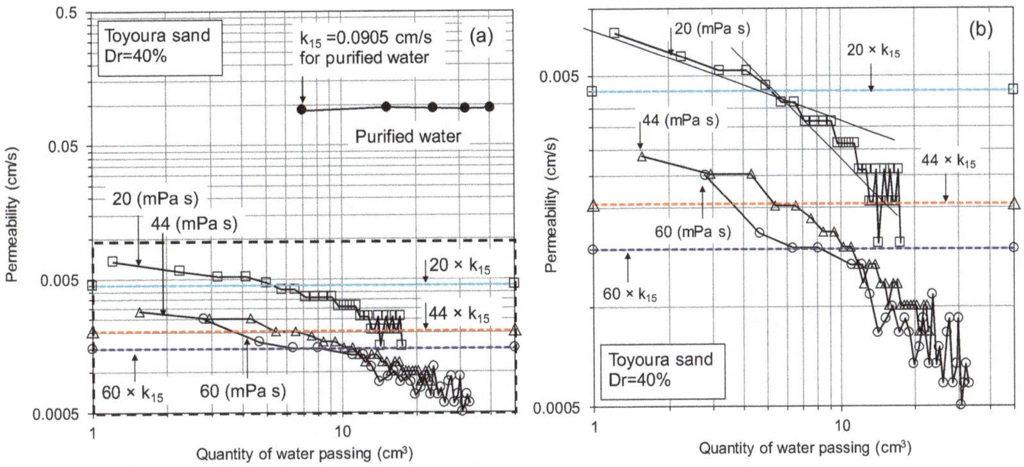


**Fig. 37.1** Variation of the permeability with quantity of passing water for Toyoura sand ($D_r = 40\%$): (**a**) showing all curves including permeability of purified water and (**b**) showing curves of viscous fluids only

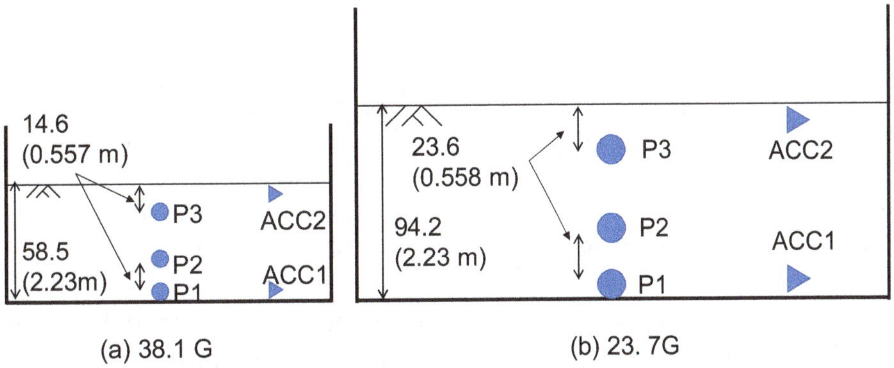


**Fig. 37.2** Model dimensions and sensors' location for "modelling of models" centrifuge tests: (**a**) 38.1G and (**b**) 23.7G

tests showed that the derived curves of permeability versus the volume of passing fluid were comprised of two segments. On the first slope, permeability continuously decreased at an almost constant rate regardless of the viscosity of the fluid until specific volumes of viscous fluid passing being reached at which point the curves reached a breakpoint and the second slope started. The rate of degradation was reduced less than 50%. The first slope of the curve may correspond to a complete coverage of pores near the surface in the sand specimen by MC fibres. From the above observation, it may be concluded that the permeability with MC solution drastically decreases with the quantity of viscous fluid passing.

Then to see the effects of this variation of permeability, centrifuge model tests were conducted under two different centrifugal accelerations: 38.1G and 23.7G (Fig. 37.2). These G levels were determined by the scaling law of time for dynamic phenomena using MC solution of 38.1 and 23.7 mPa s as a pore fluid. A centrifugal acceleration at 1/3 of ground thickness from the surface was monitored, and the G level was maintained during the experiment so that the error of confining stress due

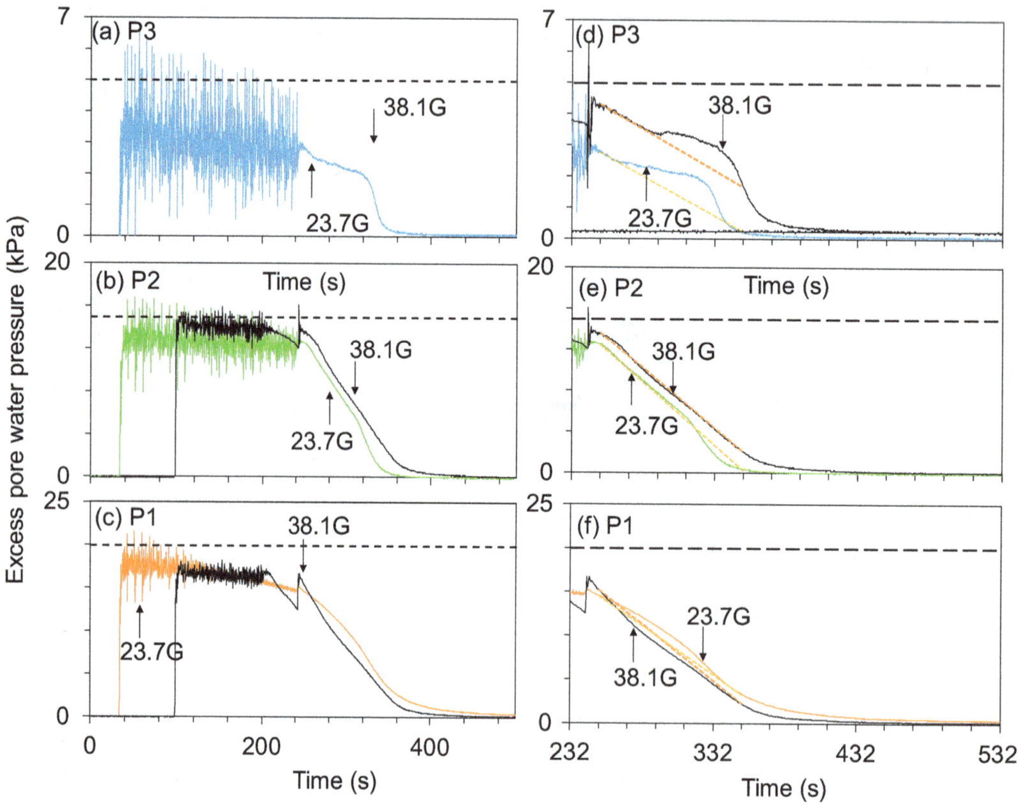


**Fig. 37.3** Time histories of excess pore water pressure: (**a**) P3, (**b**) P2, and (**c**) P1 and (**d**)–(**f**) for 232–532 s. Dotted horizontal line indicates the initial effective vertical stress

to radial acceleration field can be minimized. Attention was given to the rate of pore pressure dissipation (Fig. 37.3). What was found from the centrifuge model tests are that the curves in the dissipation phase deviates from the linear segments in both cases of 38.1G and 23.7G (Fig. 37.3d). This may indicate that as pore pressure dissipates from the ground ejected pore water may be accumulated near the ground surface because of the possible effect of clogging of MC solutions.

By assuming that the settled volume of the ground after shaking was equal to the volume of fluid ejected from the ground, the quantity of passing water was estimated to be 4.29 and 11.3 cm$^3$, respectively, for the cases of 38.1G and 23.7G. With these numbers and results of the constant head test results (Fig. 37.1), for the case of 38.1G in which the MC solution of 38.1 mPa s was used, the effect of the degradation of permeability would be minor, while for the case of 23.1G (23.1 mPa s) about 40% of reduction in permeability was expected.

We conclude that care should be taken if the duration of the diffusion process or settlements after liquefaction is of particular interest in testing a model with MC solutions. Duration times can be larger than that of a prototype because of the possible clogging effect of the MC fibres. The degree of clogging may depend on the density of MC fibres in a fluid. Thus, care should also be taken to choose a type of MC which gives the lowest density of fibres in a viscous fluid of a certain viscosity.

**Acknowledgement** The author thanks Mr. Kenta Yokoyama, a former undergraduate student at Kansai University, for conducting the constant head permeability tests and centrifuge tests as part of his graduation thesis.

## Reference

Stewart, D. P., Chen, Y.-R., & Kutter, B. L. (1998). Experience with the use of methylcellulose as a viscous pore fluid in centrifuge models. *Geotechnical Testing Journal, 21*(4), 365–369.

# Chapter 38
# Post-liquefaction Cyclic Shear Strain: Phenomenon and Mechanism

**Rui Wang, Pengcheng Fu, Jian-Min Zhang, and Yannis F. Dafalias**

**Abstract** Under undrained cyclic loading, sand experiences decrease in effective stress, which can result in liquefaction. Test results show that large cyclic shear strain is generated at zero effective stress during undrained cyclic loading. This post-liquefaction shear strain has been observed to progressively increase in amplitude with increasing number of loading cycles until it eventually stabilizes at a bounded value. However, there has been no clear explanation on why and how this cyclic shear strain is generated. The fabric mechanism behind this post-liquefaction shear strain phenomenon is briefly discussed in this study.

## 38.1 The Basic Phenomenon

Under undrained cyclic loading, a sand sample experiences decrease in effective stress $p$, which can be large enough so that $p$ approaches zero where the phenomenon of liquefaction takes place. The undrained stress path follows what is commonly known as the butterfly path shown in Fig. 38.1. Along with the reduction of $p$, some test results show that large cyclic shear strain is generated at a low but nontrivial shear stress values (Fig. 38.1a), but more recent evidence suggests that it is in fact generated at low enough shear stress values to be seen as zero (Fig. 38.1b, c). Zhang

R. Wang · J.-M. Zhang
Department of Hydraulic Engineering, Tsinghua University, Beijing, China

National Engineering Laboratory for Green and Safe Construction Technology in Urban Rail Transit, Tsinghua University, Beijing, China

P. Fu
Atmospheric, Earth, and Energy Division, Lawrence Livermore National Laboratory, Livermore, CA, USA

Y. F. Dafalias (✉)
Department of Civil and Environmental Engineering, University of California, Davis, CA, USA

Department of Mechanics, School of Applied Mathematical and Physical Sciences, National Technical University of Athens, Athens, Greece
e-mail: jfdafalias@ucdavis.edu


B. Kutter et al. (eds.), *Model Tests and Numerical Simulations of Liquefaction and Lateral Spreading*, https://doi.org/10.1007/978-3-030-22818-7_38

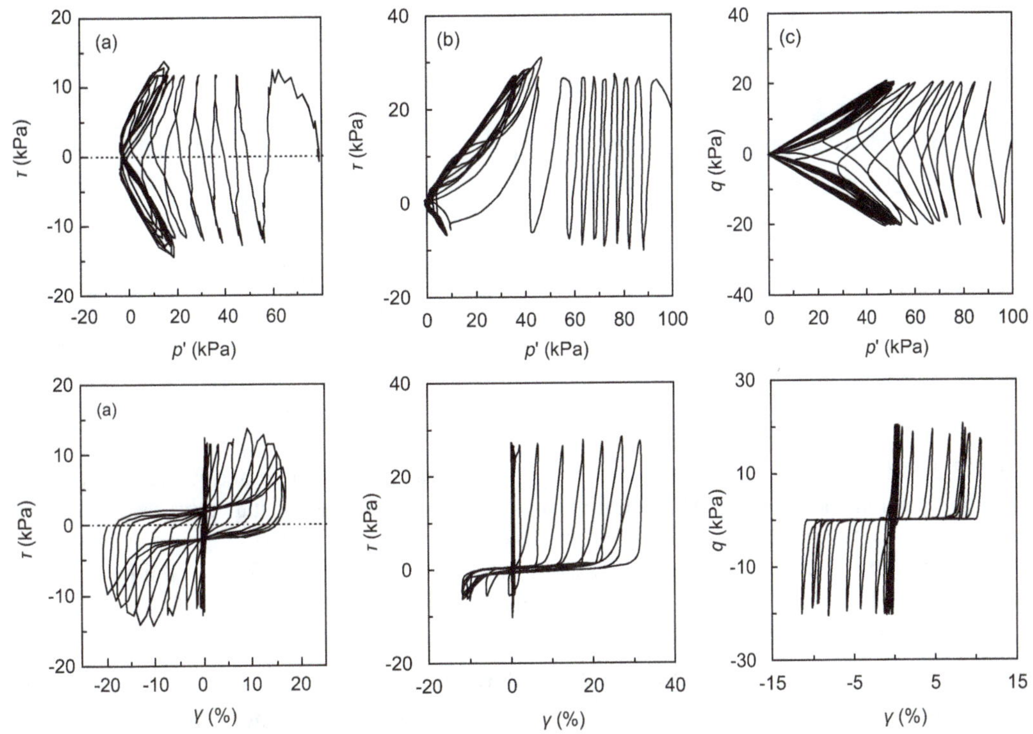


**Fig. 38.1** Typical results from undrained cyclic laboratory and numerical tests on sand: (**a**) simple shear test on Nevada sand (Arulmoli et al. 1992); (**b**) cyclic torsional test on Toyoura sand (Chiaro et al. 2013); (**c**) 2D DEM biaxial test on circular particles (Wang et al. 2016)

and Wang (2012) referred to this shear strain as post-liquefaction shear strain $\gamma_0$, considered to be generated at zero effective stress (i.e., liquefaction) state and has been observed to progressively increase in amplitude with increasing number of loading cycles until it eventually stabilizes at a bounded value. However, there has been no clear explanation on why, when and how this cyclic shear strain is generated. Without a clear understanding of this phenomenon, constitutive modelling would inevitably deviate from the actual behavior of the material, impeding the progress of validation of constitutive models by analyzing relevant BVP related to liquefaction as in the LEAP project.

## 38.2 Fabric Mechanism

Using DEM simulation, we were able to connect post-liquefaction shear deformation development to a new, theoretically measurable intrinsic fabric metric with a clear physical interpretation (Wang et al. 2016). This new fabric measurement, "Mean Neighboring Particle Distance" (*MNPD*), is the mean value over all particles of the "Neighboring Particle Distance" (*NPD*) as depicted in Fig. 38.2a, which is the mean surface-to-surface distance between a particle and its *n* closest neighbor particles,

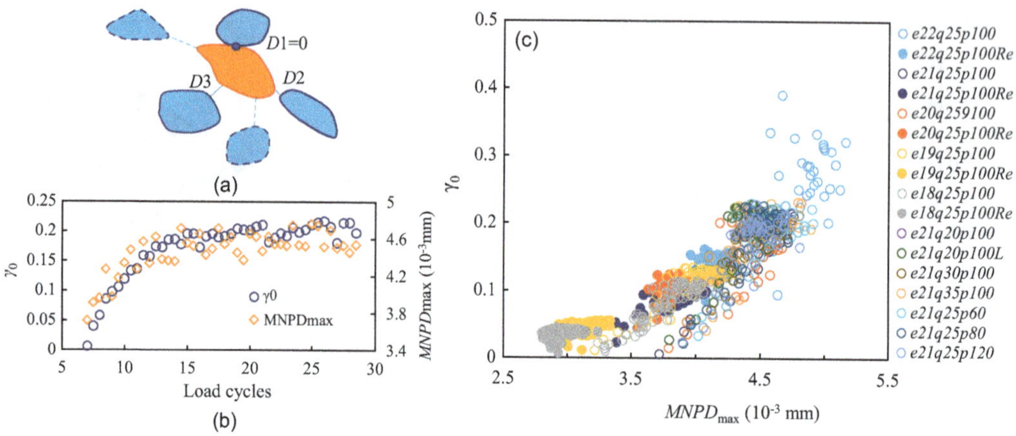


**Fig. 38.2** The concept of MNPD (Mean Neighboring Particle Distance) and its correlation with post-liquefaction shear strain: (**a**) illustration of the surface-to-surface distance between a particle and its three closest neighboring particles; (**b**) development of $\gamma_0$ and $\text{MNPD}_{\text{max}}$ in a typical DEM test; (**c**) correlation between $\gamma_0$ and $\text{MNPD}_{\text{max}}$ in each half loading cycle after initial liquefaction in 17 tests

with $n$ being the number of contacts needed to support a stable load-bearing structure ($n = 3$ in 2D and 4 in 3D). *MNPD* is formulated to capture a microstructural feature of granular materials that is closely related to deformation behavior in the post-liquefaction state by actually being a measure of "distance to establishing load-bearing contact" as opposed to a conventional measure of contact intensity, the coordination number. *MNPD*'s strong, unique correlation with $\gamma_0$ is evident from Fig. 38.2b, c. It has also been shown to influence the liquefaction resistance of sand (Wang et al. 2019). Therefore, it is expected that consideration of *MNPD* can provide a promising path to incorporating the mechanism of the post-liquefaction cyclic shear strain phenomenon into a continuum constitutive framework for practical purposes.

**Acknowledgement** Support from the European Research Council under FP7-ERC-IDEAS Advanced Grant Agreement n 290963 (SOMEF) is acknowledged.

## References

Arulmoli, K., Muraleetharan, K. K., Hossain, M. M., & Fruth, L. S. (1992). *VELACS: Verification of liquefaction analysis by centrifuge studies, laboratory testing program, soil data report*. The Earth Technology Corporation, Project No. 90-0562, Irvine, CA.

Chiaro, G., Kiyota, T., & Koseki, J. (2013). Strain localization characteristics of loose saturated Toyoura sand in undrained cyclic torsional shear tests with initial static shear. *Soils and Foundations, 53*(1), 23–34.

Wang, R., Fu, P., Zhang, J. M., & Dafalias, Y. (2016). DEM study of fabric features governing undrained post-liquefaction shear deformation of sand. *Acta Geotechnica, 11*(6), 1321–1337.

Wang, R., Fu, P., Zhang, J. M., Dafalias, Y. F. (2019). Fabric characteristics and processes influencing the liquefaction and re-liquefaction of sand. *Soil Dynamics and Earthquake Engineering, 125*, 105720.

Zhang, J. M., & Wang, G. (2012). Large post-liquefaction deformation of sand, part I: Physical mechanism, constitutive description and numerical algorithm. *Acta Geotechnica, 7*(2), 69–113.

# Index




B. Kutter et al. (eds.), *Model Tests and Numerical Simulations of Liquefaction and Lateral Spreading*, https://doi.org/10.1007/978-3-030-22818-7

www.ingramcontent.com/pod-product-compliance
Lightning Source LLC
LaVergne TN
LVHW062152200926
848484LV00007B/1904